Sponsored by the
AMERICAN SOCIETY OF LUBRICATION
ENGINEERS

CRC Handbook
of
Lubrication
(Theory and Practice of
Tribology)

Volume II
Theory and Design

Editor

E. Richard Booser, Ph.D.

Senior Engineer
Electromechanical Systems Engineering
Turbine Technology Laboratory
General Electric Company
Schenectady, New York

CRC Press, Inc.
Boca Raton, Florida

Library of Congress Cataloging in Publication Data
(Revised for Volume 2)
Main entry under title:

CRC handbook of lubrication (Tribology)

   Title of v. 2 varies: CRC handbook of lubrication
(theory and practice of tribology)
   Bibliography: v. 1, p.; v. 2, p.
   Includes index.
   Contents: v. 1. Applications and maintenance --
v. 2. Theory and design.
   1. Lubrication and lubricants--Handbooks, manuals,
etc. I. Booser, E. Richard.
TJ1075.C7   1983   621.8'9   82-4552
ISBN 0-8493-3901-4 (v. 1)
ISBN 0-8493-3902-2 (v. 2)

Direct all inquiries to CRC Press, Inc., 2000 Corporate Blvd., N.W., Boca Raton, Florida, 33431.

© 1984 by CRC Press Inc.
Second Printing, 1984

International Standard Book Number 0-8493-3901-4 (Volume I)
International Standard Book Number 0-8493-3902-2 (Volume II)

Library of Congress Card Number 82-4552
Printed in the United States

PREFACE—VOLUME II

Volume II of the *Handbook of Lubrication (Tribology)* provides coverage of basic theory involved in friction, wear, and lubrication; characteristics and application practices for lubricants; and design principles for lubricated machine elements such as bearings, gears, couplings, and seals.

Among significant developments covered in Volume II are new understandings of boundary lubrication and wear; new elastohydrodynamic theory for rolling bearings, gears and cams; extension of hydrodynamic analysis to high-speed operation in the turbulent regime and to dynamic response; and distinctive trends in the use of oils, greases, solid lubricants, additives, and synthetics.

This volume is intended to be used as a companion to Volume I with its coverage of theory and design. While construction equipment is covered in Volume I, for instance, companion coverages on the properties of oils and greases, design of bearings and gears, and lubrication fundamentals appear in Volume II.

The American Society of Lubrication Engineers has sponsored the development of the *Handbook of Lubrication.* ASLE Technical Committees and Industry Councils provided technical review, and the Handbook Advisory Committee oversaw the myriad day-to-day activities in producing the Handbook. Much of the original plan for Volume II was developed by Dr. P. M. Ku as the initial chairman of the Handbook Advisory Committee until his untimely death.

It is hoped that the Handbook will aid in achieving more effective lubrication, in control of friction and wear, and as another step to improved understanding of the complex factors involved in tribology.

E. R. BOOSER
EDITOR

THE EDITOR

Dr. E. Richard Booser has been a leader in the field of lubrication and tribology for the past 30 years. He completed his academic training in Chemical Engineering at The Pennsylvania State University in 1948 following research studies on composition, oxidation mechanisms, additives, and refining procedures for petroleum lubricants. Since that time, he has been employed by the General Electric Co. in development work on the lubrication of steam and gas turbines, electric motors and generators, nuclear plant equipment, jet engines, aircraft accessories, and household appliances.

His current assignment is Senior Engineer in Electromechanical Systems Engineering in the General Electric Turbine Technology Laboratory in Schenectady, N.Y., and he has served as leader of the Company Center of Research on Bearings and Rotor Dynamics.

He has published 60 papers covering oil oxidation, grease life in ball bearings, turbulence in high-speed oil-film bearings, selection of bearing materials, design of circulating oil systems, electric motor lubrication, and lubrication of nuclear plants. Co-author of the McGraw-Hill book *Bearing Design and Application,* he organized and taught bearing and lubrication courses for 400 engineers over the past 10 years.

Elected President of the American Society of Lubrication Engineers in 1956, he served the Society as Chairman of various activities: Lubrication Fundamentals Committee, General Technical Committee, Awards Committee, Fellows Committee and two local sections. He is also a member of the American Chemical Society, American Society of Mechanical Engineers, Sigma Xi, and is a registered professional engineer in New York State.

Dr. Booser draws on worldwide associations, and particularly on the resources and members of the American Society of Lubrication Engineers, to organize this Handbook. It is a compilation by 80 authors of developments and practices in the emerging field of tribology: the science of friction, wear, and lubrication.

CONTRIBUTORS

Frederick T. Barwell, Ph.D.
Emeritus Professor
University of Wales
and
Honorary Professorial Fellow
(Formerly Department Head)
Department of Mechanical Engineering
University College of Swansea
U.K.

E. O. Bennett, Ph.D.
Professor
Department of Biology
University of Houston
Houston, Texas

J. F. Booker, Ph.D.
Professor
School of Mechanical & Aerospace
  Engineering
Cornell University
Ithaca, New York

Donald H. Buckley, Doc. of Eng.
Chief
Tribology Branch
NASA-Lewis Research Center
Cleveland, Ohio

Michael M. Calistrat
Manager, Research & Development
Power Transmission Division
Koppers Company, Inc.
Baltimore, Maryland

Herbert S. Cheng, Ph.D.
Professor
Department of Mechanical Engineering
Technological Institute
Northwestern University
Evanston, Illinois

Horst Czichos, Ph.D.
Director and Professor
Department of "Special Fields
  of Materials Testing"
Bundesanstalt fur Materialprüfung
  (Federal Institute for Materials Research
  and Testing)
Berlin-Dahlem, West Germany

A. O. DeHart
Fluid Mechanics Department
GM Research Laboratories
GM Technical Center
Warren, Michigan

William J. Derner
Consultant
Mechanical Power Transmission
Indianapolis, Indiana

Norman S. Eiss, Jr., Ph.D.
Professor
Department of Mechanical Engineering
Virginia Polytechnic Institute and State
  University
Blacksburg, Virginia

Richard C. Elwell
Engineer — Development
Turbine Technology Laboratory
General Electric Company
Schenectady, New York

Richard S. Fein, Ph.D.
Consultant
Poughkeepsie, New York
Formerly Senior Research Associate
Texaco Inc.
Beacon, New York

Gregory Foltz
Specialist
Cimcool Technical Services
Products Division
Cincinnati Milacron
Cincinnati, Ohio

Edward J. Gesdorf
Consultant
Farval Lubricating Systems
Farval Division
Cleveland Gear Company
Cleveland, Ohio

Howard N. Kaufman
Fellow Engineer
Tribology and Experimental Mechanics
  Section
Mechanics Department
Westinghouse Research and Development
  Center
Pittsburgh, Pennsylvania

Ralph Kelly
Manager New Products
Cimcool Marketing Development
Products Division
Cincinnati Milacron
Cincinnati, Ohio

Elmer E. Klaus, Ph.D. (Retired)
Professor Emeritus
Fenske Faculty Fellow
Department of Chemical Engineering
Pennsylvania State University
University Park, Pennsylvania

John K. Lancaster, Ph.D.
Head
Materials and Structures Department
Royal Aircraft Establishment
Farnborough, Hants, U.K.

K. C. Ludema, Ph.D.
Professor
Department of Mechanical Engineers
University of Michigan
Ann Arbor, Michigan

S. Frank Murray
Senior Research Engineer
Department of Mechanical Engineering
Rensselaer Polytechnic Institute
Troy, New York

James A. O'Brien
Manager, Planning
Amoco Petroleum Additives Company
Clayton, Missouri

Eugene E. Pfaffenberger, P.E.
Manager
Engineering Analysis
Link-Belt Bearing Division
PT Components, Inc.
Indianapolis, Indiana

Ernest Rabinowicz, Ph.D.
Professor
Department of Mechanical Engineering
M.I.T.
Cambridge, Massachusetts

John L. Radovich
Senior Product Designer
Gear Division
Staff Lubrication Engineer
Farrel Company
Emhart Machinery Group
Ansonia, Connecticut

Albert A. Raimondi, Ph.D.
Manager
Tribology and Experimental Mechanics
Westinghouse R & D Center
Pittsburgh, Pennsylvania

Carleton N. Rowe, Ph.D.
Research Associate
Mobil Research and Development
  Corporation
Paulsboro, New Jersey

Irwin W. Ruge (Retired)
Product Manager
Marketing Technical Services
Union Oil Company of California
Schaumburg, Illinois

John A. Schey, Ph.D.
Professor
Department of Mechanical Engineering
University of Waterloo
Waterloo, Ontario, Canada

Milton C. Shaw, Sc.D.
Professor
Department of Mechanical and Aerospace
  Engineering
Arizona State University
Tempe, Arizona

Henry J. Sneck, Ph.D.
Professor
Department of Mechanical Engineering
Rensselaer Polytechnic Institute
Troy, New York

William K. Stair
Director
Engineering Experiment Station
and
Associate Dean
College of Engineering
University of Tennessee
Knoxville, Tennessee

Andras Z. Szeri, Ph.D.
Consultant
Westinghouse Research Laboratories
and
Professor
Department of Mechanical Engineering
University of Pittsburgh
Pittsburgh, Pennsylvania

Elmer J. Tewksbury, Ph.D. (Retired)
Professor
Department of Chemical Engineering
Pennsylvania State University
University Park, Pennsylvania

Arthur J. Twidale
Managing Director
Denco Farval Limited
Hereford, England

John H. Vohr, Ph.D.
Senior Engineer
Turbine Technology Laboratory
General Electric Company
Schenectady, New York

D. F. Wilcock, D.E.S.
President
Tribolock, Inc.
Schenectady, New York

Desmond C. J. Williams
Director
Denco Farval Limited
Hereford, England

J. Brian P. Williamson, Ph.D.
Scientific Consultant
Williamson Interface Limited
Malvern, England

TABLE OF CONTENTS

FRICTION, WEAR, AND LUBRICATION THEORY
The Shape of Surfaces 3
Properties of Surfaces 17
Friction 31
Boundary Lubrication 49
Hydrodynamic Lubrication 69
Numerical Methods in Hydrodynamic Lubrication 93
Hydrostatic Lubrication 105
Squeeze Films and Bearing Dynamics 121
Elastohydrodynamic Lubrication 139
Metallic Wear 163
Wear of Nonmetallic Materials 185
Wear Coefficients 201
Lubricated Wear 209

LUBRICANTS AND THEIR APPLICATION
Liquid Lubricants 229
Lubricating Greases—Characteristics and Selection 255
Solid Lubricants 269
Properties of Gases 291
Lubricating Oil Additives 301
Metal Processing—Deformation 317
Metal Removal 335
Cutting Fluids 357
Cutting Fluids—Microbial Action 371
Lubricant Application Methods 379
Circulating Oil Systems 395

DESIGN PRINCIPLES
Journal and Thrust Bearings 413
Sliding Bearing Materials 463
Sliding Bearing Damage 477
Rolling Element Bearings 495
Gears 539
Mechanical Shaft Couplings 565
Dynamic Seals 581
Wear Resistant Coatings and Surface Treatments 623
Systems Analysis 645

INDEX 665

CRC HANDBOOK OF LUBRICATION
(Theory and Practice of Tribology)

E. Richard Booser, Editor

Volume I
Application and Maintenance

Applications
Industrial Lubrication Practices
Maintenance
Appendixes

Volume II
Theory and Design

Friction, Wear, and Lubrication Theory
Lubricants and Their Application
Design Principles

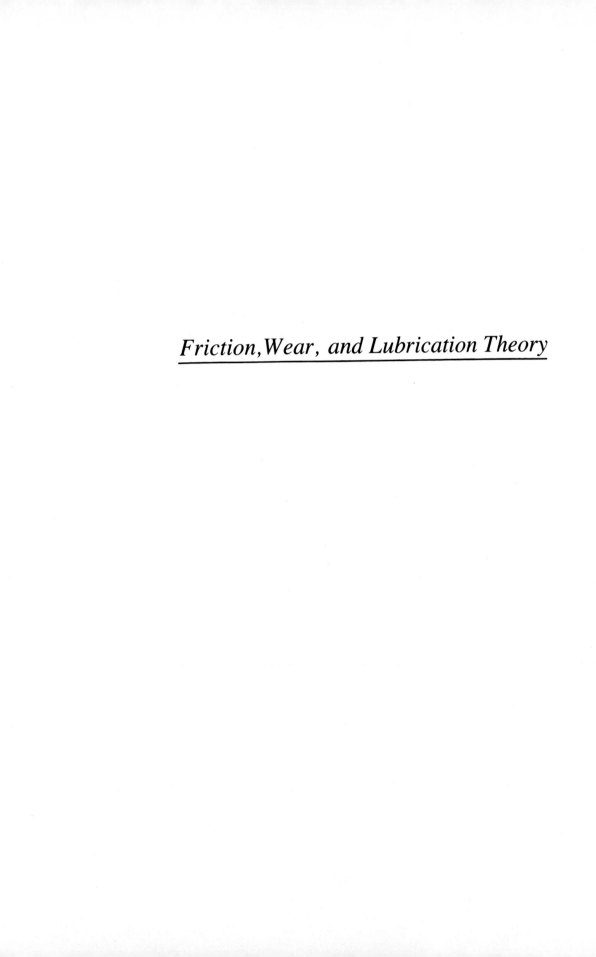

Friction,Wear, and Lubrication Theory

THE SHAPE OF SURFACES

J. B. P. Williamson

INTRODUCTION

All surfaces are rough. The world of the engineer is made of solids whose surfaces acquire their texture as the result of a great variety of processes. In some cases it is merely a by-product of forming the bulk shape, for example, in casting, molding, or cutting. More often a separate process affecting only the surface layers is applied after the part has been formed to its bulk dimensions. Some treatments remove material, as in grinding and etching. Others, such as plating, flame spraying, and sputtering, add it. Yet others merely redistribute the surface layer: peening and calendering are examples. In addition, surfaces often show the marks of unplanned treatments such as wear and corrosion.

Surface textures found in modern engineering vary widely. Figure 1A, for example, shows a mechanically polished surface, while Figure 1B shows one which has been electroplated. Such surfaces may feel smooth and give a mirror-like reflection, yet the electronmicrographs show they are covered with hills and valleys. Figure 2 places this roughness in perspective against other surface-related phenomena of interest in engineering.

Whenever two solids are brought together, they touch first where hills on one contact the surface of the other. As the hills flatten, contact areas grow and the pressure falls until it becomes too low to cause further deformation. Contact is thus limited to a relatively small area, and the rest of the surfaces are held apart. The interfacial gap formed is usually continuous, permitting gaseous and liquid access to the whole interface (Figure 3 illustrates this). Two copper surfaces were pressed together and sulfur dioxide gas was allowed to diffuse into the interfacial gap. On separation, bright areas of intimate contact (where the copper was protected from the gas) were in clear contrast to the chemically discolored surface. Areas of contact about 1 to 5 μm across and about 10 to 50 μm apart are typical of many tribological interfaces.

The texture of a surface ranges from large-scale shape deviations to tiny features such as ledges in crystal faces and steps where dislocations emerge. The scale of the world of the tribologist is essentially determined by the size of the individual contact areas between surfaces. Features which are small compared with individual contact regions are not usually significant.

MEASUREMENT OF SURFACE ROUGHNESS

The principal instruments used to study surface shape are the scanning electron microscope (SEM) and the profile analyzer. The SEM can provide micrographs with sufficient resolution to reveal individual details and, yet, has a large enough field of view that the interrelation of many such features can be seen. In practical tribology, however, it has two disadvantages: specimen size is limited and it cannot quantify roughness.

Checking a surface against a specification or measuring how texture influences perform- ance requires numerical descriptions. The profile analyzer is the most widely used instrument for this. It draws a sharp stylus lightly over the specimen and detects its movement as it follows the texture. The signal is amplified and recorded on a chart to produce a profile of the surface. Many surfaces contain flaws — unintentional, infrequent defects, such as cracks, inclusions, and scratches. Profiles should be positioned to avoid these aberrations whenever possible.

Surface profiles usually contain three major components (Figure 4):

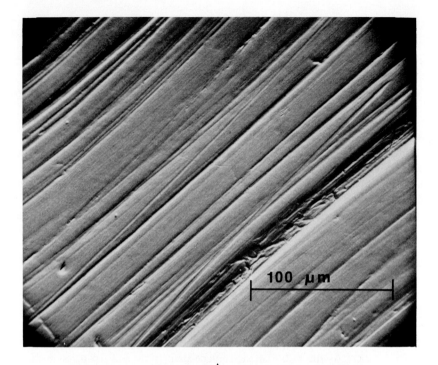

A

B

FIGURE 1.   Electron micrographs of (A) mechanically polished copper surface and (B) gold electroplate deposited on brass.

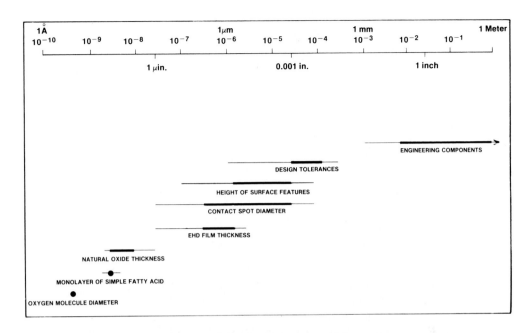

FIGURE 2. Comparative size of surface-related phenomena.

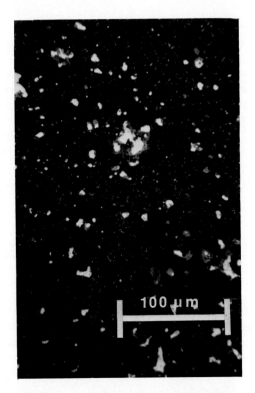

FIGURE 3. Contact areas revealed by chemical decoration. Two copper surfaces were pressed together for 1 month in an atmosphere of 1% SO_2 in air at 85% R.H., then photographed on separation.

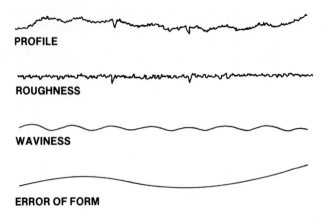

PROFILE

ROUGHNESS

WAVINESS

ERROR OF FORM

FIGURE 4.   The three components of surface texture.

1. **Roughness** — closely spaced irregularities, the height, width, and direction of which create the predominant surface pattern. Roughness includes those surface features intrinsic to the production process.
2. **Waviness** — surface irregularities of greater spacing than roughness. They are often the result of heat treatment, machine or workpiece deflections, vibrations, or warping strains.
3. **Errors of form** — gross deviations from nominal shape. They are not normally considered part of the surface texture.

It is important to analyze the profile of surface features over an appropriate length. If the function of a particular surface depends primarily on its short wavelength irregularities, for instance, a misleadingly large value of roughness would be obtained if the entire profile, including the waviness, were analyzed. The greatest spacing of surface irregularities to be included is called the "roughness-width cutoff". In most modern instruments an electronic filter removes unwanted components of the profile as illustrated in Figure 5. The top profile represents the actual movement of the stylus on a surface. The lower ones show the same profile using cutoffs of 0.8, 0.25, and 0.08 mm. The international standard roughness-width cutoffs are

Millimeters	Inches
0.08	0.003
0.25	0.010
0.80	0.030

The preferred value of 0.80 mm is assumed unless a different cutoff is specified. Roughness-width cutoff is sometimes called "roughness sampling length". The distinction between the terms is a fine technical one concerning instrument design; for practical purposes they are essentially the same.

Roughness of the filtered profile is most frequently described in terms of the arithmetic average deviation from its mean, which is called R_a. It was previously called "arithmetic average" (AA) or "center line average" (CLA). Usually the profile analyzer presents R_a as the mean of several readings taken consecutively along the profile. R_a is almost universally used on engineering drawings. Table 1 gives internationally adopted values together with the alternative "roughness grade number".

A second important measure of surface texture advocated by the International Organization

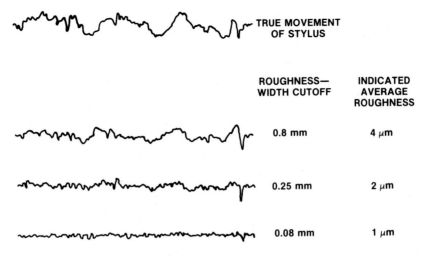

FIGURE 5.   The effect of the roughness-width cutoff is to remove all components of the total profile which have wavelengths greater than the cutoff.

Table 1
ARITHMETIC AVERAGE
ROUGHNESS GRADES

Recommended R_a values		Roughness grade number
μm	μ in.	
0.025	1	N1
0.05	2	N2
0.1	4	N3
0.2	8	N4
0.4	16	N5
0.8	32	N6
1.6	63	N7
3.2	125	N8
6.3	250	N9
12.5	500	N10
25	1000	N11

for Standardization (ISO) is the average elevation of the five highest peaks above the five lowest valleys, which is called R_z or the ISO 10-point height. It is measured over a single sampling length (roughness-width cutoff), and is particularly useful when only a small surface is available for assessment. R_a averages data from the entire height range of the profile, and is thus insensitive to occasional peaks or valleys. R_z, on the other hand, describes roughness in terms of extreme height and is valuable when performance of the surface could be impaired by excrescences or cracks.

R_a and R_z are easily determined and are widely used to monitor production consistency. They have one major disadvantage: both give equal weight to the shape of valleys and peaks. In practice, however, behavior of a surface in contact with another depends essentially on the texture of its highest strata, and hardly at all on the shape of its valleys. Frequently the highest parts of engineering surfaces differ significantly from the general texture.

Likely severity of wear between sliding surfaces is given by the Plasticity Index, which indicates whether deformation in the contact regions will be predominantly elastic or plastic. This index is given by $(E'/H)\sqrt{(\sigma/\beta)}$, where $1/E' = (1 - \nu_1^2)/E_1 + (1 - \nu_2^2)/E_2$ and E_1, E_2, ν_1, and ν_2 are Young's moduli and Poisson's ratios of the contacting solids, H is the hardness, σ is the standard deviation of the height of the hills, and β is the mean radius of their summits. A Plasticity Index of 1 or less indicates essentially elastic contact with a low probability of wear. Values above 3 indicate mainly plastic contact regions with a higher probability of wear. Ball and roller bearings and well run-in surfaces have indices around 1; almost all other engineering surfaces have 10 or more.

Some modern surface analyzers provide a voltage analog which permits a detailed computer analysis of the surface profile. In particular, height distribution of hills and the mean radius of their summits, which are components of the Plasticity Index, can be computed. Certain roughness parameters depend on the interval at which the analog signal is digitized. The number of peaks apparent in the profile, for example, and consequently the average radius of their summits, can vary widely when this interval is changed.

Departure of the surface profile from its mean may also be expressed in terms of the root mean square average deviation. This measure of roughness, called R_q, is similar to the arithmetic average R_a, although often 10 to 20% higher. It is particularly important in the theory of surface contact. In practical engineering it is frequently, through incorrectly, used interchangeably with R_a. Skewness is a useful measure of the asymmetry of the profile. A surface which is a plateau with occasional deep rifts, a "scratchy" surface, is said to have negative skewness. A plain with ridges, a "peaky" surface, has positive skewness.

Many production processes impart a directionality, or lay, to the surface. In principle, so long as the stylus of the profile analyzer traverses a representative sample of the texture, the R_a measured will be independent of the orientation of the track. However, the apparent wavelength of the surface features depends on the angle between the track and the lay. The orientation thus determines which features will be cut off by the filter. It is standard practice, therefore, to measure roughness at right angles to the lay, and this is assumed in specifications unless otherwise indicated.

The basic symbol used to designate surface texture is the checkmark.

$$\sqrt{}$$

A triangle is used when the surface is to be machined.

$$\forall$$

A circle means that the texture must be produced without any bulk removal of material.

$$\sqrt[\circ]{}$$

Various numbers and symbols may be written against the checkmark to specify features of the texture, the most common are given below.

The maximum R_a acceptable is shown in micrometers.

$$\overset{1.6}{\sqrt{}}$$

If necessary, the minimum acceptable R$_a$ is added.

A horizontal line is added to the checkmark to specify further features as in the examples given below.

The maximum acceptable waviness height and spacing (peak to valley height and peak to peak distance of the waviness) are shown in millimeters. The height is written first.

The roughness-width cutoff to be used in the measurement is shown in millimeters. When no value is given 0.8 is assumed.

The lay required is indicated by a lay symbol placed thus:

$$\sqrt{\perp}$$

Standard lay symbols are

Interpretation

Symbol

=	Approximately parallel to the line representing the surface to which the symbol is applied
⊥	Approximately perpendicular to the line representing the surface to which the symbol is applied
X	Angular in both directions to the line representing the surface to which the symbol is applied
M	Multidirectional
C	Approximately circular relative to the center of the surface to which the symbol is applied
R	Approximately radial relative to the center of the surface to which the symbol is applied
P	No lay, e.g., pitted, protuberant, particulate, or porous

CARTOGRAPHY OF SURFACE ROUGHNESS

The shape of a surface may be displayed by a computer-generated map developed from digital data derived from many closely spaced parallel profiles. Such a map shows details of individual features and also the general topography over a relatively large area. While these maps are tedious to obtain, the advent of computer-coupled profile analyzers will encourage wider use of this potent method of describing surfaces. Figure 6 shows part of such a map of a bead-blasted surface.

There is often remarkable similarity between maps of the surface of solids and ordinary contour maps of the surface of the earth. The scale factor is about 10^8 to 1. The ratio of height to spacing of the hills are similar. The slopes, in both cases, are usually between 1 and 10° and are rarely steeper than 30°. Take, for example, the mountains of New England as a model of a metal surface. A naturally occurring oxide layer would then be represented to scale by a 3-ft snowfall; an oxygen molecule by a golf ball; and a monolayer of a simple fatty acid, such as stearic acid, by a covering of 1-ft high grass. On the same scale an engineering component a few inches across would be the North American continent, and individual areas of contact would be a little larger than football stadiums.

If maps are made of two surfaces which are to be placed in contact, the gap between

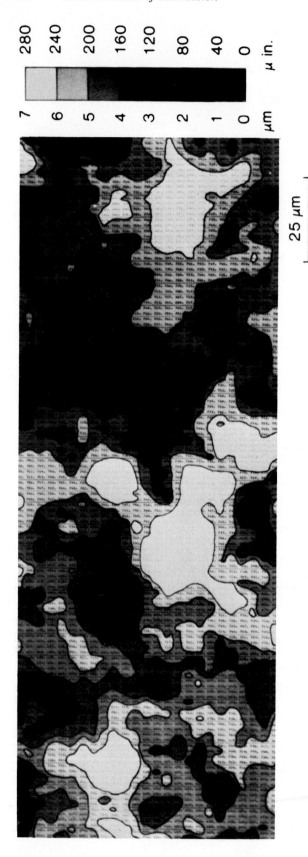

FIGURE 6. A glass surface lightly bead-blasted. Map made from height data collected from many parallel closely spaced profiles. The roughness is approximately 1 μm R_a. The white areas are plateaus of the original glass surface. The countours are every micrometer.

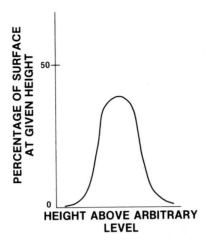

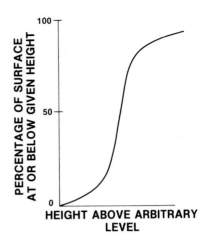

FIGURE 7. A typical height distribution (ADF) for a near-Gaussian surface.

FIGURE 8. Cumulative presentation of Figure 7: the complement of the bearing area curve (BAC).

them at each point can be determined and printed by the computer. Contours of this gap-map indicate the areas of contact between the surfaces at different loads. Analyses of gap-maps suggest normal contact is almost entirely confined to the highest 25% of each surface, and mainly occurs in the top 10%. (The percentages are of surface area, not of height.)

HEIGHT DISTRIBUTION AND BEARING AREA CURVES (BAC)

The fraction of a surface lying in each stratum can readily be obtained by tracing a profile and sampling its height at regular intervals. This gives the height distribution, sometimes called the "amplitude density function" (ADF). The profile must be long compared with the surface irregularities and include a representative sample of the texture. Figure 7 shows a typical height distribution.

It is, however, more useful to describe surfaces in terms of the integral of the distribution (Figure 8), which gives the fraction of the surface at or below each height. The well-known complementary BAC, which gives the area of contact which would exist if the hills were worn down to the given height by an ideally flat body, is the fraction of the surface at or above each height. Some modern surface analyzers provide chart or video displays of height distributions and BAC as standard features.

Gaussian Height Distribution

Many engineering surfaces have height distributions which are approximately Gaussian, i.e., they can be described by the normal probability function. A Gaussian distribution plotted on probability graph paper appears as a straight line, and deviations are readily detected. A sample of several thousand height readings will normally be needed, requiring data from many profiles (which must share a common reference level).

Figure 9 gives the BAC of a bead-blasted aluminum surface. More than 22,000 height readings show that at most only the top 0.01% of the surface was non-Gaussian. The slope of the line is a direct measure of the roughness parameter R_q. To an adequate approximation, R_q is 0.6 times the height difference between the 20th and 80th percentiles of the surface; if a longer straight line is available, R_q can be taken as 0.3 times the height difference between the 5th and 95th percentiles. For a surface with a Gaussian height distribution, R_a is approximately 0.8 R_q. In Figure 9, R_q is 1.25 μm and R_a is 1.0 μm. In general, however, there is no simple relation between R_a and R_q.

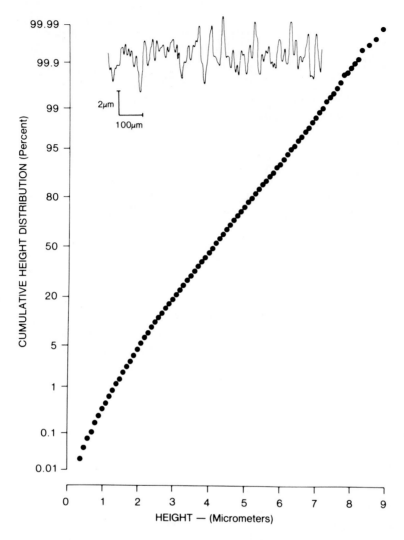

FIGURE 9.   Cumulative height distribution of a bead-blasted surface. The surface distribution is Gaussian — at the high end to at least 3.5 standard deviations from the mean.
A profile of the same surface is also shown: its vertical magnification is 50 times the
horizontal.

Several finishing processes create Gaussian height distributions. In these treatments the
texture results from a very large number of separate events each of which affects only a
small region of the surface, e.g., the impact of one bead during blasting, the deposition of
one ion in plating, or a single contact with a particle of polishing powder. If such a preparation
technique involves a large number of randomly located events, the Central Limit Theorem
of statistics says that the shape will tend towards a Gaussian height distribution. It is not
even necessary that the events are all the same — only that their effect is cumulative and
that they are numerous and randomly located. Any process satisfying these requirements
will create Gaussian surfaces, regardless of the properties of the solid.

Non-Gaussian Height Distributions

In some surface treatments the formative events, though numerous and cumulative, do
not occur randomly. Their location may depend on the microstructure of the solid, as in

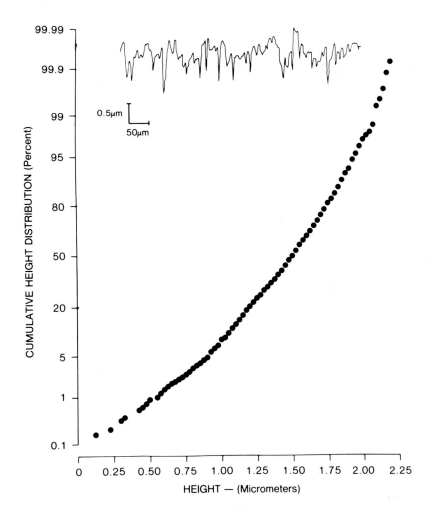

FIGURE 10.   Cumulative height distribution and typical profile of a ground surface. The distribution appears on probability paper as a gentle curve. This is characteristic of surfaces formed by extreme event processes.

etching, in which material removal is influenced by alloy phases, grain orientation, and grain boundaries. Alternatively, a process may be nonrandom because it is influenced by existing topography. Electroplating, for example, may deposit preferentially on hills.

Other surface treatments are not cumulative. In "extreme-value" processes each region of the final surface reflects only the most extreme events which occurred there. Grinding is an extreme-value process (Figure 10). When the individual events are not numerous (during turning each point on the specimen experiences only one formative event), there is again no statistical reason for a Gaussian surface. Height distribution of a turned surface is far from Gaussian (Figure 11). The distribution of *peak* heights, however, is often close to Gaussian. This reflects the randomness of the tearing where the edge of the tool cuts the wall left by the previous pass.

PURE, MIXED, AND STRATIFIED SURFACE TEXTURES

When all features on a surface result from the same treatment, the texture is said to be pure. Such textures are created only by processes which obliterate all previous treatments (milling or melting, or roughening a much smoother surface, for example).

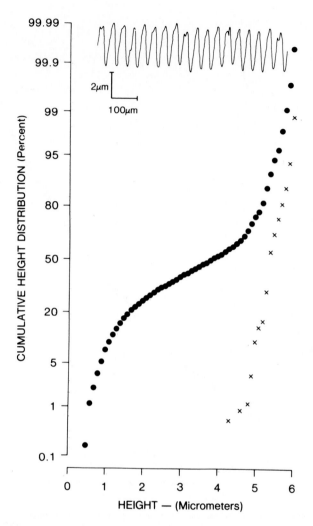

FIGURE 11.   Cumulative height distribution and typical profile of a turned surface (solid circles). The distribution is definitely non-Gaussian. The profile peaks (x's) however are much more nearly Gaussian.

Most surfaces have mixed textures as evidence of more than one preparation process. Normally features generated by the second process are distributed randomly, so that every region of the finished surface bears the marks of both treatments. However, some treatments give rise to "stratified surfaces" in which the valleys reflect one preparation process while the hills have a completely different texture from another treatment. R_a and R_z can then be very misleading; height distribution plotted on probability paper is much more useful.

Surfaces produced by running-in usually have stratified textures. Figure 12 follows such a process in which a mild steel pad lubricated with SAE 20 oil was worn against a finely ground hard steel flat. The solid circles show that before wearing the pad had a Gaussian height distribution with an R_a of 1.25 μm (given by the slope of the line). The open circles show how the texture changed as the surface wore. A typical run-in surface has a stratified texture such as that shown by curve B. The upper texture (which determines the nature of the contact) will behave as though it were part of a smooth surface. The interfacial gap (which controls lubricant access and removal of debris) depends on the lower texture.

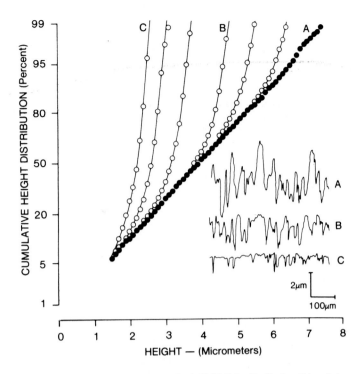

FIGURE 12.   The effect of wear. The initial height distribution (A) and six non-Gaussian distributions (open circles) of a bead-blasted surface represent, from right to left, progressive stages. Height distributions of this form are typical of those created by stratified secondary preparation processes.

Surfaces with mixed textures can be designed for particular combinations of properties which cannot be achieved with a single surface treatment. The surfaces of heavy-duty sliding bearings, for example, can be made to have wide interfacial gaps to facilitate lubricant access and debris removal; with a second process, they can also be given shallow-domed plateaus which can carry the load with very little plastic deformation, and hence little wear. Conversely, seal surfaces can be honed or lapped and then given noninterconnecting dimples to carry some lubricant to reduce wear.

REFERENCES

1. American National Standard ANSI B46.1, Surface Texture, 1978.
2. American National Standard ANSI Y14.36, Surface Texture Symbols, 1978.
3. British Standard BS 1134, Assessment of Surface Texture, 1972.
4. Canadian Standard CSA B95, 1962.
5. International Standard ISO R468, Surface Roughness, 1974.
6. **Archard, J. F.,** Surface topography and tribology, *Tribol. Int.*, 213-221, 1974.
7. **McAdams, H. T.;** Quantitative microcartography: a physico-geometric approach to surface integrity, in *Proc. Int. Conf. Surface Technology, Pittsburgh,* Society of Manufacturing Engineers, Dearborn, Mich., 1973, 96.
8. **Sayles, R. S. and Thomas, T. R.,** Mapping a small area of a surface, *J. Phys. E*, 9, 855, 1976.
9. **Tabor, D.,** The solid surface, *Phys. Bull.*, 29, 521, 1978.

10. **Whitehouse, D. J.,** Surfaces — a link between manufacture and function, *Proc. Inst. Mech. Eng. (London),* 192, 179, 1978.

11. **Whitehouse, D. J.,** Stylus techniques, in *Characterization of Solid Surfaces,* Kane, P. F. and Larrabee, G. R., Plenum Press, New York, 1974, chap. 3.

12. **Williamson, J. B. P.,** Microtopography of surfaces, *Proc. Inst. Mech. Eng. (London),* 182,(Pt. 3K), 21, 1967.

13. **Williamson, J. B. P., Pullen, J., and Hunt, R. T.,** The shape of solid surfaces, in *Surface Mechanics,* American Society of Mechanical Engineers, New York, 1969, 24.

14. **Young, R. D.,** Surface microtopography, *Phys. Today,* 42-49, November 1971.

PROPERTIES OF SURFACES

D. H. Buckley

INTRODUCTION

The first chapter showed that engineering surfaces are not smooth on a microscale but rather are rough, containing hills and valleys. The next step is to examine the chemistry, physics, metallurgy, and mechanical nature of the surfaces.

NATURE OF REAL SURFACES

Metals and alloys are, by far, the most widely used materials in practical tribological systems. If a metal is taken from the ordinary environment, placed in a vacuum, and heated mildly, the surface almost always liberates water. Various hydrocarbons are also detected if the component has been in the vicinity of operating machinery. Desorption with mild heating indicates that bonding to the surface is weak and of a physical nature.

Beneath this outer layer of absorbed water and gases, all metal surfaces (except gold) have a metallic oxide layer as indicated in Figure 1. With elemental metals, the particular oxide or oxides present depend on the environment, the amount of oxygen available to the surface, and the oxidation mechanism for the particular metal. If copper is heated in air, the outer black layer is CuO while an inner rose-colored layer is Cu_2O. Surface oxides of iron can be Fe_2O_3, Fe_3O_4, or FeO. Oxides present on an alloy surface depend upon concentration of alloying elements, affinity of alloying elements for oxygen, ability of oxygen to diffuse into surface layers, and segregation of alloy constituents to the surface. The importance of alloy element concentration is apparent, for example, in stainless steels. A certain concentration of nickel and chromium is necessary to ensure formation of nickel and chromium oxides to passivate the surface. With the relatively small concentration of 1.5% chromium in conventional 52100 ball bearing steel, presence of chromium may not even be detected in the surface oxide. Environmental conditions will vary the composition of the oxides. When 440-C bearing steel is heated in air to 500 °C, all alloying elements are present in the surface oxide layer. At 600 °C, however, iron is missing and only chromium is detected.

Chemical affinity of the metallic elements also bears on the surface oxide formed. While gold does not form a stable surface oxide in room temperature air, oxides (Cu) form on gold-copper alloys. Gold has no affinity for oxygen but copper does. Titanium reacts so readily with oxygen that it is used as an oxygen "getter" in vacuum systems. Metals such as copper (and more particularly silver) react much more slowly.

Under the absorbed gases, water, and the oxide layer, a mechanically worked layer is indicated in Figure 1. A finished mechanical component may have been machined, ground, cast, forged, extruded, or prepared by some other forming process. The energy which goes into the near surface region can result in strain hardening, recrystallization, and texturing. Grinding, for example, at high speeds is more likely to produce recrystallization than at low speeds. Further, the greater the deformation during grinding, the lower may be the recrystallization temperature. With titanium a 60% strain decreases the recrystallization temperature from 900 to 400 °C.

METHODS OF CHARACTERIZATION OF SURFACES

Microscopy

Microscopy is the most common technique employed for characterization of surfaces.

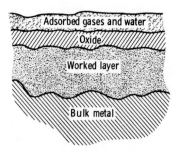

FIGURE 1.   Schematic representation of a metal surface.

The ordinary optical microscope can yield detailed surface features at magnifications up to about 500 times, 1000 times with the aid of oil immersion. Grain boundaries in metals are readily identifiable.

Development of the electron microscope permitted magnification of surfaces from 10 to 100,000 times. Atomistic features of surfaces are identifiable, and dislocation structures in metals and alloys are routinely identified and characterized.

The field ion microscope is the ultimate tool because it identifies individual atom sites on a solid surface. The photomicrograph of Figure 2a reveals the atoms and planes of a tungsten surface. The ring just to the upper right of center is the (110) plane. Each row out from the center ring represents the next nearest layer of atoms to the surface. The field ion microscope is so sensitive that it can detect the absence of a single atom from the surface or the presence of an extra or foreign atom. In Figure 2b a vacancy (absence of an atom) in the (203) plane of a platinum surface is readily discernible. The atom probe has been developed for single-atom chemistry. Used with the field ion microscope, the atom probe can characterize the chemistry of an individual atom.

Etching

Etching interacts surfaces with chemical agents such as acids or bases. With simple metals in polycrystalline form, the crystallographic orientation at the surface of each adjacent grain varies. In a crude way, one can distinguish the more atomically dense surface planes from the less dense: the more dense have lower surface energies and are not as readily attacked.

In addition to identifying orientation, grain boundaries and phases, etchants are used to identify atomistic defects such as dislocations. The proper etchant can indicate the concentration of surface dislocations, their location, and the atomic plane upon which the dislocation lies. Bibliography references identify both surface and dislocation etchants.

Simple chemical spot tests can provide insight as to the metallic elements present in the surface and detect adhesive transfer where dissimilar metals are in contact. A map of an element's distribution on a surface can be obtained if the reagents are impregnated in a porous filter paper; the paper is then pressed against the surface to be analyzed and moistened.

Analytical Surface Tools

A host of analytical tools have been developed in recent years for characterizing surfaces. Some rudiments of atomic arrangement will assist in understanding their mode of indicating surface structure.

Metals, alloys, ceramics, solid lubricants, and graphitic carbon are crystalline in nature with their atoms or molecules arranged in accordance with particular structures. Metals such as copper, nickel, silver, gold, platinum, and aluminum have a face-centered cubic structure, while iron, tantalum, niobium, vanadium, and tungsten are body-centered cubic. Zinc, cadmium, cobalt, rhenium, zirconium, and titanium have a hexagonal structure.

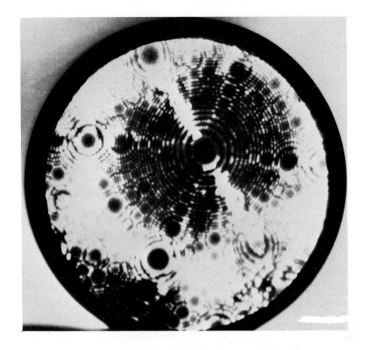

a

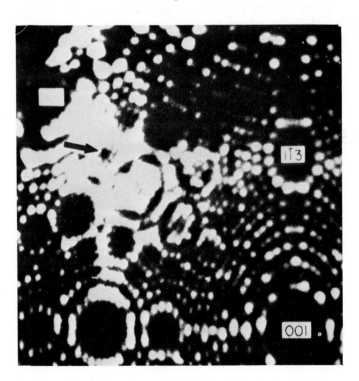

b

FIGURE 2. (a) Field ion micrograph; clean tungsten surface. (b) Concluded; vacancy in the (203) plane of a platinum surface.

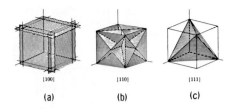

FIGURE 3.   Planes of possible slip in a cubic
crystal; (a) three (100) planes, (b) six (110) planes,
and (c) four (111) planes.

The atoms making up the faces of the cube for the face-centered cubic and the body-centered cubic are referred to as the (100) surfaces. As shown schematically in Figure 3a, these planes of atoms move relative to each other when the crystal is deformed plastically and are, therefore, referred to as slip planes. Under applied stresses, the (110) set of planes is commonly observed to slip over one another in the body-centered cubic system. The (111) planes are those upon which slip and cleavage in face-centered cubic materials is most frequent. The planes in Figure 3 are only three among these which can appear at the surface. X-ray diffraction reveals a host of different crystalline planes in a polycrystalline sample.

The atoms in the three planes of Figure 3 are arranged as indicated in Figure 4. The (111) planes have the closest atomic packing and the lowest surface energy in the face-centered cubic system and, therefore, are least likely to interact chemically with environmental constituents. The (110) planes in the face-centered cubic system are least densely packed, have higher surface energies than the (111) planes, and are, therefore, much more reactive. Because they are less densely packed, their elastic modulus and microhardness are also less. Atomic packing can vary with direction of movement. For the (111) plane in Figure 4, two basic directional packing variations are seen to exist. Surface energies vary in these two directions.

Low Energy Electron Diffraction (LEED)

LEED is widely used for characterization of the surface atomic structure on crystalline solids. Single crystals are generally studied although large grained polycrystals can also be examined. Low energy electrons in the range of 20 to 400 eV are diffracted from the surface crystal lattice to produce a reciprocal image on a phosphorus screen.

Figure 5 contains three LEED patterns from an iron (110) surface. The photograph and pattern at the left is for the iron surface with oxide removed. Upon oxide removal and heating of the iron in vacuum, the iron becomes covered with a film which produces a ring structure of diffraction spots on the surface. Auger electron spectroscopy analysis (discussed in the next section) identified the surface film to be graphitic carbon which segregates from the iron bulk. When the iron is argon ion bombarded, the carbon disappears and four diffraction spots remain in a rectangular array representative of the clean iron (110) surface. The iron diffraction spots are fuzzy and elongated from the strain in the iron surface lattice produced by the argon bombardment. Very mild heating removes the strain and produces the clean iron diffraction pattern of the right photograph.

LEED is effective in identifying a clean metal surface, its structure, condition or state, and structure of films. In a manner similar to that illustrated in Figure 5, LEED will also distinguish between organic films and their structural arrangement on an iron surface (see References for more information on LEED).

Auger Electron Spectroscopy (AES)

AES analyzes for all elements present on a surface except hydrogen and helium. It is

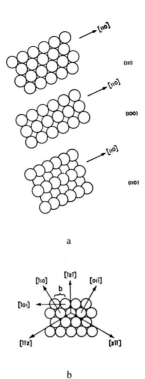

FIGURE 4. Atomic packing on various crystallographic planes and in various directions. For face-centered cubic materials. (a) Appearance of several crystal planes; (b) the (111) plane.

sensitive to an element such as oxygen to surface coverages of as little as one hundredth of a monolayer and analyzes to a depth of four or five atomic layers.

AES analysis uses a beam of electrons just as with LEED, but the electron energy is higher, usually 1500 to 3000 eV. Incident electrons strike the sample surface, penetrate the electron shells of outermost surface atoms, and cause ejection of a second electron called an Auger electron. The ejected electron carries with it an energy characteristic of the atom from which it came. Measuring the energy of the ejected electron identifies its elemental source. The electron energies detected can be recorded on a strip chart or an oscilloscope (see References for details).

An ordinary iron surface with normal surface contaminants will yield an Auger spectrum such as Figure 6a which contains peaks for sulfur, carbon, oxygen, and iron. The carbon and sulfur have two possible origins: impurities in the bulk iron which have segregated to the surface or adsorbates such as carbon monoxide, or carbon dioxide from the environment. The oxygen peak results from iron oxides present on the surface or adsorbates such as carbon compounds or water vapor. The three iron peaks originate from iron oxides and the iron metal.

If the surface of Figure 6a is bombarded with argon ions, contaminants can be knocked off leaving only iron with the spectrum in Figure 6b. The added low energy iron peak at the left end of the spectrum is easily lost when the surface is contaminated.

The shapes of Auger peaks can provide considerable information on the source of an element such as carbon, as demonstrated in Figure 7. The upper carbon Auger peak arises

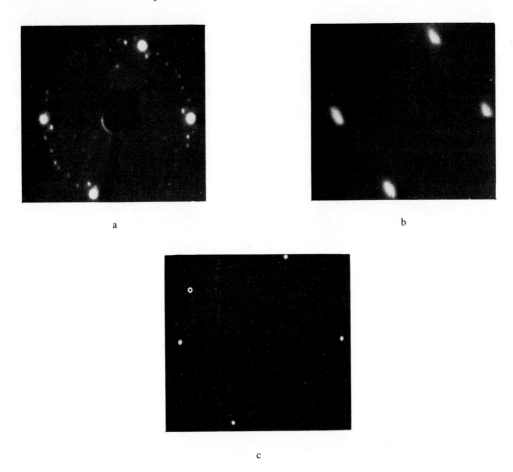

FIGURE 5. Three LEED patterns from an iron (110) surface. (a) Carbon contaminant; (b) argon bombarded; and (c) clean surface (110V).

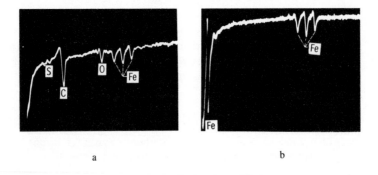

FIGURE 6. Auger spectra for an iron surface (a) before and (b) after sputter cleaning.

from carbon which segregated on the surface of the molybdenum. The second peak is from adsorbed carbon monoxide, while the third peak is from graphite.

X-Ray Photoemission Spectroscopy (XPS)

XPS is a surface tool which can determine the molecular structure from which an element came. XPS was formerly called electron spectroscopy for chemical analysis (ESCA).

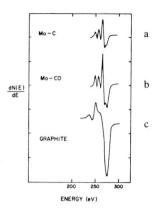

FIGURE 7. Auger electron spectra of carbon. (a) Segregated at a Mo(110) surface during initial cleaning (labeled Mo-C); (b) CO on a clean Mo(110) surface (labeled Mo-CO); and (c) in graphite.

With XPS a monochromatic X-ray beam is used as the energy source. The beam causes ejection of electrons with kinetic energies characteristic of the surface atoms. A spectrum of the elements present is obtained by plotting the total number of electrons ejected as a function of kinetic energy. XPS gives binding energies of the elements which enables identification of the compounds in which these elements exist. The binding energy of the electrons ejected from the surface is determined by their chemical environment and is roughly a function of the atomic charge.

The binding energy measured with XPS will be altered by changing the particular elements bound to the element being examined. Elemental sulfur has a characteristic binding energy of 162.5 eV. Negatively charged S^{-2} has a lower binding energy. When oxygen is bound to the sulfur, the sulfur binding energy increases. Further, the SO_4^{-2} structure has a greater binding energy than SO_3^{-2} which can be used to distinguish between sulfur bound in these two states.

Other Techniques

Over 70 surface tools have been developed for analysis and chemical characterization. A few more commonly used techniques are indicated by their acronyms in Table 1. The nondestructive techniques are nuclear back scattering spectroscopy (NBS) and electron microprobe (EM). Auger electron spectroscopy (AES), X-ray photoemission spectroscopy (XPS or ESCA), ion-scattering spectroscopy (ISS), and appearance potential spectroscopy (APS) are destructive only if sputter etching or depth profiling is used.

Two techniques which are destructive are secondary ion mass spectroscopy (SIMS) and glow discharge mass spectroscopy (GDMS). These techniques detect the species sputtered from the surface (see book by Kane and Larrabee in the References for more details). Note from Table 1 that they both detect all elements except hydrogen and helium, provide excellent chemical identification, and have sensitivities of surface elements to as little as 0.01 monolayer. Their disadvantage is that they must be operated in a vacuum system.

Probably the most versatile tool is the scanning electron microscope (SEM). It is extremely useful in obtaining a view of features on a surface such as asperities, surface irregularities, and topography where adhesion and wear have occurred. When SEM has incorporated into it X-ray energy dispersive analysis, both topography and chemistry can be determined. The X-ray analysis is not a surface analytical tool, but it can provide considerable information where material transfer takes place in adhesion or sliding. An SEM photomicrograph of an

Table 1
COMPARATIVE TABLE FOR THE VARIOUS TECHNIQUES USED FOR THE CHEMICAL CHARACTERIZATION OF SURFACES

	NBS	EM	AES	XPS	ISS	SIMS	GDMS	APS
Destructive to sample (in general)	No	No	No	No	No	Yes	Yes	No
Elements that can be detected	Heavy	$Z \geqslant 4$	$Z \geqslant 3$	$Z \geqslant 3$	$Z \geqslant 3$	All	All except He, Ne	$Z \geqslant 3$
Elemental identification[a]	F	G	E	E	E	G	G	E
Sensitivity (typical, in monolayers)	50	5	~0.01	<0.01	~0.01	<1	~1	≤0.1
Detectability (i.e., ppm)[b]	NA	100	<1	NA	NA	1	100	NA
Results are (in principle)[c]	Abs	Abs	Abs	Abs	Abs	Abs	Abs	Abs
Depth probed (in Å)	10^4	10^4—10^5	15—20	15—75	3	$\sim 5 \times 10^4$	10—10^4	~10
Depth distribution of elements[d]	Yes	Yes	Y/d	No	Yes	Yes	Yes	Y/d
Chemical (i.e., binding) information	No	Yes	Yes	Yes	No	No	No	Yes

[a] E, Excellent; G, good; F, fair.
[b] NA, Not applicable.
[c] Rel, Relative; Abs, absolute.
[d] Y/d, Yes, if destructive.

From Kane, P. F. and Larrabee, G. B., Eds., *Characterization of Solid Surfaces*, Plenum Press, New York, 1974. With permission.

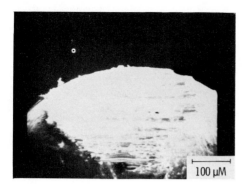
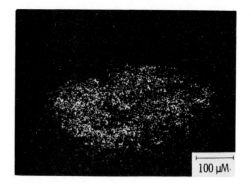

a   b

FIGURE 8.   (a) Electron image of aluminum rider wear scar; (b) iron Kα map of aluminum rider.

aluminum surface is shown in Figure 8a after sliding on an iron surface. The photomicrograph reveals surface topography while the X-ray map for iron reveals the white patch in Figure 8b where iron is detected on the aluminum wear surface.

PROPERTIES OF SURFACES

Metallurgy and Crystalline Structure

The crystal structure of ideal surfaces has already been examined in Figure 4. All engineering surfaces vary from this ideal and have grain boundaries which develop during solidification as large defects which exist in the solid and extend to the surface. They do not possess a regular structure, are highly active regions, and on the surface are very energetic. Lesser defects include subboundaries, twins, dislocations, interstitials, and vacancies.

Subboundaries are low-angle grain boundaries and usually occur where there is only a slight mismatch in orientation of adjacent grains on either side of the boundary. When the crystal lattices of adjacent grains are slightly tilted one toward the other, there is a tilt boundary. Where the lattices remain parallel but one is rotated about a simple crystallographic axis relative to the other with the boundary being normal to this axis, a twist boundary develops. The twin boundary occurs where there is only a degree or two of mismatch with the twins being mirror images. They are frequently seen on basal planes of hexagonal metals with deformation.

Dislocations are atomic line defects in crystalline solids. They may be subsurface and terminate at the surface or they may be in the surface. Edge dislocations are entirely along a line where an extra half plane of atoms exists. Screw dislocations form along a spiral dislocation line. Small angle boundaries or subboundaries are generally composed of edge dislocations. These defects in crystalline solids cause them to deviate markedly from the theoretically achievable strengths of ideal crystals.

Some of the crystalline surface defects are presented schematically in Figure 9. The vacant lattice site was seen on a real surface in the photomicrograph of Figure 2b. An interstital atom is crowded into the crystal lattice of Figure 9a. Edge and screw dislocations and a small angle boundary are also shown. Worn surfaces generally have undergone a high degree of strain and may contain large amounts of lattice distortion and defects such as dislocations. While initial dislocations cause a reduction in strength, their multiplication and interaction during deformation increase surfacial strength. Microhardness is generally higher in grain boundaries than in grains.

With plastic deformation, the strain generally produces a reduction in recrystallization

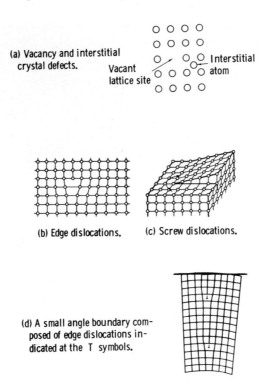

(a) Vacancy and interstitial crystal defects.

Vacant lattice site   Interstitial atom

(b) Edge dislocations.   (c) Screw dislocations.

(d) A small angle boundary composed of edge dislocations indicated at the ⊥ symbols.

FIGURE 9.   Crystalline defects in solids.

temperatures of material at the surface. The combination of strain and temperature can then bring about surface recrystallization which has an annealing effect. This process relieves lattice strain and stored energy, with a sharp reduction in the concentration of surface defects. In a dynamic, nonequilibrium system such as encountered in sliding, rolling, or rubbing, surface layers may be strained many times, recrystallized, and then strained again.

Solid State Bonding

What holds atoms or molecules in various arrangements and imparts to solids their basic cohesive strength? The answer lies in the bonding. Bonding in crystalline solids can be of four types, as shown in Figure 10: van der Waals, ionic, metallic, and covalent.

Van der Waals forces, the weakest holding solids together, are attributed to nothing more than fluctuations in the charge distributions within atoms or molecules. These forces can be represented in bonding the atoms of an inert gas together when solidified. Very little energy is required to accomplish sublimation.

An ionic bond is very strong, and some high-strength solids are held together by it. This bond is represented in Figure 10 by sodium chloride. Electrons are transferred from the metal to the nonmetal and the resulting ions are held together by the electrostatic forces developed. Aluminum and magnesium oxides are two tribological solids with this bonding.

With metals, the valence electrons are taken away from individual atoms to form a sea of electrons. This results in positively charged ions immersed in electrons. This bonding gives metals their good thermal and electrical conduction characteristics.

The fourth type of bonding is covalent where electrons are simply shared. This is indicated in Figure 10 by the overlapping of carbon atoms in diamond (the hardest material and most resistant to deformation). At the same time, the covalent bond is found in organic molecules in polymers and lubricants where it is relatively weak. No other bonding type possesses such a wide range of strengths.

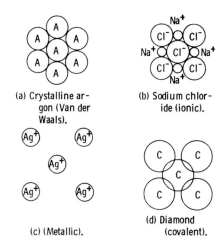

FIGURE 10. The principle types of crystalline binding forces.

CHEMISTRY OF SURFACES

Clean Surfaces

Very clean surfaces are extremely active chemically. A copper atom which lies in a (111) plane in the bulk of the solid will have a coordination number of 12: it is bonded to 12 nearest neighbors. That same copper atom at the surface will, however, have a coordination number of only 9 with only 9 nearest neighbors. The energy normally associated with bonding to three additional atoms is now available at the surface. This energy expressed over an area of many atoms is referred to as the surface energy.

Surface energy is also the energy necessary to generate a new solid surface by the separation of adjacent planes. The energy required for separation is a function of the atomic packing. For example, for copper the atomic packing density is greatest in (111) planes (greatest number of nearest neighbors within the plane). As a result, bonding forces between adjacent (111) planes is least and the surface energy of new (111) surfaces generated, say by cleavage, is less than for the (110) and (100) planes. This lesser binding strength is also a function of the distance between adjacent planes, it being greater between adjacent (111) planes than between (110) and (100) planes.

Because surface atoms have this unused energy, they can interact with each other, with other atoms from the bulk, and with species from the environment. Not bound as rigidly as atoms in the bulk, surface atoms can alter their lattice spacing by reconstruction, as depicted schematically in Figure 11. By use of LEED, this process has been found to occur in some crystalline solids but not in others.

In solids containing more than a single element, atoms from the bulk can diffuse to the surface and segregate there. In a simple binary alloy, solute atom can diffuse from near surface regions to completely cover the surface of the solvent. This has been observed for many binary systems including aluminum in copper, tin in copper, indium in copper, aluminum in iron, and silicon in iron. One hypothesis for the segregation mechanism is that the solute segregates on the surface because it reduces the surface energy. A second theory is that the solute produces a strain in the crystal lattice of the solvent, and this unnatural lattice state ejects solute atoms from the bulk.

Chemisorption

In addition to the solid interacting with itself at the surface, the surface can interact with

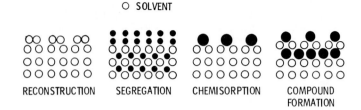

FIGURE 11.   Possible surface events.

the environment. This interaction alters the surface chemistry, physics, metallurgy, and mechanical behavior. If a metal surface is very carefully cleaned in a vacuum system and then a gas such as oxygen admitted, the gas will adsorb on the metal surface. Except with inert gases, this adsorption results in chemical bonding in a chemisorption process indicated schematically in Figure 11. Once adsorbed, these films are generally difficult to remove.

Where the species adsorbing on a clean surface is an element, adsorption is direct. Surface atoms of the solid retain their individual identity as do atoms of the adsorbate, yet each is chemically bonded to the other. When the adsorbing species is molecular, chemisorption may be a two-step process, first dissociation of the molecule upon contact with the energetic clean surface followed by adsorption of the dissociated constituents.

Chemisorption is a monolayer process. Bond strengths are a function of chemical activity of the solid surface (surface energy), degree of surface coverage of that adsorbate or another adsorbate, reactivity of the adsorbing species, and its structure. The higher the surface energy of the solid surface, the stronger the tendency to chemisorb. In general, the high-energy, low-atomic density crystallographic planes will chemisorb much more rapidly than will the high-atomic density, low-surface energy planes. Hydrogen sulfide will adsorb more readily on (110) and (100) surfaces of copper than on (111) surfaces.

The metal surface has an effect. Copper, silver, and gold are noble metals and many of their properties are similar. Yet, oxygen will chemisorb relatively strongly to copper, weakly to silver, and not at all to gold. Reactivity of the adsorbent is also important. Of the halogen family fluorine will adsorb more strongly than chlorine, chlorine than bromine, and bromine than iodine.

The structure of the adsorbing species is also significant as can be demonstrated with simple hydrocarbons. If ethane, ethylene, and acetylene are adsorbed on an iron surface, tenacity of the chemisorbed films is in direct relation to the degree of bond unsaturation. Acetylene is much more strongly bound to the surface than ethylene, which in turn is more strongly bound than ethane. The carbon to carbon bonds break on adsorption and bond to the iron. The greater the number of carbon to carbon bonds, the greater the resulting number of carbon to iron bonds.

Compound Formation

Compound formation on tribological surfaces is extremely important. The naturally occurring oxides present on metals prevents their destruction when sliding on other solids. Extreme pressure additives and many antiwear materials placed in oils perform by compound formation with the surface to be lubricated.

Once present on a surface, chemisorbed films often interact with that surface to form chemical compounds. The surface material and the adsorbate form an entirely new substance with its own characteristic properties. The process continues by diffusion of both the solid surface material and the environmental species into the film. The compound can grow in thickness on the surface if the film is porous and allows for two-way diffusion as shown in

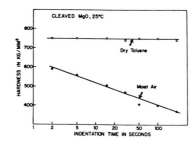

FIGURE 12.   Illustration of the effect of time on microhardness of MgO in toluene and in moist air (after Westbrook).[21]

Figure 11. An example is the oxidation of iron in moist air which continues to consume iron. In contrast, oxidation of aluminum to form aluminum oxide results in a thin, dense oxide of 120 Å which retards diffusion and film growth.

Environmental Effects

Chemical, physical, and metallurgical properties of atomically clean metal surfaces are markedly altered by foreign substances. This is extremely important because most real surfaces are not atomically clean but have film(s) present on their surface (Figure 1). The wide variations found in the literature for surface properties of materials can be attributed to the effect of these films.

Presence of oxides on metal surfaces has been observed to produce a surface hardening effect. One explanation for this hardening is that the oxygen pins dislocations which emerge at the surface, impeding their mobility.

Other surface films increase ductility. For example, water on alkali halide crystals will allow an otherwise brittle solid to deform plastically. This effect is also observed with ceramics. Magnesium oxide (MgO) is normally very brittle with a surface hardness in the clean state of about 750 kg/mm^2. Figure 12 presents the hardness of MgO as a function of indentation time in dry toluene and moist air. The increased surface ductility in the presence of water is striking, and the difference increases with increasing indentation time. This change with time makes the film effect a true surface property and not simply a lubricating effect produced by the water.

In the 1920s Rehbinder found that certain organic molecules on the surface of solids produced a softening. Such substances as oleic acid in vaseline oil were examined. This surface softening by lubricating substances can be very beneficial in certain instances such as in arresting the formation of fatigue cracks in bearing surfaces.

REFERENCES

Introduction

1. **ASTM,** Symposium on the properties of surfaces, *ASTM Mater. Sci. Ser. 4,* 1963.
2. **SCI,** *Surface Phenomena of Metals,* Monograph No. 28, Society of Chemical Industry, London, 1968.
3. **Anon.,** Conference on clean surfaces, *Ann. N.Y. Acad. Sci.,* 101, 583, 1963.
4. **Adamson, A. W.,** *Physical Chemistry of Surfaces,* 2nd ed., Interscience, New York, 1967.
5. **Gatos, H. C., Ed.,** *The Surface Chemistry of Metals and Semiconductors,* John Wiley & Sons, New York, 1960.
6. **Blakely, J. M., Ed.,** *Surface Physics of Materials,* Vols. 1 and 2, Academic Press, New York, 1975.

Method of Characterization of Surfaces

7. **Kane, P. F. and Larrabee, G. B., Eds.,** *Characterization of Solid Surfaces,* Plenum Press, New York, 1974.
8. **Bunshah, R. F., Ed.,** *Technique of Metals Research,* Vol. 2, Techniques for the Direct Observation of Structure and Imperfections, Part 2, Interscience, New York, 1969.
9. **Blakely, J. M., Ed.,** *Surface Physics of Materials,* Materials Science Series, Vols. 1 and 2, Academic Press, 1975.
10. **Somoraji, G. A.,** *Principles of Surface Chemistry,* Prentice-Hall, Englewood Cliffs, N.J., 1972.
11. Proc. 2nd Int. Conf. on Solid Surfaces. II, *Jpn. J. Appl. Phys.,* Suppl. 2, 1974.
12. **Muller, E. W. and Tsong, T. T.,** *Field Ion Microscopy,* American Elsevier, New York, 1969.

Properties of Surfaces

13. **Ehrlich, G.,** Atomistics of metal surfaces, *Surface Phenomena of Metals,* Monograph No. 28, Society of Chemical Industry, London, 1968, 13.
14. **Hayward, D. O. and Trapnell, B. M. W.,** *Chemisorption,* 2nd ed., Butterworths, Washington, D.C., 1964.
15. **Ferrante, J. and Buckley, D. H.,** A review of surface segregation, adhesion and friction studies performed on copper-aluminum, copper-tin, and iron-aluminum alloys, *ASLE Trans.,* 15(1), 18, January 1972.
16. **Burke, J. J., Reed, N. L., and Weiss, V., Eds.,** *Surfaces and Interfaces 11, Physical and Mechanical Properties,* Syracuse University Press, New York, 1968.
17. **Westwood, A. R. C. and Stolaff, N. S., Eds.,** *Environment-Sensitive Mechanical Behavior, Metallurgical Society Conference,* Vol. 35, Gordan and Bovach Science Publishers, New York, 1966.
18. **Jenkins, A. D., Ed.,** *Polymer Science, A Materials Science Handbook,* Vols. 1 and 2, North-Holland, Amsterdam, 1972.
19. **Buckley, D. H.,** Definition and Effect of Chemical Properties of Surfaces in Friction, Wear, and Lubrication, NASA TM-73806, National Aeronautics and Space Administration, Washington, D.C., 1978.
20. **Likhtman, V. I., Rehbinder, P. A., and Karpenko, G. V.,** *Effect of Surface-Active Media on the Deformation of Metals,* Chemical Publishing Company, New York, 1960.
21. **Westbrook, J. H., Ed.,** *Mechanical Properties of Intermetallic Compounds,* John Wiley & Sons, New York, 1960.

FRICTION

K. C. Ludema

DEFINITION OF FRICTION

The usual engineering definition of friction is resistance to relative motion of contacting bodies. Commonly encountered types of friction include dry, lubricated, sliding, rolling, dynamic or kinetic, static or starting or limiting, internal or hysteretic, external and viscous.

Magnitude of friction is usually expressed as a *coefficient of friction* μ, which is the ratio of the force F required to initiate or sustain relative tangential motion to the normal force (or weight) N which presses the two surfaces together. Thus, $\mu = F/N$. In the early years of technology, the value of F/N was found to be reasonably constant for each class of materials. In modern technology, μ is regarded to be widely variable, depending on operational variables, lubricants, properties of the substrate, and surface films.[1-5]

CLASSIFICATION OF FRICTIONAL CONTACTS

Friction is a phenomenon associated with mechanical components. Some are expected to slide and others are not. Four categories within which high or low friction may be desirable are given below.

1. *Force transmitting components* that are expected to operate without displacement. Examples fall in the following two classes:

    a. *Drive surfaces* or *traction surfaces* such as power belts, shoes on the floor, hose clamps, and tires and wheels on roads or rails. Some provision is made for sliding, but excessive sliding compromises the function of the surfaces. Normal operation involves little or no macroscopic slip. Static friction is often higher than the dynamic friction.

    b. *Clamped surfaces* such as press-fitted pulleys on shafts, wedge-clamped pulleys on shafts, bolted joining surfaces in machines, automobiles, household appliances, etc. To prevent movement, high normal forces must be used and the system is designed to impose a high but safe, normal (clamping) force. In some instances, pins, keys, surface steps, and other means are used to guarantee minimal motion.

    In the above examples, the application of a (friction) force frequently produces microscopic slip. Since contacting asperities are of varying heights on the original surfaces, contact pressures within clamped regions may vary. Thus, the local resistance to sliding varies and some asperities will slip when low values of friction force are applied. Slip may be referred to as microsliding as distinguished from macrosliding, where all asperities are sliding at once. The result of oscillatory sliding of asperities is a wearing mechanism, some cases of which are known as fretting.

2. *Energy absorption-controlling components* such as in braking and clutching. Efficient design usually requires rejecting materials with low coefficient of friction because such materials require large values of normal force. Large coefficients of friction would be desirable except that suitably durable materials with high friction have not been found. Thus, many braking and clutching materials have intermediate values of coefficient of friction in the range between 0.3 and 0.6. An important requirement of braking materials is constant friction, in order to prevent brake "pulling" and unexpected wheel lockup in vehicles. A secondary goal is to minimize the difference

between the static and dynamic coefficient of friction for avoiding squeal or vibrations from brakes and clutches.

3. *Quality control components* that require constant friction. Two examples may be cited, but there are many more:
    a. In knitting and weaving of textile products, the tightness of weave must be controlled and reproducible to produce uniform fabric.
    b. Sheet metal rolling mills require a well-controlled coefficient of friction in order to maintain uniformity of thickness, width, and surface finish of the sheet and, in some instances, minimize cracking of the edges of the sheet.

4. *Low friction components* that are expected to operate at maximum efficiency while a normal force is transmitted. Examples are gears in watches and other machines where limited driving power may be available or minimum power consumption is desired, bearings in motors, engines and gyroscopes where minimum losses are desired, and precision guides in machinery in which high friction may produce distortion.

SURFACE CHARACTERISTICS AND STATIC CONTACT AREA

Frequently the coefficient of friction is more dependent upon surface properties and surface finish than on substrate properties. Substrate properties, however, influence both the surface finish achieved in processing and the kinetics of adsorption of chemical species.

Surface Structure and Finish

With the exception of surfaces that solidify from the liquid (either in air, in vacuum, or in contact with a mold), most technological surfaces are formed by a cutting operation. Coarse cutting is done with a cutting tool in a lathe, drill press, milling machine, etc. Finer cutting is done with abrasives by grinding, honing, lapping, etc.

Cutting is simply localized fracture. Each individual microfracture joins another and/or extends into the substrate. The orientation of surface facets and the direction taken by subsurface cracks are often dependent upon the structure of the material. Seriousness of a substrate crack will probably depend upon the toughness of the material. For example, in cast irons and notably in white cast iron, machining often forms cracks that extend into the substrate and in fact may loosen some grains from the matrix. In more ductile materials, the cracks that extend into the substrate are less likely to be harmful and yet they may constitute a stress concentration from which fatigue cracks may emanate. Cracks may also become corrosion cells.

Many surfaces are formed by ductile fracture mechanisms with a high amount of plastic strain and residual stress remaining in the surface. All of these conditions may influence the coefficient of friction either from the beginning of sliding or as a result of surface alteration during sliding.

Adsorption on Surfaces

Material cutting operations expose atoms or molecules, formerly in the substrate, to the environment around the material. Oxygen in the air is very reactive with most metals and is usually the first to adsorb and form oxides on metal surfaces. After oxides of between 20 Å to 100 Å thick form, the rate of oxidation diminishes and other gases adsorb. In air, for example, a significant amount of water vapor adsorbs on oxides and on other materials such as gold and plastic which do not oxidize quickly. The adsorbed gases can be the same thickness as the oxide film.

Adsorption occurs very quickly. Pure oxygen gas at atmospheric pressure produces a 50% coverage by adsorption in about 1.75×10^{-9} sec.

The influence of all surface films on friction is not always the same. It might be expected that adsorbed water would act as a liquid lubricant, and that some oxides or hydroxides might act as solid lubricants. On the other hand, some oxides such as aluminum oxide (Al_2O_3) are abrasive and under some conditions greatly increase friction.

Estimating Contact Area

Explanations of friction are based upon the detailed nature of contact between two bodies. Historically the measurement of real contact area was attempted in order to decide between the two major theories of friction outlined below. The methods used include electrical resistance, heat transfer, total internal reflectance of an optical element pressed against a metal surface, phase contrast microscopy, ultrasonic transmission, electron emission phenomena, computer simulation, large-scale surface model studies, and analytical methods based on the mechanics of solids. Most methods are unsatisfactory in that either the observations are not made in real time, or the method is incapable of distinguishing between many small points of contact vs. few large regions. Results from all methods, however, produce the same conclusion: the contact area increases with normal load and when a friction force is applied.

An adequate description of the behavior of asperities may be gained by a simple analytical model. Representation as a sphere is reasonable since most asperities are reasonably rounded rather than sharp or jagged. For the simplified case of a sphere pressed against a flat surface, the radius of contact, a, may be calculated as follows:[6]

$$a = \left[\frac{3N}{4r} \left(\frac{1-\nu_1^2}{E_1} + \frac{1-\nu_2^2}{E_2} \right) \right]^{1/3}$$

where N is the normal load, r is the radius of the sphere, ν is Poisson's ratio, E is Young's modulus, and subscripts 1 and 2 refer to the two materials if the sphere and flat plate are of different materials. The pressure distribution over the area of contact is semielliptical. The average pressure is $P_m = N/\pi a^2$ and the maximum pressure q_o at the center of contact is $3/2 \ P_m$. Thus, $q_o = (3/2) \ N/\pi a^2$.

Other equations are available that give the stress state of all points in the substrate[6] and may be used to calculate the limits of elastic behavior. A principle of plasticity is that plastic flow will occur whenever the difference between the largest and smallest stresses in perpendicular directions at a point is equal to the yield strength of the material. As normal load increases, the conditions for plastic flow first occur directly under the center of the ball at a depth of 0.5a and plastic yielding will occur when $P_m = 1.1 \ Y$, where Y equals the tensile yield strength of the material.

Experimental work has shown that continued loading of the ball produces a progressively larger plastically deformed region.[1] The mean contact pressure increases and finally approaches 2.8 Y. Other experimental work on practical surfaces indicates that very many asperities are in the advanced state of plastic flow.[7] From this we may estimate the real area of contact, A, between nominally flat surfaces touching each other at asperities is approximately equal to N/3Y. For a metal with a yield strength Y = 15,000 psi, a 1-in. (2.5-cm) cube pressed with a load N as shown in the table below produces a real contact area A_r,

Load N		Contact area A_r	
lb	**kg**	**in.²**	**cm²**
10,000	4540	0.2	1.29
100	45	0.002	0.0129
1	0.45	0.00002	0.000129

The above paragraph implies that contact area increases linearly with applied load. Re-

Table 1
COEFFICIENT OF
ADHESION FOR
VARIOUS METALS

Material	Adhesion coefficient (Λ)
In air	
Steel ball on indium	1.2
Steel ball on lead	0.7
Steel ball on tin	0.4
In vacuum	
Gold on gold	0.62
Platinum on platinum	0.42
Nickel on nickel	0.30
Silver on silver	0.11

search suggests that real contact area between nominally flat surfaces increases more nearly as the 0.8 power of applied load.[7]

Adhesion and Peeling

In the above model of the elastic sphere pressing against an elastic flat plate, the radius and area of contact increase as the normal load increases. As a matter of practical experience, the area of contact also returns to 0 (point contact) as the load is decreased. From such observations it is easy to assert that there is no adhesion between surfaces. This at least has been the argument against adhesion being operative in friction. On the other hand, measurable adhesion does occur during contact between two surfaces that were vigorously cleaned in a high vacuum, which makes a total denial of adhesion untenable.

The influence of a cycle of loading and unloading of a sphere on a flat plate with and without adhesion may be seen in the illustration of a rubber ball pressed against a rigid flat surface. As each increment of load is added, a ring of larger diameter of contact forms between the ball and flat plate. The reverse occurs upon progressive removal of the load. If the flat surface were covered with a tacky substance, the increment of added load would produce increasing contact area as before, but upon decrease in load the outer ring of contact will not readily separate. A state of tension will exist across the adhesive bond. As the next increment of load reduction occurs, the second ring inward experiences higher tensile stress, etc. Finally, the normal load N may be completely removed but the ball still remains in contact with the flat surface. The stress state over the contact region is one of tension at the outer edges of contact and compression in the middle of contact to achieve static equilibrium. The compression force constitutes a recovery force and its origin is in the elastic strain field "stored" in the rubber ball.

At the outer edges of contact where the stresses are highest, there is also a sharp crack or stress concentration. Thus, the conditions are right for "peeling" or continuous fracture of adhesive bonds at the outer edge of contact. With visco-elastic materials the fracture would be time-dependent but with metals the fracture would occur progressively as the load decreases. The bonds of a ductile material do not fracture as readily as those of a brittle material, thus leaving a residual contact region. A force, $-N$, required to separate a sphere from a flat plate once N is removed, divided by N may be called the coefficient of adhesion Λ, with $\Lambda = |-N/N|$. Absolute values for various metals are shown in Table 1.

MECHANISMS OF SLIDING FRICTION

Recent Understanding

Research in the last 50 years has focused on whether friction is due to adhesion or the interlocking of asperities. The interlocking theory views surfaces as being composed of relatively rigid asperities which must follow complex paths to move around or over each other. The adhesion theory assumes that two contacting surfaces will bond or weld together and the resulting bonds must be broken for sliding to occur.

There are now two convincing arguments against the interlocking theory. First is the observation that monomolecular films of lubricants decrease the friction of the sliding pair by a factor of five or more while having a negligible effect on the size and shape of asperities. The second argument stems from the statement in the 'interlocking theory' that the coefficient of friction is related to the steepness of asperities, implying that the force to slide a body up an inclined plane has the horizontal component F. Since with continued motion the force, F, must be constantly applied, one would suppose that the upper body continues to rise and would soon be separated some distance from the lower body!

The adhesion theory has been criticized for two reasons. One is based on the belief that adhesion is a force measured normal to surfaces whereas friction is a force measured parallel to the surfaces. The second criticism arises from the common experience that surfaces are readily separated after sliding ceases, requiring no force to separate as would be required with adhesive bonding.

The modern view is that friction is primarily due to adhesion but an adhesion that is limited by the oxides and adsorbed gases found on all surfaces during sliding and destroyed by peeling when load is removed. In some instances of very rough surfaces where some of the roughness may be due to carbide particles, there may be a second component of friction due to asperity collision.

Laws of Friction

The earliest law of friction is due to Leonardo DeVinci (1452 to 1519).[8] He observed that F is proportional to N, where F is the force to initiate sliding and N is the normal force holding the surfaces together. Amontons (1663 to 1705), a French architect-engineer, in 1699 reported to the French Academy that he found F is roughly equal to N/3 and F is independent of the size of the sliding body. The specimens tested were copper, iron, lead, and wood in various combinations, and in each experiment the surfaces were coated with pork fat (suet). Amontons saw the cause of friction as the collision of surface irregularities.

Coulomb (1736 to 1806), a French physicist-engineer, supported Amontons in stating that friction is due to the interlocking of asperities. He discounted adhesion (cohesion) as a source of friction because friction was usually found to be independent of (apparent) area of contact. While Coulomb was in error in his explanation of friction and he did not improve on the findings of Amontons, yet today "dry friction" is almost universally known as "Coulomb friction". This is taken to mean simple friction, invariant with load, speed, temperature, starting rate, etc.

The investigators most commonly associated with the adhesion theory of friction are Bowden and Tabor.[1] An early model from this school began with the idea that the force of friction is the product of A_r, the summation of the microscopic areas of contact, and the shear strength, S_s, of the bond in that region; i.e., $F = A_r S_s$. To complete the model, the load, N, was thought to be borne by the tips of asperities, altogether comprising a total area of contact, A_r, multiplied by the average pressure of contact, $N = A_r P_f$, where P_f is the average pressure of contact on the asperities. Altogether, the coefficient of friction is taken as

$$\mu = F/N = A_r S_s / A_r P_f = S_s / P_f$$

S_s is usually approximately Y/2 where Y is the yield strength of the material in tension. P_f is usually no more than 3Y. Thus, the ratio S_s/P_f is about 1/6, which is not far from 0.2, a value often found in practice for "clean" metals in air. Using the best estimates for A_r and S_s, however, the closest estimate of friction is only 1/10 of the measured values. Estimation of the real area of contact is generally considered the most difficult problem in this model.

From 1938 when the above model was proposed, there have been many developments in technology, particularly in the use of vacuum equipment. In vacuum, the coefficient of friction is often seen to exceed 0.2 by a large margin and sometimes approaches 40. To explain such values and other anomalies in friction, Tabor developed a new model based on principles of biaxial stresses in metals and its influence on plastic strain of the metals.[9] Conceptually, the model of the sphere on the flat plate can be applied here. As load on the sphere increases, its contact area with the flat plate increases and the stresses pass from the elastic to the plastic regime. In the elastic regime, a superimposed shear stress on the sphere would produce an elastic shear strain in the sphere and the contact area between the sphere and flat plate would not be affected. In the plastic range, however, after a normal load is applied that produces plastic flow, a horizontal force producing a shear stress in the sphere would produce a new increment of strain in the direction of the resultant of the initial normal force and the applied shear force. Thus, the shear force causes a further normal strain in asperities with the effect of increasing the area of contact. If adhesion increases in proportion to the area of contact, the area of contact will grow in proportion to the average shear stress that can be sustained or developed at the interface between the sphere and the flat plate. The final form of the model is expressed as,

$$\mu = \frac{1}{3\sqrt{k^{-2}-1}}$$

where $k = S_i/S_s$, and S_i is the shear strength of the interface between the sphere and the flat plate. If $k = 1$ in this model, $\mu = \infty$. This corresponds to a clean surface achieved in a high vacuum. In this state, contact area increases indefinitely as a friction force is applied until the contact and adhesion area is very large. In this case, it may not be possible to separate the surfaces and this is defined as the state of seizure. Where some interruption of surface adhesion occurs, however, the value of S_i is less than S_s. The calculated values of μ for several conditions are shown in the table below.

k	μ
0.95	→1
0.8	0.45
0.6	0.25
0.1	0.03

The latest model of Tabor is not totally satisfying because of our inability to comprehend S_i in realistic terms. It may be either an average shear strength over a contact region, or the fraction of surface over which very high adhesion occurs leaving other areas to have no adhesion. Other uncertainties in the model are due to the manner in which the plastic flow properties of materials were simplified, and it does not explain the effect of surface roughness in friction. On the other hand, the interlocking theory is not aided by the frequent observation that μ increases as surface finish decreases below a roughness of 10 μin. Neither of the Tabor models or the interlocking theory explain the influence of close lateral proximity of asperities which imposes a limit on the high value of μ. This is the case in metal working where there is high-contact pressure.

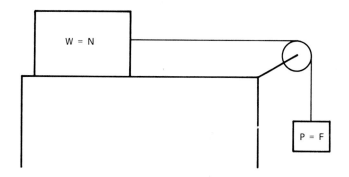

FIGURE 1. String-pulley-weight measurement of coefficient of friction.

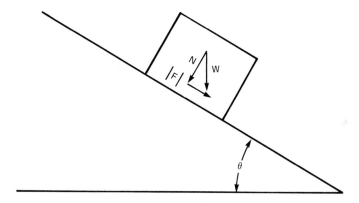

FIGURE 2. Tilting plane measurement of coefficient of friction.

COEFFICIENT OF FRICTION

Measurement of Friction

Measurement of the coefficient of friction involves two quantities, namely F, the force required to initiate and/or sustain sliding, and N, the normal force holding two surfaces together. Some of the earliest measurements of the coefficient of friction were done by an arrangement of pulleys and weights as shown in Figure 1. Weight P_s is applied to the pan until sliding begins and one obtains the static, or starting, coefficient of friction with $\mu_s = P_s/N$. If the kinetic coefficient of friction μ_k is desired, a weight is applied to the string and the slider is moved manually and released. If sliding is not sustained, more weight is applied to the string for a new trial until sustained sliding of uniform velocity is observed. In this case, the final weight P_k is used to obtain $\mu_k = P_k/N$.

A second convenient system for measuring friction is the inclined plane shown in Figure 2. The measurement of the static coefficient of friction consists simply in increasing the angle of tilt of the plane to θ when the object begins to slide down the inclined plane. By simple trigonometric relations,

$$F/N = W \sin \theta / W \cos \theta = \tan \theta = \mu$$

If the kinetic coefficient of friction is required, the plane is tilted and the slider is advanced manually. When an angle, θ, is found at which sustained sliding of uniform velocity occurs, $\tan \theta$ is the kinetic coefficient of friction.

As technology developed, it became possible to measure the coefficient of friction to a

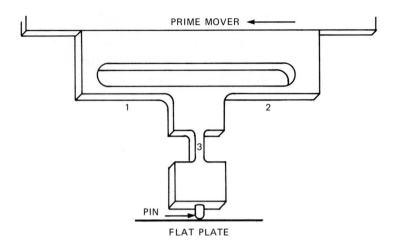

FIGURE 3. One-piece device for measuring pin-on-flat coefficient of friction. Strain gages on flexible sections 1 and 2 measure normal force; strain gage at 3 measures friction force by bending of the beam.

high accuracy under a wide range of conditions. Force measuring devices for this purpose range from the simple spring scale to devices that produce an electrical signal in proportion to an applied force. The principle of the instrumented devices is similar to the spring scale in measuring the elastic deflection of machine elements due to friction forces and normal forces on the sliding pair. The deflection can be measured by strain gages, capacitance sensors, inductance sensors, piezoelectric materials, optical interference, acoustic emission, moire fringes, light beam deflection, and several other methods. The most widely used because of its simplicity and reliability is the strain gage system.

Just as there are many sensing systems available, there are also many designs of friction measuring devices.[10] The unit shown in Figure 3 is attractive because of its simplicity. It is attached to a prime mover which moves horizontally and may be adjusted vertically to load the pin against the flat. Strain gages are attached to horizontal flexible sections 1 and 2 to measure the normal force between the pin specimen and the flat plate. Strain gages attached to vertical flexible section 3 measure friction force by bending of the beam. Designs incorporating the principle of Figure 3 are usually favored in complex, automatically controlled machinery. The chief disadvantages of this design are (1) the skill required both to calibrate the instrument and to maintain it, and (2) the inevitable interaction or ''cross talk'' between the two force-measuring signals.

A more complex system which requires less skill to operate is shown in Figure 4. It is composed of two parts. Part A can rotate about bearing G in a horizontal plane but is constrained by a wire between cantilevers x and y. Part B is attached to part A by bearing H on a horizontal axis. A slider test pin is inserted in body B. When the prime mover is moved vertically downward, the pin presses the flat plate tending to rotate body B in a clockwise direction which bends cantilever w. With strain gages attached to cantilever w, the vertical force on the pin may be measured. Motion of the prime mover to the left tends to rotate the pin about bearing G. Strain gages on cantilever x measure the force of friction of the pin against the flat plate.

The design shown in Figure 4 avoids the interaction between force signals, which plagues the design of Figure 3. The two-part design also is nearly insensitive to the amount of extension of the pin specimen, which is convenient for setup. In addition, wire z in Figure 4 can be removed and the vertical loading on the pin can be conveniently effected by dead weights. The above designs are a few of many in use. Frequently, it is more convenient to use two flat surfaces, a shaft in a bearing, or three pins instead of one.

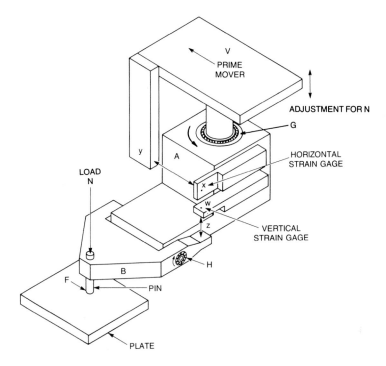

FIGURE 4.   Sketch of two-component force-measuring system with attached transducers and pin sliding on flat plate.

Data obtained from friction measuring devices are usually not easy to interpret. For some
sliding pairs a smooth force trace may be obtained on recorder strip chart but frequently the
friction force will drift or wander inexplicably. In other instances, where a flat plate rotates
for example, a repeatable behavior may be found during repeat rotations of the flat plate.
These variations may exceed 10 or 20% of the average force trace (particularly when there
is stick-slip) and have often been explained in terms of the stochastic or statistical nature of
friction. Variations are usually largest when small values of N are used and are reduced at
high values of N, where contact pressures approach the state of fully developed plastic flow.
Analysis of strip chart records is further discussed below.

General Frictional Behavior and Influence of Variables

Almost all operating parameters (speed, load, etc.) will influence the coefficient of friction.
Some of the variables and their general effects are listed below.

Sliding speed — For metals and other crystalline solids sliding on like materials, the
behavior is as shown in Figure 5.[1] The sliding speeds indicated range from imperceptibly
slow (the tip of the minute hand on a watch moves at about 10^{-3} cm/sec) to normal walking
speed ($\sim$125 cm/sec) which covers many practical conditions. At very high-sliding speed
(>2500 cm/sec) surface melting may occur to produce a low coefficient of friction. Some
polymers behave as shown in Figure 6[11] for the coefficient of friction of a steel sphere sliding
on PTFE and Nylon 6-6. Note the variation for PTFE, which is usually thought to have a
low and constant coefficient of friction. The coefficient of friction of both polymers increases
with sliding speed over a limited range of speed because sliding invokes a visco-elastic
response in the materials. Few materials slide at higher speeds in practice, producing such
a high coefficient of friction, as tire rubber on most road surfaces. Typical data are shown
in Figure 7[12] for both dry roads and roads wetted by a moderate rainfall.

Temperature — There is usually little effect on the coefficient of friction of metals until

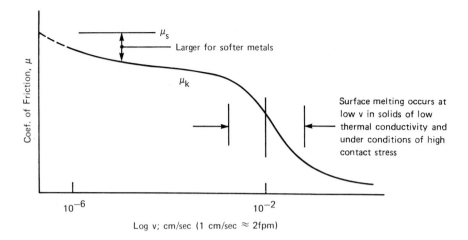

FIGURE 5.   General effect of sliding speed on coefficient of friction for metals and other crystalline solids (e.g., ice).

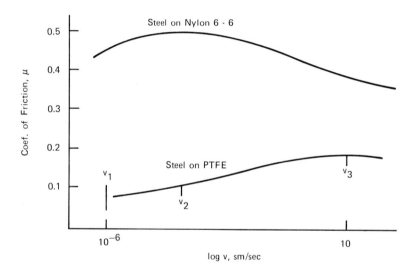

FIGURE 6.   Influence of sliding speed on coefficient of friction of a steel sphere sliding on PTFE and Nylon 6-6.

the temperature becomes high enough to increase the oxidation rate (which usually decreases μ). Increased temperature will lower the sliding speed at which surface melting occurs (see Figure 5) and increased temperature will shift the curve of coefficient of friction vs. sliding speed to a higher sliding speed in many plastics (see Figure 6).

Starting rate — Rapid starting from standstill is sometimes reported to produce a low initial coefficient of friction. In many such instances, the real coefficient of friction may be obscured by dynamic effects of the systems.

Applied load or contact pressure — In the few instances that the coefficient of friction is reported over a large range of applied load, three principles may be seen in Figure 8.[1] The first is that the coefficient of friction normally decreases as the applied load increases. For clean surfaces, as shown by curve 'a', values of μ in excess of 20 are reported at low load, decreasing to about 0.5 at high loads. An old theory suggests that the ultimate effect of increasing the contact pressure between clean surfaces is to effect adhesive bonding over

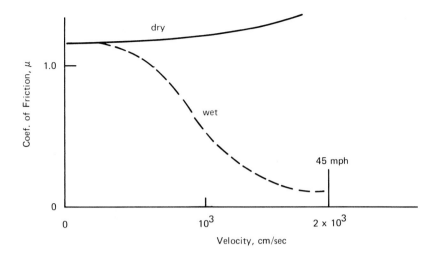

FIGURE 7. Coefficient of friction of tire rubber on asphaltic concrete road surface, both dry and when wetted by a moderate rainfall.

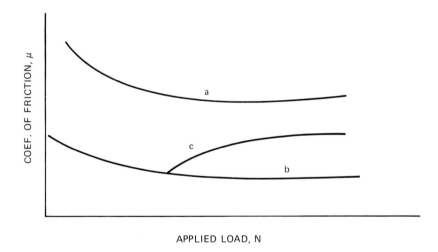

FIGURE 8. Coefficient of friction vs. applied normal load. Curve "a" applies to very clean surfaces, "b" to more practical surfaces, and "c" to materials covered with brittle oxides that flake off when their substrate is plastically deformed.

the entire macroscopic area and cause plastic shear flow at the original surface. In such a case the stress to shear the interface is the shear strength of the material which is usually taken to be about half of the compressive yield strength, thus F/N = 1/2. Practical surfaces, as represented by curve 'b', usually have a coefficient of friction less than 1/2 because surface contaminants prevent or limit adhesion. If the surface species include a brittle oxide, chipping off of oxide can expose clean substrate surfaces which increases local adhesion to cause higher coefficients of friction as shown in curve 'c'.

Surface roughness — This usually has little or no consistent effect on the coefficient of friction of clean, dry surfaces. Rough surfaces usually produce higher coefficients of friction in lubricated systems, particularly with soft metals where lubricant films are very thin as compared with asperity height.

Wear rate — One of the few consistent examples relating high coefficient of friction with surface damage is the case of scuffing. Galling and scoring also produce a high

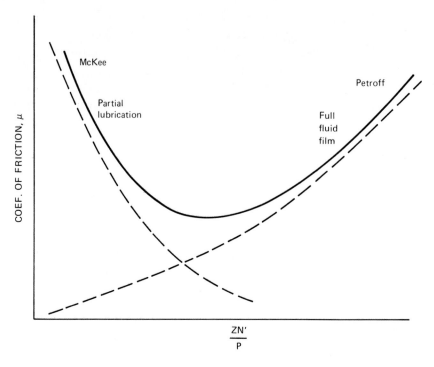

FIGURE 9.   Coefficient of friction pattern for a typical lubricated contact. Z is lubricant viscosity, N′ shaft speed, and P the unit load transferred radially by the shaft to the bearing.

coefficient of friction usually accompanied by a *severe* rearrangement of surface material with little loss of material. In most other sliding pairs there is no connection between the coefficient of friction and wear rate.

Static and Kinetic Friction

The force required to begin sliding is usually greater than the force required to sustain sliding. For dry surfaces the reason for the starting (or static) coefficient of friction being larger than the sliding (or kinetic) coefficient of friction may most simply be explained in terms of the adhesion of asperities. It is often found that the static coefficient of friction increases with time of standing. This suggests diffusion bonding of the points of contact which progresses with time. Sustained sliding could be viewed as providing a very short standing time of one asperity upon another. This should also produce a decrease in the coefficient of friction as the sliding speed increases, which is found in many systems.

When a hard sphere slides on some plastics, the frictional behavior is such as to require a new definition of static friction. For example, for a sphere of steel sliding on Nylon 6-6 the coefficient of friction at 60°C varies with sliding speed as shown in Figure 6. The "static" coefficient of friction is lower than that at v_2. Most observers would, however, measure the value of μ at v_2 as the static value of μ. The reason is that v_1 in the present example is imperceptibly slow. The coefficient of friction at the start of visible sliding at v_2 is higher than at v_3. In this case it may be useful to define the *starting* coefficient of friction as that at v_2 and the *static* coefficient of friction as that at or below v_1.

In lubricated systems the starting friction is often higher than the kinetic friction. When the surfaces slide, lubricant is dragged into the contact region and separates the surfaces. This will initially lower the coefficient of friction, but at a still higher sliding speed there is a viscous drag which again causes an increase in coefficient of friction as shown in Figure 9. This McKee-Petroff curve is typical for a shaft rotated in a sleeve bearing. The abscissa is given in units of ZN'/P where Z is the viscosity of the lubricant, N' is the shaft rotating speed, and P is the load transferred radially from the shaft to the bearing.

ROLLING FRICTION

The force required to initiate rolling motion may be larger than the force to maintain motion if the contacting surfaces are very rough. Sustained rolling motion requires very little force, usually about 0.01 times that for unlubricated sliding.

There are at least three causes for rolling resistance. The first arises from the strains within each of the solid bodies in the region of contact. During rolling a point in each body passes through complex strain cycles; since energy is lost during a cycle of strain in all materials, energy must be supplied to sustain rolling.

The second reason for rolling friction is due to differences in distortion of the contacting bodies. This can be seen by pressing the eraser of a pencil into the palm of the hand. During indentation the skin of the hand stretches, in the contact region as well as outside of it, more than the eraser increases in size. Thus, there is relative slip between the eraser and the hand. The same occurs between a ball and flat surfaces and the net effect is that energy is expended in rolling. The effect of the micro-slip can be decreased by lubrication.

A third reason for rolling friction may be that the rolling bodies are not moving in the direction of the applied force. A misaligned roller slides axially to some extent and a poorly guided ball spins about the contact region. Again, lubrication will reduce the energy loss due to slip.

Frictional resistance of ball and roller bearing assemblies is usually much greater than the rolling resistance of simple rolling elements because of the cages, grooves, and shoulders intended to control the travel of the balls or rollers.

Tapping and Jiggling to Reduce Friction

One of the practices in the use of instruments is to tap and/or jiggle to obtain accurate readings. There are two separate effects. One effect is achieved by tapping the face of a meter or gage, which may cause the sliding surfaces in the gage to separate momentarily, reducing friction resistance to zero. The sliding surfaces (shafts in bearings, or racks on gears) will advance some distance before contact between the surfaces is reestablished. Continued tapping will allow the surfaces to progress until the force to move the gage parts is reduced to zero.

Jiggling is best described by using the example of a shaft advanced axially through an O-ring. Such motion requires the application of a force to overcome friction. Rotation of the shaft also requires overcoming friction, but rotation reduces the force required to effect axial motion. In lubricated systems the mechanism may involve the formation of a thick fluid film between the shaft and the O-ring. In a dry system an explanation may be given in terms of *components* of *forces*. Frictional resistance force usually acts in the exact opposite direction as the direction of relative motion between sliding surfaces. If the shaft is rotated at a *moderate* rate, there will be very little frictional resistance to *slow* axial motion. In some apparatus the shaft is rotated in an oscillatory manner to avoid difficulties due to anisotropic (grooved) frictional behavior. Such oscillatory rotation may be referred to as jiggling, fiddling, or coaxing.

STICK-SLIP MOTION

Principles of Stick-Slip

Some sliding systems vibrate. Vibration can be a mere annoyance such as the squeal of automobile brakes or door hinges. Other vibrations serve to notify of abnormal conditions,

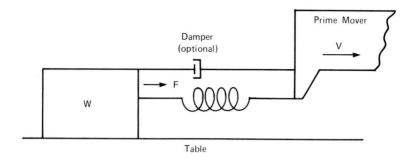

FIGURE 10.   Simplified model of vibrating sliding system.

such as in the squeal of automobile tires and unlubricated electric motors. In other instances, vibration may compromise the function of a machine. In machine tools, surface finish and shape of final parts are affected by the vibration of tool carriers.[13]

Vibration of sliding systems is usually described as stick-slip, frictional vibration, or frictional oscillation. In the simple model of Figure 10, an object of weight W is connected to a prime mover by a spring. As the prime mover moves at constant speed in the direction shown, the spring stretches until it applies a force to W that will initiate sliding. If the coefficient of friction remains constant after sliding begins, the weight W will advance at the same speed as the prime mover. If on the other hand, the coefficient of friction decreases after sliding begins, less force will be required to sustain sliding than the spring force. Weight W will, therefore, accelerate, shortening the spring, and finally overshooting the equilibrium position. The spring then exerts a force less than that required to sustain sliding so the weight decelerates and may even stop. After deceleration, its velocity is lower than before and the coefficient of friction may increase. To meet the increased force required to sustain or reinitiate sliding, the spring must stretch. This produces a never ending cycle with the weight advancing by a series of fast and slow segments of motion.

An experimental trace of the force exerted by the spring will show interesting differences depending upon the speed of the prime mover as shown in Figure 11. The upper trace for slow velocity shows true stick-slip as the force drops to nearly zero as the object comes to a standstill and the prime mover then advances to stretch the spring once more. In the second trace where the velocity is moderately high, force variations are smaller and the weight oscillates between two limiting sliding speeds in frictional oscillation.

In the simplest approach, where the coefficient of friction between the sliding object and the table is taken to be μ_k when sliding occurs and μ_s to start sliding, and where $\mu_s > \mu_k$, motion of the weight would follow simple laws of dynamics. Thus, one could reasonably expect that the frequency of frictional oscillations would be low at low speeds of the prime mover, with a large weight, and with a flexible or compliant spring. The frictional oscillations would diminish at high speeds of the prime mover, with small weights and with stiff springs, producing the lower trace of Figure 11.

It is of great commercial interest to design sliding systems to eliminate or minimize vibration. In general, the larger the difference $(\mu_s - \mu_k)$, the more likely a system will oscillate. The transition from μ_s to μ_k is influenced by the surface finish of the sliding parts and by the physical and chemical nature of the lubricant. In general, μ_k rises as velocity decreases, as lubricant viscosity increases, as chemical reactivity of lubricant with surfaces increases, and as surface finish decreases.[13]

In machine design it may be possible to stiffen the connection between the prime mover and sliding object, to reduce the weight of the sliding object, or to provide a thick fluid film. An additional design consideration is that frictional oscillation can produce pitching and yawing motion of the moving element if the driving force is applied at a different plane

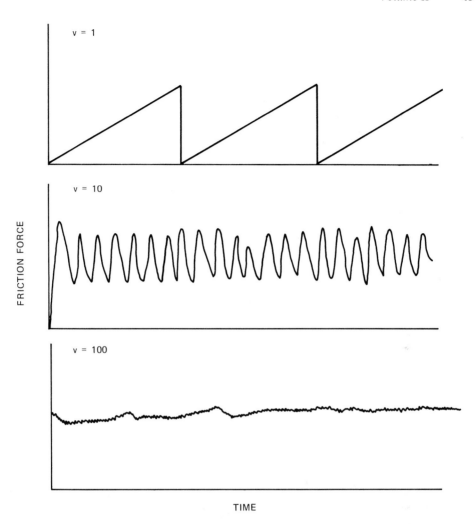

FIGURE 11.   Example of recorder strip chart traces for friction force at three sliding speeds of relative magnitude 1, 10, and 100.

from that in which the frictional resistance is located. If the sliding force cannot be applied on the same plane, long slide ways may be required to minimize pitching and yawing. In lubricated systems, additional damping capacity in the fluid film decreases oscillation amplitude and may even eliminate squeal.

Analysis of Strip Chart

When friction is measured in the laboratory, frictional oscillation is often encountered because of the tendency to construct a very flexible friction-measuring apparatus. Several traces may be obtained as shown in Figure 11. Some estimate of the coefficient of friction is often attempted from the records.

The value of μ_s may be obtained from the maximum force measured when slip starts. The shape of the curve prior to the maximum reflects only the system stiffness and speed of the prime mover. When slip begins, the "slip" portion is usually not recorded in sufficient detail to determine μ_k. It is incorrect to assume that μ_k is the average of peaks and minima in the excursions because in traces such as for $v = 1$, μ_k would be $\mu_s/2$.

A common trace is of the type given for $v = 100$ in Figure 11. In general, for small oscillations, μ_k may be taken as the average of the trace. Where excursions are greater than

Table 2
COEFFICIENTS OF STATIC AND SLIDING FRICTION

Materials	Static		Sliding	
	Dry	Greasy	Dry	Greasy
Hard steel on hard steel	0.78(1)	0.11(1,*a*)	0.42(2)	0.029(5,*h*)
		0.23(1,*b*)		0.081(5,*c*)
		0.15(1,*c*)		0.080(5,*i*)
		0.11(1,*d*)		0.058(5,*j*)
		0.0075(18,*p*)		0.084(5,*d*)
		0.0052(18,*h*)		0.105(5,*k*)
				0.096(5,*l*)
				0.108(5,*m*)
				0.12(5,*a*)
Mild steel on mild steel	0.74(19)		0.57(3)	0.09(3,*a*)
				0.19(3,*u*)
Hard steel on graphite	0.21(1)	0.09(1,*a*)		
Hard steel on Babbitt (ASTM 1)	0.70(11)	0.23(1,*b*)	0.33(6)	0.16(1,*b*)
		0.15(1,*c*)		0.06(1,*c*)
		0.08(1,*d*)		0.11(1,*d*)
		0.085(1,*e*)		
Hard steel on Babbitt (ASTM 8)	0.42(11)	0.17(1,*b*)	0.35(11)	0.14(1,*b*)
		0.11(1,*c*)		0.065(1,*c*)
		0.09(1,*d*)		0.07(1,*d*)
		0.08(1,*e*)		0.08(11,*h*)
Hard steel on Babbitt (ASTM 10)		0.25(1,*b*)		0.13(1,*b*)
		0.12(1,*c*)		0.06(1,*c*)
		0.10(1,*d*)		0.055(1,*d*)
		0.11(1,*e*)		
Mild steel on cadmium silver				0.097(2,*f*)
Mild steel on phosphor bronze			0.34(3)	0.173(2,*f*)
Mild steel on copper lead				0.145(2,*f*)
Mild steel on cast iron		0.183(15,*c*)	0.23(6)	0.133(2,*f*)
Mild steel on lead	0.95(11)	0.5(1,*f*)	0.95(11)	0.3(11,*f*)
Nickel on mild steel			0.64(3)	0.178(3,*x*)
Aluminum on mild steel	0.61(8)		0.47(3)	
Magnesium on mild steel			0.42(3)	
Magnesium on magnesium	0.6(22)	0.08(22,*y*)		
Teflon on Teflon	0.04(22)			0.04(22,*f*)
Teflon on steel	0.04(22)			0.04(22,*f*)
Tungsten carbide on tungsten carbide	0.2(22)	0.12(22,*a*)		
Tungsten carbide on steel	0.5(22)	0.08(22,*a*)		
Tungsten carbide on copper	0.35(23)			
Tungsten carbide on iron	0.8(23)			
Bonded carbide on copper	0.35(23)			
Bonded carbide on iron	0.8(23)			
Cadmium on mild steel			0.46(3)	
Copper on mild steel	0.53(8)		0.36(3)	0.18(17,*a*)
Nickel on nickel	1.10(16)		0.53(3)	0.12(3,*w*)
Brass on mild steel	0.51(8)		0.44(6)	
Brass on cast iron			0.30(6)	
Zinc on cast iron	0.85(16)		0.21(7)	
Magnesium on cast iron			0.25(7)	
Copper on cast iron	1.05(16)		0.29(7)	
Tin on cast iron			0.32(7)	
Lead on cast iron			0.43(7)	
Aluminum on aluminum	1.05(16)		1.4(3)	

Table 2 (continued)
COEFFICIENTS OF STATIC AND SLIDING FRICTION

Materials	Static		Sliding	
	Dry	Greasy	Dry	Greasy
Glass on glass	0.94(8)	0.01(10,*p*)	0.40(3)	0.09(3,*a*)
		0.005(10,*q*)		0.116(3,*v*)
Carbon on glass			0.18(3)	
Garnet on mild steel			0.39(3)	
Glass on nickel	0.78(8)		0.56(3)	
Copper on glass	0.68(8)		0.53(3)	
Cast iron on cast iron	1.10(16)		0.15(9)	0.070(9,*d*)
				0.064(9,*n*)
Bronze on cast iron			0.22(9)	0.077(9,*n*)
Oak on oak (parallel to grain)	0.62(9)		0.48(9)	0.164(9,*r*)
				0.067(9,*s*)
Oak on oak (perpendicular)	0.54(9)		0.32(9)	0.072(9,*s*)
Leather on oak (parallel)	0.61(9)		0.52(9)	
Cast iron on oak			0.49(9)	0.075(9,*n*)
Leather on cast iron			0.56(9)	0.36(9,*t*)
				0.13(9,*n*)
Laminated plastic on steel			0.35(12)	0.05(12,*t*)
Fluted rubber bearing on steel				0.05(13,*t*)

Note: Reference letters indicate the lubricant used; numbers in parentheses give sources (see References).

Key to Lubricants Used:

a	=	oleic acid	*m* =	turbine oil (medium mineral)
b	=	Atlantic spindle oil (light mineral)	*n* =	olive oil
c	=	castor oil	*p* =	palmitic acid
d	=	lard oil	*q* =	ricinoleic acid
e	=	Atlantic spindle oil plus 2% oleic acid	*r* =	dry soap
f	=	medium mineral oil	*s* =	lard
g	=	medium mineral oil plus $1/2$% oleic acid	*t* =	water
h	=	stearic acid	*u* =	rape oil
i	=	grease (zinc oxide base)	*v* =	3-in-1 oil
j	=	graphite	*w* =	octyl alcohol
k	=	turbine oil plus 1% graphite	*x* =	triolein
l	=	turbine oil plus 1% stearic acid	*y* =	1% lauric acid in paraffin oil

REFERENCES

(1) Campbell *Trans. ASME*, 1939; (2) Clarke, Lincoln, and Sterrett *Proc. API*, 1935; (3) Beare and Bowden *Phil. Trans. Roy. Soc.*, 1935; (4) Dokos, *Trans. ASME*, 1946; (5) Boyd and Robertson, *Trans. ASME*, 1945; (6) Sachs, *zeit. f. angew. Math. und Mech.*, 1924; (7) Honda and Yama 1a, *Jour. I. of M*, 1925; (8) Tomlinson, *Phil. Mag.*, 1929; (9) Morin, *Acad. Roy. des Sciences*, 1838; (10) Claypoole, *Trans. ASME*, 1943; (11) Tabor, *Jour. Applied Phys.*, 1945; (12) Eyssen, General Discussion on Lubrication, *ASME*, 1937; (13) Brazier and Holland-Bowyer, General Discussion on Lubrication, *ASME*, 1937; (14) Burwell, *Jour. SAE*, 1942; (15) Stanton, "Friction", Longmans; (16) Ernst and Merchant, Conference on Friction and Surface Finish, M.I.T., 1940; (17) Gongwer, Conference on Friction and Surface Finish, M.I.T., 1940; (18) Hardy and Bircumshaw, *Proc. Roy. Soc.*, 1925; (19) Hardy and Hardy, *Phil. Mag.*, 1919; (20) Bowden and Young, *Proc. Roy. Soc.*, 1951; (21) Hardy and Doubleday, *Proc. Roy. Soc.*, 1923; (22) Bowden and Tabor, "The Friction and Lubrication of Solids", Oxford; (23) Shooter, *Research*, 4, 1951.

From *Standard Handbook for Mechanical Engineers*, 7th ed., Baumeister, T., Ed., McGraw-Hill, New York, 1967. With permission.

about 20% of the midpoint value, averaging must be done with caution. Trace averaging can be aided by using a parallel plate (noncontacting) viscous damper to diminish oscillations during tests, as shown in Figure 10.

TABLES OF COEFFICIENT OF FRICTION

The coefficient of friction is not an intrinsic property of a material or combinations of materials. Rather the coefficient of friction varies with changes in humidity, gas pressure, temperature, sliding speed, and contact pressure. It is different for each lubricant, for each surface quality, and for each shape of contact region. Furthermore, it changes with time of rubbing, and with different duty cycles. Very few materials and combinations have been tested over a wide range of more than three or four variables, and then they are usually tested in laboratories using simple geometries. Thus, it is rarely realistic to use a general table of values of coefficient of friction as a source of design data. Information such as that in Table 2 may provide guidelines,[14] but where a significant investment will be made or high reliability must be achieved, the friction should be measured using a prototype device under design conditions.

REFERENCES

1. **Bowden, F. E. and Tabor, D.,** *The Friction and Lubrication of Solids,* Oxford University Press, Vols. I and II, London, 1954 and 1964.
2. **Buckley, D. H.,** *Surface Effects in Adhesion, Wear and Lubrication,* Elsevier, Amsterdam, 1981.
3. **Barwell, F. T.,** *Bearing Systems: Principles and Practice,* Oxford University Press, London, 1979.
4. **Ling, F. F., Klaus, E. E., and Fein, R. S., Eds.,** *Boundary Lubrication. An Appraisal of World Literature,* American Society of Mechanical Engineers, New York, 1969.
5. **Peterson, M. B., Ed.,** *Wear Control Handbook,* American Society of Mechanical Engineers, New York, 1980.
6. **Timoshenko, S. and Goodier, J. N.,** *Theory of Elasticity,* 2nd ed., McGraw-Hill, New York, 1951.
7. **Greenwood, J. A. and Williamson, J. B. P.,** Contact of nominally flat surfaces, *Proc. R. Soc. (London),* A295, 300, 1966.
8. **Dowson, D.,** An interesting account of the life and times of 23 prominent figures in the field of tribology, *J. Lubr. Technol.,* 99, 382, 1977; *J. Lubr. Technol.,* 100, 2, 1978.
9. **Tabor, D.,** Junction growth in metallic friction, *Proc. R. Soc., (London),* A251, 378, 1959.
10. **Benzing, R., Hopkins, V., Petronio, M., and Villforth, F., Jr.,** *Friction and Wear Devices,* 2nd ed., Americal Society of Lubrication Engineers, Park Ridge, Ill., 1976.
11. **Ludema, K. C. and Tabor, D.,** The friction and visco-elastic properties of polymeric solids, *Wear,* 9, 329, 1966.
12. **Yeager, R. W.,** Tire hydroplaning, in *The Physics of Tire Traction,* Hayes, D. F. and Browne, A. L., Eds., Plenum Press, New York, 1974, 25.
13. **Kato, S., Yamaguchi, K., Matsubayashi, T., and Sato, N.,** Stick-Slip motion and characteristics of friction in machine tool slideway, *Nagoya Univ.* 27, 1, 1975.
14. **Fuller, D. D.,** Friction, *Marks' Standard Handbook for Mechanical Engineers,* 8th ed., Baumeister, T., Ed., McGraw-Hill, New York, 1978.

BOUNDARY LUBRICATION

Richard S. Fein

CHARACTERISTICS OF BOUNDARY LUBRICATION

Boundary lubrication is defined by OECD as *a condition of lubrication in which the friction and wear between two surfaces in relative motion are determined by the properties of the surfaces, and the properties of the lubricant other than bulk viscosity.* Boundary lubrication also may be defined in terms of contrast with full-fluid film lubrication where load-bearing surfaces are completely separated and the load is supported entirely by pressure in the fluid film.

Usually, boundary lubrication is associated with some load support by interaction of asperities on the bearing surfaces or of the surfaces with solid particles in the fluid. In the extreme, asperity and/or particle interactions support all of the load and friction appears to be independent of bulk fluid viscosity.[1]

Friction and Wear Phenomena

Friction and wear under boundary lubrication conditions often approximately obey rather simple "laws" over considerable ranges of operating and machine configuration conditions. For friction, the Amonton-Coulomb law states that the coefficient of friction, the ratio of the friction force to the load, is independent of load and of apparent area of contact.[1]

For wear, the volume of material worn from a surface is proportional to the product of the load and sliding distance divided by the hardness of the material being worn.[2] The proportionality constant is the "wear coefficient", k,

$$k = \frac{v \cdot H}{W \cdot \ell} = \frac{d \cdot H}{p \cdot \ell} \cdot \frac{A_w}{A} \tag{1}$$

in which v is wear volume, H is hardness, W is normal load, l is distance slid, d is depth of wear, P is normal pressure, A_w is area being worn, and A is area of apparent load support. For surfaces which are not in continuous "contact" and the sliding and relative motion are parallel:

$$k = \frac{d \cdot L \cdot H}{W} \cdot \frac{1}{n} \cdot \frac{U}{|U_s|} \tag{2}$$

in which L is face width of line contact, n is number of passes through load support area, U is sweep velocity through the area, and U_s is sliding velocity.

For the more general noncontinuous contact case,

$$k = \frac{d \cdot H}{P \cdot c} \cdot \frac{1}{n} \cdot \frac{U}{|U_s|} \tag{3}$$

in which c is width of load support area in the direction of motion and P is the mean normal pressure.

Since the wear coefficient, as shown in Table 1, ranges over at least a hundred- to million-fold range, it is convenient to transform it to the AntiWear Number (AWN) defined as

$$AWN = - \log_{10}k \tag{4}$$

Table 1
COMPARISON OF RANGES OF
FRICTION AND WEAR

Lubricant/atmosphere	Approximate friction coefficient	Approximate wear coefficient	AWN
Dry argon	0.5	10^{-2}	2
Dry air	0.4	10^{-3}	3
Isoparaffin (C_8)/air	0.3	10^{-5}	5
Isoparaffin (C_{30})/air	0.12	10^{-7}	7
Aromatic oil/air	0.06	10^{-8}	8
C_{18} Fatty acid in C_{30} iso-paraffin/air	0.08	10^{-9}	9
Engine oil/air	—	$<2 \times 10^{-10}$	>9.7

Note: 52100 Steel — four-ball machine — 230 to 690 mm/sec — 20 to 50 kg.

While high wear coefficients can be tolerated for an industrial tool such as a wrench, gears and many other machine elements require quite low wear coefficients achieved by means of overt lubrication.[3] Often this depends on physical and chemical characteristics of the lubricant other than viscosity. Friction appears to be of direct importance only when (1) the heating is important, (2) energy consumption by friction is significant, (3) mechanical driving force is limited, (4) friction provides the source of mechanical drive for one of the surfaces, and/or (5) cutting or forming metals and other materials occurs. The first three cases usually require lowering of friction while the latter two cases often require an optimization of friction properties.

An example of a metal forming process made possible by lubricants providing optimized friction is shown in Figure 1. This illustrates one stage in the drawing and ironing of aluminum or steel two-piece can bodies. The side wall of a can body is being "ironed" (thinned) as the punch carries the body through the die. Friction between the punch and body must be sufficiently high to pull the metal through the die. On the other hand, friction must be sufficiently low between the die and body to prevent excessive tensile stresses from tearing the can body outside of the die-punch nip. Friction also must be sufficiently low between the punch and body to permit stripping of the body from the punch at the end of its stroke. Normal forces (σ_P and σ_D) between the punch, body, and die depend on the friction forces because plastic flow of the thinning body sidewall depends on the resultant stress field at each point in the workpiece material. Such interactions between normal forces and friction are inherent to cutting and forming operations; friction that is too low often can be as harmful as friction that is too high.

Transitions and Operational Severity

The "laws" of friction and wear commonly hold approximately over considerable ranges of operating conditions for a pair of surfaces. The ranges, however, are delimited by abrupt "transitions" in friction coefficient, wear coefficient, or surface texture. As a consequence, a small change in operating conditions sometimes leads to drastic overheating or catastrophic surface damage.

Figure 2 illustrates the transitions with gear test data for three lubricants.[4] Load is given in terms of work transmitted during a fixed period of operation at each load. Note that wear increases approximately in proportion to load with the straight mineral oil below the load equivalent to 13.4 MJ (5 hp/hr), the mild extreme pressure (EP) oil below 161 MJ (60 hp/hr), and the strong EP oil at all loads tested. This proportionality indicates that the wear

DIE–PUNCH CONJUNCTION

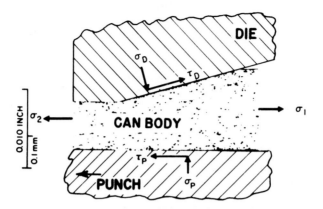

FIGURE 1.   Ironing of can wall in two-piece can manufacture.

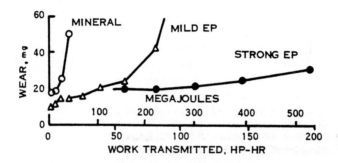

FIGURE 2.   Effect of load on wear, FZG Gear Test A, 2175 pinion speed. (Redrawn from Nieman, G., Rettig, H., and Lechner, G., *ASLE Trans.*, 4, 71, 1961.)

coefficient is relatively constant. Above 13.4 MJ for the straight mineral oil and 161 MJ for the mild EP oil, the rapid increase in wear corresponds to a wear coefficient about a hundred times greater than that at low load. Note that strong EP chemical additives increased the load-carrying capacity over 20-fold.

Operating conditions that produce a transition are commonly expressed in terms of a "severity parameter". Common parameters are load P, velocity V, PV, fPV, a measured temperature in the oil or machine, calculated conjunction temperature for the load support area, or the reciprocal of the fluid "film thickness parameter" ($1/\Lambda$). These measures all reflect mechanical or thermal stress. In addition, severity often seems to increase with decreased lubricant viscosity grade, decreased speed, increased slide-roll ratio (U_s/U), increased surface hardness, and increased asperity tip curvature ($1/r$).

Run-In

During initial sliding or rolling between a pair of surfaces, friction and wear coefficients as well as surface texture change towards a steady-state determined by operating conditions, the original surface configurations and textures, lubrication, and all other environmental and material parameters involved. Run-in is crucial since excessive friction and wear or catastrophic surface damage are much more likely initially than after the surfaces have reached a steady state. Effective run-in involves alleviation by wear of misalignment and dimensional or other errors in geometry, and generates protective surface films that shear and wear sacrificially and prevent surface damaging transitions.

During run-in, friction and wear coefficients often change approximately exponentially from beginning values f_b and k_b to steady-state values of f_{ss} and k_{ss}. Instantaneous values at any sliding distance, ℓ, tend to follow the equations

$$f = f_{ss}\left[1 + \left(\frac{f_b}{f_{ss}} - 1\right) \cdot \exp\left(\frac{\ell}{\ell_b}\right)\right] \tag{5}$$

and

$$k = k_{ss}\left[1 + \left(\frac{k_b}{k_{ss}} - 1\right) \cdot \exp\left(\frac{\ell}{\ell_b}\right)\right] \tag{6}$$

Break-in sliding distance, ℓ_b, typically may be in the order of a kilometer with an effective boundary lubricant, negligible load support by a fluid film, and negligible geometric errors in the sliding parts. This distance for the cylinder liner on a passenger car corresponds to the order of 1000 km of travel. For a typical spur gear at 30 Hz (1800 rpm) it corresponds to a day of operation.

Any factor which increases operational severity generally decreases the break-in sliding distance. Usually, shortest run-ins involve operating conditions as near as is safely possible to a wear or surface damage transition. As a consequence, run-ins often use low viscosity oils at low speed and high load. Low speed has the dual effect of increasing severity by decreasing fluid film formation and of preventing catastrophic frictional heating if a friction transition is encountered.

Common laboratory load-carrying and friction tests operate in the early stages of run-in and measure the rapidity of lubricant response to operational severity. By contrast, laboratory wear tests often are sufficiently long that steady-state conditions are approached.

BOUNDARY LUBRICANTS

Boundary films occur on almost all surfaces because they reduce the surface energy and are thus thermodynamically favored. Normally, air covers any surface with an oxidized layer plus adsorbed moisture and organic material. If surfaces slide while covered with a lubricant film, all elements in the lubricant appear in a chemically reacted boundary film along with the bearing material elements and atmospheric oxygen. Thicknesses of surface films may range from tenths of a nanometer (a few hundredths of a microinch) for single molecular layers of physically adsorbed gases to a few micrometers (several dozen microinches) for thick chemically reacted boundary films from oils with EP additives.

Gases

Inadvertent lubrication by air is the most common boundary lubrication. Among components listed in Table 2, oxygen and/or water vapor are necessary for satisfactory dry sliding of most metal surfaces. Without their adsorption and/or chemical reaction films, catastrophic levels of friction, wear, and surface damage usually occur. Hydrocarbons are also generally helpful.

A gaseous component is capable of covering the surface with a monolayer in about 1/(concentration, ppb) sec. Thus, oxygen alone could cover a clean surface in about 5 nsec ($1/0.209 \times 10^9$ ppb), while sulfur dioxide alone would require about a second. Nitrogen molecules generally do not stick well to a clean surface. With clean air, the first molecular layer formed on a clean metal surface would be expected to have a composition approximating the distribution of the strongly adsorbing/reacting components.

Table 2
ADSORBING/REACTING COMPONENTS OF
CLEAN AIR[a]

Component	Concentrations
Nitrogen	78.1% (Water-free basis)
Oxygen	20.9% (Water-free basis)
Water	0.01 to 3.5%
Carbon dioxide	330 ppm
Methane	1.5 ppm
Hydrogen	500 ppb
Carbon monoxide	~160 ppb
Ozone	~50 ppb
Hydrocarbons other than methane	~10 ppb carbon
Ammonia	~10 ppb
Iodine	~10 ppb
Sulfur dioxide	~1 ppb
Nitrogen dioxide	~1 ppb
Fine particles (<1 μm diam.)	~10 μg/m^3

Note: ~ indicates approximate concentration.

[a] At locations remote from human influence.

Carbon monoxide, nonmethane hydrocarbons, sulfur dioxide, and nitrogen dioxide in the urban U.S. atmosphere typically average ten times greater than the levels in Table 2. For short periods these components can range from a hundred to a thousand times greater.

Solid Lubricants

Solid lubricants are used alone when lubrication by gases is inadequate and liquid or grease lubricants are not tolerable or convenient. Thus, inorganic solid lubricants are used when temperatures are too high or too low, the atmosphere is too oxidizing, or oils and greases lead to contamination or maintenance problems.

Classes of solid lubricants are covered in the chapter on Solid Lubricants (Volume II). Generally, all are used as a thin film between harder substrate materials. Nylon and other polymeric organics are also commonly used as self-lubricating structural materials for light-duty applications. Sometimes, other solid and liquid lubricants are incorporated into the polymeric matrix.

Liquid Lubricants
Bulk Lubricants

Petroleum-based "mineral" oils are the most common liquid lubricants. Various esters and silicones are used as base oils in specialized applications that justify their high cost. Water is an increasingly common base lubricant for boundary lubrication because of its outstanding cooling properties, fire resistance, and environmental compatibility.

Mineral oils are composed mostly of paraffinic, naphthenic (cyclic paraffin), and aromatic hydrocarbon groups linked in a mixture of complex molecules. Some of these hydrocarbons contain more reactive "hetero-atoms" of oxygen, nitrogen, or sulfur in concentrations less than a few tenths percent by weight. Additionally, mineral oils contain small amounts of dissolved gas and water from the atmosphere. Typically, dissolved oxygen and nitrogen concentrations total about 100 mol of gas per million carbon atoms. Dissolved water concentrations range from 10 to 1000 mol per million carbon atoms (about 10 to 1000 ppm by weight) depending on ambient humidity and temperature. Although widely dispersed, many of these trace components have somewhat greater chemical activity than the hydrocarbons.

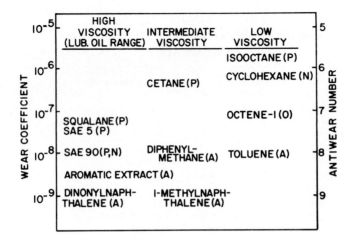

FIGURE 3.   Effect of hydrocarbon type and viscosity on wear. Four-ball machine, 50-kg load, 3.5 mm/sec, 15 to 16 hr at 38°C. P, N, O, and A indicate paraffinic, naphthenic (i.e., cycloparaffinic), olefinic, and aromatic, respectively.

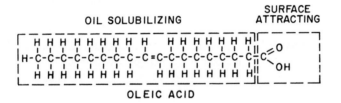

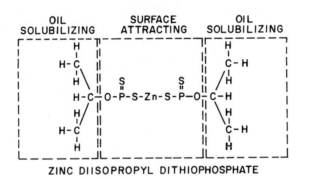

FIGURE 4.   Example of chemical compounds used as boundary lubricant additives.

Figure 3 shows that aromatic and olefinic hydrocarbons give better wear protection in air than paraffinic and naphthenic hydrocarbons of comparable viscosity (comparable molecular size). For comparable hydrocarbon type, higher viscosity (larger molecule) hydrocarbons give better wear protection even under boundary lubrication conditions.

Additives

Additives that adsorb at surfaces generally contain both base-oil and surface-attracting portions as illustrated in Figure 4. Molecules of this type cluster in "micelles" with their surface attracting portions on the interior and oil-solubilizing portions on the exterior. These

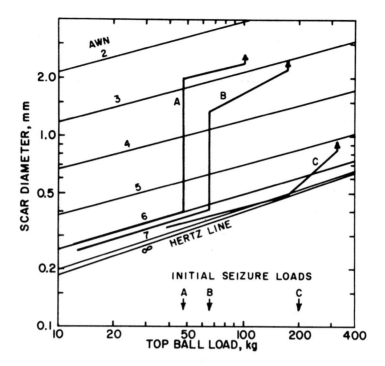

FIGURE 5. Schematic four-ball load-capacity data. Four-ball machine, 690 mm/sec, 10 sec, room temperature. Arrows on curves indicate weld load. A, B, and C indicate indifferent, moderately active EP, and active EP lubricants, respectively.

additives may be considered as "dispersed" in the oil rather than being dissolved. Inorganic solid lubricants can also be made to form stable dispersions in oil by making the particles sufficiently small and covering their surfaces with oil-solubilizing chemical groups.

Additives may be classified functionally as lubricity, antiwear, and extreme-pressure. Lubricity additives reduce and smooth friction — particularly at low sliding speeds. Antiwear additives, as the name implies, reduce wear. EP additives extend the loads that can be covered without a transition to unacceptably high friction, wear, or surface damage.

The same additive can perform more than one function. Thus, a zinc dithiophosphate, such as shown in Figure 4, is both an antiwear and an EP additive. This combination of antiwear and EP functionality is common. The combination of oiliness and antiwear also commonly occurs. By contrast, a combination of oiliness and extreme-pressure functionality is rare.

Lubricity additive molecules generally contain unbranched chains of carbon atoms that are at least ten atoms in length, plus at least one surface attracting group. The unbranched chains sometimes are only part of one or more oil-solubilizing groups. Antiwear and lubricity additives often increase load-carrying capacity over that of a neat base oil.

EP additive molecules usually contain phosphorous, sulfur, or chlorine and also may contain oxygen, lead, zinc, boron, etc. EP additive effects are most evident at high temperatures: under conditions of high sliding velocity, high and concentrated loads, and high bulk temperatures.

EP additives, as illustrated in Figure 5, usually decrease wear under conditions less severe than the lowest transition ("initial seizure" load in four-ball testing). Also, EP additives most often control wear and surface damage at loads above the lowest transition. In many cases, this permits satisfactory performance of mechanisms such as gears above the transition for part of their life. EP agents also commonly increase the severity of conditions required to produce failure by catastrophically rapid wear, surface damage, or "welding".

Boundary lubricant additives interact with base oils, with each other, with atmospheric gases and water, and with thermal and oxidative degradation products of lubricants. Table 3 illustrates interactions between the lubricity additive, behenic acid, and an antiwear/EP additive, zinc diisopropyldithiophosphate. The lubricity additive enhances the wear protection and lowest transition load over that with zinc diisopropyldithiophosphate alone. However, at the highest concentration, the additive interferes with the ability of the EP additive to prevent catastrophic wear and surface damage (scuffing) at the highest load. Incidentally, tendency for wear protection to increase with load below the lowest transition load is often observed with antiwear additives. In most cases, increasing severity by increasing surface roughness decreases load-carrying capacity. However, the reverse is true for the last two EP additives listed in Table 4.[5]

Surface Films

Cumulative evidence indicates that lubricity additives provide lubrication by covering each of the sliding surfaces with a thick film of molecules. These films can be formed from additives with molecular shapes that preclude close packing of the hydrocarbon chains. Lastly, typical lubricity additives can give chemically complex reaction products, at least for steel surfaces.[6,7]

Similarly, EP additives produce complex organic material-containing surface films in contrast to the widely held view that they give simple inorganic films (e.g., iron sulfide from an active sulfur EP additive). Films formed from neat base oils and oiliness or EP-additive oils all contain each chemical element present in the bearing metal and the lubricant as well as oxygen from ambient air.[6,7] The films generally contain organic and inorganic materials present as both chemically bound (soaps) and physical mixtures. EP films are generally solid-like, and contain the order of the same number of organic carbon and oxygen atoms as they do iron atoms.[6]

Generally, hydrocarbons give predominantly organic reaction products when wear is low and inorganic products when wear is high. The organic materials are polymeric, partially oxidized, metal-soap-containing material. The inorganic portion from steel generally contains iron oxides. Lubricity additives generally give organic complex metal-soap-containing films under low-wear conditions. Films from both hydrocarbons and additives usually appear to be liquid-like dispersions of chemical reaction products and bulk lubricant.

Films from all types of lubricants seem to become thicker, more oxidized, more inorganic, and more solid-like as operating conditions become more severe and as the sliding distance increases. Films tend to be more solid-like and inorganic close to the substrate surface and more like the bulk lubricant farthest from the substrate.

BOUNDARY LUBRICATION MECHANISMS

Asperities and Load-Bearing Characteristics

Surfaces have gently sloping hills or asperities that are large compared to molecular dimensions. These asperities range upwards in height a few orders of magnitude from approximately 10 nm (i.e., about $1/2$ μin.) for the smoothest engineering surfaces. Slopes on the sides of asperities range from the order of a degree upwards to 30° on roughly finished bearing surfaces.

When two bearing surfaces of the same material are brought together as illustrated in Figure 6, they initially "touch" at the highest asperities. The size of the real contact area may be readily calculated if the asperities are sufficiently high and steep that the pressure at contact points causes permanent deformation. In this case, the asperities yield until pressure in the contact areas just equals the asperity plastic flow strength:

Table 3
OILINESS AND EP ADDITIVE
INTERACTION IN EFFECTS ON
LOAD-CARRYING TRANSITIONS
AND AWN

Top ball load (kg)	AWN for 0.375% zinc diiso-propyldithiophosphate in cetane with varying behenic acid concentrations		
	0	**0.025%**	**0.50%**
1	6.4	6.7	7.4
10	6.5	6.7	7.6
	→[a]		
20	5.5	7.2	7.6
28	5.7	7.4	7.7
		→[a]	→[a]
35	5.5	5.9	6.7
			→[a]
50	5.5	5.5	4.4[b]

Note: Four-ball machine, 690 mm/sec, 99°C, 600 sec.

[a] Indicates transition at load between those tabulated.
[b] Scuffed wear scars.

Table 4
EFFECT OF ADDITIVE AND SURFACE
ROUGHNESS INTERACTION ON
LOAD-CARRYING CAPACITY

Additive in mineral oil	Failure load (kN)	
	0.03—0.05 μm	0.25—0.38 μm
None	0.49	0.20
7% Methyloleate (lubricity)	0.67	0.28
7% Sulfurized methyloleate (EP)	0.93	0.62
S-Cl-P type (EP)	1.13	1.76
S-Cl type (EP)	1.25	1.85

Note: SAE machine, 8.3 Hz (500 rpm).

$$W = P_p \cdot A_r \tag{7}$$

Real area A_r is directly proportional to total load W and inversely proportional to the characteristic flow strength P_p of the asperities. Approximately, the plastic flow strength may be considered as the hardness of the surface material.

When pressure at contacting asperity tips is less than the plastic flow pressure, asperities will recover their original shapes when the surfaces are separated. When such "elastic" deformation occurs, the real contact area remains approximately proportional to the load.[2] However, the plastic flow pressure, P_p, is replaced by

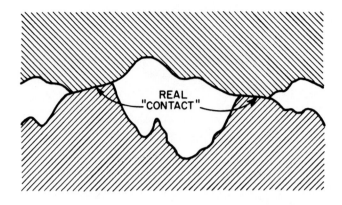

FIGURE 6.   Bearing surface "contact" at asperity tips.

$$P_e = \frac{4}{3\pi} \cdot E' \cdot \sqrt{\frac{\sigma}{r}} \tag{8}$$

Average pressure, P_e, in the real contact area increases with effective elastic modulus, E', average asperity height, σ, and the inverse of asperity tip radius, $1/r$.

An important question is whether asperities are deformed elastically or plastically.[2] Approximately, plastic support occurs when the pressure produced elastically, P_e, exceeds the plastic flow pressure, P_p. Assuming that P_p equals the indentation hardness, H, the plasticity criterion or "index",

$$\psi_c = P_e/P_p \geqslant \sim 1$$

$$\psi = \frac{1}{2.4} \frac{E'}{H} \sqrt{\frac{\sigma}{r}} > \sim 1 \tag{9}$$

Thus, the tendency toward plastic deformation depends on the ratio of two material properties, E'/H, and two surface texture properties, σ/r. Asperity shape can alternatively be expressed in terms of the autocorrelation distance, B.[2] In this case the surface texture term, $(\sigma/r)^{0.5}$ in Equation 8, becomes σ/B. To minimize plastic asperity deformation (and the wear and surface damage which accompany it), asperities should deform easily without taking a permanent set (low E'/H) and be small in height and gently rounded (low σ/r or σ/B).

The preceding paragraphs have discussed the area of real contact and asperity load support for the idealized case of one rough compliant surface bearing against a plane noncompliant surface. The equations remain valid for the case when both surfaces are compliant and rough. In this case, for surfaces 1 and 2,

P_p = Plastic flow pressure of softer surface

$E' = 2 \cdot \dfrac{E'_1 E'_2}{E'_1 + E'_2}$ = Relative asperity effective elastic modulus

H = Hardness of softer surface

$r = \dfrac{r_1 r_2}{r_1 + r_2}$ = Relative asperity tip radius

$\sigma = (\sigma_1^2 + \sigma_2^2)^{0.5}$ = Composite surface roughness

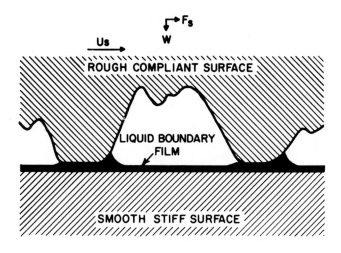

FIGURE 7.   Boundary film formation and shear.

The approximately Gaussian height distribution of asperities allows definition of the surface separations at which asperities carry load. In terms of the RMS (root mean square) surface roughness, σ, asperities must bear some load when the mean surface separation becomes less than about 2σ to 4σ. To aid understanding of boundary lubrication phenomena, the following will be restricted to the extreme case in which asperities provide all the load support.

Boundary Film Friction

In general, boundary friction consists of two components, a shear or adhesion component and a plowing or deformation component.[1] Considering Figure 7, shear component, F_s, predominates except when asperities sink too deeply into a boundary lubricant film or a soft opposing surface. When sliding occurs, the shear friction force depends on the shear resistance per unit area, S, of any "boundary film" in the real load-supporting area between asperities.

$$F_s = SA_r \tag{10}$$

Dividing by load, W, gives the shear contribution to the friction coefficient. Note that this becomes independent of total load and apparent area of contact.

$$f_s = \frac{S \cdot A_r}{W} = \frac{S}{P_p} \text{ or } \frac{S}{P_e} \tag{11}$$

Traditionally the boundary film shear resistance, S, has been assumed equal to the plastic flow shear stress, τ_p, of an ideal elastic, plastic solid.[1] Such a solid gives shear stress independent of strain (and strain rate) at strains sufficiently large to cause plastic flow. This traditional picture did not explain how Newtonian liquid lubricants give friction coefficients similar to those of organic solids.

Recent research[8] provided the unifying picture of flow properties shown in Figure 8. Materials such as mineral oils that approximate ideal liquids at atmospheric pressure and common shear rates can act like plastic solids at the high pressures and shear rates between sliding asperities. The conditions that produce the "glass transition" from liquid to plastic-like behavior are strongly dependent on the viscosity of the material at normal temperatures and pressures and the variation of viscosity with temperature and pressure. Thus, glass transition depends strongly on chemical composition.

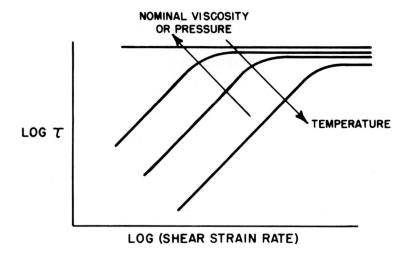

FIGURE 8.   Stress dependence on shear rate for real liquid.

These results indicate that liquid lubricants act like plastic solids in the films between asperities. Thus, $S = \tau_p$ in Equation 11 and the friction coefficient is τ_p/P_p or τ_p/P_e. Since τ_p is a relatively weak function of temperature and pressure, and P_p and P_e are independent of apparent contact area and load, the friction coefficient for a given combination of lubricant and sliding surfaces tends to be independent of operating conditions.

Rheodynamic Boundary Film Formation

Elastohydrodynamic lubrication on an asperity scale entrains film material between sliding surfaces in "micro-rheodynamic" (micro-rhd) lubrication as illustrated in Figure 7. As the upper surface slides, each asperity carries with it a pile of plowed-up film. Sufficient pressure is produced in the film to elastically deform the asperity and to force film material between the surfaces.[9,10]

In the case where the boundary film thickness is adequate to develop a full micro-rhd film, excess film flows around the asperities. For an ideal liquid (Newtonian) boundary film with pressure-viscosity coefficient, α, asperity separation from the opposing surface is[10]

$$h = 1.07 \cdot (\alpha \, \eta_o \, U_s)^{2/3} \cdot (r)^{1/3} \qquad (12)$$

For a finely ground steel surface after brief run-in ($r = 0.1$ mm), asperity separation for a boundary film of 100 P is equivalent to about 10 monolayers of stearic acid at a sliding velocity of only 1 mm/sec. For viscosities at the very high pressure in the macro-ehd conjunction between gear teeth or in the very tacky boundary films often found on wearing surfaces, much thicker micro-ehd films should occur.[11]

When the boundary film carried by the surface is less than that required for fully developed micro-rhd lubrication, asperity conjunctions will be "starved". In the extreme case, no film material is forced around the asperities and asperity separation is determined by conservation of mass considerations to be twice the thickness of the boundary film.[11] Figure 7 shows such a highly starved case. The micro-rhd lubrication mechanism is well substantiated by qualitative evidence. Thus, the influences of viscosity and velocity on friction, wear, and electrical contact resistance commonly are in accord with trends expected from Equation 12. The ehd formulas should remain valid until the sizes are reduced to the point where the lubricant film no longer acts as a continuum — probably an asperity separation of somewhat less than 100 nm. This separation is smaller than expected in most practical boundary

0.001 INCH

100μM

FIGURE 9.   Microchip wear particle from effective lubrication under sliding.

lubrication situations. For a plastic solid boundary film, the plastic flow stress, τ_p, rather than the viscous shear stress is responsible for film entrainment.

WEAR MECHANISMS

When surfaces slide under boundary lubrication conditions, wear seems to be produced by the four following principal mechanisms acting alone and in combination: corrosion, fatigue, abrasion, and adhesion.

Corrosive wear — Occurs when bearing surface material chemically reacts with its environment to form a boundary film containing some bearing material and the film material becomes detached from the surface. Examples of corrosion boundary films are (1) metal oxide formed from oxygen or water in the air or lubricant, and (2) soaps of fatty acids and other salts formed by reaction with a lubricant additive or with lubricant oxidation products. If the films are pushed aside rather than entering the real load-supporting zones between asperities or dissolve in the bulk lubricant, then corrosive wear occurs. Corrosive wear also occurs if the film spalls from the surface in the load-supporting zones.

Fatigue wear — Involves fracture of asperities or film material from repeated high stress. Commonly, this wear mechanism is evidenced by micropitting in rolling element bearings. In sliding bearings, observation of fatigue wear of asperities is usually obscured by abrasion, denting, and adhesion produced by trapping of the spalled particles between the surfaces. Asperity fatigue probably is an important mechanism at least during the early stages of run-in and severe adhesive wear.

Abrasive wear — Occurs when a hard sharp asperity (such as on a file) or a third body (such as a spalled asperity or a dust particle) plows a furrow through a surface. Evidence that it occurs is found as an occasional microchip wear particle such as shown in Figure 9. This chip was obtained under nominally well-lubricated and nonabrasive sliding (AWN above 7 and no known hard or sharp asperities or particles). Abrasion can also lead to indirect wear. As illustrated in Figure 10, plastic deformation creates ridges on either side of a plowed groove. The ridges represent new sharp asperities which tend to preferentially fatigue from the surface.

Adhesion — Is the mechanism of wear mentioned most frequently as the primary process in review articles and reference works. Particles which are removed from one surface are either permanently or temporarily attached to the other surface by solid-phase welding. It is the author's belief that adhesion is likely a secondary mechanism in boundary lubrication because of the difficulty of achieving the proper conditions.[6] High vacuum studies show that surfaces must be atomically clean and within atomic distances of each other to adhere strongly. Further, atomically clean surfaces are extremely difficult to obtain except by generation of fresh surface via plastic extension. Hence, under effective lubrication conditions, adhesion of material from one bearing surface to the other surface probably results from plowing. This would be expected when a trapped third body, such as a fatigued asperity wear particle, is deformed plastically at the same time that confining bearing surfaces are being deformed plastically. This extends the surface films on both the wear particle and the

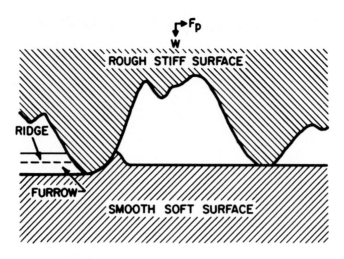

FIGURE 10. Plowing of furrow with bordering ridge.

Table 5
SIZE SCALES PERTINENT TO
BOUNDARY LUBRICATION

	Approximate size range (μm)
Monomolecular layer	$0.2—2 \times 10^{-3}$
Sliding wear debris	0.002—0.1
Boundary film	0.002—3
Ehd film	0.01—5
Asperity height	0.01—5
Rolling wear debris	0.7—10
Asperity contact	0.7—10
Hydrodynamic film	2—100
Asperity tip radius	10—1000
Concentrated contact width	30—500
Engineered counterformal radius	$1—100 \times 10^3$
Engineered conformal radius	$2—2500 \times 10^6$

confining surfaces and thereby tends to clean the surfaces and bring them together. When the surfaces adhere, sliding leads to tearing of the microwelds; this creates new and sharply edged asperities. The fresh surfaces of asperities will be particularly reactive chemically, but the sharpness of the asperities probably leads to their fatigue.

BOUNDARY LUBRICANT PROPERTIES

The physical sizes listed in Table 5 are important properties of boundary lubricants. Boundary film thickness is a significant fraction of asperity height and ehd film thickness. Thus, physical properties pertinent to asperity deformation and ehd-film thermal response may be greatly modified by boundary films. While physical properties of boundary films have not been measured directly, some can be inferred from indirect observations.

Entrainment

To affect friction and wear, the boundary film must be carried by the surfaces into macroscopic load-carrying region of the overall bearing and into the microscopic load-

Table 6
ELASTIC MODULI OF
BEARING AND FILM-LIKE
MATERIALS

Material	Young's modulus (Pa)
Steel	2.1×10^{11}
Bearing bronze	1.0×10^{11}
Inorganic glass	$\sim 0.7 \times 10^{11}$
Babbitt	0.4×10^{11}
Polymeric soap	$\sim 10^{10}$
Organic glass	$\sim 10^{9}$
Rubber	$\sim 10^{7}$

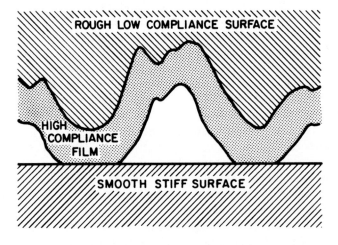

FIGURE 11.   Cushioning of asperities.

carrying zones between interacting asperities. If the film is liquid, the amount of entrainment will depend on viscosity. If the film is solid, elastic-plastic resistance to deformation will determine the entrainment. Whether liquid or solid, the film can protect surfaces only if it deforms coherently. If the film breaks and exposes substrate, those exposed areas will have substantially reduced protection.

As previously discussed, additives, many base oils, and some antiwear additives (under mild conditions) likely give boundary films that are much more viscous than the bulk lubricant.[6]

Cushioning

Table 6 gives elastic moduli for solid materials that encompass compositions of boundary films and most bearing materials. Inorganic glasses are chemically similar to metal oxides while organic glasses and rubbers are similar to low-metal-content organic boundary lubricant films found in sliding systems. Solid-like boundary films with low moduli would be expected to reduce elastic stresses on high modulus asperity tips when film thickness approaches the diameter of the load-carrying zone of an asperity contact. The size scales in Table 5 suggest that such cushioning can be important with the thickest boundary films such as those formed by EP lubricants.

Figure 11 illustrates the cushioning effect of a low-modulus (high compliance) film. Compared to Figure 7, the thick compliant film rather than the substrate surface is deformed and the stresses on the substrate are thereby reduced.

Table 7
THERMAL CONDUCTIVITY
OF STEEL AND FILM
COMPONENTS

Material	Thermal conductivity (J/m/sec/°C)
1% C steel	46.0
FeO	0.54
Fe_2O_3	0.59
Hydrocarbons	~0.2

Thermal Insulation

Table 7 indicates that boundary films consisting of iron oxides and hydrocarbons should be thermal insulators compared to substrate bearing metals.[6] With viscous heating of the hydrodynamic (or ehd) film, this insulation from the substrate metal will raise the film temperature, decrease film viscosity, and thus lower friction. Such insulating effects become important when boundary film thickness approaches the hydrodynamic (or ehd) film thickness.[12] Table 5 indicates that all but the thinnest boundary films should be capable of reducing friction of at least some ehd films. The thickest boundary films may similarly affect friction of the thinnest hydrodynamic films.

Boundary films also appear capable of reducing friction due to shearing of micro-rhd films between asperities. For this to occur, the part of the boundary film closer to the substrate must be much more shear-resistant than that farther from the surface. This often seems to be the case: the film closest to the substrate is thickened by inorganic substrate elements while that farther away is more like the bulk lubricant. Decreased frictional heating also increases the viscosity of the bulk lubricant with a consequently thicker ehd or hydrodynamic film. The reduction in substrate bearing temperatures also increases entrainment tendencies of the boundary film. The increase in surface separation reduces stresses on asperities and, thus, damage by fatigue or plastic deformation. Increased entrainment of boundary films also reduces the corrosive wear that occurs when boundary film (which contains reacted substrate metal) is removed from the surface by forces that push it aside in entrance regions to macro- and microconjunctions.

Coherence and Friability

Approach of bearing metal surfaces to atomic dimensions over sufficiently large areas of welding to occur seems to require considerable extension of the surfaces. Extension thins a boundary film and brings nascent metal to the surface-film interface. This nascent metal interacts with similar nascent metal unless its chemical bonding tendency is satisfied by the availability of boundary film material (including gases). Maximum extension before exposure of nascent metal occurs when the superposed boundary films do not fracture as they thin. Thus, the boundary films can protect against welding of the surfaces by thinning coherently in addition to reducing surface stresses. Probably this coherence is enhanced by the organic components that are an integral part of solid-like EP films.

Once a weld has been produced, the extent of surface damage will be minimized if the welded material is friable. If brittle and the resultant fragments are small, plastically deforming welds break at an early stage of deformation. Friability can be caused by surface films which inhibit emergence of dislocations from the surface. Oxide films cause embrittlement. An active sulfur EP oil has similarly been shown to embrittle steel.[11] Oxygen, water, and at least some EP additives likely reduce severe surface damage by making asperity-to-asperity and asperity-to-particle welds friable.[6]

Table 8
DIMENSIONLESS FACTORS FOR ESTIMATING
BOUNDARY LUBRICATION REQUIREMENTS

Factor	Definition
Interaction	$\dfrac{\text{Surface roughness }(\sigma)}{\text{Mean surface separation or dirt particle diameter}} = 1/\Lambda$
Thermal	Temp. $\dfrac{\text{Estimated maximum conjunction temperature (°K)}}{\text{Limiting temperature for base lubricant/material/atmosphere}}$
	PV. $\dfrac{\text{Estimated PV for application}}{\text{Limiting PV for base lubricant/material/atmosphere}}$
Wear	$\dfrac{\text{Wear coefficient required for satisfactory wear life}}{\text{Typical wear coefficient for base lubricant/material/atmosphere}}$
Friction[a]	$\dfrac{\text{Required friction coefficient}}{\text{Measured friction coefficient}}$

[a]   Significant factor only when surface separation by ehd or hydrodynamic film is less than about 5 μm (200 μ in.).

Table 9
ESTIMATED EFFECT OF FACTORS ON BOUNDARY
LUBRICATION REQUIREMENT

	Boundary lubrication requirement				
Factor	Probably none	Possible easy	Probable easy	Moderate	Difficult
Interaction	0.2	0.5	0.7	2[a]	
Thermal					
Temp.	0.8	0.95	1.4	1.7	
PV	0.7	0.9	2.5	5	
Wear	0.1	0.5	2	10	
Friction	0.8	?	?	?	

[a]   Difficult only with high thermal factor

On the other hand, lubricity additives such as fatty acids make metals less brittle because they remove dislocation-locking oxide films. This effect probably partially explains why oiliness additives often enhance severe surface damage. The effect is also a partial explanation of the beneficial effects of oiliness additives in metal processing where large surface extensions are necessary.

SELECTION OF BOUNDARY LUBRICANTS

Need for a boundary lubricant may be indicated by any of the dimensionless factors listed in Table 8. The interaction factor,[9,10] thermal factors,[13] and required wear life[14] usually can be calculated from machine geometry and operating conditions using available formulas such as Equations 1 to 4. Friction requirements generally can be calculated from design considerations. Limiting thermal parameters are available for various machine element and surface combinations.[14-17] Occasionally different lubricant/atmosphere combinations are included in these tabulations. Similarly, tabulations of representative wear coefficients are available[14] for calculation of expected wear life.

Various factors associated with the boundary lubrication requirement are given in Table 9. Values above threshold levels in the left-hand two columns lead to increasing needs for

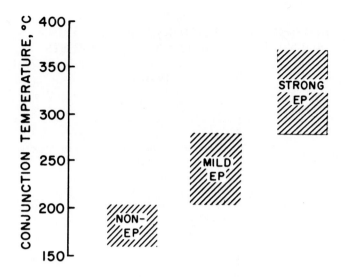

FIGURE 12.   Ranges of the maximum conjunction temperature parameter; hydrocarbon oils, steel-steel, air.

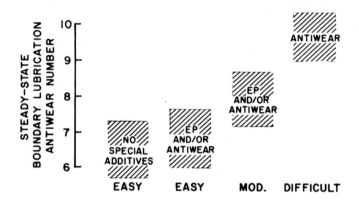

FIGURE 13.   Ranges of antiwear numbers and difficulty of achieving; hydrocarbon oils, steel-steel, air.

boundary lubrication. Problems in the easy to moderate categories can often be readily solved by selecting the proper boundary lubricant. When difficult problems are indicated, machine or operating condition modifications should be considered.

Selection of a boundary lubricant depends on experience. For steel-on-steel with hydrocarbon-based lubricants, Figure 12 summarizes this information for the maximum calculated temperatures in the load-carrying conjunctions.[15-17] Generally, the difficulty of achieving a given load-carrying capacity increases with increasing maximum conjunction temperature.

Figure 13 shows approximate ranges of AWN and the estimated difficulty of achieving them for steel-on-steel and hydrocarbon-based lubricant systems. Similar wear protection ranges and difficulties appear to exist for other common bearing material combinations such as bronze-steel. These ranges can be used to test the achievability of a given wear life.

The selection of a lubricant for boundary lubrication generally should be made from among commercially available lubricants. These lubricants economically offer a balance of boundary lubrication properties with other lubricant properties required for specific types of service.

NOMENCLATURE

AWN	=	AntiWear Number $= -\log_{10}k = \log_{10}(1/k)$
A	=	Area of apparent load support
A_r	=	Area of asperity load support
A_w	=	Area being worn
B	=	Surface profile correlation distance
c	=	Width of load support area in direction of motion
d	=	Depth of material worn from surface
E'	=	Relative effective elastic modulus, $2\,E'_1 E'_2/(E'_1 + E'_2)$
	=	Young modulus of elasticity$/(1 - \text{Poisson ratio})^2$
F_p, F_s	=	Plowing friction force, shear friction force
f, f_b, f_{ss}	=	Friction coefficient, beginning, and steady-state friction coefficients
h	=	Thickness of rheodynamic film between asperities
H	=	Hardness of surface in stress units [multiply Brinell, Knoop, or Vickers hardness numbers by 9.807 to convert to MN/m^2 (MPa) from the normal hardness number units which are kg force/mm²]
k, k_b, k_{ss}	=	Wear coefficient, beginning, and steady-state wear coefficients
L	=	Face width of line contact
ℓ, ℓ_b	=	Sliding distance, break-in sliding distance
n	=	Number of times surface passes through loaded area
P	=	Apparent pressure on load support area $= W/A$
P_e, P_p	=	Average pressure on elastically, plastically deformed asperity
r	=	Asperity tip radius
U	=	Sweep velocity of surface through load support area
U_s	=	Sliding velocity (vector difference of sweep velocities of two surfaces through load support area)
v	=	Volume of material worn from a surface
V	=	U_s = sliding speed
W	=	Load normal to load support area
η	=	Viscosity
Λ	=	Film thickness parameter: ratio of fluid film thickness to composite roughness of wearing surfaces (or to "diameter" of particles in fluid film if they are larger than three times the composite surface roughness)
σ	=	Root mean square surface roughness (for two surfaces in contact, composite roughness is square root of sum of squares of the RMS roughness of the two surfaces)
τ	=	Shear stress
ψ	=	Plasticity index

REFERENCES

1. **Bowden, F. P. and Tabor, D.**, *The Friction and Lubrication of Solids*, Oxford University Press, London, 1954.
2. **Archard, J. F.**, Wear theory and mechanisms, *ASME Wear Control Handbook*, Peterson, M. B. and Winer, W. O., American Society of Mechanical Engineers, New York, 1980.
3. **Fein, R. S.**, Boundary lubrication, *Lubrication*, 57, 1, 1971.
4. **Niemann, G., Rettig, H., and Lechner, G.**, Scuffing tests on gear oils in FZG apparatus, *ASLE Trans.*, 4, 71, 1961.

5. **Asseff, P. A.,** Study of corrosivity and correlation between chemical reactivity and load-carrying capacity of oils containing extreme pressure agents, *ASLE Trans.,* 9, 86, 1966.
6. **Fein, R. S.,** Chemistry in concentrated-conjunction lubrication, in Interdiscriplinary Approach to the Lubrication of Concentrated Contacts, NASA SP-237, Ku, P. M., Ed., U.S. Government Printing Office, Washington, D.C., 1970, 489.
7. **Fein, R. S., Rand, S. J., and Caffrey, J. M.,** Radiotracer measurements of elastohydrodynamic and boundary films, Conf. Limits of Lubr., Imperial College, London, July 1973.
8. **Blair, S. and Winer, W. O.,** A rheological model for elastohydrodynamic contacts based on primary laboratory data, ASLE/ASME Lubr. Conf., Minneapolis, Minn., October 1978.
9. **Archard, J. F. and Cowking, E. W.,** Elastohydrodynamic lubrication at point contacts, in *Elastohydrodynamic Lubrication,* Institute of Mechanical Engineers, London, 1965 and 1966, 47.
10. **Dowson, D.,** Elastohydrodynamic lubrication, in Interdisciplinary Approach to the Lubrication of Concentrated Contacts, NASA SP-237, Ku, P. M., Ed., U.S. Government Printing Office, Washington, D.C., 1970, 27.
11. **Fein, R. S. and Kreuz, K. L.,** Discussion on boundary lubrication, in Interdisciplinary Approach to Friction and Wear, NASA SP-181, Ku, P. M., Ed., U.S. Government Printing Office, Washington, D.C., 1968, 358.
12. **Fein, R. S.,** Friction effect resulting from thermal resistance of solid boundary lubricant, *Lubr. Eng.,* 27, 190, 1971.
13. **Archard, J. F.,** The temperature of rubbing surfaces, *Wear,* 2, 438, 1958-9.
14. **Peterson, M. B. and Winer, W. O., Eds.,** *ASME Wear Control Handbook,* American Society of Mechanical Engineers, New York, 1980.
15. **Anon.,** *Scoring Resistance of Bevel Gear Teeth,* Gear Engineering Standard, Gleason Works, Rochester, N.Y., 1966.
16. **Kelley, B. W. and Lemanski, A. J.,** Lubrication of involute gearing, in IME Conf. Lubr. Wear Fundam. Appl. Design, London, September 1967.
17. **Ku, P. M., Staph, H. E., and Cooper, H. J.,** On the critical contact temperature of lubricated sliding-rolling disks, *ASLE Trans.,* 21, 161, 1978.

HYDRODYNAMIC LUBRICATION

H. J. Sneck and J. H. Vohr

FLUID FILM LUBRICATION

"Lubrication theory" begins with the Navier-Stokes equations and the continuity equation. Employing simplifications consistent with hydrodynamic lubrication, the end result is a set of relatively simple working equations which describe the velocity and pressure distribution throughout the flow field.

Assumptions which are made in lubrication theory are as follows: (1) the fluid film is very thin compared to its extent, (2) effects of gravity are negligible, (3) viscosity does not vary across the fluid film thickness, (4) curvature of the surfaces in journal bearings is large compared to the fluid film thickness, (5) the fluid adheres to solid boundaries with no slip, and (6) fluid inertia is negligible.

Reynolds Equation

The following equation which governs the pressure distribution within the fluid film is named after Osborne Reynolds who first derived it. Using the notation of Figure 1:

$$\frac{\partial}{\partial x}\left(\frac{\rho h^3}{\mu}\frac{\partial P}{\partial x}\right) + \frac{\partial}{\partial y}\left(\frac{\rho h^3}{\mu}\frac{\partial P}{\partial y}\right) = 6\left\{\frac{\partial}{\partial y}\left[\rho(V_1 + V_2)h\right]\right.$$

$$\left. + \frac{\partial}{\partial x}\left[\rho(U_1 + U_2)h\right] + 2\rho(w_h - w_o) + 2\frac{\partial(\rho h)}{\partial t}\right\} \qquad (1)$$

Several simplifications can be made. It is usually possible to arrange the coordinate system so that $V_1 = V_2 = 0$, thereby eliminating the first right side term. Further, writing $U_1 + U_2 = U$ and taking x to be in the direction of U_1 and U_2 yields

$$\frac{\partial}{\partial x}\left(\frac{\rho h^3}{\mu}\frac{\partial P}{\partial x}\right) + \frac{\partial}{\partial y}\left(\frac{\rho h^3}{\mu}\frac{\partial P}{\partial y}\right) = 6\left\{U\frac{\partial(\rho h)}{\partial x} + 2\rho(w_h - w_o) + 2\frac{\partial(\rho h)}{\partial t}\right\} \qquad (2)$$

The first term on the right side of this equation indicates how the bearing surface motion combines with density gradients and the wedge action of the fluid film thickness to generate the pressure field. If the bearing walls are permeable, the net flow of fluid $(w_h - w_o)$ through the porous walls contributes to the generation of pressure by the second right side term. The last term on the right side is the squeeze film term which relates density and film thickness variation with time and the fluid film pressure.

Usually the complete Reynolds equation is not needed for a specific problem. For incompressible fluids in bearings with impermeable walls $(w_h = w_o = 0)$, for instance, density drops out and then

$$\frac{\partial}{\partial x}\left(\frac{h^3}{\mu}\frac{\partial P}{\partial x}\right) + \frac{\partial}{\partial y}\left(\frac{h^3}{\mu}\frac{\partial P}{\partial y}\right) = 6U\frac{\partial h}{\partial x} + 12\frac{\partial h}{\partial t} \qquad (3)$$

A list of further simplifying conditions and the terms affected is provided in Table 1.

Infinite Slider Bearing

A better understanding of fluid film lubrication can be gained from a few simple bearing configurations. The infinitely long slider bearing shown in Figure 2 is one of these. The

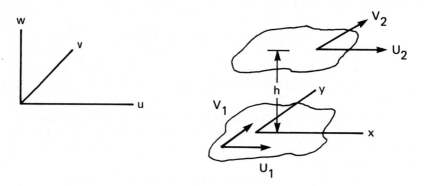

FIGURE 1.   Coordinate direction and velocity components.

Table 1
SIMPLIFICATIONS TO THE GENERAL REYNOLDS EQUATION

Simplifying condition	Term simplified
1. Gas-lubricated bearing	Density ρ replaced by pressure P
2. Incompressible lubricant	Density ρ removed from equation
3. Bearing walls impermeable	$w_n = w_o = 0$
4. Film thickness does not vary with time	$\dfrac{\partial h}{\partial t} = 0$
5. Film thickness does not vary with x or y	$\dfrac{\partial h}{\partial x} = 0$
	$\dfrac{\partial}{\partial x}\left(\dfrac{\rho h^3}{\mu}\dfrac{\partial P}{\partial x}\right) = h^3 \dfrac{\partial}{\partial x}\left(\dfrac{\rho}{\mu}\dfrac{\partial P}{\partial x}\right)$
	$\dfrac{\partial}{\partial y}\left(\dfrac{\rho h^3}{\mu}\dfrac{\partial P}{\partial y}\right) = h^3 \dfrac{\partial}{\partial y}\left(\dfrac{\rho}{\mu}\dfrac{\partial P}{\partial y}\right)$
6. Viscosity does not vary with x or y	$\dfrac{\partial}{\partial x}\left(\dfrac{\rho h^3}{\mu}\dfrac{\partial P}{\partial x}\right) = \dfrac{1}{\mu}\dfrac{\partial}{\partial x}\left(\rho h^3 \dfrac{\partial P}{\partial x}\right)$
	$\dfrac{\partial}{\partial y}\left(\dfrac{\rho h^3}{\mu}\dfrac{\partial P}{\partial y}\right) = \dfrac{1}{\mu}\dfrac{\partial}{\partial y}\left(\rho h^3 \dfrac{\partial P}{\partial y}\right)$
7. Bearing surfaces do not move	$U \dfrac{\partial(\rho h)}{\partial x} = 0$
8. Bearing very long in y direction	$\dfrac{\partial}{\partial y}\left(\dfrac{\rho h^3}{\mu}\dfrac{\partial P}{\partial y}\right) \approx 0$
9. Bearing very short in y direction	$\dfrac{\partial}{\partial x}\left(\dfrac{\rho h^3}{\mu}\dfrac{\partial P}{\partial x}\right) \approx 0$

lower surface in Figure 2 moves to the left to draw the lubricant into a convergent wedge. Reynolds equation for a steady load with $\partial P/\partial y = 0$, an incompressible lubricant, and constant viscosity becomes

$$\frac{d}{dx}\left(h^3 \frac{dP}{dx}\right) = -6\,U\mu\,\frac{dh}{dx} \qquad (4)$$

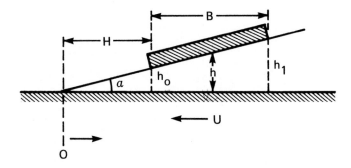

FIGURE 2. Slider bearings.

Table 2
INFINITE SLIDER BEARING PERFORMANCE

Quantity	Formula
Film thickness parameter	$K = \dfrac{h_1}{h_o} - 1$
Load capacity	$W = \dfrac{6U\mu B^2 L}{h_o^2 K^2} \left\{ \ln(1 + K) = \left(\dfrac{2K}{2 + K} \right) \right\}$
Friction force	$F_f = \dfrac{W}{2} \tan \alpha + \left(\dfrac{\mu UBL}{h_o K} \right) \ln(1 + K)$
Center of pressure	$\overline{X} = B \left\{ \dfrac{2(3 + K)(1 + K) \ln(1 + K) - K(6 + 5K)}{2K[(2 + K) \ln(1 + K) - 2K]} \right\}$

The film thickness shape is given by $(h_o/H)x$ where $H = h_o B/(h_1 - h_o)$. The first integral of this ordinary differential equation is

$$\frac{dP}{dx} = -6\,U\mu \left(\frac{h - \overline{h}}{h^3} \right) \tag{5}$$

Constant of integration $\overline{h}$ is also the film thickness at which the pressure is a maximum and where $dP/dx = 0$. If the film thickness did not converge, h would be constant throughout the film and no pressure buildup could occur. Integrating a second time yields

$$P = -\frac{6U\mu B}{h_1 - h_o} \left(\frac{\overline{h}}{2h^2} - \frac{1}{h} + c \right) \tag{6}$$

Using the boundary conditions of $P = 0$ at $h = 0$ and $h = h_1$ to evaluate integration constants $\overline{h}$ and c, the pressure in the slider becomes

$$P = \frac{6U\mu B}{h_1 - h_o} \left[\frac{1}{h} - \frac{h_o h_1}{h^2 (h_1 + h_o)} - \frac{1}{h_o + h_1} \right] \tag{7}$$

The pressure and the load it supports are directly proportional to surface velocity U and viscosity μ. Table 2 summarizes the performance characteristics of an infinitely wide slider bearing.

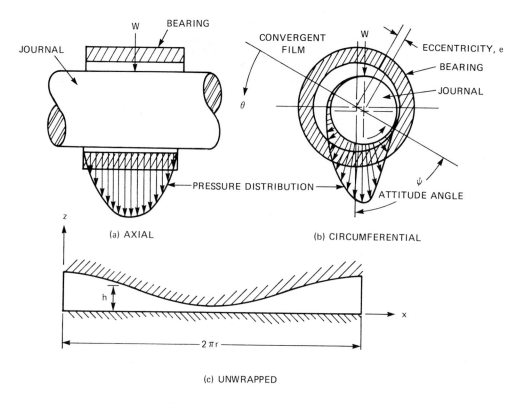

FIGURE 3.   Journal bearing oil-film relations.

Short Journal Bearing

The three factors essential for a hydrodynamic slider bearing (velocity, viscosity, and a converging film) are all contained in the right side term of the Reynolds equation 6 μU(dh/dx). The required convergent film is also formed when a shaft and bushing become eccentric as shown in Figure 3b.

The unwrapped film shape is shown in Figure 3c. A positive superambient pressure is generated in the convergent left portion of the film. Since liquids cannot withstand substantial subambient pressure, the fluid film ruptures in the divergent right section, forming discontinuous streamers which flow through this region at approximately ambient pressure. These streamers contribute nothing to load carrying capacity.

When the bearing is "narrow", i.e., its length is less than its diameter, Reynolds equation can be simplified to

$$h^3 \frac{\partial^2 P}{\partial y^2} = 6 \, U\mu \, \frac{dh}{dx} \tag{8}$$

Neglecting the pressure gradient term for the x direction attributes circumferential flow primarily to the motion of the journal surface expressed by the right side of the equation. Axial flow and leakage are due entirely to the pressure gradient term on the left side of the equation.

This "short bearing" version of Reynolds equation can be integrated twice in the y direction, because h is not a function of y, to give the pressure function

$$P = \frac{3U\mu}{h^3} \frac{dh}{dx} \left[y^2 - \left(\frac{L}{2}\right)^2 \right] \tag{9}$$

Table 3
SHORT JOURNAL BEARING

Quantity	Formula
Load capacity	$W = \dfrac{U\mu L^3 \pi}{4c^2} \left[\dfrac{\epsilon}{(1-\epsilon^2)^2} \right] (0.62\ \epsilon^2 + 1)^{1/2}$
Attitude angle	$\tan \psi = \dfrac{\pi}{4} \dfrac{(1-\epsilon^2)^{1/2}}{\epsilon}$
Friction force	$F_f = \dfrac{c\epsilon W \sin \psi}{2r} + \dfrac{2\pi\mu U r L}{c(1-\epsilon^2)^{1/2}}$
Oil flow from ends of bearing	$Q = U c L \epsilon$

where y is measured from the bearing center plane. This distribution satisfies the pressure boundary conditions of P = 0 at both y = L/2 and y = −L/2.

Axial pressure distribution is shown by this equation to approach a simple parabolic shape. Because the fluid film thickness is much smaller than the journal radius, h = c(1 − ϵ cosθ). When substituted into the pressure equation

$$P = \frac{3\ U\mu\epsilon\sin\theta}{rc^2\ (1 + \epsilon\cos\theta)^3} \left[\left(\frac{L}{2} \right)^2 - y^2 \right] \qquad (10)$$

Positive pressures are obtained in the convergent wedge portion between 0 and π. A reasonably accurate prediction of load capacity can be obtained by setting the pressure equal to zero in the divergent region between π and 2π.

Retaining the axial pressure gradient term which accounts for axial flow allows a more realistic treatment of axial pressure distribution than with the infinite slider bearing approximation. Table 3 summarizes the performance characteristics of a short journal bearing. A comprehensive table of integrals is available for use in solution of the sin-cos relations commonly encountered with journal bearings.[1]

When the shaft is lightly loaded, the eccentricity ratio, ϵ, approaches zero. The following Petroff equation results from Table 3 and is frequently used to estimate journal bearing power loss.

$$F_f = 2\pi\ \mu\ U\ rL/c \qquad (11)$$

Cylindrical Coordinates

Reynolds equation in cylindrical coordinates (used in analysis of circular thrust bearings) is

$$\frac{1}{r} \frac{\partial}{\partial r} \left(\frac{\rho r h^3}{\mu} \frac{\partial P}{\partial r} \right) + \frac{1}{r^2} \frac{\partial}{\partial \phi} \left(\frac{\rho h^3}{\mu} \frac{\partial P}{\partial \phi} \right) = \frac{6\mu}{r} \left[\frac{\partial}{\partial r} (\rho h r\ U_r) + \frac{\partial}{\partial \phi}\ (\rho h\ V_\phi) + 2\rho\ (w_h - w_o)\ r \right] \qquad (12)$$

where U_r and V_ϕ, the radical and tangential surface velocities, play the same roles as U and V in rectangular coordinates.

TURBULENCE

Very high speeds, large clearances, or low-viscosity lubricants may introduce a sufficiently

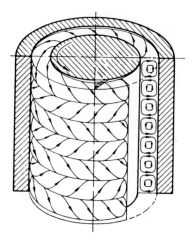

FIGURE 4. Taylor vortices between concentric rotating cylinders with the inner cylinder rotating. (From Schlichting, H., *Boundary Layer Theory,* 6th ed., McGraw-Hill, New York, 1968. With permission.)

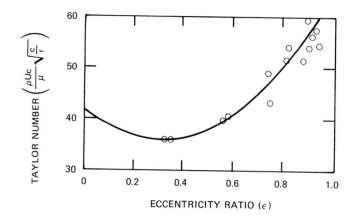

FIGURE 5. Critical Taylor number vs. eccentricity ratio. Experimental results for c/r = 0.00494. (From Frene J. and Godet, M., *Trans. ASME Ser. F,* 96, 127, 1974. With permission.)

high Reynolds number in a bearing film for departure from laminar flow velocity by either of two instabilities: Taylor vortices or turbulence.

Taylor vortex flow is characterized by ordered pairs of vortices between a rotating inner cylinder and an outer cylinder as shown in Figure 4.[2] Such vortices significantly "flatten" the velocity profile between the cylinders and increase the wall shear stress. For a concentric cylindrical journal bearing, vortices develop when the Taylor number $\rho Uc\sqrt{c/r}/\mu$ exceeds 41.1.[3] For nonconcentric (loaded) bearings the situation is less clear, although studies have recently been made.[4-8] Critical Taylor numbers are shown in Figure 5 as a function of eccentricity.[7]

Turbulence is a more familiar phenomenon. Unlike Taylor vortices, disordered flow in turbulence is not produced by centrifugal forces and will occur whether the inner or outer cylinder is rotating. In experimental studies where turbulence develops before Taylor vortices,[9] turbulence appears to set in when the Reynolds number Re = $\rho Uc/\mu$ exceeds 2000.

When vortices develop first, turbulence may begin at a lower value of Reynolds number.[4] Assuming that vortices develop when $Re\sqrt{c/r} = 41.1$, vortices will occur before turbulence for c/r ratios greater than 0.0004. In most bearings, however, turbulence sets in shortly after the development of vortices; and since random turbulent momentum transfer appears to dominate, turbulent lubrication theories have neglected the effect of vortices.

Turbulent lubricating film theories have been based on well-established empiricisms such as Prandtl mixing length or eddy diffusivity.[10-12] Most commonly employed is that due to Elrod and Ng[11] based on eddy diffusivity. Volume flows in the film are related to pressure gradient through turbulent lubrication factors G_x and G_y, i.e.,

$$\overline{uh} = q_x = -G_x (h^3/\mu) \, \partial P/\partial x + r\omega h/2 \qquad (13)$$

$$\overline{vh} = q_y = -G_y (h^3/\mu) \, \partial P/\partial y \qquad (14)$$

where x refers to the circumferential direction and y axial. Flows $\overline{uh}$ and $\overline{vh}$ are obtained from integrating velocities u and v across the film thickness. G_x and G_y depend upon the level of turbulence as a function of local Reynolds number $\overline{U}h/\nu$ where $\overline{U}$ is the local mean fluid film velocity. For Couette flow $Re_c = Uh/\nu$, where U is the bearing surface velocity, while for pressure induced flow $Re_p = |\nabla P|h^3/\mu\nu$ where $|\nabla P|$ denotes the absolute magnitude of the pressure gradient.

Using Equations 13 and 14 for lubricant flow rates, the turbulent Reynolds equation is obtained:

$$\frac{\partial}{\partial x}\left(\frac{G_x}{\mu} h^3 \frac{\partial P}{\partial x}\right) + \frac{\partial}{\partial y}\left(\frac{G_y}{\mu} h^3 \frac{\partial P}{\partial y}\right) = \frac{U}{2} \frac{\partial h}{\partial x} \qquad (15)$$

While G_x and G_y depend, in general, upon the pressure gradient, they become functions of Re_c alone at very high surface velocities and high Re_c. In Figure 6, considering the curve corresponding to $Re_p = 10^6$, G_x at $Re_c = 2 \times 10^4$ joins an envelope of curves and becomes independent of Re_p provided Re_p remains less than 10^6 and Re_c remains greater than 10^4.

For most hydrodynamic bearings, "linearized theory" with values of G_x given by the limiting envelope in Figure 6 suffices to describe turbulence.[13] Appropriate values of G_x and G_y are given by the following.[14]

$$1/G_x = 12 + 0.0136 \, (Re_c)^{0.90}$$

$$1/G_y = 12 + 0.0043 \, (Re_c)^{0.96} \qquad (16)$$

While various "fairing" procedures have been applied in the uncertain transition region between laminar and turbulent flow,[15] the writer favors the following procedure. For values of Re_c less than $41.2\sqrt{r/c}$ (onset of vortices), laminar flow theory is to be used. For values of Re_c greater than 2000, fully developed turbulent relationships such as Equation 16 are to be used. Between these critical values of Re_c, linearly interpolate between the laminar values for G_x and G_y (i.e., 1/12) and the values for G_x and G_y evaluated at $Re_c = 2000$, with Re_c being the interpolation variable.

A critical parameter affected by turbulence is the shear stress τ_s acting on the sliding member of a hydrodynamic bearing. For motion in the x direction, τ_s under laminar flow conditions is

$$\tau_s = \tau_c + h/2 \, \frac{\partial P}{\partial x} \qquad (17)$$

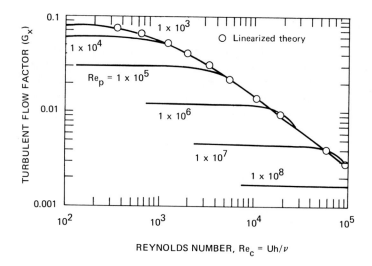

FIGURE 6.　Turbulent flow factor G_x for $\partial p/\partial x$ negative and $\partial p/\partial y = 0$.

where Couette shear stress $\tau_c = \mu U/h$. Onset of turbulence increases both τ_c and $\partial P/\partial x$. The increase in $\partial P/\partial x$ is calculated through the turbulent Reynolds Equation 15. The increase in τ_c is taken account of through turbulent Couette shear stress factor C_f, i.e.,

$$\tau_c = C_f\, \mu\, U/h \qquad\qquad (18)$$

Values of C_f are reasonably approximated from linearized theory by the following:[13,14]

$$C_f = 1 + 0.0012\,(Re_c)^{0.94} \qquad\qquad (19)$$

Qualitatively, turbulence increases apparent viscosity of the film and hence increases load capacity and power loss above that with laminar flow. If C_f is calculated to be, say, 1.25, the concentric turbulent power loss would be 25% higher than that predicted for laminar flow. Enhancement of load capacity is illustrated in Figure 7.

For turbulence in externally pressurized bearings, e.g. for cryogenic turbopumps[16] dominated by pressure-induced flow, turbulent flow factors G_x and G_y become functions only of Re_p, as shown in Figure 8. An iterative procedure is usually needed to solve for the pressure gradient. Convergence is usually rapid.

Relatively little attention has been given to transition to turbulence in thrust bearings. Pad temperatures of tilting pad bearings reported by Capitao, Gregory, and Whitford[17] show a sharp decrease as operating speed is increased beyond a certain critical value, dependent upon load. This temperature decrease is due apparently to onset of turbulence, which drastically increases heat transfer from the pads to the oil in the grooves between pads.[18] Calculations of local Reynolds numbers by Roben[19] show "transition" Reynolds numbers in the range 1000 to 1200. This is somewhat higher than the range 580 to 800 suggested by Abromovitz[20] and supported by Gregory.[21] The discrepancy may reflect that the lower range cited pertains to an average Reynolds number for the bearing pad.

Since the turbulent lubrication theories discussed above neglect centrifugal vortex effects, they apply to thrust bearings as well as to journal bearings.

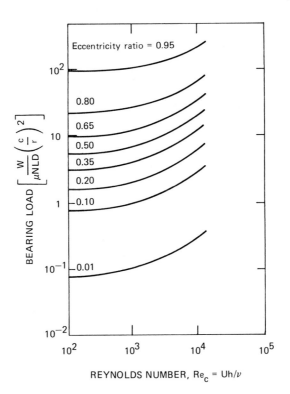

FIGURE 7. Load capacity of turbulent full journal bearing for L/D = 1. (From Ng, C. W. and Pan, C. H. T., *Trans. ASME*, 87(4), 675, 1965. With permission.)

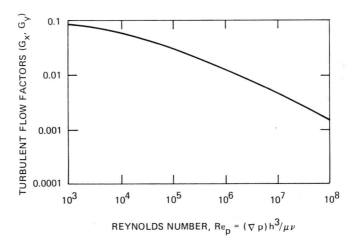

FIGURE 8. Turbulent velocity parameters G_x and G_y vs. Re_p when turbulence is dominated by pressure-induced flow. (From Reddeclif, J. M. and Vohr, J. H., *J. Lubr. Technol. Trans. ASME Ser. F*, 91(3), 557, 1969. With permission.)

ENERGY EQUATION

Temperature distribution in the lubricant film is governed by the energy equation which may be derived from the differential element of bearing film shown in Figure 9. The top bounding surface is assumed stationary while the bottom surface moves with velocity U in

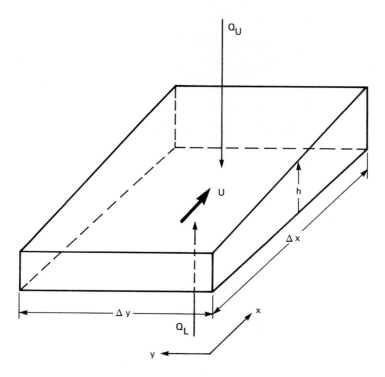

FIGURE 9.   Control volume for energy conservation.

the x direction. Considering steady flow and neglecting (1) variations in fluid properties, (2) kinetic and potential energy of the fluid, and (3) variations in temperature normal to the x-y plane, energy convected into the differential volume is given by

$$E_{in} = \rho c_p q_x (x, y + \Delta y/2) \, T(x, y + \Delta y/2) \, \Delta y$$

$$+ \rho c_p q_y (x + \Delta x/2, y) \, T(x + \Delta x/2, y) \, \Delta x \qquad (20)$$

while the energy convected out by the flow of lubricant is

$$E_{out} = \rho c_p q_x (x + \Delta x, y + \Delta y/2) \, T(x + \Delta x, y + \Delta y/2) \, \Delta y +$$

$$\rho c_p q_y (x + \Delta x/2, y + \Delta y) \, T(x + \Delta x/2, y + \Delta y) \, \Delta x \qquad (21)$$

where q_x and q_y are the volume fluxes of lubricant in $m^3/m/sec$ in the x and y directions, respectively.

Conduction of heat in the x and y directions is neglected as compared with convection. Heat conducted *into* the control volume in the z direction may be expressed simply as $Q_U \Delta x \Delta y$ and $Q_L \Delta x \Delta y$ where Q_U and Q_L represent conduction heat fluxes through the upper and lower bearing surfaces in W/m^2.

Pressure work done on the lubricant entering and by the lubricant leaving is as follows:

$$W_{in} = P(x, y + \Delta y/2) \, q_x (x, y + \Delta y/2) \, \Delta y +$$

$$P(x + \Delta x/2, y) \, q_y (x + \Delta x/2, y) \, \Delta x$$

$$W_{out} = P(x + \Delta x, y + \Delta y/2) \, q_x (x + \Delta x, y + \Delta y/2) \Delta y +$$

$$P(x + \Delta x/2, y + \Delta y) \, q_y (x + \Delta x/2, y + \Delta y) \, \Delta x \qquad (22)$$

Work done on the lubricant by shear stress τ_s at the moving surface is given by $\tau_s U \Delta x \Delta y$. Equating net heat flow convected and conducted out of the control volume to work done on the fluid, dividing through by Δx and Δy, taking the limit as Δx and Δy go to zero, and applying the following continuity equation

$$\partial q_x / \partial x + \partial q_y / \partial y = 0 \tag{23}$$

results in the following overall energy balance

$$\rho c_p \left[q_x \frac{\partial T}{\partial x} + q_y \frac{\partial T}{\partial y} \right] = Q_u + Q_L - q_x \frac{\partial P}{\partial x} - q_y \frac{\partial P}{\partial y} + \tau_s U \tag{24}$$

For laminar flow, the lubricant fluxes q_x and q_y are given by:

$$q_x = \frac{-h^3}{12\mu} \frac{\partial P}{\partial x} + \frac{Uh}{2}$$

$$q_y = \frac{-h^3}{12\mu} \frac{\partial P}{\partial y} \tag{25}$$

These expressions may be modified to take account of turbulence by means of the turbulent flow factors G_x and G_y described earlier.

$$q_x = \frac{-G_x h^3}{\mu} \frac{\partial P}{\partial x} + \frac{Uh}{2}$$

$$q_y = \frac{-G_y h^3}{\mu} \frac{\partial P}{\partial y} \tag{26}$$

Stress τ_s at the moving surface is given for laminar flow by the first of the following relations, and is modified in the second case by turbulence factor C_f discussed earlier.

$$\tau_s = \frac{\mu U}{h} + \frac{h}{2} \frac{\partial P}{\partial x}$$

$$\tau_s = C_f \mu \frac{U}{h} + \frac{h}{2} \frac{\partial P}{\partial x} \tag{27}$$

Since Equation 24 depends upon the pressure gradient, it must be solved simultaneously with the Reynolds equation. The usual procedure assumes an initially uniform temperature (and viscosity) distribution to solve Reynolds equation for $P(x,y)$; q_x, q_y, and τ_s are then determined from Equations 26 and 27. Equation 24 is then solved for $T(x,y)$ and hence $\mu(T)$. Reynolds equation is then resolved for pressure using the new distribution for viscosity. This iterative procedure typically converges quickly.

Equation 24 requires specification of a bearing film inlet temperature. This temperature is usually higher than that of the oil supplied to the bearing as a result of (1) heating as the oil comes into contact with bearing metal parts, and (2) hot oil recirculation in and around the bearing.

DYNAMICS

Three important concerns are commonly associated with the dynamic behavior of bearings: (1) avoiding any bearing-rotor system natural frequencies, or "critical speeds", near op-

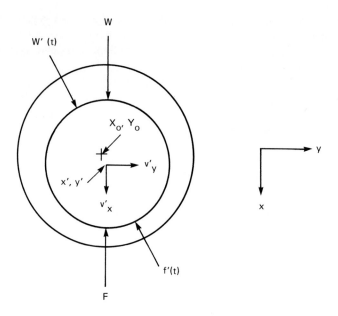

FIGURE 10.   Dynamic motion of shaft.

erating speeds, (2) limiting forced vibrations of the rotor-bearing system to acceptable levels, and (3) freedom from self-excited "half frequency whirl" or "oil whip" instability.

Dynamic motion of a shaft in a fluid film bearing is illustrated in Figure 10. In general, steady load W is balanced by steady film force F. Superposed dynamic loading $W'(t)$ produces dynamic shaft displacement $x'(t)$ and $y'(t)$ away from steady state position x_o, y_o. Corresponding dynamic bearing film forces $f_x'(t)$ and $f_y'(t)$ may be calculated from Reynolds equation since film thickness h and the time rate of change of h can be determined if x', y', and v'_x, and v'_y are known.

From instantaneous film forces $f_x'(t)$ and $f_y'(t)$, dynamic motions of a rigid shaft of mass M can be determined from the following equations of motion:

$$M \frac{d^2x'}{dt^2} = f_x'(t) + W_x'(t)$$

$$M \frac{d^2y'}{dt^2} = f_y'(t) + W_y'(t) \tag{28}$$

where $W_x'(t)$ and $W_y'(t)$ represent the x and y components of the externally imposed dynamic loading on the shaft.

Since $f_x'(t)$ and $f_y'(t)$ are, in general, determined by a numerical solution of Reynolds equations, Equations 28 may be solved numerically by a time integration procedure. This "nonlinear transient analysis" does not involve the common simplifying assumption that a linear relationship exists between shaft motions and the bearing film forces f_x' and f_y'. This transient approach is usually cumbersome and expensive since Reynolds equation must be solved at each time step, and many cycles of shaft motion may need to be analyzed to define rotor-bearing critical speeds, response to forced vibrations, and rotor bearing stability.

When using linearized stiffness and damping coefficients, shaft motions are assumed to be sufficiently small that dynamic bearing forces f_x' and f_y' are linearly proportional to journal displacements x' and y', and to journal velocities v_x' and v_y'.

$$-f_x' = K_{xx}x' + K_{xy}y' + B_{xx}v_x' + B_{xy}v_y' \tag{29}$$

$$-f_y' = K_{yx}x' + K_{yy}y' + B_{yx}v_x' + B_{yy}v_y' \qquad (30)$$

Proportionality constants K_{xx}, K_{xy}, K_{yx}, and K_{yy} are referred to as bearing spring coefficients while B_{xx}, B_{xy}, B_{yx}, and B_{yy} are bearing damping coefficients.

The usual computational procedure for determining the stiffness coefficients is to solve Reynolds equation for bearing force components $F_x(x_o y_o)$ and $F_y(x_o y_o)$ with the journal steady-state position x_o, y_o. For a very small shift, δ, of the journal in the x direction, bearing forces $F_x(x_o + \delta, y_o)$ and $F_y(x_o + \delta, y_o)$ are then recalculated. Stiffness coefficients K_{xx} and K_{yx} are given approximately by:

$$K_{xx} = \frac{F_x(x_o + \delta, y_o) - F_x(x_o, y_o)}{\delta}$$

$$K_{yx} = \frac{F_y(x_o + \delta, y_o) - F_y(x_o, y_o)}{\delta} \qquad (31)$$

K_{xy} and K_{yy} are similarly computed from a small journal displacement in the y direction.

Damping coefficients B_{xx}, B_{xy}, etc. are calculated by solving Reynolds equation to determine the bearing squeeze film forces that arise due to journal velocities v_x' and v_y'. For incompressible lubricants, squeeze film forces are linearly proportional to squeezing velocity provided boundary conditions for the pressure solution do not change.

Substituting Equations 29 and 30 into Equation 28 gives the equations of motion of a rigid rotor supported in a fluid film bearing. Velocities v_x' and v_y' in Equations 29 and 30 are replaced by their equivalents dx'/dt and dy'/dt.

$$M \frac{d^2x'}{dt^2} = W_x' - K_{xx}x' - K_{xy}y' - B_{xx}\frac{dx'}{dt} - B_{xy}\frac{dy'}{dt} \qquad (32)$$

$$M \frac{d^2y'}{dt^2} = W_y' - K_{yx}x' - K_{yy}y' - B_{yx}\frac{dx'}{dt} - B_{yy}\frac{dy'}{dt} \qquad (33)$$

Coupled linear Equations 32 and 33 relate bearing dynamic motions x' and y' to the imposed dynamic load components $W_x'(t)$ and $W_y'(t)$. Following four solutions of the associated Reynolds equations to determine the eight stiffness and damping coefficients, Equations 32 and 33 are then solved for "bearing response" and "bearing stability".

Bearing Natural Frequency

Natural frequencies of a rigid rotor are those at which the rotor would tend to vibrate if it were initially disturbed and then left free to return to steady state equilibrium. Natural frequencies of rotor-bearing systems are referred to as critical speeds because large vibration response is often experienced when running speed is the same as the natural frequency. Mathematically, these natural frequencies for a single bearing are determined by equations of motion 32 and 33 with dynamic forces $W_x'(t)$ and $W_y'(t)$ set equal to zero. The solution to this set of equations may be expressed in the form:

$$x'(t) = \text{Real} \{ x^* e^{st} \}$$
$$y'(t) = \text{Real} \{ y^* e^{st} \} \qquad (34)$$

where x* and y* are complex amplitudes which are not functions of time and where $s = \lambda + i\nu$ is a complex "frequency". Acutal motion $x'(t)$ is given by the real part of the complex number formed by multiplying x* and e^{st}. The last terms can be given as

$$e^{st} = e^\lambda e^{i\nu t} = e^{\lambda \tau} (\cos \nu t + i \sin \nu t) \qquad (35)$$

If we denote $x^* = \alpha + i\beta$, then

$$x'(t) = e^{\lambda t} (\alpha \cos \nu t - \beta \sin \nu t) \qquad (36)$$

Differentiating Equation 34 we find that

$$\frac{dx'}{dt} = \text{Real} \{ s\, x^* e^{st} \}$$

$$\frac{d^2 x'}{dt^2} = \text{Real} \{ s^2\, x^* e^{st} \} \qquad (37)$$

Thus we obtain from Equations 32 and 33

$$\text{Real} \{ [Ms^2\, x^* + B_{xx} s\, x^* + B_{xy} s\, y^* + K_{xx} x^* + K_{xy} y^*]\, e^{st} \} = 0 \qquad (38)$$

$$\text{Real} \{ [Ms^2\, y^* + B_{yx} s\, x^* + B_{yy} s\, y^* + K_{yx} x^* + K_{yy} y^*]\, e^{st} \} = 0 \qquad (39)$$

With the real part of the complex number inside the brackets forced to be zero, canceling out the common term e^{st} gives the following simultaneous equations for x^* and y^* in matrix form.

$$\begin{bmatrix} Ms^2 + B_{xx} s + K_{xx} & B_{xy} s + K_{xy} \\ B_{yx} s + K_{yx} & Ms^2 + B_{yy} s + K_{yy} \end{bmatrix} \begin{Bmatrix} x^* \\ y^* \end{Bmatrix} = 0 \qquad (40)$$

According to linear equation theory, nonzero solutions for natural vibration amplitudes x^* and y^* will be obtained if and only if the determinant of the left side matrix in Equation 40 is zero. This leads to the following characteristic equation for s:

$$Ms^4 + M(B_{xx} + B_{yy})\, s^3 + [M(K_{xx} + K_{yy}) + (B_{xx} B_{yy} - B_{yx} B_{xy})]\, s^2$$

$$+ (B_{xx} K_{yy} + B_{yy} K_{xx} - B_{yx} K_{xy} - B_{xy} K_{yx})\, s + K_{xx} K_{yy} - K_{yx} K_{xy} = 0 \qquad (41)$$

Many "canned" computer routines are available for ready solution of this fourth order polynomial equation for complex number s. Solutions for s occur in complex conjugate pairs, i.e., the four solutions for s are

$$s_1 = \lambda_1 + i\nu_1$$

$$s_2 = \lambda_1 - i\nu_1$$

$$s_3 = \lambda_2 + i\nu_2$$

$$s_4 = \lambda_2 - i\nu_2 \qquad (42)$$

That is, dynamic vibration $x'(t)$, in the absence of an external forcing function, will decay back to an equilibrium state at a rate given by the exponential decay factor λ_1 and λ_2, and at frequency ν_1 or ν_2. If the damping factor roots λ_1 and λ_2 are both negative, the bearing is stable: free vibrations will tend to die out. On the other hand, if either λ_1 or λ_2 is positive, then vibration will tend to grow without external excitation, an unstable condition.

For a simplified calculation of bearing natural frequencies, consider a bearing in which cross-coupling stiffness terms K_{xy} and K_{yx} are zero. Tilting pad bearings satisfy this condition.

As a further simplification, neglect damping in the solution for ν_1 and ν_2. This simplification is valid for bearings having negligible cross-coupling stiffness and damping coefficients, but is often not valid for fixed-arc bearings where the cross-coupling terms are significant.

With these simplifying assumptions, the bearing coefficient matrix becomes simply

$$\begin{bmatrix} K_{xx} + Ms^2 & \\ & K_{yy} + Ms^2 \end{bmatrix}$$

The characteristic equation for s is

$$(K_{xx} + Ms^2)(K_{yy} + Ms^2) = 0 \tag{43}$$

where solutions are

$$s_1 = i\sqrt{\frac{K_{xx}}{M}}$$

$$s_2 = -i\sqrt{\frac{K_{xx}}{M}}$$

$$s_3 = i\sqrt{\frac{K_{yy}}{M}}$$

$$s_4 = -i\sqrt{\frac{K_{yy}}{M}} \tag{44}$$

Referring to Equation 42 we see that our two natural frequencies are

$$\nu_1 = \sqrt{\frac{K_{xx}}{M}}$$

$$\nu_2 = \sqrt{\frac{K_{yy}}{M}} \tag{45}$$

which are the natural frequencies of two simple, uncoupled, spring-mass systems.

Bearing Response

When a rotor supported in a fluid film bearing is subjected to an arbitrary dynamic loading $W'(t)$, linearized response motions $x'(t)$ and $y'(t)$ would require numerical integration of Equations 32 and 33. Frequently, however, dynamic loading is harmonic in nature and is most commonly due to shaft unbalance. Such a rotating dynamic loading can be described by the following equations:

$$W_x' = A \cos \omega t$$

$$W_y' = A \sin \omega t \tag{46}$$

where A is the amplitude of the loading and ω is the frequency. Substituting into Equations 32 and 33 we obtain

$$M \frac{d^2x'}{dt^2} + B_{xx} \frac{dx'}{dt} + B_{xy} \frac{dy'}{dt} + K_{xx}x' + K_{xy}y' = A \cos \omega t \quad (47)$$

$$M \frac{d^2y'}{dt^2} + B_{yx} \frac{dx'}{dt} + B_{yy} \frac{dy'}{dt} + K_{yx}x' + K_{yy}y' = A \sin \omega t \quad (48)$$

The following particular solution for x' and y' satisfies the right-hand sides of Equations 47 and 48:

$$x'(t) = A \alpha_x \cos \omega t - A \beta_x \sin \omega t \qquad (49)$$

$$y'(t) = A \alpha_y \cos \omega t - A \beta_y \sin \omega t \qquad (50)$$

where:

$$\alpha_x = \frac{\psi(K_{yy} - M\omega^2) + \phi\omega B_{yy} - \psi\omega B_{xy} + \phi K_{xy}}{\psi^2 + \phi^2}$$

$$\beta_x = \frac{\psi\omega B_{yy} - \phi(K_{yy} - M\omega^2) + \psi K_{xy} + \phi\omega B_{xy}}{\psi^2 + \phi^2}$$

$$\alpha_y = \frac{\psi\omega B_{xx} - \phi(K_{xx} - M\omega^2) - \psi K_{yx} - \phi\omega B_{yx}}{\psi^2 + \phi^2}$$

$$\beta_y = \frac{-\psi(K_{xx} - M\omega^2) - \phi\omega B_{xx} - \psi\omega B_{yx} + \phi K_{yx}}{\psi^2 + \phi^2}$$

$$\psi = \omega^4 M^2 + \omega^2 [-B_{xx}B_{yy} + B_{xy}B_{yx} - M(K_{xx} + K_{yy})] + K_{xx}K_{yy} - K_{xy}K_{yx}$$

$$\phi = -\omega^3 M(B_{xx} + B_{yy}) + \omega(B_{xx}K_{yy} + B_{yy}K_{xx} - B_{xy}K_{yx} - B_{yx}K_{xy})$$

Equations 49 and 50 give the forced vibration response solution to Equations 47 and 48 as distinct from the natural frequency or free vibration solutions discussed earlier. Both forced and free vibration frequencies may be present in the total frequency spectrum of a vibrating rotor. Unless the natural frequency vibrations are excited by random forces, however, they will decay exponentially with time for a stable bearing. Forced vibrations, however, will be sustained at the stable amplitudes given by Equations 49 and 50.

For some insight into bearing response, consider a simplified bearing for which cross-coupling stiffness terms are zero and damping may be neglected. Equations 49 and 50 reduce to:

$$x'(*) = \frac{A}{(K_{xx} - M\omega^2)} \cos \omega t$$

$$y'(*) = \frac{A}{(K_{yy} - M\omega^2)} \sin \omega t \qquad (51)$$

When the dynamic loading frequency ω becomes equal to one of the natural frequencies as given by Equation 45, the response amplitude becomes infinitely large in either the x or y direction.

For purpose of discussion, assume that $K_{yy} < K_{xx}$. For values of ω less than $(K_{yy}/M)^{1/2}$, i.e., below the critical speed, Equations 51 yield the response orbit shown schematically in Figure 11a. This ellipse is traced out by the center of the rotor whirling in the same direction as shaft rotation. In this region, the y amplitude will always be greater than the x.

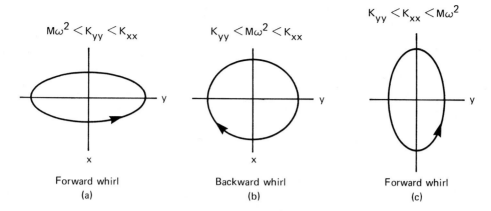

FIGURE 11.   Typical response ellipses for undamped bearing with no cross-coupling coefficients.

For values of ω between ν_1 and ν_2, the response orbit will be backward whirling as shown in Figure 11b. As ω increases from ν_1 to ν_2, the y amplitude will decrease and the x amplitude will increase. For values of ω greater than ν_2 in Figure 11c, the response orbit again reverts to forward whirl with the x amplitude greater. With damping and cross-coupling terms included, the axis of response ellipses will, in general, be inclined at an angle to the horizontal. Damping also limits response amplitude as the rotor passes through critical speeds.

Bearing Stability

As noted earlier, a bearing is unstable if the real part, λ, of any roots of characteristic Equation 41 are positive. A simpler criterion for stability is the "threshold of stability", i.e., the point where λ passes through zero going from negative to positive and the corresponding value for $s = \lambda + i\nu$ becomes purely imaginary.

Operating conditions for this threshold are determined by setting $s = i\nu$ and then solving Equation 41 to determine what critical values of shaft mass M_c and frequency ν_c are required in order that the rotor-bearing system be at the threshold of stability.

$$\nu_c^2 = \frac{(K_{xx} - K)(K_{yy} - K) - K_{xy}K_{yx}}{B_{xx}B_{yy} - B_{xy}B_{yx}} \tag{52}$$

$$M_c = K/\nu_c^2 \tag{53}$$

where

$$K = \frac{K_{xx}B_{yy} + K_{yy}B_{xx} - K_{xy}B_{yx} - K_{yx}B_{xy}}{B_{xx} + B_{yy}} \tag{54}$$

ν_c is the instability whirl frequency while M_c is the critical rotor mass for onset of instability. If the actual shaft mass is less than M_c, the system will be stable and conversely.

These stability formulas may be extended to a simple, symmetrical, flexible rotor of effective shaft stiffness K_R, supported on two equal bearings, by modifying the expression for M_c as follows:

$$M_c = \frac{1}{\nu_c^2} \frac{2KK_R}{2K + K_R} \tag{55}$$

Greater shaft flexibility (lower K_R) tends to decrease M_c, i.e., decrease stability. Shaft

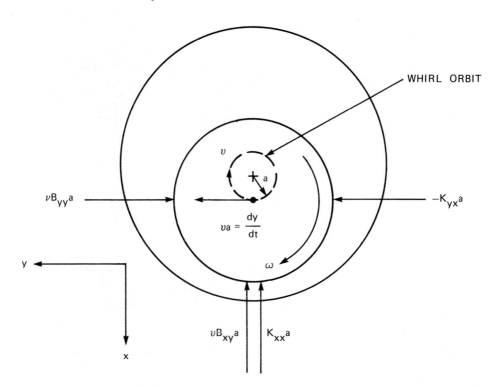

FIGURE 12.   Dynamic forces acting on whirling journal.

stiffness, K_R, can be determined approximately by dividing the weight of the rotor by the mid-span deflection resulting from that weight.

Figure 12 indicates the role played by the various bearing coefficients in stability. The journal is shown whirling at the frequency ν in a circular orbit of radius a about its equilibrium position. At the instant depicted, the journal center is displaced in the x direction while the velocity of the bearing center is in the y direction and given by $dy/dt = \nu a$. If dynamic forces acting on the journal in both the x and y directions are assumed to be in equilibrium, the journal is just at the "threshold" of instability and the whirl orbit is neither growing or decaying. Then

$$-K_{yx}a = B_{yy}\nu a \text{ (y direction)}$$

$$K_{xx}a + B_{xy}\nu a = M\nu^2 a \text{ (x direction)} \qquad (56)$$

In the y direction, cross-coupling force $-K_{yx}a$ tending to drive the journal in whirl must be balanced by damping term $B_{yy}\nu a$. In the x direction, restoring force, $K_{xx}a + B_{xy}\nu a$, tending to increase the orbit, must balance centrifugal force $M\nu^2 a$ which tends to increase the orbit. The first equation determines the whirl frequency

$$\nu = -K_{yx}/B_{yy} \qquad (57)$$

Typically, for fixed arc bearings operating at light loads, ν is very close to half the journal rotational speed. As loading increases, ν decreases as the bearing becomes more stable.

The equation for the x direction gives the "critical" shaft mass M_c necessary for the system to be at the threshold of instability:

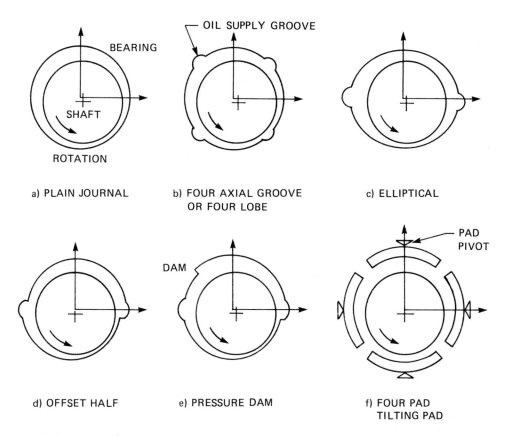

FIGURE 13. Geometry for various bearing types. (From Allaire, P. E., in *Fundamentals of the Design of Fluid Film Bearings,* Rhode, S. M., Maday, C. J., and Allaire, P. E., Eds., American Society of Mechanical Engineers, New York, 1979. With permission.)

$$M_c = (K_{xx} + B_{xy}\nu)/\nu^2 \qquad (58)$$

where ν would be given by the preceding equation. If the actual shaft mass is less than M_c, the whirl orbit will diminish in amplitude (system is stable) while if the actual shaft mass is larger, the system will be unstable.

The above analytical exercise is not valid because it considers only one point on the orbit. In general, forces change as the shaft moves about the orbit and what happens "on average" determines system stability. The exercise does identify, however, the role of cross-coupling terms in promoting whirl, the type of balance which determines the whirl frequency ratio, and a critical mass criterion for stability.

Design of Bearings For Optimum Dynamic Performance

Preceding equations provide tools for evaluating bearing dynamic behavior from stiffness and damping coefficients. Fortunately, such data is being published for a variety of bearing geometries (References 22 to 25). Some bearing types in common use are shown schematically in Figure 13.[26] The first two bearings, the plain journal and the four-axial groove, are circular in shape. The arcs of elliptical and four-lobe bearings are displaced inwards some fraction of the bearing clearance to introduce preload. In bearing (d), the bearing arcs are displaced sideways for preload. Bearing (e) is usually circular but with a groove machined over a portion of one of the bearing arcs, the circumferential end of the groove being a dam.

Tilting pad bearing (f) is the most stable of all and is commonly employed in high-speed

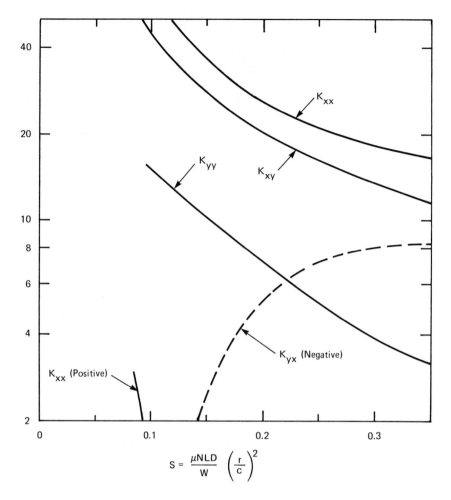

FIGURE 14.   Elliptical bearing stiffness coefficients. L/D = 0.5; preload = 0.5. (From Allaire, P. E., Nicholas, J. C., Gunter, E. J., and Pan, C. H. T., Incompressible Fluid Film Bearings, U.S.A.F. Tech. Rep. AFAPL-TR-78-6, Air Force Aero Propulsion Laboratory, Wright-Patterson Air Force Base, Ohio, March 1980.)

machinery. Since each arc is free to pivot, the center of pressure generated in each bearing arc must be opposite the pivot point. Thus, if the shaft in Figure 13f moves directly down toward the bottom pad, the resultant force acts directly up, eliminating the cross-coupling force components which contribute to instability.

Stiffness and damping for fluid film bearings are usually presented in terms of a dimensionless stiffness coefficient Kc/W and a dimensionless damping coefficient ωBc/W plotted or tabulated as a function of either Sommerfeld Number S = (μNLD/W)(r/c)^2 or eccentricity ratio ϵ = e/c. A typical plot of $\overline{K}$ vs. S for an elliptical bearing is shown in Figure 14.[22] As load increases in Figure 14, S decreases and the cross-coupling coefficient K_{yx} decreases to provide increased stability. A common cure for unstable bearing designs is to increase unit load W/LD by decreasing the bearing length.

Relative stability of some bearing types in Figure 15[26] is based on threshold of stability analysis represented by Equations 52 to 54. In this case, stability results are plotted in terms of a dimensionless critical mass $\overline{M}_c$ defined as $\omega\sqrt{cM_c/W}$. Figure 16 shows the corresponding whirl ratio ν_c/ω at the threshold of instability.

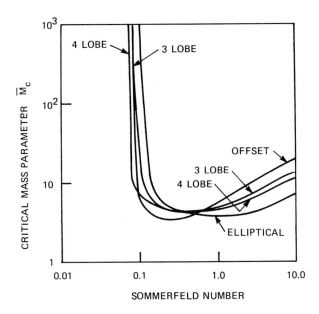

FIGURE 15. Linearized bearing stability for a rigid rotor (elliptical, 3-lobe and 4-lobe bearings have preload = 0.5). Offset bearing has 0.5 offset. (From Allaire, P. E., in *Fundamentals of the Design of Fluid Film Bearings,* Rohde, S. M., Maday, C. J., and Allaire, P. E., Eds., American Society of Mechanical Engineers, New York, 1979. With permission.)

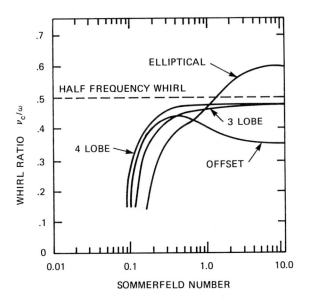

FIGURE 16. Whirl speed ratio vs. Sommerfeld number. (From Allaire, P. E., in *Fundamentals of the Design of Fluid Film Bearings,* Rohde, S. M., Maday, C. J., and Allaire, P. E., Eds., American Society of Mechanical Engineers, New York, 1979. With permission.)

NOMENCLATURE

B	=	Bearing length
c	=	Journal bearing radial clearance
D	=	Journal diameter
e	=	Eccentricity
F_f	=	Friction force
h_o	=	Slider minimum film thickness
h_1	=	Slider maximum film thickness
h	=	Film thickness
L	=	Bearing width
N	=	Rotational speed, cycles per second
P	=	Pressure
Q	=	Volumetric flow rate
q_x	=	x-Component of lubricant flux
q_y	=	y-Component of lubricant flux
r	=	Radius, radial coordinate
R_e	=	Reynolds number, cUp/μ
t	=	Time
T	=	Temperature
u	=	x-Component of fluid velocity
U, U_1, U_2	=	x-Components of surface velocities
U_r	=	Radial component of surface velocity in cylindrical coordinates
v	=	y-Component of fluid velocity
V, V_1, V_2	=	y-Components of surface velocities
V	=	Tangential component of velocity in cylindrical coordinates
w	=	z-Component of fluid velocity
W	=	Load capacity
$\overline{X}$	=	Center of pressure coordinates
ϵ	=	Eccentricity ratio e/c
θ	=	Journal bearing angular coordinate
μ	=	Dynamic viscosity
ρ	=	Density
δ	=	Thrust bearing angular coordinate
ψ	=	Journal bearing attitude angle
ω	=	Angular velocity
ν	=	Kinematic viscosity μ/ρ, vibration frequency

REFERENCES

1. **Booker, J. F.,** A table of the journal-bearing integral, *Trans. ASME Ser. D,* 87, 533, 1965.
2. **Schlichting, H.,** *Boundary Layer Theory,* 6th ed., McGraw-Hill, New York, 1968.
3. **Taylor, G. I.,** Stability of a viscous liquid contained between two rotating cylinders, *Philos. Trans. R. Soc. London Ser. A,* 223, 289, 1923.
4. **Vohr, J. H.,** An experimental study of Taylor vortices and turbulence in flow between eccentric rotating cyclinders, *Trans. ASME Ser. F,* 90, 285, 1968.
5. **DiPrima, R. C. and Stuart, J. T.,** Non-local effects in the stability of flow between eccentric rotating cylinders, *J. Fluid Mech.,* 54, 393, 1972.
6. **Mobbs, F. R. and Younes, M. A.,** The Taylor vortex regime in the flow between eccentric rotating cylinders, *Trans. ASME Ser. F,* 96, 127, 1974.

7. **Frene, J. and Godet, M.,** Flow transition criteria in a journal bearing, *Trans. ASME Ser. F,* 96, 135, 1974.

8. **Gross, W. et al.,** *Fluid Film Lubrication,* John Wiley & Sons, New York, 1980.

9. **Taylor, G. I.,** Fluid friction between rotating cylinders. I. Torque measurements, *Proc. R. Soc. A.,* 157, 546, 1936.

10. **Constantinescu, V. N.,** On turbulent lubrication, *Proc. Inst. Mech. Eng.,* 173(38), 881, 1959.

11. **Elrod, H. G. and Ng, C. W.,** A theory for turbulent fluid films and its application to bearings, *J. Lubr. Technol., Trans. ASME Ser. F,* 89(3), 346, 1967.

12. **Hirs, G. G.,** A bulk theory for turbulence in lubricant films, *J. Lubr. Technol., Trans. ASME,* 95(2), 137, 1973.

13. **Ng, C. W. and Pan, C. H. T.,** A linearized turbulent lubrication theory, *J. Basic Eng., Trans. ASME,* 87(4), 675, 1965.

14. **Constantinescu, V. N.,** Basic relationships in turbulent lubrication and their extension to include thermal effects, *J. Lubr. Technol., Trans. ASME,* 95, 147, 1973.

15. **Constantinescu, V. N., Pan, C. H. T., and Hsing, F. C.,** A procedure for the analysis of bearings operating in the transition range between laminar and fully developed turbulent flow, *Rev. Roum, Sci. Techn.-Mech. Appl.,* 16(5), 945, 1971.

16. **Reddeclif, J. M. and Vohr, J. H.,** Hydrostatic bearings for cryogenic rocket engine turbopumps, *J. Lubr. Technol., Trans. ASME Ser. F,* 91(3), 557, 1969.

17. **Capitao, J. W., Gregory, R. S., and Whitford, R. P.,** Effects of high operating speeds on tilting pad thrust bearing performance, *Trans. ASME Ser. F,* 98, 73, 1976.

18. **Vohr, J. H.,** Prediction of the operating temperature of thrust bearings, *J. Lubr. Technol., Trans. ASME,* 103(1), 97, 1981.

19. **Roben, G. D.,** private communication, General Electric Company, 1981.

20. **Abramovitz, S.,** Turbulence in a tilting-pad thrust bearing, *Trans. ASME,* 78, 7, 1956.

21. **Gregory, R. S.,** Performance of thrust bearings at high operating speeds, *Trans. ASME Ser. F,* 96, 7, 1974.

22. **Allaire, P. E., Nicholas, J. C., Gunter, E. J., and Pan, C. H. T.,** Rotor Bearing Dynamics Technology Design Guide, V. Dynamic Analysis of Incompressible Fluid Film Bearings, U.S.A.F. Tech. Rep. AFAPL-TR-78-6, Air Force Aero Propulsion Laboratory Wright-Patterson Air Force Base, Ohio, March 1980.

23. **Nicholas, J. C., Gunter, E. J., and Allaire, P. E.,** Stiffness and damping coefficients for the five-pad tilting pad bearing, *ASLE Trans.,* 22(2), 113, 1979.

24. **Nicholas, J. C., Allaire, P. E., and Lewis, D. W.,** Stiffness and damping coefficients for finite length step journal bearings, *ASLE Trans.,* 23(4), 353, 1980.

25. **Lund, J. W. and Thomsen, K. K.,** A calculation method and data for the dynamic coefficients of oil-lubricated journal bearings, in *Topics in Fluid Film Bearing and Rotor Bearing System Design and Optimization,* American Society of Mechanical Engineers, New York, 1978, 1.

26. **Allaire, P. E.,** Design of journal bearings for high speed rotating machinery, in *Fundamentals of the Design of Fluid Film Bearings,* Rohde, S. M., Maday, C. J., and Allaire, P. E., Eds., American Society of Mechanical Engineers, New York, 1979.

NUMERICAL METHODS IN HYDRODYNAMIC LUBRICATION

J. H. Vohr

INTRODUCTION

Analysis of the performance of fluid-film bearings involves the solution of Reynolds equation, derived previously in the chapter on Hydrodynamic Lubrication (Volume II). For an incompressible lubricant, the steady state form of this equation is

$$\frac{\partial}{\partial x}\left(\frac{h^3}{12\mu}\frac{\partial P}{\partial x}\right) + \frac{\partial}{\partial y}\left(\frac{h^3}{12\mu}\frac{\partial P}{\partial y}\right) = \frac{U}{2}\frac{\partial h}{\partial x} \qquad (1)$$

While there are many analytical solutions to this equation for relatively simple geometries, the usual method of solution is by numerical analysis, employing either finite difference or finite element techniques.[1-3] The following provides a brief description of these procedures.

FINITE DIFFERENCE SOLUTIONS

Assuming constant lubricant viscosity, Equation 1 can be rewritten

$$3h^2\frac{\partial h}{\partial x}\frac{\partial P}{\partial x} + h^3\frac{\partial^2 P}{\partial x^2} + 3h^2\frac{\partial h}{\partial y}\frac{\partial P}{\partial y} + h^3\frac{\partial^2 P}{\partial y^2} = 6\mu U\frac{\partial h}{\partial x} \qquad (2)$$

The purpose of finite difference approximations is to reduce a continuous differential equation, such as Equation 2, to a series of algebraic equations that can be solved for the pressure distribution $P(x,y)$ for a known distribution of $h(x,y)$ and $\mu(x,y)$.

First, the distribution of all functions in all coordinates concerned are replaced by samplings at several discrete points. Usually the spacing between points along each coordinate axis is chosen to be constant, but uneven spacing may be more suitable for some special applications. The governing differential equations are then written at each point of the coordinate grid with suitable approximations for the functions and their derivatives. Most common of such approximations are central difference formulas applicable to a uniform grid spacing as shown in Figure 1.

$$x = (i - 1)\,\Delta x \qquad (3)$$

$$y = (j - 1)\,\Delta y \qquad (4)$$

$$f(x,y) = f(i,j) \qquad (5)$$

$$\partial P/\partial x = [P(i+1,j) - P(i-1,j)]/2\Delta x \qquad (6)$$

$$\partial P/\partial y = [P(i,j+1) - P(i,j-1)]/2\Delta y \qquad (7)$$

$$\partial^2 P/\partial x^2 = [P(i+1,j) - 2P(i,j) + P(i-1,j)]/(\Delta x)^2 \qquad (8)$$

$$\partial^2 P/\partial y^2 = [P(i,j+1) - 2P(i,j) + P(i,j-1)]/(\Delta y)^2 \qquad (9)$$

$$\partial^2 P/\partial x\partial y = [P(i+1,j+1) - P(i-1,j+1) + P(i-1,j-1) - P(i+1,j-1)]/4\Delta x\Delta y \qquad (10)$$

Substituting these approximations in Equation 2, one obtains the following algebraic equation in the unknown pressure $P(i,j)$, $P(i+1,j)$; $P(i-1,j)$, $P(i,j+1)$, and $P(i,j-1)$ at each i,j grid point.

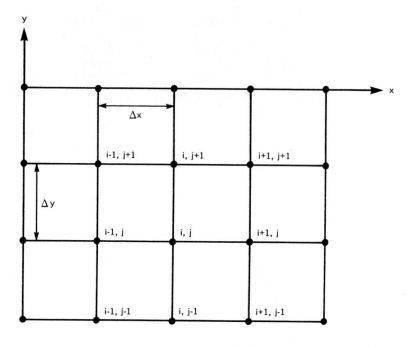

FIGURE 1.   Uniform grid spacing for central difference formulas.

$$3h^2 \, \partial h/\partial x \left[\frac{P(i+1,j) - P(i-1,j)}{2\Delta y} \right] + h^3 \left[\frac{P(i+1,j) - 2P(i,j) + P(i-1,j)}{(\Delta x)^2} \right] \tag{11}$$

$$+ 3h^2 \, \partial h/\partial y \left[\frac{P(i,j+1) - P(i,j-1)}{2\Delta y} \right] + h^3 \left[\frac{P(i,j+1) - 2P(i,j) + P(i,j-1)}{(\Delta y)^2} \right] = 6\mu \, U \, \partial h/\partial x$$

Analytical expressions are usually known for the clearance functions, h, $\partial h/\partial x$, and $\partial h/\partial y$. Their value is then obtained at the x,y coordinate locations.

By writing difference equations of the form of Equation 11 at each point within the lubricant film where Reynolds' equation applies, one obtains a set of algebraic equations for the discrete unknown pressure P(i,j). The set of equations for P(i,j) is completed by the various boundary conditions which the pressure P must satisfy. For locations at the edges of the lubricant film, one simply sets the pressure equal to the known boundary pressure P_a:

$$P(i,j) = P_a \tag{12}$$

A symmetry boundary condition at, say, j = N is handled by writing the normal field difference equations Equations 11 along the line j = N, N being the total number of grid lines used in the y direction. These equations will involve pressures P(i,N + 1) defined along the nonexistent grid line j = N + 1. These pressures are replaced by the pressures P(i, N − 1) reflecting that by symmetry

$$P(i, N + 1) = P(i, N - 1) \tag{13}$$

Difference Equations 11, together with the boundary relationships described above, provide a set of M × N independent linear algebraic equations for the M × N discretized pressures defined at each i,j grid point. This set of equations can be solved by several methods, some of which are described in linear algebra and numerical analysis texts. Three

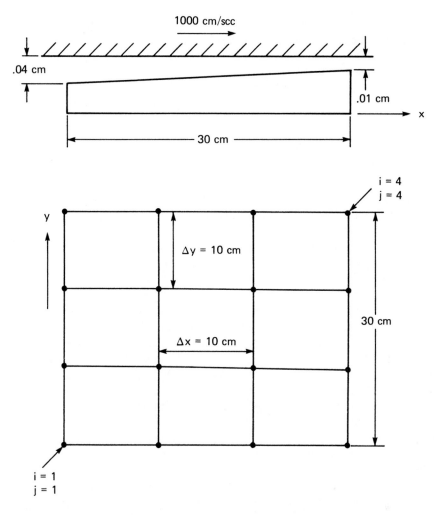

FIGURE 2.   Notation for inclined slider pad.

major methods will be discussed here: relaxation, direct matrix inversion, and columnwise influence coefficients.

Relaxation

This method reduces all equations in the system to equations in only one unknown only which can be solved immediately. All other unknowns in each equation are considered to be known and equal to some previously computed value. Therefore, an initial distribution is guessed and then successively improved by solving the system of equations in one unknown. As an example, consider the square inclined slider thrust pad shown in Figure 2. For the given geometry, $\partial h/\partial x = \Delta h/\Delta x = 0.001$ and $\partial h/\partial y = 0$ throughout the film. Applying Equation 11 at the $i = 2$, $j = 2$ coordinate point and assuming $\mu = 0.01$ Pa-sec and $U = 1000$ cm/sec.

$$3(0.03)^2.001 \left[\frac{P(3,2) - P(1,2)}{20} \right] + (0.03)^3 \left[\frac{P(3,2) - 2P(2,2) + P(1,2)}{(10)^2} \right] \tag{14}$$

$$+ (0.03)^3 \left[\frac{P(2,3) - 2P(2,2) + P(2,1)}{(10)^2} \right] = 6(0.01)1000(0.001)$$

Writing Equation 11 at the three remaining interior points yields four equations such as Equation 14 in which $P(i,j)$ is expressed in terms of its neighbors. Pressures on the boundaries $(i = 1, j = 1, i = 4,$ and $j = 4)$ are set equal to zero. With this boundary condition imposed, Equation 14 may be rewritten

$$P(2,2) = 1/4 \ [P(3,2) + P(2,3)] \ + 3/8 \ 10/0.03 \ P(3,2) - 6/4 \ 1/(0.03)^3 \tag{15}$$

Similarly, at the $i = 2, j = 3$ grid point

$$P(2,3) = 1/4 \ [P(3,3) + P(2,2)] \ + 3/8 \ 10/0.03 \ P(3,3) - 6/4 \ 1/(0.03)^3 \tag{16}$$

and so forth for $i = 3, j = 2$ and $i = 3, j = 3$.

Starting with some assumed initial distribution for the interior pressures $P(i,j)$ [$P(i,j) = 0$ everywhere is an adequate choice], one can solve Equation 15 for $P(2,2)$. This "updated" value for $P(2,2)$ may now be used in Equation 16, together with the initially assumed zero for the rest of the pressures, to obtain an "updated" value for $P(2,3)$. One proceeds in this fashion up the $i = 3$ grid columns, solving for $P(3,2)$ and $P(3,3)$, always using the most recently updated values of surrounding pressures when solving for $P(i,j)$. Successive sweeps through the finite difference grid continue until successive interior pressures $P(i,j)$ converge to a final distribution. Typically, 30 sweeps may be required for convergence.

Several improved procedures of applying relaxation methods exist such as "overrelaxation", which strives to accelerate the convergence process by multiplying the change in pressure from one sweep to the next by a factor such as 1.2. A drawback of relaxation procedures is their proneness to numerical instability. This phenomenon sets in depending on the value of the coefficients, and of the first guess distribution.

As seen in the example above, the equation written for point (i,j) is solved for $P(i,j)$. Therefore, the value of $P(i,j)$ is corrected at each iteration by the influence of its immediate neighbors. As a consequence, the influence of each point propagates one grid interval each iteration and the number of iterations involved in achieving steady-state solutions is proportional to the number of points on the longest side of the grid. In addition, the number of operations is proportional to the total number of grid points $N \times M$. Therefore, the total time needed for converging to a steady-state solution is proportional to $N^2 \times M$, where N has been assumed to be larger than M. For computation speed, the relaxation method has advantages over the others for large values of N and M. As a general rule, however, the stability of a relaxation solution decreases when the number of grid points is increased. Slowing the pace of the solution ("underrelaxation") usually circumvents the problem but obviates the speed advantages for large grids. Since relaxation requires minimum internal computer capacity, this method still finds occasional use with a desk-top computer.

Direct Matrix Inversion

In this simple method, all equations are written together as a system.
Letting $L = N \times M$,

$$a_{11}q_1 + a_{12}q_2 + \cdots a_{1L}q_L = r_1$$

$$a_{21}q_1 + a_{22}q_2 + \cdots a_{2L}q_L = r_2$$

$$\cdot \ \cdot \ \cdot \ \cdot \ \cdot \ \cdot \ \cdot$$

$$a_{L1}q_1 + a_{L2}q_2 + \cdots a_{LL}q_L = r_L \tag{17}$$

where

$$q_1 = P(1,1); q_2 = P(1,2); \cdots q_N = P(1,N);$$

$$q_{N+1} = P(2,1); \cdot q_{2N} = P(2,N) \cdots q_L = P(M,N)$$

Obviously, most of the "a's" are zero. The system (Equations 17) may be written as in matrix notation as

$$[a]\{q\} = \{r\} \tag{18}$$

By methods such as Gaussian inversion, the problem can be solved as

$$\{q\} = [a]^{-1}\{r\} \tag{19}$$

Inversion of matrix [a] is performed by routines available at any computer installation. If this method were economical in computation time, it would be the most desirable since it solves the linear system of algebraic equations with a minimum of bookkeeping complications. Unfortunately, the computation time involved in performing the key matrix inversion step grows proportionately to L^3 or $M^3 \times N^3$ which is exorbitant except for very small grids.

Columnwise Influence Coefficients

This method is similar to matrix inversion, but takes advantage of some features of the system of equations. If we denote the jth column of unknown pressure as

$$\{P(i)\}_j \quad i = 1,M$$

then we can see by Equation 11 that the equations involving this j^{th} column vector also involve only the adjacent column vectors $\langle P(i)\rangle_{j-1}$ and $\langle P(i)\rangle_{j+1}$. These equations can be written in matrix form as

$$[A]_j\{P\}_j + [B]_j\{P\}_{j-1} + [C]_j\{P\}_{j+1} = \{R\}_j \tag{20}$$

where $[A]_j$, $[B]_j$, and $[C]_j$ are matrices which are $M \times M$ in size, M being the total number of grid points in the i direction, i.e., the "height" of each column vector $\langle P\rangle$ j. Matrices $[A]_j$, $[B]_j$, and $[C]_j$ can be determined from Equation 11.

At the start of the bearing film, i.e., at $j = 1$, pressure $\langle P\rangle$ is known and equal to the boundary pressure P_a. Consequently, Equation 20 at $j = 2$, gives

$$[A]_2\{P\}_2 + [B]_2\{P_a\} + [C]_2\{P\}_3 = \{R\}_2 \tag{21}$$

Since $\langle P_a\rangle$ is a column vector, each element of which is the known boundary pressure P_a, this equation can be written in the form

$$\{P\}_2 = [E]_3\{P\}_3 + \{F\}_3 \tag{22}$$

where

$$[E]_3 = -[A]_2^{-1}[C]_2$$
$$\{F\}_3 = -[A]_2^{-1}([B]_2\{P_a\} - \{R\}_2) \tag{23}$$

If next written at $j = 3$, Equation 20 will involve unknown column pressure $\langle P\rangle_2$,

$\langle P \rangle_3$, and $\langle P \rangle_4$. By using Equation 22 to eliminate $\langle P \rangle_2$ from this equation, one can obtain an equation containing only the unknown column pressures $\langle P \rangle_3$ and $\langle P \rangle_4$. We can write this equation formally like Equation 22, i.e.,

$$\{P\}_3 = [E]_4 \{P\}_4 + \{F\}_4 \tag{24}$$

By continuing this procedure, we can successively write equations of the form

$$\{P\}_{j-1} = [E]_j \{P\}_j + \{F\}_j \tag{25}$$

for all values of j from j = 3 to j = N. Moreover, Equation 26 also holds at j = 2 if we define $[E]_2 = 0$ and $\langle F \rangle_2 = \langle P_a \rangle$.

Substituting Equation 25 into Equation 20 we obtain

$$([A]_j + [B]_j[E]_j) \{P\}_j = [C]_j \{P\}_{j+1} + \{R\}_j - [B]_j \{F\}_j \tag{26}$$

Comparing Equation 26 with Equation 25 we can deduce the recursion relationship for determining $[E]_j$ and $\langle F \rangle_j$.

$$[E]_{j+1} = -[T]_j[C]_j$$

$$\{F\}_{j+1} = [T]_j(\{R\}_j - [B]_j\{F\}_j)$$

$$[T]_j = ([A]_j + [B]_j[E]_j)^{-1} \tag{27}$$

The solution procedure for determining the unknown column pressures $\langle P \rangle_j$ goes as follows:

1. Starting with the initial values $[E]_2 = 0$ and $\langle F \rangle_2 = \langle P_a \rangle$, use the recursion relationships (Equation 27) to determine $[E]_j$ and $\langle F \rangle_j$ for all values of j from 2 to N.
2. At the end of the bearing (j = N), set $\langle P \rangle_N$ equal to the known boundary pressure $\langle P_a \rangle$.
3. Use Equation 25 successively from j = N to j = 2 to determine all column pressures $\langle P \rangle_j$.

While the matrices which must be inverted here are of the order M × M, in the direct inversion method the matrix is of the order M × N. If M is chosen to be significantly smaller than N, computation cost of the columnwise method will be significantly less than that of the direct method.

A more detailed description of the columnwise solution method, including a discussion of cyclical boundary conditions, is given in Reference 1.

Calculation of Flow and Power Loss

Once the discretized pressure field is obtained, quantities such as bearing flow and power loss may be obtained from approximate finite difference expressions. For example, lubricant flow per unit width in the direction of sliding at the i,j grid point would be evaluated by the following central difference expression

$$q_x = -h^3 (i, j) \frac{[P(i+1, j) - P(i-1, j)]}{2\Delta x} + \frac{U\,h(i, j)}{2} \tag{28}$$

At the leading edge, i.e., at i = 1, a forward difference formula must be used

$$q_x = \frac{-h^3(1, j)}{12\mu} \frac{[P(2, j) - P(1, j)]}{\Delta x} + \frac{U h(1, j)}{2} \qquad (29)$$

Similarly, a backwards difference formula must be used at the trailing edge.

Power loss is obtained by multiplying surface shear stress τ_s acting on the surface by surface velocity. In finite difference form

$$\tau_s U = \frac{\mu U^2}{h(i, j)} + \frac{U[P(i + 1, j) - P(i - 1, j)] h(i, j)}{2\Delta x} \qquad (30)$$

Multiplying $\tau_s U$ by $\Delta x \Delta y$ and summing over the entire i,j grid, will yield the bearing power loss.

FINITE ELEMENT ANALYSIS

An alternate approach for solving Reynolds' equation is use of finite element analysis. Developed originally for use in structural analysis, this method has been finding increasing favor in the analysis of fluid film bearings. The method is particularly advantageous for complex film geometries since the elements into which the solution field is broken do not have to be uniform or follow coordinate lines.

Descriptions of the finite element method are given by Reddi,[4] Booker and Huebner,[5] Allaire et al.,[6] and Hays.[7] In this present discussion, the method will be described in its simplest form by solving for the pressure in a one-dimensional, step slider bearing following the sequence presented by Reddi.[4]

Reddi presents a proof that if a functional $\phi(P)$ is given by

$$\phi(P) = \int_R -\left[h^3/12\mu \, \vec{\nabla} P \cdot \vec{\nabla} P + h \, \vec{U} \cdot \vec{\nabla} P - 2\dot{h}P \right] dA + 2 \int_{C_2} q \, P \, dS \qquad (31)$$

integrated over the region R, comprising a lubricant film, then

$$\delta \phi(P) = 0 \qquad (32)$$

if and only if P is a solution of the incompressible lubrication problem. In Equation 31, μ, h, and P are the lubricant film viscosity, thickness, and pressure, respectively, $\vec{U}$ is the vector velocity of the bearing surface, and q represents the outflow flux of lubricant from the region R in in.^{3}/sec/in. (m^2/sec/m) across the boundary C_2.

In Equations 32 and 33 pressure P is described by a set of functions which are continuous over R and which satisfy prescribed boundary conditions. In the finite element method, region R is broken up into small elements and the distribution of pressure P over each element is described by convenient interpolation functions particular to the element. For example, triangular elements are convenient for two-dimensional solutions as shown in Figure 3a.

The most convenient interpolation function is a linear one, i.e., the pressure P_m over the mth triangular element is assumed to vary as

$$P_m(x,y) = a_{m_1} + a_{m_2} x + a_{m_3} y \qquad (33)$$

By evaluating pressure P_m at the three vertices or nodes of each element, one can solve for constants a_{m1}, a_{m2}, and a_{m3} in terms of the three unknown node pressures P_{m1}, P_{m2}, P_{m3} at

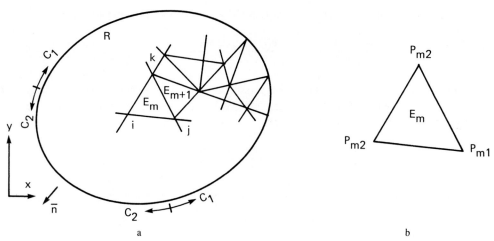

FIGURE 3.   Incompressible lubricant domain and finite element idealization (a), and single triangular element with node pressures (b).

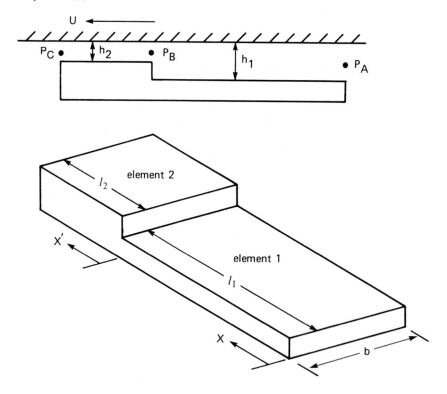

FIGURE 4.   One-dimensional step bearing with linear elements.

the boundaries of the mth element (see Figure 3b). Hence, the distribution of pressure P_m over each element and also the gradient of the pressure can be described in terms of the unknown node pressures at the boundaries of each element. Substitution of these functions for P_m and $\overrightarrow{\nabla}P_m$ for each discrete element m into Equation 31, integrating, and applying Equation 32, yields a set of linear algebraic equations in the unknown node pressures which may be solved for these pressures.

For a simple, one-dimensional problem, consider the infinitely wide-step bearing shown in Figure 4 in which the bearing film is broken into two finite elements of arbitrary width b and lengths ℓ_1 and ℓ_2, respectively. Since P is a function of x only, the finite elements are linear rather than triangular. The pressures of the nodes of element 1 are P_A and P_B while the pressures at the nodes of element 2 are P_B and P_C. Using a linear interpolation function for the pressure distributions over elements 1 and 2

$$P_1(x) = P_A + x/l_1 (P_B - P_A) \tag{34}$$

$$P_2(x') = P_B + x'/l_2 (P_C - P_B) \tag{35}$$

We can write these equations in the general matrix form used in Reference 4,

$$P_m(x) = T_m P \tag{36}$$

where $P_m(x)$ represents the variation of P over element m. P represents the column vector of the unknown node pressures

$$P = \begin{Bmatrix} P_A \\ P_B \\ P_C \end{Bmatrix} \tag{37}$$

and T_m represents the appropriate row vector relating $P_m(x)$ to P. In our simple example

$$T_1 = \{ \ (1 - x/l_1) \quad x/l_1 \quad 0 \ \}$$

$$T_2 = \{ \quad 0 \quad (1 - x'/l_2) \quad x'/l_2 \ \} \tag{38}$$

By differentiating Equation 36, one obtains the gradient of $P_m(x)$. For two-dimensional problems, the gradient would be a column vector, i.e.,

$$\vec{\nabla} P_m(x) = \begin{Bmatrix} \dfrac{\partial P_m}{\partial x} \\ \dfrac{\partial P_m}{\partial y} \end{Bmatrix} = \begin{Bmatrix} \dfrac{\partial T_m}{\partial x} \\ \dfrac{\partial T_m}{\partial y} \end{Bmatrix} P = R_m P \tag{39}$$

The matrix formed from differentiating the row vector T_m first with respect to x and then with respect to y is denoted as R_m. In our case, $\partial T_m/\partial y$ is zero so R_m reduces to a row vector, where

$$R_1 = \{ \ -1/l_1 \quad 1/l_1 \quad 0 \ \}$$

$$R_2 = \{ \quad 0 \quad -1/l_2 \quad 1/l_2 \ \} \tag{40}$$

Since we now have expressions for P and $\vec{\nabla}P$ over each element m, the integral over R in Equation 31 can be written as a summation of integrals over each element m. Using the general notation of Reference 4, Equation 31 becomes

$$\phi(P) = \sum_{m=1}^{N} \left[\int_m \left(P^T R_m^T C_m R_m - h_m U_m R_m + 2\dot{h}_m T_m \right) dA_m + 2 \int_{C_{2m}} q_m T_m dS_m \right] P \tag{41}$$

In our simple one-dimensional case, the element property term C_m is purely the scalar quantity $h_m^3/12\mu$ and the velocity U_m is simply the velocity U in the x direction. In the more general two-dimensional case described in Reference 4, C_m is a matrix and U_m is a vector. Equation 41 can be restated as

$$\phi(P) = (P^T K - V + H + Q)P \tag{42}$$

where K is the volumetric fluidity matrix, and V, H, and Q are the element flow, squeeze film, and boundary flow column vectors, respectively, given by:

$$K = \sum_{m=1}^{N} \int_m R_m^T C_m R_m dA_m$$

$$V = \sum_{m=1}^{N} \int_m h_m U_m R_m dA_m$$

$$H = \sum_{m=1}^{N} 2 \int_m \dot{h}_m T_m dA_m$$

$$Q = \sum_{m=1}^{N} 2 \int_{C_{2m}} q_m T_m dS_m \tag{43}$$

Let us now evaluate K, V, and Q for our example (H is zero for this steady-state case). For elements 1 and 2, from Equation 41

$$R_1^T C_1 R_1 = \frac{h_1^3}{12\mu} \begin{Bmatrix} -1/l_1 \\ 1/l_1 \\ 0 \end{Bmatrix} \{ -1/l_1 \quad 1/l_1 \quad 0 \} = \frac{h_1^3}{12\mu} \begin{bmatrix} 1/l_1^2 & -1/l_1^2 & 0 \\ -1/l_1^2 & 1/l_1^2 & 0 \\ 0 & 0 & 0 \end{bmatrix} \tag{44}$$

$$R_2^T C_2 R_2 = \frac{h_2^3}{12\mu} \begin{Bmatrix} 0 \\ -1/l_2 \\ 1/l_2 \end{Bmatrix} \{ 0 \quad -1/l_2 \quad 1/l_2 \} = \frac{h_2^3}{12\mu} \begin{bmatrix} 0 & 0 & 0 \\ 0 & 1/l_2^2 & -1/l_2^2 \\ 0 & -1/l_2^2 & 1/l_2^2 \end{bmatrix} \tag{45}$$

Substituting these into the expression for K, integrating over each element, and then summing over the elements:

$$K = \frac{b}{12\mu} \begin{bmatrix} h_1^3/l_1 & -h_1^3/l_1 & 0 \\ -h_1^3/l_1 & (h_1^3/l_1 + h_2^3/l_2) & -h_2^3/l_2 \\ 0 & -h_2^3/l_2 & h_2^3/l_2 \end{bmatrix} \tag{46}$$

For the element flow vector V,

$$h_1 U_1 R_1 = h_1 U \{ -1/l_1 \quad 1/l_1 \quad 0 \}$$

$$h_2 U_2 R_2 = h_2 U \{ 0 \quad -1/l_2 \quad -1/l_2 \} \tag{47}$$

Substituting these into the expression for V, integrating over each element, and summing

$$V = \{-b\,U\,h_1 \quad b\,U\,(h_1 - h_2) \quad b\,U\,h_2\} \tag{48}$$

Finally, for Q, flows q_1 and q_2 *leaving* the bearing per unit width at the leading and trailing edge boundaries, respectively, are

$$q_1 = -U\,h_1/2 + (h_1^3/12\mu)\,(P_B - P_A)/l_1$$

$$q_2 = U\,h_2/2 - (h_2^3/12\mu)\,(P_C - P_B)/l_2 \tag{49}$$

Using expressions (Equation 38) for T_1 and T_2, $x = 0$ at the leading edge boundary and $x' = \ell_2$ at the trailing edge:

$$q_1 T_1 = \{\,[-U\,h_1/2 + (h_1^3/12\mu)\,(P_B - P_A)/l_1\,] \quad 0 \qquad\qquad 0 \qquad\qquad\}$$

$$q_2 T_2 = \{\qquad\qquad 0 \qquad\qquad 0 \quad [U\,h_2/2 + (h_2^3/12\mu)\,(P_C - P_B)/l_2\,]\,\} \tag{50}$$

Substituting these expressions into Equation 45 for Q, integrating along the leading and trailing edge boundaries, and summing

$$Q = 2b\left\{\left(\frac{-U\,h_1}{2} + \frac{h_1^3}{12\mu}\frac{(P_B - P_A)}{l_1}\right)\; 0 \;\left(\frac{U\,h_2}{2} - \frac{h_2^3}{12\mu}\frac{(P_C - P_B)}{l_2}\right)\right\} \tag{51}$$

Now, Equation 32 states that if the pressure P is a solution of the incompressible lubrication problem, then $\delta\phi(P) = 0$ over the region R

$$\delta\,\phi\,(P) = \sum_i \frac{\partial\phi}{\partial P_i}\,\delta P_i = 0 \tag{52}$$

where P_i are the unknown pressures at the nodes of the elements. Since δP_i is an independent variation, it follows that $\partial\phi/\partial P_i = 0$ for each i. From Equation 42

$$\left\{\frac{\partial\phi}{\partial P_i}\right\} = 2\,P^T K - V + H + Q = 0 \tag{53}$$

or

$$KP = 1/2(V^T - H^T - Q^T) \tag{54}$$

Equation 54, in general, yields a system of simultaneous equations for the nodal pressures P which may be solved by any of a number of established techniques. In our simple example, Equation 54 becomes

$$\frac{b}{12\mu}\begin{bmatrix} h_1^3/l_1 & -h_1^3/l_1 & 0 \\ -h_1^3/l_1 & (h_1^3/l_1 + h_2^3/l_2) & -h_2^3/l_2 \\ 0 & -h_2^3/l_2 & h_2^3/l_2 \end{bmatrix}\begin{Bmatrix} P_A \\ P_B \\ P_C \end{Bmatrix}$$

$$= \frac{1}{2}\begin{Bmatrix} -b\,U\,h_1 \\ b\,U(h_1 - h_2) \\ b\,U\,h_2 \end{Bmatrix} - b\begin{Bmatrix} \dfrac{-U\,h_1}{2} + \dfrac{h_1^3}{12\mu}\dfrac{(P_B - P_A)}{l_1} \\ 0 \\ \dfrac{U\,h_2}{2} - \dfrac{h_2^3}{12\mu}\dfrac{(P_C - P_B)}{l_2} \end{Bmatrix} \tag{55}$$

To solve Equation 55, conditions must be supplied for boundary pressures P_A and P_C. For the simplest case where $P_A = P_C = 0$, the set of three equations represented by Equation 55 reduces to the following single equation for P_B.

$$\frac{b}{12\mu} \left(\frac{h_1^3}{l_1} + \frac{h_2^3}{l_2} \right) P_B = \frac{1}{2} \, b \, U \, (h_1 - h_2) \qquad (56)$$

The above example illustrates the steps involved in setting up the matrices and vectors associated with a finite element solution. As such, it is intended to provide only a "flavor" of how the method works. In this simple case, the answer could have been obtained more easily by a less involved procedure, so our example does not illustrate the power of the method. Numerical advantages the finite element method has over finite difference techniques are discussed in Reference 4.

The numerical techniques described in the preceding pages may also be applied to solution of the energy equation derived in the chapter on Hydrodynamic Lubrication (Volume II). Application of the columnwise influence coefficient finite difference method to solving both the pressure and energy equations for a tilting pad thrust bearing are given in Reference 8. Application of the finite element method to solving for both the static and dynamic characteristics of tilting pad and step journal bearings is provided in References 9 and 10.

REFERENCES

1. **Castelli, V. and Shapiro, W.,** Improved method for numerical solutions of the general incompressible fluid film lubrication problem, *J. Lubr. Technol., Trans. ASME,* 89(2), 211, April 1969.
2. **Castelli, V. and Pirvics, J.,** Review of numerical methods in gas bearing film analysis, *J. Lubr. Technol., Trans. ASME,* 90, October 1968.
3. **Castelli, V.,** Numerical methods, in *MTI Gas Bearing Design Manual,* Wilcock, D. F., Ed., Mechanical Technology Inc., Latham, N.Y., 1972.
4. **Reddi, M. M.,** Finite element solution of the incompressible lubrication problem, *J. Lubr. Technol., Trans. ASME, Ser. F,* 91(3), 524, July 1969.
5. **Booker, J. F. and Huebner, K. H.,** Application of finite elements to lubrication; an engineering approach, *J. Lubr. Technol.,* Trans. ASME, Ser. F, 24(4), 313, October 1972.
6. **Allaire, P. E., Nicholas, J. C., and Gunter, E. J.,** Systems of finite elements for finite bearings, *J. Lubr. Technol., Trans. ASME, Ser. F,* 99(2), 187, April 1977.
7. **Hays, D. F.,** A variational approach to lubrication problems and the solution of the finite journal bearing, *J. Basic Eng., Trans. ASME,* 81(1), 13, March 1959.
8. **Castelli, V. and Malanoski, S. B.,** Method for solution of lubrication problems with temperature and elasticity effects: application to sector, tilting-pad bearings, *J. Lubr. Technol., Trans. ASME,* 91, 634, October 1969.
9. **Nicholas, J. C., Gunter, E. J., and Allaire, P. E.,** Stiffness and damping coefficients for the five-pad tilting pad bearing, *ASLE Trans.,* 22(2), 113, April 1979.
10. **Nicholas, J. C., Allaire, P. E., and Lewis, D. W.,** Stiffness and damping coefficients for finite length step journal bearings, *ASLE Trans.,* 23(4), 353, October 1980.

HYDROSTATIC LUBRICATION

R. C. Elwell

INTRODUCTION

In contrast to the *hydrodynamic* bearings described earlier, *hydrostatic* bearings support loads on fluid supplied from an external source, usually a pump, which feeds pressurized fluid to the film. For this reason, these bearings are often called "externally pressurized".

Hydrostatic bearings are used over a wide range of applications, from miniature high-precision devices to large heavy masses, for the following reasons:

1. Friction is theoretically zero at zero rotational speed. Heavy objects can be moved with small forces because the weight is "floated" on a fluid film.
2. Friction is repeatable. There is no mechanical contact between the parts, hence no wear or burnishing of surfaces. This is important in machine tools and other precision equipment, e.g., radar antennas.
3. Fluids with poor lubricating qualities can be used. Gases (nitrogen, helium, air), water, and liquid metals (sodium) have all been used.
4. Large loads can be lifted on small bearing areas. In many machines, e.g., ore crushers, axial space for journal bearings is limited. A compact bearing can be used here, with fluid pressures of the order of 56 MPa (8000 psi). Reference 1 describes such a bearing, designed to accommodate local structural deflections.
5. High startup loads on hydrodynamic bearings can be relieved by hydrostatic pockets built into their surfaces. These "oil lifts" are common in hydrogenerators and large steam turbines.

In recent years, there have been a number of large-scale, low-pressure applications of the basic principle. Among these are "air pallets" for handling materials, and Mile High Stadium in Denver in which an entire grandstand with 21,000 seats slides back on hydrostatic water films to convert the stadium from football to baseball.

BASIC THEORY

Almost any fluid may be used in a hydrostatic bearing but this chapter will discuss only incompressible fluids (liquids) in laminar flow. Fluid compressibility and turbulence introduce special problems in gas bearings and high speed applications. The literature on these topics is abundant.

Load

Load is the product of pressure and bearing area:

$$W = k(P_p - P_o) A \tag{1}$$

The constant k is less than 1 and is a function only of the bearing geometry. One major method of analysis[2] uses a "load coefficient" (a_f) for the same purpose. The constant may be thought of as the efficiency of the design in using the recess pressure to support the load. For the geometry in Figure 1 for instance, k is approximately equal to $(a - b)^2/a^2$ and $A = a^2$. The load as a function of supply pressure for this bearing is therefore:

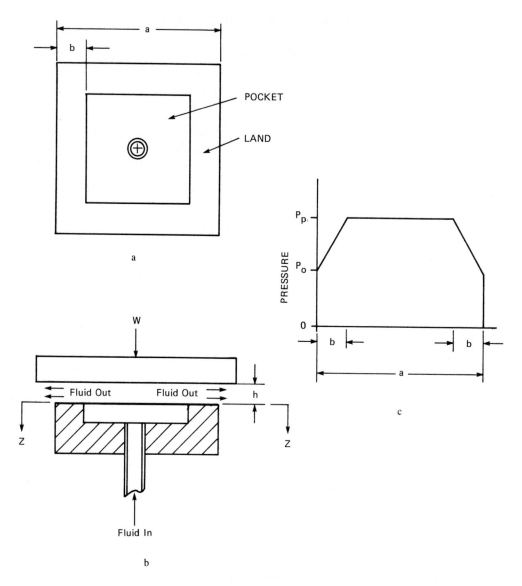

FIGURE 1.   A square hydrostatic bearing. Pocket is sometimes called "recess"; another name for land is "sill". (a) Cross-section; (b) view Z-Z; and (c) pressure distribution.

$$W \simeq (P_p - P_o)(a - b)^2 \qquad (2)$$

The height (h) the load is lifted is determined by the flow rate.

Flow

Pressure P_p is assumed to be uniform across a recess, or "pocket", in the bearing surface. (This will not necessarily be so for a very shallow pocket, or in a gas bearing.) The pressure is sealed by the lands, or "sills", over which the fluid must flow in all directions from the pocket. As fluid flows through the gaps over these lands, pressure drops in accordance with the following equation for laminar flow through a slot:

$$Q = \frac{h^3 w(P_p - P_o)}{12\mu\ell} \qquad (3)$$

For the bearing in Figure 1, w $\simeq$ (a − b) and ℓ = b. In most applications, P_o will be atmospheric ambient pressure; some nuclear coolant pumps, however, have hydrostatic journal bearings in which P_o is of the order of 14 MPa (2030 psi), and P_p is only slightly higher. Calculating performance of such a bearing requires that the fluid viscosity be corrected for pressure.

If Equation 3 is applied to a rectangular slot, for example the gap over one of the lands in Figure 1a, the pressure drops linearly along the distance b (Figure 1c). Integration of the pressure distribution over the area A gives the load capacity W. An especially important aspect of Equation 3 is its sensitivity to the film thickness (h). As a practical matter, this limits the separation between the moving and fixed parts of most bearings. The flow rate usually becomes prohibitively high when the film is thicker than about 0.1 mm (0.004 in.).

Power

Power required to supply fluid to the bearing is given by:

$$H = QP_s/\eta \qquad\qquad (4)$$

Ordinarily the power consumption of a hydrostatic bearing system is not large enough to be important. Occasionally it becomes a problem because most of the pump power is converted to heat when the fluid pressure is expended in the bearing. This can be serious in high-precision applications such as astronomical telescopes and machine tools.

In the hydrostatic steadyrest of a large lathe, for example, a bearing with a geometry resembling Figure 1 supports a heavy steel forging near midspan. The dimensions are a = 100 mm (3.94 in.) and b = 25 mm (0.98 in.). Oil is supplied at a rate of 0.025 ℓ/sec (0.4 gpm) and a supply pressure of 56 MPa (8000 psi) by a fixed-displacement pump with an efficiency of 0.5. The oil has a specific heat of 890 J/ℓ (3.2 Btu/gal). From Equation 4, the input power is calculated in SI units as follows:

$$H = 0.025 \ (56 \times 10^6)/1000 \ (0.5) = 2800 \ \text{watts}$$

In customary U.S. units, the power calculation would be:

$$H = 0.4(231)(8000)/60(12)(550)(0.5) = 3.73 \ \text{hp}$$

If all this power is converted to heat in the small oil flow, the resulting temperature rise of the fluid would be 69°C (124°F). With a reservoir temperature of 38°C (100°F) this would result in an oil temperature of 107°C (224°F) leaving the bearing. This local flow of hot oil can cause a "hot spot" on the object being supported. In the lathe application, much of the heat is absorbed by the rotating forging and is distributed by conduction and radiation. In stationary applications local heating can cause distortions which may interfere with supporting the load.

It is possible to minimize the power consumption with a rotating load because friction (shear) losses come down and pump losses go up as the film thickness increases. The simple case of a flat thrust bearing is given in Reference 6. A more general discussion of optimizing for minimum loss is given in Reference 3, based upon the parameter S_H developed by Rowe et al.[4] In applications sensitive to bearing heating, either of the above analytical methods may be applied.

Flow Restrictors

Most hydrostatic thrust bearings use more than one pocket in order to resist misalignment or off-center loads (Figure 2). (Journal bearings have many pockets distributed around their

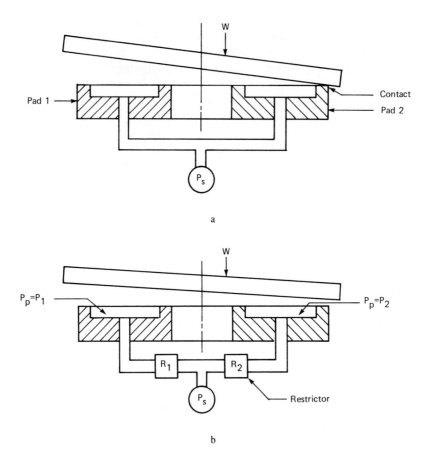

FIGURE 2. Use of restrictors to support offset load on a multipocket bearing. (a) Unrestricted flow and (b) restricted flow, $P_2 > P_1$.

bore and can be ''unwrapped'' into developed surfaces resembling thrust bearings with off-center loads.)

When the entire bearing flow is supplied from a single source, pressure differences must be maintained between the pockets in order to support asymmetric loading. In Figure 2a, an off-center load calls for Pad 2 to generate a larger force than Pad 1. Because their pockets directly communicate however, their pressures remain the same. Now, with the larger gaps over Pad 1 more fluid will flow out of it than out of Pad 2. By Equation 3 the film thickness over Pad 2 is therefore reduced. If the floating member is free to tilt, contact will occur as shown.

Putting a hydraulic resistance (a ''restrictor'') in each supply line (Figure 2b) isolates them from each other at the cost of a line pressure drop ($P_s - P_p$). The flow increase through Pad 1 causes a larger pressure drop across restrictor R_1 than across R_2. The resulting pressure distribution is now asymmetric ($P_2 > P_1$) and supports the load without contact.

The three dominant types of restrictors used in hydrostatic bearings are discussed below in their approximate order of practicality. Other less widely used restrictors include thin feed slots, servo valves,[3] and porous metals in gas bearings.[5]

Orifice

Orifices are often used in the form of simple threaded plugs. The equation for flow is as follows:

$$Q_o = CA_o\sqrt{2(P_s - P_p)/\rho}$$ (5)

Constant Flow

$$Q_k = \text{a constant}$$ (6)

Multiple gear pumps, mechanically coupled together, are commonly used for supplying constant flow to several pockets at once. Each pump supplies a single pocket. Single-pump systems with a constant flow valve in each line are also used.

Capillary Tube

$$Q_c = \frac{\pi d_c^4 (P_s - P_p)}{128\mu L_c}$$ (7)

Equation 7 emphasizes one of the difficulties in using capillary tubes: slight variations in the bore anywhere along the length are serious because of the fourth power exponent. Furthermore, d_c cannot be measured except at the ends.

Restricted bearings are analyzed by equating the inflow through the restrictor (Equations 5 to 7) to the outflow from the bearing (Equation 3). The resulting equation is solved for the pocket pressure as illustrated in the next section.

THRUST BEARINGS

Single Acting

Analysis of the circular thrust bearing with a central pocket (Figure 3) is available from several sources[6,7] and has been verified in experiments.[8] Integration of the pressure distribution in this bearing gives the load as a function of pocket pressure:

$$W = \frac{\pi(P_p - P_o)}{2} \left[\frac{R^2 - R_o^2}{\ln (R/R_o)} \right]$$ (8)

Equation 3 written for this geometry gives the flow.

$$Q = \frac{\pi h^3 (P_p - P_o)}{6\mu \ln (R/R_o)}$$ (9)

A restrictor in the supply line forces the bearing flow to assume the characteristic of the restrictor. Equations 5 to 7 result in the flow-vs.-load relations illustrated in Figure 4 since the pocket pressure in these equations is a linear function of load (Equation 8). By equating each restrictor flow equation with the flow equation for this bearing (Equation 9), load capacity expressions can be derived as below.

Orifice Restriction

$$W = \frac{\pi K_2 K_3^2}{4h^6} \left[-1 + \sqrt{1 + \frac{4P_s h^6}{K_3^2}} \right]$$ (10)

Constant Flow

$$W = 3\mu Q_K (R^2 - R_o^2)/h^3$$ (11)

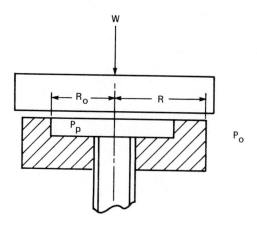

FIGURE 3.   Circular hydrostatic bearing.

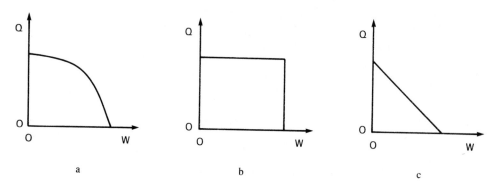

FIGURE 4.   Flow vs. load for a thrust bearing with three types of restrictors. (a) Orifice; (b) constant flow; and (c) capillary.

Capillary Restriction

$$W = \frac{\pi(R^2 - R_0^2)P_s}{2\left[\dfrac{64L_c h^3}{3d_c^4} + \ell n \dfrac{R}{R_0}\right]}$$ (12)

Note that the load-film thickness relation for the capillary-restricted bearing is not dependent upon viscosity as it is for the other two types. This characteristic may be useful in sensitive applications.

When the three types of restrictors are compared at a common operating point, differences in the film stiffnesses are apparent (Figure 5). The constant-flow bearing is stiffest, followed by the orifice-restricted and capillary-restricted. The constant flow bearing has a maximum load, set by the maximum supply pressure.

Double-Acting

Two thrust bearings are often used in opposition. Each bearing is calculated independently over the possible range of film thickness, and the load, flow, and power losses are combined for the assembly.

The following example for two unequal-size bearings supporting a rotor (Figure 6) will illustrate the method of Reference 2. Supply pressure (P_s) is 0.8 MPa (116 psi) and end clearance is 0.5 mm (0.0197 in.). Oil viscosity is 0.025 Pa·sec (25 cP). Geometry is as follows:

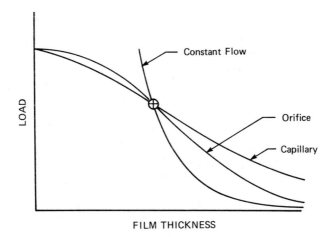

FIGURE 5. Effect of restrictor type on stiffness of a thrust bearing (Reference 8).

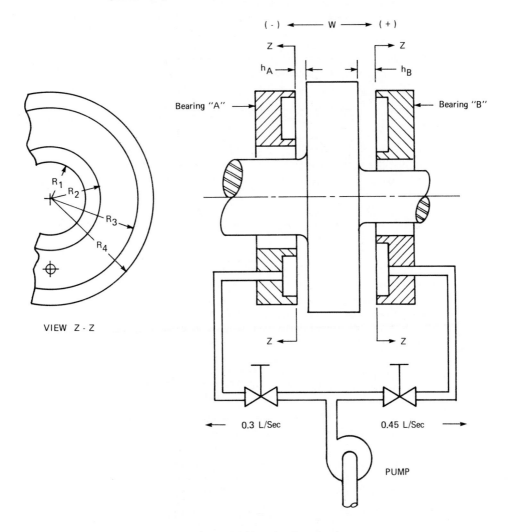

FIGURE 6. Double-acting thrust bearing.

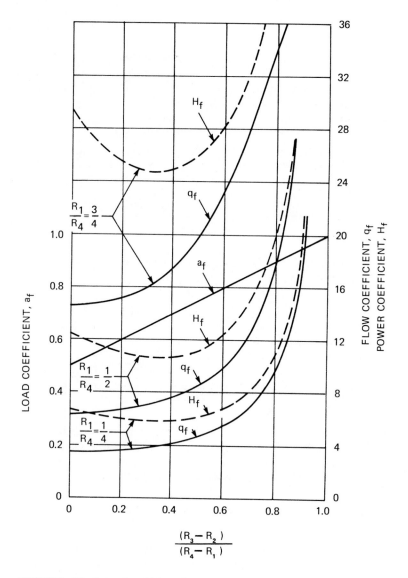

FIGURE 7. Bearing pad coefficients for annular thrust pad bearing. Annular recess is centrally located within bearing width ($R_1 + R_4 = R_2 + R_3$). Curve for a_f applies to all R_1/R_4 ratios. (Reproduced from Rippel, H. C., *Cast Bronze Hydrostatic Bearing Design Manual*, Courtesy of Cast Bronze Bearing Institute, Chicago.)

	Bearing A		Bearing B	
	mm	(in.)	mm	(in.)
R_1	100	(3.94)	50	(1.97)
R_2	120	(4.72)	70	(2.76)
R_3	180	(7.09)	180	(7.09)
R_4	200	(7.87)	200	(7.87)
k_v = Flow, ℓ/sec (gpm)	0.30	(4.76)	0.45	(7.13)
R_1/R_4	1/2		1/4	
$(R_3 - R_2)/R_4 - R_1)$	0.60		0.73	

From Figure 7 for annular thrust pad bearings:

	Bearing A	Bearing B
a_f (Load coeff.)	0.80	0.87
q_f (Flow coeff.)	10.0	7.5
H_f (Power coeff.)	12.4	8.6
$\beta = W/a_f A_p P_s$	16.58×10^{-6} W	12.2×10^{-6} W

Values of β (Equation 21 of Reference 2) are calculated over a range of loads for each bearing. Performance coefficients for film thickness, pump power, stiffness and flow are then taken from Figure 8. In a constant-flow bearing, pump power is proportional to load so k_H equals β. Film thickness coefficients over the range of loads in this example are as follows:

	Bearing A		Bearing B	
W, N	β	k_h	β	k_h
82,000	—	—	1.00	1.00
70,000	—	—	.85	1.06
60,300	1.00	1.00	.74	1.11
50,000	.83	1.07	.61	1.19
30,000	.50	1.26	.37	1.41
10,000	.17	1.80	.12	2.05

Film thickness is calculated from the following relation from Reference 2:

$$h = k_h \sqrt{\frac{k_v \mu}{a_f q_f P_s}}$$

The load capacity for each bearing is an inverse function of the film thickness cubed. When the load increases to the value at which the pocket pressure reaches the supply pressure, the calculated film thickness becomes indeterminate. This happens at 60.3 kN on bearing A and 82 kN on bearing B. Pump power (neglecting efficiency) is calculated from:

$$H = k_H k_v P_s$$

To find the total power required, the calculated values must be divided by the pump efficiency (of the order of 0.5 to 0.9 for most gear, vane, and piston-type pumps).

Combining the individual bearing performances curves gives the results shown in Figure 9 for the overall assembly. Figure 9a shows that the total usable load capacity is about +70 to −50 kN and that the rotor would move a distance of 0.26 mm (0.010 in.) between these loads. Power loss of the combined bearing varies from 70 to 360 W (Figure 9b).

JOURNAL BEARINGS

A common arrangement for a full journal bearing with six pockets is shown in Figure 10. The bearing is in a housing with an annular groove supplying fluid to all orifices. With the load downward as in Figure 10, the pressure in the bottom pocket (P_L) rises, by Equation 1. The orifice feeding the top pocket causes the pressure there (P_u) to go down as its flow increases. Analysis of hydrostatic journal bearings is more complex than for thrust bearings because the film thickness is not generally uniform over the individual pads and the direction of load is usually not limited.

A wide range of geometries is possible. Reference 9 presents design charts for four-pocket journal bearings loaded in the direction of a pocket only. The same method was extended

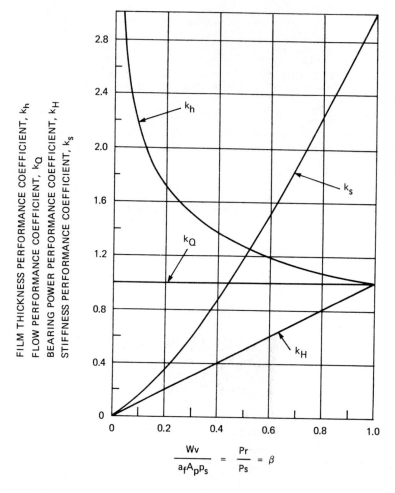

FIGURE 8.   Performance coefficients of flow-control valve-compensated bearings. (Reproduced from Rippel, H. C., *Cast Bronze Hydrostatic Bearing Design Manual*, Courtesy of Cast Bronze Bearing Institute, Chicago.)

to six-pocket bearings in Reference 10 which compared analytical results to experiments on 41.3 mm (1.625 in.)-diameter bearings using light turbine oil. Although maximum test speed was only 1000 rpm this caused substantial equalization of pocket pressures and a loss of load capacity. In designs with different proportions, moderate rotational speeds may enhance overall load capacity due to hydrodynamic pressure generation. Without rotation, pressure differences between pockets cause circumferential flow as internal leakage (Figure 10b). This reduces the net overall force.

Reference 11 analyzes circumferential flow in bearings with and without axial drain slots between pockets. This is a good reference for basic equations applicable to bearings with four different flow controls: orifice, capillary, constant flow, and "diaphragm valve", a form of proportional control which accentuates the fluid film stiffness. Many practical design examples and questions are considered, including some effects of rotational speed.

Reference 3 is part of a useful series on externally pressurized journal bearings. Six pockets are compared to four, and are recommended as optimum. A unique geometry is described in which thin slits in the wall replace the more conventional orifice or capillary restrictors. Charts permit selection of bearing variables to minimize power consumption and maximize load capacity. A four-pocket bearing design is worked out for the following specifications:

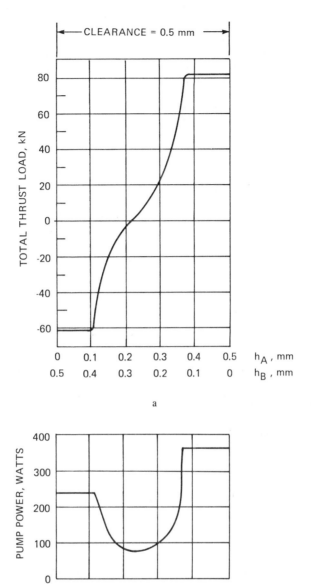

FIGURE 9. Performance of bearing assembly shown in Figure 6. (a) Load capacity; (b) pump power.

Concentric pressure ratio (β = pocket pressure ÷ supply pressure) = 0.4
Load = 4000 N (900 lb) at 50% eccentricity (max)
Load parameter (W) at design load ≃ 0.25
Bore diameter = 64 mm (2.54 in.)
Diametral clearance = 0.080 mm (0.0031 in.)
Max supply pressure = 5 MPa (725 psi)
Viscosity = 0.014 Pa·sec (14 cP)
Number of pockets = 4
L/D = 1
Axial land length = L/4

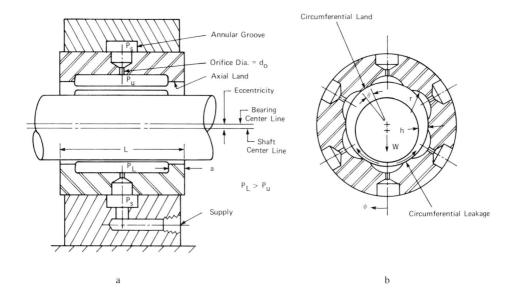

FIGURE 10.   Section through hydrostatic journal bearing with downward load on a pocket. (a) Longitudinal section through bearing and housing; (b) radial section through bearing.

Supply pressure at the design load is calculated from another form of Equation 1:

$$P_s = W/(D^2 \overline{W}) = 4000/((0.064)^2 \times 0.25) = 3.9 \text{ MPa (round off to 4.0)}$$

From Figure 11a, the load at four different eccentricity ratios is obtained for $\beta = 0.4$, with the following results:

ϵ	W, N (lb)
0.25	2460 (553)
0.50	4710 (1060)
0.75	7370 (1660)
1.00	9010 (2026)

The design load is carried at less than 50% eccentricity, as specified.

From Figure 11b, the flow rate parameter is 0.8 with the journal centered in the bearing ($\beta = 0.4$, L/D = 1). This is the maximum because increasing eccentricity reduces the flow. Converting the flow rate parameter gives the oil flow, as follows:

$$Q = \overline{Q}P_s C_D^3/8\mu = 0.8(4)(0.08)^3/8(0.014)(10^6) = 0.015 \text{ ℓ/sec}$$

Reference 2 gives other data for single- and multiple-pad journal bearings which may be assembled into an approximate analysis of a complete cylindrical bearing. In Chapter 10 of Reference 2, however, the author concludes it is not possible to achieve sufficient accuracy without using a computer. Effects of fluid heating and circumferential leakage are described, and an approximate method is given for estimating the inception of half-frequency rotor whirl.

OIL LIFTS

To reduce starting friction, wear, and low speed vibration problems in large rotating

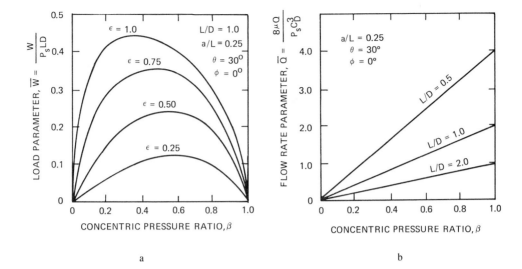

FIGURE 11. 4-Pocket recessed bearings. (a) Variation of load parameter with pressure ratio; (b) variation of flow parameter with pressure ratio. (Reproduced from Stout, K. J. and Rowe, W. B., *Tribol. Int.,* 7(5), 195, 1974. Courtesy of IPC Science and Technology Press Ltd.)

machines (hydrogenerators, steam turbines, electric motors), pressurized oil pockets called "oil lifts" are sometimes incorporated in their bearings. The machines are started (and usually stopped) with the lift pumps on. At some speed above which the bearings are operating hydrodynamically, the pumps are shut off. Reliable check valves are essential in the oil lift supply lines, otherwise flow will leak back through them from the hydrodynamic films, endangering the bearings at operating speed. This backflow can also drive the pumps backwards, leading to excessive wear in them and their drives.

Lift pockets are made as small as possible, i.e., a few percent of projected bearing area, in order to minimize damage from dirt pumped in with the lift oil. If possible the pockets are made elliptical, with the long axis in the direction of rotation to further reduce shaft damage. The juncture between the pocket and the bearing surface should be flared to make it easier for the pressurized oil to get between the bearing and the rotor surface. When the pockets are pressurized in large machines the oil takes several seconds to wend its way through the tiny local gaps formed by the surface finishes of the parts, until the pressure times the area equals the load and the rotor lifts.

Oil flow is very low during this process and the pocket pressure is as high as the pump can deliver. The pump must be protected (usually by a relief valve) from damage during this "breakaway" process. Attention must also be paid to the bond line in cast iron bearings to prevent the high-pressure oil from penetrating behind the babbitt and lifting it off. Usually a threaded fitting is screwed in beforehand and babbitt is cast around it. In steel-backed bearings the babbitt is metallurgically bonded and a hole is simply drilled through (Figure 12).

In large journal bearings with one or more lift pockets (Figure 12 shows a typical arrangement), the breakaway pressure has been found to be predictable by the following empirical relation:

$$P_{BA} = \frac{K_{BA}W}{Dn\sqrt{A_L}}$$ (13)

The constant K_{BA} is determined by test for a given type of machine. It may also vary

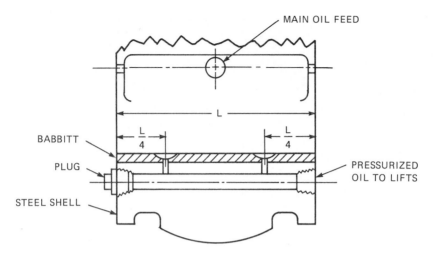

FIGURE 12. Two elliptical oil lift pockets located at quarter points along a journal bearing axis. Removing the plug allows the supply passages to be flushed.

with size. A_L, the area of a single bearing pocket, is usually made about 1.5% of A in a bearing of this type.

Once the load is lifted, the pocket pressure falls because pressurized oil is distributed over the whole bearing area. The flow, oil film thickness, and other quantities can be calculated for this condition as a pure hydrostatic bearing (before rotation starts). The lift is the following function of the pump flow:

$$Q = \frac{WC_D}{12\mu} \left(\frac{C_D}{D}\right)^2 \frac{(1-\epsilon)^2}{(2-\epsilon)} \qquad (14)$$

This is the corrected form of Equation (12–19) in Reference 12.

A numerical example will illustrate the use of the above equations, where Q = 0.07 ℓ/sec (1.1 gpm), D = 533.4 mm (21 in.), L = 304.8 mm (12 in.), C_D = 0.711 mm (0.028 in.), W = 382,500 N (86,000 lb), n = 2 pockets, A_L = 2580 mm² (4 in.²), and μ = 0.058 Pa·sec (8.4 μreyn).

Solving Equation 13, assuming K_{BA} = 3, gives a breakaway pressure of 21.1 MPa (3072 psi). The oil pump should be sized to give at least this much pressure with some margin (say, 50 to 100%) to allow for performance deterioration over a period of time. Employing appropriate units, Equation 14 gives an eccentricity ratio of 0.63. Assuming that the shaft moves straight up, the lift equals the minimum film thickness, $(1 - \epsilon)(C_D/2)$, or 0.132 mm (0.005 in.). This is a realistic value which will allow for some misalignment or shaft deflection.

In thrust bearings somewhat more pocket area has historically been used than in journal bearings, about 5% of the bearing area. The pad in Figure 13 is from a bearing with an outside diameter of 2286 mm (90 in.) with a design load of 4.1 MPa (600 psi). At breakaway the pocket pressure rises to 12.4 MPa (1800 psi). It then falls back to a steady-state value of 4.8 MPa (700 psi).

The lift (h) may be estimated by assuming a circular pressure distribution similar to the bearing in Figure 3. For the bearing pad in Figure 13 the following values may be assigned to the variables: Q = 0.025 ℓ/sec (0.4 gpm), P_p = 4.8 MPa (700 psi), μ = 0.056 Pa·sec (56 cP), (ISO VG 68 oil at 38°C), R = 286 mm (11.3 in.), and R_o = 51 mm (2.0 in.).

Solving Equation 9 for h, using consistent units, gives a lift of 0.099 mm (0.004 in.),

FIGURE 13.   Large thrust bearing pad with oil lift pocket. (Photo courtesy of General Electric Co., Schenectady, N.Y.)

comparable to the film thickness value computed above for the large journal bearing. Oil is metered to individual tilting thrust bearing pads, usually by constant-flow valves or fixed-displacement pumps, to prevent the supply from being vented out of any unloaded pad.

NOMENCLATURE

a   = Length of side, square pad (Figure 1)
A   = Total bearing area (LD for journal bearing)
A_L = Area of a single oil-lift pocket
A_o = Orifice area
b   = Length of land
C   = Orifice coefficient, see Equation 5
C_D = Diametral clearance in journal bearing
D   = Diameter of journal bearing
d_c = Capillary bore diameter
d_o = Orifice bore diameter
e   = Eccentricity in journal bearing
g   = Acceleration due to gravity
h   = Film thickness
H   = Power
K_1 = $1/\ell n\ (R/R_o)$
K_2 = $(R^2 - R_0^2)/\ell n(R/R_o)$
K_3 = $3Cd_o^2\mu/K_1\sqrt{2\rho}$
K_{BA} = Constant in Equation 13
k   = Load coefficient, Equation 1
ℓ = Length of land in direction of flow
L   = Length of journal bearing (overall), see Figure 10a

L_c = Length of capillary tube
n = Number of oil-lift pockets in journal bearing
P_{BA} = Breakaway pressure in oil lift
P_o = Ambient pressure around bearing sealing land
P_p = Pressure in bearing pocket
P_s = Supply pressure
Q = Flow
Q_c = Capillary tube flow
Q_k = Constant flow
Q_o = Orifice flow
$\overline{Q}$ = Flow rate parameter, Reference 3
R = Outside radius of thrust bearing (Figure 3)
R_o = Pocket radius of thrust bearing (Figure 3)
w = Width of land normal to direction of flow
W = Load
ϵ = Eccentricity ratio = e ÷ radial clearance
ρ = Fluid density
θ = Angle subtended by circumferential land in journal bearing (Figure 10)
ϕ = Direction of loading in journal bearing (zero is toward a pocket) (Figure 10)
μ = Fluid viscosity
η = Efficiency

REFERENCES

1. **Anon.,** Floating shoes form big bearings, *Mach. Design,* 49(27), 37, 1977.
2. **Rippel, H. C.,** *Cast Bronze Hydrostatic Bearing Design Manual,* Cast Bronze Bearing Institute, Chicago, 1975.
3. **Stout, K. J. and Rowe, W. B.,** Externally pressurized bearings — design for manufacture. III. Design of liquid externally pressurized bearings for manufacture including tolerancing procedures, *Tribol. Int.,* 7(5), 195, October 1974.
4. **Rowe, W. B., O'Donoghue, J. P., and Cameron, A.,** Optimization of externally pressurized bearings for minimum power and low temperature rise, *Tribology,* 3(4), 153, August 1970.
5. **Sneck, H. J.,** A survey of gas-lubricated porous bearings, *Trans. ASME, Ser. F,* 90(4), 804, October 1968.
6. **Fuller, D. D.,** Hydrostatic lubrication, in *Standard Handbook of Lubrication Engineering,* O'Connor, J. J., Boyd, J., and Avallone, E. A., Eds., McGraw-Hill, New York, 1968, 3-17.
7. **Szeri, A. Z.,** Hydrostatic bearings, in *Tribology: Friction, Lubrication, and Wear,* Szeri, A. Z., Ed., McGraw-Hill, New York, 1980, 47.
8. **Elwell, R. C. and Sternlicht, B.,** Theoretical and experimental analysis of hydrostatic thrust bearings, *Trans. ASME, Ser. D,* 82(3), 505, September 1960.
9. **Raimondi, A. A. and Boyd, J.,** Hydrostatic journal bearings (compensated), in *Standard Handbook of Lubrication,* O'Connor, J. J., Boyd, J., and Avallone, E. A., Eds., McGraw-Hill, New York, 1968, 5-66.
10. **Hunt, J. B. and Ahmed, K. M.,** Load capacity, stiffness and flow characteristics of a hydrostatically lubricated six-pocket journal bearing supporting a rotary spindle, Part 3N, *Proc. I.M.E.,* 182, 53, 1967-8.
11. **O'Donoghue, J. P. and Rowe, W. B.,** Hydrostatic bearing design, *Tribology,* 2(1), 25, February 1969.
12. **Wilcock, D. F. and Booser, E. R.,** *Bearing Design and Application,* McGraw-Hill, New York, 1957.

SQUEEZE FILMS AND BEARING DYNAMICS

J. F. Booker

INTRODUCTION

This chapter covers *transient* behavior of viscous lubricant films under loads which may be fixed or variable in magnitude and/or direction. Since it takes time for such films to be squeezed out from between surfaces, bearings can often carry surprisingly high *peak* loads as compared to those they might sustain in steady-state operation.

"Squeeze-film" action is often of interest because of the damping it provides. Occasionally such special devices as dampers for turbomachinery are involved; more often, as in reciprocating machinery, the damping action is provided by conventional bearings.

The following analysis begins with treatment of the normal approach of planar bearings. It proceeds with examination of cylindrical bearings in one- and two-dimensional translation, both without and with accompanying rotation. Finally, by way of an example for connecting-rod bearings, analysis is supplemented by a parametric design study and correlation of a failure criterion with field experience.

GENERAL REYNOLDS EQUATION

In its general form the incompressible Reynolds equation derived in an earlier chapter can be written in rectangular coordinates x, y

$$\frac{\partial}{\partial x}\left(\frac{h^3}{12\mu}\frac{\partial p}{\partial x}\right) + \frac{\partial}{\partial y}\left(\frac{h^3}{12\mu}\frac{\partial p}{\partial y}\right) = \frac{\partial}{\partial x}\left(h\bar{U}^x\right) + \frac{\partial}{\partial y}\left(h\bar{U}^y\right) + \frac{\partial h}{\partial t}$$

or in polar coordinates, r, θ

$$\frac{1}{r}\frac{\partial}{\partial r}\left(r\frac{h^3}{12\mu}\frac{\partial p}{\partial r}\right) + \frac{1}{r^2}\frac{\partial}{\partial \theta}\left(\frac{h^3}{12\mu}\frac{\partial p}{\partial \theta}\right) = \frac{1}{r}\frac{\partial}{\partial r}\left(rh\bar{U}^r\right) + \frac{1}{r}\frac{\partial}{\partial \theta}\left(h\bar{U}^\theta\right) + \frac{\partial h}{\partial t}$$

For an important class of normal approach "squeeze film" problems, the average tangential surface velocity $\bar{U}$ has negligible effect, leaving only the squeeze rate $\partial h/\partial t$ as an effective driving term.

PLANAR BEARINGS IN NORMAL APPROACH

For isoviscous planar normal approach with uniform film thickness, the Reynolds equation simplifies in rectangular coordinates to

$$\frac{\partial^2 p}{\partial x^2} + \frac{\partial^2 p}{\partial y^2} = 12\mu\dot{h}/h^3$$

or in polar coordinates

$$\frac{1}{r}\frac{\partial}{\partial r}\left(r\frac{\partial p}{\partial r}\right) + \frac{1}{r^2}\frac{\partial^2 p}{\partial \theta^2} = 12\mu\dot{h}/h^3$$

Special Formulation for Circular Section

As an example, consider Figure 1 in which a film is squeezed by the normal approach

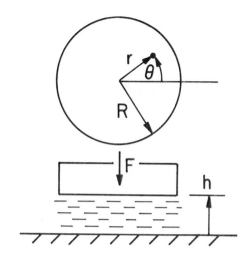

FIGURE 1.   Planar circular section in normal approach.

of a circular plate. Fully flooded boundary conditions at ambient pressure simply require p to vanish at radius R. These boundary conditions and the Reynolds equation give

$$p = -3(R^2 - r^2)\,\mu\dot{h}/h^3$$

with the maximum (central) value

$$p^* = -3R^2\mu\dot{h}/h^3$$

Resultant volumetric outflow rate is then

$$q = \frac{-\pi R h^3}{6\mu}\,\frac{\partial p}{\partial r}\bigg|_{r=R} = \pi R^2 \dot{h}$$

while the normal force applied to the lubricant film by the moving plate is

$$F = \int_A p\,dA = \int_0^{2\pi}\int_0^R p\,r\,dr\,d\theta$$

$$= \frac{-3\pi R^4}{2}\,\frac{\mu\dot{h}}{h^3}$$

Inverting gives the equation of motion

$$\dot{h} = -\frac{2\,h^3}{3\pi R^4 \mu}\,F$$

While numerical integration over time (e.g., by Euler's linear extrapolation) is straightforward, formal integration over some interval of approach gives the *general* solution

$$\int_{t_1}^{t_2} F\,dt = -\frac{3\pi R^4 \mu}{2}\int_{h_1}^{h_2}\frac{dh}{h^3} = \frac{3\pi R^4 \mu}{4}\,(h_2^{-2} - h_1^{-2})$$

where, for a *constant* load

$$\int_{t_1}^{t_2} F \, dt = F (t_2 - t_1)$$

For particular fixed values μ, R, and F, the above relations give an approach rate slowing asymptotically as final closure is approached. This qualitative behavior is typical of all "squeeze films" in response to time integral (impulse), *not* instantaneous values of loading.

General Formulation

The relations derived for the circular section are also valid for general geometries if expressed in terms of area A and dimensionless shape factors P and K. Thus,

$$p^* = P F/A$$

$$q = -A\dot{h} = - \frac{h^3}{KA\mu} F$$

$$F = - \frac{KA^2 \mu}{h^3} \dot{h}$$

$$\dot{h} = - \frac{h^3}{KA^2 \mu} F$$

$$\int_{t_1}^{t_2} F \, dt = \frac{KA^2 \mu}{2} (h_2^{-2} - h_1^{-2})$$

Shape factor P is a measure of the sharpness or nonuniformity of the pressure distribution, K the dynamic stiffness or damping rate of the lubricant film as a whole.

Circular Section

The circular section in the example has area $A = \pi R^2$ and shape factors

$$P = 2 \quad \text{and} \quad K = \frac{3}{2\pi} = 0.477$$

Elliptical Section

An elliptical section with major and minor diameters L and B has area $A = \pi LB/4$ and shape factors as shown in Figure 2

$$P = 2 \quad \text{and} \quad 1/K = (B/L + L/B) \pi/3$$

Note the reduction to the circular section result as slenderness ratio $B/L \rightarrow 1$.

Rectangular Section

A rectangular section with sides L and B has area $A = LB$ and shape factors P and K as shown in Figure 3. Though these results have been computed from an exact series solution,[1] they are quite accurately fit by the optimum approximate Warner solution[2,3] expressions

$$K \approx B/L[1 - \tanh(\Lambda L/B)/(\Lambda L/B)]$$

$$P \approx 3/2[1 - 1/\cosh (\Lambda L/B)]/[1 - \tanh (\Lambda L/B)/(\Lambda L/B)]$$

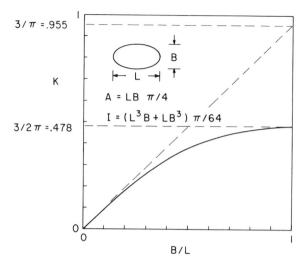

FIGURE 2. Shape factors for elliptical section.

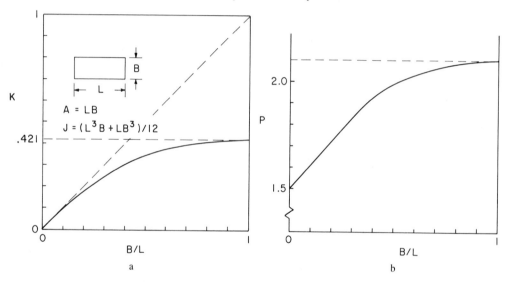

a

b

FIGURE 3. Shape factors for rectangular section.

where $\qquad\qquad \Lambda^2 = 5/2 \qquad$ and $\qquad B/L \leqslant 1$

which also show the exact asymptotic behavior of Figure 3 as slenderness ratio $B/L \to 0$.

Approximate Relations

Figures 2 and 3 show a general insensitivity to slenderness ratio, which suggests wide applicability of rough "rule-of-thumb" approximations

$$P \approx 2 \qquad \text{and} \qquad K \approx 1/2$$

Two somewhat more elegant approximations follow.

"Narrow-Section" Formulas

The previous results for rectangular sections show the asymptotic behavior

$$K \to L/B \quad \text{and} \quad P \to 3/2$$

while holding

$$A = LB \quad \text{as} \quad B/L \to 0$$

These relations, which correspond to a one-dimensional parabolic pressure distribution (usually attributed to Sommerfeld), are applicable to any narrow section. For example, they hold in the limit for an annular ring with relatively similar inner and outer radii, corresponding to many simple thrust bearings.

"Broad-Section" Formulas

The previous results for elliptical sections can be expressed as

$$P = 2 \quad \text{and} \quad K = \frac{3}{4\pi^2} \frac{A^2}{J}$$

in terms of area and polar moment

$$A \equiv \int_A dA \quad \text{and} \quad I \equiv \int_A r^2 \, dA$$

These relations, which hold exactly for elliptical (and circular) sections, are also applicable approximately to any broad section.

Application of this approximation (usually attributed to Saint Venant) to the *rectangular* section studied previously gives

$$P \approx 2 \quad \text{and} \quad 1/K \approx (B/L + L/B) \, (\pi/3)^2$$

so that for a *square* section

$$P \approx 2 \quad \text{and} \quad K \approx (1/2) \, (3/\pi)^2 = 0.456$$

as compared to the approximate values computed from the Warner solution above

$$P \approx 2.167 \quad \text{and} \quad K \approx 0.419$$

and the numerically exact series values plotted in Figure 3

$$P = 2.100 \quad \text{and} \quad K = 0.421$$

Similarly, application of the approximations to an *equilateral triangular* section gives

$$P \approx 2 \quad \text{and} \quad K \approx \frac{9\sqrt{3}}{4\pi^2} = 0.395$$

as compared to exact values

$$P = 20/9 = 2.222 \quad \text{and} \quad K = \sqrt{3}/5 = 0.346$$

Other Sections and Surfaces

Though the literature[1,4-6] contains exact formulas for normal approach of many other special planar sections (including complete and annular circles and sectors), the results given here should be entirely adequate for most purposes. The literature[1,4-8] also contains results for normal approach of a variety of nonplanar surfaces, including plates with small curvature (single and double), cones (complete and truncated), and spheres of various extents.

CYLINDRICAL JOURNAL BEARINGS[9-13]

The "squeeze-film" behavior of nonrotating cylindrical bearings in one-dimensional radial motion is *qualitatively* quite similar to that for planar bearings in normal approach, and generalization to two-dimensional motion is conceptually straightforward. Remarkably, even the addition of journal rotation causes no real difficulties. Thus solution of general cylindrical journal bearing dynamics problems rests on an understanding of "squeeze-film" behavior in simpler nonrotating cases.

One-Dimensional Motion Without Rotation

Figure 4 shows a nonrotating journal moving radially downward into a cylindrical half-sleeve. As before, rigorous analysis proceeds from the general Reynolds equation in rectangular coordinates wrapped around the journal circumference, a procedure justified by the clearance ratio $h/R \ll 1$. Tangential surface velocities are neglected. Fully flooded ambient boundary conditions assumed at the axial and circumferential ends of the bearing film complete specification of the problem.

Solution for pressure, etc., can be numerical or semianalytical.[11] In the latter case, computations are facilitated by special tables[23] for the "journal bearing integrals" which arise.

General Formulation

Relations analogous to previous ones can be expressed in terms of dimensional geometrical and material factors μ, L, D, R, and C and dimensionless *functions* P, Q, W, M, and J of dimensionless slenderness ratio L/D and dimensionless eccentricity ratio $\epsilon < 1$. (Recall that previous dimensionless quantities for planar bearings were *constants*.)

Thus,

$$e = C\epsilon = C - h^*$$

$$p^* = \frac{F}{LD} P$$

$$q = LD\dot{e} = \frac{R(C/R)^3(D/L)^2}{\mu} FQ$$

$$F = \frac{2\mu L}{(C/R)^3} \dot{e}W$$

$$\dot{e} = \frac{(C/R)^3}{2\mu L} FM$$

$$\int_{t_1}^{t_2} Fdt = \frac{\mu LD}{(C/R)^2} (J_2 - J_1)$$

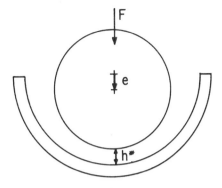

FIGURE 4. Cylindrical bearing in one-dimensional motion.

Evidently,

$$Q = (L/D)^2 M$$

while the last three factors are related by

$$MW = 1 \quad \text{and} \quad \frac{dJ}{d\epsilon} = \frac{1}{M}$$

so that

$$(J_2 - J_1) = \int_{\epsilon_1}^{\epsilon_2} \frac{d\epsilon}{M}$$

Pressure ratio[9] $P(\epsilon,L/D)$ is a measure of the sharpness or nonuniformity of the pressure distribution; flow factor $Q(\epsilon,L/D)$ the outward flow responsiveness to loading; impedance[10] $W(\epsilon,L/D)$ the stiffness or damping rate; mobility[11-15] $M(\epsilon,L/D)$ the velocity responsiveness to loading; and impulse[15,16] $J(\epsilon,L/D)$ the displacement responsiveness to loading over an interval.

Short-Bearing Relations

While numerically exact solutions are available elsewhere,[17,18] the results given here are obtained by the qualitatively correct and widely used short-bearing approximation discussed in previous chapters. For this particular solution (with its parabolic axial pressure distribution), factors P and Q are entirely independent of ratio L/D, while factors J and W (or M) vary with its square. Figure 5 shows data from several sources[9-11,19] for L/D = 1; adjustments are easily made for other slenderness ratios.

Though computed from closed-form expressions for the short-bearing model, these data are quite accurately fit by the approximations

$$P \approx -\frac{8}{9\pi} \frac{(1-\epsilon)^{5/2}}{\epsilon} \quad , -1 \leqslant \epsilon \leqslant -1/2$$

$$\approx \frac{6}{\pi} \frac{1}{(1-\epsilon)^{1/2}} \quad , -1/2 \leqslant \epsilon < 1$$

$$Q \approx (1-\epsilon)^{5/2}/\pi$$

$$W \approx \pi(L/D)^2/(1-\epsilon)^{5/2}$$

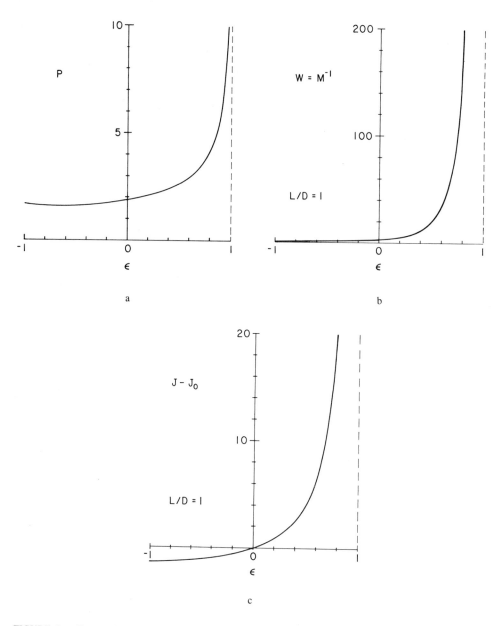

FIGURE 5. Characteristics for cylindrical bearings in one-dimensional motion (short-bearing film model). (a) Pressure vs. eccentricity, (b) impedance and mobility vs. eccentricity, and (c) impulse vs. eccentricity.

so

$$M = 1/W \approx (D/L)^2 (1 - \epsilon)^{5/2}/\pi$$

and

$$J - J_0 = \int_0^\epsilon d\epsilon/M \approx 2\pi (L/D)^2 [1/(1 - \epsilon)^{3/2} - 1]/3$$

For liquid films, which will not support significant negative pressures without rupturing, the short-bearing results given here for the *half*-sleeve bearing of Figure 4 apply equally well to radial motion of the *full*-sleeve bearing of Figure 6.

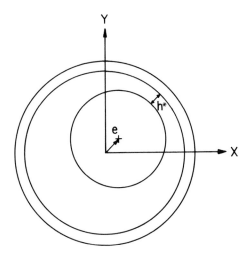

FIGURE 6. Cylindrical bearing in two-dimensional motion.

Two-Dimensional Motion

General Formulation — Without Rotation

Figure 6 shows a non-rotating journal moving in an arbitrary direction within a cylindrical sleeve. Fully flooded conditions assumed at the axial ends of the film would appear to complete specification of the problem. For liquid films, however, special analytical arrangements must be made to avoid negative pressures in the bearing half with normally *receding* surfaces.

The journal motion in the clearance space is now two-dimensional. Since journal eccentricity, squeeze velocity, maximum film pressure, and resultant film force (applied *by* the journal *to* the lubricant film) all have magnitude and direction, previous *scalar* relations must be replaced by the *vector* forms

$$\mathbf{e} = C\boldsymbol{\epsilon}$$

$$\mathbf{p}^* = \frac{1}{LD} \, |F| \mathbf{P}$$

$$\mathbf{F} = \frac{2\mu L}{(C/R)^3} \, |\dot{e}| \mathbf{W}$$

$$\dot{\mathbf{e}} = \frac{(C/R)^3}{2\mu L} \, |F| \mathbf{M}$$

so that

$$|\mathbf{W}| |\mathbf{M}| = 1$$

involving dimensionless eccentricity ratio, pressure ratio, force (impedance), and velocity (mobility) *vectors* $\boldsymbol{\epsilon}$, **P**, **W**, and **M**. *Scalar* impulse relations for the two-dimensional problem and their relationship to mobility relations are discussed elsewhere.[15,16]

The various vectors can be displayed in "fixed" coordinates X, Y or in "moving" coordinates x, y or x′,y′ referenced, respectively, to velocity or force directions as shown in Figure 7. Using the two "moving" frames, bearing data can be displayed in maps of

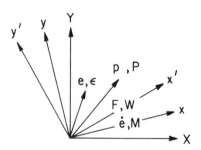

FIGURE 7.   Coordinate axes and vectors for two-dimensional motion.

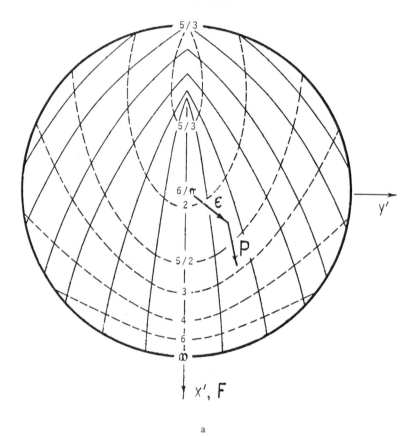

a

FIGURE 8.   Characteristics for cylindrical bearing in two-dimensional motion (short-bearing film model).[9-13] (a) Pressure vs. eccentricity, (b) mobility vs. eccentricity, and (c) impedance vs. eccentricity.

vector or scalar quantities plotted over the clearance space of all possible eccentricity ratios. Figure 8 allows a comparison of typical maps[9-13] for the liquid-film short-bearing model (in which film pressure is positive throughout the bearing half with normally approaching surfaces and vanishes in the other). The maps are oriented to velocity or force directions as shown. Dashed/solid curvilinear families indicate magnitude/direction of pressure ratio, mobility, and impedance vectors in Figures 8a, b, and c, respectively. Though the same basic data are displayed in both impedance and mobility maps, each point on one map corresponds to a (different) point on the other. In particular, the sample points indicated in Figures 8b and c do not correspond.

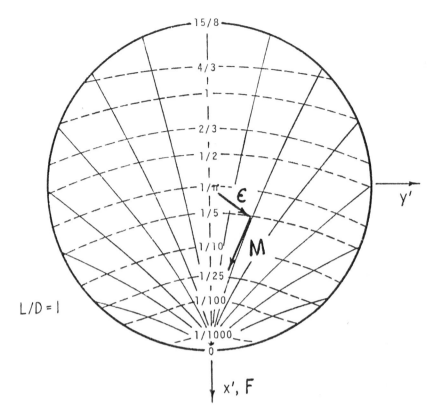

FIGURE 8b

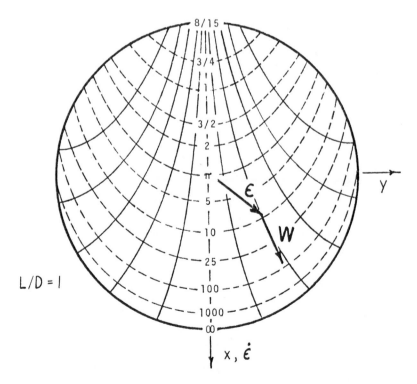

FIGURE 8c

Generally, such maps are specific to a particular slenderness ratio; for the *short-bearing* film model vector **P** is entirely independent of ratio L/D, while vector **W** (or **M**) varies with its square.

One-dimensional Figures 5a and b correspond to the midlines of two-dimensional Figures 8a, b, and c. Similarly, the two-dimensional short-bearing approximations

$$M^{x'} \approx (D/L)^2 (1 - \epsilon^{x'})^{5/2} / \pi$$

$$M^{y'} \approx -4 (D/L)^2 \epsilon^{y'} (1 - \epsilon^{x'})^{3/2} / \pi^2$$

are generalizations of the one-dimensional approximations given earlier. More exact map data are available elsewhere.[9-13,24,25]

Application of the map data to nonrotating bearings is straightforward: specification of **e** and **ė** allows direct determination of **F** via **W**; specification of **e** and **F** allows direct determination of **p*** (or **ė**) via **P** (or **M**). Transformations are required if (as is often the case) the map frames x, y and/or x′,y′ do not coincide instantaneously with the computation frame X, Y. (Graphically, this simply requires rotating maps.)

General Formulation — With Rotation

For extension of these procedures to problems involving rotation of journal and/or sleeve, consider an "observer" fixed to the sleeve center but rotating at the average angular velocity $\overline{\omega}$ of journal and sleeve (positive CCW). The absolute journal center velocity $\dot{e}_{abs}$ seen in the "fixed" computation frame, X, Y and the relative velocity $\dot{e}_{rel}$ apparent to the observer are related to journal eccentricity **e** and $\overline{\omega}$ by the simple kinematic expression

$$\dot{e}_{abs} - \dot{e}_{rel} = \overline{\omega} \times e$$

Since the average angular velocity of journal and sleeve (fluid entrainment velocity) apparent to this observer would vanish, maximum pressure **p*** and resultant force **F** would seem to be related solely to the relative (squeeze) velocity $\dot{e}_{rel}$ *in exactly the same way as for the nonrotating bearings considered previously.*

Thus extension of the previous procedures to general problems requires only use of the kinematic relation above *before* the impedance procedure for finding force from (relative) velocity and/or *after* the mobility procedure for finding (relative) velocity from force; the procedure for finding maximum pressure from force requires no modification, however.

The impedance and mobility methods are perfect complements. Both provide for efficient storage of bearing characteristics based on *any* suitable film model. Because pressure distributions are not calculated, both methods permit efficient computation. In appropriate applications the resulting equations of motion are in explicit form, and iterative calculations can thus be avoided in most system simulation studies.

Since the impedance formulation is appropriate to cases in which instantaneous force is *desired,* it seems most suited to problems in rotating machinery, particularly with damper bearings.

Since the mobility formulation is appropriate when instantaneous force is *known*, it has found widest application in reciprocating machinery. By giving instantaneous journal center velocity, the mobility method provides a basis for predicting an entire journal center path by numerical extrapolation (while allowing simultaneous prediction of maximum film pressure). Numerical implementation of the mobility method is straightforward; simplified versions require as few as 50 steps on programmable calculators. A digital computer program which accepts tabulated duty cycle data can be compiled from about 200 FORTRAN statements.[12,13]

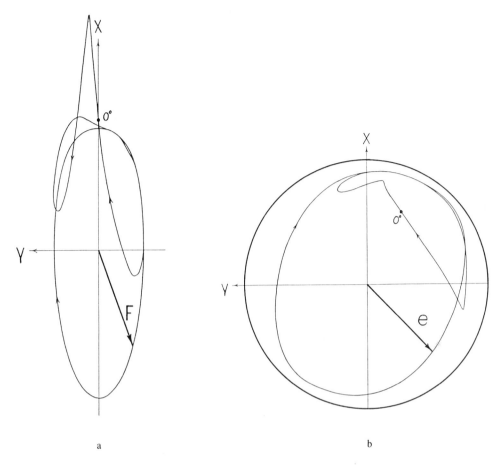

FIGURE 9.   Connecting-rod bearing polar diagrams.[12,13,15] (a) Journal loading cycle (four-stroke combustion), and (b) journal displacement cycle (short-bearing film model).

Connecting-Rod Bearing Example[13]

Orbit Computation by the Mobility Method

Connecting-rod bearings have complex loading as well as unsteady angular motion. Engine, bearing, and lubricant parameters for one such bearing are given elsewhere,[12,15] together with load components in a coordinate system X, Y fixed to the (moving) connecting rod. Figure 9a shows a full cycle of such loading; Figure 9b shows the corresponding steady-state displacement response computed using the short bearing model data represented in Figure 8b.

Computations using essentially the same program with data for more accurate film models show qualitatively similar results.[12,13,25,26] Extensive comparisons[20] support continued use of the short-bearing model in correlation studies. Further applications of orbit analysis, parametric studies, corresponding design criteria (minimum film thickness, maximum film pressure, power dissipation, etc.), and their correlation with failure modes and field experience are discussed elsewhere.[13,20-22,26,27]

Parametric Studies

Mobility method results for four-stroke engines (automotive Otto and Diesel, and industrial Diesel) are summarized elsewhere.[20] In all cases, firing loads have a minor (less than 20%) effect on predicted minimum film thicknesses so long as the maximum bearing load due to

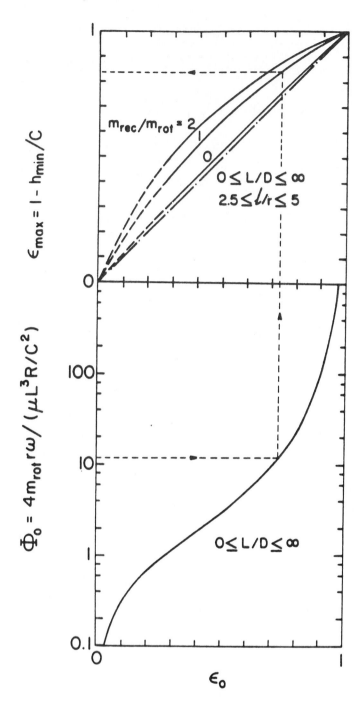

FIGURE 10.   Connecting-rod bearing under inertia loading: minimum film thickness/maximum journal displacement.[13]

firing alone is no more than about seven times the maximum bearing load due to inertia alone. For a particular medium-speed Diesel this means that firing loads have negligible effect on predicted film thicknesses above about 300 rev/min.

Thus for many higher-speed engines, firing loads have very little effect on minimum film thickness and can be reasonably ignored in preliminary design computations. Figure 10[13,21] summarizes 120 different minimum film thickness predictions for connecting-rod bearings

Table 1
DANGER LEVELS FOR FILM THICKNESSES
PREDICTED BY SHORT BEARING FILM MODEL
FOR CONNECTING-ROD BEARINGS[13]

	D(typical) mm(in.)	h(dangerous) μm(μin.)
Automotive (Otto)	50 (2)	1.0 (40)
Automotive (Diesel)	75—100 (3—4)	1.75 (70)
Industrial (Diesel)	250 (10)	2.5 (100)

loaded by inertia forces alone. The omission of firing loads allows convenient characterization of results by the dimensionless parameters shown. (For film models other than the short-bearing, both upper and lower graphs will be weakly dependent on slenderness ratio L/D.) Thus preliminary design guidance is available through Figure 10 without resort to computation.

The intermediate value ϵ_o given by the lower graph of Figure 10 can be interpreted as the steady-state response to a steadily rotating inertia load applied to a steadily rotating journal in a nonrotating sleeve. The final value ϵ_{max} given by the upper graph of Figure 10 reflects the further effects of reciprocating inertia and engine geometry on extremes of periodic response.

Field Experience

Film thickness predictions and field experience for about 60 practical connecting-rod bearings,[20] together with experiences from several other sources,[22] suggest the danger levels in Table 1. Main bearings are understood to be a bit more tolerant in smaller sizes.

Noting the uncertainties involved, exceeding these values in no sense guarantees success; though believed to be representative, they are offered for information only. It is also interesting to compare these limiting values of predicted oil film thickness with peak-to-valley estimates of surface finish.

NOMENCLATURE

A	Area	$[L^2]$
I	Area polar moment	$[L^4]$
L	Length	$[L]$
B	Breadth	$[L]$
D	Diameter	$[L]$
R	Radius	$[L]$
C	Radial clearance	$[L]$
U	Surface tangential velocity	$[LT^{-1}]$
ω	Surface angular velocity	$[T^{-1}]$
μ	Film viscosity	$[FL^{-2}T]$
h	Film thickness	$[L]$
p	Film pressure	$[FL^{-2}]$
P	Dimensionless pressure ratio	$[-]$
q	Outflow rate	$[L^3T^{-1}]$
Q	Dimensionless outflow ratio	$[-]$
F	Film force	$[F]$
K	Dimensionless film force (stiffness)	$[-]$
W	Dimensionless film force (impedance)	$[-]$
M	Dimensionless velocity (mobility)	$[-]$
J	Dimensionless impulse	$[-]$
e	Eccentricity	$[L]$
ε	Dimensionless eccentricity ratio	$[-]$
r	Crank radius (throw)	$[L]$
ℓ	Rod length	$[L]$
m_{rec}	Reciprocating mass	$[M]$
m_{rot}	Rotating mass	$[M]$
t	Time	$[T]$
r,θ	Coordinates	$[L,-]$
x,y	Coordinates	$[L,L]$
x′,y′	Coordinates	$[L,L]$
X,Y	Coordinates	$[L,L]$
·	Time derivative	$[T^{-1}]$
*	Special value	
—	Average value	

REFERENCES

1. **Gross, W. A., Matsch, L. A., Castelli, V., Eshel, A., Vohr, J. H., and Wildmann, M.,** *Fluid Film Lubrication,* John Wiley & Sons, New York, 1980, chap. 8.
2. **Warner, P. C.,** Static and dynamic properties of partial journal bearings, *J. Basic Eng.,* 85, 247, 1963.
3. **Booker, J. F. and Rohde, S. M.,** *Toward the Optimum Side Leakage Correction Factor, Research Rep.,* General Motors Research Laboratories, Warren, Mich., in press.
4. **Archibald, F. R.,** Squeeze films, in *Standard Handbook of Lubrication Engineering,* O'Connor, J. J., Ed., McGraw-Hill, New York, 1968, chap. 7.
5. **Moore, D. F.,** A review of squeeze films, *Wear,* 8, 245, 1965.
6. **Moore, D. F.,** *Principles and Applications of Tribology,* Pergamon Press, Oxford, 1975, 113.
7. **Hays, D. F.,** Squeeze films for rectangular plates, *J. Basic Eng.,* 83, 579, 1961.
8. **Goenka, P. K. and Booker, J. F.,** Spherical bearings: static and dynamic analysis via the finite element method, *J. Lubr. Technol.,* 102, 308, 1980.
9. **Booker, J. F.,** Dynamically loaded journal bearings: maximum film pressure, *J. Lubr. Technol.,* 91, 534, 1969.
10. **Childs, D., Moes, H., and van Leeuwen, H.,** Journal bearing impedance descriptions for rotordynamic applications (with discussion by Booker, J. F.), *J. Lubr. Technol.,* 99, 198, 1977.
11. **Booker, J. F.,** Dynamically loaded journal bearings: mobility method of solution, *J. Basic Eng.,* 87, 537, 1965.
12. **Booker, J. F.,** Dynamically loaded journal bearings: numerical application of the mobility method, *J. Lubr. Technol.,* 93, 168 and 315, 1971.
13. **Booker, J. F.,** Design of dynamically loaded journal bearings, in *Fundamentals of the Design of Fluid Film Bearings,* Rohde, S. M., Maday, C. J., and Allaire, P. E., Eds., American Society of Mechanical Engineers, New York, 1979, 31.
14. **Barwell, F. T.,** *Bearing Systems,* Oxford University Press, Oxford, 1979, 261.
15. **Campbell, J., Love, P. P., Martin, F. A., and Rafique, S. O.,** Bearings for reciprocating machinery: a review of the present state of theoretical, experimental and service knowledge (with discussion by Booker, J. F.), *Proc. Inst. Mech. Eng.,* 182(3A), 51, 1967.
16. **Blok, H.,** Full journal bearings under dynamic duty: impulse method of solution and flapping action (with discussion by Booker, J. F.), *J. Lubr. Technol.,* 97, 168, 1975.
17. **Hays, D. F.,** Squeeze films: a finite journal bearing with a fluctuating load (with discussion by Phelan, R. M.), *J. Basic Eng.,* 83, 579, 1961.
18. **Donaldson, R. R.,** Minimum squeeze film thickness in a periodically loaded journal bearing, *J. Lubr. Technol.,* 93, 130, 1971.
19. **Booker, J. F.** Analysis of Dynamically Loaded Journal Bearings: The Squeeze Film Considering Cavitation, Ph.D. thesis, Cornell University, Ithaca, N.Y., 1961.
20. **New, N. H.,** The use of computer design techniques applied to IC engines, presented at Int. Symp. on Plain Bearings, Štrbské Pleso, Czechoslovakia, October 24 to 26, 1972.
21. **Martin, F. A. and Booker, J. F.,** Influence of engine inertia forces on minimum film thickness in con-rod big-end bearings, *Proc. Inst. Mech. Eng.,* 181, 30, 1967.
22. **Warriner, J. F.,** Factors affecting the design and operation of thin shell bearings for the modern diesel engine, *Diesel Engines for the World,* Whitehall Press, England, 1977/78, 13-23.
23. **Booker, J. F.,** A table of the journal-bearing integral, *J. Lubr. Technol.,* 87, 533, 1965.
24. **Moes, H. and Bosma, R.,** Mobility and impedance definitions for plain journal bearings, *J. Lubr. Technol.,* 103, 468, 1981.
25. **Goenka, P. K.,** Analytical curve fits for solution parameters of dynamically loaded journal bearings, ASME PaperNo. 83-Lub-33, *J. Tribology,* in press.
26. **Martin, F. A.,** Developments in engine bearings design, in *Tribology International,* 16, 147, 1983, from *Tribology of Reciprocating Engines,* (Proc. 9thLeeds-Lyon Symp. on Tribology, Leeds, England, September 1982), Dowson, D., Taylor, C. M., Godet, M., and Berth, D., Eds., Butterworths, 1983, 9.
27. **Hollander, M. and Bryda, K. A.,** Interpretation of engine bearing performance by journal orbit analysis, Paper 830062, presented at SAE International Congress, Detroit, Mich., February 28 to March 4, 1983.

ELASTOHYDRODYNAMIC LUBRICATION

Herbert S. Cheng

INTRODUCTION

Elastohydrodynamic lubrication (EHL) applies to conditions where surface deformation and hydrodynamic action both play important roles. The most prominent example of EHL is the Hertzian contact in rolling element bearings, gears, and cams where surface deformation often exceeds the mean film thickness.

In elastohydrodynamic contacts, lubrication effectiveness is often measured by the average film thickness separating the asperities between two surfaces. Good lubrication performance is usually achieved when the mean separation approaches or exceeds three times the composite rms value of surface roughness. In this full-film region, overall performance can usually be predicted by EHL theories considering smooth boundaries. While full-film EHL theories are useful in predicting performance, partial-film EHL (Figure 1) relations are required in forecasting conditions for failure by pitting, scuffing, or wear.

Effects of surface deformation in sliding journal and thrust bearings can also be regarded as EHL problems and will be discussed briefly. Finally, application of EHL theories in roller and ball bearings, spur gears, helical gears, and cams will be reviewed.

FULL-FILM EHL

After two decades of intensive efforts, characteristics of full-film EHL are now reasonably well understood. The principal features were first revealed in two-dimensional analyses for line contacts. Later, three-dimensional analyses showed additional features in point or elliptical contacts.

Line Contacts

Line contact geometry applies to cases which correspond to two cylinders in contact: rolling-element bearings, uncrowned spur and helical gears, etc. Most EHL analyses have used this geometry both because it is easiest to analyze and because it reveals basic features common in more complicated contacts.

Film Thickness

Classical lubrication theories for rigid rollers are of interest to show their inadequacy in predicting film thickness for EHL contacts. Two rigid rollers are equivalent to a rigid roller on a flat plate, shown in Figure 2. The lubricant transported at any point along the film is

$$\left(\frac{u_1 + u_2}{2}\right) h - \frac{h^3}{12\mu} \left(\frac{dp}{dx}\right) \tag{1}$$

where p, h, u_1, u_2, and μ are pressure, film thickness, surface velocity of roller 1 and 2, and viscosity, respectively.

Let the lubricant flow rate be $1/2 (u_1 + u_2)h^*$, where h^* is the gap where dp/dx is zero. Continuity of flow gives

$$\frac{dp}{dx} = 6\mu(u_1 + u_2) \left(\frac{h - h^*}{h^3}\right) \tag{2}$$

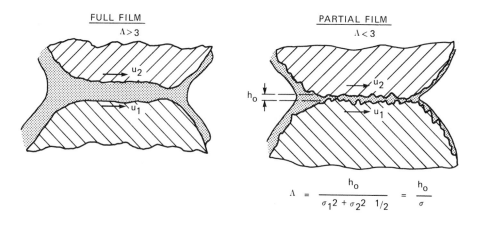

FIGURE 1. Full- and partial-film EHL contacts.

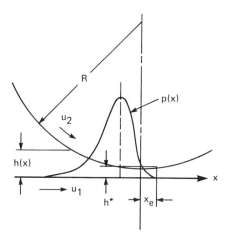

FIGURE 2. Lubrication of rigid roller on flat plate.

The boundary conditions are

$$p = 0 \text{ at } x = -\infty$$

$$p = 0, \frac{dp}{dx} = 0; \ h = h^* \text{ at } x = x_e$$

On the left side of h^*, dp/dx is positive, resulting in a hydrodynamic pumping action. On the right side of h^*, dp/dx is negative, resulting in a sealing action. By prescribing h^*, one can solve Equation 2 for the pressure distribution, and hence the load and film thickness relationship.

After Martin[1] showed that this simple theory could not adequately predict a lubricating film for rolling contacts, many believed that pressure-viscosity dependence could account for a greater film thickness. Including the exponential relation $\mu = \mu_o e^{\alpha p}$ in Equation 2, one obtains:

$$\frac{1}{e^{\alpha p}} \frac{dp}{dx} = 6\mu_o(u_1 + u_2) \frac{h - h^*}{h^3} \tag{3}$$

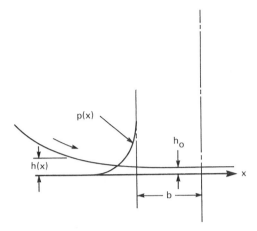

FIGURE 3. Inlet geometry of Grubin-type EHL line contact.

where μ_o is the ambient viscosity and α the pressure-viscosity coefficient. By introducing a "reduced pressure" $q = (1 - e^{-\alpha p})/\alpha$, Equation 3 can be simplified to

$$\frac{dq}{dx} = 6\mu_o (u_1 + u_2) \frac{h - h^*}{h^3} \tag{4}$$

Equation 4 can be solved for q, which lead to p and the load capacity. Improvement in film thickness was still far from enough to suggest a full-film hypothesis.

Ertel-Grubin[2] provided the first convincing evidence of full-film EHL by solving Equation 4 for a Hertzian shape shown in Figure 3. They determined h_o, the inlet film thickness, for the reduced lubricant pressure q to reach $1/\alpha$ (or $p \to \infty$) at $x \to -b$, the entrance of the Hertzian conjunction. Once $e^{-\alpha p}$ approaches zero, μ becomes very large in the conjunction. A very minute change in conjunction film thickness would cause an enormous change in pressure. Therefore, for heavily loaded contacts, the gap in the conjunction is practically uniform. Equation 4 was solved using the Hertzian inlet film profile:

$$h(x) = h_o + \frac{4}{\Pi} \frac{w}{E} \left[|\bar{x}| \sqrt{\bar{x}^2 - 1} - \ell n \left(|\bar{x}| + \sqrt{\bar{x}^2 - 1} \right) \right] \tag{5}$$

where $1/E = 1/2 [(1 - \nu_1^2)/E\ 1 + (1 - \nu_2^2)/E_2]\ \bar{x} = x/b$, and w = load per unit width. Empirical fitting yields the following equation which gives film thicknesses one or two orders of magnitude higher than rigid roller theories:

$$\frac{h_o}{R} = 1.95 \frac{(GU)^{8/11}}{(W)^{1/11}} \tag{6}$$

where $G = \alpha E$, $U = \mu_o u/ER$, $W = w/ER$, and $R = R_1 R_2/(R_1 + R_2)$.

Ertel-Grubin's formula for predicting h_o holds well for a reasonably wide range of conditions. It becomes less accurate when $G < 1000$ (lubricant with a small pressure-viscosity dependence or material of low modulus) and when W becomes small ($<10^{-5}$). Solution for h_o for the full range of G from 0 to 5000 can be found in References 3, 4, and 5. Lubricant compressibility was included.

When U is very large, reduction of viscosity due to inlet shear heating requires solving the energy equation simultaneously with Equation 4 in the inlet region.[6-8] Results led to a

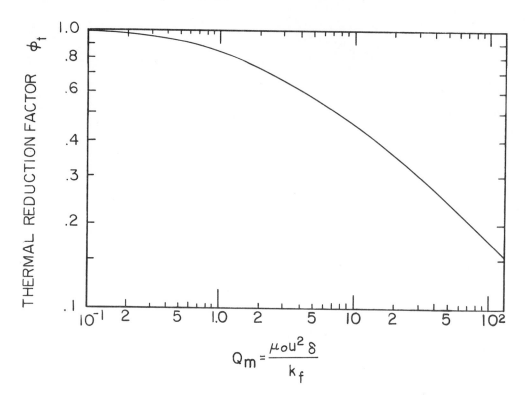

FIGURE 4.   Thermal reduction factor, where Q_m = thermal parameter, μ_c = ambient lubricant viscosity, u = rolling velocity, T_o = ambient temperature, k_f = lubricant thermal conductivity, and δ = temperature viscosity coefficient in the empirical relation $\mu = \mu_o\, e^{-\delta(T - T_o)}$.

thermal reduction factor, ϕ_T, which can be multiplied by the Ertel-Grubin film thickness for the actual h_o. With ϕ_T varying slightly with W and lubricant pressure G, Figure 4 gives a first approximation.[7]

Starvation introduced another reduction factor which is a function of the distance between the inlet film meniscus and the entrance edge of the Hertzian conjunction as indicated in Figure 5.[9,10] For rollers lubricated with an initial charge only, film thickness would be gradually reduced to about 0.71 of the fully flooded initial value.[10]

Measurements of film thickness in line contracts have been reviewed by Archard.[11] The first confirmation of Ertel-Grubin's film prediction was due to Crook,[12] who measured the capacitance between two discs. The second significant film measurement method was developed by Sibley and Orcutt[13] using X-ray techniques. Optical measurements were made by Wymer and Cameron.[14]

Film Shape and Pressure Distribution

For a detailed film thickness and pressure distribution, the following coupled Reynolds and deformation equations must be solved:

$$\frac{dp}{dx} = 6\mu(u_1 + u_2)\,\frac{h - h^*}{h^3} \tag{7}$$

$$h = h^* + \left(\frac{x^2 - x^{*2}}{2R}\right) - \frac{4}{\Pi E}\int_{-\infty}^{x^*}\ell n\,\frac{|\xi - x|}{|\xi - x^*|}\,p(\xi)d\xi \tag{8}$$

where x^* is the coordinate at which the lubricant film terminates. At $x = x^*$, $h = h^*$, $dp/dx = 0$.

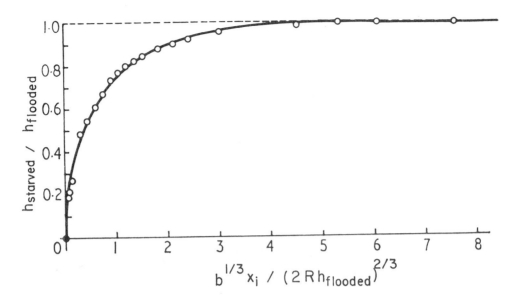

FIGURE 5. Effect of starved inlet boundary on film thickness, where b = half Hertzian width, x_i = distance from the inlet oil meniscus to the inlet edge of Hertzian boundary, R = effective radius $R_1R_2/(R_1 + R_2)$, $h_{starved}$ = starved film thickness, and $h_{flooded}$ = flooded film thickness. The circle points are derived from the computer solution by Orcutt and Cheng.[83]

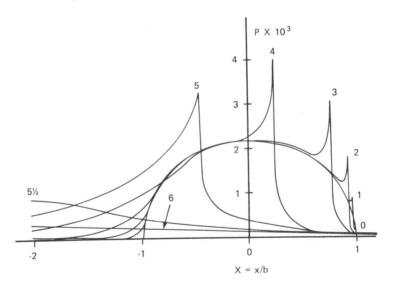

FIGURE 6. Pressure distributions for a compressible lubricant, where W = 3×10^{-5}, G = 5000, U = (0) 0 (dry contact), (1) 10^{-13}, (2) 10^{-12}, (3) 10^{-11}, (4) 10^{-10}, (5) 10^{-9}, ($5\frac{1}{2}$) $10^{-8.5}$, and (6) 10^{-8} (for definition of W, G, and U, see Equation 6).

The first full EHL numerical solution revealed a sharp pressure spike accompanied by a film constriction at the exit, as shown in Figure 6.[15] As speed decreases or load increases, the pressure spike moves further towards the exit and eventually disappears for very heavily loaded contacts.[17] Similar features were found in later isothermal solutions.[18-20]

Inlet and central film thickness obtained in Reference 16 agrees well with the Ertel-Grubin solution. Minimum film thickness at the constriction was found to be approximately 70 to 75% of the central film thickness. This provided the following widely used minimum film thickness for line contacts.[21]

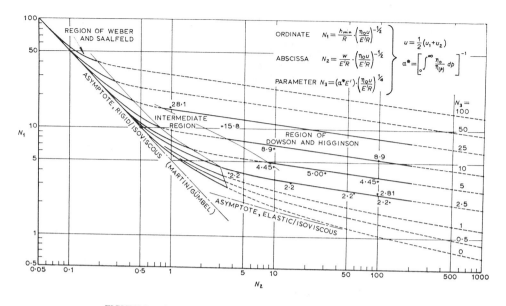

FIGURE 7.   Survey diagram for incompressible and isothermal EHL.

$$\frac{h_{min}}{R} = 2.65 \, \frac{G^{0.54} U^{0.7}}{W^{0.13}} \tag{9}$$

The exit construction was first seen experimentally in the circumferential film profile measured by the X-ray technique.[22] This was followed by optical interferometry.[14,22] While the measured nominal film shows good correlation with EHL theories, the measured ratio of minimum film to nominal film appears to be considerably smaller than 0.7 to 0.75, as predicted analytically. Film pressure measurements by means of a vapor-deposited manganin strip confirmed analytical predictions on effect of load on location of the pressure spike.[18,24,25]

Film Thickness Chart

Minimum film thickness of EHL line contacts may be determined from the Moes diagram[26] of Figure 7. Film thickness parameter h_{min}/R, speed parameter $\eta_o u/E'R$, load parameter, and lubricant parameter $\alpha^* E'$ are regrouped to form an implicit relation among only three independent parameters. Only one family of curves is needed to relate the film thickness parameter with the other two parameters over a wide range of loads, speeds, and lubricant parameters. Martin[1] results are an asymptote for the rigid/isoviscous case, and Herrebrugh[27] for the elastic/isoviscous case.

Point Contacts

Film Thickness

Figure 8 shows a point contact which is characterized by principal radii R_{x1}, R_{y1} for body 1 and R_{x2}, R_{y2} for body 2. In general, the principal planes containing R_{x1} and R_{x2} may not coincide; however, for most EHL contacts such as rolling bearings and gears, principal radii R_{x1} and R_{x2} do lie in the same plane. These surfaces can be convex, concave, or saddle-shape, depending on whether R_x and R_y are both positive, both negative, or mixed.

Ertel-Grubin type of analysis can also be carried out for spherical contacts with a circular conjection. Archard and Cowking[28] solved the two-dimensional Reynolds equation outside the circular conjunction region for a film thickness distribution compatible to the Hertzian solution for an unlubricated contact. An Ertel-Grubin type boundary condition, $q = 1/\alpha$, around the circumference of the circular conjunction gave:

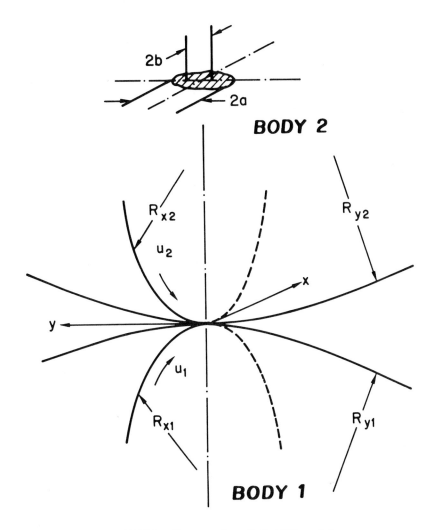

FIGURE 8. Geometry of an elliptical EHL contact.

$$\frac{h_o}{R} = 1.37\,(GU)^{0.74}\left(\frac{p_o}{E}\right)^{-0.22} \tag{10}$$

where p_o is the maximum Hertzian pressure. For point contacts with an elliptical conjunction, analytical solution is not feasible and numerical solutions have been obtained.[23,29] Results in Reference 29 gave:

$$\frac{h_o}{R} = C\,(GU)^{n_1}\left(\frac{p_o}{E}\right)^{n_2} \tag{11}$$

where C, n_1, n_2 in Table 1 are shown for four values of ellipticity ratios of the conjunction, a/b (b = semi-axis in the direction of motion, a = semi-axis normal to the motion).

Predictions of h_o using Equations 10 and 11 become inaccurate at high rolling speed where both inlet heating and starvation cause a reduction in h_o. While inlet heating for elliptical point contact has not been investigated, the thermal reduction factor for line contacts in Figure 4 may be applicable.

Table 1
VALUES OF C, n_1, AND n_2
FOR EQUATION 11

a/b	C	n_1	n_2
5	1.625	0.74	−0.22
2	1.560	0.736	−0.209
1	1.415	0.725	−0.174
0.5	1.145	0.688	−0.066

Full computer solutions for elliptical contacts were made for flooded as well as for starved contacts.[30] Film thickness formulas for the flooded contacts appear as:

$$H_{c,F} = 2.69 \ U^{0.67} \ G^{0.53} \ W^{-0.067} \left(1 - 0.61e^{-0.73k}\right) \qquad (12)$$

$$H_{min,F} = 3.63 \ U^{0.68} \ G^{0.49} \ W^{-0.073} \left(1 - e^{-0.68k}\right) \qquad (13)$$

where $H_{c,F} = h_{c,F}/R_x$, $h_{c,F}$ = central film thickness for flooded contacts, $H_{min,F} = h_{min,F}/R_x$, $h_{min,F}$ = minimum film thickness for flooded contacts, k = elliptical parameter, k = 1.03 $(R_y/R_x)^{0.64}$, $R_x = R_{x1}R_{x2}/(R_{x1} + R_{x2})$, $R_y = R_{y1}R_{y2}/(R_{y1} + R_{y2})$, $W = w/ER_x^2$, w = total load, $U = \mu_o (u_1 + u_2)/2ER_x$, and $G = \alpha E$.

For the starved contacts, the formulas are

$$H_{c,s} = H_{c,F} \left(\frac{m-1}{m^* - 1}\right)^{0.29} \qquad (14)$$

$$H_{min,s} = H_{min,F} \left(\frac{m-1}{m^* - 1}\right)^{0.25} \qquad (15)$$

where subscript s refers to starved contacts, m is the distance of the inlet meniscus from the center of the contact, and m^* is the inlet distance required for achieving flooded conditions; m^* can be expressed as

$$m^* = 1 + 3.06 \left[\left(\frac{R_x}{b}\right)^2 H_{c,F}\right]^{0.58} \qquad (16)$$

where b is the semiminor axis of the elliptical conjunction in the rolling direction.

In most starvation analyses, location of the inlet meniscus is not known beforehand and is dependent on lubricant supply rate and system configuration. Reduction in film thickness due to starvation in most ball bearings is considerably greater than from inlet heating.[31,32]

Optical techniques enabled extensive point-contact film thickness measurements and correlations with analysis.[33-36] Film thickness data by X-ray transmission[37] with crowned rollers showed a much stronger load dependence than that predicted by EHL analysis at maximum Hertzian pressures beyond 1 GPa (145,000 psi). This disagreement was explained by Gentle et al.[38] as possibly due to a combination of thermal and surface roughness effects. The point contact EHL film theory at extreme pressures [up to a maximum Hertzian pressure of 2 GPa (290,000 psi)] was validated by using a sapphire disk and tungsten carbide ball.[38] Film thickness measurements using tungsten carbide disks confirm EHL film theories for loads as high as 2.5 GPa (362,500 psi).[39]

Point Contact Film Shape and Pressure Distribution

Film shape in a circular point contact was experimentally revealed by the interferometric map between a highly polished steel ball and a transparent plate.[23,34,35,36,40] By identifying successive fringes, constant thickness contours can be mapped. The effect of speed on film shaped constriction which is very narrow and very close to the trailing edge. As speed increases or load decreases, the constriction becomes wider and less distinctive. Except for extremely low loads and high speeds, the minimum film thickness is found at the two sides rather than at the center of the trailing edge, and it is more sensitive to load variation than the minimum film thickness for a line contact. Analytical confirmation of the horseshoe-shaped constriction in a numerical solution of two-dimensional EHL equations[41] with a solid-like lubricant was followed by a series of full EHL solutions for circular and elliptical contacts[30,42] with a Newtonian lubricant.

Temperature

For a line contact, a detailed study of thermal effects required solution of the energy equation in the lubricant film considering heat generation by shearing and compressing the lubricant, heat convected away by the lubricant, and heat conducted into the solids.[43,44]

Film thickness level is influenced only by temperature rise in the inlet region discussed earlier; subsequent large temperature rise in the Hertzian conjunction has little influence. The predicted temperature field within the Hertzian conjunction depends strongly on the lubricant rheological model used in evaluating the heat generated by sliding. For a Newtonian model[44] for steel contacts with a maximum Hertzian pressure up to 0.5 GPa, the principal feature of exit film constriction and pressure peaks are unaltered when thermal effects are included. However, for loads higher than 0.5 GPa, the Newtonian lubricant model predicts a sliding frictional coefficient almost an order of magnitude higher than measured. Since practical sliding EHL contacts such as gears and cams involve pressures greater than 0.5 GPa, a non-Newtonian lubricant model is needed for the frictional heat. Successful non-Newtonian models are discussed in the next section.

Early measurement of surface temperature profiles were made with a platinum wire temperature transducer[24] at moderate loads. Later, an improved transducer[45,46] with a titanium wire deposited over a silica layer gave good results at much higher loads. While vapor-deposited probes are yet to perfected, a promising infrared technique was developed by Nagaraj et al.[47] for measuring the surface as well as film temperature in circular contacts. Figure 10 shows surface temperature along the center strip of the circular contact. The measured temperature show good agreement with that predicted from the Jaeger-Archard[48] formula.

Friction

In rolling and sliding EHD contacts, frictional force has two components: one due to rolling and the other due to slip between the surfaces. Except for nearly pure rolling conditions, sliding friction is always much larger than rolling friction. Basic features of sliding friction are revealed in Figure 11 by data from a two-disk machine.[49] In the low-slip region, friction increases linearly with slip. As slip increases, friction gradually tapers off in a nonlinear region where the stress is no longer governed by linear constitutive relations of the lubricant. In the high-slip region, friction decreases with slip because of thermal influence on lubricant properties at high-sliding speeds.

Low-Slip Friction

For low-slip frictions, sliding friction can be predicted from the Maxwell viscoelastic model. If equilibrium viscosity and shear modulus are used, however, predicted friction is much greater than measured. This disagreement led to the argument that the viscosity, when

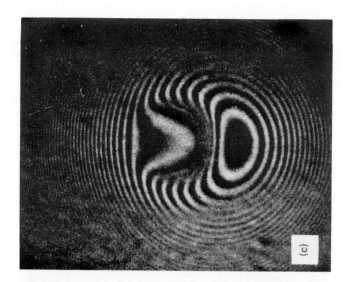
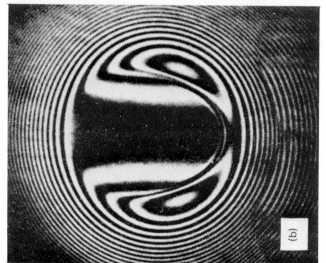
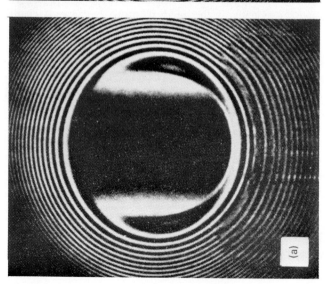

FIGURE 9. Effect of load and speed on film shape. (a) $W = 10$ lb, $U = \mu_0 u/ER = 3 \times 10^{-11}$; (b) 10 lb, 1.8×10^{-10}; and (c) 5 lb, 1.5×10^{-9}.

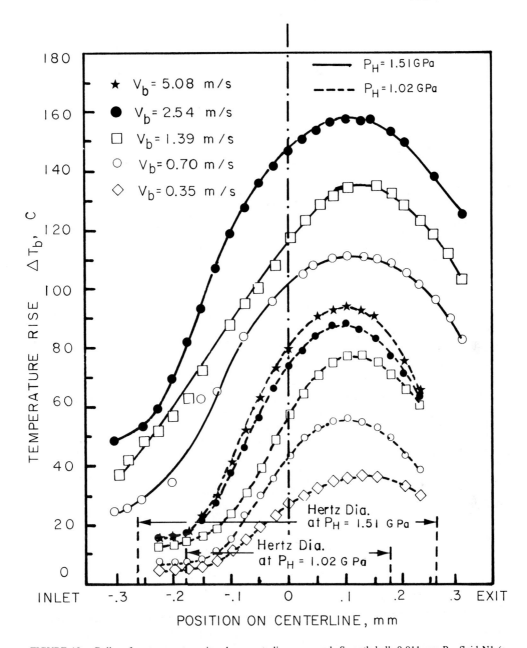

FIGURE 10. Ball surface temperature rise along centerline vs. speed. Smooth ball: 0.011 μm R_a, fluid N1 (a naphthenic oil), $T_{bath} = 40°C$, $V_{sapphire} = 0$, V_b = speed of ball surface, P_h = maximum Hertzian pressure, and T_b = ball surface temperature rise.

subjected to a sudden, higher pressure, cannot reach its equilibrium immediately. A time delay in the viscosity model[50] can account for the observed low friction[51] in the low-slip region even without considering shear elastic strain.

In a spin roll point contact, Johnson and Tevaariverk[52] have shown that oil under high-pressure and a short-transient time tends to deform more as an elastic solid than as a viscous liquid. However, values of shear modulus deduced from the low spin tests are still considerably lower than the known equilibrium shear modulus under high pressure.[53] They attributed this discrepancy in shear modulus to transient effects under a suddenly applied pressure.

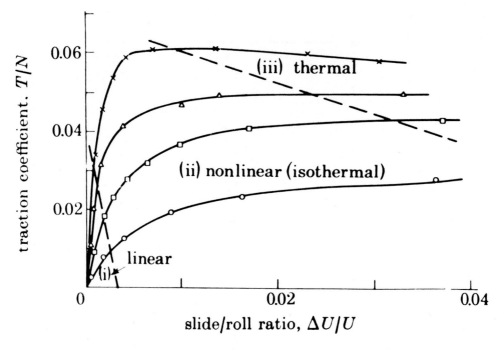

FIGURE 11.   Typical traction curves measured on a two-disc machine in EHL line contact at varying mean contact pressure, p̄ : x, 1.03; △, 0.68; □, 0.51; ○, 0.40 GPa. (From Johnson, K. L. and Cameron, R., *Proc. Inst. Mech. Eng.*, 182, 307, 1967. With permission.)

Montrose et al.[54] developed perhaps the most complete and promising viscoelastic model to describe low-slip traction behavior. Shear elastic strain is considered in the same manner as the conventional viscoelastic theory. However, instead of actual local pressure, a fictive pressure accounts for the transient viscosity effects similar to that considered in Reference 50. The fictive pressure is determined from an exponentially decaying structural relaxation function.

High-Slip Friction

As sliding velocity increases, slope of the friction curve gradually decreases to zero, and in many cases becomes slightly negative at very high speed. Under these conditions, the fluid undergoes a large shear strain usually in the midplane of the lubricant film.

The Newtonian model was shown to be inadequate for predicting friction for heavily loaded contacts. It yields a friction force almost one order of magnitude higher than the measured value.[49] Non-Newtonian models for predicting friction in the high-slip region include a shear-dependent Ree-Eyring model,[55] a limiting shear liquid model,[56] a plastic solid model,[57] and a promising nonlinear viscous and plastic model.[52,58] This final model is capable of describing linear and nonlinear viscous, linear and nonlinear viscoelastic, as well as elastic plastic behavior in a single equation. Depicted in Figure 12, it is expressed as

$$\dot{\gamma} = \frac{1}{G}\frac{d\tau}{dt} + F(\tau) = \dot{\gamma}_e + \dot{\gamma}_e + \dot{\gamma}_v \qquad (17)$$

$$F(\tau_e) = \left(\frac{\tau_o}{\eta}\right)\sinh\left(\frac{\tau_e}{\tau_o}\right) \qquad (18)$$

where $\dot{\gamma}$ = total strain rate, $\dot{\gamma}_e$ = elastic strain rate, $\dot{\gamma}_v$ = viscous strain rate, τ_e = equivalent

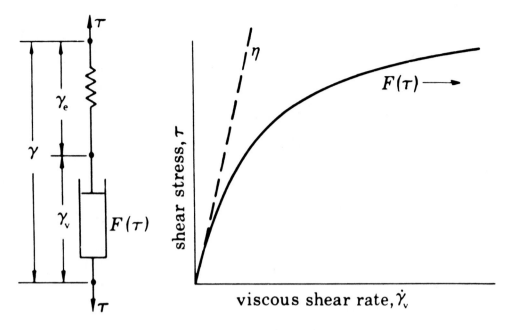

FIGURE 12. Nonlinear Maxwell fluid with zero-shear-rate viscosity and infinite-rate shear modulus G. F(τ) denotes the nonlinear viscous function.

stress = $(1/2\tau_{ij}\tau_{ij})^{1/2}$, τ_o = representative stress, a fluid property, η = viscosity, and G = shear modulus. τ_o, η, and G are fluid properties deduced from traction tests. Limited values are given for five oils.[52] A similar nonlinear viscous and plastic model introduced recently by Bair and Winer[58] used rheological constants from tests totally independent of any data from the EHL contact itself.

The Bair and Winer[58] model in dimensional form can be written as

$$\dot{\gamma} = \frac{1}{G_\infty} \frac{d\tau}{dt} - \frac{\tau_L}{\mu_o} \ell n \left(1 - \frac{\tau}{\tau_L} \right) \tag{19}$$

In dimensionless form, it is

$$\hat{\dot{\gamma}} = \hat{\dot{\tau}} - \ell n \, (1 - \hat{\tau}) \tag{20}$$

where

$$\hat{\dot{\gamma}} = \frac{\dot{\gamma}\mu_o}{\tau_L}$$

$$\hat{\dot{\tau}} = \frac{\mu_o}{\tau_L G_\infty} \frac{d\tau}{dt}$$

$$\hat{\tau} = \frac{\tau}{\tau_L}$$

Only three primary physical properties are required: low shear stress viscosity, μ_o, limiting elastic shear modulus, G_∞, and limiting yield shear stress, τ_L, all as functions of temperature and pressure. The behavior of the dimensionless equation is shown in Figure 13. Agreement between theory and experiment is good.

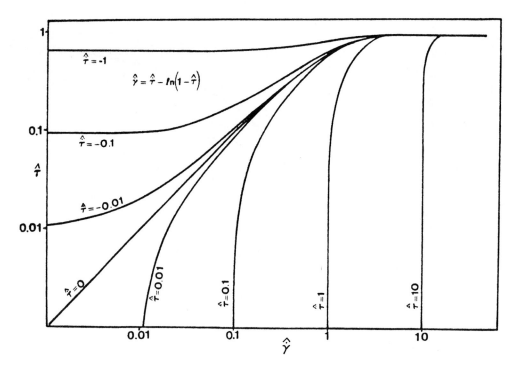

FIGURE 13.   Nondimensionalized shear stress vs. nondimensionalized shear strain for fluid model.

PARTIAL-FILM EHL

Partial EHL is the regime where average film thickness becomes less than three times the composite surface roughness and the local film thickness is interrupted at the tip of tall asperities.[59] A majority of Hertzian contacts in gears, cams, and rolling element bearings operate in this regime.

For years, roughness of bearing surfaces has been characterized by a single parameter such as σ, the rms standard deviation of roughness amplitude from a mean plane. To supplement this indication of roughness, surface roughness texture can be described by two statistical parameters: height distribution and autocorrelation function, acf. For surfaces finished by an abrasive process, height distribution is approximately Gaussian.[60] For running-in surfaces, the height distribution is slightly skewed from Gaussian.

The acf measures the wave length structure of a surface profile in a given direction. It is defined as

$$R_x(\lambda) = \frac{1}{\ell} \int_0^\ell \delta(x) \cdot \delta(x + \lambda) dx \tag{20a}$$

where λ is the correlation length, δ is the height function along the x-direction, and $R_x(\lambda)$ is the acf in the x-direction. The acf for most engineering surfaces is an exponentially decaying function.[61] A linear function has been used as an approximation.[62]

Many engineering surfaces contain long wave lengths in one direction and short wave lengths in the normal direction. The pattern can be described by parameter γ defined as the ratio of x and y correlation lengths:[63]

$$\gamma = \frac{\lambda_{0.5\,x}}{\lambda_{0.5\,y}} \tag{21}$$

where $\lambda_{0.5x}$ is the correlation length at which the acf of the profile is 50% of the value at the origin; γ may be interpreted as the length-to-width ratio of a representative asperity contact. Purely transverse, isotropic, and purely longitudinal roughness patterns have $\gamma = 0,1,\infty$, respectively. Surfaces with $\gamma > 1$ are longitudinally oriented.

For determining partial EHL performance, surface roughness parameters required for each surface include: (1) σ — rms surface roughness, (2) height distribution function, (3) $\lambda_{0.5x}$, $\lambda_{0.5y}$ — 50% correlation lengths in x and y directions, and (4) acf (autocorrelation function).

Average Film Thickness

Pure longitudinal or transverse roughness was first explored by Johnson et al.[64] for pure rolling contact based on Christensen's stochastic theory.[65] They developed a Grubin type solution for $\sigma \ll h$ and concluded that the effect of roughness on average film thickness is minimal.

For $h/\sigma > 3$ and for rolling and sliding contacts, Berthè[66] and Chow and Cheng[67] showed that:

1. For pure rolling contact with pure transverse roughness, average film thickness is higher than predicted by smooth surface EHL theory. This effect is greatly enhanced as h/σ approaches three. For sliding contacts with one surface smoother than the other, the roughness effect is enhanced if the smoother surface is faster and retarded if the smoother surface is slower.
2. For pure longitudinal roughness, average film thickness is lower than predicted by the smooth surface theory. Superimposing of sliding on rolling has little influence on the roughness effect for pure longitudinal surfaces.

Patir and Cheng[62] developed an average flow model to handle roughnesses of an arbitrary surface pattern parameter γ and extended the results to h/σ below three where part of the load is shared by asperity contacts. Figure 14 depicts the flow pattern for longitudinally oriented ($\gamma > 1$), trasversely oriented ($\gamma < 1$), and isotropic roughness ($\gamma = 1$). In Figure 15, the ratio of actual film thickness to the smooth surface film thickness is plotted against film parameters $\Lambda = h_{smooth}/\sigma$.

Asperity Load to EHL Load Ratio

Average asperity contact pressure in partial EHL is a function of the ratio of compliance to composite surface roughness h/σ. Here, compliance is the distance between the two mean planes based on the undeformed surfaces. For Gaussian surfaces, Tallian[59] has derived the asperity load as a function of h/σ for both plastically or elastically deformed asperities.

The load sharing ratios in circumferential ground EHL contacts (longitudinal roughness) can be obtained by a full numerical solution for disks with known surface roughness characteristics.[68]

Average Friction

Once the ratio of asperity load to fluid pressure load is determined, total friction force in partial EHL can be estimated by:

$$F = \mu_a Q_a + \mu_{EHL} Q_{EHL} \tag{22}$$

where F = total frictional force, μ_a, μ_{EHL} = coefficient of friction for asperity load and hydrodynamic load, respectively, and Q_a, Q_{EHL} = asperity, hydrodynamic load. For most partial EHL contacts, μ_a is believed to be between 0.1 and 0.2. The value of μ_{EHL} can be taken from frictional coefficients for full-film EHL contacts.

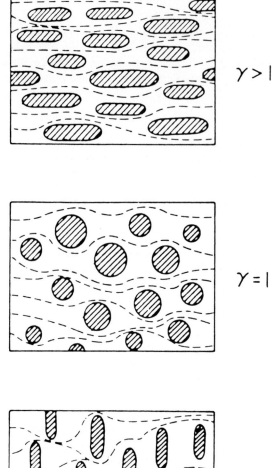

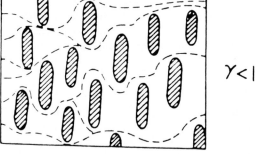

FIGURE 14. Typical contact area patterns for longitudinally oriented, isotropic, and transversely oriented rough surfaces.

Micro-Elastohydrodynamic Lubrication

Micro-EHL[69] deals with local pressure and film fluctuations around asperities or furrows within a macro-EHL conjunction. For pure rolling, local pressure and film thickness distributions at asperities are governed mainly by the normal approach action. As an ellipsoidal asperity approaches the opposing surface, the lubricant at the asperity center becomes highly pressurized and entrapped to form a central pocket as it travels through the Hertzian conjunction.[69,70]

For pure transverse ridges, the normal approach action in pure rolling causes the lubricant in the Hertzian region ahead of the asperity to become extremely viscous. Subsequently, the asperity becomes frozen together with the lubricant and is transported through the Hertzian conjunction as an integral unit.[71]

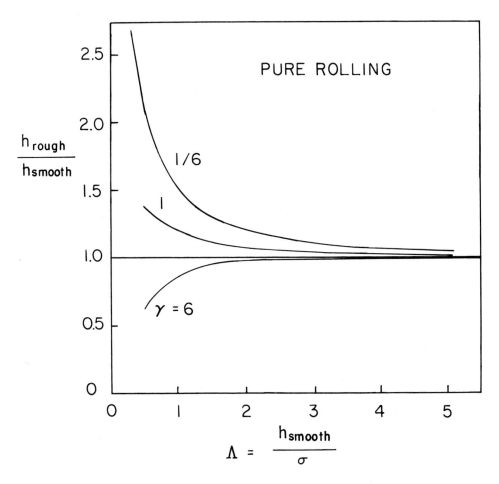

FIGURE 15. Effect of surface roughness on the average film thickness of EHL contacts: $P_o/E = 0.003$, pure rolling, $\alpha E = 3333$, and $\sigma/R = 1.8 \times 10^{-5}$.

For sliding EHL contacts, micro-EHL film thickness is controlled largely by entrainment of lubricant at the inlet of an asperity. For tranverse asperities, a lower limit of film thickness can be estimated by applying the classical EHL film thickness formulas to a sliding asperity in a low pressure ambient. For longitudinal asperities, very little is available to estimate minimum film thickness.

For a pair of transverse asperities colliding in a lubricant of low ambient pressure, micro-EHL film thickness can be estimated with existing theories.[72,73] If collision takes place in a high-pressure ambient, micro-EHL film is expected to increase considerably but cannot be predicted quantitatively.

COMPLIANT HYDRODYNAMIC JOURNAL AND THRUST BEARINGS

In hydrodynamic journal and thrust bearings, EHL effects can become significant if deformation of the bearing surfaces is of the same order as the film thickness. This occurs in heavily loaded journal bearings for large diesel engines, in high-pressure thrust bearings for hydroelectric turbines, and in elastomeric bearings used to tolerate dirt. Reference 74 gives a detailed review of compliant hydrodynamic bearings.

In journal bearings, minimum film thickness is increased slightly and peak film pressure is reduced when elastic effects are included. Surface deformations caused by local compres-

Table 2
VALUES OF C
FOR BEARING RACEWAYS

Bearing type	Inner race	Outer race
Ball	8.65×10^{-4}	9.43×10^{-4}
Spherical and cylindrical	8.37×10^{-4}	8.99×10^{-4}
Tapered and needle	8.01×10^{-4}	8.48×10^{-4}

Table 3
TYPICAL VALUES OF σ
FOR BEARINGS

Bearing type	Composite roughness (μm)	(μin.)
Ball	0.178	7
Spherical and cylindrical	0.356	14
Tapered and needle	0.229	9

sion, bending of pads, and thermal distortion can significantly affect performance of large, high-speed thrust bearings.[75,76] Because deformation effects are sensitive to detailed pad geometry, they can only be determined by elaborate computer codes.[77]

APPLICATION TO MACHINE COMPONENTS

Based on EHL theories, effectiveness of lubrication in rolling element bearings,[3,78-81] gears,[3,78] and cams[82] can be calculated through the film parameter Λ, the ratio of film thickness to the composite surface roughness. In this section, formulas are taken mostly from an EHL guide book.[78]

Rolling Element Bearings

Roller bearings usually have line contacts and Equation 9 should be used to calculate film thickness. For ball bearings, contacts are elliptical with semimajor axis normal to the direction of rolling and Equations 10 through 13 should be used; to evaluate the speed and load parameter, rolling speed and contact dimensions must be determined from the geometry and kinematics of the system. Reference 78 gives formulas for all common commercial rolling bearings. A simplified film thickness formula, which does not involve detailed bearing geometry and yet gives an adequate prediction of film thickness, is given below:[78]

$$\Lambda = \frac{1}{\sigma} \text{ CD } [LP \cdot N]^{0.74} \qquad (23)$$

where $\Lambda = h/\sigma$, D = bearing outside diameter, m or in., C = a constant given in Table 2, dimensionless, LP = $\mu_o \alpha \cdot 10^{11}$, sec, μ_o = viscosity, N-sec/m² or lb-sec/in.², α = pressure-viscosity coefficient, m²/N or in.²/lb, N = difference between the inner and outer race speeds, rpm, h = film thickness in microns if D is in meters or in microinches if D is in inches, and σ = composite roughness, μm or μin. Typical values of σ for bearings are given in Table 3.

An adequate Λ for protecting bearing surfaces against early surface fatigue was shown to be greater than 1.5. Typical values of lubricant parameter, LP, for motor oils can be found in Figure 16.

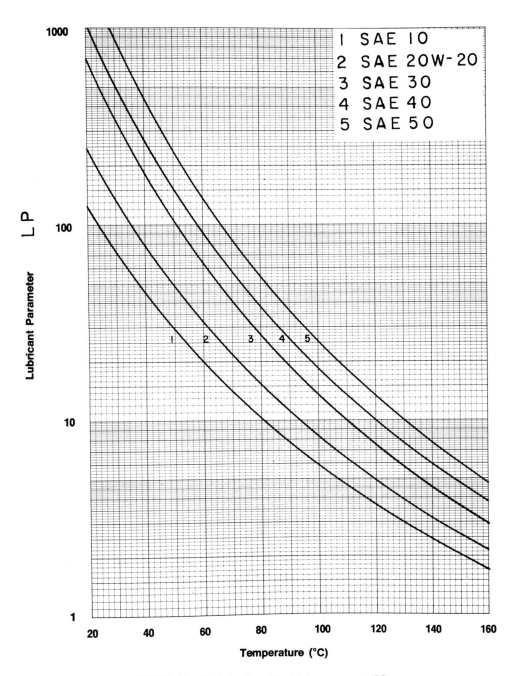

FIGURE 16. Typical value of the lubricant parameter LP.

Gears

Equation 24 is given here for calculating film parameter Λ in involute gears of the following types:

1. Internal and external, parallel, fixed axis spur, and helical gears.
2. Straight, zerol, and spiral bevel gears with any shaft angle.
3. Simple planetary gear trains.

Table 4
GEAR EQUATIONS

	Gear Type	N	G	W_T/ℓ	V	Notes
FIXED AXIS	Parallel Axis External	N_G	$\dfrac{3.4\times10^{-4}(m_G C\sin\phi_n)^{1.5} E_D^{0.148}}{(m_G+1)^2}$	$\dfrac{T_G(m_G+1)}{m_G CF\cos\phi_n\cos^2\psi}$	$\dfrac{2\pi m_G CN_G}{60(m_G+1)}$	Helical gears: Helix Angle = ψ Spur Gears: $\psi=0$
	Parallel Axis Internal	N_R	$\dfrac{3.4\times10^{-4}(m_G C\sin\phi_n)^{1.5} E_D^{0.148}}{(m_G-1)^2}$	$\dfrac{T_R(m_G-1)}{m_G CF\cos\phi_n\cos^2\psi}$	$\dfrac{2\pi m_G CN_R}{60(m_G-1)}$	
	Bevel Gear 90° Shaft Angle	N_G	$\dfrac{3.4\times10^{-4}(R_{Gm}\sin\phi_n)^{1.5} E_D^{0.148}}{(1+m_G^2)^{0.25}}$	$\dfrac{T_G}{R_{Gm}F\cos\phi_n\cos^2\psi_m}$	$\dfrac{2\pi R_{Gm}N_G}{60}$	Spiral bevels; Spiral angle=ψ_m
	Bevel with Shaft Angle Other than 90°	N_G	$\dfrac{3.4\times10^{-4}(R_{Gm}\sin\phi_n)^{1.5} E_D^{0.148}}{(\cos\gamma_G+m_G\cos\gamma_P)^{0.5}}$	$\dfrac{T_G}{R_{Gm}F\cos\phi_n\cos^2\psi_m}$	$\dfrac{2\pi R_{Gm}N_G}{60}$	Straight, Zerol: $\psi_m = 0$
PLANETARY	Sun-Planet	$\lvert N_S-N_C\rvert$	$3.4\times10^{-4}(R_S\sin\phi_n)^{1.5}\left(\dfrac{R_R-R_S}{R_R+R_S}\right)^{0.5} E_D^{0.148}$	$\dfrac{\lvert T_S\rvert}{nR_S F\cos\phi_n\cos^2\psi}$	$\dfrac{2\pi R_S}{60}\lvert N_S-N_C\rvert$	Helical gears: Helix Angle = ψ
	Ring-Planet	$\lvert N_R-N_C\rvert$	$3.4\times10^{-4}(R_R\sin\phi_n)^{1.5}\left(\dfrac{R_R-R_S}{R_R+R_S}\right)^{0.5} E_D^{0.148}$	$\dfrac{\lvert T_R\rvert}{nR_R F\cos\phi_n\cos^2\psi}$	$\dfrac{2\pi R_R}{60}\lvert N_R-N_C\rvert$	Spur gears: $\psi = 0$

Note: Where:

$\lVert$	= Absolute (positive) value	N_g	= gear wheel speed, rpm	T_s	= sun gear torque
C	= Center distance	N_R	= ring gear speed, rpm	T_R	= ring gear torque
E_D	= reduced modulus (equation 2)	N_S	= sun gear speed, rpm	γ_G	= gear cone angle
F	= face width	R_{Gm}	= midface pitch radius	γ_P	= pinion cone angle
m_G	= gear ratio	R_R	= ring gear radius	ϕ_n	= normal pressure angle
n	= Number of planets	R_S	= sun gear radius	ψ	= helix angle
N_C	= Carrier speed, rpm	T_G	= gear wheel torque	ψ_m	= midface spiral angle

Table 5
TYPICAL VALUES OF COMPOSITE ROUGHNESS, σ

	Initial value		Run-In value	
Tooth finish	μm	μ.in.	μm	μ.in.
Hobbed	1.78	70	1.02	40
Shaved	1.27	50	1.02	40
Ground soft	0.89	35	—	—
Ground hard	0.51	20	—	—
Polished	0.18	7	—	—

$$\Lambda = \frac{1}{\sigma}\left[G\,LP\,N\left(\frac{W_T}{\ell}\right)^{-0.148}\right]^{0.74} \tag{24}$$

where G = geometrical parameter from Table 4, LP = $\mu_o\alpha\cdot 10^{11}$, sec, N = gear rotational speed, rpm, W_T/ℓ = load per unit length of contact from Table 4, and σ = composition roughness, see Table 5.

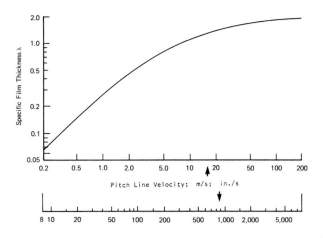

FIGURE 17. Adjusted specific film thickness vs. pitch line velocity (5% probability of distress).

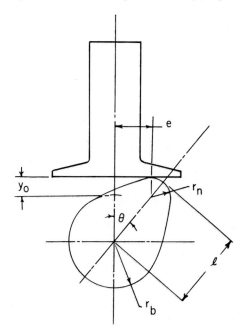

FIGURE 18. Geometry of a cam-follower contact.

The critical value of Λ at which a 5% probability of surface distress is expected is an empirical function of pitch line velocity V as shown in Figure 17. Equations for V for different types of gears are given in Table 4.

Cam-Follower Systems

The film parameter Λ for a cam-flat follower Figure 18 system can be calculated by Equation 25:

$$\Lambda = \frac{1}{\sigma} \, 4.35 \times 10^{-3} \left[f_N \, LP \, N \right]^{0.74} R^{0.26} \qquad (25)$$

where N = cam shaft speed, rpm, LP = lubricant parameter, sec, $f_N = |2r_n - \ell|$, where ℓ is the distance from the nose tip to the shaft axis and r_n is the nose radius (see Figure 18), m or in., $R = (1/r_n + 1/r_f)^{-1}$, m or in., r_n = nose radius, m or in., r_f = follower radius, m or in., and σ = composite roughness, μm or μin.

In general, Λ in cam systems is well below one. In this regime, EHL is ineffective and one must rely heavily on surface film or boundary lubrication to protect surfaces against scuffing and wear.

REFERENCES

1. **Martin, H. M.**, Lubrication of gear teeth, *Engineering (London)*, 102, 199, 1916.
2. **Grubin, A. N.**, Contact stresses in toothed gears and worm gears, *Central Scientific Research Institute for Technology and Mechanical Engineering*, Book No. 30,, Moscow. (D..S.I..R. English Translation No. 337. As communicated by Prof. M. M. Krushchov to Prof. A. Cameron, Grubin's contribution was originally studied by A. M. Ertel and after his death was seen through the press by his co-worker Grubin and is thus often known as Grubin's name alone.)
3. **Dowson, D. and Higginson, G. R.**, *Elastohydrodynamic Lubrication*, Pergamon Press, Oxford, 1977.
4. **Cheng, H. S.**, Isothermal EHD theory for the full range of pressure-viscosity coefficient, *J. Lubr. Technol. Trans. ASME*, 94(1), 35, 1972.
5. **Ford, R. A. J.**, The Lubrication of High Speed Gas Turbine Roller Bearings, Ph.D. thesis, University of London, March, 1975.
6. **Greenwood, J. and Kanzlarich, J.**, Inlet shear heating in elastohydrodynamic lubrication, *J. Lubr. Technol. Trans. ASME*, 95(4), 417, 1973.
7. **Cheng, H. S.**, *Calculation of Elastohydrodynamic Film Thickness in High-Speed Rolling and Sliding Contacts*, Rep. No. MTI-67TR24, Mechanical Technology Inc., Latham, N.Y., May 1967.
8. **Murch, L. E. and Wilson, W. R. D.**, A thermal elastohydrodynamic inlet zone analysis, *J. Lubr. Technol., Trans. ASME*, 97(2), 212, 1975.
9. **Wolveridge, P. E., Baglin, K. P., and Archard, J. F.**, The starved lubrication of cylinders in line contact, Proc. Inst. Mech. Eng., 185, 1159, 1970.
10. **Dowson, D., Saman, W. Y., and Toyoda, S.**, A study of starved elastohydrodynamic line contacts, *Proc. 5th Leeds-Lyon Symp. Tribology*, Leeds, England, 1979.
11. **Archard, J. F.**, Experimental studies of elastohydrodynamic lubrication, *Proc. Inst. Mech. Eng.*, 180(38), 17, 1965.
12. **Crook, A. W.**, The lubrication of rollers. II. Film thickness with relation to viscosity and speed, *R. Soc. London Philos. Trans. Ser. A*, 254, 223, 1961.
13. **Sibley, L. B. and Orcutt, F. K.**, Elastohydrodynamic lubrication of rolling contact surfaces, *Am. Soc. Lubr. Eng. Trans.*, 4(2), 234, 1961.
14. **Wymer, D. G. and Cameron, A.**, EHD lubrication of a line contact, *Proc. Inst. Mech. Eng.*, 188, 221, 1974.
15. **Dowson, D. and Higginson, G. R.**, A numerical solution to the elastohydrodynamic problem, *J. Mech. Eng. Sci.*, 1(1), 6, 1959.
16. **Dowson, D., Higginson, G. R., and Whitaker, A. V.**, Elastohydrodynamic Lubrication — a survey of isothermal solutions, *J. Mech. Eng. Sci.*, 4(2), 121, 1962.
17. **Archard, G. D., Gair, F. C., and Hirst, W.**, The elastohydrodynamic lubrication of rollers, *Proc. R. Soc. London Ser. A*, 262, 51, 1961.
18. **Hamilton, G. M. and Moore, S. L.**, Deformation and pressure in an EHD contact, *Proc. R. Soc. London Ser. A*, 322, 313, 1971.
19. **Rodkiewicz, C. M. and Srinivanasan, V.**, EHD lubrication in rolling and sliding contacts, *J. Lubr. Technol. Trans. ASME*, 94(4), 324, 1972.
20. **Rohde, S. M.**, A unified treatment of thick and thin film EHD problems by using high order element methods, *Proc. R. Soc. London Ser. A*, 343, 315, 1975.
21. **Dowson, D.**, Elastohydrodynamic Lubrication, Interdisciplinary Approach to the Lubrication of Concentrated Contacts, Spec. Publ. No. NASA SP-237, National Aeronautics and Space Administration, Washington, D.C., 1970, 34.
22. **Kannel, J. W. et al.**, A Study of the Influence of Lubricants on High-Speed Rolling-Contact Bearing Performance, Part IV, Tech. Rep. No. ASD-TR-61-643, Air Force Aero Propulsion Laboratory, Dayton, Ohio, 1964.

23. **Gohar, R. and Cameron, A.,** The mapping of EHD contacts, *ASLE Trans.,* 10, 214, 1967.
24. **Orcutt, F. K.,** Experimental study of elastohydrodynamic lubrication, *ASLE Trans.,* 8, 381, 1965.
25. **Kannel, J. W.,** The measurement of pressure in rolling contacts, *Proc. Inst. Mech. Eng.,* 180(3B), 135, 1965.
26. **Moes, I. H.,** Communications to EHL symposium held at Leeds University, *Proc. Inst. Mech. Eng.,* 180(3B), 244, 1965.
27. **Herrebrugh, K.,** Solving the incompressible and isothermal problem in elastohydrodynamic lubrication through an integral equation, *J. Lubr. Technol., Trans. ASME,* 90(1), 262, 1968.
28. **Archard, J. F. and Cowking, E. W.,** A simplified treatment of elastohydrodynamic lubrication theory for a point contact, *Proc. Inst. Mech. Eng.,* 180(3B), 47, 1965.
29. **Cheng, H. S.,** A numerical solution of the elastohydrodynamic film thickness in an elliptical contact, *J. Lubr. Technol., Trans. ASME,* 92(1), 155, 1970.
30. **Hamrock, B. J. and Dowson, D.,** Isothermal elastohydrodynamic lubrication of point contacts, *J. Lubr. Technol.,* 98(2), 223, 1976; 98(3), 1976; 99(2), 264, 1977; 99(1), 15, 1977.
31. **Chiu, Y. P.,** An analysis and prediction of lubricant film starvation in rolling contact systems, *ASLE Trans.,* 17, 22, 1974.
32. **Chiu, Y. P. et al.,** *Exploratory Analysis of EHD Properties of Lubricants,* Rep. No. AL72P10, SKF Industries, King of Prussia, Pa., 1972.
33. **Snidle, R. W. and Archard, J. F.,** Experimental investigation of elastohydrodynamic lubrication at point contacts, *Proc. 1972 Symp. Elastohydrodynamic Lubrication,* Paper C2/72, Institute of Mechanical Engineers, London, 1972, 5.
34. **Wedevan, L. D.,** Optical Measurements in EHD Rolling-Contact Bearings, Ph.D. thesis, University of London, March 1970.
35. **Westlake, F. J. and Cameron, A.,** Interferomatric study of point contact lubrication, *Proc. 1972 Symp. Elastohydrodynamic Lubrication,* Paper C39/72, Institute of Mechanical Engineers, London, 1972, 153.
36. **Sanborn, D. M. and Winer, W. O.,** Fluid rheological effects in sliding elastohydrodynamic point contacts with transient loading. I. Film thickness, *J. Lubr. Technol., Trans. ASME,* 93(2), 262, 1971.
37. **Parker, R. J. and Kannel, J. W.,** EHD Film Thickness Between Rolling Discs with a Synthetic Paraffinic Oil to 589 K, NASA Tech. Note D-6411, National Aeronautics and Space Administration, Washington, D.C., 1970.
38. **Gentle, C. R., Duckworth, R. R., and Cameron, A.,** EHD Film thickness at extra pressures, *J. Lubr. Technol., Trans. ASME,* 97, 383, 1975.
39. **Hirst, W. and Moore, A. J.,** Elastohydrodynamic Lubrication at High Pressures, Tech. Rep., University of Reading, Reading, U.K., 1977.
40. **Foord, C. A. et al.,** Optical elastohydrodynamics, *Proc. Inst. Mech. Eng.,* 184(1), 487, 1969.
41. **Jacobson, B.,** On the lubrication of heavily loaded spherical surfaces considering surface deformations and solidification of the lubricant, *Acta Polytechn. Scand. Mech. Eng. Ser.,* 54, 1970.
42. **Ranger, A. P.,** Numerical Solutions to the EHD Problems, Ph.D. thesis, University of London, March, 1974.
43. **Cheng, H. S. and Sternlicht, B.,** A numerical solution for pressure, temperature and film thickness between two infinitely long rolling and sliding cylinders under heavy load, *J. Basic Eng., Trans. ASME,* 87(3), 695, 1965.
44. **Dowson, D. and Whittaker, B. A.,** A numerical procedure for the solution of the elastohydrodynamic problem of rolling and sliding contacts lubricated by a newtonian fluid, *Proc. Inst. Mech. Eng.,* 180(3B), 57, 1965.
45. **Kannel, J. W. and Bell, J. C.,** A method for estimating of temperature in lubricated rolling-sliding gear or bearing EHD contacts, Paper C24/72, *Proc. 1972 Symp. Elastohydrodynamic Lubrication,* Institute of Mechanical Engineers, London, 1972, 118.
46. **Kannel, J. W., Zugaro, F. F., and Dow, T. A.,** A method for measuring surface temperature between rolling/sliding steel cylinders, *J. Lubr. Technol. Trans. ASME,* 100(1), 100, 1978.
47. **Nagaraj, H. S., Sanborn, D. M., and Winer, W. O.,** Direct surface temperature measurement by infrared radiation in elastohydrodynamic contacts and the correlation with the Block temperature theory, *Wear,* 49, 1, 1978.
48. **Jaeger, J. C.,** Moving sources of heat and the temperature at sliding contacts, *Proc. R. Soc. N.S.W.,* 56, 203, 1942.
49. **Johnson, K. L. and Cameron, R.,** Shear behavior of elastohydrodynamic oil film at high rolling contact pressures, *Proc. Inst. Mech. Eng.,* 182, 307, 1967.
50. **Harrison, G. and Trachman, E. G.,** The role of compressional viscoelasticity in the lubrication of rolling contacts, *J. Lubr. Technol., Trans. ASME,* 95, 306, 1972.
51. **Dyson, A.,** Frictional traction and lubricant rheology in elastohydrodynamic lubrication, *Philos. Trans. R. Soc. London,* 266, 1170, 1970.

52. **Johnson, K. L. and Tevaariverk, J. L.,** Shear behavior of EHD oil films, *Proc. R. Soc. London,* A356, 215, 1977.

53. **Barlow, A. J. et al.,** The effect of pressure on the viscoelastic properties of liquids, *R. Soc. London Proc.,* A327, 403, 1972.

54. **Montrose, C. J., Moynihan, C. T., and Sasake, H.,** Dynamic Shear and Structural Viscoelasticity in EHD Lubrication, Vitreous State Laboratory Tech. Rep., July 1977.

55. **Bell, J. C.,** Lubrication of rolling surfaces by a Ree-Eyring fluid, *ASLE Trans.,* 5, 160, 1962.

56. **Trachman, E. and Cheng, H. S.,** Thermal and non-Newtonian effects on traction in elastohydrodynamic lubrication, Paper C37/72, *Proc. 1972 Symp. Elastohydrodynamic Lubrication,* Institute of Mechanical Engineers, London, 1972, 142.

57. **Smith, F. W.,** Rolling contact lubrication — the application of elastohydrodynamic theory, *J. Lubr. Technol., Trans. ASME Ser. D,* 87, 170, 1965.

58. **Bair, S. and Winer, W. O.,** A rheological model for EHD contacts based on primary laboratory data, *J. Lubr. Technol., Trans. ASME,* 101(3), 258, 1979.

59. **Tallian, T. E.,** The theory of partial elastohydrodynamic contacts, *Wear,* 21, 49, 1972.

60. **Williamson, J. B. P.,** Topography of solid surfaces, in *Interdisciplinary Approach to Friction and Wear,* NASA SP-181, National Aeronautics and Space Administration, Washington, D.C., 1968, 143.

61. **Whitehouse, D. J. and Archard, J. F.,** The properties of random surfaces of significance in their contact, *Proc. R. Soc. London,* A316, 97, 1970.

62. **Patir, N. and Cheng, H. S.,** An average flow model for determining effects of three dimensional roughness on partial hydrodynamic lubrication, *J. Lubr. Technol., Trans. of ASME,* 100(1), 12, 1978.

63. **Peblenik, J.,** New developments in surface characterization and measurement by means of random process analysis, *Proc. Inst. Mech. Eng.,* 182(3K), 108, 1967.

64. **Johnson, K. L., Greenwood, J. A., and Poon, S. Y.,** A simple theory of asperity contact in elastohydrodynamic lubrication, *Wear,* 19, 1972, 91.

65. **Christensen, H.,** Stochastic models for hydrodynamic lubrication of rough surfaces, *Proc. Inst. Mech. Eng. Tribology Group,* 184(1,55), 1013, 1969.

66. **Berthè, D.,** Les Effects Hydrodynamiques Sur La Fatigue Des Surfaces Dans Les Contacts Hertziens, D. Sc. thesis, University of Lyon, France, 1974.

67. **Chow, L. S. H. and Cheng, H. S.,** The effect of surface roughness on the average film thickness between lubricated rollers, *J. Lubr. Technol., Trans. ASME,* 98(1), 117, 1976.

68. **Cheng, H. S. and Dyson, A.,** Elastohydrodynamic lubrication of circumferentially ground disks, *ASLE Trans.,* 21(1), 25, 1978.

69. **Cheng, H. S.,** On some aspects of micro-elastohydrodynamic lubrication, *Proc. 4th Leeds-Lyon Symp. Lubr.,* April 1977.

70. **Christensen, H.,** Elastohydrodynamic theory of spherical bodies in normal approach motion, *J. Lubr. Tech., Trans. ASME,* 92, 145, 1970.

71. **Lee, K. M. and Cheng, H. S.,** The Effect of Surface Asperity on the Elastohydrodynamic Lubrication, NASA CR-2195, National Aeronautics and Space Administration, Washington, D.C., 1973.

72. **Fowles, P. E.,** The application of elastohydrodynamic theory to individual asperity-asperity collisions, *J. Lubr. Tech., Trans. ASME,* 91, 464, 1969.

73. **Fowles, P. E.,** A thermal elastohydrodynamic theory for individual asperity-asperity collision, *J. Lubr. Tech., Trans. ASME,* 93, 383, 1971.

74. **Rohde, S.,** Thick film and transient elastohydrodynamic lubrication problems, *Proceedings on Fundamentals of Tribology,* MIT Press, Cambridge, Mass., 1979.

75. **Castelli, V. and Malanowski, S. B.,** Method for solution of lubrication problems with temperature and elasticity effects: Application to sector, tilting-pad bearings, *J. Lubr. Technol. Trans. ASME,* 91(4), 634, 1969.

76. **Taniguichi, S. and Ettles, C.,** A thermal elastic analysis of the parallel surface thrust washer, *ASCE Trans.,* 18(4), 299, 1975.

77. **Ettles, C.,** The development of a generalized computer analysis for sector shaped tilting pad thrust bearings, *ASLE Trans.,* 19(2), 153, 1976.

78. **Anon.,** *EHL Guidebook,* Mobile Oil Corporation, New York, 1979.

79. **McGrew, J. M. et al.,** Elastohydrodynamic Lubrication — Preliminary Design Manual, Tech. Rep. AFAPL-TR-70-27, Air Force Propulsion Laboratory, Dayton, Ohio, 1970.

80. **Cheng, H. S.,** *Application of Elastohydrodynamics of Rolling Element Bearings,* ASME Paper 74-DE-32, American Society of Mechanical Engineers, New York, 1974.

81. **Anon.,** *SKF Engineering Data,* SKF Industries, Inc., King of Prussia, Pa., 1968.

82. **Dyson, A.,** Discussion of "Elastohydrodynamic Lubrication" by D. Dowson, Spec. Publ. SP-237, National Aeronautics and Space Administration, Washington, D.C., 1970.

83. **Orcutt, F. K. and Cheng, H. S.,** Lubrication of rolling contact instrument bearings, gyro spin-axis, *Hydrodynamic Bearing Symp.,* Vol. 2, M.I.T. Instrument Laboratory, Cambridge, Mass., 1966.

METALLIC WEAR

F. T. Barwell

INTRODUCTION

Nature of Wear

Wear of material from machine elements may occur as the result of direct overstressing of surface material, by fatigue of subsurface material, melting, evaporation, chemical attack, or by electrical or electrolytic action. Because various mechanisms may act either singly or in combination, the rate of wear may sometimes be determined by competition and sometimes by mutual reinforcement of two or more effects. There are, therefore, no simple laws to enable wear rates to be calculated without reference to specific environmental and operational conditions relating to the actual machine under consideration. For example, the expression

$$\gamma = kWV/H \tag{1}$$

where γ is the wear rate, W the applied load, V the sliding speed, and H the hardness of the material is only applicable over a very limited range of variables.[28]

Lubrication is essential in most machines to reduce friction and the rate of wear to tolerable values. This introductory section will, however, concentrate on the wear of metal without deliberate lubrication.

Conformal and Counterformal Surfaces

The most important consideration governing tribological interaction of two solid objects is their shape, because this determines both the nature of the stress system and the thermal regime. Two broad categories are as follows:

1   *Conformal* surfaces wherein stress is distributed over a comparatively wide nominal area.
2.   *Counterformal* surfaces which produce either "point" or "line" contact. The surfaces deform either elastically or plastically so as to provide an adequate area of contact.[56] Compressive stress at the surface of such a Hertzian contact is distributed in accordance with a parabolic law with the highest stress being at the center. Shear stress in the absence of tangential loading reaches a maximum at a depth within the surface of about 1/6th of the breadth of the contact zone. While detailed methods enable calculating the stresses in bodies of various shapes,[4,12,56] the following simple cases will enable the nature of Hertzian stress to be appreciated.

Spheres in Contact

Radius of circle of contact = a

$$a = \left| \frac{3}{4} W \left\{ \frac{1 - \nu_1^2}{E_1} + \frac{1 - \nu_2^2}{E_2} \right\} \frac{r_1 r_2}{r_1 + r_2} \right|^{1/3} \tag{2}$$

where W = load, ν_1 and ν_2 = Poisson's ratio of material of spheres 1 and 2, respectively, E_1 and E_2 = Young's modulus of elasticity of spheres 1 and 2, respectively, and r_1 and r_2 = radii of spheres.

When both spheres are made from material having the same modulus of elasticity and when Poisson's ratio equals 0.3,

$$a = 1.109 \left| \frac{W}{E} \frac{r_1 r_2}{r_1 + r_2} \right|^{1/3} \tag{3}$$

$$p_o = 0.338 \, (W^{1/3}) \left(E \frac{r_1 + r_2}{r_1 r_2} \right)^{2/3} \tag{4}$$

where p_o, the maximum compressive stress, is 2/3 times the average value. The maximum shearing stress is approximately at a depth of 1/2 a below the surface. The maximum tensile stress occurs at the edge of the contact zone. Its magnitude is given by

$$p_t = p_o \, (1 - 2\nu)/3 \tag{5}$$

Parallel Cylinders in Contact

Breadth of contact strip $= 2a$

$$a = \left| \frac{4P}{\pi} \frac{r_1 r_2}{r_1 + r_2} \times \left\{ \frac{1 - \nu_1^2}{E_1} + \frac{1 - \nu_2^2}{E_2} \right\} \right|^{1/2} \tag{6}$$

where P is load per unit length of the cylinders in contact.

$$p_o = 0.418 \left| PE \frac{r_1 + r_2}{r_1 r_2} \right|^{1/2} \tag{7}$$

Maximum shearing stress is 0.304 p_o and occurs at a depth of 0.78a below the surface. The elastic limit is reached when $p_o = 1.7 \times$ the yield strength in compression.[5]

The fact that maximum shear stress is situated, a definite distance within the interacting bodies causes crack formation to occur by fatigue action on an internal plane. Eventually, cracks will join this plane to the surface giving rise to "pitting"-type failure.

Temperature of Interacting Surfaces

The amount of heat liberated at contact will be determined by the product of the force acting between the surfaces, the velocity of relative motion, and the coefficient of friction. The temperature of the contact will depend on the thermal diffusivity of the material and the rate of supply of fresh material into the contact zone. A good estimate of the "flash temperature", that is the excess of temperature caused by friction over the bulk temperature of the machine elements, is given by Blok[7] as

$$\theta_f = \frac{1.11 \, \mu P |U_1 - U_2|}{(K_1 \rho_1 C_1 U_1)^{1/2} + (K_2 \rho_2 C_2 U_2)^{1/2}} \frac{1}{(2a)^{1/2}} \tag{8}$$

where P is the load per unit width (measured at right angles to direction of relative motion), 2a is the width at the Hertzian contact band (measured in direction of relative motion), U_1 and U_2 are the surface velocities, K_1 and K_2 are the thermal conductivities, ρ_1 and ρ_2 are the densities, and c_1 and c_2 are specific heats per unit mass relating to the two bodies. An appropriate value for μ, the coefficient of friction, may vary widely from about 0.15 to 1 according to the effectiveness of any lubrication or any surface films. Oxygen or water vapor, for example, may cause films to be generated, which either mitigate the effects of metal-to-metal contact or have harmful abrasive effects.

CHARACTERISTIC MODES OF DAMAGE

Terminology

For the purpose of this handbook, wear will be divided into the following categories:

smooth sliding (mild), severe, adhesive, fatigue, fretting, impact, gouging, firecracking, corrosion, cavitation, and electrical effects.

When two ferrous materials are rubbed together under moderate load, a certain degree of protection may be offered by the generation of oxide films and the sliding will occur smoothly and without gross damage. This has been described as ''mild wear'' but the writer proposes the substitution of ''smooth sliding'' because the rate of wear can in certain circumstances be greater during ''mild'' than during ''severe'' conditions.

Transition Between ''Smooth Sliding'' and ''Severe'' Wear

The transition between different modes of wear may be abrupt and may be reversed even when the severity of the applied conditions is apparently increased. The complexity of the interaction is illustrated by two classical experiments.

Kerridge[22] loaded a flat-ended pin of tool steel against the peripheral surface of a rotating ring within an enclosure which could be evacuated. The pin was made radioactive and was softer than the ring, the hardness values being 270 and 860 HD_{30}, respectively. Material was rapidly transferred from the pin to the ring which soon attained a constant value of radioactivity and the wear rate was constant throughout the test. When the radioactive pin was replaced by an inactive pin the activity of the ring soon ceased. These results showed that metal was first transferred from the soft pin to the harder ring. This metal rapidly formed an oxide film which resisted further transfer and thus introduced a ''rate limiting'' action. As this oxide was gradually removed by rubbing or fatigue, further transfer of material could take place at a controlled rate.

Kragelskii[26] found that during some experiments on the sliding of Armco iron, the wear rate fell off by a factor of about 600 when the rate of sliding increased beyond a certain value. When the interfacial region was cooled by liquid nitrogen there was a very high-wear rate. When it was electrically heated, wear was reduced one-thousand fold. Kragelskii explains the contrasting behavior by a hardness or strength gradient within the material. When the interface between rubbing surfaces is composed of weak material, sliding will take place there with relatively little damage; when the bond between the surfaces is stronger than the underlying layers, failure will occur within the bulk of the material causing considerable roughening and superficial damage.

There are, therefore, two fundamentally different ways in which rubbing surfaces may react. During ''external'' friction, contact between the surfaces is dispersed and the true area of interaction depends on the applied load and the strength properties of the weaker material. With ''internal'' friction the surface of action is continuous, is independent of load and the zone undergoing deformation occupies a considerable volume.

Effect of Environment

Clarification of the effect of environment is provided by Soda and Sasada[50] who studied the wear of pure metals in air under pressure ranging from 10^{-6} to 760 torr. For most metal combinations the wear was ''cohesive'', but for the transition metals (Ni, Fe, Pt, Mo, and W) ''noncohesive'' wear occurred except under high loads (Figure 1).

This transition was explained by the ''mean free time'' of a contact point. A small surface of bare metal would be formed every time a contact bridge was broken. Gas molecules would attack the clean spot forming a chemisorbed layer which would be thereby protected from welding. When either the speed or the load was high, the time between events would be too short for an effective protective layer to form and severe or cohesive wear would occur. When air is replaced by nitrogen, Sasada and Kando[43] showed that the mean size of particles was increased by over one hundred times. The powerful effect of oxygen prevented surface adhesion and subsequent particle growth.

Figure 2 shows the variation of wear rate with speed for Ni on Ni in air and vacuum.

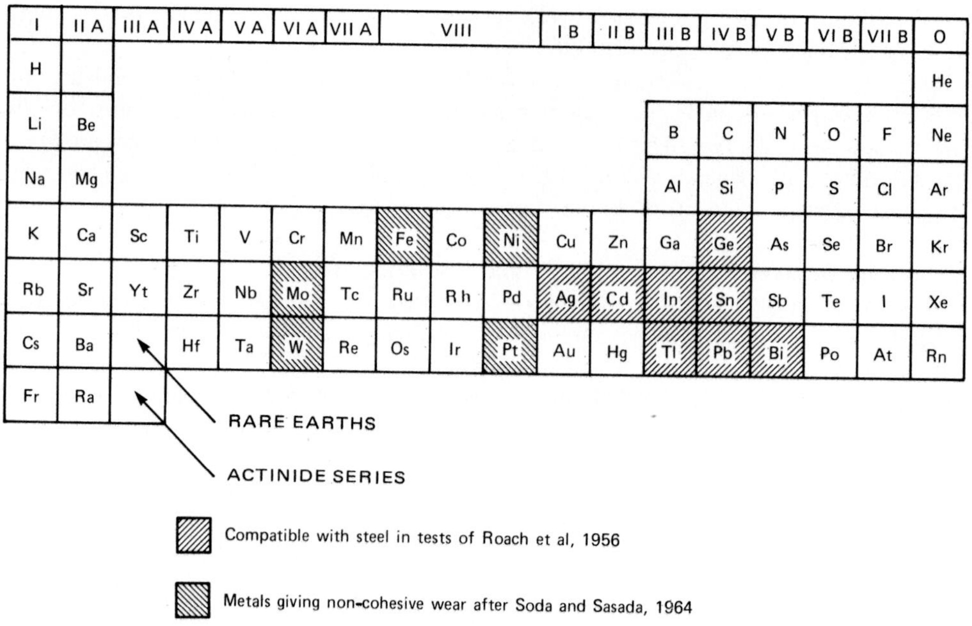

I | II A | III A | IV A | V A | VI A | VII A | VIII | I B | II B | III B | IV B | V B | VI B | VII B | O
H He
Li Be B C N O F Ne
Na Mg Al Si P S Cl Ar
K Ca Sc Ti V Cr Mn Fe Co Ni Cu Zn Ga Ge As Se Br Kr
Rb Sr Yt Zr Nb Mo Tc Ru Rh Pd Ag Cd In Sn Sb Te I Xe
Cs Ba Hf Ta W Re Os Ir Pt Au Hg Tl Pb Bi Po At Rn
Fr Ra
RARE EARTHS
ACTINIDE SERIES
Compatible with steel in tests of Roach et al, 1956
Metals giving non-cohesive wear after Soda and Sasada, 1964

FIGURE 1. Periodic chart of the elements showing metals compatible with Fe as reported by Roach et al. (1956) and metals which gave "noncohesive" wear in the tests of Soda and Sasada.[50]

0.10
0.08
0.06
0.04
0.02
Wear W, mm3
2.0
1.6
1.2
0.8
0.4
0
Coefficient of Friction, f
f (vac.)
f (air)
W (air)
W (vac.)
0
100
200
300
400
Sliding speed v, mm/s

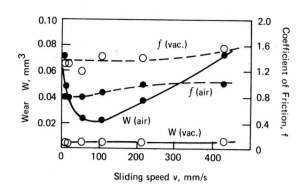

FIGURE 2. Effect of environment on the wear of Ni on Ni after Soda and Sasada.[50] Ni/Ni wear and friction, total sliding distance Z = 20 m, load P = 2N.

The actual rate of removal of metal is greater with the so called "mild" wear in air than the "severe" wear at higher speeds in vacuum.

Another environment of importance in power technology is water at high temperatures and pressures. Roberts[42] carried out tests at temperatures up to 350°C and pressures up to 1.8 MPa (260 lb/in.^{2}). When identical surfaces of soft nickel-chromium alloys were in contact, severe wear occurred with the coefficient of friction approaching unity. Satisfactory operation was obtained with a particular nickel alloy which was harder. Stainless steel sliding on hard-chromium plating was also satisfactory.

Severe Sliding Wear

The mode of degradation of surfaces subject to high stresses depends very much upon the circumstances of the application. Extensively studied, the railway wheel on rail can

Table 1
RELATIVE WEAR RESISTANCE OF RAIL STEELS IN CURVED TRACK

Type of rail steel	Wt %				UTS (N/mm²)[a]	Relative wear resistance
	C	Mn	Si	Cr		
BS11 (oxygen)	0.5/0.6	0.95/1.25	0.25 max		700	1
UIC Regular Grade	0.4/0.6	0.8/1.2	0.35 max		700	1
UIC wear resistant grade A (UICA)	0.6/0.75	0.8/1.3	0.5 max		880	2
UIC wear resistant grade B (UICB)	0.5/0.7	1.3/1.7	0.5 max		880	2
UIC wear resistant grade C (UICC)	0.45/0.65	1.7/2.1	0.5 max		880	2
1% Cr (Germany)	0.65/0.80	0.8/1.3	0.07 max	0.8/1.3	1 100	4
1% Cr (Canada)	0.83	0.51	0.30	1.2	1 100	4
1% Cr (USSR)	0.69/0.82	0.7/1.0	0.04 max	0.5/1.0	1 100	4
Standard AREA rail (US)	0.67/0.80	0.7/1.0	0.1/0.23		900	2
Standard USSR rail	0.6/0.8	0.6/1.0				2
Hadfield's manganese	1.0/1.4	10.0/14.0	0.2/0.6		900	4—10
Fracture tough (IRS) cast 1511	0.30	1.28	1.06			≤1
Fracture tough (IRS) cast 692	0.37	1.43	0.24			1

[a] Ultimate tensile strength.

From Clayton, P., in *Tribology 1978 Materials Performance and Conservation,* Institute of Mechanical Engineers Conference Publication, Swansea, 1978, 83. With permission.

provide an example of the reaction of steel to intense Hertzian stresses. Clayton[9] describes a type of wear where intermittent high-stress dry-rubbing contact induces severe surface deformation which leads to the formation of small deformed metal wear platelets.

Table 1 shows the relative wear resistance of different steels under operating conditions. Laboratory tests produced results which were consistent with track observations and Clayton concluded that the wear-rate was a function of cumulative plastic strain ϵ. If the equation for a monotonic stress-strain curve is given by

$$\sigma = K(\epsilon)^n \tag{9}$$

where K and n are constants, then plastic strain is given by

$$\epsilon = \left(\frac{\sigma}{K}\right)^{1/n} \tag{10}$$

Hence, wear rate is a function of

$$\left(\frac{J}{K}\right)^{1/n}$$

He considers that wear is more likely to be stress-related than strain controlled and produces the

$$\text{Wear rate, mm}^3/\text{cm} = 0.2605 \left(\frac{1000}{K}\right)^{1/n} + 0.0243 \tag{11}$$

For Hadfields steel with K taken as 2800 N/mm² and n as 0.31, a wear rate of 0.0332 mm³/cm is predicted which corresponds to the relative value obtained.

Jamieson[18] investigated mechanical wear of wheels and rails due to negotiation of curves. Two mechanisms differed in the rate of metal loss by factors of 10 to 100. Low-rate "flow-

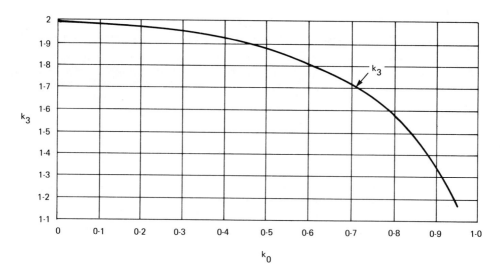

FIGURE 3.   Chart for calculating Hertzian deflection.

$$k_o = \left[\left(\frac{1}{r_1}\right)^2 + \left(\frac{1}{r_2}\right)^2 - 2\left(\frac{1}{r_1} \cdot \frac{1}{r_2}\right)\right]^{1/2} \Bigg/ \left(\frac{1}{r_1} + \frac{1}{r_2}\right)$$

where r_1 = wheel radius and r_2 = rail crown radius.

fatigue'' wear occurred when normal forces between wheel and rail produced subsurface stresses which exceeded 2.3 times the yield strength of the material. High-rate ''smearing wear'' occurred when tractive forces were superimposed onto the normal force so that the subsurface plastic flow boundary intercepted the surface. Laboratory experiments indicated that high humidity could sufficiently lower traction forces so that the flow-fatigue type of wear occurred, rather than smearing. It is suggested that rail lubrication will produce the same effect in practice.

Analysis indicated that contact stresses between wheels and rail could exceed 350,000 lb/in.[2] (2.4 GPa) sufficient for severe smearing wear. Examination of wear debris from a high wear region of track in Canada confirmed that smearing wear predominated.

Corrugation

Deflection of the wheel and rail in the contact zone in the normal direction due to normal loading W is given by Equation 12.[61]

$$\delta = \text{Deflection} = 1.39 \times 10^{-8} \times K_3 \left(\frac{1}{r_1} + \frac{1}{r_2}\right)^{1/3} \times W^{1/3} \qquad (12)$$

where r_1 is the radius of the wheel rim and r_2 is the radius of the wheel head. Values of K_3 are given in Figure 3. If the mass of the wheels and other directly coupled masses such as axle boxes is denoted by M, the system can resonate as follows

$$M\ddot{y} + Ky = 0 \qquad (13)$$

where K is the stiffness. The frequency of vibration will be

$$\sqrt{K/M} \text{ rad/sec or } \frac{1}{2\pi} \sqrt{\frac{K}{M}} \text{ Hz}$$

If M is taken as 800 kg the frequency will be 216 Hz. Nayak[32] has shown that random surface characteristics can excite oscillation at preferred frequencies.

Johnson and Gray[20] demonstrated that such vibration can cause the interacting surfaces to develop "corrugations". Some evidence of plastic flow has been found in the crests of corrugations but none in the troughs. The position is so serious that a number of railways periodically reprofile rail surface by grinding.

Adhesion

When perfectly clean, flat metal surfaces are brought into close proximity (less than 1/5 nm) they unite chemically. When separation is wider, they are attracted to each other by van der Waals' forces. At small separations (less than 10^{-8} m) these are governed by a square law and at greater separations (greater than 10^{-7} m) by a cube law. Practical surfaces are usually covered by oxide films and are so rough on the atomic scale that when bodies are brought together, only a tiny fraction of the contacting area (about 1/1000) at the peaks of asperities is subjected to powerful adhesive forces. Experimental measurements of adhesive forces are available with soft materials which conform when pressed together,[19] with mica which has been cleaved to produce an exceptionally smooth surface,[17] and with hard spheres of very small diameters.[25,36] Adhesive forces were sometimes two to three orders of magnitude higher than those applied initially to force the spheres together.[25]

Buckley[8] measured the force to rupture junctions made within a vacuum system evacuated to 10^{-11} torr. Crystals of copper, gold, silver, nickel, platinum, lead, tantalum, aluminum, and cobalt were cleaned by argon ion bombardment before being forced against a clean iron (011) surface by a force of 20 dyn. When iron was pressed against iron, a separating force greater than the 400 dyn was required. In the case of other metals this force varied from 50 to 250 dyn. In every case the strength of the junction was greater than the force used to promote it. Even in the case of lead (which is insoluble in iron), Auger analysis indicated transfer of lead to the iron surface. Thus the adhesive bonds of lead to iron were stronger than the cohesive bonds within the lead. In general, the cohesively weaker metals adhered and transferred to the cohesively stronger.

The adhesion theory of wear is based on the assumption that a similar welding action occurs between a limited number of asperities and that the welds are ruptured when the solids slide one relative to the other.[54]

The actual process of formation of wear particles has been studied by Sasada and Kando[43] with a pin and disc machine. They concluded that an initial metal-to-metal junction is sheared by the frictional motion and a small fragment of either surface becomes attached to the other surface. As sliding continues this fragment constitutes a new asperity becoming attached once more to the original surface. This "transfer element" is repeatedly transferred from one surface to the other continually increasing in size and being flattened by the force between the pin and the disc. Once a flattened particle attached to the disc grows to such a size that it supports the load, it becomes the only contact between pin and disc. It then grows quickly to a large size, absorbing many of the transfer elements dotted over the disc surface so as to form a flake-like particle from materials of both rubbing elements. Unstable thermal and dynamic conditions brought about by rapid growth of this transfer element finally account for its removal as a wear particle. These authors experimented with the following materials: Mo, Fe, Mi, Cu, Ag, Zn, and Al.

The combination of Al disc and pins of Mo, Fe, Ni, and Cu produced violent ploughing. The following combinations produced smooth sliding: Mo/Mo, Fe/Fe, Ni/Ni, Ni/Fe, Fe/Cu, Cu/Fe, Fe/Mo, Mo/Fe, Ni/Mo, and Mo/Ni. Metal transfer was scarcely observed and the wear rate was very low in the case of Mo-Cu, Mo-Ag, Fe-Ag or Ni-Ag where the metals have poor mutual solubility.[44]

Relative importance of adhesion and plastic flow is covered by Andarelli et al.[1] who observed the occurrence of dislocations by transmission electron-microscopy. Glass fibers were slid against aluminum specimens 10^{-5} m thick, and normal and tangential forces were determined from the shape assumed by the loaded fiber. Further tests[31] employed a cold-rolled tungsten wire with a hemispherical tip of radius 2.5 10^{-6} m as the stylus. The load ranged from 1 to 100 μN as compared with values of 1.2 to 2.4 μN calculated on the basis of an interfacial energy of 100 to 200 mJ/m². This indicates that van der Waals' forces between metals shielded by absorbed gases were responsible for the adhesion. Load had to exceed a critical value before the stylus suddenly penetrated the surface. Measurements of friction were consistent with this, nearly zero at low loads as long as deformation remained elastic.

These results emphasize that plastic deformation rather than adhesion was the important agency determining friction and wear. Comparison of the dislocation density based on tensile tests showed that 90% of the frictional energy was dissipated as heat with only a minor proportion being stored within the material.

Fatigue
Sliding Wear
In all machinery there is a periodic variation of stress. An element of metal at the surface of a rotating shaft will be subject to reversal of bending stresses, the race of a rolling contact bearing will experience continual application and release of Hertzian stress, and the surface of a conformal bearing will experience repeating stresses on a micro scale due to the passage of asperities on the rotating surface. All these repeating stresses can give rise to fatigue action. Tsuya et al.[57] and Quinn and Sullivan[39] have provided evidence of changes in the substrate of a wearing part due to relative motion.

Because it provides a more direct account of the formation of a wear particle than the adhesion theory, the fatigue theory of wear warrants close attention. Soda et al.[51] reported a series of experiments on the face-centered-cubic metals Ni, Cu, and Au. When atmospheric pressure was reduced, wear of Ni and Cu decreased but that of Au remained unchanged. This was shown to affect the rate of wear fragment formation in contrast to mechanical factors which affected wear by changing the volume of fragments. Mean thickness of the wear fragments was about one fourth of that of the plastically deformed substrate layer. Correlation with direct fatigue tests indicated that the number of wear fragments was governed by the resistance of the materials to fatigue. Environmental factors such as atmospheric pressure had similar effect on wear rate as on fatigue strength. Kimura[23] produces additional evidence of a correlation between the thickness of the deformed layer and that of the wear fragments.

A particularly comprehensive test program was carried out by Tsuya[58] who used a variety of test arrangements and ambient conditions. Plastic working of the subsurface regions of materials in contact led to the formation of micronized crystals and cracks which originated in the boundary region between the micronized crystals and those nearer to the surface which had been simply distorted. These cracks tend to develop in the direction of material flow until particles are released.

The Delamination Theory of Wear
Koba and Cook[24] studied the wear of leaded bronze running against steel and demonstrated by scanning electron electron micrographs that metal flowed freely at the surface, smoothing out hills and valleys. Some metal transfer was observed but did not appear to be an essential part of the wear process.

Suh[52] investigated a number of wearing systems and put forward the "Delamination Theory of Wear" which can be summarized as follows:

1. When two sliding surfaces come into contact, asperities on the softer surface are deformed by repeated loading to generate a relatively smooth surface. Eventually asperity-to-asperity contact is replaced by asperity-plane contacts and the softer surface experiences cyclic loading as the asperities of the harder surface plough through it.
2. Surface traction by the harder asperities on the softer surface induces plastic shear deformation.
3. As the subsurface deformation continues, cracks are nucleated below the surface. Crack nucleation very near to the surface is inhibited by the triaxial compressive stress existing just below the contact region.
4. Further loading causes the cracks to propagate parallel to the surface.
5. When these cracks finally intercept the surface, long-thin wear sheets "delaminate" giving rise to plate-like particles.

Figure 4 shows the initiation of subsurface cracks which then spread to release laminae. The preponderance of plate-like particles under conditions of lubricated smooth sliding provides powerful evidence for some mechanism of the type proposed in the delamination theory.

Pitting

As indicated earlier, counterformal contact between solids leads to shearing stresses which attain their maximum value a short distance within a surface. When there is relative motion, either rolling, sliding, or a combination of both, a band of material will be stressed repeatedly and cracks will form at points of stress concentration such as nonmetallic inclusions. These subsurface cracks will eventually reach the surface and release a flake of metal to create a pit.

A number of investigators have observed spherical particles about 1 μm in diameter which appear to be formed within the growing crack (see Figure 11). These particles are found during the early stages of crack formation and provide an early warning of impending surface pitting.

Materials are often evaluated for resistance to pitting using disc machines, but the life of actual gears may be considerably less than would be expected from these results.[35] The difference may be due to dynamic loading or the transient nature of the film forming process. Berthe[6] distinguishes between pitting failure from the action of Hertzian stress and micropitting which may be related to the stresses arising from interaction of surface asperities.

Pitting failure may be minimized by using hard-clean steel. When parts are case-hardened, the hardened case must be sufficiently thick to embrace the zone of maximum Hertzian shearing stress.

Abrasion

Definition

Abrasive wear may be defined as damage to a surface by a harder material. This hard material may be introduced between two interacting surfaces from outside, it may be formed *in situ* by oxidation, or it may be the material forming the second surface.

The action of granular abrasive particles has been simulated by Sakamoto and Tsukizoe[46] who used cones of mild steel sliding on copper under a normal load of 9.8 N. Figure 5 shows front-ridges of displaced material formed by a steel cone having an apex angle of 160°. Although the hardness of the steel rider was about twice that of the copper, the depth of the groove diminishes with distance of sliding. This indicates that the harder of two bodies also suffers plastic deformation during the confrontation.

Test Methods for Abrasion Resistance

Results of many tests of the resistance of materials to abrasive wear using abrasive paper

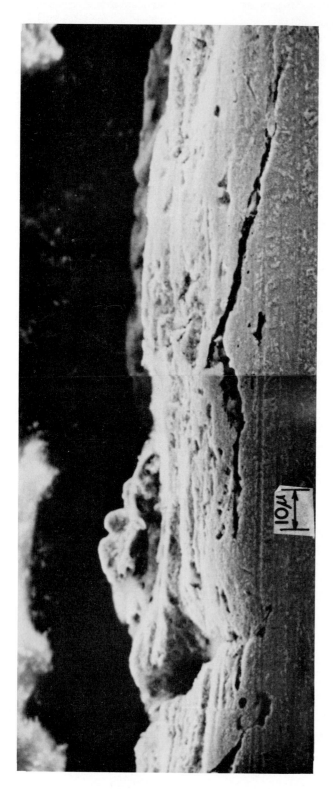

FIGURE 4.   The delamination theory of wear development of subsurface cracks.[52]

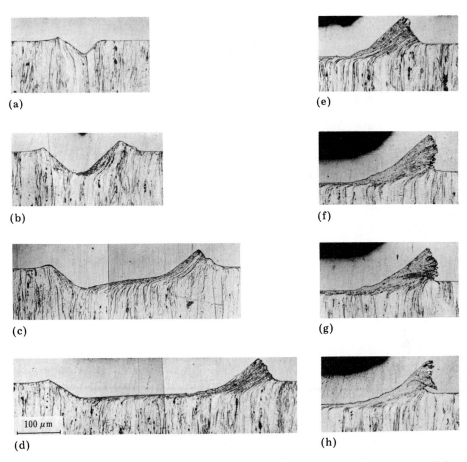

FIGURE 5.   Friction grooves and front ridges formed by 100° cone of steel sliding on copper. Sliding distance in mm (a) 0; (b) 0.15; (c) 0.3; (d) 0.45; (e) 0.6; (f) 3; (g) 6; and (h) 7.2. (From Sakamoto, T. and Tsukizoe, T., *Wear*, 48, 93, 1978. With permission.)

or cloth are confusing, chiefly because the abrading material changes during the test. Kruschov[27] arranged for the test piece undergoing abrasion to traverse across a rotating disc of abrasive material in a spiral path. Direct proportionality was observed between the resistance to wear of a series of pure metals in the annealed condition when abraded with electro-carburundum cloth having an average grain size of 80 μm and the material hardness as measured by a 136° diamond pyramid. With steels hardened by heat treatment, however, the increase in wear resistance was not commensurate with that of a material having the same hardness in the annealed condition.

Richardson[41] reported a large measure of correlation between field trials of the wear of metals when cutting soil (as in ploughing) and laboratory tests carried out with Kruschov's method. Some results are presented in Table 2. It will be noted that hardness as measured before test is not a very good guide to the wear resistance of materials. Thus a 0.74% carbon steel was shown by all tests to be superior to 0.37% carbon steel, although the hardness was identical. Austenitic manganese steel was more wear resistant than carbon steel having twice its initial hardness although, of course, the enhanced wear resistance is accompanied by extensive work-hardening. The difference between results in different soils may depend on the proportion of wear caused by stones which exert a "gouging" action.

The rate of abrasive wear appears to be effected by the relative hardnesses of the abrasive and the material under attack. The hardness of some abrasives is given in Table 3.

Table 2

RELATIVE WEAR RESISTANCES RECORDED IN THE FIELD AND IN LABORATORY TESTS

Material		Nearest Equivalent AISI steel	H (kg/mm²)	Relative wear resistance	Field results			
				Flint paper (40 grit)	Light soil iron stone	Light soil flint	Quartz sand soil (Kenya)	Flint paper (180 grit)
Carbon and low-alloy steels (B.S. 970:1955)								
En42	0.74% C steel	1074	500	—	1.20	1.14	—	1.14
	0.74% C steel	—	650	1.23	1.42	1.37	—	1.58
	0.74% C steel	—	820	1.80	1.76	1.95	2.32	2.06
En8	0.43% C steel	1040	500	—	1.05	1.02	—	1.26
	0.43% C steel	—	600	1.17	1.34	1.34	1.45	—
En24	0.37% C, Ni-Cr-Mo steel	4341	350	—	—	0.72	0.86	0.94
	0.37% C, Ni-Cr-Mo steel	Reference	500	1.00	1.00	1.00	1.00	1.00
Alloy steels								
A.Mn	Austenitic manganese steel	—	220	1.27	1.09	1.08	1.39	1.60
KE275	0.4% C, 10% W, 3% Cr hot-die steel	H21	600	1.37	1.66	1.67	2.48	2.89
CCr	2% C, 14% Cr die steel	D3	700	1.78	1.94	2.07	—	11.7
	2% C, 14% Cr die steel	—	860/900	3.50	2.93	3.34	—	32.6
Cast hardfacing alloys								
Delcrome	3% C, 30% Cr, Fe base	—	~610	2.25	2.28	3.32	9.60	129
Stellite 1	2.5% base C, 33% Cr, 13% W, Co	—	~630	2.29	2.49	4.26	10.3	26.9
White cast irons								
Ni-Hard	3% C, 1.7% Cr, 3% Ni	—	~700	1.50	1.71	2.50	—	5.95
W.I.	3.6% C	—	~700	1.59	2.32	3.81	—	4.32

From Richardson, R. C. D., Proc. Inst. Mech. Eng. (London), 182(3A), 29, 1967. With permission.

Table 3
HARDNESS OF ABRASIVES[41]

Material	Hardness (MNm^{-2})
Glass	5 790
Quartz	10 390
Garnet	13 370
Corundum	21 180
Silicon carbide	29 420

The laboratory tests using 180 grit paper gave wear resistances for some materials which were much higher than those recorded in soils. When larger grit was used (40 or 36) the high resistance was not repeated and there was good correlation with field results. A distinction is, therefore, drawn between hard abrasive wear which is little effected by particle size and soft abrasive wear. Transition from hard to soft abrasive wear appears to occur when the ratio of the hardness of the metal in the fully work hardened condition to the hardness of the abrasive drops below 0.8. During the soft abrasive wear of heterogenous marterials (these having some phase harder than the abrasive and some softer) particle size is particularly significant. Large carbide particles may obstruct wear whereas features which are small compared with a chip of wear debris are ineffective. Suh et al.[53] conclude that the increasing particle size causes a transition from the cutting type of wear to the sliding mode (see Section on Gouging).

Impact Wear

Percussive Impact

A number of tools, notably rock drills, are used in the percussive mode: they strike the work at right angles to its surface. This gives rise to both Hertzian and oscillatory stresses governed by the speed of sound within the material. In an investigation using a ballistic impact test machine, Engel[11] showed that there was an induction period involving hardly any change followed by roughening and general deterioration of the surface. A typical induction period was 10 cycles for air-hardening tool steel.

Percussive wear on elements stressed beyond their elastic limit has been studied by Wellinger and Brechel.[59] They reported good experimental agreement between the logarithmic slopes of impact wear and impact velocity.

Abrasion by Impacting Particles

The limited applicability of erosion abrasion models based on mechanical properties has resulted in recent investigations into the thermal nature of the impact zone. It was recognized that only 5% of the expended energy was used in mechanical work, the remaining energy dissipated in heating and melting the target material.[49] Ascarelli[2] proposed a thermodynamic parameter "thermal pressure", the product of coefficient of linear expansion of the metal, bulk modulus and the difference between the target material temperature and its melting point. The erosion resistance of metals ranging from tin to tungsten was shown to be proportional to this function. Hutchings[16] suggested that lip formation on the impact crater edge was the main source of erosion damage and concluded that the erosion resistance of a metal was proportional to the product of specific heat, density, and the difference between the metal temperature and melting point.

The parameters above successfully predict the erosion behavior of most ductile metals, notable exceptions are, however, alloy steels. Jones and Lewis[21] have shown for a range of alloy steels considered for use in gun barrels that erosion resistance was inversely proportional to the linear expansion coefficient, Figure 6.

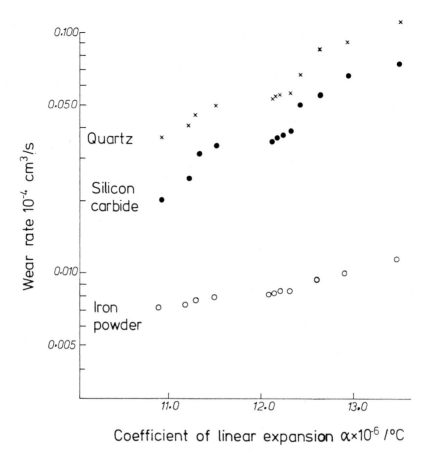

FIGURE 6.   Effect of coefficient of thermal expansion on wear rate for alloy steels using three abrasive grits (quartz, silicon carbide, and iron powder).

Gouging Type Wear

The wear which occurs in some soils is frequently determined by the number and size of stones and it is difficult to distinguish grooves formed by cutting and those formed by plastic flow of the surface. Where loads are high, such as where gravel or flints embedded in the ground strike a digger blade or plough share, plastic flow predominates and the wear mode is referred to as gouging wear as illustrated in Figure 7. Gouging is presumably the smooth-sliding mode of wear referred to by Suh et al.[53]

Firecracking

Although abrasive wear is a factor in the life of steel rolls in rolling mills, breakdown of the surface due to thermal shock and fatigue is also important.[10] The immediate surface layer of a hot-mill roll can be subjected to severe thermal stress gradients which being repetitive, rapidly lead to fatigue. Figure 8 shows a surface which has been effected by firecracking. This particular roll has been heat treated to provide a tough core and then the surface region has been reaustenitized, quenched, and tempered to provide a wear resistant surface. Table 4 relates metallurgical features to wear resistance of roll steels.

Fretting

When components are subject to very limited relative vibratory movement, an interactive form of wear takes place which is initiated by adhesion, amplified by corrosion, and has its

FIGURE 7.   Gouging type wear of grey cast iron. (Material extracted from Swansea Tribology Centre, Rep. No. 76/331; *Guide to the Selection of Materials to Resist Wear*, by permission of Swansea Tribology Centre.)

FIGURE 8.   Fine firecrack pattern typical of a differentially hardened steel. (From Dickinson, W. A. and Porthouse, D., in *Tribology 1978 Materials Performance and Conservation,* Proc. Inst. Mech. Eng. Convention, 1978, 71. With permission.)

main effect by abrasion. It frequently occurs between components which are not intended to move, press fits, for example. While an increase in hardness sometimes reduces fretting, hardening of interacting components does not prevent it.

At temperatures above 200°C the fretting of mild steel diminished with temperatures until a second transition was reached between 500 and 600°C above which the wear rate increased.[14,15] This variation is attributed to the different nature of oxide formed at different temperatures. Above 380°C the proportion of Fe_3O_4 to Fe_2O_3 increased with a corresponding reduction in wear rate. FeO appeared to be the most harmful oxide.

Environment has an important effect on fretting. Wright[62] found that dry conditions produced very rapid wear which was reduced as the relative humidity was increased to 30%. Further increase in humidity up to 100% resulted in increased wear.

There appears to be no complete cure for fretting apart from eliminating relative motion. Phosphating the surfaces may be a palliative and ion plating can delay the incidence of fretting.[34] Ion-plated chromium, cadmium, and zirconium were effective in preventing fretting. Best results were obtained from ion-plated boron carbide film which (at small amplitudes) prevented fretting up to 5×10^5 cycles of movement.

Corrosive Wear

Some chemical attack is likely on any exposed surface, and in normal atmospheric conditions this likely takes the form of oxidation. Quinn and Sullivan[39] described a system where oxidation occurs on virgin metal formed by the dislodgement of a wear fragment. This oxidation will proceed at an increasing rate until a critical oxide thickness is reached.

Table 4
EFFECT OF METALLURGICAL
FEATURES ON WEAR OF ROLL STEELS

Phase	Roll properties
Ferrite	Resistance to breakage, good grip, good firecrack resistance, poor wear and low hardness; high toughness
Lamellar Pearlite	Good strength and firecrack resistance; reasonable wear; finer pearlite improves strength and wear but at the expense of ductility and firecrack resistance; good grip
Spheroidized Pearlite	Combines high toughness with reasonable wear resistance; good firecrack resistance; relatively soft; good grip
Bainite	High strength and hardness together with good wear resistance; upper bainite structures have good firecrack resistance
Martensite	Very high hardness, good surface finish; very good wear resistance; poor firecrack resistance; low toughness
Carbide	Extremely good wear resistance; high-carbide contents cause embrittlement; this effect can be alleviated by suitable heat treatment with carbon contents of up to 1.4%
Graphite	Improves firecrack resistance and spall resistance; lowers strength due to internal notch effect; this effect can be largely avoided by producing nodular graphite; improves grip

From Dickinson, W. A. and Porterhouse, D., in *Tribology 1978 Materials Performance and Conservation,* Proc. Inst. Mech. Eng. Convention, University College of Swansea, 1978, 71.

The film then cracks up due to such factors as differences in thermal expansion of the metal and its oxide.

Oxidative wear may occur by spalling of oxide flakes from a substrate which itself shows little evidence of deformation. Shivaneth et al.[48] investigated the transition from oxidative wear to severe mechanical wear of binary aluminium-silicon alloys.

Cavitation Erosion

When material is subjected to a hydrodynamic situation wherein bubbles are formed and then collapse due to violent changes in pressure, the surfaces become damaged by pitting sometimes followed by gross removal of material.[55]

Lord Rayleigh[40] related the instantaneous pressure developed in a liquid due to collapse of a cavity (bubble) with the compressibility and density of the liquid and the speed of collapse. Wilson and Graham[60] reported that weight loss of silver surfaces correlated well with this concept and that erosion damage may be related to the product of the density and the speed of sound in a liquid. Cavitation erosion has been shown to be characterized by a delay period in which little or no damage occurs followed by a period of wear at a constant rate.[30] Duration of the delay period is determined by the initial surface state and is closely related to the endurance limit in mechanical fatigue tests. Once cavitation has commenced,

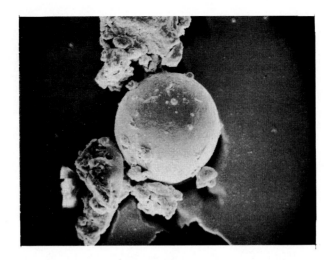

FIGURE 9.   Particles arising from cavitation erosion test. (From Thiruvengadam, A., *Trans. ASLE,* 21, 344, 1978. With permission.)

the rate of removal of material is broadly related to its strength as measured by diamond hardness or ultimate tensile strength. The effect may not be entirely mechanical because the nature of the liquid, i.e., whether or not an electrolyte, markedly affects test results.

Thiruvengadam[55] has observed the formation of spherical particles of the type illustrated in Figure 9. Size of the spheroids varied from 0.5 to 30 μm. Origin of the particles is attributed to collapse of cavitation bubbles. This collapse produces indentations at very high rates of strain causing metal to melt and to splash into the surrounding fluid.

Electrical Wear

Electrical switchgear embodies contacting members which function in accordance with the following sequence: (1) to close the circuit, (2) to allow current to flow when required, and (3) to open the circuit and suppress the current. Repeated operation results in surface deterioration from electrical effects as well as ordinary mechanical wear.

When two charged conductors approach each other, intense electro-static forces are set up at microscopic protuberances so that conduction can commence even before physical contact is made. Once the circuit has been completed, contaminating films or rough surfaces will limit areas of true contact so as to concentrate the current and melt the metal locally. As the contact begins to open, the current becomes concentrated at fewer and fewer points of contact until finally restricted to a single microscopic area. A molten globule of metal is formed and the temperature can reach the boiling point of the metal when it evaporates or even explode. Detailed analysis of the rupture of a micro-bridge by Llewellyn Jones[29] indicates the following progression:

1.   A small gap (10^{-6} m) is set up between the two electrodes
2.   Each contact spot reaches a high temperature becoming an intense thermionic emitter
3.   The gaseous atmosphere becomes mixed with metal vapor

Self-inductance of the local circuit can set up a pulse of voltage sufficient to produce ionization of the gas or metal vapor. This will generate a micro-arc which may be the primary cause of electrical wear.

RESIDUAL STRESS

Stress can become ''locked in'' to a solid as the result of combinations of plastic and

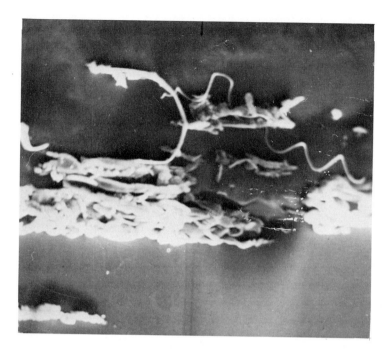

FIGURE 10.   Cutting wear particles formed during running in period of operation.

elastic strain arising from mechanical working, thermal stresses, or volume changes arising from phase or chemical transformations. While such stresses are generally undesirable, superficial compressive stresses may be beneficial in countering the formation of fatigue cracks.

A method of evaluating residual stress in Hertzian contacts has been developed.[37,38] This consisted of cutting two mutually orthogonal strip specimens from the surface of each body to be examined. Resistance strain gauges were attached to each strip and the assembly placed in an automatically controlled electropolishing bath. The degree of stress relief associated with the removal of a given quantity of material from the surface was automatically recorded.

A series of tests on hardened steel under Hertzian conditions revealed residual strains in the region of maximum shearing stress and the superposition of traction forces produced tensile strain at the surface.

PARTICLE FORMATION

Loose particles generated by wear provide a useful history of the process. One method of dealing with wear particles which are suspended in a fluid is known as "Ferrography".[47] Particles are precipitated for examination by pumping the fluid containing the sample at a slow-steady rate ($\frac{1}{4}$ cc/min) between the poles of a magnet. The fluid runs down an inclined microscope slide which has been treated so that the oil is confined to a central strip and so that the particles will adhere to the slide surface on removal of the fluid. The viscous, gravitational and magnetic forces acting on ferrous particles sort them by size, the larger particles being deposited first and the smaller particles and oxides of iron, magnetite and hematite, being deposited lower down.

In addition to the number and size of particles, much useful information is available from microscopic observations of their nature and shape. Smooth sliding is characterized by the formation of plate-like particles as predicted by the delamination theory of wear. The coiled particles of Figure 10 provide evidence of a cutting form of wear. They are often present

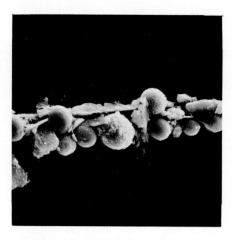

FIGURE 11.   Spherical particles formed during
fatigue in rolling element bearing.

Table 5
WEAR DEBRIS FROM DISC MACHINE TESTS

Load (N)	Speed (m/sec)			
	0.6	**1.27**	**1.9**	**3.2**
445	Mainly rubbing wear particles	Rubbing wear platelets with a few spheres and cutting wear particles	Rubbing wear platelets	Cutting wear particles and oxide
890	Rubbing wear platelets with few spherical particles	Larger size platelets + spherical particles	Rubbing wear platelets + cutting wear and some oxide	Rubbing wear particles but more oxide
1780	Rubbing wear and cutting wear particles	Large platelets also oxide	Rubbing wear + oxide and few spherical wear particles	Mainly rubbing wear particles and oxides

Note: Hardened steel discs, 300 mm diam, 9.4-mm wide.

From Odi-Owei, S., Barwell, F. T., and Roylance, B. J., *Trans. ASLE,* 20, 177, 1977. With permission.

when a plant is being commissioned and can sometimes originate during manufacture of a machine. If, however, running-in proceeds satisfactorily, the particles will cease to be formed. Their reappearance is indicative of the inception of a severe wear condition.

Spherical particles of the type shown in Figure 11 are related to pitting failure of rolling elements. It is possible to relate the characteristics of wear particles to the condition of the machine from which the sample was taken. Table 5 indicates the effect on wear particles of speed and load in a laboratory test.[33]

REFERENCES

1. **Andarelli, G., Maugis, D., and Courtel, R.,** Observations of dislocations created by friction on aluminium thin films, *Wear,* 23, 21, 1973.
2. **Ascarelli, P.,** Relation Between the Erosion by Solid Particles and the Physical Properties of Materials, Rep. 71-47, U.S. Army Materials and Mechanics Research Center, 1971.
3. **Barwell, F. T.,** Some further thoughts on the nature of boundary lubrication, *Rev. Roum. Sci. Tech. Ser. Mec. Appl.,* 11(3), 683, 1968.
4. **Barwell, F. T.,** *Bearing Systems, Principles and Practice,* Clarendon Press, Oxford, 1979.
5. **Beeching, R. and Nicholls, W.,** A theoretical discussion of pitting failure in gears, *Proc. Inst. Mech. Eng. (London),* 158A, 317, 1968.
6. **Berthe, D.,** Dissertation thesis, No. 216, 'L' University Claud Bernard, Lyon, France, 1974.
7. **Blok, H.,** The Postulate About the Constancy of Scoring Temperature. Interdisciplinary Approach to the Lubrication of Concentrated Contacts, NASA SP 237, National Aeronautics and Space Administration, Washington, D.C., 1970, 153.
8. **Buckley, D. H.,** Metal-to-metal interface and its effect on adhesion and friction, *J. Colloid Interface Sci.,* 58, 36, 1977.
9. **Clayton, P.,** Lateral wear of rails on curves, in *Tribology 1978 Materials Performance and Conservation,* Institute of Mechanical Engineers Conference Publication, Swansea, 1978, 83.
10. **Dickinson, W. A. and Porthouse, D.,** Influence of recent cast roll developments on roll wear, in *Tribology 1978 Materials Performance and Conservation,* Proc. Inst. Mech. Eng. Convention, University College of Swansea, 1978, 71.
11. **Engel, P. A.,** *Impact wear of Materials,* Elsevier, Amsterdam, 1978.
12. Stresses Deflections and Contact Dimensions for Normally Loaded Unlubricated Elastic Components, Engineering Sciences Data Item No. 78035, Contact phenomena 1, London, 1979; **Hertz, H.** (1896), as cited in **Timoshenko, S.** and **Goodier, J. N.,** *Theory of Elasticity,* McGraw-Hill, New York, 1934, 366.
13. **Hisakado, T. and Tsukizoe, T.,** Deformation mechanism at solid interfaces subjected to tangential loads, *Mec. Mater. Electr.,* No. 337, 38, January 1978.
14. **Hurricks, P. L.,** The fretting of mild steel from room temperature to 200°C, *Wear,* 19, 207, 1972.
15. **Hurricks, P. L.,** The fretting of mild steel from 200°C to 500°C, *Wear,* 30, 189, 1974.
16. **Hutchings, I. M.,** Prediction of the resistance of metals to erosion by solid particles, *Wear,* 35, 371, 1975.
17. **Israelachvili, J. N. and Tabor, D.,** The measurement of van der Waals dispersion forces in the range 1.5 to 150 mm, *Proc. R. Soc. London Ser. A,* 331, 19, 1972.
18. **Jamieson, W. E.,** The wear of railroad freight car wheels and rails, *Lubr. Eng.,* 36, 401, 1980.
19. **Johnson, K. L., Kendall, K., and Roberts, A. D.,** Surface energy and the contact of elastic solids, *Proc. R. Soc. London Ser. A,* 324, 301, 1971.
20. **Johnson, K. L. and Gray, C. G.,** Development of corrugations on surfaces in rolling contact, *Proc. Inst. Mech. Eng. (London),* 189, 567, 1975.
21. **Jones, M. H. and Lewis, R.,** Solid Particle Erosion of a Selection of Alloy Steels, Cambridge Conference ELSI.v., September 1979.
22. **Kerridge, M.,** Metal transfer and wear process, *Proc. Phys. Soc. London,* 68B, 400, 1955.
23. **Kimura, Y.,** An Interpretation of Wear as a Fatigue Process, *Proc. J.S.L.E.—ASME Int. Lubrication Conf.,* Tokyo, 1975, Elsevier, Amsterdam, 1976.
24. **Koba, H. and Cook, N. H.,** *Wear Particle Formation Mechanics,* Massachusetts Institute of Technology, Cambridge, 1974.
25. **Kohno, A. and Hyodo, S.,** The effect of surface energy on the micro-adhesion between hard solids, *J. Phys. D,* 7, 1243, 1974.
26. **Kragelskii, I. V.,** *Friction and Wear,* Butterworths, London, 1965.
27. **Kruschov, M. H.,** Resistance of metal to wear by abrasion as related to hardness, *in Proc. Conf. Lubrication and Wear,* Institute of Mechanical Engineers, London, 1967, 655.
28. **Ludema, K. C.,** Editorial, *Wear,* 60, 5, 1980.
29. **Llewellyn Jones, F.,** *4th Int. Conf. Gas Discharges,* Swansea, 1976, 429.
30. **Mathuson, R. and Hobbs, J. M.,** Cavitation erosion comparative tests, *Engineering,* 189, 136, 1960.
31. **Maugis, D., Desalos-Andaralli, G., Heutel, A., and Courtel, R.,** Adhesion and friction on Al thin foils related to observed dislocation density, *Trans. ASLE,* 21, 1, 1976.
32. **Nayak, P. R.,** Contact vibrations of rolling discs, *J. Sound Vib.,* 22, 297, 1972.
33. **Odi-Owei, S., Barwell, F. T., and Roylance, B. J.,** Some implications of surface texture in partial elasto hydrodynamic lubrication, *Trans. ASLE,* 20, 177, 1977.
34. **Ohmae, N., Tsukizoe, T., and Nakai, T.,** Ion-plated thin films for anti-wear applications, *Trans. ASME J. Lubr. Technol.,* 100, 129, 1978.
35. **Onions, R. A. and Archard, J. F.,** Pitting of gears and discs, *Proc. Inst. Mech. Eng.,* 188, 673, 1974.

36. **Pollock, H. M., Shufflebottom, P., and Skinner, J.,** Contact-adhesion between surfaces in vacuum, deformation and surfaces energy, *J. Phys. D.,* 10, 127, 1977.
37. **Pomeroy, R. S. and Johnson, K. L.,** Residual stresses in rolling contact, *J. Strain Anal.,* 4, 208, 1969.
38. **Pomeroy, R. S.,** Measurement of residual stresses in contact, *Wear,* 16, 393, 1971.
39. **Quinn, T. F. J. and Sullivan, J. L.,** A review of oxidational wear, in *Trans. ASME Wear of Materials,* American Society of Mechanical Engineers, New York, 1977, 110.
40. **Rayleigh (Lord),** On the pressure developed in a liquid during the collapse of a spherical cavity, *Philos. Mag.,* 34, 94, 1917.
41. **Richardson, R. C. D.,** Laboratory simulation of abrasive wear such as that imposed by soil. The abrasive wear of metals and alloys, *Proc. Inst. Mech. Eng. (London),* 182(3A), 29 and 410, 1967.
42. **Roberts, W. H.,** Measurement of sliding friction and wear in high temperatures and high pressure water environments, *Proc. Inst. Mech. Eng. (London) Part 3K,* 186, 37, 1966.
43. **Sasada, T. and Kando, H.,** Formation of wear particles by the mutual transfer and growth process, Proc. 17th Japan Congr. on Materials Research, 1973, 33.
44. **Sasada, T., Norose, S., and Mishina, H.,** Effect of mutual solubility on metallic wear, Proc. 20th Japan Congr. on Materials Research, 1977, 99.
45. **Sasada, T., Norose, S., and Mishina, H.,** The behaviour of adherend fragments interposed between sliding surfaces and the formation process of wear particles, in *Proc. Int. Conf. Wear of Materials,* Dearborn, Mich., 1979, 72.
46. **Sakamoto, T. and Tsukizoe, T.,** Deformation and wear behaviour of the junction in quasi-scratch friction, *Wear,* 48, 93, 1978.
47. **Seifert, W. W. and Westcott, V. C.,** A method for the study of wear particles in lubricating oil, *Wear,* 21, 27, 1972.
48. **Shivaneth, R., Sengupta, P. K., and Eyre, T. S.,** Wear of aluminium — silicon alloys, in *Trans. ASME Wear of Materials,* American Society of Mechanical Engineers, New York, 1977, 120.
49. **Smeltzer, C. E., Gulden, M. E., McElmury, S. S., and Cromption, W. A.,** Mechanism of Sand and Dust Erosion in Gas Turbine Engines, U.S.A. ALABS Tech. Rep. 70-36, 1970.
50. **Soda, N. and Sasada, T.,** Studies in adhesive wear, effect of gas-absorbed films on metallic wear (abstract); see also Mechanism of lubrication by surrounding gas molecules in adhesive wear, *Trans. ASME, J. Lubr. Technol.,* 100, 492, 1978.
51. **Soda, N., Kimura, Y., and Tanada, A.,** Wear of some f.c.c. metals during unlubricated sliding. I. Effects of load, velocity and atmospheric pressure on wear, *Wear,* 33, 1, 1975a; II. Effects of normal load, sliding velocity and atmospheric pressure on wear fragments, *Wear,* 35, 331, 1975b; III. A mechanical aspect of wear, *Wear,* 40, 23, 1976; IV. Effects of atmospheric pressure on wear, *Wear,* 43, 165, 1977.
52. **Suh, N. P.,** The delamination theory of wear, *Wear,* 25, 111, 1973.
53. **Suh, N. P., Saka, N., and Sin, H. C.,** Effect of Abrasive Grit Size on Abrasive Wear, Prog. Rep. Advanced Research Projects Agency, U.S. Department of Defense, Washington, D.C., June 1978.
54. **Tabor, D.,** Junction growth in metallic friction: the role of combined stresses and surface contamination, *Proc. R. Soc. London Ser. A,* 229, 198, 1959.
55. **Thiruvengadam, A.,** Cavitation erosion, *Appl. Mech. Rev.,* 215, 1971; Mechanism of spheroids produced by cavitation erosion, *Trans. ASLE,* 21, 344, 1978.
56. **Timoshenko, S. and Goodier, J. N.,** *Theory of Elasticity,* McGraw-Hill, New York, 1934, 366.
57. **Tsuya, Y., Yamada, Y., and Takagi, R.,** Damage and internal deformation near the surface caused by friction, *J. Mater. Sci. Soc. Jpn.,* 1, 35, 1964.
58. **Tsuya, Y.,** Microstructure of wear, friction and solid lubrication, Tech. Rep. Mechanical Engineering Laboratory, No. 81, Tokyo, Japan, 1976.
59. **Wellinger, K. and Brechel, H.,** Kenngrössen und vershleiss beim stoss metallischer werkstoffe, *Wear,* 13, 257, 1969.
60. **Wilson, R. W. and Graham, R.,** Cavitation of metal surfaces in contact with lubricants, in *Proc. Conf. Lubrication and Wear,* Institute of Mechanical Engineers, London, 1957, 707.
61. **Whittlemore, H. L. and Petrenko, S. N.,** Friction and Carrying Capacity of Ball and Roller Bearings, Tech. Paper No. 191, National Bureau of Standards, Washington, D.C., 1921.
62. **Wright, K. A. R.,** An investigation of fretting corrosion, *Proc. Inst. Mech. Eng.,* lB, 556, 1952.

WEAR OF NONMETALLIC MATERIALS

Norman S. Eiss, Jr.

INTRODUCTION

Substitution of a nonmetal for a metal in one of the components of a sliding system will usually result in a change in the dominant wear mechanism. For example, when two metals are in sliding contact, the disruption of surface films permits metallic contact to occur and adhesive wear is the predominant wear mode. When a nonmetal replaces one of the metals, the metallic bond no longer dominates at the interface. The interfacial bonds are made weaker and the dominant wear mechanism becomes abrasion. The change in the dominant wear mechanism is caused by the difference in properties between metals and nonmetals.

Properties of metals result from the metallic bond. The metallic bond is responsible for good thermal and electrical conductivity. High strengths of metals, their ductility, and their capability for alloying and being welded all result from the metallic bond. Nonmetals are bonded by ionic, covalent, molecular, and hydrogen bonds. Polymers are characterized by large molecular weights. The bonding within the molecule is covalent while the bonding between molecules is by molecular van der Waals bonds. Because the molecular bond is weak, much effort has been devoted to strengthening the intermolecular bonds of polymers to improve mechanical properties. Strengthening has been accomplished by crystallization, cross-linking, and stiffening the polymer chain.

Ceramics consist of a combination of metals with a nonmetallic element, usually oxygen. The ionic and covalent bonds involved are the primary cause for the stability and strength of ceramic materials. Ceramics are more brittle than metals and more resistant to chemical attack because they are highly oxidized. While the properties of metals and nonmetals can be linked to the nature of the atomic bond, the wear of materials is a function of their properties, the conditions of sliding and the environment.

The emphasis in this discussion will be on wear mechanisms rather than on reporting wear test results for specific systems. However, some data and precautions in their use are presented at the end of this chapter. This discussion will be restricted to the wear of polymers and elastomers, and one specific example of filled polymers, i.e., mineral-filled epoxies. The discussion of the wear of filled polymers (and polymers used as fillers), carbon, graphite, ceramics, and cermets appears in the chapter on Solid Lubricants (Volume II). The wear of ceramics is also discussed in the chapter on Wear-Resistant Coatings and Surface Treatments (Volume II). A more comprehensive review of the wear of nonmetallic materials is given in several excellent reviews.[1-3] The reader is also referred to the published papers of several recent conferences on wear.[1,2,4,5]

WEAR OF UNFILLED POLYMERS

Wear of polymers is a complex phenomenon which is often discussed in simplified terms. For example, Briscoe and Tabor[3] discussed polymer wear under two main headings: deformation wear and interfacial wear. Deformation wear included abrasive and fatigue mechanisms and interfacial wear included adhesive or transfer wear. When a polymer slides against a hard surface, the roughness of the surface dictates the dominant wear mechanism. Hence, deformation wear occurs when surfaces are rough and interfacial wear dominates when surfaces are smooth.

There is no agreement in the literature on the roughness at which the dominance of the two wear modes changes. Buckley[6] found that the wear of polyethylene on stainless steel is a minimum at an RMS roughness of 0.38 μm. Dowson et al[7] measured the minimum wear of ultrahigh molecular weight polyethylene (UHMWPE) on stainless steel at an arithmetic average (R_a) roughness of 0.03 μm. Eiss and Warren studied the wear of polychlorotrifluoroethylene (PCTFE)[8] and low density polyethylene (LDPE)[9] sliding on mild steel and found that wear monotonically decreased as the R_a roughness decreased to 0.06 μm. In the following discussion, the term "rough" surface will refer to one on which deformation wear predominates and "smooth" surface when interfacial wear predominates.

Deformation Wear on Rough Surfaces

Abrasive wear occurs when a hard, sharp particle cuts or displaces material from the polymer. The simplest form of abrasive wear occurs during single traversal sliding (the slider is continuously exposed to new surface). In this case, wear of the polymer is a direct response to the interaction of the hard surface topography and the polymer properties. The wear is not complicated by modification of the surface by polymer transferred on previous traversals. However, Lancaster[10] has shown that even on single traversal sliding, wear is a function of polymer transfer. Material transferred from the leading edge of a polymer slider influences the polymer transferred from the rear of the slider. For rectangular sliders, higher wear was measured when the sliding direction was parallel to the short side than when parallel to the long side.

Most single traversal sliding experiments are performed at sliding speeds less than 1 cm/sec to avoid heating the polymer and changing its mechanical properties. Investigators have correlated wear in single traversal sliding with polymer properties and with certain topographical features of the rough surface. Positive correlation of wear has been found with the inverse of the product of the stress and elongation at rupture.[11-13] Giltrow has correlated single traversal wear of thermoplastic polymers with their cohesive energies, provided that plastic deformation predominated during the wear process.[14] Lontz and Kumnick[15] found that the wear rate of polytetrafluoroethylene (PTFE) was directly proportional to the flexure modulus and inversely proportional to the yield strain.

Several surface topography features have been correlated with single-pass abrasive wear. One of the simplest models for abrasive wear[3,11] is expressed by the equation

$$V = \frac{KW \tan\theta}{H} \tag{1}$$

where V is the wear volume per unit sliding distance, K the abrasive wear coefficient, W the normal load, θ the average slope of the asperities, and H the hardness of the polymer.

Single traversal sliding of PCTFE on surfaces produced by bead blasting, grinding, and lapping results in a positive correlation between the mass of polymer transferred and the average value of the asperity slopes.[8] The transferred mass also correlated positively with the arithmetic average roughness R_a of the surfaces. Lancaster[12] also showed that single-traversal polymer wear correlated positively with R_a and the average of the asperity slopes. However, in neither of these studies was the linear relationship predicted by Equation 1 found. In general, wear was proportional to the average slope to a power greater than 1.

Hollander and Lancaster[17] found a positive correlation between the ratio of the standard deviation of asperity heights to the average radius of the asperities and the wear of polymers sliding on abraded mild steel surfaces. Warren and Eiss[13] found that polymers transfer to a rough surface by shearing of the polymer slider. The transferred material was deposited at an angle which was significantly different for each polymer. The angle correlated with the inverse of the product of stress and elongation at rupture.

These angles were used in a model[18] to predict the transfer of polyvinylchloride (PVC), PCTFE, and LDPE to rough surfaces. It was assumed that each asperity that penetrated the polymer removed a wedge of material where the wedge angle was that found in Reference

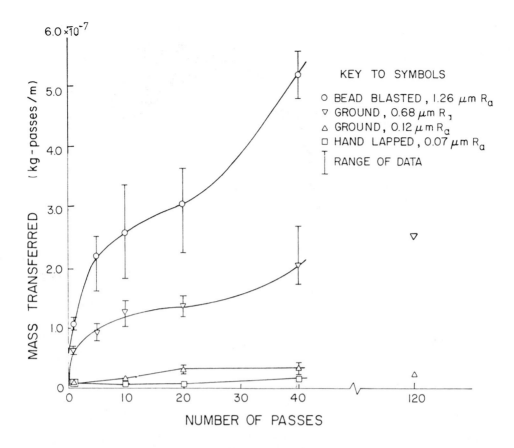

FIGURE 1.   Transfer of PCTFE to a mild steel disk as a function of surface roughness and number of passes. Normal load — 9.8 N, sliding speed — 0.2 cm/sec. (From Eiss, N. S., Jr. and Warren, J. H., *The Effect of Surface Finish on the Friction and Wear of PCTFE Plastic on Mild Steel,* Paper No. IQ75-125, Society of Manufacturing Engineers, Detroit, Mich., 1975. With permission.)

13. The model predicted the transfer in single traversal sliding to within a factor of five for the three polymers. The predictions were most accurate when the combination of loads and polymer properties produced a ratio of a calculated real area to apparent area in the range of 0.1 to 0.3. Below 0.1, too few asperities penetrated the polymer to obtain accurate data on penetration. For values above 0.3 the predicted volume of the wedge of polymer was usually greater than the void space available in the valleys of the rough surface. While this model has a limited range of validity, it does show that given sufficient information about the polymer, mechanical properties, the surface topography, and the normal load, transfer can be predicted.

In multiple-pass sliding the initial wear rate decreases as the polymer transfers until some steady-state wear rate is achieved. The number of passes to reach steady-state wear depends on the surface topography, direction of sliding relative to the lay of the surface, and the polymer properties. Figure 1 shows that on the smoother surfaces steady-state wear was achieved in a fewer number of passes and the wear rates tended to be lower.

On rougher surfaces with a pronounced lay, the highest wear was measured[19] when the direction of sliding was at an angle to the lay. The wear particles which collected in the grooves between the asperities (ridges) were moved to the edge of the wear track by the component of the friction force parallel to the lay. Hence, the grooves never became loaded with debris and the wear rate remained high. Sliding perpendicular to the lay produced the

FIGURE 2.   Rolls of LDPE debris in the valleys of a steel surface after 38 400 passes at 1.28 m/sec R_a = 1.16 µm, arrow indicates the direction of sliding. (From Eiss, N. S., Jr. and Bayraktaroglu, M. M., The effect of surface roughness on the wear of low density polyethylene, *ASLE Trans.*, 23, 269, 1980. With permission.)

next highest wear. The debris particles built up in the grooves and eventually supported some of the load, thereby reducing the abrasive wear. Some transport of debris occurred by friction forces pulling the polymer over the asperity ridges. When sliding was parallel to the lay, the wear rate was lowest. Abrasive action was minimized and other mechanisms such as a thin-film transfer dominated the wear process.

While single- and multiple-traversal experiments at sliding speeds below 1 cm/sec have indicated that the lowest wear rates occur on smoother surfaces, little experimental work at high sliding speeds on rough surfaces has been reported on the literature. Studies on wear of LDPE sliding at 1.3 m/sec on mild steel surfaces of 1.2 µm R_a roughness indicated the nature of the polymer wear debris.[9] Figure 2 shows rolls of debris lying in the valleys

Table 1
ABRASION LOSS OF
POLYETHYLENES OF
DIFFERENT MOLECULAR
WEIGHT (MELT INDEX) AND
CRYSTALLINITY (DENSITY)

Melt index	Density	Abrasion loss (g/5000 cycles)[a]
22.0	0.925	0.24
7.0	0.935	0.08
30.0	0.965	0.05
3.0	0.919	0.07
3.0	0.934	0.03
1.0	0.960	0.02
1.0	0.924	0.11
1.0	0.931	0.10
0.9	0.938	0.03
0.15	0.960	0.02
0[b]	High	0.005
0[b]	High	0.005

[a]   ASTM D1044, CS-17 wheel, 1000 g load.
[b]   UHMWPE.

between ridges on the steel surface. It is not clear whether the rolls are formed as the LDPE is removed by a ridge or formed as a result of multiple interactions between the polymer pin and the previously transferred material. It is postulated that the rolls are indicative of the ductility of the polymer.

When ductile polymers are slid on abrasive surfaces, a complex interaction occurs between surface asperities and polymer properties. Deanin and Patel[20] studied the abrasive wear of polyethylene as a function of molecular weight and degree of crystallinity. They concluded:

The mechanism by which an abrasive wheel produces wear on the surface of polyethylene can thus be visualized as a series of simpler mechanical processes which are relatively well understood. 1. The abrasive particles have high hardness and modulus and sharp edges. 2. The hard high-modulus abrasive particles indent the soft low-modulus polyethylene surface, deforming it. 3. The sharp edges . . . cut the polyethylene surface . . . . 4. The moving abrasive particles snag and catch in these cuts, and pull the surface layers of the polyethylene along with them. 5. Such microscopic surface strips of polyethylene are stretched beyond the yield point . . . 6. . . . until they actually break or tear away from the massive polyethylene sample.

In these terms, it is easy to understand why high molecular weight, which increased tear strength, also increased tear resistance. Similarly, high crystallinity increased indentation resistance, tensile modulus, tensile yield strength, ultimate tensile strength, and tear strength; and these in turn increased abrasion resistance.

Practically, the combined benefits of high molecular weight and high crystallinity are best seen in the use of ultra-high-molecular weight high-density polyethylene for extremely abrasion resistant applications.

The abrasion loss for the polyethylenes tested are shown in Table 1.

Interfacial Wear on Smooth Surfaces

When the asperities which cause abrasive transfer and the formation of wear particles are removed from the surface, thin film transfer becomes the predominant wear mode. There are two modes of thin film transfer; one mode involves films on the order of 10 nm to 50 nm thick and the other involves films 0.1 to 1.0 μm thick. The only two polymers which have been found to transfer on the first mode are PTFE and high-density polyethylene at low sliding speeds and intermediate temperatures.[21]

The mechanism postulated for this very thin transfer film is based on the smooth molecular profile of these two polymers and the ability of the molecular chains to reorient in the surface layers prior to the transfer. Hence, the transferred films consist of molecular chains oriented in the direction of sliding. At higher sliding speeds, thicker films form. For further speed increases, frictional heating causes melting of the HDPE and a two-order-of-magnitude increase in wear rate.[22] Sections through the polymer slider show clear evidence of a melted zone. Conflicting evidence exists on the melting of PTFE at high sliding speeds.

The second mode of film transfer is observed for semicrystalline polymers which do not have a smooth molecular profile when the sliding temperature is above their glass transition temperature. Such polymers as LDPE, polypropylene, and nylon 6/6 transfer in this mode. This is the more prevalent of the two modes, primarily because it occurs over a wider range of sliding speeds, up to speeds where the temperatures developed at the interface cause softening or melting of the polymer. Scanning electron microphotographs of the films (Figure 3a) confirmed that they were formed by an initial adhesion of the polymer to the smooth steel surface followed by growth in area and thickness.

Thickness of the film appears to be limited by removal of patches of the film and eventually larger regions (Figure 3b). Fatigue of the film followed by delamination[25] could explain the break up. Likewise, elastic-stored energy in the film exceeding the adhesion energy at the interface could also explain a limiting film thickness.[26] Debris formed during thin film wear consists of conglomerates of sheet-like particles, again confirming the breakup configuration (Figure 4).

Whether the transfer film is 10-nm or 0.1-μm thick, no models are available which predict the quantity of transfer, the thickness of the film before breakup, or the wear rate once loose particles have started to form.

Wear of brittle polymers (below their glass transition temperatures), does not involve thin film transfer. None of these polymers have the ability to reorient the molecular segments, a property found to be necessary for film formation. Thus, on smooth surfaces, these polymers would tend toward true interfacial sliding with little transfer and debris produced in the process.

ELASTOMERS

The wear of elastomers is a result of three possible mechanisms: abrasive wear, fatigue wear, and wear by roll formations.[27] Abrasive wear is dominant on rough surfaces with sharp asperities, fatigue wear is dominant on rough surfaces with rounded asperities, and the formation of rolls occurs on smooth surfaces. While predominant modes of wear have been identified, no models exist for predicting the rate of wear from independently measured fundamental strength properties.

Removal of rubber by abrasive wear has been attributed to a tensile failure at right angles to the direction of sliding.[28] The lips of rubber formed on the surface by these tensile failures eventually become detached. More recently, the abrasive wear of rubber has been related to the growth rate of cracks into the rubber.[29] In tests performed with a razor blade perpendicular to the rubber surface and sliding in a direction perpendicular to the plane of the blade, good agreement was found between crack growth and abrasion data for noncrystallizing rubbers. Strain crystallizing natural rubber abraded more than was expected on the basis of its crack growth behavior, indicating that crystallization was inhibited or ineffective in the razor-blade abrasion test.

Fatigue wear on blunt asperities is caused primarily by surface deformations resulting from frictional traction.[30] The blunt asperity pushes up a surface section of the rubber, compressing the rubber in front and stretching that behind. The rubber either tears and subsequently recovers or it overcomes the friction by its elastic stress and then returns to

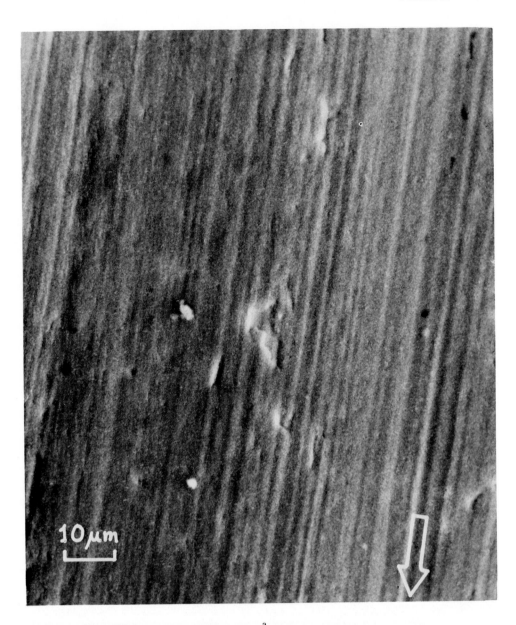

a

FIGURE 3.   Transfer of LDPE to a steel surface, R_a = 0.065 μm at 1.28 m/sec, arrow indicates sliding direction. (a) Uniform film with small broken areas after 57 600 passes; (b) break up of film after 76 800 passes. (From Eiss, N. S., Jr. and Bayraktaroglu, M. M., The effect of surface roughness on the wear of low density polyethylene, *ASLE Trans.*, 23, 269, 1980. With permission.)

its former position as the protrusion moves on. Subsequent passes eventually separate the torn piece. Thus, the wear is a direct function of the friction between the rubber and the blunt asperities. Support for this mechanism was based on positive correlation between the distance of sliding until surface failure and the number of cycles to tensile failure.[31]

On very smooth surfaces, highly elastic and soft materials can form rolled shreds of material at the sliding interface.[32] These shreds are eventually torn off to become loose debris. Moore[27] has associated wear by roll formation to the Schallamack waves of detach-

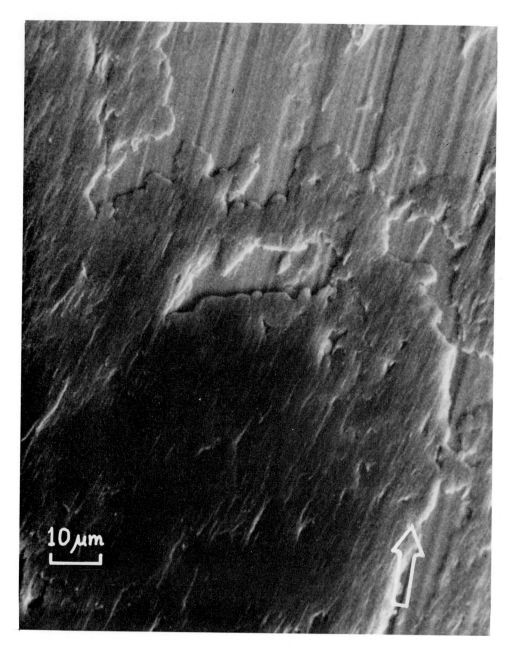

FIGURE 3b

ment which arise from surface buckling of the rubber-like material within the contact area as a result of frictional stress. The more severe forms of elastomer wear are caused by the abrasive and roll formation mechanisms while fatigue wear on blunt asperities is the cause of mild wear. Severity of wear is controlled by the critical shear stress for each elastomer. Above the critical shear stress, abrasive wear or roll formations result; below the critical shear stress, fatigue wear. Since shear stresses are directly related to the friction force, each material combination has a critical friction coefficient[33] above which high wear rates are observed involving abrasive or roll formation.

In contrast to other polymers, nonmetals, and metals, the wear of elastomers has been

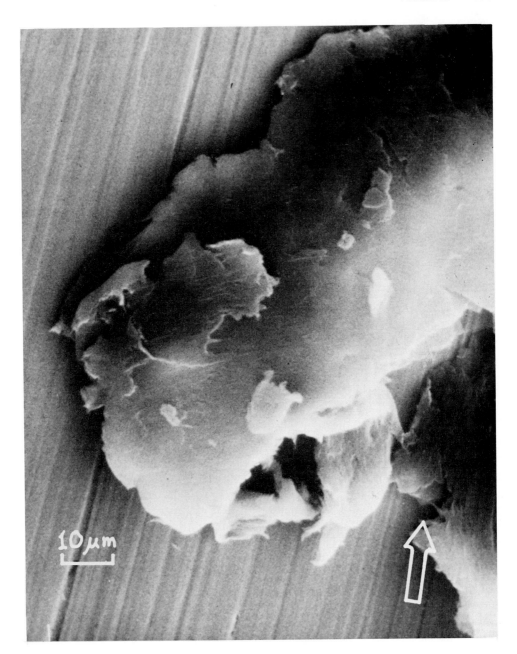

FIGURE 4.   Wear particle of LDPE with a multilayered structure after 76 800 passes on a steel surface, R_a = 0.065 μm at 1.28 m/sec, arrow indicates direction of sliding. (From Eiss, N. S., Jr. and Bayraktaroglu, M. M., The effect of surface roughness on the wear of low density polyethylene, *ASLE Trans.*, 23, 269, 1980. With permission.)

found to be related to interfacial friction. Figure 5 shows the proportionality between wear and the logarithm of the coefficient of friction. As noted above each of the wear mechanisms depends on the adhesion at the interface to transmit the forces from one body to another. Moore[27] notes that in spite of the relationship demonstrated between friction and wear, wear is much more complex. The effects of bond formation, stretching, and fracture may appear as shredding, tearing, buckling, plucking, and rolling into a whorl. In addition, local irreversible changes occur in elastomeric structure and properties.

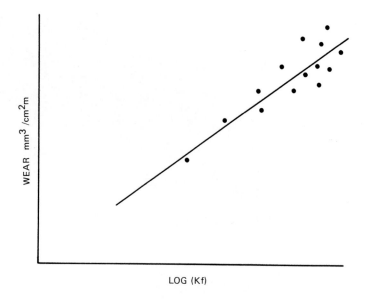

FIGURE 5.   Proportionality between volumetric wear and coefficient of friction, f. (From Moore, D. F., in *The Wear of Non-Metallic Materials,* Dowson, D., Godet, M., and Taylor, C. M., Eds., Mechanical Engineering Publ., London, 1976, 141. With permission.)

MINERAL- AND CERAMIC-FILLED RESINS

Mineral-filled resins have been used in a variety of applications to control the viscosity of the uncured mixture and the hardness, differential thermal expansion, and isothermal shrinkage of the cured material. These applications include dental restorative materials and dielectric materials in a multisegmented switch. The dental restorative materials must resist the abrasive action of food and dirt particles trapped between the teeth during mastication, resist the grinding of teeth, and resist the mild abrasives in dentifrices. The dielectric material must be capable of being polished by abrasive action so that the smooth surface will not cause excessive wear of the graphite composite brushes sliding over the surface.

A study of the polishing of mica-filled epoxy[34] identified two mechanisms of wear. When the abrasive particles were larger than the mica particles, wear occurred by crushing and fracturing the mica particles by the rolling and sliding motion. Subsequent abrasive particles removed the fractured mica and the resin surrounding them. When the abrasive particles were smaller than the mica particles, wear occurred by erosion of the resin surrounding the mica. As the support for the mica gradually wore away, the mica particles were removed by the polishing motions.

The erosive wear mechanism has also been proposed for the wear of dental restorative materials.[35] These materials were quartz or glass filled BIS/GMA resins. Wear data suggest the filler volume fraction and the particle size are the most significant parameters affecting wear resistance. Interaction of these two parameters as they affect packing density is also important. Scanning electron micrographs of the worn surfaces confirmed the erosion mechanism in a simulated tooth-brushing wear test. The study showed that an impact sliding wear test gave better correlation with in vivo wear of the composites in rabbits than the simulated tooth brushing.

Wear rates of poly (methacrylate) and BIS/GMA, unfilled and filled with coupled quartz, when sliding on 180-grit SiC abrasive cloth were all similar in magnitude.[36] These results suggest that the resin fracture properties govern the wear rates. The negligible correlation of the wear results with hardness of the composites tends to confirm resin wear and erosion around the filler as the wear mechanism.

Clinical observations also support the erosion mechanism.[37] In a study of the wear characteristics of five experimental resins[38] sliding on SiC, aluminum, and quartz papers, the highest wear was caused by the SiC and the lowest by quartz. Tests in which a diamond stylus was slid on the resins showed three failure modes. Ductile failure (as evidenced by a smooth wear track) was found at low loads; brittle failure (surface cracks, chevron-shaped) occurred at intermediate loads. Catastrophic failure (gross disruption of the surface) was observed at the highest loads.

In summary, wear of mineral-filled epoxies occurs by erosion of the resin around the filler for abrasive particle sizes smaller than the filler. For large abrasive particles, stresses are high enough to fracture the resin and filler particles, removing both at a rapid rate.

POLYMER SELECTION USING PUBLISHED WEAR DATA

Selection of polymers for wear resistance is usually based on wear data and on measured load and sliding speed at which the wear rate becomes catastrophic. The latter data are usually referred to as the PV limit for the polymer, where P is the interface pressure and V is the sliding velocity.[39] The PV limit is a measure of the energy input to the sliding interface which is sufficient to cause the polymer to soften or melt. This softening results in high wear rates which are unacceptable in most applications.

Published wear rate data cannot, however, be used to predict absolute wear rates for applications in which the conditions are different from those used to obtain the data. In some circumstances, published data cannot even be used to predict relative wear rates of the polymers in a different application. One of the major reasons for the poor predictions of wear based on published data is the strong influence that surface roughness has on wear rate.[40] Figure 6 shows that an order of magnitude increase in the R_a of a surface can result in wear rate increases that range from a factor of 3 to 1000. In addition, relative wear rates can completely reverse as the roughness changes. For example, at a roughness of 0.1 μm, the wear rate of polyacetal is about one ninth that of polyethylene. At roughness of 1.0 μm, the wear rate of polyacetal is about six times that of polyethylene.

A second reason why published data may not predict wear performance of polymers is the different wear rates that result from single traversal and multiple traversal tests. In the latter, the ability of the polymer to form transfer films on the counterface surface can reduce the wear rates experienced on single traversals over the same surface.[40] Table 2 shows that the ratio of steady-state wear to single traversal wear can vary widely. The polymers with the lower ratios tend to be more ductile than those with the higher ratios.

A third reason why wear performance may not be predicted from published data is that different test geometries produce different rankings of polymers. In Figure 6 and Table 2, Nylon 6/6 has a lower wear rate than polyacetal for both single-traversal and steady-state wear for a cylinder-on-ring geometry. However, Nylon 6/6 is reported to have a higher wear rate than that for polyacetal in steady state sliding in a thrust-washer configuration, 0.2 m/sec sliding speed, 0.20 μm R_a roughness, and an interfacial pressure of 2.8×10^5 Pa.[41]

The general conclusion can be drawn that wear rate data in the literature can be useful in predicting performance of polymers only if the conditions of the test and the application are very similar.

Published PV limits of polymers must also be used with caution in predicting performance in a given application. Often a single number is published for the PV limit. The implication is that any combination of P and V less than the limit will be satisfactory. At low velocities, however, the pressure is limited by the flow characteristics of the polymer. At low pressures, the velocity is limited by frictional heating which causes softening of the surface layers and

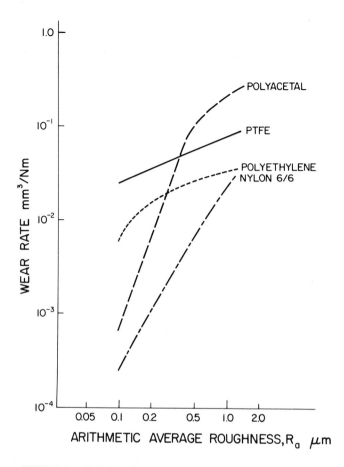

FIGURE 6.   Variation of wear rate with surface roughness of the steel counterface, cylinder-on-ring, 0.6 m/sec sliding speed, 20°C, helical wear track, silicone lubricant. (Adapted from Lancaster, J. K., *Proc Inst. Mech. Eng.*, 183(3P), 98, 1969. With permission.)

increases the wear rate. Some attempts have been made to predict PV limits by calculating interfacial temperature rise as a function of load, sliding speed, material properties, and measured coefficients of friction.[42]

The PV limit is determined in a test where the velocity is fixed and the load is increased in steps until the temperature is no longer stabilized. The most useful data includes PV limits measured at several different sliding velocities. While the PV limit is usually measured in a thrust-washer configuration, some literature data do not include the configuration of the experiment or the roughness of the steel counterface. An example of some PV data is given in Table 3.[43]

The previous discussion has centered on the availability and usefulness of wear rate data and PV limits for polymers. However, the designer may want no wear in his application. In this case, the engineering model for wear may be used. In this model the load is calculated which will result in wear on the order of the initial surface finish of the material (called zero wear) for a given contact geometry, life, and materials.[44] One important parameter in this model is the ratio of the maximum shear stress τ_{max} in the interface to the shear strength of the material τ_y'. For polymers, the zero wear condition is satisfied if this ratio is 0.54.[45] Therefore, polymers with higher shear strengths will be able to support larger loads and satisfy the zero wear condition. The ratio, τ_{max}/τ_y' for several polymers is shown in Figure 7.

Table 2
RATIO OF STEADY-STATE WEAR TO
SINGLE-TRAVERSAL WEAR

Polymer	Ratio	Elongation (%)
Polyethylene	0.06	600
Polyacetal	0.08	35
Polypropylene	0.26	320
Nylon 6/6	0.28	100
Polytetrafluorethylene	0.33	500
Nylon 11	0.44	100
Polycarbonate	0.58	10
Acrylonitride-butadiene-styrene	1.0	20
Polytrifluorochloroethylene	3.5	10
Polymethylmethacrylate	16.7	12
Polyester (17449)	20	10
Polyvinylidenechloride	27	5
Polystyrene	33	6
Epoxy (828)	250	5

Note: Test conditions: counterface-steel with 0.15 μm R_a, 9.8 N load, 0.6 m/sec sliding speed. Steady-state wear was unlubricated, single-traversal wear lubricated with silicone fluid.

From Theberge, J. E., *Mach. Design,* 42, 114, 1970. With permission.

Table 3
PV DATA FOR SEVERAL POLYMERS

Polymer	Sliding velocity (m/sec)		
	0.05	0.5	5.0
Polycarbonate	0.03	0.01	<0.01
Acetal (homopolymer)	0.14	0.12	0.09
Nylon 6/6	0.11	0.09	<0.09
Polyimide	4.0	—	—
PTFE	0.04	0.06	0.09
PTFE filled 15% weight glass	0.33	0.39	0.5

Note: PV (MPa × m/sec).

From Lancaster, J. K., *Tribology,* 6, 219, 1973. With permission.

In the model, τ_{max} is assumed proportional to the maximum Hertz pressure in the contact, and the normal load is proportional to the cube of τ_{max}. In Figure 7, for example, the acetal homopolymer can tolerate a τ_{max} which is eight times that which medium density polyethylene can tolerate for the zero wear condition. In terms of load, the acetal can carry a load $8^3 = 512$ times that which polyethylene can support and maintain zero wear.

In summary, the designer who tries to select polymers for tribological applications based on published data must know all of the significant test parameters to determine if the data are valid for his application. If the data are not applicable, a test program in which the

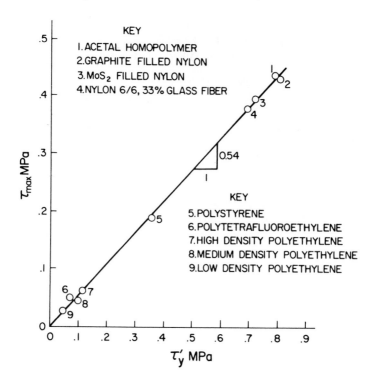

FIGURE 7.   Experimentally determined values of τ_{max} at zero wear vs. measured values of τ'_y for different plastics. Dry sliding of 302 stainless steel sphere on platens of plastics for 2000 passes. (From Clinton, W. C., Ku, T. C., and Schumacker, R. A., *Wear*, 7, 354, 1964. With permission.)

conditions of the application are matched as closely as possible must be performed to measure polymer wear. For applications where minimal wear is desired the engineering model for wear can be used. Experience with this model has shown that the contact may take larger loads than calculated and still satisfy the zero wear criterion.

REFERENCES

1. **Tabor, D.,** The wear of non-metallic materials: a brief review, in *The Wear of Non-Metallic Materials*, Dowson, D., Godet, M., and Taylor, C. M., Eds., Mechanical Engineering Publ., London, 1976, 3.
2. **Tabor, D.,** Wear, a critical synoptic view, in *Wear of Materials — 1977*, Glaeser, W. A., Ludema, K. C., and Rhee, S. K., Eds., American Society of Mechanical Engineers, New York, 1977.
3. **Briscoe, B. J. and Tabor, D.,** The sliding wear of polymers: a brief review, in *Fundamentals of Tribology*, Suh, N. P. and Saka, N., Eds., MIT Press, Cambridge, Mass., 1980, 733.
4. **Ludema, K. C., Glaeser, W. A., and Rhee, S. K., Eds.,** *Wear of Materials — 1979*, American Society of Mechanical Engineers, New York, 1979.
5. **Lee, L. H., Ed.,** *Advances in Polymer Friction and Wear*, Plenum Press, New York, 1974.
6. **Buckley, D. H.,** Introductory remarks — friction and wear of polymeric composites, in *Advances in Polymer Friction and Wear*, Lee, L. H. Ed., Plenum Press, New York, 1974, 601.
7. **Dowson, D., Challen, J. M., Holmes, K., and Atkinson, J. R.,** The influence of counterface roughness on the wear rate of polyethylene, in *The Wear of Non-Metallic Materials*, Dowson, D., Godet, M., and Taylor, C. M., Eds., Mechanical Engineering Publ., London, 1976, 99.
8. **Eiss, N. S., Jr. and Warren, J. H.,** *The Effect of Surface Finish on the Friction and Wear of PCTFE Plastic on Mild Steel*, Paper No. IQ75-125, Society of Manufacturing Engineers, Detroit, Mich., 1975.

9. **Eiss, N. S., Jr. and Bayraktaroglu, M. M.,** The effect of surface roughness on the wear of low density polyethylene, *ASLE Trans.*, 23, 269, 1980.

10. **Lancaster, J. K.,** Geometrical effect on the wear of polymers and carbons, *J. Lubr. Technol., Trans. ASME*, 97, 187, 1975.

11. **Ratner, S. B., Farberova, I. I., Radyerkevich, O. V., and Lur'e, E. G.,** Connection between wear-resistance of plastics and other mechanical properties, in *Abrasion of Rubber*, James, D. S., Ed., MacLaren, London, 1967, 145.

12. **Lancaster, J.,** Abrasive wear of polymers, *Wear*, 14, 233, 1969.

13. **Warren, J. H. and Eiss, N. S., Jr.,** Depth of penetration as a predictor of the wear of polymers on hard, rough surfaces, *J. Lubr. Technol., Trans. ASME*, 100, 92, 1978.

14. **Giltrow, J. P.,** A relation between abrasive wear and the cohesive energy of materials, *Wear*, 15, 71, 1970.

15. **Lontz, J. F. and Kumnick, M. C.,** Wear studies on moldings of polytetrafluoroethylene resin, considerations of crystallinity and graphite content, *ASLE Trans.*, 16, 276, 1973.

16. **Rabinowicz, E.,** *Friction and Wear of Materials*, John Wiley & Sons, New York, 1965, 168.

17. **Hollander, A.E. and Lancaster, J. K.,** An application of topographical analysis to the wear of polymers, *Wear*, 25, 155, 1973.

18. **Eiss, N. S., Jr., Wood, K. C., Smyth, K. A., and Herold, J. H.,** Model for the transfer of polymers on hard, rough surfaces, *J. Lubr. Technol., Trans. ASME*, 101, 212, 1979.

19. **Eiss, N. S., Jr., Warren, J. H., and Quinn, T. F. J.,** On the influence of the degree of crystallinity of PCTFE on its transfer to steel surfaces of different roughnesses, in *Wear of Non-Metallic Materials*, Dowson, D., Godet, M., and Taylor, C., Eds., Mechanical Engineering Publ. Ltd., London, 1976, 18.

20. **Deanin, R. D. and Patel, L. B.,** Structure, properties, and wear resistance of polyethylene, in *Advances in Polymer Friction and Wear*, Lee, L. H., Ed., Plenum Press, New York, 1974, 569.

21. **Pooley, C. M. and Tabor, D.,** Friction and molecular structure: the behavior of some thermoplastics, *Proc. R. Soc. London, Ser. A*, 329, 251, 1972.

22. **Tanaka, K. and Uchiyama, Y.,** Friction, wear, and surface melting of crystalline polymers, in *Advances in Polymer Friction and Wear*, Lee, L. H., Ed., Plenum Press, New York, 1974, 499.

23. **Kar, M. K. and Bahadur, S.,** Micromechanism of wear at polymer-metal sliding interface, in *Wear of Materials — 1977*, Glaeser, W. A., Ludema, K. C., and Rhee, S. K., Eds., American Society of Mechanical Engineers, New York, 1977, 501.

24. **Tanaka, K., Uchiyama, Y., and Toyooka, S.,** The mechanism of wear of polytetrafluoroethylene, *Wear*, 23, 153, 1973.

25. **Suh, N. P.,** The delamination theory of wear, *Wear*, 25, 111, 1973.

26. **Rabinowicz, E.,** *Friction and Wear of Materials*, John Wiley & Sons, New York, 1965, 151.

27. **Moore, D. F.,** Some observations on the interrelationship of friction and wear in elastomers, in *The Wear of Non-Metallic Materials*, Dowson, D., Godet, M., and Taylor, C. M., Eds., Mechanical Engineering Publ., London, 1976, 141.

28. **Schallamack, A.,** Abrasion of rubber by a needle, *J. Polym. Sci.*, 9, 385, 1952.

29. **Southern, E. and Thomas, A. G.,** Some recent studies in rubber abrasions, in *The Wear of Non-Metallic Materials*, Dowson, D., Godet, M., and Taylor, C. M., Eds., Mechanical Engineering Publ., London, 1976, 157.

30. **Ratner, S. B. and Farberova, I. I.,** Mechanical testing of plastics — wear, in *Abrasion of Rubber*, James, D. S., Ed., MacLaren, London, 1967, 297.

31. **Kraghelsky, I. V. and Nepomnyashaki, E. F.,** Fatigue wear under elastic contact conditions, *Wear*, 8, 303, 1965.

32. **Aharoni, S. M.,** The wear of polymers by roll formation, *Wear*, 25, 309, 1973.

33. **Hurricks, P. L.,** The wear and friction of elastomers sliding against paper, in *The Wear of Non-Metallic Materials*, Dowson, D., Godet, M., and Taylor, C. M., Eds., Mechanical Engineering Publ., London, 1976, 145.

34. **Eiss, N. S., Jr., Lewis, N. E., and Reed, C. W.,** Polishing of mica-filled epoxy, in *Wear of Materials — 1979*, Ludema, K. C., Glaeser, W. A., and Rhee, S. K., Eds., American Society of Mechanical Engineers, New York, 1979, 589.

35. **Lee, H. L., Orlowshi, J. A., Kidd, P. D., and Glace, R. W.,** Evaluation of wear resistance of dental restorative materials, in *Advances in Polymer Friction and Wear*, Lee, L. H., Ed., Plenum Press, New York, 1974, 705.

36. **Wright, K. H. R. and Burton, A. W.,** Wear of dental tissues and restorative materials, in *The Wear of Non-Metallic Materials*, Dowson, D., Godet, M., and Taylor, C. M., Eds., Mechanical Engineering Publ., London, 1976, 116.

37. **Kusy, R. P. and Leinfelder, K. F.,** Pattern of wear in posterior composite restorations, *J. Dental Res.* 56, 544, 1977.

38. **Powers, J. M., Douglas, W. H., and Craig, R. G.,** Wear of dimethacrylate resins used in dental composites, in *Wear of Materials — 1979,* Ludema, K. C., Glaeser, W. A., and Rhee, S. K., Eds., American Society of Mechanical Engineers, New York, 1979, 605.
39. **Lewis, R. B.,** Predicting the wear of sliding plastic surfaces, *Mech. Eng.,* 86, 32, 1964.
40. **Lancaster, J. K.,** Relationships between the wear of polymers and their mechanical properties, *Proc. Inst. Mech. Eng.,* 183(3P), 98, 1969.
41. **Theberge, J. E.,** A guide to the design of plastic gears and bearings, *Mach. Design,* 42, 114, 1970.
42. **Lancaster, J. K.,** Estimation of the limiting PV relationships for thermoplastic bearing materials, *Tribology,* 4, 81, 1971.
43. **Lancaster, J. K.,** Dry bearings: a survey of materials and factors affecting their performance, *Tribology,* 6, 219, 1973.
44. **Bayer, R. G., Clinton, W. C., Nelson, C. W., and Schumacker, R. A.,** Engineering model for wear, *Wear,* 5, 378, 1962.
45. **Clinton, W. C., Ku, T. C., and Schumacker, R. A.,** Extension of the engineering model for wear of plastics, sintered metals, and platings, *Wear,* 7, 354, 1964.

WEAR COEFFICIENTS

Ernest Rabinowicz

INTRODUCTION

Other chapters have revealed that wear is extremely complicated. For example the temperature can rise, thus increasing the thickness or changing the nature of the oxide layer. Alternatively, plastic deformation or fatigue can damage the material, increasing its tendency to wear. Thus, prediction of the wear rate to any high degree of accuracy (say ± 10%) has so far defied all attempts, even in cases where the sliding system is well understood and well controlled.

However, the task is much easier when we are content with an estimate of the wear rate within a factor of four or five. In this case the exact contributions of some of the complicating factors are less significant and leave the order of magnitude of the wear rate unaffected.

Wear predictions, even though imperfect, can be used in a number of ways besides estimating the wear rate. First, an equation for wear indicates the relative influence of various parameters, such as load, hardness, velocity, surface roughness and grain size, and suggests the change in wear that might result if the sliding system is changed. Second, computation of the wear is also important in failure analysis, or in the study of any worn component of a system.

Quantitative analysis of wear starts with the concept that, while a sliding system may be losing material in more than one way, one mechanism will dominate the overall wear rate. This dominant mechanism is generally identified as one of the following:

1. Adhesive wear — considered as 'mild' or 'severe' depending on the rate of wear and the size of the wear debris. In this case wear results from adhesion and pulling out of regions of one sliding surface by the other (see the chapter on Metallic Wear for further details of wear mechanisms).
2. Abrasive wear — results from a hard sharp object, which may be a loose abrasive particle or a sharp projection on one of the sliding surfaces, scratching out a groove in a sliding surface.
3. Corrosive wear — caused by mechanical removal of a surface layer formed by a corrosion process.
4. Surface fatigue wear — removal of particles loosened by a growth of surface of subsurface fatigue cracks arising from stress variations during continued sliding.

ADHESIVE WEAR

This principal form of wear is invariably present when two surfaces slide together, and adhesive wear is predominant in many cases where wear occurs by more than one mechanism. Adhesive wear has been investigated most intensely, probably because the same interatomic forces at the interface which cause adhesive wear also cause friction.

Although a number of studies have investigated the mechanism accounting for the formation of adhesive wear particles[1,2,3,4] only one simple quantitative relationship has been developed for predicting the wear rate. This is the equation derived by Holm[5] and refined by Archard.[1]

$$\text{Wear volume V} = \frac{k \times \text{normal load L} \times \text{distance slid X}}{\text{hardness p}} \qquad (1)$$

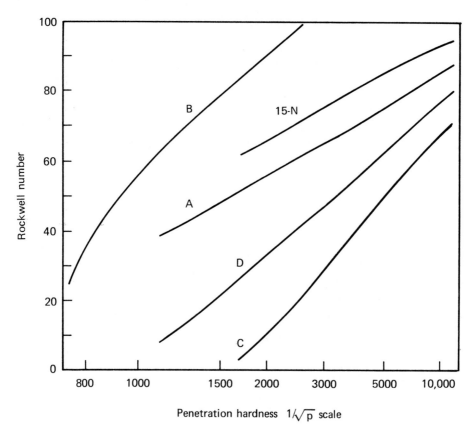

FIGURE 1.   Diagram to help convert various Rockwell arbitrary hardness numbers into hardness stresses, p, in units of N/mm².

Here p represents indentation hardness, i.e., the ratio of load applied to area of indentation produced by plastic yielding, of the softer material being worn. Parameter k is a non-dimensional constant, the wear coefficient. In Archard's original derivation with a factor of three in the denominator of Equation 1, the wear coefficient physically represented the probability that a sizeable wear particle was produced during the contact of the two surfaces at an asperity. Data presented in this section are based on Equation 1 (i.e., the wear equation without the factor of three) for simplicity. Note that such parameters as surface roughness, grain size, sliding velocity, and apparent pressure do not appear in Equation 1. The wear is independent of them, except insofar as they influence other parameters (e.g., the distance slid is proportional to velocity).

The indentation hardness is best measured by a Vickers, Knoop, or Brinell hardness test. The number given in one of these tests is the hardness in units of kg/mm², and must be multiplied by 9.8 to convert to N/mm². The various Rockwell scales are arbitrary. Conversion to N/mm² may be made using Figure 1. The hardness of a material is about 3.2 times the yield stress in uniaxial tension or compression.

All terms in the wear equation, except k, are readily available parameters like load or material hardness. Thus, the key factor in determining the adhesive wear to be expected in any sliding situation is a knowledge of the only unknown, the wear coefficient k.

VALUES OF WEAR COEFFICIENT k

Very few investigations have been carried out with the primary aim of generating wear

Table 1
WEAR COEFFICIENTS FOR
UNLUBRICATED SURFACES

Material combination	Wear coefficient (k)
Low carbon steel on low carbon steel	70×10^{-4}
60/40 Brass on tool steel	6
Teflon® on tool steel	0.25
70/30 Brass on tool steel	1.7
Lucite on tool steel	0.07
Molded bakelite on tool steel	0.024
Silver steel on tool steel	0.6
Beryllium copper on tool steel	0.37
Tool steel on tool steel	1.3
Stellite #1 on tool steel	0.55
Ferritic stainless steel on tool steel	0.17
Laminated bakelite on tool steel	0.0067
Tungsten carbide on low carbon steel	0.04
Polyethylene on tool steel	0.0013
Tungsten carbide on tungsten carbide	0.01

From Archard, J. F. and Hirst, W., *Proc. R. Soc. London Ser. A*, 236, 397, 1956. With permission.

Table 2
WEAR COEFFICIENTS FOR ADHESIVE WEAR

Lubrication	Metal-on-metal				Metal-on-nonmetal Nonmetal-on-nonmetal
	Identical	Soluble	Intermediate	Insoluble	
None	1500×10^{-6}	500×10^{-6}	100×10^{-6}	15×10^{-6}	3×10^{-6}
Poor	300	100	20	3	1.5
Good	30	10	2	0.3	1
Excellent	1	0.3	0.1	0.03	0.5

coefficient data. One experimental study was that of Archard and Hirst[6] and the results are shown in Table 1. The second[7] consisted of gathering the few wear coefficient values available from published papers, and this led to the information in Table 2. The utility of these wear coefficients is leading more and more to their use in reporting specific machine element and material test results.[8]

The two tables represent two different approaches to compiling wear coefficient data. Table 1 lists typical data, and then a user must find a value that is representative of his sliding conditions. In view of the endless variety of sliding, such a tabulation would have to be extremely extensive for general use. The other approach divides all sliding systems into a limited number of categories, and then gives appropriate wear coefficient data for each category. This approach is used in the present chapter.

For sliding metals, the two factors which mainly determine the value of the wear coefficient are the degree of lubrication and the metallurgical compatibility as indicated by the mutual solubility. The metallurgical compatibility represents the degree of intrinsic attraction of the atoms of the contacting metals for each other. Such compatibility is best determined from binary metal phase diagrams, which show the extent of mutual solubility or insolubility in the liquid or solid states.

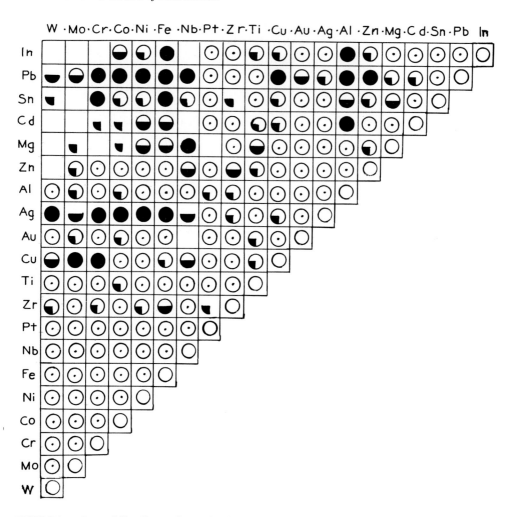

FIGURE 2.   Compatibility diagram for metal pairs. The significance of the various circles is shown in Table 3. Partial circles and blank squares are due to insufficient information.

Table 3
COMPATIBILITY RELATIONSHIPS FOR METALS

Symbol	Metallurgical solubility	Metallurgical compatibility	Sliding compatibility	Anticipated wear
○	100%	Identical	Very poor	Very high
⊙	Above 1%	Soluble	Poor	High
◓	0.1—1%	Intermediate soluble	Intermediate	Intermediate
◒	Below 0.1%	Intermediate insoluble	Intermediate or good	Intermediate or low
●	Two liquid phases	Insoluble	Very good	Very low

The compatibility of a large number of metal pairs is shown in Figure 2.[9] The significance of the various circles, in terms of metallurgical solubility at room temperature, liquid miscibility, metallurgical compatibility, sliding compatibility, and anticipated wear are shown in Table 3. The general rule is that the blacker the circle, the better the sliding characteristics and the lower the adhesive wear coefficient.

Four degrees of lubrication are considered in Table 2. The 'unlubricated' case assumes that the sliding surfaces are clean and slide without the presence of any introduced lubricant or contaminant. The second condition, labeled 'poor lubrication', assumes that the surfaces are covered by a poor lubricant, for example an inert fully fluorinated hydrocarbon, water, or gasoline. The third condition assumes the presence of a good lubricant such as mineral oil with lubricity additives. The fourth condition, excellent lubrication, is one in which excellent lubricants such as those containing zinc dialkyl dithiophosphate are used at moderate pressures and temperatures, so that surface burnishing occurs rather than wear particle formation on a large scale.

The wear coefficient values of Table 2 are based on experimental work[10] designed to generate wear data, as well as on published and unpublished wear data. Note that nonmetals sliding against metals or against nonmetals give similar wear coefficients, and these are indicated in the last column.

In applying the data of Table 2 to practical situations, two points come up. First, the solubility of an alloy is generally that of its major constituent. Thus, the rating of an aluminum bronze (copper 75%, aluminum 25% by volume) sliding against a stainless steel (iron 74%, chromium 18%, nickel 8%, by volume) is the same as that of copper against iron, namely intermediate. Second, when a liquid is heated its performance as a lubricant deteriorates above a characteristic transition temperature. For mineral oils this temperature is around 160°C. If the interfacial temperature is above 160°C, we use the wear value for the next higher line (e.g., a good lubricant above 160°C acts like a poor lubricant at room temperature).

To assist in using Equation 1, we show a simple example.

Illustrative Example 1
Problem

A person is writing with a pencil containing a 'lead' of Vickers hardness 5 kg/mm². If the writing force pressing the pencil against the paper is 0.5 N, estimate the volume of wear in writing a five-page letter (assuming that each page corresponds to 20 m of pencil-paper sliding).

Solution

The wear coefficient for unlubricated nonmetals is 3×10^{-6} (Table 2). The total sliding distance is $5 \times 20 \times 1000 = 10^5$ mm. The hardness of the pencil is 5×9.8 or 49 N/mm². The normal force is given as 0.49 N. On substitution in Equation 1,

$$V = \frac{3 \times 10^{-6} \times 0.5 \times 10^5}{49} = 3.1 \times 10^{-3} \, mm^3$$

Table 2 demonstrates the very large range of wear coefficients encountered in practice, ranging over about five orders of magnitude. Changes of the compatibilities of the surfaces may change the wear by two orders of magnitude, while changing the state of lubrication can affect the wear by as much as three orders of magnitude.

The condition described as excellent lubrication, in which the surfaces experience very little wear and become burnished, is not readily attainable. For like or compatible metals it is hardly possible to achieve excellent lubrication except when there is combined rolling and sliding, as in spur gears. With intermediate and insoluble metals, burnishing is generally possible if the mean interfacial pressure is less than 0.05 of the hardness stress of the softer material.

Lubricants have much less effect on the wear coefficient of nonmetals than on metals. Nonmetals when unlubricated give lower wear coefficients than metals, but the reverse is true in the presence of excellent lubricants.

Table 4
ABRASIVE WEAR COEFFICIENTS

	Coefficient of wear (k)			
	File (new)	Abrasive paper (new)	Loose abrasive grains	Coarse polishing
Dry surfaces	50×10^{-3}	10×10^{-3}	1×10^{-3}	0.1×10^{-3}
Lubricated	100	20	2	0.2

Note: These are maximum rates for sharp fresh abrasive surfaces. After wear and clogging, abrasive wear rates are generally reduced by up to a factor of 10.

ADHESIVE WEAR OF THE HARDER MATERIAL

For materials of different hardness sliding against each other, Equation 1 gives the wear of the softer one. There has been relatively little study of the wear of the harder material, but a good estimate may be obtained by the following procedure. If the harder surface is harder by less than a factor of three, its wear volume will be less than that of the softer surface as the hardness ratio squared. If the harder surface is more than a factor of three harder, its wear will be less by three times the hardness ratio.

Illustrative Example 2
Problem

In a certain system an aluminum alloy (hardness 60 kg/mm^2) slides against tungsten (hardness 480 kg/mm^2). After a certain period of time 10 mg of aluminum are worn away. Estimate the wear of the tungsten.

Solution

Since the density of aluminum is 2.7 mg/mm^3, wear of the aluminum is 3.70 mm^3. Tungsten is 8 times harder than aluminum, and hence its wear is less by a factor of 3 × 8, or 24. Thus, the volume wear of the tungsten is 0.154 mm^3. Since the density of tungsten is 18.8 mg/mm^3, expected weight loss of the tungsten will be 2.9 mg.

ABRASIVE WEAR

Abrasive wear, the removal of material by a plowing, cutting, or scratching process, also obeys Equation 1. The wear coefficient is determined mainly by the abrasive geometry, i.e., the effective sharpness of the abrasive, and to a smaller extent by the lubrication, which determines the ease with which wear debris can be removed from the sliding interface.

Typical wear coefficient values are shown in Table 4. Abrasive wear only occurs when the sharp material, present as protuberances or as loose grains, has a higher hardness than the surface subject to abrasive wear. Abrasive wear coefficients tend to be higher than adhesive wear coefficients, certainly in the case of lubricated surfaces. Thus, when abrasive wear and adhesive wear occur together, abrasive wear is generally predominant.

CORROSIVE WEAR

This is a very complicated form of wear, in that corrosion occurs and then the products of corrosion are worn away. Many corrosion products have some lubricating ability which also affects the wear rate. Quantitative analyses of corrosive wear have generally resulted in very complex expressions,[11] and nothing as simple as the wear equation can adequately characterize corrosive wear.

The least wear coefficient resulting from the combined effects of corrosion and adhesive wear is about 10^{-6}, but as temperature is raised and corrosion becomes rampant, the wear coefficient increases to 10^{-2} and even higher. For well-designed combinations of lubricant, surfaces, and temperature, corrosive wear coefficient values in the range of 10^{-4} to 10^{-5} are frequently observed.

SURFACE FATIGUE WEAR

This is generally observed only in systems undergoing combined sliding and rolling, such as ball bearings, gears, cams, and automotive valve lifters. It is characterized by an induction period during which microcracks grow on and under the sliding surfaces without any readily detectable wear. Then suddenly one or more large wear particles are produced. This process is not amenable to treatment by expressions such as Equation 1 which assume that wear is a steady continuous process.

APPENDIX: SOME TYPICAL USES OF WEAR COEFFICIENTS

Three examples from Reference 12 indicate ways in which wear coefficients can be used in evaluation of worn sliding surfaces. A knowledge of the wear coefficient value can be used to eliminate one or more possible modes of wear, and this can frequently be done even though typically there is an uncertainty factor of ± 4 in the value of wear coefficient.

Example 1: Wear of Hollow Jet Engine Turbine Blades
Certain jet engine turbine blades are cast hollow, a soft metal tube is fitted inside each blade, and cooling air is passed through the space between the tube and the blade. The tube was wearing away rather rapidly. The first hypothesis was that slip between the tube and the turbine blade occurred because the turbine blade was vibrating, and consequent sliding action led to the observed wear. The likely loads, amplitudes, and frequencies suggested a wear coefficient of 2.4×10^{-4}. The second theory was that slip occurred as result of a thermal expansion mismatch between the tube and blade. Knowing the number of thermal cycles the engine had undergone, one could compute a wear coefficient of 1.1.

Since adhesive wear coefficients of this magnitude do not exist, the thermal expansion theory was eliminated. If the other theory is correct, then Table 2 suggests that the blade and tube (which in terms of metallurgical compatibility represented a 'soluble' combination) operated in a region of poor lubrication, or else rampant corrosive wear was occurring. Actually, the latter was a correct description. The blades and tubes were at very high temperatures, and the protective oxide layers were being continually worn off.

Example 2: Unusually Severe Wear of Railroad Rails
Excessive wear of railroad track carrying heavily loaded ore cars in Canada was noted. On a sharply curved section of track, the rail wore down to an unsafe condition in only about a year.

Approximation of the amount of slip or sliding between the train wheels and the track (pure rolling causes essentially no adhesive wear) indicated a wear coefficient of 2×10^{-4}. Table 2 suggests that for identical metals (in terms of composition, railroad rails and wheels are very similar steels), this wear coefficient is characteristic of "poor lubrication". This seems not unreasonable for marginal lubrication from rainfall and contamination. Thus, it was not necessary to postulate any new and unusual phenomenon; simply the amount of traffic and the slippage associated with the sharp curves had produced this large amount of adhesive wear.

Example 3: Excessive Wear of the Ways of a Cold Deformation Processing Machine

Sliding occurred between a high-carbon steel die and an aluminum bronze way at reasonably high speeds and loads. A good lubricant was provided. The aluminum bronze was wearing rather rapidly, the wear coefficient being 5×10^{-5}.

The combination of iron against copper is of intermediate solubility and the use of a steel as against iron, and of aluminum bronze as against copper, does not change the rating. For a combination of intermediate solubility a wear coefficient of 5×10^{-5} indicates poor lubrication. In this case the machine design was such that the sliding surfaces became overheated, and consequently the 'good' lubricant behaved like a 'poor' one. The remedy: either improve the cooling, or use a lubricant with better performance at high temperatures.

REFERENCES

1. **Archard, J. F.,** Contact and rubbing of flat surfaces, *J. Appl. Phys,* 24, 981, 1953.
2. **Endo, K. and Fukada, Y.,** The role of fatigue in wear of metals, *Proc. 8th Japan Cong. Testing Materials,* Kyoto, 1965, 69.
3. **Suh, N. P.,** The delamination theory of wear, *Wear,* 25, 111, 1973.
4. **Halling, J.,** A contribution to the theory of mechanical wear, *Wear,* 34, 239, 1975.
5. **Holm, R.,** *Electric Contacts,* Almqvist & Wiksells, Stockholm, 1946, sect. 40.
6. **Archard, J. F. and Hirst, W.,** The wear of metals under unlubricated conditions, *Proc. R. Soc. London Ser. A,* 236, 397, 1956.
7. **Rabinowicz, E.,** New coefficients predict wear of metal parts, *Product Eng.,* 29(25), 71, 1958.
8. **Peterson, M. B. and Winer, W. O., Eds.,** *Wear Control Handbook,* American Society of Mechanical Engineers, New York, 1980.
9. **Rabinowicz, E.,** The determination of the compatibility of metals through static friction tests, *ASLE Trans.,* 14, 198, 1971.
10. **Rabinowicz, E.,** The dependence of the adhesive wear coefficient on the surface energy of adhesion, *Wear of Materials — 1977,* American Society of Mechanical Engineers, New York, 1977, 36.
11. **Uhlig, H. H.,** Mechanism of fretting corrosion, *J. Appl. Mech., Trans. ASME,* 76, 401, 1954.
12. **Rabinowicz, R.,** The wear coefficient — magnitude, scatter, uses, *J. Lubr. Technol., Trans. ASME,* 103, 188, 1981.

LUBRICATED WEAR

Carleton N. Rowe

INTRODUCTION

The prime function of a lubricant is to reduce friction, wear, and general surface damage by prevention of solid-solid contact of asperities on opposing surfaces. When the surfaces are metals, the lubricant must inhibit the formation of any strong metallic junctions that would lead to adhesive wear. By definition, any gas (vapor), liquid, or solid can serve as a lubricant as long as it reaches the surface and physically keeps the asperities separated.

Liquid lubricants scavenge wear debris and remove heat from contacting surfaces. The latter reduces operating temperature, resulting in the formation of thicker oil films and/or a lower demand on the lubricant additives. Scavenging of wear debris lessens the chance of interaction of small wear particles to form larger, work-hardened particles which can cause abrasive wear, higher surface temperatures, and mechanical destruction of the surface antiwear film.

ADHESIVE WEAR

Adhesive wear occurs under lubricated conditions when the hydrodynamic or elastohydrodynamic (EHL) film is so thin that surface asperities penetrate the film. Despite the presence of intervening lubricant between two asperities on a collision path, high hydrodynamic pressures elastically deform the metal and squeeze out the lubricant until only a very thin surface (boundary) film separates the surfaces.[1] Concurrently, asperity contact area grows due to high normal and tangential stresses on the metal (plastic as well as elastic deformation can occur) so that the trapped boundary film may be stretched until it ruptures, thereby allowing the formation of a metal-metal junction. In addition, high-shear stresses cause considerable local heating, which weakens the adsorption forces of the surface film. Mineral oil molecules and simple polar compounds desorb under these high temperatures, again allowing metal-metal contact.

A number of mathematical models for lubricated wear are based on modifications of the equation for unlubricated adhesive wear. The usual mathematical model for adhesive wear rate (V/d) may be modified to:[2]

$$\frac{V}{d} = K_m A_m = K_m \alpha A = K_m \alpha \frac{W}{H} \tag{1}$$

where V is the volume of wear, d is the sliding distance, K_m is the dimensionless adhesive wear coefficient for the particular sliding couple free of any surface contamination, A_m is the true area of metallic contact within the total true area of contact A, W is the load, H is the hardness, and α is the fractional defect of the surface film. Lubricant effectiveness in mitigating wear is expressed entirely by α, although the metal is indirectly involved through lubricant-metal interactions. Even under nominally dry sliding conditions, adsorbed gas molecules or oxide film function to some extent as a lubricant so that αK_m is reduced to a typical value of 1×10^{-3}. K_m for most metals is of the order of 0.1 to 0.2.[3]

A model for a defect in the surface film assumes that the energy of adsorption-desorption is critical to the effectiveness of the lubricant molecule on the surface.[2] Parameter α may be expressed as

$$\alpha = \frac{X}{Ut_o} e^{-E/RT_s} \tag{2}$$

where X is the diameter of the area associated with an adsorbed lubricant molecule, t_o is the fundamental time of oscillation of the molecule in the adsorbed state, U is the sliding velocity, E is the energy of adsorption, T_s is the surface temperature, and R is the gas constant. Combining with Equation 1,

$$\frac{V}{d} = K_m \frac{X}{t_o H} \cdot \frac{W}{U} \cdot e^{-E/RT_s} \tag{3}$$

A modification of the equation for lubrication by gases correlates the wear results of graphite in the presence of hydrocarbon vapors at varying vapor pressure, and the vapor pressure for a given wear rate correlates with the heats of adsorption of hydrocarbon, alcohol, carbon tetrachloride and water vapors.[4]

The following expression was derived for the presence of an antiwear additive in a base oil:[5]

$$\left(\frac{V}{d}\right)_c = \frac{e^{-\frac{\Delta E}{RT_s}}}{e^{\frac{\Delta S^\circ}{R}}} \left[\frac{\left(\frac{V}{d}\right)_b - \left(\frac{V}{d}\right)_c}{C}\right] + \left(\frac{V}{d}\right)_b \frac{e^{-\frac{\Delta E}{RT_s}}}{t_o'} \tag{4}$$

where $(V/d)_c$ and $(V/d)_b$ are the wear rates for additive concentration C and for base oil, respectively, ΔS° is the entropy change for the system, and the t_o' is the ratio of the fundamental oscillation times of additive and solvent on the surface. Figure 1 plots the wear rate of a copper pin-on-steel against the quantity in brackets of Equation 4 for stearic acid solutions at 77°C. From the slope and intercept, the adsorption-desorption equilibrium constant K can be calculated from the expression

$$K = e^{\Delta S^\circ/R} e^{\Delta E/RT_s} \tag{5}$$

Figure 2 is a plot of the calculated values of the equilibrium constant and $\Delta E (\Delta E = E_a - E_b)$ from wear tests of *n*-octadecanol, *n*-octadecanoic acid, and *n*-octadecylamine solutions in *n*-hexadecane using AISI-1020 steel surfaces in a pin-on-disk machine. The values are plotted against independently published surface potential measurements of the same additives on steel surfaces.[6] The order amine > acid > alcohol agrees with friction measurements of monolayers and wetting measurements of adsorbed films from *n*-hexadecane on metal[7] and with four-ball wear results.[8]

Collectively, these results support the concept that adsorbed surface films of long-chain polar compounds reduce adhesive wear and that the magnitude of the energy of adsorption is a controlling factor. For antiwear additives that function by chemical reaction with the metal substrate, reaction rates are probably fast compared to desorption of the reaction product, so that desorption energy of the reaction product is controlling and the model remains applicable.

CORROSIVE CHEMICAL WEAR

Antiwear and extreme pressure additives generally function by chemical reaction with surfaces. For example, organic disulfides form an iron mercaptide on steel surfaces under mild conditions[9] and iron sulfide under more severe operating conditions.[10] Likewise, organophosphates, such as tricresyl phosphate, react with iron surfaces to produce iron organophosphate under mild operating conditions and iron phosphate under severe conditions of high-load and high-surface temperature.[11]

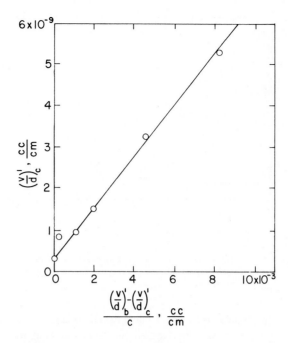

FIGURE 1. Wear rate of copper pin as a function of stearic acid concentration at 77°C. (From Rowe, C. N., *ASLE Trans.*, 13, 179, 1970. With permission.)

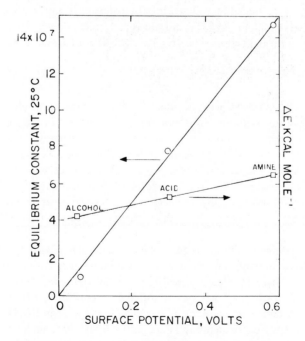

FIGURE 2. Correlation of calculated adsorption parameters from wear tests with surface potential measurements on steel surfaces. (From Rowe, C. N., *ASLE Trans.*, 13, 179, 1970. With permission.)

During relative motion of surfaces, some reaction product is lost by shearing or mechanical action, giving rise to chemical wear. Generally, the term corrosive wear is used when chemical wear is the dominant wear mode and it exceeds by a wide margin (>10 to 1) the

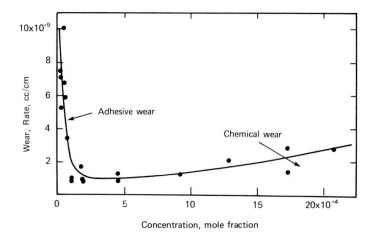

FIGURE 3.   Effect of concentration on wear rate of a copper pin with zinc *0,0*-diisopropylphosphorodithioate additive in cetane. (From Rowe, C. N., *ASLE Trans.*, 13, 179, 1970. With permission.)

amount required to control adhesive wear. Under normal circumstances, however, the amount of corrosive wear is only a small fraction of the adhesive wear that would occur if the antiwear or EP additive were not present.

Corrosive wear may occur by selective chemical reaction with one component in a bearing alloy followed by the reaction product dissolving in the oil. An example is the selective loss of lead in copper-lead connecting rod bearings by halide scavengers from the leaded fuel.[12] Where corrosion is a problem, additives are added to the oil to prevent the chemical reactions, either by inhibiting oil oxidation or by surface adsorption of metal deactivators.

ADHESIVE WEAR-CHEMICAL WEAR BALANCE

A balance may be obtained between adhesive and chemical wear under optimum conditions. Under operating conditions in which chemical reactivity of the lubricant is very low the adhesive wear process will dominate; alternatively, corrosive wear will dominate when chemical reactivity is excessive. An optimum balance is found as additive concentration is varied. Figure 3 shows the wear rate of a copper pin as a function of zinc *0,0*-diisopropylphosphorodithioate concentration in *n*-hexadecane.[5] At extremely low-additive concentrations, the surface is only partially protected and adhesive wear predominates. At high-additive concentrations, an excessively thick chemical reaction film is formed and worn during sliding.

More severe sliding conditions require greater chemical reactivity or higher additive concentrations for achieving the optimum balance between adhesive and chemical wear. Consequently, the curve in Figure 3 will shift to the right, as shown in Figure 4. The observed wear with reactivity A will be controlled by chemical wear for severity I, optimum balance of adhesive and chemical wear in II, and adhesive wear in III. This figure illustrates the difficulty frequently encountered in rating antiwear additives among various bench wear tests and field applications arising from the variation in severity at the rubbing interfaces.

Definition and measurement of chemical reactivity at a metal surface is difficult because reaction rate is exceedingly low under normal ambient conditions. Figure 5 plots wear vs. corrosion rate of an electrically heated iron wire for a variety of organic phosphites and phosphates in a refined bright stock oil.[13] Others measured the rate of decomposition of zinc *0,0*-dialkylphosphorodithioates and found correlation between the wear rate of a copper pin sliding against steel vs. the rate of hydrogen sulfide formation.[14]

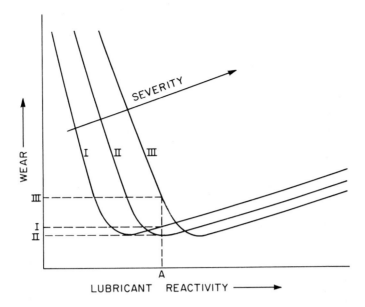

FIGURE 4. Schematic plot illustrating effect of severity of sliding on adhesive-chemical wear balance.

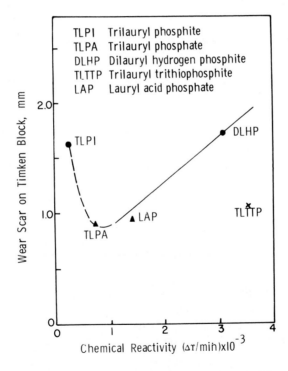

FIGURE 5. Relation between wear and chemical reactivity for phosphorus compounds. Chemical reactivity defined as rate of change of radius (Δr/min) in hot wire method. (From Sakurai, T. and Sato, K., *ASLE Trans.*, 13, 252, 1970. With permission.)

Figure 6 correlates load-carrying capacity, expressed as ratio of Mean Hertz Load with and without EP additive, with chemical reactivity as measured by the hot wire method for two types of EP additives.[13] Chemical reactivity is expressed as the ratio of rate constants

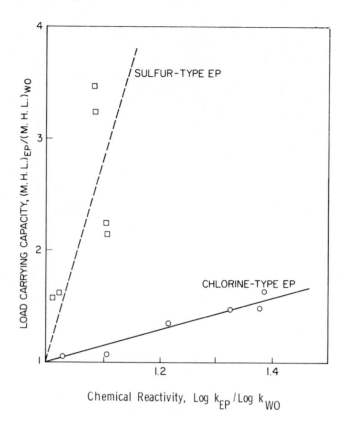

FIGURE 6.   Dependence of load-carrying capacity on chemical reactivity. (From Sakurai, T. and Sato, K., *ASLE Trans.*, 13, 252, 1970. With permission.)

k with and without EP additive, and k is the rate of change of radius of the electrically heated wire. The separate curves for the sulfur- and chlorine-containing compounds suggest that the effectiveness of the surface reaction product is also an important parameter.

Unfortunately, comprehensive guidelines for achieving maximum wear life are often not available due to a lack of understanding of many parameters including the following:

1.   Local surface temperatures and pressures
2.   Catalytic properties of deformed metal and newly created surface
3.   Chemical reactions between the additive and the metal
4.   Possible role of oxygen in the chemical reactions
5.   Surface reaction and diffusion kinetics
6.   Lubricant residence time in the contact zone
7.   Optimum surface film thickness for maximum antiwear response
8.   Properties of the surface film that control its rheology, shear strength, and adherence to the metal

With a completely formulated lubricant, additive interactions can occur in the bulk phase as well as in competing reactions with the surface, making the composition of the boundary film and its properties even more complex. In addition, a given lubricant is frequently required to lubricate different types of contacts with varying levels of severity and different metallurgies that can give entirely different surface reaction products. Nevertheless, various load-carrying and wear test results have led to valuable guidelines for producing effective additives to satisfy increasingly severe performance demands.

Table 1
HEAT OF ADSORPTION OF
ADDITIVES ON IRON OXIDE

Additive	cal $\times$ 10^{-3}/g Fe$_2$O$_3$
[CH$_3$(CH$_2$)$_{11}$–S–]$_2$	4.9
[–(CH$_2$)$_4$COOC$_2$H$_5$]$_2$	7.0
[–S–(CH$_2$)$_2$COOC$_2$H$_5$]$_2$	11.4

From Forbes, E. S., Allum, K. G., Neustadter,
E. L., and Reid, A. J. D., *Wear*, 15, 341, 1970.
With permission.

Kinetics of Surface Reactions

Kinetics of chemical reactions of antiwear and EP additives with sliding surfaces are important for maximum effectiveness. Reaction of radioactive S^{35} EP additives with rubbing surfaces in a pin-on-disk machine has been investigated, modeled mathematically, and various rate constants calculated. References 15 to 18 constitute a formidable beginning to sort out the kinetics of adsorption, chemical reactions, and wear of reaction products in contact areas.

The adsorption and the configuration of the adsorbed molecule before chemical reaction takes place are important.[19] Diester sulfides having the structure

$$[—S—(CH_2)_nCOOC_2H_5]_2$$

where n = 1, 2, or 3 were found to be effective antiwear compounds at low concentrations. At high concentrations, however, they possess poor EP properties, so that wear is adhesive. Electron probe microanalysis showed appreciable sulfur present on the wear scar at low concentrations but very little at high. Heats of adsorption in Table 1 show that the ester groups contribute more to adsorption than do the sulfur atoms. Evidently, at low concentrations both sulfur and ester groups lie on the surface and the sulfur is available for forming iron sulfide. At high concentrations, the more strongly adsorbed ester group displaces the sulfur atoms, thereby eliminating sulfur from chemical reaction with the surface.

The importance of the kinetics of additive-metal chemical reactions may be illustrated in comparison of phosphorus and sulfur EP additives in CRC L-37 and L-42 hypoid gear tests. Sulfur compounds generally function best at high-speed/low-load test (L-42) conditions while phosphorus compounds function best at low-speed/high-load test (L-37) conditions. High-speed operation results in short time intervals between repeated contacts so that the reaction kinetics of sulfur compounds may be more favorable for the L-42 test.

Temperature

Temperature is the key parameter in many lubrication mechanisms for controlling wear. Temperature determines the viscosity of the lubricant and its ability to generate a viscous film for keeping the surfaces separated. As temperature increases, the chance of asperity interactions increases and greater demand is placed on the surface film.

Surface temperature determines the reaction kinetics of antiwear and EP additives with rubbing surfaces. Many EP additives require some minimum threshold temperature to become effective. Providing the reaction is endothermic, a further increase in surface temperature can lead to excessive chemical reaction and possible chemical wear (Figure 4).

At excessively high surface temperatures, however, the following factors can lead to a loss in additive response: (1) a decrease in adhesion to the metal substrate, (2) increased desorption due to greater oil solubility, and (3) a modification of the rheology or shear properties of the film. As the surface film becomes less effective, the increased metallic contact leads to higher friction, even higher surface temperatures, and the onset of catastrophic wear. This has become known as the transition temperature phenomenon.[20]

Load

Wear rate is proportional to load in Equation 1 because the true area of contact is proportional to load under thin film conditions. As the number of asperity contacts increases with increasing load, a point is eventually reached where the diminishing lubricant film can no longer support the load. This is the point of transition from mild (acceptable) wear to ineffective lubrication and catastrophic adhesive wear.

Wear transitions with increasing load are common in the four-ball wear test. The value of the dimensionless wear coefficient K,

$$K = \alpha k_m = \frac{VH}{dW} \tag{6}$$

for a paraffinic oil at 15-kg load and 1 hr was 7.8×10^{-8}, while at 60- and 70-kg load and 1 min tests the K values are 6.6×10^{-4} and 6.9×10^{-4}, respectively.[21] Wear rate above the transition is about 10,000 times higher.

A plausible mechanism for the abrupt transition in wear may be found in the behavior of wear debris. Below the load transition, generated wear particles are able to escape the geometrical area of contact without extensive interaction with each other. As the number of metallic contacts and resultant wear particles increase, the rate of generating wear particles eventually exceeds the rate at which they can escape without interaction and aggregation. Wear particle interaction then causes a rapid deterioration of the lubricant surface film, probably by both a mechanical action and a temperature rise at the surfaces.

Sliding Velocity

Sliding velocity can have opposing effects. At low sliding velocity an increase in velocity decreases wear because of the generation of micro-EHL films[1] and the reduced chance of squeezing the surface film from between colliding asperities.[22] Equation 3 suggests that wear is inversely proportional to sliding velocity. At high sliding velocities, high surface temperatures may be experienced since temperature rise is proportional to sliding velocity. Comments made previously on temperature apply.

Wear Coefficients

The value of K ($K = \alpha K_m$) for unlubricated AISI-52100 steel is of the order of 1×10^{-3}.[23] For antiwear and EP additives, K values are of the order of 10^{-9} to 10^{-7} in four-ball wear tests.[24] Table 2 gives K values for zinc dialkylphosphorodithioates and typical phosphorus and sulfur compounds in the four-ball wear test and shows that the K ranking is roughly

$$\text{Zinc dialkylphosphorodithioates} < \frac{\text{phosphorus}}{\text{compounds}} < \frac{\text{sulfur}}{\text{compounds}}$$

These results show that effective lubricants can reduce wear to about one millionth of that under unlubricated sliding conditions.

ABRASIVE WEAR

Abrasive wear results from a cutting action by (1) a rough, hard surface sliding against

Table 2
WEAR COEFFICIENT K

Additive	Wear scar diameter (mm)	Wear coefficient K ($\times 10^{-8}$)
Zinc Dialkylphosphorodithioates		

$$Zn \left[SP \overset{S}{\underset{OR}{-OR}} \right]_2$$

Additive	Wear scar diameter (mm)	Wear coefficient K ($\times 10^{-8}$)
R = Isopropyl	0.258	0.17
R = 2-Ethylhexyl	0.296	0.43
R = n-Dodecyl	0.290	0.37
Phosphonates		
$(R_1O)_2P(O)R_2$		
R_1; R_2		
Butyl; H	0.64	14.1
2-Ethylhexyl; H	0.36	1.2
Butyl; phenyl	0.48	4.2
Phosphates		
Tributyl	0.47	3.9
Tricresyl	0.28	0.29
Disulfides		
Dibenzyl	0.560	8.1
Di-n-butyl	0.855	46.0
Di-n-dodecyl	0.623	12.6

Note: Four-ball test, 15 kg, 1500 r/min, 50°C, 60 min.

From Rowe, C. N., *Wear Control Handbook*, Peterson, M. B. and Winer, W. O., Eds., American Society of Mechanical Engineers, New York, 1980, chap. 6. With permission.

a softer surface, or (2) contaminant hard particles trapped between sliding surfaces. Hydrodynamic, or elastohydrodynamic, lubricant films either sufficiently thick to separate the surfaces in (1) or thicker than the largest hard particles in (2) will greatly reduce abrasive wear.[25] Abrasive wear in lubricated systems is controlled either by providing a lubricant of the appropriate viscosity for generating a sufficiently thick lubricant film, or by using adequate filters for removal of unavoidable hard particle contamination.

If adhesive wear particles remain in the contact area, chances are good for the formation of larger particle aggregates which can undergo work-hardening and cause abrasive wear. Liquid lubricants can flush the adhesive wear particles out of the contact area and prevent the series of steps that lead to abrasive wear. With operating conditions and systems where it is impossible to have thick lubricant films, antiwear and EP additives may offer some protection through the generated surface film, both by its thickness and its reduction of adhesive wear.

CONTACT FATIGUE

Fatigue is associated with cyclic loading as encountered with rolling element bearings, gears, and valve trains. Cyclic deformation of the contacting surface leads to the initiation

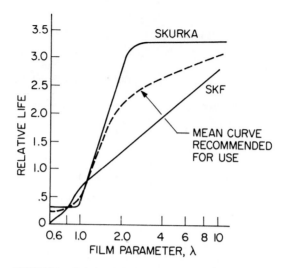

FIGURE 7. Relative fatigue life as a function of film param-
eter, λ. (From Bamberger, E. N., Harris, T. A., Kacmarsky,
W. M., Moyer, C. A., Parker, R. J., Sherlock, J. J., and
Zaretsky, E. V., *Life Adjustment Factors for Ball and Roller
Bearings — An Engineering Guide,* American Society of Me-
chanical Engineers, New York, 1971. With permission.)

and propagation of microcracks. Subsurface crack initiation is generally in regions of maximum shear stress and is often associated with inclusions in the metal. Surface imperfections,
imbedded debris, scratch marks, dents, and contaminant hard particles in the lubricant can
act as stress initiators for surface microcracks. Development of a network of cracks via crack
propagation results in spalling or pitting, which are forms of fatigue wear.

Physical and chemical properties of the lubricant play a significant role in fatigue wear.[26]
Specific film thickness, λ, which is the ratio of EHL film thickness to the composite surface
roughness, is considered to be an important parameter affecting the fatigue life of bearings.[27-29] Since λ is a function of lubricant viscosity and viscosity-pressure coefficient, fatigue
can often be controlled by selecting lubricants of proper viscosity. Figure 7 plots relative
fatigue life as a function of the film parameter, or specific film thickness.[30]

Effect of Water

Small amounts of water added to lubricants may significantly reduce the fatigue life of
ball and roller bearings as summarized in Table 3. With varying water content in an SAE-
20 rust and oxidation-inhibited mineral oil in tapered roller bearing tests at 2.03 GPa (290,000
psi) Hertz stress, relative life (L_{Rel}) follows the expression:

$$L_{Rel} = \left(\frac{100}{C}\right)^{0.6} \tag{7}$$

where C is water concentration in parts per million (ppm) and the relative life is one at 100
ppm.[31] The effect of added seawater to commercial hydraulic fluids on the fatigue life of
angular contact ball bearings at 2.27 GPa (330,000 psi) Hertzian stress is found in Reference
32. Figure 8 is the present writer's plot of the log of life/EHL film thickness ratio against
the log of water content in the oils as received. A least square analysis of the data shows
that

$$\frac{L_{10}}{h} = \text{constant } C^{-0.42}$$

$$\frac{L_{50}}{h} = \text{constant } C^{-0.68} \tag{8}$$

Table 3
EFFECT OF WATER IN LUBRICANTS ON ROLLING
CONTACT FATIGUE LIFE

Lubricant description[a]	Water content of oil (%)	Fatigue life reduction (%)	Test equipment and Hertzian stress	Ref.
SAE 20 rust and oxidation inhibited mineral oil (0.01%)	0.04	48	Tapered roller bearing 2.03 GPa (0.29 × 10⁶ psi)	31
Mineral oil-based	0.1	45	Angular contact bearing	
Emulsifying hydraulic oil (0.02%)	0.5	56	2.27 GPa (0.33 × 10⁶ psi)	32
SAE-10-based mineral oil (0.15%)	0.5	45	Unisteel bearing fatigue test 3.90 GPa (0.57 × 10⁶ psi)	33
SAE-10-based mineral oil (≈0.05%)	0.5	17	Rolling 4-ball 7.52 GPa (1.09 × 10⁶ psi)	34
Emulsifying hydraulic oil (purged with Argon)	1.0 (seawater)	45	Rolling 4-ball 6.89 GPa (1.00 × 10⁶ psi)	35

[a]   Water content of oil without added water given in parentheses.

where C is the water content in ppm. The exponents of C are in good agreement with the 0.6 in Equation 7. It follows that the Weibull slope increases with increasing water content, suggesting that the lives of those bearings which run the longest before failure are affected most since the water will have a longer time to affect the crack initiation and crack propagation processes.

Most industrial oils contain dissolved water in the range of 50 to 500 ppm and may become further contaminated with moisture from the operating environment. Water molecules, being extremely small compared to the base lubricant and additive molecules, readily diffuse to the tips of microcracks and decompose on the highly reactive newly formed surface to produce atomic hydrogen. The hydrogen diffuses into the metal ahead of the crack, causing hydrogen embrittlement of the steel. The embrittlement not only allows the crack to propagate more readily, but promotes crack branching.[26]

Water-accelerated fatigue and its prevention by lubricant additives has been investigated using a rotating-beam fatigue apparatus.[36] The additives are considered to function by proton neutralization, formation of hydrophobic film, or water sequestration.

Lubricant Type and Lubricant Additives

Lubricant type can significantly affect fatigue life. Fire-resistant fluids lead to an appreciable reduction in life.[37,38] Recent comparative results for several fluids with tapered roller bearings and the Unisteel fatigue test are given in Table 4.

Extreme pressure and antiwear lubricant additives can have a significant influence on

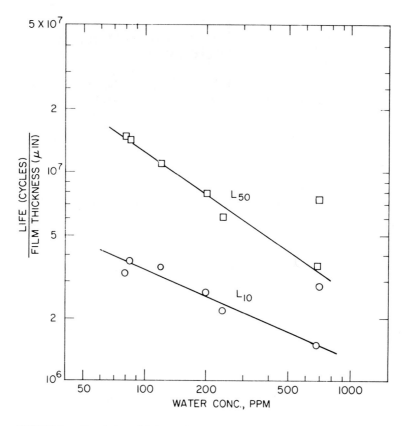

FIGURE 8.   Correlation of fatigue life/EHL film thickness ratio with inherent water content for a series of hydraulic fluids.

Table 4
COMPARISON OF RELATIVE FATIGUE LIFE FOR
FIRE-RESISTANT FLUIDS

	Tapered roller bearing 1.97 GPa (0.285 × 10⁶ psi)[37]	Unisteel 3.88 GPa (0.563 × 10⁶ psi)[38]
Mineral oil	1.0	1.0
Phosphate ester	0.49—0.80	0.79
Water-in-oil emulsion	0.31—0.53	0.21
Water glycol	0.14—0.20	0.17
Dilute emulsion, 5/95 oil/water	—	0.06

fatigue life.[26] A sulfur-phosphorus EP additive reduced fatigue life at high values of the specific film thickness, but increased life at low values.[39] Additive response, positive or negative, has been found to be sensitive to oil base stock, additive concentration, and metallurgy.[40] Oxidation of the oil can lead to reduced fatigue life,[41] and the magnitude of reduction is influenced by acid strength.[42] Under combined rolling/sliding conditions, a zinc dialkylphosphorodithioate additive reduced fatigue life to a greater extent at lower stress level than at a higher stress level.[43] Since many bench fatigue tests operate at stress levels that significantly exceed those in field applications, it is important to know the influence of stress level on additive response.

Table 5
EFFECT OF REAR AXLE LUBRICANTS ON
FATIGUE LIFE OF TAPERED ROLLER
BEARINGS

SAE Grade	Type of EP additive package	L_{50} life, normalized
90	None	1.00
80	Lead-sulfur	0.24
80	Lead-sulfur	0.34
80	Phosphorus-sulfur	0.32
90	Zinc-phosphorus-sulfur	0.17
90	Zinc-phosphorus-sulfur-chlorine	0.40
90	Phosphorus-sulfur	0.67

Note: 1.03 GPa (150,000 psi) maximum Hertz stress.

From Kepple, R. K. and Johnson, M. F., *Effect of Rear Axle Lubricants on Fatigue Life of Tapered Roller Bearings,* Paper No. 760329, Society of Automotive Engineers, Warrendale, Pa., 1976. With permission.

As a general conclusion, many EP and antiwear additives reduce fatigue life. Table 5 compares the results for a series of rear axle lubricants with different types of EP additives in a tapered roller bearing test.[44] Recently, however, an organophosphonate was reported to provide an increase in fatigue life via the generation of a film on rolling element surfaces.[45]

EROSIVE WEAR

Erosive wear involves loss of material from a solid surface due to the abrasive action, or impingement, of fluids or solid particles suspended in the fluids. If the fluid is corrosive, the phenomenon may be called erosion corrosion. Erosive wear of plain bearings can occur from the rapid flow of a viscous fluid over sharp edges of bearing features.[46] Erosion by solids under lubricated conditions can occur when particles are carried by the oil stream. In the case of plain bearings, erosive wear occurs in the vicinity of the inlet ports.[47]

CAVITATION WEAR

Wear can occur by the high impact pressure resulting from the collapse of vapor and gas bubbles in a liquid. The phenomenon is generally found where the hydrodynamic condition is characterized by a sudden and gross change in pressure resulting in the formation and the collapse of bubbles. Wear of ship propellers constituted one of the first field problems of cavitation damage.

In lubricating films two types of disruption of the continuous liquid phase are recognized: gaseous cavitation and vapor cavitation. Gaseous cavitation results when the pressure to which the lubricant is subjected falls below the saturation pressure of dissolved gases. Vapor cavitation results when the pressure in a liquid falls below the vapor pressure of the liquid causing local boiling and the formation of bubbles which then collapse as the pressure is increased. Gaseous cavities form and collapse more slowly than vapor cavities, thereby generally causing less surface damage.[46,48]

Film rupture and the fundamental aspects of the formation of cavities or bubbles in plain bearings are found in a number of papers in the Proceedings of the First Leeds-Lyon Symposium on Tribology.[49] Cavitation can influence the performance of dynamically loaded

Film rupture and the fundamental aspects of the formation of cavities or bubbles in plain bearings are found in a number of papers in the Proceedings of the First Leeds-Lyon Symposium on Tribology.[49] Cavitation can influence the performance of dynamically loaded journal bearings[50] and externally pressurized journal bearings.[51] The effect is one of lower load capacity, lower film thickness, and the onset of surface damage. The side plates in gear pumps may undergo surface damage and wear due to cavitation.

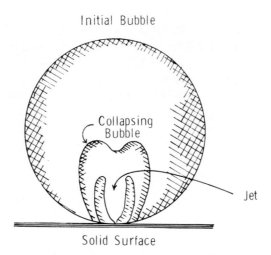

FIGURE 9.   Collapse of a bubble. (From Plesset, M. S. and Chapman, R. B., *Collapse of an Initially Spherical Vapor Cavity in the Neighborhood of a Solid Boundary*, California Institute of Technology, Pasadena, 1970. With permission.)

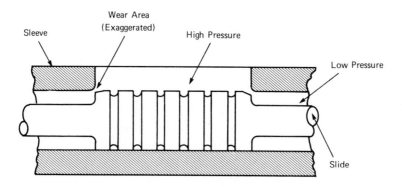

FIGURE 10.   Electrochemical wear in aircraft servo valves. (From Beck, T. R., Mahaffey, D. W., and Olsen, J. H., *J. Basic Eng.*, 92, 782, 1970. With permission.)

Cavitation may be important in wear of rolling contact bearings.[52] Resistance to fatigue in a four-ball rolling contact test correlated with the resistance to pitting in a cavitation erosion test. A similar conclusion has been reached that cavitation erosion is a fatigue process due to the related impacts of liquid jets as a result of bubble collapse.[53].

The detailed mechanism of cavitation erosion is still unknown. Figure 9 shows the collapse of a bubble with the formation of a high-velocity microjet in the center.[54] The similarity of these jets to liquid impingement suggests that cavitation wear and fluid erosion are similar. Surface tension increases the rate of collapse, while compressibility, viscosity, and the presence of noncondensable gases tend to decrease the rate.[55]

ELECTROCHEMICAL WEAR

A less frequent type of wear is that caused by electrochemical reactions. Wear of aircraft hydraulic servo valves in the presence of a phosphate ester fluid has been ascribed to the generation of an electrokinetic or streaming current.[56] As shown in Figure 10 the wear occurs on the upstream side of the valve, and it occurs most rapidly when the valve is in the null position with a large pressure drop (3000 psi) across the small orifice. Theory hypothesizes that fluid flow will sweep free charges in the diffuse outer regions of the electrical double layer, thereby setting up an electrokinetic potential between the solid and the fluid outside the diffuse layer. The conservation of charge imposes a current flow between the metal and fluid, known as the wall current, which causes electrochemical reactions.

For this type of wear to occur, the fluid must meet two conditions: the electrical double layer is thin compared to the hydrodynamic boundary layer, and the conductivity of the fluid is relatively low compared to that of the metal. Low-conductivity fluids violate the first condition, while high-conductivity fluids fail the second. The phosphate ester evidently met both conditions.

Wear can be investigated in water-based emulsion lubricants by electrochemical methods.[57] As an example, the potential in a cationic emulsion was used to distinguish between adhesive and corrosive wear of a chromium-plated yarn guide.[58]

REFERENCES

1. **Fowles, P. E.,** The application of elastohydrodynamic lubrication theory to individual asperity-asperity collisions, *J. Lubr. Tech., Trans. ASME,* 91F, 464, 1969.
2. **Rowe, C. N.,** Some aspects of the heat of adsorption in the function of a boundary lubricant, *ASLE Trans.,* 9, 100, 1966.
3. **Rowe, C. N.,** Discussion to a chapter on ''Wear'' by Archard, J. F., Interdisciplinary Approach to Friction and Wear, NASA SP-181, Ku, P. M., Ed., National Aeronautics and Space Administration, Washington, D.C., 1968, 308.
4. **Rowe, C. N.,** A relation between adhesive wear and heat of adsorption for the vapor lubrication of graphite, *ASLE Trans.,* 10, 10, 1967.
5. **Rowe, C.N.,** Role of additive adsorption in the mitigation of wear, *ASLE Trans.,* 13, 179, 1970.
6. **Martin, P.,** Surface Potentials of Adsorbed Organic Monolayers on Metals, Rock Island Arsenal Lab. Rep. No. 67-1754, Research and Engineering Division, July 1967.
7. **Zisman, W. A.,** Friction and wear, *Proc. Symp. Friction and Wear, Detroit, 1957,* Elsevier, Amsterdam, 1959, 143.
8. **Groszek, A. J.,** Heat of preferential adsorption of surfactants on porous solids and its relation to wear of sliding steel surfaces, *ASLE Trans.,* 5, 105, 1962.
9. **Allum, K. G. and Forbes, E. S.,** The load carrying properties of organic sulfur compounds. II. The influence of chemical structure on the anti-wear properties of organic disulphides, *J. Inst. Petrol.,* 53, 173, 1967.
10. **Allum, K. G. and Forbes, E. S.,** The load carrying properties of organic sulfur compounds, I., *J. Inst. Petrol.,* 51, 145, 1965.
11. **Forbes, E. S., Upsdell, N. T., and Battersby, J.,** *Current Thoughts on the Mechanism of Action of Tricresyl Phosphate as a Load Carrying Additive,* Institute of Mechanical Engineers, London, 1973, 7.
12. **Weetman, D. G., Kreuz, K. L., Hellmuth, W. W., and Becker, H. C.,** *Scanning Electron Microscope Studies of Copper-Load Bearing Corrosion,* Paper 760559, Society of Automotive Engineers, Warrendale, Pa., 1976.
13. **Sakurai, T. and Sato, K.,** Chemical reactivity and load-carrying capacity of lubricating oils containing organic phosphorus compounds, *ASLE Trans.,* 13, 252, 1970.
14. **Rowe, C. N. and Dickert, J. J.,** The relation of antiwear function to thermal stability and structure for metal 0,0-dialkylphosphorodithioates, *ASLE Trans.,* 10, 86, 1967.
15. **Sakurai, T., Sato, K., and Ishida, K.,** Reaction between sulfur compounds and metal surfaces at high temperatures, *Bull. Jpn. Petrol. Inst.,* 6, 40, 1964.
16. **Sakurai, T., Ikeda, S., and Okabe, H.,** The mechanism of reaction of sulfur compounds with steel surfaces during boundary lubrication using S^{35} as a tracer, *ASLE Trans.,* 5, 67, 1962.
16a. **Sakurai, T., Ikeda, S., and Okabe, H.,** A kinetic study on the reaction of labelled sulfur compounds with steel surfaces during boundary lubrication, *ASLE Trans.,* 8, 39, 1965.

17. **Sakurai, T., Okabe, H., and Takahashi, Y.,** A kinetic study of the reaction of labelled sulfur compounds in binary additive systems during boundary lubrication, *ASLE Trans.,* 10, 91, 1967.
18. **Okabe, H., Nishio, H., and Masuko, M.,** Tribochemical surface reaction and lubricating oil film, *ASLE Trans.,* 22, 67, 1979.
19. **Forbes, E. S., Allum, K. G., Neustadter, E. L., and Reid, A. J. D.,** The load carrying properties of diester disulphides, *Wear,* 15, 341, 1970.
20. **Fein, R. S.,** Effects of lubricants on transition temperatures, *ASLE Trans.,* 8, 59, 1965.
21. **Allum, K. G. and Forbes, E. S.,** The load-carrying properties of metal dialkylphosphorodithioates: the effect of chemical structure, *Proc. Inst. Mech. Eng.,* 183(3P), 7, 1969.
22. **Kingsbury, E. P.,** The heat of adsorption of a boundary lubricant, *ASLE Trans.,* 3, 30, 1966.
23. **Fein, R. S.,** AWN — a proposed quantitative measure of wear protection, *Lubr. Eng.,* 31, 581, 1975.
24. **Rowe, C. N.,** Lubricated wear, in *Wear Control Handbook,* Peterson, M. B. and Winer, W. O., Eds., American Society of Mechanical Engineers, New York, 1980, chap. 6.
25. **Ronen, A., Malkin, S., and Loewy, L.,** Wear of dynamically loaded hydrodynamic bearings by contaminated particles, in *Wear of Materials,* Ludema, K. C., Glaeser, W. A., and Rhee, S. K., Eds., American Society of Mechanical Engineers, New York, 1979, 319.
26. **Scott, D.,** Lubricant effects on rolling contact fatigue — a brief review, in *Rolling Contact Fatigue; Performance Testing of Lubricants,* Tourret, R. and Wright, E. P., Eds., Heyden and Son Ltd., London 1977, chap. 1.
27. **Skurka, J. C.,** Elastohydrodynamic lubrication of bearings, J. Lubr. Tech., *Trans. ASME,* 93(F), 281, 1971.
28. **Danner, C. H.,** Fatigue life of tapered roller bearings under minimal lubricant films, *ASLE Trans.,* 13, 241, 1970.
29. **Liu, J. Y., Tallian, T. E., and McCool, J. I.,** Dependence of bearing fatigue life on film thickness to surface roughness ratio, *ASLE Trans.,* 18, 144, 1975.
30. **Bamberger, E. N., Harris, T. A., Kacmarsky, W. M., Moyer, C. A., Parker, R. J., Sherlock, J. J., and Zeretsky, E. V.,** *Life Adjustment Factors for Ball and Roller Bearings — An Engineering Guide,* American Society of Mechanical Engineers, New York, 1971.
31. **Cantley, R. E.,** The effect of water in lubricating oil on bearing fatigue life, *ASLE Trans.,* 20, 244, 1977.
32. **Felsen, I. M., McQuaid, R. W., and Marzani, J. A.,** Effect of seawater on the fatigue life and failure distribution of flood-lubricated angular-contact ball bearings, *ASLE Trans.,* 15, 8, 1972.
33. **Murphy, W. R., Armstrong, E. L., and Wooding, P. S.,** Lubricant performance testing for water-accelerated bearing fatigue, in *Rolling Contact Fatigue; Performance Testing of Lubricants,* Tourret, R. and Wright, E. P., Eds., Heyden and Son, London, 1977, chap. 16.
34. **Armstrong, E. L., Leonardi, S. J., Murphy, W. R., and Wooding, P. S.,** Evaluation of water-accelerated bearing fatigue in oil-lubricated ball bearings, *Lubr. Eng.,* 34, 15, 1978.
35. **Schatzberg, P.,** Inhibition of water-accelerated rolling contact fatigue, *J. Lubr. Tech. Trans. ASME,* 93(F), 231, 1971.
36. **Murphy, W. R., Polk, C. J., and Rowe, C. N.,** Effect of lubricant additives on water-accelerated fatigue, *ASLE Trans.,* 21, 63, 1978.
37. **Culp, D. V. and Widner, R. L.,** *The Effect of Fire-Resistant Hydraulic Fluids on Tapered Roller Bearing Fatigue Life,* Paper No. 770748, Society of Automotive Engineers, Warrendale, Pa., 1977.
38. **March, C. N.,** The evaluation of fire-resistant fluids using the Unisteel Rolling Contact Fatigue machine, in *Rolling Contact Fatigue; Performance Testing of Lubricants,* Tourret, R. and Wright, E. P., Eds., Heyden and Son Ltd., London, 1977, chap. 13.
39. **Phillips, M. R. and Quinn, T. F. J.,** The effect of surface roughness and lubricant film thickness on the contact fatigue of steel surfaces lubricated with a sulfur-phosphorus type of extreme pressure additive, *Wear,* 51, 11, 1978.
40. **Rounds, F. G.,** Some effects of additives on rolling contact fatigue, *ASLE Trans.,* 10, 243, 1967.
40a. **Rounds, F. G.,** Lubricant and ball steel effects on fatigue life, *J. Lubr. Tech., Trans. ASME,* 93(F), 236, 1971.
41. **Mould, R. W. and Silver, H. B.,** The effect of oil deterioration on the fatigue life on EN 31 steel using the rolling four-ball machine, *Wear,* 31, 295, 1975.
42. **Mould, R. W. and Silver, H. B.,** A study of the effects of acids on the fatigue life on EN 31 steel balls, *Wear,* 37, 333, 1976.
43. **Littman, W. E., Kelley, B. W., Anderson, W. J., Fein, R. S., Klaus, E. E., Sibley, L. B., and Winer, W. O.,** Chemical effects of lubrication in contact fatigue. III. Load-life exponent, life scatter and overall analysis, *J. Lubr. Technol., Trans. ASME,* 98(F), 308, 1976.
44. **Kepple, R. K. and Johnson, M. F.,** *Effect of Rear Axle Lubricants on the Fatigue Life of Tapered Roller Bearings,* Paper No. 760329, Society of Automotive Engineers, Warrendale, Pa., 1976.

45. **Fowles, P. E., Jackson, A., and Murphy, R. W.,** Lubricant chemistry in rolling contact fatigue — the performance and mechanism of one anti-fatigue additive, ASLE Paper 79-LC-4A-1, ASLE/ASME Lubr. Conf., Dayton, Ohio, October 16 to 18, 1979.
46. **James, R. D.,** Erosion damage in engine bearings, *Tribology Int.,* 8, 161, 1975.
47. **Love, P. P.,** Diagnosis and analysis of plain bearing failures, *Wear,* 1, 196, 1958.
48. **Wilson, R. W.,** Cavitation damage in plain bearings, in *Cavitation and Related Phenomena,* Dowson, D., Godet, M., and Taylor, C.M., Eds., Institute of Mechanical Engineers, London, 1975, Paper 7.
49. **Dowson, D., Godet, M., and Taylor, C. M., Eds.,** *Cavitation and Related Phenomena,* Institute of Mechanical Engineers, London, 1975.
50. **Marsh, H.,** Cavitation in dynamically loaded journal bearings, in *Cavitation and Related Phenomena,* Dowson, D., Godet, M., and Taylor, C. M., Eds., Institute of Mechanical Engineers, London, 1975, Paper 4 (ii).
51. **Davies, P. B.,** Cavitation in dynamically loaded hydrostatic journal bearings, in *Cavitation and Related Phenomena,* Dowson, D., Godet, M., and Taylor, C. M., Eds., Institute of Mechanical Engineers, London, 1975, Paper 4 (iii).
52. **Tichler, J. W. and Scott, D.,** A note on the correlation between cavitation erosion and rolling contact fatigue resistance of ball bearings, *Wear,* 16, 229, 1970.
53. **Tao, F. F. and Appledoorn, J. K.,** Cavitation erosion in a thin film as affected by the liquid properties, *J. Lubr. Technol., Trans. ASME,* 93(F), 470, 1971.
54. **Plesset, M. S. and Chapman, R. B.,** *Collapse of an Initially Spherical Vapor Cavity in the Neighborhood of a Solid Boundary,* Rep. 85-49, Div. Eng. Appl. Sci., California Institute of Technology, Pasadena, 1970.
55. **Knapp, R. T., Daily, J. W., and Hammitt, F. G.,** *Cavitation,* McGraw-Hill, New York, 1970.
56. **Beck, T. R., Mahaffey, D. W., and Olsen, J. H.,** Wear of small orifices by streaming current driven corrosion, *J. Basic Eng.,* 92, 782, 1970.
57. **Waterhouse, R. B.,** Tribology and electrochemistry, *Tribology,* 3, 158, 1970.
58. **Ijzermans, A. B.,** Corrosive wear of chromium steel in textile machinery, *Wear,* 14, 397, 1969.

Lubricants and Their Application

LIQUID LUBRICANTS

E. E. Klaus and E. J. Tewksbury

INTRODUCTION

Liquid lubricants available for the 1980s include petroleum fractions, synthetic liquids and mixtures of two or more of these materials. Various additives are used to improve specific properties. A partial list of the liquid lubricant types currently available include the following:

Type	Principal attribute
Mineral oil fractions	Low cost
Synthetic hydrocarbons	Low temperature fluidity
Organic esters	Low temperature fluidity
Polyglycol ethers	Good viscosity-temperature properties
Water base lubricants	Less flammable
Oil-water emulsions	Less flammable
Phosphate esters	Less flammable
Silicones	Excellent viscosity-temperature properties
Polyphenylethers	Thermal stability
Perfluoropolyethers	Oxidation resistant
Halocarbons	Nonflammable

The physical properties of lubricants are attributable primarily to the structure of the lubricant base stock. Chemical properties of the finished or formulated lubricants are due primarily to the additive package and response of base stocks to the additive package. Properties considered in this chapter include the following:

Viscosity	Thermal properties
Viscosity-temperature relationships	Surface tension
Viscosity-pressure relationships	Gas solubility
Viscosity-shear properties	Foaming
Viscosity-volatility relationships	Electrical
Vapor pressure	Thermal stability
Density	Oxidation stability
Bulk modulus	Lubrication specifications

VISCOSITY

The viscosity values most frequently reported for a lubricant are at 40 and 100°C (previously 100 and 210°F) at atmospheric pressure and low-shear rates. The following sections will deal in turn with viscosity measurement and correlations available for viscosity-temperature, viscosity-pressure and viscosity-shear for the extension of usual viscosity information to conditions commonly encountered in lubrication systems and machine elements. Viscosity is a measure of resistance to flow; the basic unit is the pascal-second (10 P). The poise is equivalent to the force of one dyne per centimeter shearing a liquid at a rate of one centimeter per second per centimeter. The common unit of absolute viscosity is the centipoise (0.001 Pa·sec).

The most common method of viscosity measurement is described in ASTM D445. Viscometers commonly depend on the force of gravity on the fluid head to drive the fluid through a capillary. The direct viscosity measurement from this procedure gives a kinematic viscosity in stokes (St) or centistokes (0.01 St). A stoke in SI units is equal to 1 cm²/sec or

10^{-4} m²/sec. Absolute viscosity (η) in centipoise is equal to kinematic viscosity (ν) in centistokes multiplied by density (ρ) in kg/dm³.

The viscosity of a fluid in a capillary viscometer at a given temperature can be defined by Poiseuille's law:

$$\nu = \frac{\pi \, P r^4 t}{8 L V} \tag{1}$$

where P = pressure drop across the capillary, r = radius of the capillary, L = length of the capillary, V = volume of fluid flowing through the capillary, and t = time of flow.

Two types of rotational viscometers are also commonly used. One type consists of two concentric cylinders with their annulus containing the fluid for viscosity measurement. Viscosity is related to the angular velocity of the moving cylinder and the torque generated between the two cylinders according to the following equation:

$$\eta = \frac{\tau(r_1^2 - r_2^2)}{4 \pi \ell \omega \, r_1^2 \, r_2^2} \tag{2}$$

where τ = torque measured, r_1 = radius of the larger cylinder, r_2 = radius of the smaller cylinder, ℓ = depth of fluid in the annular space, and ω = angular velocity. A common viscometer of this type is the Brookfield (ASTM D2669 and ASTM D2983).

The other type of rotational viscometer places the fluid between a flat circular plate and a circular cone with only a slight taper from planar. As either the cone or plate is rotated with a thin fluid film between, torque is measured and viscosity is related to viscometer geometry in the following manner:

$$\eta = \frac{3 \tau \theta}{2 \omega \pi r^3} \tag{3}$$

where τ = torque, θ = angle between cone and plate, ω = angular velocity, and r = radius of cone.

This cone and plate viscometer can be used to measure a normal force at right angles to the direction of flow, which tends to force the cone and plate apart. This is a manifestation of the Weisenberg effect[1] or the viscoelasticity of the fluid. Polymer solutions tend to show this effect at high molecular weights coupled with high molecular volumes in solution. These viscoelastic effects may become significant with drastic changes in flow path in short time periods. Greases tend to show viscoelasticity if the time for flow change is of the order of 0.1 to 0.01 sec. Conventional VI improvers show the same behavior in 10^{-4} to 10^{-6} sec and in severe EHD lubrication events mineral oils may show a similar behavior due to high transient loading.

Viscosities of mineral oil fractions and true solutions of molecules of low molecular weights are Newtonian. That is, the viscosity of the fluid is independent of shear rate or shear stress. Shear stress equals viscosity (η) multiplied by shear rate or velocity gradient (γ), and the viscosity determined by all three viscometers at a common pressure and temperature should be the same. In the case of polymer solutions, viscosity may vary with shear rate and shear stress at the same temperature and pressure. The three viscometer types must, therefore, be evaluated for the measurement of viscosity as a function of shear. The capillary viscometer provides the most complex problem for shear rate on the flowing fluid. Liquid molecules on the capillary wall remain fixed and the fluid flows in concentric layers producing a parabolic flow profile with a maximum shear rate at the capillary wall and zero shear rate in the capillary center. Maximum shear rate at the wall is

$$\gamma = \frac{4 V}{\pi r^3 t} \tag{4}$$

Values ranging from 0.4 to 0.7 times this maximum shear rate can be found in the literature to represent the average shear in the capillary. Excellent agreement between high shear viscosity values has been obtained for 0.5 times the maximum shear rate at the capillary wall and the shear rate in a rotational (tapered plug) viscometer.[2]

Shear rate in the concentric cylinder rotational viscometer is given simply by dividing the linear surface velocity of the cylinder by the film thickness. In this case, the liquid layer on each cylinder is stationary with respect to the wall and shear rate across the small annular space is constant under given operating conditions.

The cone and plate viscometer presents a similar relationship to a concentric cylinder for shear rate measurements as given in the following equation:

$$\gamma = \frac{\text{Linear velocity}}{\text{Film thickness}} = \frac{2\pi r N}{r \tan \theta} = \frac{2\pi N}{\tan \theta} \tag{5}$$

The angle θ in the cone and plate viscometer makes the shear rate independent of the radius. The cone and plate viscometer provides one unique feature. Alignment of the polymer molecules in the direction of flow resulting in reduced viscosity in that direction produces a normal force at right angles to this flow.

One problem common to all three types of viscometers is temperature control at high shear rates. As an example of this temperature effect in a capillary of 0.0856 cm diameter by 10.4 cm long with a flow rate that produced a pressure drop of 2000 psi (13.9 MPa), the temperature rose along the capillary wall from an inlet of 21°C (70°F) to 198°C (420°F) at the outlet. The temperature profile across the capillary flow at the exit showed a temperature of 70°F (21.1°C) at the center and 420°F (198°C) next to the capillary wall.[3] Despite these temperature problems, capillary viscometers have been used successfully for measuring viscosities of Newtonian and non-Newtonian fluids at high shear rates. The advantage of rotational viscometers for viscosity-shear measurements is constant shear rate across the viscous film; a disadvantage is difficult temperature control of the overall unit and temperature rise in the sheared film. Temperature rise is minimized by short test time.

VISCOSITY-TEMPERATURE RELATIONSHIPS

Eyring and Ewell[4] derived the following viscosity-temperature relationship from fundamental principles:

$$\eta = \frac{h\overline{N}}{\overline{V}} \; e^{\Delta E/RT} \tag{6}$$

where h = Planck's constant, $\overline{N}$ = Avagadro's no., $\overline{V}$ = molecular volume = molecular weight/density, ΔE = energy of activation for viscous flow per mole, R = gas constant, and T = temperature, K. A simplified version of this relationship has been derived by Andrade using constants A and b:

$$\mu = A \, e^{b/T} \tag{7}$$

These equations predict a straight line relationship between logarithm of the viscosity and the reciprocal of the temperature (1/T). This approximate relationship can be used only over very limited temperature ranges. The widely used Walther equation was derived from a large number of viscosities of mineral oil fractions:

$$\log \log (\nu + c) = a - b \log T \tag{8}$$

where a and b are constants for a given fluid and c varies with viscosity level (ASTM D341).

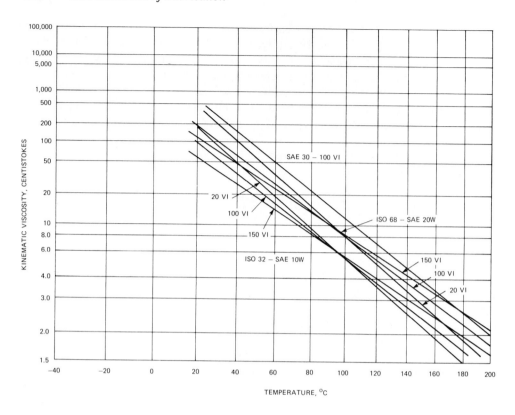

FIGURE 1. Viscosity-temperature properties of some commonly used oils.

For viscosities above 2.0 cSt, c is 0.7 for all viscosity-temperature charts available from ASTM. These charts cover the viscosity range from 1.5 to 20×10^6 cSt and temperatures from -70 to $370°C$ (-94 to $689°F$). Mathematical relationships for the viscosity-temperature charts are presented as appendixes to ASTM D341. Figure 1 shows Walther plots for a series of typical ISO and automotive lubricants.

Viscosities of mineral oils can be predicted from the Walther equation (ASTM chart) and two measured values over the temperature range of $350°F$ ($177°C$) down to $20°F$ ($11°C$) above the cloud point. In general, synthetics follow the Walther relationship over the range of -18 to $175°C$ (0 to $347°F$) to within 5%. Measured high temperature viscosities at 254 and $375°C$ (490 and $707°F$) in Table 1 show most synthetics to have a substantially lower viscosity than that predicted by the Walther equation. Aromatic structures with poor viscosity-temperature characteristics tend to show higher than predicted viscosity at high temperatures.

Low-temperature viscosities have been a problem to obtain by extrapolation of the Walther equation. Mineral oils at and below the cloud point show viscosity values substantially above the predicted value in Table 2. Addition of polymer to improve the viscosity-temperature properties results in higher than predicted viscosities at low temperatures. Silicones and polyglycolethers also show higher than predicted low temperature viscosities. Many esters, synthetic hydrocarbons, and low pour-point mineral oils exhibit low temperature viscosities substantially below Walther equation predictions. In general, these esters and hydrocarbons consist of molecules containing three or four alkyl groups attached to a central carbon atom or grouping.

The widely used expression for viscosity-temperature properties is the viscosity index. The 100 VI reference points consist of the viscosity-temperature properties of a series of refined Pennsylvania oil fractions starting with a neutral fraction of 4 cSt at $210°F$ ($98.9°C$).

Table 1
EXTRAPOLATED VISCOSITIES
AT ELEVATED TEMPERATURES

Test fluid	Measured centistoke viscosity at			% Dev. from extrapolated viscosity[a]	
	40°C	254°C	375°C	254°C	375°C
Methylphenyl silicone	44.4	3.25	1.46—1.37	−17	−36
Chlorinated silicone	59.9	4.52	2.2—2.0	−14	−25
Tetra-2-ethylhexyl silicate	6.79	0.59	—	−10	—
Pentaerythritol ester	22.4	0.83	0.42—0.59	−15	−27
Naphthenic bright stock	201	1.16	0.57—0.66	−4	−8
Naphthenic neutral	44.6	0.87	0.45	−11	−18
Chlorinated aromatic hydrocarbon	42.5	0.55	—	+15	—
Tricresyl phosphate	38.3	0.76	—	+3	—

[a]   Based on 40 and 100°C viscosities.

Table 2
EXTRAPOLATED VISCOSITIES AT LOW TEMPERATURES

Test fluid	ASTM cloud pt.(°C)	ASTM pour pt. (°C)	Measured centistokes viscosity at			% Deviation from extrapolation[a]	
			40°C	−18°C	−40°C	−18°C	−40°C
Naphthenic min. oil	<−23	−23	201	156,000	—	−2	—
Pa. neutral + pour depr.	−4	−23	40.4	9,550	—	+300	—
Dewaxed Pa. neutral + pour depr.	−14	−18	45.1	10,800	—	+210	—
	−14	−29	46.3	4,010	—	+38	—
Naphthenic min. oil + polymethacrylate	<−54	—	33.0	655	7,880	+61	+250
+ polybutene	<−54	—	33.3	817	10,700	+49	+170
Spec. MIL-H-5606 hydraulic fluid	<−54	—	14.2	100	483	+29	+110
Dicapryl phthalate	<−40	—	26.8	—	31,800	—	−56
Pentaerythritol ester	<−54	—	22.4	—	5,310	—	−21
Tri-N-butyl phosphate	<−46	—	2.66	—	46.9	—	−42
Di-N-hexyl-carbonate	<−40	—	2.93	—	59.1	—	−26
Tri-2-ethylhexyl borate	<−40	—	6.28	—	433	—	−12
Poly-α-olefin	<−54	—	20.9	—	44,000[b]	—	−45[b]
Superrefined min. oil	<−54	—	14.4	—	21,000[b]	—	−53[b]

[a]   Based on 40 and 100°C viscosities.
[b]   Values at −54°C.

Zero is assigned to a similar series of refined fractions taken from a Gulf coast crude oil. A change in the reference temperatures from 210°F (98.9°C) and 100°F (37.8°C) to 100°C and 40°C has had little or no effect on the VI of an oil. VIs for oils having values under 100 are calculated by the general formula:

$$VI = \frac{L - U}{L - H} \times 100 \qquad (9)$$

where U = 40°C viscosity of the unknown, L = 40°C viscosity of an oil of 0 VI with the same 100°C viscosity as U, and H = 40°C viscosity of an oil of 100 VI with the same 100°C viscosity as U. Values of H and L for viscosities of 2 cSt and above at 100°C are provided in ASTM D2270.

This method gives confusing results for VI values above 100. To make increasing VI values represent improved viscosity-temperature properties at high viscosity levels, an empirical fit was developed as shown in the following equation:

$$VI = [((antilog\ N) - 1)/0.00715] + 100 \qquad (10)$$

where N = (log H − log U)/log Y, and Y = viscosity in cSt at 100°C for the fluid of interest.

Several aromatic type materials, polyphenyl ethers, aryl phosphate esters and halogenated aromatic hydrocarbons have negative VI values.

The ASTM viscosity-temperature chart has another useful feature. The section between 0 and 100°C can be used as a blending chart for two components of different viscosities by considering the 0 to 100 scale to be weight percent of the viscous component. The viscosity of the less viscous component is plotted on the 0 line and the viscosity of the more viscous component on the 100 line; the straight line connection is a good representation of the viscosity values of any mixture. This chart cannot be used for blending with VI improvers.

VISCOSITY-PRESSURE RELATIONSHIPS

Viscosity of a liquid increases with decreasing temperature and with increasing pressure. In order to appreciate the range of viscosity-temperature and viscosity-pressure conditions, typical values should be considered. In lubrication and hydraulic systems, bulk conditions tend to range between atmospheric pressure and 10,000 psi (0.1 and 69 MPa) with temperatures from −65 to 300°F (−54 to 149°C). Hydrodynamic bearings tend to exhibit temperature rises of 100°F (55°C) or less and pressures of 10,000 psi (69 MPa) or less over bulk system values. For elastohydrodynamic (EHD) contacts in gears, cams and roller and ball bearings, the temperature may be 100 to 300°F (55 to 167°C) over bulk and pressures may be in the 50,000 to 500,000 psi (345 to 3450 MPa) range. Boundary lubrication implies temperatures of the order of 650°F (343°C) or higher and pressures in the same range as EHD contacts.

Viscosity-temperature and viscosity-pressure properties of synthetics provide a far wider spectrum, as shown in Table 3, than are available in mineral oil lubricants. The viscosity-pressure coefficient that determines the amount of fluid in an elastohydrodynamic bearing film (see Elastohydrodynamic Lubrication) is the value of α at or below 10,000 psi (69 MPa). Coefficient α is essentially constant over the range of 0 to 10,000 psi (0.1 to 69 MPa) for a wide variety of fluids.[5]

Four major classes of viscometers are used to measure viscosity as a function of pressure. Falling weight viscometers have been used to measure viscosities at high ambient pressures and low shear rates.[6,7] Capillary viscosities have been measured at low shear rates and high

Table 3
VISCOSITY-TEMPERATURE AND
VISCOSITY-PRESSURE PROPERTIES

All Fluids — 20 cSt at 40°C and 0.1 MPa

Lubricant type	cSt Viscosity at		cSt Viscosity (40°C) at pressure (MPa)			Viscosity index
	100°C	− 40°C	138	275.9	551.7	
Fluorolube	2.9	500,000	2,700	200,000	>1,000,000	− 132
Hydrocarbon	3.4	50,000	800	36,000	>1,000,000	0
Hydrocarbon	3.9	14,000	340	12,000	270,000	100
Ester	4.4	3,600	110	500	4,900	151
Polyglycol ether	4.6	7,000	120	570	8,800	164
Phosphate-base[a]	4.6	8,000	—	—	—	164
Ester-base[a]	6.3	1,000	—	—	—	197
Silicone	9.5	150	160	700	48,000	195

[a] Contains polymeric additive.

precision at pressures up to 10,000 psi (69 MPa).[5,8-10] High-pressure, high-shear capillary viscosities have been measured.[11-13] Vibrating crystal viscometers have been used for pressure-shear viscosities at moderate viscosity levels.[14] Optical viscometers designed as elasto-hydrodynamic bearings[15] have been used to measure relative viscosity and determine the temperature in the EHD film.[16,17]

A number of correlative equations have been suggested for viscosity-pressure properties.[18] Fresco et al.[5] and Klaus and So[10] correlations are applicable to all fluid types; and the Kim correlation[9] is designed specifically for polymer solutions. Johnston[19] reviewed the literature for viscosity-pressure relations and proposed a fundamental expression relating the viscosity-pressure coefficient to viscosity-temperature properties, compressibility, and thermal expansion coefficient. The best correlation from readily available lubricant properties appears to be given by the following:[10]

$$\alpha = 1.216 + 4.143 \, (\log \nu_o)^{3.0627} + 2.848 \times 10^{-4} \, b^{5.1903} \, (\log \nu_o)^{1.5976}$$
$$- 3.999 \, (\log \nu_o)^{3.0975} \rho^{0.1162} \tag{11}$$

where α = pressure-viscosity coefficient, $kPa^{-1} \times 10^{-5} = \partial(\log \nu)/\partial P$ at P = 0, $\log \nu_o$ = base 10 logarithm of atmospheric kinematic viscosity in centistoke at temperature of interest, b = viscosity-temperature property based on atmospheric kinematic viscosities at 37.8°C (100°F) and 98.9°C (210°F) in Equation 8, and ρ = atmospheric density in g/mℓ at temperature of interest.

Predicted viscosities at the Hertzian pressures existing in EHD and boundary lubrication indicate that the fluid film is very viscous and may in fact be a plastic or viscoelastic solid-like film. It is not clear how well developed the thermal and pressure effects are on the fluid in the milliseconds contact time. Winer et al.[20,21] has shown a correlation between glass and pseudoglass transition behavior based on lubricant viscosity in EHD bearing situations and at low temperatures at atmospheric pressure. Lubricant behavior in EHD and boundary lubrication is currently under extensive investigation.

VISCOSITY-SHEAR RELATIONSHIPS

Many premium hydraulic fluids and lubricants are mineral oil base stocks containing polymeric VI improvers. In addition, some synthetics are polymeric in nature, e.g., silicones, polyglycol ethers, polyesters, polyperfluoro ethers, etc. The presence of a polymer in the lubricant raises the possibility of two viscosity-related effects. First, under high shear rates and streamlined flow, the viscosity of the solution may be reduced reversibly. In the case of turbulent flow and extremely high shear rates, the polymer can be mechanically degraded and the viscosity of the solution lowered permanently. Characteristics of flow in the lubricant system determine the severity of mechanical degradation and, thereby, limits the size and effectiveness of polymeric additives that can be used.

Shear rates in lubricant applications range from low values to the order of 10^6 reciprocal seconds in various common lubricating systems. Another important area is the shear rate or shear stress for cold starting. In an operating automotive engine, oils of the order of 3 to 15 cP (0.003 to 0.015 Pa·sec) are subjected to 5×10^5/sec shear rate. At cold starting, oils of 3000 to 50,000 cP (3 to 50 Pa·sec) are subjected to shear rates of the order of 10^3 to 10^4/sec.

A typical polymer solution gives a characteristic behavior as shown in Figure 2. Viscosity remains constant up to a critical shear rate after which the viscosity falls linearly to a stable or second Newtonian zone. The greater this slope, $(n-1)$, the higher the molecular weight of the polymer in these "power law" fluids. Relative polymer size can be judged by the power law index η in the Ostwald de Waele equation $\eta = K \gamma^{n-1}$ where K is a constant typical of the polymer system. A plot of percent temporary viscosity loss vs. log of the shear rate also provides a straight line which, when extrapolated to lower shear rates, predicts the shear rate at which non-Newtonian behavior begins. With polymer molecular weights limited by mechanical viscosity loss, the second Newtonian zone appears to be greater than 10^6/sec.

Non-Newtonian lubricants may provide some advantages in a journal bearing. Studies have shown that a non-Newtonian lubricant can maintain the film thickness predicted by the low shear viscosity and show as much as a 40% friction reduction.[22] Non-Newtonian lubricants also tend to give lower than predicted friction in EHD bearings, possibly by partial starvation of the EHD contact. While polymer-containing lubricants lower friction and improve gas mileage in automotive engines, the specific mechanism responsible is not well defined.

Polymer Degradation

Polymers undergo permanent size reduction under the turbulence and cavitation involved in the valving system in pumps, relief valves in hydraulic systems, and all types of EHD and boundary lubrication.[23] Three types of test devices to evaluate mechanical breakdown in polymer solutions are (1) a pump system with an orifice or needle valve in the discharge line to create a pressure drop and severe cavitation, (2) an ultrasonic oscillator, and (3) a roller bearing rig to provide severe mechanical degradation with an EHD contact. A large number of cycles are required to achieve a final breakdown value. For a given pressure drop across an orifice, viscosity reduction will approach an asymptote after 5000 to 10,000 c. The breakdown is also a function of severity, but is surprisingly independent of the characteristics of all but the most severe unit in a system.

Recent studies have shown that the amount of mechanical degradation of a given polymer is a function of the initial molecular weight and either the pressure drop across an orifice type loading device or Hertzian pressure in EHD contacts, as illustrated in Figure 3.[24] The roller bearing data were determined in a tester comprising two loaded tapered roller bearings running at 3500 rpm with 15 mℓ of lubricant.[23] The mechanical breakdown appeared to be stepwise with an estimated nine successive molecular scissions for a 5000 nm molecule to the stable size of 5 to 8 nm. Polar polymers exhibit a lower initial rate of breakdown than do nonpolar polymers, indicating some reduction in mixing rate near the bearing surface.

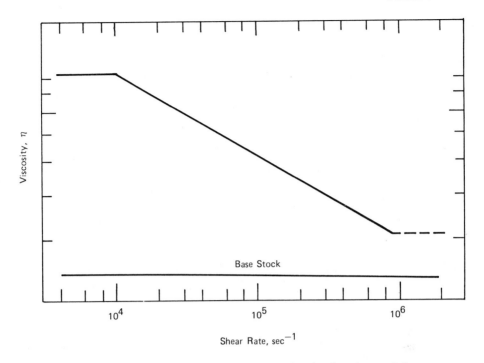

FIGURE 2.   Typical effect of shear rate on the viscosity of a polymer solution.

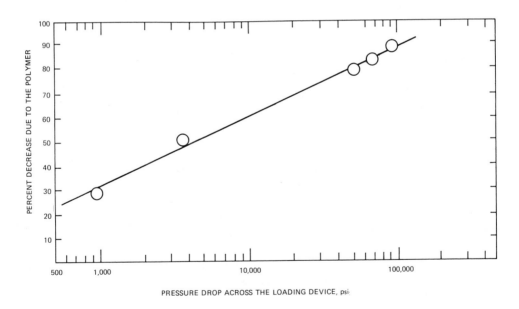

FIGURE 3.   Polymer breakdown of a polybutene solution as a function of pressure drop across the loading mechanism. Data below 5,000 are for orifice system. Data above 40,000 are for tapered roller bearings.

VAPOR PRESSURE

In the case of automotive crankcase oils, vapor pressure or volatility is significant in determining the rate of oil consumption and thereby limits the viscosity level of the base oil. The most commonly available properties of mineral oil fractions relating to vapor pressure

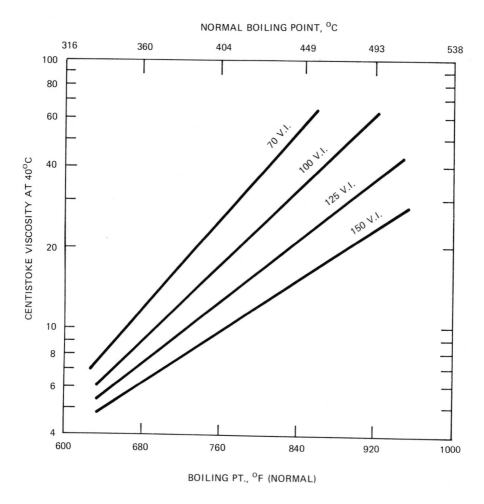

FIGURE 4.   Viscosity-volatility relationship.

are the Cleveland Open Cup flash and fire points (ASTM D92). The flash and fire points of a well-distilled petroleum fraction should differ by about 10°F (5.5°C)/100°F (55.5°C) of fire point. Thus, a flash point of 400°F (204°C) and a fire point of 440°F (227°C) would be expected of a typical lubricating oil fraction. A larger spread would indicate a relatively poor separation by distillation. As a rule of thumb, the fire point of a typical mineral oil fraction is approximately equal to the 20% boiling point at 10 mmHg (1.33 kPa) pressure. A careful measure of the boiling points for a typical mineral oil neutral fraction by temperature-programed gas chromatography indicates that a boiling range from the 5 to 95% boiling points of 150 to 170°C (302 to 338°F) is typical. Base oils for most industrial and automotive lubricants exhibit this range.

For a variety of synthetic compounds or narrow boiling range (30°C) mineral oil fractions, viscosity-boiling point properties are correlated with viscosity-temperature properties in Figure 4. Evaporation losses from a relatively thin film evaporation test give another useful measure of volatility.

The boiling point or boiling range of lubricant fractions or components can be measured using temperature programed gas chromatography (ASTM D2887) for boiling points up to 1000°F (538°C). One convenient method of converting the boiling point to vapor pressure or going from a vacuum fractionation to normal boiling points is the vapor pressure chart

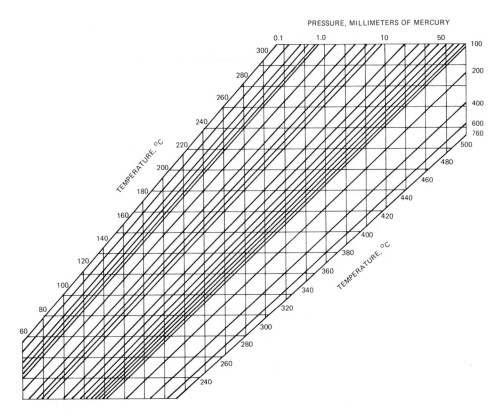

FIGURE 5.   Vapor pressure chart for hydrocarbons.

for hydrocarbons shown in Figure 5.[25] This relationship was generated using hydrocarbons and esters of organic acids as single compounds. The figure can also be used to convert a 10 or 50% normal boiling point to reduced boiling points or vapor pressures. A good method of predicting vapor pressures of lubricants down to 10^{-6} mmHg (1.33×10^{-4} Pa) is given by Beerbower and Zudkevitch.[26]

Vapor pressure of a typical mineral oil lubricant is influenced strongly by its more volatile components. Thus, in a lubricant with a 150°C (302°F) boiling range, the 5 to 20% boiling points are the most important in establishing a vapor pressure for the system.

In addition to oil consumption, evaporation, and safety (flammability), volatility plays a role in boundary lubrication. There is evidence[27,28] that lubricants with high volatilities cause higher wear in systems than do lubricants with matched viscosities and fluid types of lower volatility levels.

DENSITY

Specific gravity is defined as the ratio of the weight of a given volume of product at 60°F (15.6°C) to the weight of an equal volume of water at the same temperature. The petroleum industry has modified the Baume scale to provide an API gravity defined by the equation:

$$\text{degrees API} = \frac{141.5}{\text{specific gravity } 60/60°F\ (15.6/15.6°C)} - 131.5 \quad (12)$$

A high API gravity value matches a low specific gravity and vice versa. Tables are available for conversion of density or gravity measurements at any temperature between 0°F ($-17.8°C$)

Table 4
COEFFICIENT OF EXPANSION
FOR MINERAL OIL LUBRICANTS
ESTIMATED FROM ASTM TABLES

Specific gravity at 60°F (15.6°C)	API gravity at 60°F (15.6°C)	Coefficient of expansion	
		Per °F	Per °C
1.076—0.967	0—14.9	0.00035	0.00063
0.966—0.850	15—34.9	0.00040	0.00072
0.850—0.776	35—50.9	0.00050	0.00090
0.775—0.742	51—63.9	0.00060	0.00108

and 500°F (260°C) to the standard conditions of 60°F (15.6°C) (ASTM D1250). Density change with temperature (coefficient of thermal expansion) is more sensitive to the boiling point of the hydrocarbon fraction than to its density, although both independent variables are necessary to correlate the data properly.[29] For mineral oil lubricants an engineering approximation for the coefficient of expansion is summarized in Table 4. The chapter on ''Lubricant Properties and Test Methods'' in Volume I gives typical densities of commercial lubricants.

A similar straight line relationship exists between temperature and density over the range of 0 to 500°F (−17.8 to 260°C) for high-boiling synthetic lubricants. In addition to its usual engineering applications, density often offers a simple way of identifying specific lubricants. In petroleum and hydrocarbon-based lubricants, gravity can aid in distinguishing among paraffinic, naphthenic, and aromatic structures in the lubricant base oil (ASTM D3238).

Lubricant compressibility is usually expressed as bulk modulus which is defined by the equation:

$$\bar{B} = \frac{(P - P_o)V_o}{V_o - V} \tag{13}$$

where $\bar{B}$ = isothermal secant bulk modulus, psi (Pa), P = pressure of measurement, psi (Pa), P_o = atmospheric pressure, 0 psi (101.3 kPa), V_o = relative volume at P_o, and V = relative volume at P.

BULK MODULUS

Bulk modulus expresses the resistance of a fluid to compression (reciprocal of compressibility). This property, which varies with pressure, temperature, and molecular structure, is significant in (1) hydraulic and servosystem efficiencies and response time, (2) resonance and water-hammer effects in pressurized-fluid systems, (3) explanation of viscosity-pressure properties in hydrodynamic and EHD lubricants, and (4) in thermodynamic considerations of liquids.

Two general methods used to measure bulk modulus are (1) pressure-volume-temperature determination of density or density change directly, and (2) velocity of sound in a liquid at the desired temperature and pressure. The former method provides isothermal secant bulk modulus or average values over a pressure range. Tangent bulk modulus or bulk modulus for a specific pressure is obtained by differentiation from the secant data. Velocity of sound measurements provide adiabatic tangent bulk modulus values.

Klaus and O'Brien[30] measured the isothermal secant bulk modulus for a variety of lubricants over the range of 0 to 10,000 psi (0.101 to 69 MPa). For engineering accuracy, the isothermal secant bulk modulus, $\overline{B}$, can be converted to an isothermal tangent bulk modulus, B, in accordance with the relationship:

$$\tan B_p \cong \text{secant } \overline{B}_{2p} \tag{14}$$

Isothermal and adiabatic tangent bulk modulus are related by the equation:

$$\frac{B_s}{B_r} = \gamma = \frac{C_p}{C_v} \tag{15}$$

where B_s = adiabatic tangent bulk modulus, B_r = isothermal tangent bulk modulus, γ = ratio of bulk moduli or specific heats, C_p = specific heat at constant pressure, and C_v = specific heat at constant volume.

Wright[31] proposed a useful method for predicting isothermal secant bulk modulus values for mineral oils based on Figures 6 and 7. Figure 6 shows the relationship between $\overline{B}$ and temperature at 20,000 psi (138 MPa) as a function of fluid density at atmospheric pressure. Figure 7 shows a relationship between isothermal secant bulk modulus and pressure. These relationships work well for mineral oil base stocks and formulated lubricants, organic acid esters, synthetic hydrocarbons, and phenyl ethers. Both silicones and perfluoropolyethers show a relatively low bulk modulus (high compressibility) based on a density correlation. Bulk modulus is a physical property of the base fluid which cannot be changed significantly by additives.

Entrained air (or other gas) in a hydraulic system being pumped at high pressure shows two deleterious effects on system response. First, any entrained air dissolves upon raising the pressure, causing a greater volume reduction than the compressibility of the original fluid. Secondly, the gas-saturated fluid is somewhat more compressible than the same fluid with only air saturation at atmospheric pressure. Air saturation at atmospheric pressure does not measurably change $\overline{B}$ over that of a degassed fluid.

GAS SOLUBILITY

Solubility of gases in lubricants is a physical property that in turn affects related lubricant properties such as viscosity, foaming, bulk modulus, cavitation, heat transfer, oxidation, and boundary lubrication. In many cases, gas is entrained at low pressures and then dissolved in the high-pressure portion of lubrication and hydraulic systems. As the pressure is again reduced to that in the reservoir or sump, the gas comes out of solution to produce foam or just entrained gas bubbles. The dissolved oxygen, in the case of air, can also react with the lubricant as the temperature in bearings or hot portions of the system reaches the threshold of the oxidation reaction.

Gas solubility can be measured with precision at temperatures up to 260°C in a gas chromatograph (GC) with a precolumn of solid adsorbent to remove the liquid which contains the gas.[32] The experimental data can be plotted as a straight line of log gas dissolved vs. 1/temperature K. As the molecular weight of the gas increases, the rate of increase in gas solubility with temperature rise drops off. At a molecular weight of about 32 (oxygen), change in gas solubility with temperature is small. At higher molecular weights, e.g., CO_2, gas solubility decreases with increasing temperature. At fluid temperatures where the vapor pressure of the liquid is 60 mmHg (8 kPa) or above, gas solubility falls below levels predicted from lower temperatures. At the normal boiling point, gas solubility drops to zero. Small amounts of volatile products in the lubricant can have the same effect as a more volatile base oil and result in reduced gas solubility. With gas mixtures, solubility of individual gases follows the partial pressure of the gas in the mixture.

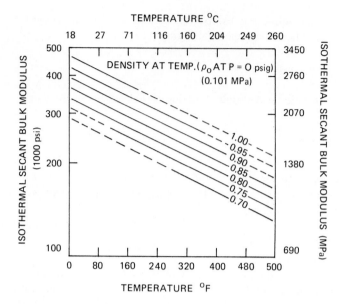

FIGURE 6.   Secant bulk modulus vs. temperature for petroleum oils at 20,000 psi (138 MPa).

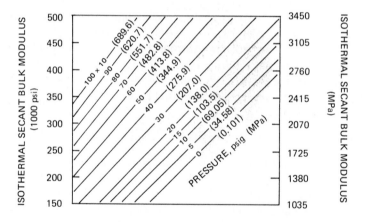

FIGURE 7.   Secant bulk modulus vs. pressure for petroleum oils. Enter chart with modulus at 20,000 psi, at desired temperature. A vertical line at the intersection with 20,000 psi line gives modulus at any other pressure and the selected temperature.

Solubility of gases at atmospheric pressure follows a semiempirical correlation proposed by Beerbower.[33]

$$\ln L = [0.0395\,(S_1 - S_2)^2 - 2.66]\,(1 - 273/T) + 0.303 S_1$$
$$- 0.0241\,(8.60 - S_2)^2 + 5.731 \tag{16}$$

where L = Ostwald coefficient at desired temperature T (degrees K) in desired liquid, and S_1 = solubility parameter of liquid at 298 K (MPa$^{1/2}$). For mineral oils $S_1 = 8.63\, n_D^{20} + 0.96$ where n_D^{20} is the refractive index of the D sodium line at 20°C. S_2 = solubility parameter of the hypothetical liquefied gas at 298 K (MPa$^{1/2}$).

Table 5
GAS SOLUBILITY PARAMETERS

Gas	Ostwald coefficient L at 0°C[a]	Solubility parameter S_2 for Equation 16, MPa$^{1/2}$
Helium	0.012	3.35
Neon	0.018	3.87
Hydrogen	0.040	5.52
Nitrogen	0.069	6.04
Air	0.098	6.69
Carbon monoxide	0.12	7.47
Oxygen	0.16	7.75
Argon	0.18	7.77
Methane	0.31	9.10
Krypton	0.60	10.34
Carbon dioxide	1.45	14.81
Ammonia	1.7	
Ethylene	2.0	
Xenon	3.3	
Hydrogen sulfide	5.0	

[a] L applied only to petroleum liquids of 0.85 kg/dm^3 density, d, at 15°C. To correct the other densities, L_c = 7.70L(0.980 − d) (see ASTM D2779 for details).

Gas parameters to use in this equation are given on Table 5. The Ostwald coefficient is the equilibrium volume of gas dissolved in a unit volume of oil. This coefficient can be used directly for many engineering approximations below 5 atm pressure and 373 K (100°C). Solubility of air is, for instance, about 9.8% by volume in petroleum oils under conditions encountered in lubrication systems. The weight solubility of air at 2 atm is then double the solubility at 1 atm for a given temperature. Liquid solubility parameter, S_1, is approximately 18.0 for diesters commonly used in aircraft fluids, 18.5 to 19.0 for higher esters, 18.41 for methyl phenyl silicone, 15.14 for dimethyl silicone, 18.29 for tri-2-ethylhexyl phosphate, and 18.82 for tricresyl phosphate.[33]

In cases where thin films of lubricant are exposed to gases at high pressures, the gases dissolve rapidly. The resulting fluid can show a dramatic reduction in viscosity. Typical viscosity effects are shown on Table 6.[18] In general, the effectiveness of dissolved nitrogen in reducing viscosity negates the normal augmenting effect of pressure on viscosity.

FOAMING AND AIR ENTRAINMENT

Tendency to foam generally increases with increasing fluid molecular size, increasing viscosity, or decreasing temperature. Foaming is caused by the escape of insoluble gases or the physical mixing of excess gas with the fluid. The best way to minimize foam is with mechanical design. The chemical approach to reducing foaming is the use of a silicone additive that tends to lower surface tension at gas-liquid interfaces.

Air entrainment is similar to the problems of foam. In hydraulic systems, air entrainment can result in response problems, while in gear systems air entrainment can result in reduced heat transfer and higher operating temperatures. Antifoam additives are not necessarily helpful; several commercial additives are available to improve air entrainment characteristics.

THERMAL PROPERTIES

Thermal properties of lubricants are involved in considering heat transfer, temperature

Table 6
EFFECT OF DISSOLVED GASES ON VISCOSITY OF LUBRICANTS AT
1000 PSI (7.0 MPa) PRESSURE

	Gas-free viscosity, cSt (mm²/sec) at 100°F (38°C)	Gas-saturated viscosity at 100°F (38°C)		
Fluid		**He saturated**	**N₂ saturated**	**CO₂ saturated**[a]
Diester	14.9		12.9	
Paraffinic neutral	15.4	14.0	13.1	4.0
Paraffinic bright stock	625		470	
Polyester	630		515	
Naphthenic oil	3980		2620	
Paraffinic resin	5400	4600	3800	1500

[a]   Saturated at 500 psi (3.55 MPa).

Table 7
THERMAL CONDUCTIVITIES AND SPECIFIC
HEAT VALUES FOR SEVERAL MATERIALS

Material	**Thermal conductivity Btu/hr/ft²/F/Ft at 212°F (W/m·K at 373 K)**	**Specific heat Btu/lb °F·68°F (J/kg·K 293 K)**
Steel	27 (46.7)	0.11 (460)
Oil	0.08 (0.14)	0.47 (1966)
Water	0.39 (0.67)	1.0 (4184)

rise, and related thermal performance factors for machine elements, coolers, heaters, and lubrication systems. Typical thermal conductivities and specific heats for the mineral oil, water and steel commonly encountered in lubrication systems are given in Table 7. More specific values for mineral oils can be found in the *API Data Book for Hydrocarbons*.

Thermal Conductivity

Thermal conductivity changes linearly with temperature. Based on the data presented by Maxwell,[29] thermal conductivity of liquid hydrocarbons ranges from 0.14 W/m·K at 0°C (273 K) to 0.11 W/m·K at 400°C (673 K). These values should hold for essentially all mineral oil and synthetic hydrocarbon base lubricants. Polarity and hydrogen bonding affect thermal conductivity. For example, ethylene glycol has a thermal conductivity value at 100°C of 0.31 W/m·K.

Specific Heat

Like thermal conductivity, specific heat varies linearly with temperature and shows significant increases with increased polarity or hydrogen bonding of the molecules. Specific heat of water is about twice that for oil at 100°C in Table 7. Ethylene glycol has a specific heat of 2384 J/kg·K at 273 K. For mineral oil and synthetic hydrocarbon base lubricants, specific heat values range from 0.45 Btu/lb·F (1882 J/kg·K) at 0°C (273 K) to 0.78 Btu/lb·F (3263 J/kg·K) at 400°C (673 K).

Heat of Vaporization

Heat of vaporization is the difference in enthalpy or energy content between the saturated vapor and saturated liquid of a material at constant temperature. Heat of vaporization is a function of pressure and the molecular weight or boiling point of a mineral oil. Most conventional mineral oil base lubricants have latent heats of vaporization between 60 and 90 Btu/lb (140 to 209 kJ/kg) at atmospheric pressure. The heat of vaporization at the boiling temperature decreases with the increasing pressure (and increasing boiling point). A typical mineral oil, e.g., ISO grade 68, exhibits the following values.

Pressure		Latent heat of vaporization	
atm	kPa	Btu/lb	kJ/kg
0.013	1.3	100	233.0
1.0	101.3	75	174.0
10.0	1013.0	5	11.6

By comparison, heat of vaporization of water is 969.7 Btu/lb (2255 kJ/kg) at atmospheric pressure. Much less heat is required for evaporation of oil compared with water under conditions where these fluids exhibit the same vapor pressure.

Electrical Conductivity

Electrical conductivity can generally be related to relative amounts of ions and ion-forming materials in the lubricant. In water base lubricants, the suppression of electrical conductivity is desirable since metallic corrosion is related to electrical activity. Metal corrosion products also tend to add to further degradation of the lubricant.

Electrical conductivity of well-refined and dry mineral oil and most synthetic lubricant base stocks is extremely small, in the 10^{-14} mho/cm^2 range. In some hydraulic control systems, streaming potential or zeta potential has been related to corrosion in servo valves. It appears that alteration of the electrical conductivity of the base oil is primarily due to the ionic nature of many additives, impurities such as water and chlorides, or oxidative or thermal degradation of the base stock. Electrical conductance of the unused lubricant is considered critical primarily in such applications as electrical equipment and in some aircraft and industrial control systems where streaming currents have caused damage.

Surface Tension

Surface and interfacial tension are related to free energy at a surface. Surface tension is the manifestation of this surface free energy at a gas-liquid interface, while interfacial tension exists at an interface between two immiscible liquids (ASTM D971). Surface tension can be measured by the du Nouy ring method. In this procedure a platinum wire ring is placed in contact with the clean surface of the liquid and the force F required to pull the ring away from the surface is measured.

$$F = 4\pi r\sigma \tag{17}$$

where F = force in dyn/cm^2 (N/m^2), r = radius of the platinum ring, and σ = surface tension in dyn/cm (N/m).

The surface tension of several base oils is shown in Table 8. Surface tension for the finished lubricant is surprisingly sensitive to additives. For example, less than 0.1 wt% of a silicone in a mineral oil will reduce the surface tension to essentially that of the silicone. The use of additives to lower surface tension does improve surface wetting of the bearing.

Interfacial tension between two immiscible liquids is approximately the difference between the surface tensions of the two liquids. Additives that create stable emulsions and micro-emulsions are capable of reducing the interfacial tension between the two phases to very low values approaching zero. The preferred method of measuring low interfacial tensions

Table 8
SURFACE TENSION OF
SEVERAL BASE OILS

Liquid	Surface tension dyn/cm (N/m)
Water	72 ($\times$ 10^{-3})
Mineral oils	30—35 ($\times$ 10^{-3})
Esters	30—35 ($\times$ 10^{-3})
Methylsilicone	20—22 ($\times$ 10^{-3})
Fluorochloro compounds	15—18 ($\times$ 10^{-3})

spinning drop apparatus.[34] Relationships in this unit are expressed in the following equation.

$$\sigma = K(\rho_1 - \rho_2)(d)^3 / \theta^2 \tag{18}$$

where ρ_1 = density of the heavy phase, ρ_2 = density of the light phase, d = drop width, θ = rotational speed, and K = constant characteristic of the test unit.

With this unit interfacial tensions down to 10^{-4} to 10^{-5} dyn/cm (10^{-7} to 10^{-8} N/m) can be measured. The detergents and dispersants in many automotive lubricants are so effective at reducing interfacial tension that used crankcase oils contaminanted with 10 to 15% water form a stable emulsion which defies separation by techniques which do not involve distillation. These general techniques can be used to suspend graphite in motor oil, disperse calcium carbonate in over-based diesel lubes, or prepare a 95% water invert emulsion. Both surface tension and interfacial tension are altered by additives and by lubricant degradation.

THERMAL STABILITY

Thermal stability is the resistance of the lubricant to either molecular breakdown or rearrangement at elevated temperatures in the absence of oxygen. Stability in an ordinary air environment (oxidation stability) is covered in the next section.

One method of measuring thermal stability involves the isoteniscope, a closed vessel with a manometer for measuring the rate of pressure increase at a specified heating rate. Thermal gravimetric and differential thermal analyses can also be used to evaluate thermal stability. Several thermal stability tests are described in Federal Specifications.[35-37] The test should allow for decomposition of a significant portion of the test sample and provide an analysis of the liquid and solid decomposition products as well as the gases formed.[37]

Fluids such as mineral oils with a substantial percentage of C–C single bonds as the most vulnerable point for breakdown exhibit a thermal stability of about 650 to 700°F (343 to 371°C). Synthetic hydrocarbons prepared by a polymerization or aligomerization process and then hydrogenated involve the same basic structures as mineral oils, but exhibit a thermal stability of 50°F (28°C) or more below that of a mineral oil. In thermal breakdown, a mineral oil produces more moles of methane than of ethane and ethylene. That is, the molar quantities of the thermal decomposition product tend to decrease continuously with increasing molecular weight. A synthetic hydrocarbon made by polymerization will produce a significant quantity of the monomer from which it was made as a telltale fingerprint.

Molecules containing only aromatic linkages or aromatic linkages with methyl groups as side chains show a thermal threshold of the order of 850 to 900°F (454 to 482°C). Polyphenyl ethers, chlorinated biphenyls, and condensed ring aromatic hydrocarbons fall in this category.

With organic acid esters the functional group is the weak link in the molecule, and thermal stabilities range from 500 to 600°F (260 to 316°F). The presence of metals such as iron in

Table 9
SPECTRAL DATA OBTAINED FOR VARIOUS LUBRICANTS

Lubricant	Wavelength λ(nm)	Extinction coefficient (ℓ/g-cm)	Wavelength λ(nm)	Extinction coefficient (ℓ/g-cm)
DEHS (orig.)	~280	NA	~220	NA
DEHS (HMW)	277	7.19	219	13.69
MLO 7558 (HMW)	278	11.65	225	17.62
MLO 7219 (HMW)	275	48.47	223	73.18
MLO 7828 (HMW)	277	14.00	226	17.14
TMPTH (HMW)	~280	A	~220	A
TDP (HMW)	~280	A	~220	A

Note: DEHS — di-2-ethylhexyl sebacate, HMW — high molecular weight oxidation product, NA — no absorption at this wavelength, A — absorbs at this wavelength, but extinction coefficient not reported, MLO 7558 — paraffinic white oil, MLO 7828 — naphthenic white oil, and MLO 7219 — partially hydrogenated aromatic stock.

a nitrogen atmosphere tend to push the thermal stability limit of the common dibasic acid esters and polyol esters toward the low end of this range. An all-glass system[35] produces a thermal stability advantage for the polyol esters that is probably not reflected in use in a lubrication system. Methyl esters have thermal stability levels about the same as those of mineral oil.

Polymers used as VI improvers tend to have thermal stability thresholds that are lower than smaller molecules of the same general structure. Polymethacrylates show thermal breakdown at 450°F (232°C) and polybutenes at 550°F (288°C). In both cases, thermal breakdown is distinctly different from mechanical degradation.[38]

Additives used for lubrication improvement tend to have thermal stability limits below those of base oils. Zinc dialkyldithiophosphates used to improve boundary lubrication properties show thermal degradation at 400 to 500°F (204 to 260°C). Generally, the more active the EP additive, the lower the thermal stability threshold.

OXIDATION STABILITY

Stability of a lubricant in the presence of air or oxygen is commonly its most important chemical property. Unlike thermal stability, oxidation stability can be altered significantly. Additives control oxidation by attacking the hydroperoxides formed in the initial oxidation step or by breaking the chain reaction mechanism. Aromatic amines, hindered phenols, and alkyl sulfides are compounds that provide oxidation protection by one of these mechanisms. A third type of oxidation control involves metal deactivators that can keep metal surfaces and soluble metal salts from catalyzing the condensation polymerization reactions of oxidized products to produce sludge and varnish.

A number of bulk oxidation tests are described in the ASTM (D2272, D1313) and Federal Test Method Standards No. 791, Method No. 5308. These tests are good for measuring stable life or the effectiveness of oxidation inhibitors. Oxygen diffusion limits the value of these tests in correlations with many actual lubrication systems.

The first step in oxidation of hydrocarbons is formation of a peroxide at the most vulnerable carbon-hydrogen bonds. This initiates a free radical chain mechanism which propagates formation of hydroperoxides. Further oxidation leads to other oxygen-containing molecules such as aldehydes, ketones, alcohols, acids, and esters. A similar peroxide path of oxidation has been shown for dibasic acid esters and polyol esters.

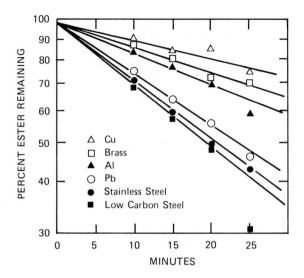

FIGURE 8. Oxidation of trimethylolpropane triheptanoate at 498 K.

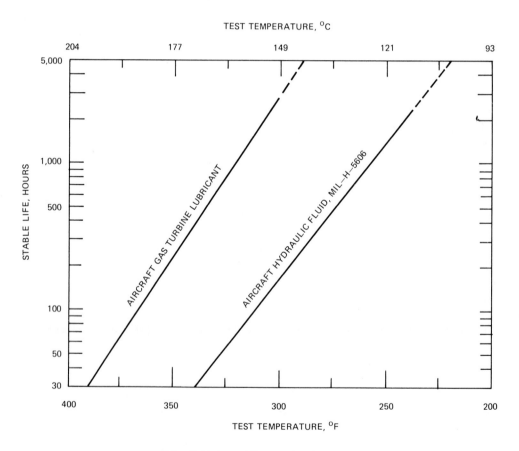

FIGURE 9. Oxidation stability as a function of temperature.

To monitor the oxidation process, a microoxidation test has been developed along with analytical procedures based on gel permeation chromatography (GPC) and atomic absorption spectroscopy (AAS).[39] In these tests, oxidations were carried out until 50% or more of the

Table 10
INTERNATIONAL
ORGANIZATION FOR
STANDARDIZATION (ISO)
VISCOSITY
CLASSIFICATION SYSTEM
FOR INDUSTRIAL FLUID
LUBRICANTS

ISO viscosity grade numbers	Viscosity grade ranges (cSt at 40°C)	
	Min	Max
2	1.98	2.42
3	2.88	3.52
5	4.14	5.06
7	6.12	7.48
10	9.00	11.0
15	13.5	16.5
22	19.8	24.2
32	28.8	35.2
46	41.4	50.6
68	61.2	74.8
100	90.0	110
150	135	165
220	198	242
320	288	252
460	414	506
680	612	748
1,000	900	1,100
1,500	1,350	1,650

Note: The viscosity grade numbers for the ISO System are identical to those shown for the ANSI/ASTM system (ASTM D 2422, ISO 3448 — 1975).

original base oil was oxidized. The large molecules separated by GPC are found to be rich in metal corrosion products. These large molecular size products appear to be condensation polymers with a characteristic beta keto conjugated unsaturation (−C=C−C−) which can be found in oxidation products from dibasic acid esters, polyol esters, monoesters, and mineral oils. These fluids all show oxidation products with the same general UV absorption patterns as shown in Table 9. In Figure 8 the rates of oxidation for the same polyol ester show that a copper catalyst has an inhibiting effect, while lead and iron accelerate the primary oxidation rate.

The effect of temperatures on stable life of lubricants is illustrated in Figure 9. This extrapolation system relating log of life to temperature provides a design guideline for the limiting bulk lubricant temperatures in a system.

LUBRICATION SPECIFICATIONS

Several widely used specifications include SAE engine oil grades, SAE gear lubrication grades, ASTM/International Organization for Standardization (ISO) grades for industrial

Table 11
TYPICAL MILITARY SPECIFICATIONS FOR HYDRAULIC FLUIDS AND LUBRICANTS

Properties	Specification designation					
	MIL-H-27601	MIL-H-83282	MIL-L-6387	MIL-L-7808	MIL-L-23699
cSt viscosity at					
98.9°C (219°F)	3.2 (min)	3.5 (min)	4.5 (min)	3.0 (min)	5.0—5.5 (min-max)
54.4°C (130°F)	—	—	10.0 (min)	—	
37.8°C (100°F)	—	16.5 (min)	—	11.0 (min)	25.0 (min)
−40°C (−40°F)	4,000 (max)	2,800 (max)	1,500 (max)	—	13,000 (max)
−54°C (−65°F)	—	—	7,500 (max)	13,000 (max)	—
Viscosity index	89 (min)	—	—	140	130
COC flash point (°C)	182 (min)	202 (min)	177 (min)	205 (min)	246 (min)
Pour point (°C)	−54 (max)	−54 (max)	−60 (max)	−60 (max)	−54 (max)
Total acid no.	0.20 (max)	0.10 (max)	0.2 (max)	—	0.05 (max)

Note: MIL-H-27601 — Hydraulic fluid, petroleum base, high temperature, flight vehicle, MIL-H-83282 — Hydraulic fluid, fire resistant synthetic hydrocarbon base, aircraft, MIL-L-6387 — Lubricating oil, synthetic base, MIL-L-7808 — Lubricating oil, gas turbine, aircraft, and MIL-L-23699 — Lubricating oil, aircraft turbine engine, synthetic base.

Table 12
PHYSICAL PROPERTIES OF SEVERAL FLUIDS

Fluid	cSt Viscosity at 40°C	Viscosity index	Density (g/cm³ at 15°C)	Pour point (°C)	Flash point (°C)
Mineral oils					
Naphthenic	17	70	0.876	−43	177
Paraffinic	17	100	0.861	−18	190
Superrefined	14	92	0.842	−60	204
Polymer-thickened	15	226	0.827	<−62	107
Polyolefin	16	121	0.816	<−62	204
Esters					
Dibasic acid	13	153	0.915	<−54	227
Polyol	15	124	0.960	<−54	252
Tricresyl phosphate	38	−67	1.180	−23	235
Tributyl phosphate	2.7	90	0.970	<−46	171
Compounded phosphate	11	239	1.070	<−60	182
Polyglycol ether	35	148	0.980	−46	210
Silicate	6.8	151	0.910	<−60	188
Silicones					
Dimethyl	79	400	0.970	<−60	260
Phenyl methyl	74	360	0.990	<−60	260
Chlorophenyl methyl	56	164	1.030	<−60	288
Alkyl methyl	80	200	0.910	<−60	260
Perfluoro methyl	44	158	1.210	<−60	—
Phenyl ether					
4P-3E	75	−20	1.180	9.5	263
5P-4E	355	−74	1.210	4.4	343
Perfluoro polyether					
Low viscosity	36	90	1.880	−48	—
High viscosity	287	115	1.870	−26	—

Table 13
PROPERTIES OF TYPICAL SAE GRADE LUBRICANTS

SAE grade	ISO grade	cSt Viscosity 100°C	cSt Viscosity 40°C	VI	Density (g/cm³ at 15°C)	Visc-press coeff. at 40°C Psi⁻¹ × 10⁴	Visc-press coeff. at 40°C KPa⁻¹ × 10⁵
10W	32	5.57	32.6	107	0.871	1.58	2.29
20W	68	8.81	62.3	118	0.873	1.71	2.48
30	100	11.9	100	110	0.875	1.85	2.68
40	150	14.7	140	102	0.901	1.84	2.67
5W—20	46	6.92	38.0	140	0.864	1.50	2.17
10W—30	68	10.2	66.4	135	0.870	1.63	2.36
10W—40	100	14.4	77.1	193	0.864	1.55	2.25
10W—50	—	20.5	117	194	0.874	1.61	2.34

Note: For automotive oil specifications, see "Automobile Engines" and subsequent chapters in Volume I.

fluid lubricants, and military specifications. Examples of these standards and classifications are shown in Tables 10 and 11 and in pertinent chapters of Volume I. These specifications define the lubricants in terms of physical properties and in some cases, particularly the military specifications, with respect to oxidation stability, thermal behavior, and wear characteristics.

General specifications for a fluid type do not imply that all fluids meeting the requirements are of equal quality. Relative quality must be determined by the ultimate user in his particular application. A summary of some properites for several classes of fluids with potential use in the formulation of lubricants is shown in Table 12. Properties of some typical SAE grade lubricants are shown in Table 13. Characteristics of a variety of commercial lubricants are also provided in the chapter on ''Lubricant Properties and Test Methods'' in Volume I.

NOMENCLATURE

$\overline{B}$ = Isothermal secant bulk modulus
B_s = Adiabatic bulk modulus
B_r = Isothermal tangent bulk modulus
ΔE = Energy of activation
F = Force
h = Planck's constant
L = Length
ℓ = Depth
N = Rotational speed
$\overline{N}$ = Avagadro's No.
n = Power law index
n_D^{20} = Refractive index
P = Pressure
R = Gas constant
r = Radius
T = Temperature
t = Time
V = Volume
$\overline{V}$ = Molecular volume
VI = Viscosity index
α = Viscosity-pressure coefficient
γ = Shear rate
η = Viscosity in centipoise
θ = Angle
ν = Viscosity in centistokes
ρ = Fluid density
σ = Surface tension; interfacial tension
τ = Torque
ω = Angular velocity

REFERENCES

1. **Fredrickson, A. G.,** *Principles and Applications of Rheology,* Prentice-Hall, Englewood Cliffs, N.J., 1964, 118.
2. **Fenske, M. R., Klaus, E. E., and Dannenbrink, R. W.,** The comparison of viscosity-shear data obtained with the Kingsbury tapered plug viscometer and the PRL high shear capillary viscometer, Special Tech. Publ. No. 111, *Symposium on Methods of Measuring Viscosity at High Rates of Shear,* Tech. Publ. 111, American Society for Testing and Materials, Philadelphia, Pa., 1950, 45.
3. **Gerrard, J. E., Steidler, F. E., and Appeldoorn, J. K.,** Viscous heating in capillaries, *Ind. Eng. Chem. Found.,* 4, 332, 1965; 5, 260, 1966.
4. **Ewell, R. H. and Eyring, H. J.,** *Chem. Phys.,* 5, 726, 1937.
5. **Fresco, G. P., Klaus, E. E., and Tewksburg, E. J.,** Measurement and prediction of viscosity-pressure characteristics of liquids, *J. Lubr. Tech., Trans. ASME,* 91, 454, 1969.
6. **Kuss, E.,** The Viscosities of 50 Lubricating Oils Under Pressures up to 2000 Atmospheres, Rep. No. 17 on Sponsored Res., (Germany), Department of Scientific and Industrial Research, London, 1951.
7. **ASME,** *Pressure-Viscosity Report,* American Society of Mechanical Engineers, New York, 1953.
8. **Klaus, E. E., Johnson, R. H., and Fresco, G. P.,** Development of a precision capillary-type pressure viscometer, *ASLE Trans.,* 9, 113, 1966.
9. **Kim, H. W.,** Viscosity-Pressure Studies of Polymer Solutions, Ph.D. thesis, Pennslyvania State University, University Park, Pa., 1970.
10. **So, B. Y. C. and Klaus, E. E.,** Viscosity-pressure correlation of liquids, *ASLE Trans.,* 23, 409, 1980.
11. **Jones, W. R., Johnson, R. L., Sanborn, D. M., and Winer, W. O.,** Viscosity-pressure measurements for several lubricants to 5.5×10^8 N/m² (8×10^4 psi), and 149°C (300°F), *Trans. ASLE,* 18, 249, 1975.
12. **Novak, J. and Winer, W. O.,** Some measurements of high pressure lubricant rheology, *J. Lubr. Technol. Trans. ASME,* 90, 580, 1968.
13. **Jakobsen, J., Sanborn, D. M., and Winer, W. O.,** Pressure-viscosity characteristics of a series of siloxanes, *J. Lubr. Technol., Trans. ASME,* 96, 410, 1974.
14. **Appledoorn, J. K., Okrent, E. H., and Philippoff, W.,** Viscosity and elasticity at high pressures and high shear rates, *Proc. Am. Pet. Inst.,* 42(3), 1962.
15. **Foord, C. A., Wedeven, L. D., Westlake, F. J., and Cameron, A.,** Optical elastohydrodynamics, *Proc. Inst. Mech. Eng.,* 184, 487, 1969/1970.
16. **Nagaraj, H. S., Sanborn, D. M., and Winer, W. O.,** Surface temperature measurements in rolling and sliding EHD contacts, *ASLE Trans.,* 22, 277, 1979.
17. **Nagaraj, H. S., Sanborn, D. M., and Winer, W. O.,** Direct surface temperature measurements by infrared radiation in EHD, and the correlation of the Blok flash temperature theory, *Wear,* 49, 43, 1978.
18. **API,** *Technical Data Book — Petroleum Refining,* 3rd ed., American Petroleum Institute, Washington, D.C., 1977.
19. **Johnston, W. G.,** A method to calculate the pressure-viscosity coefficient from bulk properties of lubricants, *ASLE Trans.,* 24, 232, 1981.
20. **Alsaad, M., Bair, S., Sanborn, D. M., and Winer, W. O.,** Glass transitions in lubricants: its relation to EHD lubrication, *J. Lubr. Technol., Trans. ASME,* 100, 404, 1978.
21. **Bair, S. and Winer, W. O.,** Shear strength measurements of lubricants at high pressure, *J. Lubr. Technol., Trans. ASME,* 101, 251, 1979.
22. **Dubois, G. B., Ocvirk, F. W., and Wehe, R. L.,** Natl. Advisory Committee for Aeronautics, Contract No. NAw6197, Prog. Rep. 9 (revised), August 1953.
23. **Klaus, E. E. and Duda, J. L.,** Effect of Cavitation on Fluid Stability in Polymer-Thickened Fluids and Lubricants, Sp. Publ. 394, U.S. National Bureau of Standards, Washington, D.C., 1974, 88.
24. **Bhatia, R.,** Mechanical Shear Stability and Blending Efficiency of Polymers in Lubricant Formulations, M.S. thesis, Pennsylvania State University, University Park, Pa., 1978.
25. **Myers, H. S., Jr.,** Volatility Characteristics of High-Boiling Hydrocarbons, Ph.D. thesis, Pennsylvania State University, University Park, Pa., 1952.
26. **Beerbower, A. and Zudkevitch, D.,** Predicting the evaporation behavior of lubricants in the space environment, ACS Meet. 8, C-99, Div. Pet. Chem., American Chemical Society, Los Angeles, April 1963, preprint.
27. **Klaus, E. E. and Bieber, H. E.,** Effects of some physical and chemical properties of lubricants on boundary lubrication, *ASLE Trans.,* 7, 1, 1964.
28. **Fein, R. S.,** Chemistry in concentrated-conjunction lubrication, in An Interdisciplinary Approach to the Lubrication of Concentrated Contacts, National Aeronautics and Space Administration, Washington, D.C., 1970, chap. 12.
29. **Maxwell, J. B.,** *Data Book on Hydrocarbons,* D Van Nostrand, New York, 1950.

30. **Klaus, E. E. and O'Brien, J. A.,** Precision measurement and prediction of bulk-modulus values for fluids and lubricants, *J. Basic Eng., ASME Trans.,* 86 (D-3), 469, 1964.
31. **Wright, W. A.,** Prediction of bulk moduli and pressure-volume-temperature data for petroleum oils, *ASLE Trans.,* 10, 349, 1967.
32. **Wilkinson, E. L., Jr.,** Measurement and Prediction of Gas Solubilities in Liquids, M.S. thesis, Pennslyvania State University, University Park, Pa., 1971.
33. **Beerbower, A.,** Estimating the solubility of gases in petroleum and synthetic lubricants, *ASLE Trans.,* 23, 335, 1980.
34. **Cayias, J. L., Wade, W. H., and Schecter, R. S.,** The measurement of low interfacial tension via the spinning drop techniques, *Adsorption at Interfaces,* ACS Symp. Ser. No. 8, American Chemical Society, Washington, D.C., 1975.
35. Military Specification, MIL-L-23699B, Lubricating Oil, Aircraft Turbine Engine, Synthetic Base, U.S. Department of Defense, Washington, D.C., 1969.
36. Federal Test Method Standards No. 791, Lubricants, Liquid Fuel, and Related Products; Methods of Testing, U.S. Bureau of Standards, Washington, D.C., 1974.
37. Military Specification MIL-H-27601A (USAF), Hydraulic Fluid, Petroleum Base, High Temperature, Flight Vehicle, U.S. Department of Defense, Washington, D.C., 1966.
38. **Klaus, E. E., Tweksbury, E. J., Jolie, R. M., Lloyd, W. A., and Manning, R. E.,** *Effect of Some High Energy Sources on Polymer-Thickened Lubricants,* Spec. Tech. Publ. No. 382, American Society for Testing and Materials, Philadelphia, Pa., 1965, 45.
39. **Lockwood, F. E. and Klaus, E. E.,** Ester oxidation under simulated boundary lubrication conditions, *ASLE Trans.,* 24, 276, 1981.

LUBRICATING GREASES — CHARACTERISTICS AND SELECTION

I. W. Ruge

GREASES

Grease is a semisolid lubricant consisting essentially of a liquid mixed with a thickener; the liquid does the lubricating, the thickener primarily holds the oil in place and provides varying resistance to flow. It may be hard enough to cut into blocks, or soft enough to pour through a funnel.

THICKENERS

Variations in grease characteristics are largely determined by the material used to thicken it. If the thickener can withstand heat, the grease will be usable at high temperatures. If the thickener is unaffected by water, the grease will be also. The many different kinds of thickeners used in commercial greases can be divided into two primary classes: soap and nonsoap. Table 1 describes properties of greases with various thickeners.

A soap is a metallic element reacted with fat or fatty acid. Metallic elements include lithium, calcium, sodium, aluminum, barium, and others. Fats and fatty oils may be animal or vegetable in origin, ranging from cattle, hog, fish, castor bean, coconut, cottonseed, or flaxseed. Choice of these and reaction conditions offer a wide variety of soaps, and control the characteristics of grease.

Among soap type greases, lithium accounts for the majority used in the U.S., followed by calcium, aluminum, sodium, and others (mainly barium). Soap-type grease production in 1981 were approximately 80% of the total, nonsoap thickeners accounted for about 10% (see Table 2).

Lithium Soap Greases

Development of lithium greases on a large scale got its start just before and during World War II. They can be made by virtually any conventional compounding procedure with no unusual problems. Current products may be divided into those using 12-hydroxystearate and those using organic acid radicals.

Lithium 12-hydroxystearate soaps can generally be dispersed at temperatures around 200°F (93°C), while most of the other lithium soaps require temperatures in the range of 400°F (204°C) or more. A large variation in fiber structure and grease properties results from using soaps derived from organic acid compounds.

After soap dispersion, lithium greases are cooled in various ways. They can be transferred to shallow pans, but more common methods include circulating through coolers, or slow cooling and stirring in the kettle. Slow cooling, with or without stirring, will form long fibers, increase mechanical stability and increase the bleeding tendency. Rapid, or quench cooling yields fine fibers which result in poor mechanical stability but good oil retention. Ideally, a grease should combine a variety of fibers to give a good compromise in stability and bleeding. Milling (subjecting the product to very high shear) as a finishing touch is desirable because it gives a more uniform smooth texture which is less likely to change in consistency during service.

When properly formulated, lithium-base greases are very acceptable as multipurpose products for automotive and industrial requirements. High-quality products have no serious deficiencies in any of these applications except for very severe extremes of temperature, speed, loads, and pressures. They have given good service in journal and antifriction bearings

Table 1
PROPERTIES OF GREASES WITH VARIOUS THICKENERS

	Soap						Nonsoap		
	Lithium	Calcium	Calcium complex	Sodium	Aluminum	Barium	Inorganic	Organic	Synthetic
Texture	Smooth feathery	Buttery	Buttery	Smooth or fibrous	Smooth and Clear	Buttery	Buttery	Smooth	Smooth
Dropping point °F(°C)	375(191)	180(82)	500(260)	340(171)	170(77)[a]	370(188)	Nonmelt	500(260)	>480(249)
Highest usable temp °F(°C)	300(149)[b]	170(77)	300(149)[b]	250(121)	180(82)	225(107)	250(121)	325(163)[b]	325(163)[b]
Water Resistance	Fair to good	Good	Good	Poor to fair	Good	Good	Fair to good	Good to excellent	Excellent
Work softening	Good to excellent	Fair	Fair	Poor to fair	Poor	Fair to poor	Excellent	Excellent	Good to excellent
Slumpability	Good	Poor	Fair	Smooth to Poor Fibrous to good	Poor	Fair	Good	Good	Good
Compatibility	Advice about compatibility ranges from "don't mix" to "possibly okay"; switching from one type of grease to another can usually be satisfactorily accomplished by flushing the new grease through the system with vent plugs removed, lines open, and making the first few relubrication intervals shorter								

[a] Softening point
[b] With effective inhibitors.

Table 2
PERCENTAGE OF GREASE
PRODUCTION BY
THICKENER TYPE — 1981

Thickener	%
Soap	
Lithium	59
Calcium	16
Aluminum	8
Sodium	4
Others (mainly barium)	3
Nonsoap	
Inorganic	7
Organic	3
	100

and are used extensively in prepacked and lubricated-for-life rolling contact bearings. However, the smooth buttery texture may not have proper adhesion and retention in some sliding and reciprocating devices.

Calcium (Lime) Soap Greases

Mineral oils thickened with calcium soap were among the first lubricating greases available. Several important types will be discussed individually.

The oldest and probably most widely used types of calcium greases require water as a stabilizing agent to maintain the grease structure. This explains, in part, the limited temperature capability and good water resistance of the product.

In the manufacture of calcium soap greases, the reaction of animal fat and lime to make soap typically takes place at temperatures far above the boiling point of water. Therefore, water is not introduced until soap structure has been formed and an initial amount of oil has been stirred into the mixture. About twice as much water as needed in the finished product is introduced before final blending to grade to allow for evaporation during final stages of manufacture. Depending on many factors, the finished product may have from less than 0.5% to more than 1.5% water by weight. If this water is driven out of the grease by short exposure to temperatures over about 180°F (82°C), consistency is destroyed.

Evaporation of the water in an operating environment is influenced by time, temperature, and air circulation. In order to assure long life, operating temperatures of 125 to 150°F (52 to 66°C) are considered satisfactory. Exceeding 175°F (79°C) is not advisable. Generally, calcium greases are replaced at frequent intervals; service conditions are mild with little exposure to high temperatures.

Since calcium soap greases are water resistant, they find many uses in food plants, water pumps, wet industrial and sewage plant machinery, equipment exposed to weather, harvesting equipment for damp crops, marine hardware, and chassis lubrication. A harder grade such as NLGI #4 would be preferred in water pumps; a softer, more easily pumpable grease would be used to lubricate through fittings of farm equipment, some industrial applications, and other light-duty specialized requirements such as tractor track rollers, mine cars, and textile machinery. For many of these applications, calcium greases are being replaced by more versatile multipurpose products.

Calcium 12-Hydroxystearate Greases

There is objection to some calcium soap thickeners where there is a possibility of incidental contact with food. To meet this need, calcium 12-hydroxystearate was developed. It does not use water to stabilize the structure, so the maximum operating temperature ranges up to 250°F (121°C). Some satisfactory services have been reported at temperatures near the dropping point, which varies from 280 to 300°F (138 to 149°C), depending on formulation.

These products approach multipurpose greases in their temperature capability, but adhesion and retention can be poor in some types of service because of the extremely smooth buttery texture. Special formulations which provide unique characteristics and advantages have been developed. Zinc oxide is frequently included when the grease is used in the food industry, but the base oil must also be acceptable for incidental contact with food.

Calcium Complex Greases

Calcium soaps modified by a small quantity of calcium acetate provide characteristics quite different from the normal water-stabilized product. Stability is usually improved, and extreme pressure (EP) characteristics are enhanced without the use of additives. Other complex calcium soaps have been developed, but the salts of acetic acid have been the preferred complexing agent for performance improvement. Calcium complex greases frequently perform under conditions that cannot be handled by any other soap-thickened product. There have been, however, unsuccessful incidents. Bearing failures can result from excessive thickening in use.

Sodium Soap Greases

These were originally developed to provide a higher service temperature limit than was possible with conventional calcium soap. Even with the poor water resistance of most sodium greases, they are still popular for some demanding applications in lubricating electric motor bearings, particularly those with effective seals. Lighter grades are used in textile plants since leaked product is easily removed from cloth in the normal washing process. Although sodium base greases were the accepted "standard" in automotive wheel bearings for many years, they are now being replaced by high-quality multipurpose greases with much better water resistance.

Sodium greases are manufactured in the same type of equipment used for most types of soap-thickened grease, but adequate kettle volume should be provided because of a tendency to foam during early processing steps. The type of fatty acid or glyceride used is important in determining characteristics of the finished product. Glycerine from the glyceride stabilizes the grease, but reduces oxidation stability. Small amounts of water and/or glycerine are common modifiers to control fiber structure and stability. The goal is to produce fibers as thin as possible for good shear stability.

Base oil selected depends on the end use. As usual, viscosity is the important consideration. With a high-viscosity index, medium to light viscosity base oil, sodium base greases can equal or surpass low-temperature performance of calcium or lithium greases. However, high-viscosity oils give rise to variations in other grease characteristics.

Sodium soap greases for ball and roller bearings usually contain oxidation inhibitors. EP additives such as sulfurized fatty oil, organic sulfur, and/or chlorine compounds will improve load-carrying ability. Tackiness additives are seldom needed, and rust inhibition is not normally required.

When sodium greases are mixed with other type greases, particularly with calcium type, the combination can become soupy. Despite this apparent incompatibility, there are some anomalies. Sodium-calcium soap mixed base greases are manufactured by special reactions and used in wheel bearings and in ball and roller bearing service. The calcium is claimed to shorten the fiber. Sodium-aluminum greases have been used in many industrial applications. The aluminum modifies the structure to give a smooth texture and a dropping point of 375 to 425°F (191 to 218°C). At lower temperatures this grease has reasonably good water resistance.

Aluminum Soap Greases

These are manufactured by mixing dry, powdered aluminum stearate in the base oil as the temperature is increased to the range of 240 to 350°F (116 to 177°C). Cooling is followed by mild working prior to packaging. Care must be taken to avoid air bubbles which would spoil the clear, translucent appearance of the grease. Various modifiers may be added during the manufacture to change the structure, increase plasticity, or increase stringiness.

Aluminum base greases have a smooth texture and are water resistant. Their shear stability is generally poor, however, and when exposed to temperatures above about 170°F (77°C) the normal smooth structure becomes rubbery. In the rubber-like state, the grease pulls away from metal and ceases to lubricate. They also tend to aerate badly when severely agitated or churned.

Barium Soap Greases

These are among the least understood soap-oil systems, but they have unique and valuable characteristics. Even impurities can affect the structure and performance. Typically, soap content to achieve a given consistency is very high.

Barium soap greases were among the first multipurpose products with both high temperature capability and good water resistance. They are frequently used to lubricate electrical cables for power transmission lines. Wind, temperature, and current surges cause flexing and stretching between the conductors and the sheath.

Nonsoap Thickeners

These fall into separate classifications: inorganic, organic, and "synthetic" materials. Inorganic thickeners are very fine powders which have enough surface area and porosity to thicken by absorbing oil. Silica and bentone, a process bentonite, have been the most successful commercially. Both types are very sensitive to water unless the thickener particles are protected by a coating, which may break down at 300°F (149°C).

While clay-thickened greases usually are manufactured without the need for the hot cooking or reaction cycle required for soaps, high-shear mechanical mixing or milling is needed. Frequently dispersing aids are required during formulation; these aids are volatile and evaporate during processing.

Since these greases have no melting point, maximum continuous service temperatures depend on the oxidation stability of the base oil and its inhibitor treatment. Rapid deterioration is sometimes encountered with continued exposure of mineral oil-based grease to temperatures of 250°F (121°C). When properly inhibited against oxidation, however, inorganic-thickened greases are successfully used in many high-temperature applications.

Greases of this type are considered multipurpose lubricants. Industrially, they have been widely used in rolling contact bearings operating at moderate speeds and temperatures. In automotive service they have been used as general purpose greases, but performance in wheel bearings has varied widely.

Organic thickeners such as amides, anilides, arylureas, dyes and synthetic products used in commercial products are also superior to soap based greases in high temperature applications. These thickeners are chemically and structurally stable over a wide temperature range and do not act as catalysts for the oxidation of the base oil. Current use is mainly for high temperature ball bearing greases and special synthetic greases for military and aerospace use.

Generally, they have dropping points above 500°F (260°C) and are serviceable down to very low temperatures. The oxidation rate at all temperatures is typically lower than for greases prepared with other thickeners. Oxidation is commonly limited by the base fluid. While rust protection is poor, they are water resistant and in manufacture respond well to inhibition.

OIL PHASE

In grease, the lubricating petroleum oil or synthetic fluid is the main component and makes a most important contribution in structure, performance, and stability.

Petroleum Oils

Petroleum oils used as lubricants in grease vary widely in type. Various crude sources and refining methods result in differing oil characteristics relating to hydrocarbon types. Naphthenic oils are most common despite their relatively low viscosity index. Their low-temperature fluidity and ability to combine readily with soap contributes to their wide use. A base oil viscosity range of 65 to 175 cSt at 40°C (approximately 300 to 800 SUS at 100°F) is the most widely used. Greases for low-temperature or high-speed use may have lower viscosity base oils while greases used for particularly low speed, high loads, and shock loading will be higher in viscosity.

Viscosity, viscosity index, and chemical characteristics are each important. Low-temperature pumpability and handling ease are mostly a result of properly selected base oil viscosity. Viscosity is usually more significant than pour point. Viscosity-temperature relationship of the base oil is important where service conditions involve a wide temperature range. Load-carrying ability at moderate to high shear rates is mainly due to base oil viscosity, particularly in the absence of EP additives.

The two main classes of petroleum base oils, paraffinic and naphthenic, have different effects on thickeners. With some soaps, the gel structure is weaker with paraffinic oils. Due to the relative stability of paraffinic oils, they are less likely to react chemically during grease formation. Paraffinic oils are poorer solvents for many additives used in greases. Naphthenic oils, particularly when some unsaturates are present, can function chemically during manufacture. When controlled, this can be advantageous and explains in part why naphthenic base oils are popular.

Synthetic Fluids

These have proven to be particularly well suited for extreme conditions. Among soap-type greases, probably more synthetic fluids are thickened with lithium soap than any other. Tailoring the desired characteristics of the grease can only be achieved by careful selection of both thickener and fluid. Synthetic fluid greases are normally designed for improved performance in some extreme temperature range, either high or low. By suitable compromise, reasonable service performance can be achieved over wide ranges of temperature. Products of this type find their greatest application in high-performance aircraft, missiles, and space vehicles. When thickeners and fluids are both synthetic, use is almost exclusively in such high-performance equipment. For some missile applications, a service life of minutes or less might be adequate. A variety of synthetic military specification greases and their temperature ranges are indicated in Table 3.

ADDITIVES AND FILLERS

Additives of the types indicated in Table 4 are often needed to augment or improve performance, or meet special needs. Some modify soaps; others enhance natural characteristics of the oil, give it longer life, or improve its ability to protect equipment.

Antioxidants

Antioxidants (oxidation inhibitors) are the most common additives and must be selected to match the individual grease. The original objective of inhibitors was to protect the grease during storage prior to use. Most multipurpose greases and greases designed for high tem-

Table 3
SYNTHETIC GREASES USED BY MILITARY[a]

Military specification	Oil type[a]	Thickener[a]	Temp range (°F)	Use
MIL-G-4343	Diester/silicone	Lithium	−65 +250	Pneumatic systems as a lubricant between rubber seals and metal parts (under dynamic conditions)
MIL-G-6032	Animal, vegetable, ester, and or silicone, etc.	Soap or nonsoap	+200	Tapered plug valves, gaskets, or seals and other plug valve service in systems where resistance to gasoline, oil, alcohol or water is required.
MIL-G-21164 (5% MoS$_2$)	Diester	Lithium soap or clay	−100 +250	Accessory splines, heavily loaded sliding steel surfaces or for anti-friction bearings carrying high loads and operating through wide temperature ranges
MIL-G-23827	Diester	Lithium soap or clay	−100 +250	Ball, roller, and needle bearings, gears and sliding and rolling surfaces of instruments, cameras, electronic gear; also general use on aircraft gears, actuator screws, and control systems
MIL-G-25013	Silicone	Nonsoap	−100 +450	Ball and roller bearings operating in extreme high and/or low temperature range, when speed factor (DN value) does not exceed 200,000 and main action does not involve sliding of metal-on-metal
MIL-G-27617	Perfluoroalkylpolyether	Inorganic	−30 +400	Taper plug valves, gaskets, and bearings in fuel systems. Also for valves, threads and bearings in the presence of liquid oxygen systems, or extreme high temperatures
MIL-G-81322	Synthetic hydrocarbon	Clay	−65 +350	Aircraft wheel bearings, antifriction bearings, gear boxes operating at high speeds and over wide temperature range; also instrument bearings
MIL-G-81827 (5% MoS$_2$)	Synthetic hydrocarbon	Clay	−65 +350	Heavily loaded accessory splines, sliding surface, and antifriction bearings operating through a wide temperature range
MIL-G-81937	Diester	Lithium soap	−65 +250	Wide temperature range high precision miniature and instrument bearings requiring ultra-clean lubrication
MIL-G-83261	Fluorinated polysiloxane	F&P and/or: PTFE	−100 +450	Aircraft actuators, gear boxes, gimbel rings, oscillating bearings and other applications involving heavy loads and extreme temperature ranges
MIL-G-83363	Polyol aliphatic ester and fluorinated polysiloxane	FEP-clay	−65 +300	Helicopter tail rotor and intermediate transmissions and gear boxes
DOD-G-24508	Synthetic hydrocarbon only	Clay	−65 +300	Multipurpose ball and roller bearings

[a] Oil type and thickener specified only in MIL-G-83261. Other MIL-G numbers are performance specs and do not demand oil and thickener type.

Table 4
TYPICAL ADDITIVES USED IN LUBRICATING GREASES

Function or designation	Types of chemicals	Specific illustrations[a]	Common proportions (%)
Antioxidants or oxidation inhibitors	Amines	Phenyl alpha naphthylamine	0.1—1.0
	Phenols	Di-*tert*-butyl-*para*-cresol	0.05—1.0
	Sulfur	Zinc dibutyl dithiocarbonate	0.1—1.0
	Selenium compounds	Di-lauryl selenide	2.0—5.0
Corrosion inhibitors	Sulfonates	Sodium sulfonate	0.2—3.0
	Sorbitan esters	Sorbitan monooleate	1.0
Color stabilizers	Similar to oxidation inhibitors	Substituted hydroquinones	0.01—0.10
		Furfural azine	0.01—0.10
Dyes	Oil-soluble aniline colors	Oil-soluble red or green	0.01—0.03
EP or film-strength agents[b]	Chlorine, phosphorus, or sulfur compounds	Chlorinated paraffin	5.0—15
		Zinc dithiophosphate	
		Dibenzyl disulfide	
Metal deactivators	Sulfur compounds or phosphites	Mercaptobenzothiazole	0.01—0.05
Rust inhibitors	See corrosion inhibitors	Amine of tallow fatty acids	0.01—6
	Amines or oxidized petroleum fractions	Butyl stearate	
Stringiness additives	Polymers	Polymerized isobutylene	0.02—1.0
Structure modifiers for soap-oil systems	Glycols, higher alcohols, monoglycerides	Propylene glycol	0.1—1.0
		Glycerol	
		Glycerol monooleate	

[a]   Illustrations are typical but not at all complete.
[b]   Film-strength additives most often consist of mixtures of compounds of two or more active elements; that is, a chlorine compound will be used in conjunction with a sulfur compound, or phosphorus and sulfur will be used in the same mixture.

From Boner, C. J., *Standard Handbook of Lubricating Engineering*, O'Connor, J. J. and Boyd, J., Eds., McGraw Hill, New York, 1968. With permission.

peratures contain inhibitors to assure extended service life and enable longer regreasing intervals. Even at normal atmospheric temperatures, oxidation inhibitors can extend service life. If glycerides are used in saponification, the finished product will normally contain some glycerine which can oxidize rapidly and make inhibition more difficult.

In order to combat catalytic action of copper and other metals that may aggravate oxidation, metal deactivators and passivators are added.

Rust and Corrosion Inhibitors

These are somewhat related. Most properly formulated greases are noncorrosive to bearing materials in machinery. However, corrosion can be caused by breakdown of the grease, or by contamination.

Simple sodium grease will give rust protection where traces of water are present, but unfortunately sodium soap loses its structure completely when exposed to only a moderate amount of water. Under extremely wet or corrosive conditions the performance of most greases can be improved by a rust inhibitor. In such industrial applications as food preparation and handling, suitable inhibition is a necessity. As an example, chicken soup can be extremely corrosive to bearings and equipment. Another example is boric acid which is suitable at low concentration as an eye wash, but can be extremely corrosive, even to stainless steel. Most high-quality greases recommended for multipurpose uses contain rust and corrosion inhibitors.

Tackiness Additives

Products such as Paratac* are sometimes added to increase adhesion and cohesion, impart stringy texture, and give better bearing retention. The latter characteristic is also occasionally improved by polyethylene. When used with lower viscosity base oils (which may be desired for low temperature performance), tackiness agents give the appearance and structure of higher viscosity base oils.

Extreme Pressure (EP) Additives

EP additives provide improved load-carrying ability and give added protection under shock loads. In most rolling contact bearings other than tapered rollers, EP characteristics have little beneficial effect. In tapered roller bearings where thrust loads are high, EP additives are often needed to prevent scoring and wear of roller ends. In other extreme pressure conditions, the EP agents react with steel surfaces to form a surface or interface which will act like a solid lubricant and prevent metal-to-metal contact or welding.

EP additives for greases generally contain various combinations of sulfur and phosphorus. They are similar, if not identical, to those used in industrial oils and gear lubes.

Fillers

Fillers are usually, but not always, fine micron-size solids employed for special functions. As a rule of thumb, particles less than one half the size of the bearing clearance give little trouble with wear, but rolling element bearing clearance may be as low as 0.0001 in. Nevertheless, a proper filler can improve grease performance under adverse conditions. In food processing and food canning, an alkaline filler helps protect bearings against the corrosive action of food acids. In some applications, the cosmetic effect of the color or pleasing appearance may be more important than any change in performance.

There are decreasing numbers of fillers in common use. Ranging from popular to almost extinct are graphite, molybdenum disulfide, metal oxides, metal flakes, carbon black, talc, and mica. Another material that might be classed as a filler is yarn.

If enough is present, graphite can minimize metal-to-metal contact and wear of sliding surface bearings. Artificial graphite is most often used because it is free of abrasive impurities. But even pure graphite deposits tend to promote wear in rolling element bearings, so graphite grease should usually be avoided in them. Molybdenum disulfide (Moly) is popular as a filler in lubricating greases. While the advantages at low concentrations are open to question, higher concentrations of 3% and over can be demonstrated to provide a protective film when grease lubrication is difficult to maintain.

Zinc and magnesium oxides have been used in food processing industries. The light color and ability to neutralize acid are the main advantages. For best performance, these oxides must be dispersed in the grease by milling.

Metal flakes and powdered soft metals such as lead, tin, zinc, and aluminum are used in pipe thread and antiseize compounds. These compounds typically are greases with a high-filler content. At one time, talc was widely used as a die and drawing lubricant, and in roll neck lubricants for older type mill stands. Carbon black, because of its fineness, also must be vigorously milled to get complete dispersion. Although usually acting only as fillers, some carbon blacks have thickening characteristics.

Wool yarn is still used, mostly in large open bearing boxes. It is exactly what its name implies: wool yarn cut into short lengths to help give structural depth to grease. It is occasionally blended with curled horsehair, but avoid synthetic fibers in blends.

Dyes

Dyes are nonperformance additives that may be used to identify grease, improve color,

* Trademark for high molecular weight iso-butylene polymer.

and mask slight variations between batches. They may camouflage changes in color that do not detract from performance.

SELECTION FACTORS

Performance characteristics used in selecting and specifying greases are best described by commonly used test procedures. These tests use finished greases, but base oil viscosities are often included in specifications because of their obvious importance. A brief description of the American Society for Testing and Materials (ASTM) test procedures for grease follows.

Penetration (ASTM D-217, ASTM D-1403)

This test is an arbitrary measure of hardness, consistency, or shear strength. It is defined as the depth in tenths of a millimeter that a precisely standardized cone penetrates the grease in a specified cup under prescribed conditions of time and temperature.

Greases are graded by the National Lubricating Grease Institute (NLGI) classifications based on ASTM penetration, see Table 5. Note that there is only a 30-point spread for each NLGI grade, with a 15-point gap between each grade. The 30-point spread in a given grade may often make minor variations in penetration appear unduly important. In selecting a suitable grease, penetration is but one of the many factors which must be considered. The correlation between penetration and pumpability is so poor that penetration cannot be used for predicting the behavior of a grease under various temperature conditions. As Table 5 indicates, the grading system separates the lubricants into two groups, then classifies them into two semifluid grades and from soft to hard in six grades.

Grades 000 and 00 are most frequently used in relatively low-speed gear boxes, usually larger ones with rudimentary or inefficient seals. These grades are not appropriate for grease-lubricated antifriction bearings because they would leak. Antifriction bearings are most frequently lubricated with #2 grease because of ideal oil release and good feeding ability.

Grades 0 and 1 are most suited for central lube systems with lengthy tubing runs. Number 3 grease may be better for large rolling element bearings which hold substantial quantities of grease. Harder greases may be used on large open gears or large shaft bushings in which a block of grease is resting on the rotating element. Greases harder than #3 constitute a very minor proportion of lubricating greases.

Dropping Point (ASTM D-566)

This is the temperature at which a drop of fluid forms and falls from a grease under test conditions established by ASTM. Although dropping points are poor predictors of service performance of lubricating greases, they do represent a limiting temperature for most applications. The limiting temperature for prolonged exposure is well below the dropping point. This is particularly true of products containing volatile components or additives. A common example is straight calcium grease that is stabilized by a small amount of water. While the dropping point is below 200°F (93°C), water loss can become a problem at about 125°F (52°C) and a serious one above 150°F (66°C).

Bleeding (ASTM D-1742)

Bleeding is a characteristic tested mainly for evaluation, not as a batch or shipment acceptance. Practically all greases will separate oil under certain conditions. In most applications, the separation of a limited amount of oil is not harmful. If none separated, lubricant starvation could result. In some cases separation permits lubricant to creep into narrow clearances by capillary action where soap won't go. Pressure and vibration promotes separation; this may cause trouble with soap plugging in central lube systems and pressure grease cups.

Table 5
NATIONAL LUBRICATING GREASE
INSTITUTE (NLGI) SPECIFICATIONS

NLGI no.	ASTM worked penetration [25°C (77°F) mm/10]	
000	445—475	Semifluid
00	w/modified cone	
	400—430	
	w/modified cone	
0	355—385	Soft
1	310—340	
2	265—295	
3	220—250	
4	175—205	
5	130—160	
6	85—115	Hard

Oil separation on the top of grease in a container, as long as it is not excessive, may be remixed with no sacrifice in performance. Because of the possibility of further changes in storage and use, however, excessive separation or severe cracking on the surface should be investigated before using the product.

ASTM D-1742 correlates with oil separation in a 35-lb (16-kg) pail during storage. It consists of buttering a sample of grease on to a screen to a specific depth controlled by a ring, weighing it, and sealing it over a funnel in a cylinder. A slight pressure is applied above the grease. The funnel directs any free oil into a small beaker. The bleeding tendency is the weight percent of oil that separates from the grease in 24 hr. at 77°F (25°C).

Oxidation Stability (ASTM D-942)

Oxidation stability is the ability of grease to resist oxidation at elevated temperatures. In the ASTM bench test five small dishes containing grease are encased in a bomb charged with oxygen under pressure and placed in a bath at high temperature. Pressure drop accompanying absorption of oxygen is recorded at time intervals. The length of the test usually mitigates its use as an acceptance procedure.

Evaporation (ASTM D-972)

This characteristic can also be of value in choosing a grease where high temperatures are involved and evaporation loss is a concern. A sample of grease is placed in a small cell, which in turn is immersed in a bath. Heated air is passed over the surface at a controlled rate. Evaporation is calculated from the weight loss of the sample, expressed in percent.

Apparent Viscosity (ASTM 1092)

This is expressed in poises at desired temperatures and shear rates. It is of concern, particularly when grease is required to perform at low temperatures. Grease is forced through a capillary tube at a designated temperature. The rate of forcing the grease is governed by hydraulic pressure on a floating piston separating hydraulic oil from the grease sample in a cylinder. Apparent viscosity is the ratio of shear stress to shear rate. The viscosity of a grease changes with the rate of shear. When at rest, or at low rates of shear, apparent viscosity of the grease can be very high. Once motion begins and the rate of shear is increased, the apparent viscosity of grease approaches, but never reaches, the viscosity of the fluid component.

In industrial applications, the apparent viscosity is useful in predicting:

1. How grease might actually perform in a bearing
2. Leakage tendency from a journal bearing
3. Performance at low temperature
4. Pumpability and flowability

Unfortunately, apparent viscosity data from a pressure viscosimeter are of little value in predicting performance of the grease at high temperatures.

Mechanical Stability

These tests also find their way into specifications. The Grease Worker used in conjunction with penetration testing (ASTM D-217) involves the penetration cup which contains about a pound of grease. A sturdy disc with 51 holes is forced through the grease in the closed cup for a specified number of strokes. Changes in penetration from unworked grease are recorded. Some may soften, some harden.

Wheel Bearing Test (ASTM D-1263)

This test involves placing a weighed amount of grease, properly packed into an automotive type front wheel hub and spindle. Instead of installing a seal, the large end of the hub is encircled by a trough or collector ring. The assembly is placed in an enclosed box equipped with heaters. It is belt driven by an electric motor outside the box. Temperature, time, and speed are specified. The grease that leaves the hub is slung into the ring; its weight is recorded for comparison with performance parameters.

Roll Stability Test (ASTM D-1831)

This test, sometimes referred to as the Shell roll test, is also used in conjunction with the penetration test, which is run before and after a prescribed number of hours. A small sample of grease is weighed into a cylinder. A weighted steel roller whose outside diameter is about two thirds the inside diameter of the cylinder is inserted after the grease is distributed in the cylinder. The cylinder is rolled on a horizontal axis for a timed interval (usually 2 hr), starting at essentially room temperature. The difference in penetration before and after rolling is considered to be a measure of shear stability.

Extreme Pressure

These tests are frequently seen in specifications, although correlation with actual service is sometimes lacking.

Timken Test (ASTM D-2509)

The Timken test involves rotating a bearing race against a fixed block at a prescribed speed under increasing load while a constant flow of grease is maintained. Results, expressed as pounds lever load, will differentiate between low, medium, or high levels of extreme pressure performance.

Four-Ball Methods (ASTM D-2596, ASTM D-2783)

These tests, likewise, do not necessarily correlate with service results. In these procedures, one steel ball is rotated at a controlled speed against a nest of three stationary steel balls of identical size. Load is increased incrementally until seizure occurs. This is known as the weld-point.

A variation in these methods utilizing similar specimens (one ball rotating against a nest of three), but not under the high load conditions, is the Four-Ball Wear Tester (ASTM D-2266). It measures relative wear performance of grease on the steel-on-steel balls by measuring scar diameters rather than measuring the load at seizure.

Falex Tests (ASTM D-3233)

These tests consist of determining load carrying limits as indicated by shear pin breakage. A small journal is rotated between two vee blocks which are clamped against the journal and immersed in the lubricant. Load is steadily increased by clamp levers. Performance is measured in pounds force when the shear pin fails. A variation of the method consists of increasing the load incrementally, maintaining a constant load for 1 min at each step. Load at the time the shear pin breaks is also the criteria in this procedure.

Other Selection Factors

Numerous in-house and special tests are performed for qualification and product acceptance. Among better known performance tests is the Lincoln ventmeter which measures pumpability and ventability of greases in a central system. It consists of a coil of $^1/_4$ in. tubing with a pressure gage at one end, a grease fitting at the other, and valves at both ends. The tubing is filled with grease at a high pressure, and the entire system is allowed to stabilize to a desired temperature. The venting valve is opened for 30 sec and the pressure is read. This gives a clue about relative pumpability, and helps to determine either NLGI grade or type of grease that will function, or to select pipe size for the central system.

COMMENTS

All the expertise that goes into the development, manufacture, selection, and application of the grease does not necessarily assure performance. One other most important consideration is cleanliness. Contaminated grease is faulty grease. Keep containers intact, closed, stored in a clean, dry place. Be sure fittings, plugs, and nozzles are clean before application.

REFERENCES

1. **Boner, C. J.**, Additives used in lubricating greases, in *Standard Handbook of Lubricating Engineering*, O'Connor, J. J. and Boyd, J., McGraw-Hill, New York, 1968, 11.
2. **Rebuck, N. D.**, Naval Air Development Center, private communication, March, 1981.

ADDITIONAL REFERENCES

3. **Boner, C. J.**, *Manufacture and Application of Lubricating Greases*, Hafner Publ. Co., New York.
4. **Bailey, C. A. and Aarons, J. S.**, *The Lubrication Engineers Manual*, 1st ed., United States Steel, Pittsburgh, Pa.,
5. Technical Bulletin, *Lubrication*, Texaco Inc., Houston.
6. **Quigg, J. S.**, Grease survey, paper presented at 49th Ann. Meet. Natl. Lubr. Grease Inst., Hilton Head, South Carolina, 1982.
7. Witco, Inc., *Target Talk and Technical Letter*, Southwest Petro-Chem., Inc., Wichita, Kan.

SOLID LUBRICANTS

J. K. Lancaster

INTRODUCTION

A solid lubricant is "any material used as a powder or a thin film on a surface to provide protection from damage during relative movement and to reduce friction and wear." This chapter describes the various materials currently being used as solid lubricants, indicates areas of application in Table 1, discusses their particular advantages and limitations, and shows how they can be used in the four main ways which are outlined in Table 2.

Reviews of solid lubricants are also given in References 1 to 5. Discussions of the nature and influence of surface films, boundary lubrication, and wear mechanisms are covered in earlier handbook chapters. Many factors influence friction and wear of polymers and have been reviewed elsewhere.[6,7]

MATERIALS

Important features needed in a solid lubricant are tabulated in Table 3. The extent to which all of these can be realized in various groups of materials is covered in the following discussion.

Lamellar Solids

These materials crystallize with a layered structure in which interatomic bonding between the layers is weaker than that within them. Graphite and MoS_2 are the best known examples.

Graphite and MoS_2

Bragg first suggested in 1923 that the low friction of graphite, typically 0.05 to 0.1, might be a consequence of its lamellar structure with easy shear occurring between basal planes. Subsequently, low friction was found only in environments containing water or other condensable vapors and interlamellar bonding in vacuum was greater than that attributable to weak Van der Waals forces alone. These forces are now believed to be supplemented by π-electron interactions and the intrinsic interlamellar shear strength of graphite is thus not particularly low. Most recent ideas attribute the low friction of graphite to its basal planes being low energy surfaces.[8] During sliding the basal planes orient themselves almost parallel to the surface and adhesion between them is very low. However, once a basal plane becomes damaged, high-energy edge sites are exposed and adhesion and friction increase appreciably unless these edge sites can be neutralized by adsorbed vapors.

Thermal stability of graphite is extremely high (>2000°C), but its use in ordinary environments is limited by onset of oxidation in the 500 to 600°C range; the greater the degree of graphitic order in the crystallites, the higher the oxidation temperature. The necessity for adsorbed vapors to maintain low friction can restrict the use of graphite to much lower temperatures. In air, the amount of physically adsorbed water may decrease at around 100°C to such an extent that low friction can no longer be maintained. Organic vapors are very effective substitutes for water and may be available as contaminants in the surrounding environment, or be derived from organic material introduced deliberately or accidentally into the graphite itself. Some added inorganic compounds are also able to extend the temperature range over which low friction occurs. PbO, CdO, Na_2SO_4, and $CdSO_4$ have all been shown to be effective on nickel alloy substrates to around 550°C.[9]

Although graphite continues to play a major role in metal working lubrication and in

Table 1

AREAS OF APPLICATION OF SOLID LUBRICANTS

Fluid Lubricants Undesirable

Contaminate product or environment	Prolonged storage	Maintenance unlikely or impossible	Abrasive contamination
Food processing machines	Missile components	Nuclear reactors	Agricultural and mining equipment
Optical equipment		Space	
		Consumer "durables"	

Fluid Lubricants Ineffective

High temperatures	Cryogenic temperatures	Vacuum	Ionizing radiation	High pressures	Fretting
Metal working	Space	Satellites	Nuclear reactors	Metalworking	General
Compressors	Refrigeration plant	X-ray equipment	Space	Bridge supports	

Table 2
METHODS OF USING SOLID LUBRICANTS

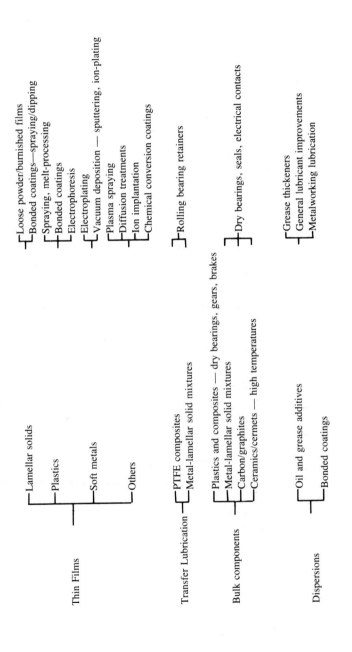

Table 3
PROPERTIES REQUIRED BY SOLID LUBRICANTS

To Provide Low-Friction and Wear

Thin films	Self-lubricating materials
Good film-forming ability (powders)	Ability to form transfer films
Ductility	Low-moderate elastic modulus
Good adhesion to substrate	Adequate strength for required load capacity
Film continuity	

Low-Shear Strength

General

High thermal/oxidative/hydrolytic stabilities
High softening/melting points
Chemically inert
High-thermal conductivity/diffusivity
Corrosion protection of substrate
Appropriate electrical conductivity
No abrasive impurities
Low toxicity/environmental compatibility
Low-thermal expansion

electrical contacts, it is being increasingly supplanted by MoS_2 for three reasons. First is the wide variability in graphites from different sources; MoS_2 quality is more rigidly controlled by specifications. Second, the low friction of MoS_2 does not depend on adsorbed vapors and is, in general, lower in vacuum than in air. Finally, the load-carrying capacity of MoS_2 is generally superior.

MoS_2 has a lamellar structure, but with interlamellar bonding being between adjacent layers of S atoms. The bonding is relatively weak, via Van der Waals forces only, and MoS_2 is therefore an *intrinsic* solid lubricant. Adsorbed vapors usually increase friction but the effects are comparatively small. The thermal stability of MoS_2 in nonoxidizing environments is of the order of 1100°C, but in air oxidation begins to become significant at around 350 to 400°C. The normal air-oxidation product is MoO_3, once believed to be abrasive but now known to be virtually innocuous.[10]

A major concern with MoS_2 is the presence of abrasive impurities.[11] The reasons for concern are twofold. First and foremost, chemical analyses provide no information about the form of the impurity; abrasion by hard particles, such as SiO_2, depends greatly on their shape and size. Second, other factors in addition to impurities can play a role in abrasiveness, e.g., crystallite modifications or anisotropy in hardness.

Other Lamellar Solids

The only dichalcogenides, other than MoS_2, with real promise appear to be sulfides and selenides of Mo, Ta, W, and Nb. Since these are synthesized directly from the elements, the compositions are not always stoichiometric and the crystal structure not wholly hexagonal. Some compounds are nevertheless superior to MoS_2 in two main areas. TaS_2, $TaSe_2$, and WS_2 have greater oxidation stability while TaS_2, $TaSe_2$, and $NbSe_2$ have much greater electrical conductivity.[12] Experimental determinations of frictional properties and endurance of surface films are somewhat conflicting, but no synthetic dichalcogenides appear to be consistently superior to MoS_2. Together with uncertainties about composition and high expense, this has precluded their widespread use. WS_2 and $NbSe_2$ have found limited application in self-lubricating composites.

The most recent synthetic solid lubricant to receive serious attention is poly (carbon monofluoride), usually referred to as graphite fluoride, $(CF_x)_n$.[13] Prepared by direct elemental synthesis, differing reaction temperatures lead to variation in x from about 0.25 to 1.1. Friction is largely independent of composition for x >0.6, but film endurance increases monotonically until x reaches about 1.0.[14] Thermal stability depends markedly on composition and ranges from 200 to over 500°C.[15] In comparison to graphite, graphite fluoride is distinctly superior in a number of respects. For burnished films of powder, the load-carrying capacity of $(CF_x)_n$ is greater,[16] wear life is longer,[13] and effective lubrication occurs in both vacuum[17] and inert gases.[18] Comparisons with MoS_2, however, are less favorable; $(CF_x)_n$ is variously reported as being superior[14] or not,[15] depending on the test method and on the formulation of the lubricant film. As a very general summary, $(CF_x)_n$ appears to offer little over MoS_2 in most applications.

A number of other lamellar solids with crystal structures of the CdI_2 or $CdCl_2$ type also form coherent surface films from powder and exhibit low friction. However, their thermal, oxidative, and hydrolytic stabilities are generally much inferior to those of MoS_2. BN, similar in crystal structure to graphite, seems to be largely ineffective as a high-temperature lubricant due to its inability to form surface films.

Metal Salts

Numerous other inorganic salts with low shear strength and film-forming ability have shown promise as solid lubricants. The main interest is in their high-temperature potential, and PbO and CaF_2 are particularly important. PbO provides effective thin film lubrication from room temperature to about 350°C, and again from 500°C upwards. Between these temperatures, however, it oxidizes in air to Pb_3O_4, which has poor lubricating properties. Attempts to bridge this gap have been made by addition of SiO_2 to form a silicate phase containing excess PbO which is then protected against oxidation.[19] With this mixture, lubrication is possible between 250 and 700°C, but below 250°C friction becomes high, and film endurance low. CaF_2 and eutectic mixtures of CaF_2/BaF_2 also provide effective lubrication in the range 250 to 1000°C; high friction (f>0.3) below 150°C can be partially alleviated by the addition of Ag.[20] A series of metal oxides, tungstates, and molybdates also show promise as high-temperature lubricants, with reasonably low-friction coefficients at 700°C, e.g., f ≃ 0.2 (MoO_3, K_2MoO_4) and f ≃ 0.3 (Co_2O_3, $NiMoO_4$). None, however, are effective at room temperature.

Synthetic mixed metal sulfides, e.g., $AsSbS_4$, $Ce_2(MoS_4)_3$, are claimed to increase the load-carrying capacity of greases more than comparable additions of MoS_2.[21] Their performance as solid films, however, is inferior to that of MoS_2.

Reaction Films

The ability of oxide and other reaction films on metals to prevent intermetallic contact and reduce wear, and sometimes friction, is well known. Coefficients of friction of oxide films are not particularly low (0.4 to 0.8), but during continuous sliding at high temperature, increased rates of oxidation can combine with substrate softening and plastic flow to generate a complex, oxide-rich, surface layer which may greatly reduce wear.[22] Deliberate introduction of readily oxidizable alloying elements, e.g., Si or Fe, into Ni-alloys enhances the production of such layers.

Soft Metal Films

Several low shear strength metals can be deposited as lubricating films on harder substrates by conventional electroplating or by newer techniques of vacuum deposition — evaporation, sputtering, ion-plating. Most metals of interest — In, Pb, Sn, Ag, Au, Cu, Zn, Tl, Ba, and Bi have low-solid solubility in Fe. Thin metal-film lubrication is most relevant to high

Table 4
SOME SURFACE TREATMENTS TO REDUCE FRICTION AND/OR WEAR OF METALS

Diffusion Treatments

Carburizing	
Nitriding	
"Tuftride"	
"Sulfinuz"	C, N, and/or S into ferrous alloys
"Sursulf"	
"Sulf-BT"	
"Noskuff"	
"Forez"	Sn-Cu electroplate + duffusion into ferrous alloys
Siliconizing	Si into ferrous alloys
Boriding	B into ferrous alloys
"Tiduran"	
"Tifran"	C, N, and O into titanium alloys
"Tiodize"	
"Delsun"	Sb-Sn-Cd electroplate + diffusion into copper alloys
"Zinal"	Zn-Cu-In electroplate + diffusion into alumunium alloys
"Metalliding"	Many elements into most metals from molten fluoride baths

Chemical Conversion Coatings

Phosphate	Ferrous alloys
Anodize	Aluminium and titanium alloys
Oxalate	Copper alloys
Dichromate	Magnesium alloys
Phosphate-fluoride	Titanium alloys

Ion-Implantation

Most elements into most metals via ion-bombardment

temperatures or to applications where sliding is limited, e.g., rolling element bearings. Ag-Pd films have been used at temperatures up to 1000°C, and Pb films have been very successful for long-term rolling bearing lubrication in space mechanisms.[23] Au is also of interest in the latter application, but test results have proved extremely variable. Vacuum sputtering and ion-plating permit close control of film composition and thickness and can provide outstanding adhesion to the substrate. Optimum film thickness for maximum wear life is generally very similar to that required to give minimum coefficient of friction, 0.1 to 1 μm.

Diffusion Coatings

An alternative to deposition of a surface film for reducing friction and wear of metals is the thermal diffusion of foreign atoms into a surface. Some commonly available treatments of this type, listed in Table 4, have different objectives: to increase wear resistance by increasing surface hardness (C,N in steels), to produce a low-shear strength surface to inhibit scuffing or seizure (S in steels), or to provide either of the above in conjunction with increased corrosion-resistance (Sn-Cu in steels).

Analogous to diffusion treatments, although not involving high bulk temperatures, is the recently developed "ion-implantation" in which surfaces are bombarded with ions of the element of interest accelerated to high energies. The surface usually increases in hardness and also develops a compressive stress which improves fatigue resistance. Although depth of penetration is small, ≃ 100 nm or less, beneficial effects on wear appear to persist long after removal of material to this depth.[24]

Table 5
PLASTICS AND FILLERS FOR SELF-LUBRICATING COMPOSITES

Thermoplastics	Max useful temp (°C)	Reinforcements
		Glass ⎫
		Asbestos ⎬ fibers
		Carbon ⎭
Polyethylene (high MW and UHMW)	80	Textiles (polyester, "Nomex", cotton)
Polyacetal (homo- and co-polymer)	125	Mica
Nylons (types 6,6.6, 11)	130	**Friction and Wear Reducing Additives**
Poly (phenylene sulfide)	~200	Graphite
Poly (tetrafluoroethylene)	275	MoS$_2$
Poly (*p*-oxybenzoate)	300	PTFE
		Metal oxides
Thermosetting		Silicone fluid
Phenolics	~150	**To Increase Thermal Conductivity**
Cresylics	~150	
Epoxies	~200	Bronze
Silicones	~250	Graphite
Polyimides	~300	Silver

Chemical conversion coatings listed in Table 4 comprise "built-up" films produced by reactions in salt solutions. Thicknesses are typically 2 to 25 μm. The films are porous and are most important in the present context as substrates on which to deposit lubricating solids. Without additional lubrication, solid, or liquid, they are of little value for long-term reduction in friction and wear of metals.

Polymers

Polymers are used for solid lubrication in three main ways: as thin films, as self-lubricating materials, or as binders for lamellar solids. PTFE is outstanding in this group and, in thin film form, can exhibit lower friction than any other known polymer (~0.03 to 0.1). Its other main advantages are effectiveness over a wide temperature range, −200 to +250°C, and general lack of chemical reactivity. The low friction of PTFE is attributed to the smooth molecular profile of the polymer chains which, after orientation in early stages of sliding, can then slip easily over each other.[25]

PTFE films are conventionally produced by spraying followed by sintering at temperatures above 325°C. Coating formulations are also available in which PTFE particles are bonded with a synthetic resin curing at a lower temperature. A recent technique of radio-frequency sputtering can produce very uniform, thin films with excellent adhesion to metals.[26] Since load-carrying capacity and endurance of PTFE films on metals are generally inferior to those of the best MoS$_2$ coatings, and low thermal conductivity limits the maximum speed, they tend to be used mainly in moderate conditions of sliding or where contamination by MoS$_2$ might create problems. Anti-stick coatings in food processing equipment and in plastics molding are major areas. The only other polymers widely used as thin-film lubricants are the polyimides.[27] Their maximum useful temperature for long-term use, ≃300°C, exceed that of PTFE but the frictional properties are inferior, f ≃ 0.13 to 0.3.

By far the greatest use of polymers in solid lubrication is in self-lubricating composites as direct replacements for lubricated metals.[28,29] Of the hundreds of polymers commercially available, the few finding widespread use as self-lubricating materials are listed in Table 5. Reinforcing fibers, fillers, and additives commonly incorporated to improve particular prop-

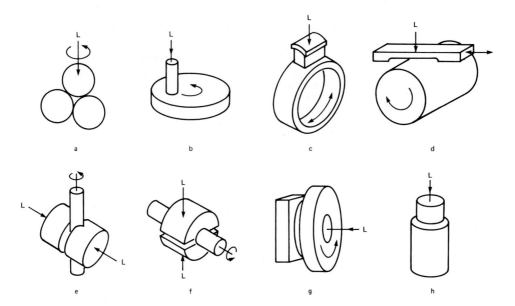

FIGURE 1.   Wear-testing apparatus for solid lubricants. Initial point contact: (a) four-ball; (b) hemisphere on disc (may be 3 pins). Initial line contact: (c) block on ring (Timken, LFW1); (d) Reciprocating pad on ring; and (e) Falex. Conforming contact: (f) journal bearing (Almen-Wieland); (g) thrust bearing (LFW3); and (h) Press-fit (LFW4).

erties are also given. PTFE almost invariably requires reinforcement when used in bulk as it is extremely susceptible to viscoelastic deformation under load. Reinforcements are also commonly used with some thermosetting resins, e.g., phenolics, to increase toughness. Friction and wear properties of the latter are improved by addition of lamellar solids such as MoS_2, or PTFE powder or flock. Some additives can also be multifunctional. A good example is graphite which, particularly in fiber form, not only reduces friction and wear but also increases the strength, stiffness and thermal conductivity of polymer composites.

FRICTION AND WEAR TESTING

Three separate objectives are involved in performance-testing solid lubricants and self-lubricating materials: provision of design data, selection or development of materials, and quality control. Unfortunately, reliable design information is available only from tests either in the intended application or in a very close laboratory simulation. For materials selection, development and quality control, however, a variety of accelerated test procedures can be used,[30] and the most common are illustrated in Figure 1. With tests involving nonconformal geometry (Figures 1a to e), thin-film solid lubricants are usually applied to the larger, rotating surface because this makes the greatest contribution to the total wear life. With conformal geometries, both surfaces are usually coated. The wear life of thin-film lubricants is obtained by determining the time or sliding distance before the coefficient of friction rises to some arbitrarily fixed value such as 0.2. Amount of wear is seldom measured per se, although an average wear rate can be inferred from the film thickness and time to failure. Load-carrying capacity is frequently found by increasing the applied load in increments until failure occurs, either by increased friction (thin films), by greatly increased wear, or by excessive temperature rise (self-lubricating composites).

Relative ratings of different materials may vary significantly between one test and another. One attempt to provide a basis for comparison suggests that the fundamental parameter affecting wear life is the number of cycles of compression/flexure to which each element of the film is subjected.[31] Even in very carefully controlled conditions, repeat determinations of wear life can show considerable scatter. With the Timken apparatus, Figure 1c, scatter in wear life determinations can exceed ± 100%. With Falex tests, Figure 1e, scatter is usually less than ± 50%. Falex tests are commonly incorporated into specification requirements for thin film lubricants.

The four-ball machine, Figure 1a, is widely used for evaluating solid lubricant additives in oils; the pin/disc and pin/ring arrangements (Figures 1b to d) are used for wear testing self-lubricating composites as well as thin film lubricants; reciprocating line-contact arrangements (Figure 1d) show promise for wear testing thin, self-lubricating, bearing-liner materials;[32] the press-fit test (Figure 1h) is used for dry powders and rubbed films and the journal and thrust-bearing configurations (Figures 1f and g) simulate bearing applications for both thin films and self-lubricating composites.

OPERATIONAL PERFORMANCE

Thin Film Lubricants
Rubbed Films

The simplest way to coat a solid lubricant on a metal surface is by burnishing of dry powder (MoS_2, graphite, etc.) with a soft tissue. MoS_2 films produced in this way range from 0.1 to 10 μm thick, depending on rubbing time. Film thickness also increases with increasing humidity.[33] Bonding of lamellar solids to the substrate appears to involve three mechanisms: (1) particles can be physically trapped within surface depresssions, (2) crystallites may be mechanically embedded into the substrate and act as nuclei around which film growth occurs via intercrystallite cohesion, and (3) the lubricant may interact chemically with the substrate. The importance of the last component is supported by observations that effectiveness of MoS_2 film formation on different metals correlates with the strength of the metal-sulfur bond.[34]

Behavior of rubbed MoS_2 films shows some general trends with operational parameters. Friction rises with increasing relative humidity,[35] possibly as a result of increased hydrogen bonding between adsorbed water molecules. Initial reduction in friction with increasing temperature can be attributed to desorption of water vapor, but reduction in wear life as temperatures rise above 200°C is more probably a consequence of increasing oxidation of the MoS_2. Effects of substrate roughness on wear life are consistent with the idea that mechanical entrapment of particles plays a major role in film formation; if the topography is very smooth, little lubricant is contained within the surface depressions, but if the surface is very rough metal peaks may protrude through the lubricant film. Relation of wear life to substrate hardness involves an uncertain trend.[36,37]

The possibility that MoS_2 might induce corrosion of ferrous substrates in humid environments has been the subject of much controversy. Oxidation of MoS_2 is accelerated by moisture, and after prolonged storage of powder in air at room temperature, MoO_3, adsorbed H_2O, and H_2SO_4 can all be present as surface contaminants. For this reason, pH limits of aqueous extracts from MoS_2 powder are required by most specifications,[38] or a direct corrosion test.[39] MoS_2 powder is commonly protected against oxidation during storage either by adsorption of long chain organic inhibitors or by enclosure in an inert gas atmosphere.

Bonded Coatings

To overcome the dependence of burnished film thickness on relative humidity, and to obtain greater film thickness and wear lives, lamellar solids are often incorporated within a synthetic resin binder to produce a "bonded coating". An enormous number of coating formulations has been developed[40] and some of the more widely used constituents are listed in Table 6. MoS_2 is by far the most common. Relevant specifications are given in Table 7.

Table 6
CONSTITUENTS OF BONDED-FILM
FORMULATIONS

Binders	Lubricants	Other Components
Organic-air drying	MoS_2	Corrosion inhibitors
	WS_2	
Cellulosics	Graphite	Sodium phosphite
Acrylics	$(CF_x)_n$	Stannous chloride
		Lead phosphite
Organic-heat curing		
	PTFE	
Alkyd	Phthalocyanine	Solvents
Phenolic		
Vinyl-butyral		Water
Epoxy	CaF_2/BaF_2	Isopropyl alcohol
Silicone	PbO	Toluene
Polyimide	Pbs	Amyl acetate
	Sb_2O_3	Ethyl acetate
		Naphtha
Inorganic salts		
	Au	
Sodium silicate	Ag	
Aluminium phosphate	In	
Sodium phosphate	Pb	
Potassium silicate		
Sodium borate		
Titanates		
CaF_2/BaF_2		
Inorganic ceramics		
Silica		
B_2O_3		
Hydrated Al_2O_3		
Metals		
Ag		
Cu		
Au		

Table 7
SPECIFICATIONS FOR SOLID FILM BONDED COATINGS

US-MIL-L23398	Lubricant, solid film, air-drying
UK-DEF-STAN 91-19/1 }	Lubricant, solid film, heat-curing
US-MIL-L-8937	
US-MIL-L-46010	Lubricant, solid film, heat-cured, corrosion inhibiting
US-MIL-L-81329	Lubricant, solid film, extreme environment

Most organic resins are only stable below about 300°C. Inorganic binders required for higher temperatures fall into two groups: (1) salts, such as Na_2SiO_3 and $AlPO_4$ which are used in solution with dispersed MoS_2 and (2) oxides/ceramics which are applied together with MoS_2 by spraying and subsequently sintered or fused. The choice of binder is also influenced by mechanical properties, environmental compatibility, and ease of processing. For convenience, many coatings for moderate-duty applications incorporate room-temperature-curing acrylic binders and are applied from pressurized aerosol containers. Thermosetting resin binders requiring heat-curing tend to provide longer wear lives than those curing at room temperature.

With the possible exception of polyimides, most binders have intrinsically poor frictional properties and the optimum lubricant to binder ratio usually ranges from 1:1 to 4:1. High ratios minimize friction while low ratios maximize wear life. Other additives can also be included in the coating. Sb_2O_3 generally increases the wear life of MoS_2 coatings when added at a concentration of around 30% by weight, and is believed to function as a sacrificial antioxidant. Inhibitors, such as dibasic lead phosphite, reduce substrate corrosion and other metal sulfides can increase wear life. Graphite additions increase wear life but are falling into disfavor because of possible electrochemical corrosion.

Bonded coatings are generally applied from dispersions in a volatile solvent by spraying, brushing, or dipping. Spraying is usually the most consistent, but dipping is widely used because of low cost. Recommended thicknesses range from 5 to 25 μm, but even thicker coatings may be useful in low-stress applications. Surface pretreatment is essential both to remove organic contamination and to provide a suitable topography for mechanical "keying". Optimum roughness depends on the finishing process used: abrasion 0.5 μm Ra, grit-blasting 0.75 μm Ra, grinding 1.0 μm and turning 1.25 μm Ra. An alternative, or additional, pretreatment is phosphating for steels and analogous chemical conversion treatments for other metals.

It is more difficult to generalize performance trends for bonded coatings than for rubbed films of lamellar solids because their properties depend on the type of binder and on the test method. In low stress conditions wear life usually increases with film thickness but at high stresses the reverse may occur.[41] Sliding speed usually has little effect on either friction or wear until it becomes so high that frictional heating begins to soften or degrade organic resin binders. The most important variable is temperature. With organic binders, wear life tends to decrease with increasing temperature but with inorganic binders the converse is sometimes observed because of low-temperature brittleness. Probably best all-round performance over the widest temperature range is given by formulations incorporating high-temperature resin binders such as polyimides. Binder properties may also affect the way in which wear life depends on relative humidity.

Significant reductions in both wear life and load-carrying capacity of solid lubricant films occur in the presence of conventional oils.[42] In some cases the reduction in performance is a consequence of the resin binder being attacked by certain fluids, e.g., acrylics by chlorinated organic solvents. More generally, fluids tend to cause adhesion failures at the substrate interface and also impede reaggregation of lubricant debris produced during wear. Despite these reductions in performance, some MoS_2-bonded coatings persist sufficiently long in the presence of oils to facilitate running-in,[43] and to reduce tool wear during machining operations.[44]

The most promising high-temperature coatings are those incorporating CaF_2/BaF_2 eutectic. These may be applied by spraying from dispersions, followed by fusing at around 1000°C, or bonded with metal salts such as monoaluminum phosphate.[45] Thicker coatings, 0.1 mm upwards, can be produced by plasma-spraying mixtures of CaF_2/BaF_2 with metals, oxides, or graphite, followed by machining and a final heat treatment to enrich the lubricant phase in the surface.[46] Applications include seals for gas turbine regenerators and high-temperature air-frame bearings. Thin coatings of mixed fluorides have also been used on retainers of ball bearings for hostile environments.[47] For cryogenic applications, bonded coatings containing either MoS_2 or PTFE are generally satisfactory, although some resin binders can become rather brittle. PTFE films tend to lose adhesion to metal substrates on cooling to low temperatures as a result of their high thermal expansion coefficients; this may be offset by low expansion fillers in the coatings, e.g., lithium aluminum silicate.

Self-Lubricating Composites

The main applications of self-lubricating composites are for dry bearings, gears, seals, sliding electrical contacts, and retainers in rolling element bearings. This section concentrates on the influence of composition and sliding conditions on wear.

Polymer Composites

Because low thermal conductivity inhibits dissipation of frictional heat, thermoplastics undergo large increases in wear above critical loads and speeds as a consequence of surface melting. Effects on thermosetting resins are less dramatic because oxidative degradation, leading to surface embrittlement, is a function of exposure time as well as temperature. Thermal conductivity of the counterface is also relevant and at high sliding speeds can become more important than the conductivity of the polymer composite itself. Limiting speeds for polymers sliding against themselves are, in general, several hundred times lower than those for polymers sliding against metals.[48]

Wear rates of polymer composites depend strongly on the surface roughness of metal counterfaces. In early stages of sliding, wear rate varies typically with initial Ra roughness raised to a power of 2 to 4;[49] for this reason smooth counterfaces are always recommended for applications such as dry bearings. During running-in, however, the initial counterface roughness is frequently reduced, either by transfer of the polymer and/or fillers or by polishing/abrasive action of fillers, leading to a reduction in wear rate. Steady-state roughness and steady-state rate of wear depend both on the composite composition and on relative hardness of the fillers and counterface.[50] Relationships between steady-state rate of wear and initial counterface roughness thus become very variable and examples are shown in Figure 2. Although an optimum counterface roughness for minimum wear is sometimes suggested, experimental results are conflicting.

For PTFE composites and other polymers incorporating solid lubricants which rely on transfer film formation on the counterface to achieve low wear, wear behavior is strongly influenced by environmental factors. Relative humidity is particularly important and increasing humidity can either reduce or increase wear depending on the type of filler; there are no systematic trends.[51] Liquid water, however, increases wear by inhibiting transfer film formation and the aggregation of wear debris. Other fluids, including conventional hydrocarbon lubricants, produce similar effects although to a smaller extent. For polymer composites which do not rely on transfer film formation, e.g., nylons and acetals, hydrocarbon lubricants usually reduce wear[52] and are often effective in extremely small amounts. Small pockets of fluid within the bulk structure can provide a continuous source of lubricant.[53]

Applications of polymer composites are extremely diverse. For dry bearings, some of the most successful composites are of complex construction, e.g., a layer of sintered bronze of graded porosity on a steel backing and filled with PTFE/Pb,[3] or a fabric liner of interwoven PTFE and glass fibers impregnated with synthetic resin and adhesively bonded to a steel backing.[54] Composites of the latter type are widely used in aerospace applications; a typical modern aircraft may contain several hundred. For transfer lubrication of rolling-element bearings, a particularly successful composite for retainers is PTFE/glass fiber/MoS_2.[55,56]

Metal-Lamellar Solid Composites

A wide variety of metal-solid lubricant mixtures have been developed and some examples are listed in Table 8. With those containing lamellar solids, low friction is achieved via transfer. Since transfer film formation is an inefficient process, a high proportion of solid lubricant, 25% or more, is usually needed. Since such composites are mechanically weak, low friction tends to be associated with high wear and vice versa, as shown in Figure 3. For any given materials, however, conditions which reduce friction, such as increased temperature with fluoride or oxide films, usually reduce wear rate also.

A great deal of effort has been devoted to material combinations and/or composite fabrication to obtain both low friction and wear. Incorporation of PTFE in lamellar solid-metal composites appears to facilitate transfer film formation, and carbides in Ta-Mo-MoS_2 improve strength.[57] Fabrication techniques use conventional powder metallurgy, infiltration of porous metals, electrochemical codeposition, plasma spraying, and machining of holes or recesses

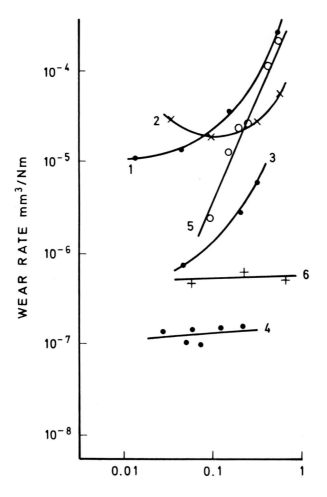

FIGURE 2.   Variation of steady-state wear rate with initial counterface roughness for some PTFE composites.

1.⎫
2.⎪PTFE fiber/glass fiber/resin ⎫Line contact tests
3.⎬liners on a steel backing   ⎪Journal bearings
4.⎭   ⎬Pad on track
                                  ⎭Partial journal bearings

5. PTFE + 25 % graphite fiber⎫
6. PTFE + 25% glass fiber   ⎬ Pin and disc tests

in metals which are subsequently filled with solid lubricant formulations. Porous Ni-Cr impregnated with BaF_2/CaF_2 eutectic has been used for high-temperature piston rings, etc. in hostile environments, but optimum performance necessitates careful surface preparation and heat treatment.[58] Some multicomponent, plasma-sprayed coatings, e.g., 30% Ag, 30% Ni-Cr, 25% CaF_2, and 15% glass avoid the latter problems and give low friction and wear over a very wide temperature range, -100 to $+870°C$ in both air and vacuum.[59]

A major application of metal-solid lubricant composites is for self-lubricating rolling bearing retainers in vacuum, or at elevated temperatures up to 400°C. Retainers based on Ga-In-WSe_2, Ta-Mo-MoS_2, and metals with recesses containing MoS_2-graphite-sodium silicate are all capable of providing useful bearing lives at 400°C in air or vacuum. At more

Table 8
SOME METAL-SOLID LUBRICANT COMPOSITES AND THEIR APPLICATIONS

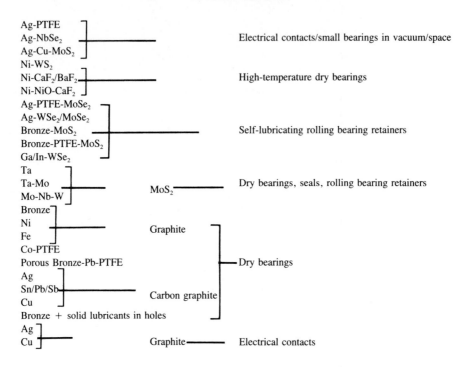

moderate temperatures, however, lives are generally very much shorter than those obtainable for conventional retainers with oil or grease lubrication. Bearings with self-lubricating retainers have also proved useful for "fail-safe" applications where a limited bearing life is required following failure of a conventional lubricant supply.

Carbons and Graphites

Materials falling into this category are listed in Table 9. During repeated sliding over the same contact area, even carbons of low graphiticity undergo structural degradation and produce low shear strength surface films with many of the properties of graphite, including low friction. The main limitations of carbons for use in bulk are low tensile strength and lack of ductility, but high thermal and oxidative stabilities make them particularly attractive for use at high temperatures and/or high sliding speeds. The oxidation limit, typically 500 to 600°C, can be further increased by additives, particularly compounds of B or P. Patent literature on this topic is extensive.[60]

Wear of carbon is very sensitive to counterface roughness before its topography is modified by sliding. For graphitic carbons, transfer is the main mechanism of counterface modification, but for harder, less graphitic carbons, abrasion is more important and hardened counterfaces are usually required. Fluid lubrication sometimes increases the wear rates of graphitic carbons both by inhibiting transfer film formation and by inducing hydrostatic stresses within the porosity leading to surface disruption.[61] Wear rates in dry conditions tend to increase with increasing temperature, partly as a result of localized oxidation if sliding speed is high,[62] and partly as a consequence of loss of physically adsorbed water vapor inhibiting debris aggregation and transfer-film formation.[63] The greatest uncertainty at high temperatures or high speeds is possible catastrophic "dusting" wear in the absence of sufficient water vapor in the environment. This was originally encountered on the carbon brushes of aircraft

Table 9

CLASSIFICATION OF CARBONS AND GRAPHITES

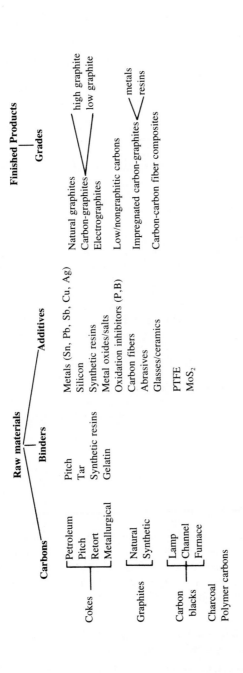

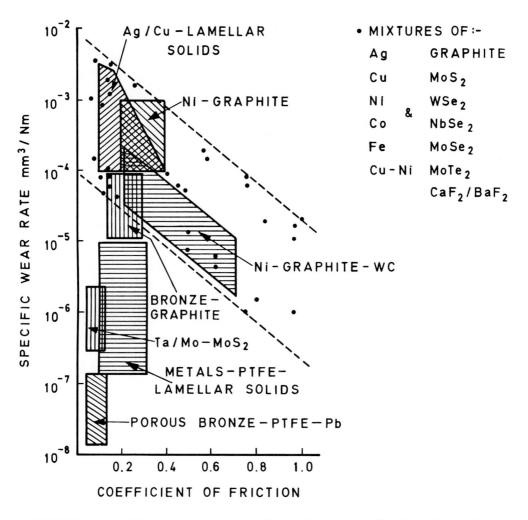

FIGURE 3.   Relationship between wear rate and coefficient of friction for metal-lamellar solid composites.

Note: $1 \dfrac{mm^3}{Nm} \simeq 10^{-9} \dfrac{m^2}{N} \simeq 8 \times 3.10^{-5} \dfrac{in^3}{ft\text{-}lbf} \simeq 5.2 \times 10^{-3} \dfrac{in^3 - min}{ft\text{-}lbf\text{-}hr}$

electrical generators at high altitudes but has subsequently been observed in vacuum, inert gases, and dry air.[64] Dusting can now be inhibited satisfactorily by a wide range of additives including halides of Pb, Ba, and Sr, organic resins, MoS_2, and PTFE.[60] Even without such additives, sufficient organic contamination is present in most industrial conditions to act as an effective substitute for water vapor. Use of carbons as electrical contacts is reviewed by Shobert,[65] while seal applications are discussed by Mayer.[66]

Ceramics and Cermets

High temperature solid film lubricants based on CaF_2/BaF_2 or PbO/SiO_2 have already been described, but for applications where low friction is less important than low wear rate, several types of ceramics and cermets can be used up to temperatures of the order of 1000°C. Some of the most widely used compositions are listed in Table 10. While it is economically attractive to use these materials as plasma-sprayed coatings on temperature-resistant substrates, adhesion may then present problems during thermal cycling. The advantages of

Table 10
SOME CERAMICS AND CERMETS FOR
HIGH-TEMPERATURE USE

Ceramics	Cermets	Sprayed coatings
Al$_2$O	TiC ⎤	⎡Ni	Al$_2$O$_3$ (+ TiO$_2$)

Ceramics: Al$_2$O, B$_4$C, Si$_3$N$_4$, SiC

Cermets: TiC, Cr$_3$C$_2$, WC + Ni, Co, Mo, Cr; Al$_2$O$_3$-Cr-Mo; NiO-Ni

Sprayed coatings: Al$_2$O$_3$ (+ TiO$_2$); Cr$_3$C$_2$ (+ Ni, Cr, Co); WC (+ Ni, Fe, Cr, Co); Cr$_2$O$_3$(+ Cr + Al$_2$O$_3$)

Table 11
SELECTION OF BEARING MATERIALS FOR
VARIOUS CONDITIONS

Requirement		Order of decreasing suitability →			
Low wear/long life	5	3	6	7	4
Low friction	11	9	5	3	4
High temperatures	10	9	11	7	8
Low temperatures	3	11	9	4	5
High loads	9	10	6	5	4
High speeds	11	9	8	5	7
High stiffness	11	9	6	4	5
Dimensional stability	10	11	9	7	8
Compatibility with fluid lubricants	7	10	4	8	2
Corrosive environments	10	7	3	4	2
Compatibility with abrasives	1	3	2	4	5
Tolerance to soft counterfaces	1	9	2	3	4
Compatibility with radiation	7	4	9	10	2
Space/vacuum	11	9	4	3	2
Minimum cost	1	2	3	4	5

Note: Key: 1 = unfilled thermoplastics, 2 = filled/reinforced thermoplastics, 3 = filled/reinforced PTFE, 4 = filled/reinforced thermosetting resins, 5 = PTFE impregnated porous metals, 6 = woven PTFE/glass fiber, 7 = carbons-graphites, 8 = metal-graphite mixtures, 9 = solid film lubricants, 10 = ceramics, cermets, hard metals, and 11 = rolling bearings with self-lubricating cages.

cermets over ceramics are greater toughness and ductility, but the metal content, usually Co or Ni, reduces the maximum temperature.

Few general guidelines are available to predict the wear behavior of ceramics, particularly coatings where properties depend as much on method of deposition as on composition. Friction coefficients tend to be very variable but can be as low as 0.2 to 0.25 at high temperatures, e.g., Cr$_2$C$_3$-Ni-Cr or Cr$_2$O$_3$ sliding against themselves.[5] Attempts to incorporate solid lubricants into bulk ceramics to reduce friction have met with little success, except when confining them to machined holes and recesses.[67]

Selection of Materials for Dry Sliding

Various attempts have been made to provide general guidelines for selection of materials for specific applications. For dry bearings, one approach is to identify major application requirements as listed down the left hand side of Table 11, and then select the group of

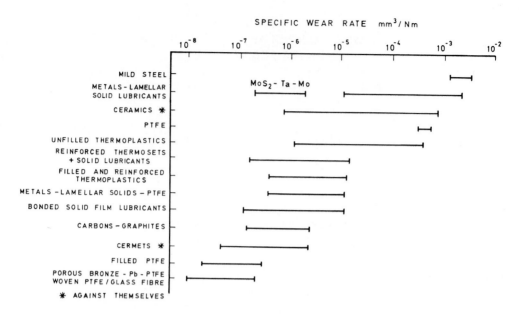

FIGURE 4.   Order-of-magnitude wear rates of self-lubricating composites sliding against steel at room temperature, light loads, and low speeds

materials which offers the best compromise solution. Published wear rates of the selected materials obtained in low-duty sliding conditions where frictional heating is negligible are then modified to take into account sliding conditions appropriate to the intended application. Figure 4 illustrates the range of wear rates typical of various groups of self-lubricating composites, and approximate wear rate correction factors are listed in Table 12. A more complete account of this procedure, together with information about individual materials, is given elsewhere.[68] Unfortunately, a similar approach is not yet available for self-lubricating components other than dry bearings, e.g., gears, seals, or thin-film solid lubricant coatings.

Dispersions in Oils and Greases

Graphite and MoS are extensively used as additives in conventional oils and greases to reduce friction and wear when full-film hydrodynamic or elastohydrodynamic lubrication cannot be achieved. The concentrations added vary widely, from 0.1 to 60% by weight, the higher values producing pastes used primarily for component assembly purposes. Relevant specifications are listed in Table 13. Numerous rig tests have demonstrated that MoS_2 can provide increases in load-carrying capacity, reductions in wear, and increased life of rolling bearings. The optimum concentrations depend on the type of carrier fluid and the sliding conditions but are typically around 3% by weight in oils and 20% by weight in greases. Automotive experience has confirmed the beneficial effects of MoS_2 additions to oils in reducing both wear and fuel consumption (friction).[69] Two cautionary comments are in order. First, detergent additives in automotive oils can inhibit the wear-reducing ability of MoS_2 and graphite, and some anti-wear additives can even increase wear rates slightly.[70] Second, solid lubricant additions can affect the oxidation stability of oils and greases, and this may influence the concentration of oxidation inhibitors required; smaller particles have a greater effect on oxidation stability than larger ones.

The influence of solid lubricant particle size on performance in oils and greases is confused.[71] Particle shape can be important, and significant improvements in performance have been reported when using dispersions of "oleophilic" graphite and MoS_2.[72] These materials are produced as very thin, plate-like particles by grinding in hydrocarbon media, and can

Table 12
APPROXIMATE CORRECTION FACTORS FOR WEAR RATES

Geometrical Factor a	Continuous motion	Rotating Load	0.5
	Oscillatory motion	Unidirectional Load	1
			2

Heat Dissipation Factor b	Metal Housing, intermittent	0.5
	Metal housing, continuous	1
	Nonmetal housing, continuous	2

Temperature Factor c			
	20°C	1	1
	100°C	2	3
	200°C	5	6
		PTFE composites	Carbons/graphites, thermosets

Counterface Material Factor d	Stainless steels, chromium	0.5
	Steels	1
	Soft, nonferrous alloys(Cu,Al)	2—5

Counterface Roughness Factor e	0.1—0.2 μm cla	1
	0.2—0.4 μm cla	2—3
	0.4—0.8 μm cla	4—10

Note: Wear depth/unit time = a b c d e K P V, (K = specific wear rate of material, m^2/N).

Table 13
SPECIFICATIONS FOR SOLID LUBRICANT DISPERSIONS IN OILS AND GREASES

Paste

UK-DTD-392B US-MIL-T-5544	Antiseize compound, high temperatures (50% graphite in petrolatum)
UK-DTD-5617	Antiseize compound, MoS_2 (50% MoS_2 in mineral oil)
US-MIL-A-13881	Antiseize compound, mica base (40% mica in mineral oil)
US-MIL-L-25681C	Lubricant, MoS_2, silicone (50% MoS_2 — antiseize compound)

Grease

US-MIL-G-23549A	Grease, general purpose (5% MoS_2, mineral oil base)
UK-DTD-5527A US-MIL-G-21164C	Grease, MoS_2, low and high temperature (5% MoS_2, synthetic oil base)
US-MIL-G-81827	Grease, MoS_2, high load, wide temperature range (5% MoS_2)
UK-DEF-STAN 91-18/1	Grease, graphite, medium (5% in mineral oil base)
UK-DEF-STAN 91-8/1	Grease, graphite (40% in mineral oil base)

Oil

UK-DEF-STAN 91-30/1 US-MIL-L3572	Lubricating oil, colloidal graphite (10% in mineral oil)

be made with sufficiently high surface areas to act as grease thickeners in their own right. Graphite dispersions are widely used in high-temperature metal-working operations such as extrusion and drop forging, and as parting agents in casting. Performance can be greatly enhanced by additives. Effects of additions of metal oxides and salts to graphite-oil pastes during high-temperature extrusion have been surveyed by Cook.[73]

Solid lubricants other than graphite and MoS_2 which have been used as additives to conventional fluid lubricants are various phosphates, oxides, and hydroxides such as $Zn_2P_2O_7$ and $Ca(OH)_2$, and PTFE. The former groups are of interest where the black color of MoS_2 or graphite is a disadvantage, e.g., in textile machinery. PTFE may also be used for this purpose, but its special properties are more fully exploited in PTFE-thickened fluorocarbon greases, which can provide effective lubrication in oxidizing environments over a wide temperature range.[74] Typical applications are in rocket motors and space components.

REFERENCES

1. **Campbell, W. E.,** Solid lubricants, in *Boundary Lubrication: An Appraisal of World Literature,* Ling, F. F., Klaus, E. E., and Fein, R. S., Eds., American Society of Mechanical Engineers, New York, 1969, 197.
2. **Lansdown, A. R.,** Molybdenum disulphide: a survey of the present state of the art, *Swansea Tribol. Cent. Rep.,* 74, 279, 1974.
3. **Pratt, G. C.,** Plastic-based bearings, in *Lubrication and Lubricants,* Braithewaite, E. R., Ed., Elsevier, Amsterdam, 1967, 377.
4. **Claus, F. J.,** *Solid Lubricants and Self-Lubricating Solids,* Academic Press, New York, 1972.
5. **Lancaster, J. K.,** Dry bearings: a survey of materials and factors affecting their performance, *Tribology,* 6, 219, 1973.
6. **Lancaster, J. K.,** Friction and wear (of polymers), in *Polymer Science,* Jenkins, A. D., Ed., North Holland, Amsterdam, 1972, 960.
7. **Tabor, D.,** Friction, adhesion and boundary lubrication of polymers, in *Advances in Polymer Friction and Wear,* Lee, L.-H., Ed., Plenum Press, New York, 1974, 1.
8. **Roselman, I. C. and Tabor, D.,** The friction of carbon fibres, *J. Phys. D.,* 9, 2517, 1976.
9. **Peterson, M. B. and Johnson, R. L.,** Friction Studies of Graphite and Mixtures of Graphite With Several Metallic Oxides and Salts at Temperatures to 1000°F, TN-3657, National Aeronautics and Space Administration, Washington, D.C., 1956.
10. **Grattan, P. A. and Lancaster, J. K.,** Abrasion by lamellar solid lubricants, *Wear,* 10, 453, 1967.
11. **Giltrow, J. P. and Lancaster, J. K.,** The role of impurities in the abrasiveness of MoS_2, *Wear,* 20, 137, 1972.
12. **Magie, P. M.,** A review of the properties and potentials of the new heavy metal derivative solid lubricants, *Lubr. Eng.,* 22, 262, 1966.
13. **Fusaro, R. L. and Sliney, H. E.,** Graphite fluoride, $(CF_x)_n$ — a new solid lubricant, *ASLE Trans.,* 13, 56, 1970.
14. **Play, D. and Godet, M.,** Study of the Lubricating Properties of $(CF_x)_n$, *Coll. Int. CNRS,* 233, 441, 1975; NASA Rep. TM 75191, National Aeronautics and Space Administration, Washington, D.C., 1975.
15. **McConnell, B. D., Snyder, C. E., and Strang, J. R.,** Analytical evaluation of graphite fluoride and its lubrication performance under heavy loads, paper 76-AM-5C-3, *ASLE Trans.,* 1976. preprint.
16. **Gisser, H., Petronic, M., and Shapiro, A.,** Graphite fluoride as a solid lubricant, *Lubr. Eng.,* 28, 161, 1972.
17. **Martin, C., Sailleau, J., and Roussel, M.,** The ultra-high vacuum behavior of graphite-fluoride filled self-lubricating materials, *Wear,* 34, 215, 1975.
18. **Fusaro, R. L.,** Effect of Fluorine Content, Atmosphere and Burnishing Technique on the Lubricating Properties of Graphite Fluoride, TN-D-7574, National Aeronautics and Space Administration, Washington, D.C., 1974.
19. **Bisson, E. E.,** Non-conventional lubricants, in Advanced Bearing Technology, SP-38 Bisson, E. E. and Anderson, W. J., Eds., National Aeronautics and Space Administration, Washington, D.C., 1964, 203.
20. **Olsen, K. M. and Sliney, H. E.,** Additions to Fused Fluoride Lubricant Coatings for Reduction of Low Temperature Friction, TN-D-3793, National Aeronautics and Space Administration, Washington, D.C., 1967.
21. **Devine, M. J., Cerini, J. P., Chappell, W. H., and Soulen, J. R.,** New sulphide addition agents for lubricant materials, *ASLE Trans.,* 11, 283, 1968.

22. **Stott, F. H., Lin, D. S., Wood, G. C., and Stevenson, C. W.,** The tribological behavior of nickel and nickel-chromium alloys at temperatures from 20° to 800°C, *Wear,* 36, 147, 1976.

23. **Todd, M. J. and Bentall, R. H.,** Lead film lubrication in vacuum, *Proc. ASLE 2nd Int. Conf. Solid Lubr.,* SP-6, American Society of Lubrication Engineers, Park Ridge, Ill., 1978, 1948.

24. **Dearnaley, G. and Hartley, N. E. W.,** Ion implantation of engineering materials, *Proc. Conf. Ion Plating and Allied Techniques,* CEP Consultants Ltd., Edinburgh, 1977, 187.

25. **Pooley, C. M. and Tabor, D.,** Friction and molecular structure: the behavior of some thermoplastics, *Proc. R. Soc. London Ser. A,* 239, 251, 1972.

26. **Spalvins, T.,** Sputtering technology in solid film lubrication, *Proc. ASLE 2nd Int. Conf. on Solid Lubr.,* SP-6, American Society of Lubrication Engineers, Park Ridge, Ill., 1978, 109.

27. **Fusaro, R. L.,** Friction and Wear Life Properties of Polyimide Thin Films, TN-D-6914, National Aeronautics and Space Administration, Washington, D.C., 1972.

28. **Brydson, J. A.,** *Plastic Materials,* 3rd. ed., Butterworths, London, 1975.

29. **Theberge, J. E.,** Properties of internally lubricated, glass-fortified thermoplastics for gears and bearings, *Proc. ASLE Int. Conf. Solid Lubr.,* SP-3, American Society of Lubrication Engineers, Park Ridge, Ill., 1971, 106.

30. **Benzing, R. J., Goldblatt, I., Hopkins, V., Jamison, W., Mecklenburg, K., and Peterson, M.,** *Friction and Wear Devices,* 2nd ed., American Society of Lubrication Engineers, Park Ridge, Ill., 1976.

31. **McCain, J. W.,** A theory and tester measurement correlation about MoS_2 dry film lubricant wear, *SAMPE J.,* February/March, 1970, 17.

32. **Lancaster, J. K.,** Accelerated wear testing of PTFE composite bearing materials, *Tribol. Int.,* 12, 65, 1979.

33. **Johnston, R. R. M. and Moore, A. J. W.,** The burnishing of molybdenum disulphide on to metal surfaces, *Wear,* 19, 329, 1972.

34. **Stupian, G. W., Feuerstein, S., Chase, A. B., and Slade, R. A.,** Adhesion of MoS_2 powder burnished on to metal substrates, *J. Vac. Sci. Technol.,* 13, 684, 1976.

35. **Pritchard, C. and Midgley, J. W.,** The effect of humidity on the friction and life on unbonded molybdenum disulphide films, *Wear,* 13, 39, 1969.

36. **Tsuya, Y.,** Microstructure of wear, friction and solid lubrication, *Tech. Rep. Mech. Eng. Lab. Tokyo,* 81, 1975.

37. **Lancaster, J. K.,** The influence of substrate hardness on the friction and endurance of molybdenum disulphide films, *Wear,* 10, 103, 1967.

38. Military specifications, Molybdenum Disulphide Powder, Lubricating, U.K.: DEF-STAN 68-62/1; France: AIR 4223; W. Germany: VTL - 6810-015; Canada: 3-GP-806a.

39. Military specifications, Molybdenum Disulphide, Technical, Lubrication Grade, U.S.: MIL-M-7866B.

40. **Campbell, M. E. and Thompson, M. B.,** Lubrication Handbook for Use in the Space Industry, Part A — Solid Lubricants, CR-120490, National Aeronautics and Space Administration, Washington, D.C., 1972.

41. **Hopkins, V. and Campbell, M. E.,** Film thickness effect on the wear life of a bonded solid lubricant film, *Lubr. Eng.,* 25, 15, 1969.

42. **Hopkins, V. and Campbell, M. E.,** Important considerations in the use of solid film lubricants, *Lubr. Eng.,* 27, 396, 1971.

43. **Kawamura, M., Hoshida, K., and Acki, I.,** Running-in effect of bonded solid film lubricants on conventional oil lubrication, *Proc. ASLE 2nd Int. Conf. Solid Lubr.,* SP-6, American Society of Lubrication Engineers, Park Ridge, Ill., 1978, 101.

44. **Harley, D. and Wainwright, P.,** Development of a dry film tool lubricant, *Proc ASLE 2nd Int. Conf. on Solid Lubr.,* SP-6, American Society of Lubrication Engineers, Park Ridge, Ill., 1978, 281.

45. **Lavik, M. T., McConnell, B. D., and Moore, G. D.,** The friction and wear of thin, sintered, fluoride films, *J. Lubr. Technol.,* Trans. ASME, 95, 12, 1972.

46. **Sliney, H. E.,** Self-Lubricating Plasma-Sprayed Composites for Sliding-Contact Bearings to 900°C, TN D-7556, National Aeronautics and Space Administration, Washington, D.C., 1974.

47. **Sliney, H. E.,** A Calcium Fluoride-Lithium Fluoride Solid Lubricant Coating for Cages of Ball-Bearings to be Used in Liquid Fluorine, TMX-2033, National Aeronautics and Space Administration, Washington, D.C., 1970.

48. **Evans, D. C. and Lancaster, J. K.,** The wear of polymers, in *Treatise on Materials Science and Technology,* Vol. 13, Scott, D., Ed., Academic Press, New York, 1979, 85.

49. **Lancaster, J. K.,** Relationships between the wear of polymers and their mechanical properties, *Proc. Inst. Mech. Eng.,* 183 (3P)), 98, 1969.

50. **Lancaster, J. K.,** Polymer-based bearing materials: the role of fillers and fibre reinforcement, *Tribology,* 5, 249, 1972.

51. **Arkles, B. C., Gerakaris, S., and Goodhue, R.,** Wear characteristics of fluoropolymer composites, *Advances in Polymer Friction and Wear,* Plenum Press, New York, 1974, 663.

52. **Evans, D. C.,** Fluid-polymer interactions in relation to wear, *Proc. 3rd Leeds-Lyon Symp. Wear of Non-Metallic Materials,* Mechanical Engineering Publication, London 1978, 47.
53. **Ikeda, H.,** Plastic-Based Anti-Friction Materials, Japanese Patent, 75101441, 1975.
54. **Williams, F. J.,** Teflon airframe bearings — their advantages and limitations, *SAMPE Quart.,* 8, 30, 1977.
55. **Sitch, D.,** Self-lubricating rolling element bearings with PTFE-composite cages, *Tribology,* 6, 262, 1973.
56. **Anon.,** *Self-Lubricating Bearings — A Performance Guide,* U.K. Natl. Center of Tribology, Risley, Warrington, 1977.
57. **McConnell, B. D. and Mecklenburg, K. R.,** Solid lubricant compacts — an approach to long-term lubrication in space, 76-AM-2E-1, *ASLE Trans.,* 1976, preprint.
58. **Gardos, M. N.,** Some Topographical and Tribological Characteristics of a CaF_2/BaF_2, Eutectic-Containing Porous Nichrome Alloy Self-Lubricating Composite, 74LC-2C-2, *ASLE Trans.,* 1974, preprint.
59. **Sliney, H. E.,** Wide-Temperature-Spectrum Self-Lubricating Coatings Prepared by Plasma Spraying, TM-79113, National Aeronautics and Space Administration, Washington, D.C., 1979.
60. **Paxton, R. R.,** Carbon and graphite materials for seals, bearings, and brushes, *Electrochem. Tech.,* 5, 1974, 1967.
61. **Strugala, E. W.,** The nature and cause of seal carbon blistering, *Lubr. Eng.,* 28, 333, 1972.
62. **McKee, D. W., Savage, R. H., and Gunnoe, G.,** Chemical factors in carbon brush wear, *Wear,* 22, 193, 1972.
63. **Giltrow, J. P.,** The influence of temperature on the wear of carbon fibre-reinforced resins, *ASLE Trans.,* 16, 83, 1973.
64. **Lancaster, J. K.,** The wear of carbons and graphites, in *Treatise on Materials Science and Technology,* Vol. 13, Scott, D., Ed., Academic Press, New York, 1979, 141.
65. **Shobert, E. I.,** *Carbon Brushes: The Physics and Chemistry of Sliding Contacts,* Chemical Publishing Co., New York, 1965.
66. **Mayer, E.,** *Mechanical Seals,* 2nd ed., Illiffe, London, 1972.
67. **Van Wyk, J. W.,** Ceramic Airframe Bearings, 75-AM-7A-3, *ASLE Trans.,* 1975, preprint.
68. **Anon.,** A Guide on the Design and Selection of Dry Rubbing Bearings, Item 76029, Engineering Sciences Data Unit, London, 1976.
69. **Braithewaite, E. R. and Greene, A. B.,** A critical analysis of the performance of molybdenum compounds in motor vehicles, *Wear,* 46, 405, 1978.
70. **Bartz, W. J. and Oppelt, J.,** Lubricating effectiveness of oil soluble additions and graphite dispersed in mineral oil, *Proc. 2nd ASLE Int. Conf. on Solid Lubr.,* SP-6, American Society of Lubrication Engineers, Park Ridge, Ill., 1978, 51.
71. **Bartz, W. J.,** Solid lubricant additives — effect of concentration and other additives on anti-wear performance, *Wear,* 17, 421, 1971.
72. **Groszek, A. J. and Witheredge, R. E.,** Surface properties and lubricating action of graphite and MoS_2, *Proc. ASLE Conf. Solid Lubr.,* SP-3, American Society of Lubrication Engineers, Park Ridge, Ill., 971, 371.
73. **Cook, C. R.,** Lubricants for high temperature extrusion, *Proc. ASLE Conf. Solid Lubr.,* SP-3, American Society of Lubrication Engineers, Park Ridge, Ill., 1971, 13.
74. **Messina, J.,** Rust-inhibited, non-reactive perfluorinated polymer greases, *Proc. ASLE Conf. Solid Lubr.,* SP-3, American Society of Lubrication Engineers, Park Ridge, Ill., 1971, 326.

PROPERTIES OF GASES

Donald F. Wilcock

INTRODUCTION

Increasing interest in and application of gas bearings requires knowledge of a number of gas properties which are not as readily available as the properties of common liquid lubricants. This is particularly true in process fluid lubrication where gases other than air are involved.

This section provides as much as possible of the information required in the design of a wide variety of gas bearings. Some brief background is followed by property data and by discussions on a number of typical applications.

NATURE OF A GAS

In the gaseous state of matter, individual atoms or molecules are in constant motion and are separated from each other by distances of several times their diameter. The gas particles collide with each other frequently and travel in straight lines between collisions. The average velocity of the particles is an expression of the gas temperature, increasing with temperature.

When a gas particle hits a solid surface and bounces off, the change in momentum of the particle exerts a force on the surface. The sum of the countless surface collisions is the pressure the gas exerts on the surface. If one of a pair of parallel surfaces is moving, it will impart an additional component of velocity to each gas particle hitting it. This additional velocity is transmitted to other particles in the course of collisions and eventually to the other surface. The result is a force on the other surface expressed as the product of the area of the surface, the rate of shear, and the viscosity. The rate of shear is defined as the velocity difference between the surfaces divided by the distance between them.

If a volume of gas is compressed, more particles must hit each unit of surface, and the pressure increases. If the temperature is increased, average particle velocity is increased, momentum change in each surface collision increases, and again the pressure is increased. This behavior is expressed in the "perfect gas" law $PV = nRT$ where R is the "gas constant" and n the mass (moles) of the volume of gas involved.

Almost all gases are "perfect" at low pressures, usually one atmosphere or less. Deviation occurs at very low pressures when not enough particles are present to provide many collisions between impacts with the surface. When the pressure is high, the gas particles are forced more closely together, molecular attractions between particles begin to exert an influence, and deviations from the perfect gas law are observed.

Mixtures of gases behave as if each were alone in the total volume. Each exerts a partial pressure equal to the pressure it would exert if it were alone in the volume. The total pressure is then the sum of the partial pressures of the gases that are mixed in the volume.

PROPERTIES OF A GAS

In designing gas bearings, viscosity is usually the property of prime interest. A number of other physical properties may also be required, however, and are described in this section. Chemical properties of any particular gas may influence mixing of the gas with fluids in the system, reactions with other gases, or reaction with bearings or other surfaces. The designer should, therefore, ascertain from other sources the chemical reactivity of the gas.

Boiling point — T_B, is the absolute temperature in degrees Kelvin at which a gas will condense into a liquid. Boiling point increases with pressure.

Density—ρ, also termed mass density, is the mass of gas in kilograms in a volume of one cubic meter.

Absolute viscosity—μ, is used in determining flow of a gas in both hydrostatic and hydrodynamic designs. It is the force in Newtons on an area of one square meter that is exerted by the gas when that area is moved at a velocity of one meter per second parallel to a second surface one meter distant. The units are Newtons per square meter $\times$ seconds, or Pascal-seconds (Pa-sec).

Kinematic viscosity—ν, is the absolute viscosity divided by the mass density. Units are Newton-meter-seconds per kilogram. Since the Newton is the force required to accelerate one kilogram by one meter per second squared, kinematic viscosity has units of m^2/sec.

Temperature— The relation between absolute temperature T in degrees Kelvin and relative temperature in degrees Celsius C is C = T − 273.1.

Specific heat—C_p and C_v, is the energy required to raise the temperature of a unit quantity of gas by one degree. When the process is carried out at constant pressure, the quantity is C_p. At constant volume, the quantity is C_v. The units are kilo-Joules per kilogram per degree.

Sonic velocity—Φ, is the speed with which a pressure wave is transmitted through a gas. Since an increase in pressure can be transmitted only through particle collisions, the speed of transmission will be related to the particle velocities, and hence to gas temperature.

Mean free path—λ, is the average distance traveled by a gas particle between collisions. This quantity is of interest in bearings operating with very close clearances or at very low pressures where the mean free path approaches the surface separation distance. Its calculation is treated in a following section.

Equation of state— Relates the physical properties of a perfect gas to each other and to the quantity of gas present. It is PV = nRT, where V is the volume in m^3 occupied by n kg-mol of gas at an absolute temperature T. Gas constant R = 8.3143 kJ/kg-mol·K.

PHYSICAL DATA

Data in Table 1 are abstracted from an extensive listing of thermophysical properties of liquids and gases.[1] The first three columns give the common name of the gas, its chemical formula, and its molecular weight. Column four gives the boiling point in K at a pressure of 760 mmHg or 1.01 bar. Also, given are specific volume in m^3/kg, heat capacity C_p in kJ/kg·K, speed of sound in m/sec, viscosity in Pa·sec, and the viscosity-temperature exponent in Equation 1.

Viscosity

The viscosity of a gas is nearly independent of pressure over a wide range of lower pressures, but at higher pressures it will increase significantly. Figure 1 illustrates this point for nitrogen, the principal component of air: the viscosity is 18×10^{-6}Pa·sec up to 40 atm pressure, 20×10^{-6} at 100 bar, and 53×10^{-6} at 1000 bar. The viscosities of air at several pressures from 1 to 100 bar are shown in Figure 2 as a function of absolute temperature. This shows that the effect of pressure increases at lower temperatures.

Viscosities of a number of common gases at 1 bar are shown in Figure 3 to increase rapidly with absolute temperature, contrary to the behavior of liquids. The low viscosity of hydrogen is striking, as is the deviation of water vapor from the general trend. The water vapor curve terminates at its boiling point of 373 K. Data for air at a number of temperatures and pressures are shown in Table 2.

In determining viscosity as a function of temperature, two equations are often used. As can be seen from Figure 2, log (gas viscosity) is nearly linear with log (temperature) and can be represented by:

$$\mu = \mu_o (T/T_o)^n \tag{1}$$

Table 1
PROPERTIES OF SELECTED GASES[1]

Gas	Chemical formula	Molecular weight	Boiling point at P — 760 mmH₈ K (1.01 bar)	Heat capacity, C (kJ/Kg-deg.)		Viscosity (msec/m² × 10⁶)		Speed of sound (m/sec)ᵃ		Specific volume (m³/Kg)ᵃ		Viscosity exponent n in Equation 1
				300 K	400 K	300 K	400 K	300 K	400 K	300 K	400 K	
Air	$N_2H_7O_2$	28.96	78.8	1.007	1.014	18.5	23.0			0.861	1.148	0.757
Ammonia	NH_3	17.03	239.6	2.158	2.287	15.76ᵇ	16.35ᵇ			1.451	1.947	0.128
Argon	Ar	39.94	87.3	0.522	0.521	22.7	28.9	322.6	372.6	0.624	0.833	0.839
Carbon dioxide	CO_2	44.01	194.6ᶜ	0.851	0.942	14.9	19.4	269.4	307.3	0.564	0.754	0.918
Carbon monoxide	CO	28.01	81.6	1.043	1.049	17.7	21.8	353.1	407.3	0.890	1.188	0.724
Fluorine	F_2	38.00	85.2			24.0	30.0			0.656	0.876	0.776
Freon 12	CF_2Cl_2	120.92	243.3			12.4	15.3			0.203	0.273	0.731
Freon 13	CF_3Cl	104.47	191.6							0.237	0.318	
Freon 21	$CHF\,Cl_2$	102.92	282.0			11.4				0.236	0.320	
Freon 11	$CF\,Cl_3$	137.39	296.8			11.5	13.7			0.176	0.239	0.608
Freon 113	$C_2F_3Cl_3$	187.39	320.8			10.3						
Helium	He	4.003	4.2	5.192	5.192	19.81	24.02			6.23	8.31	0.670
Hydrogen	H_2	2.016	20.3	14.31	14.48	8.96	10.8	1319.3	1519.6	12.38	16.50	0.634
Krypton	Kr	83.8	119.8	0.2492	0.2486	25.6	33.1			0.297	0.397	0.893
Nitrogen	N_2	28.02	77.4	1.041	1.045	17.9	22.1	353.1	407.4	0.890	1.187	0.733
Neon	Ne	20.183	27.09			31.8	38.8			1.24		0.692
Oxygen	O_2	32.00	90.2	9.920	9.942	20.7	25.8	329.6	379.0	0.779	1.039	0.766
Sulfur dioxide	SO_2	64.07	263.0	0.622	0.675	12.8ᵈ	17.2			0.369	0.495	1.027
Water	H_2O	18.02	373.0	4.20ᵈ	2.00	858.°	13.2					
Xenon	Xe	131.3	160.05	0.160	0.159	23.3	30.8			0.189	0.253	0.970

ᵃ At 1 bar.
ᵇ At 30 bar.
ᶜ Sublimes.
ᵈ Liquid.

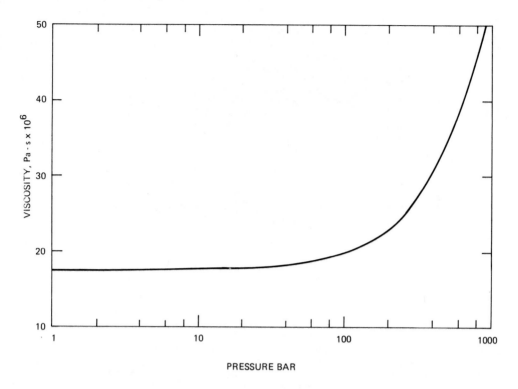

FIGURE 1.   Viscosity of nitrogen at 300 K.

where μ_o is the viscosity at a known absolute temperature, $T_o(K)$. Exponent n is usually below one and is given in Table 1. The Sutherland formula fits the data slightly better than Equation 1 for hydrogen and helium:

$$\mu = \mu_o \left[\frac{T_o + C}{T + C} \right] (T/T_o) \tag{2}$$

For $T_o = 300$ K and $T = 400$ K, the constant C is 61 for hydrogen and 70.5 for helium based on the viscosity data in Table 1. The differences between Equations 1 and 2 are commonly so small that Equation 1 with the exponent n from Table 1 can normally be used.

When other data are not available, viscosity of a gas may be estimated from its pressure and temperature at the critical point. The critical values for selected pure gases are given in Table 3, abstracted from Reference 1. In order to estimate the viscosity, first calculate the reduced temperature and pressure for the T and P for which the viscosity is desired.

$$P_r = P/P_{cr} \text{ and } T_r = T/T_{cr} \tag{3}$$

These values may then be used in Figure 4 to estimate the reduced viscosity. To obtain the actual viscosity, one also needs the viscosity at the critical point. If the viscosity is not known at the critical point, it may be estimated from a known viscosity at some other P_o and T_o. By calculating P_{cr} and T_{cr} from Equation 3, Figure 4 may be used to estimate μ_{cr}, and then

$$\mu_{cr} = \mu_o/\mu_{cr} \tag{4}$$

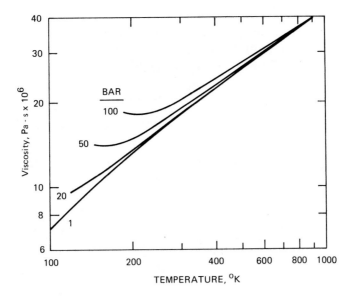

FIGURE 2.   Viscosity of air at several pressures.

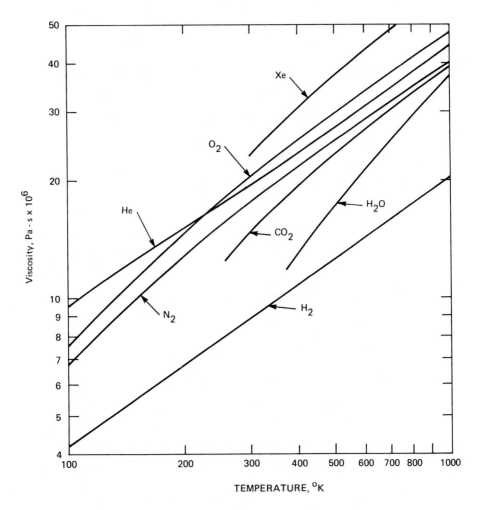

FIGURE 3.   Viscosity of several gases at P = 1 bar.

Table 2

**VISCOSITY OF AIR AS A FUNCTION OF
TEMPERATURE AND PRESSURE,
(PA·SEC × 10⁷, GASEOUS AIR)[1]**

Temp (K)	Pressure (bar)						
	1	10	20	50	100	200	400
100	71.1						
150	103.4	106.4	110.8	143.4	315.2	481.2	706.4
200	132.5	134.6	137.2	148.6	181.2	279.1	445.8
250	159.6	161.2	163.1	170.4	187.9	237.6	351.2
270	169.6	171.1	172.8	179.3	194.1	235.0	332.2
300	184.6	185.9	187.4	192.9	205.0	275.5	315.9
330	199.2	200.4	201.7	206.5	216.7	243.3	308.7
350	208.2	209.3	210.4	215.0	224.2	248.0	306.6
400	230.1	231.1	232.1	235.8	243.4	262.1	308.3
450	250.7	251.6	252.5	255.7	262.0	277.5	315.4
500	270.1	270.9	271.7	274.5	279.9	293.1	325.1
600	305.8	306.4	307.1	309.4	313.6	323.7	348.0

From Vargaftile, N. B., *Tables on the Thermophysical Properties of Liquids and
Gases,* John Wiley & Sons, New York, 1975. With permission.

Table 3

**CRITICAL POINT VALUES FOR SELECTED
GASES[1]**

Gas	$T_{cr}(K)$	$P_{cr}(bar)$	$\rho_{cr}(kg/m^3)$
Ammonia	405.5	112.9	235.0
Argon	150.9	50.0	536.0
Carbon dioxide	304.2	73.8	468.0
Carbon monoxide	133.1	35.0	301.0
Fluorine	144.0	53.2	535.0
Freon 12	384.9	41.3	562.0
Freon 13	302.2	39.0	571.0
Freon 21	451.3	51.8	525.0
Freon 11	471.1	[a]	554.0
Freon 113	487.2	34.1	576.0
Helium	5.26	2.29	69.3
Hydrogen	32.23	13.16	31.6
Krypton	209.4	54.9	911.0
Nitrogen	126.3	34.0	304.0
Neon	44.4	26.5	483.0
Oxygen	154.8	50.9	405.0
Sulfur dioxide	430.6	78.8	525.0
Water	647.2	221.2	318.0
Xenon	289.7	58.3	1110.0

[a]   Not given in Reference 1.

The viscosity at the desired point is obtained by determining μ_r at the desired P_r and T_r from
Figure 4, and then

$$\mu = \mu_r \mu_{cr} \qquad\qquad (5)$$

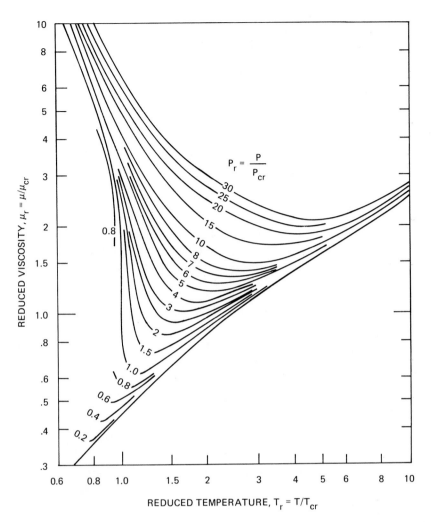

FIGURE 4. Generalized reduced viscosity of gases. (From Hougen, O. A., Watson, K. M., and Ragatz, R. A., *C.P.P. Charts*, 2nd ed., John Wiley & Sons, New York, 1960. With permission.)

Table 4
COMPARISON OF ESTIMATED NITROGEN VISCOSITIES AT 1 BAR

T	T_r	μ_r (from Figure 4)	μ (based on μ at 300 K)	μ (based on $\mu_{cr} = 19.1 \times 10^{-6}$)	μ actual (Reference 1)
200	1.58	0.69	12.4	13.2	12.9
300	2.38	1.00	17.9	19.1	17.9
400	3.17	1.25	22.4	23.9	22.1
500	3.96	1.42	25.4	27.1	25.8

If no viscosities are known, the critical viscosity may be estimated from the values of P_{cr}, T_{cr}, and molecular weight M, as follows:

$$\mu_{cr} = 7.70 \times 10^{-7} M^{0.5} P_{cr}^{0.667} T_{cr}^{-0.167} \tag{6}$$

Table 4 illustrates the application of the two methods to the calculation of the viscosity of nitrogen. For each temperature, T_r is calculated from $T_{cr} = 126.3$ K. Next are listed the values of μ_r estimated from Figure 4 from the low density limit curve. Assuming we know the viscosity is 17.9×10^{-6}Pa·sec at 300 K where μ_r is estimated to be 1.00, the value of μ_{cr} is the same and the viscosities at the other temperatures are directly calculated as shown. If no viscosities are known, Equation 6 is used:

$$\mu_{cr} = 7.70 \times 10^{-7} (28.02)^{0.5} (34.0)^{0.667} (126.3)^{-0.167} = 19.1 \times 10^{-6}$$

Values for the other temperatures are calculated directly from the estimated values of μ_r. The final column shows the actual viscosities, indicating a reasonable check.

For convenience, the following are conversions to the SI system: 1 reyn $= 1.45 \times 10^{-10}$ Pa·sec and 1 P $= 0.1$ Pa·sec.

Specific Volume

The specific volumes listed in Table 1 indicate the degree of "perfection" of a gas. At 273.1 K and 1 bar pressure, 1 g-mol of a perfect gas occupies a volume of 22.4 ℓ. Adjusting this to 300 K gives 24.6 ℓ. If the specific volumes in Table 1 are multiplied by the molecular weight for each gas, the result is liters per gram mole (ℓ/g-mol) and also cubic meters per kilogram-mole (m³/kg-mol). Values for air, nitrogen, and oxygen are 24.9. Freon 21, Freon 11, and sulfur dioxide are below the perfect gas figure, indicating some degree of association between molecules.

Pressure is given here in bars or atmospheres. For use in the SI system, 1 bar is equivalent to 101,300 Pa.

APPLICATION OF DATA

Hydrodynamic bearings principally require knowledge of the viscosity of the gas at the temperature and pressure involved. When the pressures are very low or the spacing between surfaces is very small, consideration must be given to the mean free path. Hydrostatic bearings involve feeding of gas at an elevated pressure into the bearing film area. Viscosity is required in calculating the flow through the film; thermal properties are required in calculating flow through feed orifices or ports. Sonic velocity sets a limit on flow rate in these situations.

Hydrodynamic and Hydrostatic Designs

The viscosity enters directly in hydrodynamic calculations through the principal terms in Reynolds equation which are of the form:

$$\frac{d}{dx} \left(\frac{h^3}{12\mu} \right) \frac{dP}{dx}$$

Because of the usually good heat transfer to the surfaces, hydrodynamic films are treated as isothermal and at the bearing surface temperature. Ambient temperature and pressure conditions are adequate for the determination of operating viscosity from the data in the previous section.

In hydrostatic design computations, one is concerned with mass flow through thin slots obeying the equation:

$$\dot{m} = \frac{h^3 \rho w}{12\mu} \frac{dp}{dx} = \frac{h^3 w}{12\nu} \frac{dp}{dx} \tag{7}$$

where the kinematic viscosity $\nu = \mu/\rho$ enters. Here h is the slot thickness, ρ is the mass density, and w is the slot width. Where turbulence may be involved, the kinematic viscosity enters in the Reynolds number:

$$\text{Re} = \frac{\rho U h}{\mu} = \frac{U h}{\nu} \qquad (8)$$

where U is the velocity of one surface relative to another a distance h away.

Feed restrictors to a hydrostatic gas bearing are more commonly of the orifice type rather than of the laminar flow type. For discharge of a perfect gas through an orifice,

$$(P_1/P_2)^\alpha = T_1/T_2 \qquad (9)$$

where T is the absolute temperature. The exponent α is usually expressed in terms of the ratio of specific heats, $k = C_p/C_v$, but can also be expressed in terms of the gas constant R, C_p, and the molecular weight M:

$$\alpha = (k - 1)/k = R/C_p M \qquad (10)$$

The limiting pressure ratio at which the throat velocity equals the speed of sound is given by:

$$r_c = (P_1/P_2)_c = 2k/(k + 1)(k - 1) = \left(\frac{1 - \alpha}{1 - \alpha/2}\right)^{1/\alpha} \qquad (11)$$

The key constant α can be calculated directly from Equation 10.

As an example, consider the monomolecular gas argon with a molecular weight of 39.94 and a heat capacity at 300 K of $C_p = 0.522$ kJ/kg·deg. For argon at 300 K, $\alpha = 8.314/(0.522 \times 39.94) = 0.399$, and $r_c = (0.601/0.8005)^{1/0.399} = 0.488$. This is equal to the theoretical value for a perfect gas of 0.49.

Mean Free Path

The mean free path is a measure of the average distance between collisions of the gas molecules. It is a function of the volume density of the gas and is given by:

$$\lambda = 1/(1.414\pi \sigma^2 n) \qquad (12)$$

where σ is the molecular diameter and n is the molecular density in molecules per cubic centimeter. The number of molecules per gram-mole is Avogadro's Number, 6.02×10^{23}.

As an example, the molecular density of argon at 273 K is

$$n = 6.02 \times 10^{23}/22,400 = 2.69 \times 10^{19}$$

The molecular diameter is approximately 2.9×10^{-8} cm. Applying Equation 12:

$$\lambda = 1/[1.414\pi (2.9 \times 10^{-8})^2 \times 2.69 \times 10^{19}] = 9.9 \times 10^{-6} \text{ cm}$$

The value estimated in the *Handbook of Chemistry and Physics* is 9.0. This accuracy is quite sufficient for low pressure or very thin film bearing design.

Speed of Sound

Speed of sound in a gas is a function of temperature, molecular weight, heat capacity at constant pressure, and the gas constant:

$$\Phi = \left[\frac{RT}{M(1-\alpha)}\right]^{1/2} = \left[\frac{RT}{M - R/C_p}\right]^{1/2} \tag{13}$$

Applying this data to oxygen, Table 1 lists $M = 32.00$, and $C_p = 0.920$ kJ/kg·K. Using this data in Equation 13 yield 329 m/sec, as compared with the value of 353 m/sec listed in Table 1.

NOMENCLATURE

C = Temperature, C
C_p = Specific heat at constant pressure, kJ/kg·K
C_v = Specific heat at constant volume, kJ/kg·K
P = Pressure, N·m^{-2}
T = Absolute temperature, K
T_B = Boiling point, K
U = Surface velocity, m/sec
h = Film thickness, m
M = Molecular weight
w = Width of leakage path, m
ν = Kinematic viscosity, m^2/sec
λ = Mean free path, m
μ = Absolute viscosity, Pa·sec
ρ = Mass density, kg/m^3
Φ = Sonic velocity, m/sec

REFERENCES

1. **Vargaftile, N. B.,** *Tables on the Thermophysical Properties of Liquids and Gases,* 2nd ed., John Wiley & Sons, New York, 1975.
2. **Bird, R. B., Steward, W. E., and Lightfoot, E. N.,** *Transport Phenomena,,* John Wiley & Sons, New York, 1960.
3. **Uyehara, O. A. and Watson, K. M.,** *Natl. Pet. News,* 36, 764, 1944.
4. **Hougen, O. A., Watson, K. M., and Ragatz, R. A.,** *C.P.P. Charts,* 2nd ed., John Wiley & Sons, New York, 1960.

LUBRICATING OIL ADDITIVES

J. A. O'Brien

INTRODUCTION

The modern history of lubricant additives began in the early 20th century with the use of fatty oils and sulfur in mineral oils to improve lubrication under high loads. World War II provided a major impetus to the development of lubricant additives as the military, engine builders, and machine manufacturers demanded more performance from their equipment.

Consumption of lubricant additives in the U.S. increased from 127 thousand metric tons in 1950 to 710 thousand metric tons in 1978.[1] Lubricants for internal combustion (IC) engines account for 72% of the market. Total free-world consumption of lubricant additives is estimated to be about three times that of the U.S.

The unique feature of IC engine lubricants is their exposure to combustion products from blow-by, fuel combustion products which leak past piston rings and contact the lubricant. Blow-by contains unburned fuel, reactive intermediates of fuel oxidation, fixed nitrogen in the forms of nitrogen oxides, and their fuel reaction products, soot, products of fuel additives, sulfur oxides, carbon monoxide, carbon dioxide, and water. IC engine lubricants require extensive additive treatment to counteract the effects of blow-by, such as internal engine rust, bearing corrosion, surface deposits which interfere with engine clearances, sludge formation which blocks lubricant passages, and lubricant decomposition.

Some industrial lubricants, such as those used in a steel or paper mill, also encounter severe environments and contamination. External and internal environments may subject lubricants to severe oxidizing conditions, extreme pressures and temperatures, water, dust, metal catalysts, and active chemicals.

ADDITIVE FUNCTIONS

Many minerals are used as lubricant additives, far too many to list in detail. Ramney published three texts[2,3,4] listing recent additive patents. This chapter discusses some of the more widely used additives, emphasizing their primary performance characteristics. Chemical structure and manufacture of some major lubricant additives are included in the Appendix.

Boundary Lubrication Additives

In boundary lubrication, surface asperities contact each other even though the lubricant supports much of the load. Friction depends mainly on the shearing forces necessary to cleave these adhering asperities and wear and friction can be reduced by certain additives. Table 1 lists common boundary lubrication additives.

Wear inhibitors and *lubricity agents* are polar materials that adsorb on a metal and provide a film that reduces metal-to-metal contact. *Extreme pressure* (EP) *additives* are a special class of boundary lubrication additive which react with the metal surface to form compounds with lower shear strength than the metal. This low-shear compound provides the lubrication. *Friction modifiers* can either adsorb or react with the surface to reduce friction by forming a very low shear-strength film. For example, Figure 1 demonstrates the effect of a friction modifier in an automatic transmission fluid. Without a friction modifier, the friction coefficient in the transmission would increase at low-sliding velocity where surface asperities make contact. This would result in rough shifting and lead to high-transmission loading and driver discomfort in vehicles. The friction modified fluid now in common use permits smooth shifting at low speeds while it maintains adequate friction under normal driving to prevent

Table 1
BOUNDARY LUBRICATION ADDITIVES

Metal dialkyl dithiophosphates[a]
Metal diaryl dithiophosphates[b]
Alkyl phosphates
Phosphorized fats and olefins
Phospho-sulfurized fats and olefins
Sulfurized olefins and paraffins
Sulfurized fats, fat derivatives and carboxylic acids
Chlorinated fats, fat derivatives and carboxylic acids
Fatty acids, other carboxylic acids and their metal salts
Esters of fatty acids
Oxidized paraffins and oils
Organic molybdenum compounds
Molybdenum disulfide (solid)
Graphite (solid)
Borate dispersions

[a]   With the metal zinc it is commonly called alkyl zinc dithiophosphate
[b]   With the metal zinc it is commonly called aryl zinc dithiophosphate

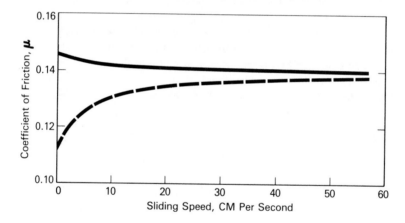

FIGURE 1.   Effect of friction modifier in automatic transmission fluid.

clutch slippage. Friction modifiers also reduce wet-brake chatter in tractors by reducing the difference between the static and dynamic coefficients of friction.

Friction modifiers conserve energy when used in automotive engine and drive-train lubricants where they are now being widely applied. Passut and Kollman[5] demonstrated the effect of friction modifiers on gasoline engine friction losses in a motored engine as demonstrated in Figure 2. At low-crankcase temperature, high-lubricant viscosity reduces boundary lubrication and the friction modifier had little effect. At high-crankcase temperature, low-lubricant viscosity increases boundary lubrication and the friction modifier reduced engine friction losses. Haviland and Goodwin[6] reported an average of 4.1% improvement in fuel economy for several friction modified engine lubricants in passenger car road tests. Fuel savings from friction modifiers in automotive drive-train gear oils and bearing greases are also being investigated.[7]

Corrosion Inhibitors

Rust, the formation of hydrated iron oxide, is a particularly important form of corrosion. Corrosion inhibitors (Table 2) can (1) react with metal surfaces to form a protective coating

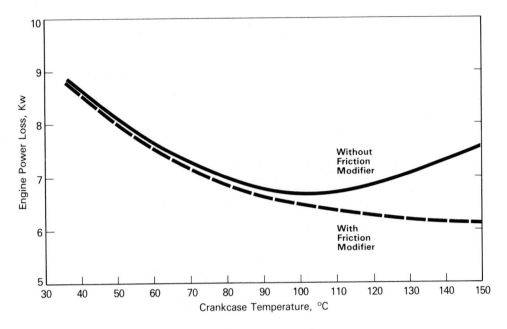

FIGURE 2.   Effect of friction modifier in engine crankcase lubricant.

Table 2
CORROSION INHIBITORS

Zinc dithiophosphates
Other dithiophosphates
Metal sulfonates
Overbased metal sulfonates
Metal phenate sulfides
Overbased metal phenate sulfides
Fatty acids
Acid phosphate esters
Chlorinated wax
Amines
2,4-Bis (alkyldithio)-1,3,4-thiadiazoles
Alkyl succinic acids

or (2) deactivate corrosive contaminants in the lubricant. Certain additives that inhibit corrosion in some environments can actually cause corrosion in others. For example, zinc dithiophosphates inhibit copper-lead bearing corrosion in an oxidative environment, yet cause silver bearing distress from sulfidation. When used in high concentrations, zinc dialkyldithiophosphates can also pit some ferrous metals.

Oxidation Inhibitors

Oxidation, the most common form of lubricant deterioration, proceeds through free-radical reactions which are accelerated by heat and catalyzed by metals. In hydrocarbon lubricants, free-radicals react with oxygen to form peroxy free-radicals which attack hydrocarbons to form new free-radicals and hydroperoxides. Free-radicals are formed faster than they are used and the rate of oxidation increases.

Some hydroperoxides decompose into aldehydes, ketones, carboxylic acids, and other oxygen-containing hydrocarbons. The oxygen compounds polymerize to form viscous soluble materials (lubricant thickening) and insoluble materials (sludge and deposits.) Some of the oxygen compounds are active, polar materials that accelerate rust and corrosion.

Table 3
OXIDATION INHIBITORS

Zinc dithiophosphates
Metal dithiocarbamates
Hindered phenols
Phenol sulfides
Metal phenol sulfides
Metal salicylates
Aromatic amines
Phospho-sulfurized fats and olefins
Sulfurized olefins, fats, fat derivatives, paraffins,
  and carboxylic acids
Disalieylal-1,2-propane diamine
2,4-bis (Alkyldithio)-1,3,4-thiadiazoles
Dilauryl selenide

Table 4
DETERGENTS AND DISPERSANTS

Detergents	Dispersants
Metal sulfonates	Polyamine succinimides
Overbased metal sulfonates	Hydroxy benzyl polamines
Metal phenate sulfides	Polyamine succinamides
Overbased metal phenate sulfides	Polyhydroxy succinic esters
Metal salicylates	Polyamine amide imidazolines
Metal thiophosphonates	

Table 5
VI IMPROVERS

Ethylene-propylene copolymers
Polymethacrylates
Styrene isoprene copolymers
Styrene butadiene copolymers
Styrene maleic ester copolymers
Polyisobutylenes

Oxidation inhibitors (Table 3) generally function by one or more of three mechanisms: free radical inhibition, peroxide decomposition, or metal deactivation. Each mechanism inhibits oxidation at a different link in the chain reaction. Hindered phenols, such as 2,6-di-*t*-butyl-*para*-cresol (DBPC), are effective free-radical oxidation inhibitors because they react with free-radicals to form nonfree-radical compounds. Some sulfur compounds decompose peroxides into stable compounds. Some selective polar additives react with metal ions and surfaces to inhibit their catalytic activity and are known as metal deactivators.

Detergents and Dispersants

Both detergent and dispersant additives (Table 4) are polar materials which provide a cleaning function. Detergency is a surface phenomenon of cleaning surface deposits. Dispersancy is a bulk lubricant phenomenon of keeping contaminants suspended in the lubricant. Detergent and dispersant additives each perform both of the above functions and differ in their relative ability to function at the machine surface or in the bulk of the lubricant.

Viscosity Modifiers

Viscosity index (VI) improvers (Table 5) are polymers that cause minimal increase in

Table 6
EFFECT OF VI IMPROVER

SAE viscosity grade	VI Improver	Viscosity		VI
		cSt at 100°C	cP at −18°C	
40	No	14.0	15000	100
10W	No	6.5	2000	100
10W40	Yes	14.0	2350	145

lubricant viscosity at low temperature, but considerable increase at high temperature. Table 6 demonstrates the effects of a VI improver. A typical 100 VI SAE 40 viscosity grade oil has proper viscosity (14 cSt) for engine lubrication at 100°C, but is too viscous (15000 cP) to permit engine starting at −18°C. A typical 100 VI SAE 10W oil would permit engine starting at −19°C, but is not viscous enough to protect the engine from wear at 100°C. By adding a VI improver to the SAE 10W oil, the product can permit engine starting at −18°C and still protect the engine from wear at 100°C.

Many of the same chemicals are also used as thickeners to increase the viscosity of products for special applications such as gear oils. Molecular weight may be varied to optimize specific performance characteristics. Viscous petroleum fractions, such as bright stocks, are also used as thickeners but are not considered additives.

Pour Point Depressants

Petroleum oils contain paraffinic wax which crystallizes in a lattice-like structure as the lubricant cools and prevents the lubricant from flowing. The lowest temperature at which the lubricant flows is called the pour point. Pour point depressants co-crystallize with the paraffinic wax, modify growth of the lattice-like structure, and permit flow at temperatures below the pour point of the unmodified lubricant. Common pour point depressants include: polymethacrylates, wax alkylated naphthalene polymers, wax alkylated phenol polymers, and chlorinated polymers.

Emulsion Modifiers

Emulsifiers give stable emulsions of water-in-oil or oil-in-water. They are used where high amounts of water improve cooling due to the high specific heat and thermal conductivity of water. Lubricants which contain water are increasingly being used to conserve petroleum base stocks.

Demulsifiers make emulsions unstable, which permits separation of water and lubricant. They are particularly important where water contamination can damage the lubricant in marine or industrial applications. Emulsion modifiers change the interfacial tension of oil and water. Low-interfacial tension permits stable emulsions. Table 7 lists emulsion modifiers.

Foam Decomposers

Excessive lubricant foaming can cause an overflow of the lubricating system, displace lubricant in pumps, increase response time of hydraulic systems, and disrupt the lubricant supply. Two theories predominate on the function of foam decomposers. The first is that they increase gas-lubricant interfacial tension to the point where the bubbles collapse. The second is that these partially soluble compounds with low-surface tension cause openings in the bubbles which allow the gas to escape.

Foam decomposers function at concentrations from 1 to 50 ppm. High concentrations can lead to excessive foaming, more than the original lubricant, and increased air entrainment. Common foam decomposers include: polysiloxanes (silicones), polyacrylates, organic copolymers, and candellilla wax.

Table 7
EMULSION MODIFIERS

Emulsifiers
    Soaps of fats and fatty acids
    Low-molecular weight Na and Ca sulfonates
    Low-molecular weight Na and Ca naphthenates

Demulsifiers
    High-molecular weight Ca and Mg sulfonates
    Alkylene oxide derivatives
    Heavy metal soaps

Tackiness Agents

Tackiness agents help the lubricant adhere to machine surfaces, or to itself, rather than flow, splatter, leak, mist, or otherwise contaminate the surrounding area. Tackiness agents function through physical, viscoelastic phenomena by increasing low-shear viscosity and providing the lubricant with the ability to stretch into fibers. Common additives of this type include viscosity modifiers (Table 7), aluminum soaps of unsaturated fatty acids, and other soaps.

Seal Swell Agents

Lubricants frequently contact elastomer seals, for example, in automatic transmissions. Shrinkage of seals results in leaks while excessive swelling and softening cause wear or extrusion from the seal seat — again resulting in leaks. Most lubricants are intended to cause a minor amount of seal swell to ensure sealing without excessive softening. Seal swell characteristics usually depend on the base stock. If the base stock causes excessive seal swell, additives can do very little to correct it. However, if the base stock shrinks the seals, the following seal swell additives can correct the problem: aromatics, aldehydes, ketones, and esters.

Dyes

Dyes are occasionally used in lubricants to provide a distinctive, attractive, or uniform color. Dyes must be soluble in the lubricant, have high-coloring power and not be detrimental to other lubricant properties.

LUBRICANT FORMULATION

Problems with Interactions

Most modern lubricants require more than one additive to meet all performance demands. In some cases, individual additives are blended directly into the base oil. In other cases, a group of additives are blended into an additive "package," which is subsequently blended into the base oil. Since most additives are active chemicals, they can interact in the package or in the lubricant to form new compounds. These interactions can decrease additive effectiveness and lead to insoluble or otherwise undesirable by-products.

Additive functions frequently depend on their limited solubility in the lubricant. For example, zinc dithiophosphate (ZDTP) must be able to leave the bulk of the lubricant and adhere to the machine surface to function as a wear inhibitor. When a dispersant is used in the same lubricant, the dispersant can hold the ZDTP in solution and prevent the latter from functioning. Many lubricants require both ZDTP and dispersant. Dispersants are frequently manufactured to minimize their ability to disperse ZDPT. Moreover, ZDPT's for these applications are selected to perform in the presence of a dispersant.

Surface active additives can also compete with each other. Both wear inhibitors and some

Table 8
SF ENGINE LUBRICANT TESTS[8]

Test sequence	Performance aspect
ASTM IID[10]	Engine rust under short-trip, low-temperature operation
ASTM IIID[10]	Oxidation, wear, and deposits under high-speed, high-temperature operation
ASTM VD[11]	Overhead cam wear and sludge and varnish deposits under stop-and go operation
CRC L-38[12]	Oxidation and bearing corrosion under high-temperature operation

Table 9
CD ENGINE LUBRICANT TESTS[9]

Test sequence	Performance aspect
Caterpillar 1-G2[12]	High-temperature diesel piston deposits
Caterpillar 1-D2[a]	High-temperature diesel piston deposits with high-sulfur fuel
CRC L-38[12]	Oxidation and bearing corrosion under high-temperature operation

[a] Proposed designation of a new test to replace the obsolete Caterpillar 1-D[12] test.

rust inhibitors function by adsorbing on metal surfaces and they compete for the same surface. The wear inhibitor can displace the rust inhibitor on the surface and be detrimental to rust inhibition. Likewise, the rust inhibitor can displace the wear inhibitor.

Meeting Performance Requirements

Universal engine lubricants meet the performance requirements of passenger car gasoline engines, as well as turbocharged two-stroke cycle and four-stroke cycle truck diesel engines. The additive package requires a very careful balance because it deals with diverse, complex quality requirements. Passenger car engine lubricant requirements in the U.S. are defined by a series of laboratory engine tests designated SF[8] by the American Petroleum Institute (API). The SF designation signifies that the lubricant passed the tests shown in Table 8.

Truck diesel engine lubricant requirements in the U.S. are designated CD.[9] To obtain the CD designation, the lubricant must pass the tests shown in Table 9. Most engine lubricant requirements outside the U.S. also require SF or CD performance plus additional local performance requirements.

Each SF and CD test stresses certain performance aspects of the lubricant and has been correlated with field experience. The IID test simulates short-trip winter driving, which is the most severe rust-forming condition. The IIID test simulates high-speed, high-load, and high-temperature driving conditions, such as towing a camper-trailer across the desert in the summer at high speed, which are severe for oxidative thickening and wear. The VD test simulates continuous stop-and-go city driving with an overhead cam engine — a severe condition for sludge and varnish formation and cam wear. The 1-G2 test simulates a heavily loaded, turbocharged, four-stroke diesel using high-sulfur fuel. The L-38 test stresses protection of copper-lead bearings from corrosion. In addition to passing all SF and CD tests, a universal engine oil must also have less than 1% sulfated ash to be compatible with two-stroke cycle diesel engine performance. Lubricant formulation involves handling all of these conditions with the same lubricant.

Table 10
EXAMPLE OF ADDITIVE EFFECTS IN UNIVERSAL ENGINE
LUBRICANT

Performance aspect	Dispersant	ZDTP	Overbased sulfonate	Low-based sulfonate	Phenate
Gasoline engine					
Rust	−	+	+ + +	+	+
Lubricant oxidation	=	+ + +	−	−	+ +
Deposits	+ + +	=	+	−	+
Wear	−	+ + +	−	−	=
Bearing corrosion	−	+ + +	=	−	+
Diesel Engine					
Piston deposits	+	− −	+	+ + +	+ + +
Fuel sulfur	=	=	+ + +	+	+ + +
Low ash	+ +	−	− −	+	− −

Note: Key: + + + , very beneficial; + + , beneficial; + , slightly beneficial; = , no effect; − , detrimental; and − − , very detrimental.

Table 10 shows how the additives commonly used in universal engine oils perform in the major aspects of the SF and CD sequence tests. Additive formulation for most lubricants involves similar consideration of beneficial and detrimental additive effects.

Relationship of Additives and Base Stocks

Lubricant base stocks influence additive performance through two main functions; solubility and response. For example, performance of surface active additives depends largely on their ability to adsorb on the machine surface at the proper time and place. Base stocks with poor solubility characteristics may allow these additives to separate before they can fulfill their intended functions. Conversely, base stocks with very high-solubility characteristics may keep the additives in solution, not allowing them to adsorb.

Additive response depends on base stock composition. Natural sulfur, nitrogen, and phenolic inhibitors are removed along with undesirable materials during base stock refining. Removal of these natural inhibitors often results in reduced oxidation inhibition relative to unrefined stocks. However, the natural inhibitors, as well as the undesirable materials removed during base stock refining, often interfere with additive performance.

Synthetic base oils, depending on their chemical structure, exhibit very specific solubility characteristics, additive response, and additive compatibility that are sometimes different from mineral oils. The most common synthetic base oils are synthetic hydrocarbons, such as polyalpha olefins, and esters, such as adipate, azelate and pentaerythritol esters.

Synthetic hydrocarbons exhibit excellent additive response but are poor additive solvents. Esters vary in additive response and are excellent solvents except for additives with which they react to form precipitates. Synthetic oils can be blended with each other or with mineral oil to provide the optimum balance of solubility and additive response.

Formulation

Additive formulation requires consideration of interaction and competition among additives as well as individual additive performance and solubility. Undesirable additive side-effects must be overcome and the total additive formulation balanced to achieve optimum performance in the finished lubricant.

Often the solution to a lubricant formulation problem lies not in changing the relative concentration of an additive, but in changing the additives themselves. Molecular weight, molecular weight distribution, reaction conditions, purity, and many other variables influence these parameters. Small amounts of chemicals added during additive manufacture often induce large changes in additive interaction performance. These "additive additives" are considered highly proprietary by additive manufacturers.

ACKNOWLEDGMENT

The author expresses sincere appreciation to Roger Watson of Batavia, Illinois, who consulted on chemical accuracy, authored the appendix, and provided constructive suggestions for the other aspects of the chapter. Assistance received from many members of the Amoco Research Center, Naperville, Illinois, is also gratefully acknowledged.

APPENDIX: EXAMPLES OF LUBRICANT ADDITIVES

A description of representative additives follows. Some additives are relatively pure chemicals, for example, the zinc dithiophosphates. Many additives, however, are reaction products of industrial grade chemicals using reaction conditions to control the product quality and performance. In these cases, the complex mixture is not extensively separated and the structures shown represent the major component of a generic group. A vast literature describes additive properties and preparation in more details.[2,3,4,13,14]

1. Metal Dialkyldithiophosphates (Metal Dialkyl Phosphorodithioates)

Variations — M is usually zinc but may also be molybdenum, tungsten, or other metals. The R–O– groups are derived from primary and secondary alcohols and alkyl phenols and may be single or mixed.

Manufacture — Alcohols or alkyl phenols are reacted with P_4S_{10} to form the dialkyl-dithiophosphoric acids. The zinc salts are made by reacting the acids with zinc oxide.

Application — Antiwear, anticorrosion, and antioxidant used almost universally in lubricants.

2. Tricresyl Phosphate

Manufacture — Mixed cresols (except ortho) reacted with phosphorous oxychloride.

Application — Wear inhibitor for synthetic oils and greases.

3. Sulfurized Fats, Olefins, Hydrocarbons

Variations — The R groups are residues from olefin polymers, petroleum stocks, unsaturated fats, terminal olefins, and terpenes.

Manufacture — Selected fats and hydrocarbons are heated with sulfur, sometimes in the presence of an alkaline catalyst.

Application — Antioxidants, antiwear, and antifriction are widely used in industrial lubricants including cutting oils and gear oils.

4. Benzotriazole

Manufacture — Nitrous acid on *o*-phenylene diamine.

Application — Inhibitor for sulfur corrosion of silver and copper in industrial lubricants.

5. 2-Alkyl-4-Mercapto-1,3,4-Thiadiazole

Variations — 2,4-bis (alkyldithio)-1,3,4-thiadiazoles. R is -octyl or -dodecyl.

Manufacture — Hydrazine and carbon disulfide are condensed to form the dimercapto thiadiazole, which is then coupled with tertiary mercaptan using hydrogen peroxide.

Application — Inhibitors for sulfur corrosion of silver and copper; wear and oxidation inhibitors in industrial lubricants.

6. Metal Dialkyldithiocarbamates

Variations — M may be any of a variety of metals, including zinc and molybdenum. R is C_4 to C_{10}. x is a function of the metal valence.

Manufacture — Secondary amines are reacted with carbon disulfide and caustic solution. The sodium dithiocarbamate is then reacted with the metal chloride.

Application — Antioxidants and antifriction additives.

7. 2,4-Ditertiarybutyl-*p*-Cresol (DBPC)

Variations — This is the most important member of a class called hindered phenols. Several variations are formed by coupling two phenolic groups through the ortho or para positions by methylene groups, sulfur, or nitrogen.

Manufacture —*p*-Cresol is ortho alkylated using aluminum alkyl catalyst.

Application — Antioxidants widely used in many kinds of lubricants.

8. Phenothiazine

Variations — Ring alkyl groups.

Manufacture — Reaction of diphenylamine with sulfur.

Application — Antioxidant in synthetic oils for jet engines and in greases.

9. Phenyl Alpha Naphthylamine (PAN)

Variations — Ring substitutions.

Manufacture — 1-Napthyl amine heated with phenol.

Applications — Antioxidants for greases and synthetic oils.

10. Dialkyldiphenylamine

Variations — R may be phenyl or alkyl derived from olefins.

Manufacture — Aniline is heated with strong acid to yield diphenylamine which is then alkylated with chlorobenzene or olefin polymers.

Applications — Antioxidants for greases, mineral oils, and synthetic oils.

11. Phosphosulfurized Pinene

$$R-P \underset{S}{\overset{S}{\parallel}} \overset{S}{\diagdown} \underset{S}{\overset{S}{\diagup}} \overset{S}{\parallel} P-R$$

Variations — R is derived from either alpha or beta pinene or a turpentine mixture.
Manufacture — Pinene is reacted with P_4S_{10}.
Application — Antioxidant and anticorrosion additives.

12. Dilauryl Selenide

$$C_{12}H_{25}-Se-C_{12}H_{25}$$

Manufacture — Lauryl chloride heated with dimethyl selenide.
Application — Antioxidant for greases and synthetic oils.

13. Neutral and Basic Metal Sulfonates

$$R-\underset{O}{\overset{O}{S}}-O-M-O-\underset{O}{\overset{O}{S}}-R$$

$$R-\underset{O}{\overset{O}{S}}-O-M-O-H$$

Variations — Sulfonates are one of the oldest lubricant additives to be used on a large scale; they have evolved into a large variety of similar materials. The R groups have come from many different sources including by-products of lube oil refining by sulfuric acid treating, heavy ends of alkylbenzenes from laundry detergent manufacture, alkylation of benzene and naphthalene with olefin polymers, and alkylation of benzene with various chlorinated petroleum fractions. Many different metals are used, but the most important are sodium and the alkaline earths, magnesium, calcium and barium (see also *Overbased Metal Sulfonates*).
Manufacture — The alkyl aromatics are prepared as indicated above. Sulfonation is done with gaseous SO_3 or oleum. The soaps are prepared by direct neutralization with the metal oxide or hydroxide, or by metathesis of the sodium sulfonate with a metal halide.
Application — Emulsifiers and rust inhibitors are widely used in industrial lubricants.

14. Overbased Metal Sulfonates

$$\left[R-\underset{O}{\overset{O}{S}}-O-\left(M-CO_3\right)_x-M-O-\underset{O}{\overset{O}{S}}-R \right]_y$$

Variations — For R, see *Neutral and Basic Metal Sulfonates*. x = 1 to 15. Occasionally, the CO_3 group is partially replaced by a hydroxyl or another anion. y = 10 to 30. M is usually magnesium, calcium, or barium. A micellar structure is postulated.
Manufacture — Several overbasing processes are used. In one, metal cellosolve carbonate is decomposed with water in the presence of a sulfonic acid. In other processes the basic or neutral soap, together with a suspension of the metal oxide or hydroxide, methanol, water, and a promoter such as ammonia or amine, is blown with carbon dioxide.

Application — Detergents, alkaline agents, and rust inhibitors widely used in crankcase motor oils and marine cylinder oils.

15. Metal Alkylphenate Sulfides

Variations — Phenates are generally produced in the form of complex basic or overbased (see Overbased Metal Sulfonates) soaps of magnesium, calcium, or barium. A represents a weak acid anion other than phenate, such as hydroxyl or glycolate. x is generally from 1 to 2. The metal-free phenol sulfides are also produced. R is ordinarily in the range of C_8 to C_{12}.

Manufacture — Alkylphenol is sulfurized with one of the sulfur chlorides or with sulfur and a base catalyst. The phenol sulfides are then reacted with the metal oxide or hydroxide in the presence of a promoter such as ethylene glycol. Overbasing may be accomplished by contacting with CO_2.

Application — Phenates are widely used as detergents, antioxidants, and alkaline agents in crankcase motor oils and in marine cylinder oils. Metal-free phenol sulfides are used as antioxidants in industrial lubricants.

16. Metal Alkylsalicylates

Variations — R is derived from olefin polymers of 300 to 1000 mol wt. M may be any of a variety of metals, but usually calcium or barium.

Manufacture — Alkylphenols are heated with carbon dioxide under pressure with alkaline catalyst. The alkysalicylic acid is then neutralized with metal oxide or hydroxide.

Application — Detergents and antioxidants mainly used for crankcase motor oils.

17. Alkenyl Polyamine Succinimides

Variations — R is alkenyl from olefin polymer, e.g., polybutene, mol wt is 500 to 2000, x = 1 to 4. Stoichiometry may be varied to yield 2 succinimides: 1 polyamine. Other types of polyamines may be used.

Manufacture — Maleic anhydride is condensed with olefin polymers. The resulting alkenyl succinic anhydrides and acids are then reacted with polyamines.

Applications — Dispersants are widely used in crankcase motor oil. The alkyl succinic anhydrides and acids are used as rust inhibitors.

18. Alkyl Hydroxyl Benzyl Polyamine

Variations — R is from an olefin polymer, e.g., polybutene, mol wt 500 to 2000. x is 2 to 5. Stoichiometry may be varied to yield 2 phenols: 1 amine or 2 amines: 1 phenol or oligomers.

Manufacture — Phenol is alkylated with olefin polymer then reacted with formaldehyde and polyamine.

Application — Dispersants are widely used in crankcase motor oils.

19. Polyamine Amide Imidazoline

Variations — R–C≡ is derived mainly from commercial isostearic acid, but may come from other carboxylic acids, e.g., naphthenic. Other polyamines may be used.

Manufacture — Polyamine is condensed with carboxylic acid.

Application — Detergents and inhibitors are used in two-stroke cycle oils and industrial oils.

20. Esters of Polymethacrylic Acid

Variations — R is a mixture of normal and iso-paraffin groups. Mol wt is 10,000 to 50,000.

Manufacture — Polymerization of mixed methacrylic esters.

Application — Viscosity modifiers (VI improvers) and wax crystallization modifiers (pour point depressants) for petroleum lubricants, especially crankcase motor oils.

21. Polyisobutylene

Variations — Copolymers with butene-1. Mol wt vary from 1,000 to 1,000,000.

Manufacture — Polymerization of pure isobutylene or mixed butenes via Friedel Crafts catalysts.

Application — Viscosity modifiers are tackiness agents.

22. Alkylene Oxide Derivatives

$$R\left(\!CH_2-CH_2-O\!\right)_x\!\!\left(\!CH_2-\overset{\overset{\textstyle CH_3}{|}}{CH}-O\!\right)_y\!\!A$$

Variations — A large variety of materials have been made by condensing alkylene oxides, mainly ethylene oxide and propylene oxide on active hydrogen compounds such as alcohols, phenols, and amines. In the structure above, ethylene and propylene oxide are condensed singly, together, or blockwise where x and y may vary from 0 to 50 or more and A may be hydrogen or an alkyl or acyl group.

Manufacture — Active hydrogen compounds with a strong base catalyst are condensed with alkylene oxides under pressure in a single operation or sequentially to form block polymers.

Application — Dispersants, emulsifiers, demulsifiers, and ancillary rust inhibitors are used in a wide variety of lubricants.

REFERENCES

1. *Lubricating Oil Additives,* SRI International, Menlo Park, Calif., 1979.
2. **Ramney, M. W.,** *Lubricant Additives,* Noyes Data Corporation, Park Ridge, N. J., 1973.
3. **Ramney, M. W.,** *Lubricant Additives Recent Developments,* Noyes Data Corporation, Park Ridge, N. J., 1978.
4. **Ramney, M. W.,** *Synthetic Oils and Additives for Lubricants: Advances Since 1977,* Noyes Data Corporation, Park Ridge, N. J., 1980.
5. **Passut, C. A. and Kollman, R. E.,** Laboratory Techniques for Evaluation of Engine Oil Effects on Fuel Economy, Paper Number 780601, Society of Automotive Engineers, Inc., Warrendale, Pennsylvania (1978).
6. **Haviland, M. L. and Goodwin, M. C.,** *Fuel Economy Improvements with Friction Modified Engine Oils in Environmental Protection Agency and Road Tests,* Paper No. 790945, Society of Automotive Engineers, Warrendale, Pa., 1979.
7. **Davis, B. T. et al.,** *Fuel Economy benefits from Modified Crankcase Lubricants,* Paper presented at American Society of Lubrication Engineers, 34th Annual Meeting, St. Louis, Mo., 1979.
8. *SAE Handbook 1981,* SAE J183 preprint, Society of Automotive Engineers, Warrendale, Pa, February 1980.
9. *SAE Handbook 1979,* Society of Automotive Engineers, Warrendale, Pa., 1979, 13.02.
10. ASTM Special Tech. Publ. 315G, *Multicylinder Test Sequences for Evaluating Automotive Engine Oils,* American Society for Testing and Materials, Philadelphia.
11. ASTM Special Tech. Publ. 315H, Part III, American Society of Testing and Materials, Philadelphia, in press.
12. ASTM Special Tech. Publ. 509, *Single Cylinder Engine Tests for Evaluating Performance of Crankcase Lubricants,* American Society for Testing and Materials, Philadelphia.
13. **Smalheer, C. V. and Smith, R. K.,** *Lubricant Additives,* The Lezius-Hiles Co., Cleveland, Ohio, 1967.
14. **Smalheer, C. V.,** Additives, in *Interdisciplinary Approach to Liquid Lubricant Technology,* NASA SP-318, NTIS N74-12219-12230, Ku, P. M., Ed., 1973, 433.

METAL PROCESSING — DEFORMATION

John A. Schey

INTRODUCTION

In manufacturing, the desired shape of individual parts is often obtained by plastic deformation. Some 85 to 90% of all steel and other technically important metals and alloys are subject to rolling, forging, extrusion, or drawing, often repeatedly.

The workpiece surface frequently undergoes a very substantial extension; fresh metal surfaces are exposed, and lubricants must protect not only the old but also these new surfaces. Success or failure of such lubrication is important in determining the magnitude of pressures, forces, energy requirements, and often the very possibility of plastic deformation itself. For this reason, friction, lubrication, and wear in metalworking have been of great interest.[1-13]

EFFECTS OF FRICTION IN METAL DEFORMATION

Mathematical Representation of Friction

In the absence of plastic deformation, the coefficient of friction is

$$\mu = F/P = \tau_i/p \tag{1}$$

where P is normal force; F, frictional force; p, interface pressure; and τ_i, interface shear stress; all referred to apparent total area of contact, A. In plastic deformation, the interface pressure is usually high, and physical sliding at the tool/workpiece interface is possible only if $\tau_i < \tau_f$, the shear flow stress of the material. When $\tau_i > \tau_f$ the workpiece deforms by internal shear, and sliding at the interface ceases in so-called "sticking friction" (even though no actual adhesion needs to occur). Thus, $\tau_{i\ max} = \tau_f$ (Figure 1) and, since τ_f is approximately one half of the uniaxial flow strength σ_f, for plastic deformation the maximum possible value of μ is $\mu_{max} = \tau_f/\sigma_f = 1/2$. This is true only at $p = \sigma_f$; at higher interface pressures μ_{max} actually drops (Figure 1).

An alternative description of the interface is $\tau_i = m\tau_f$, where m is the interface shear factor, with a value ranging from 0 (absence of friction) to 1 (at sticking). In many instances, real properties of the interface can be described only by τ_i.

Forging

Upsetting a Cylinder

Axial upsetting of a cylindrical workpiece between flat, overhanging dies reveals many effects of friction. In the absence of friction, the diameter of the workpiece increases uniformly so as to maintain a constant volume. Since there is no restraint to sliding on the interface, the die pressure p is everywhere the same and equals the flow stress in compression σ_f (Figure 2a).

In the presence of slight friction, free expansion of the end face is hindered and die pressure must increase from edge to center (friction hill, Figure 2b). Assuming that the cylinder remains a cylinder (deformation is homogeneous), the pressure distribution can be calculated. Using a constant coefficient of friction μ, the friction hill assumes an exponential shape; with a constant shear factor m, the shape is linear. In the latter case

$$p = \sigma_f \left[1 + \frac{2m}{\sqrt{3h}} \left(\frac{d}{2} - x \right) \right] \tag{2}$$

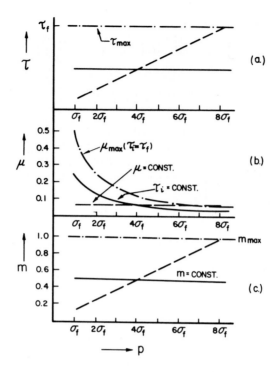

FIGURE 1. Examples of the variation of frictional stress with normal pressure. (a) Variation of shear stress; (b) coefficient of friction; and (c) interface shear strength factor. (From **Schey, J. A.,** Ed., *Metal Deformation Processes: Friction and Lubrication,* Marcel Dekker, New York, 1970. With permission.)

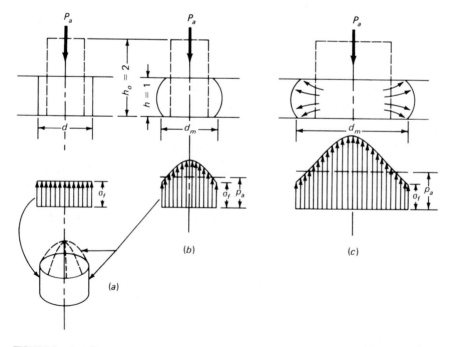

FIGURE 2. Interface pressures in upsetting a cylinder with (a) no friction, (b) high friction, and (c) high friction and larger d/h ratio. (From **Schey, J. A.,** *Introduction to Manufacturing Processes,* McGraw-Hill, New York, 1977. With permission.)

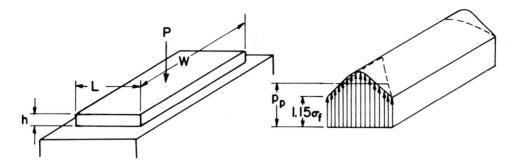

FIGURE 3. Interface pressure in upsetting a flat, rectangular workpiece with friction. (From **Schey, J. A.,** *Introduction to Manufacturing Processes,* McGraw-Hill, New York, 1977. With permission.)

where x is the distance from the edge, d is workpiece diameter, and h is the workpiece height. Both maximum and average pressures are a function of the d/h ratio (Figures 2b and c).

Stresses set limitations: interface pressures cause an elastic deformation of the tool, of significant proportions when precise forgings are to be made; maximum die pressure may exceed the pressure rating of the tooling; total force required may be too high for any press or hammer of reasonable size. All these limitations are a function of τ_i and of process geometry, characterized by the d/h ratio. For this reason, forging of relatively thin workpieces can become extremely difficult unless friction is kept very low with a suitable lubricant.

Friction also affects the deformation process. With low friction, resistance to sliding at the interface results in barrelling (Figure 2b). When friction is high enough to immobilize part of the end face, deformation becomes highly inhomogeneous, and some of the end face is actually formed by a folding over of the original side surfaces (Figure 2c). Barreling and folding over generate tensile stresses on the barrel surface. These ''secondary tensile stresses'' may lead to surface cracking in moderately ductile materials.

Upsetting a Slab

In forging flat slabs, major material flow takes place in the direction of minimum resistance, i.e., in direction L in Figure 3. If the workpiece is very wide or material flow in the w direction is hindered by a die element, pressure distribution remains constant along the entire width and, in the absence of friction, equals the plane-strain flow stress. With friction, a friction hill develops which is the same for identical L/h and d/h ratios. Materials flows away from the neutral center line. When friction is high enough to cause sticking over part of the interface, the neutral line broadens into a neutral zone and the workpiece changes shape by the folding out of side surfaces (as in Figure 2c). In a workpiece of finite width or unconstrained by die elements, some spread takes place and this increases with higher friction.

Ring Upsetting

In the absence of friction, the hole (and outer diameter) of a ring expand as though the workpiece were solid. Friction hinders free expansion and some material now flows towards the center; a flow-dividing ''neutral circle'' develops (Figure 4). With increasing friction this neutral circle moves towards the outer diameter and the internal diameter actually shrinks. This offers one of the easiest methods for studying friction and lubricant evaluation: higher friction results in a greater decrease of internal diameter.

Impression Die Forging

When forging in shaped dies, a flash is generated which contributes to die filling by preventing the free escape of material from the die cavity. Therefore, high friction in the

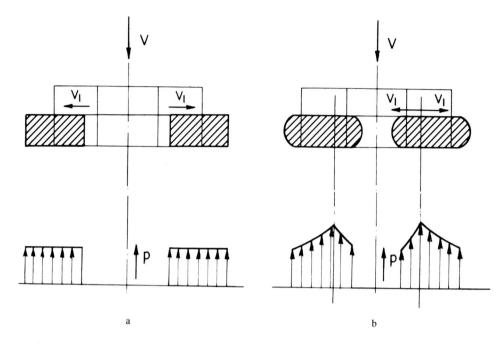

FIGURE 4.   Interface sliding velocities and pressures in ring upsetting with (a) no friction and (b) sliding friction. (From **Schey, J. A., Ed.,** *Metal Deformation Processes: Friction and Lubrication,* Marcel Dekker, New York, 1970. With permission.)

flash area is desirable; at the same time, low friction in the cavity would aid die filling. These contradictory requirements dictate a compromise, usually governed by considerations such as die wear and prevention of metal transfer unto the die surface.

Indentation

When the tool is smaller than the workpiece itself, and L/h > 1, the process is similar to that of upsetting a slab (except L remains constant) and friction still determines interface pressures and forces. When h/L > 1 (Figure 5), localized deformation takes place, and the interface pressure rises to overcome the resistance exerted by the nondeforming material between the two deformation zones. Lubrication makes no difference, and the pressure is a function of only the h/L ratio, reaching a limiting value of $p = 3\sigma_f$ at h/L = 8. As the h/L ratio drops, the deformation zones begin to interact and the pressure drops to the plane-strain flow stress 1.15 σ_f at h/L = 1. Lubrication serves only to reduce wear.

Rolling

Deformation takes place between two rolls (Figure 6) which contact the slab over a projected length L. When h/L > 1, the process is similar to indentation and lubrication does not affect rolling forces. When L/h > 1, effects of the friction hill (Figure 3) predominate, and at large L/h ratios lubrication assumes a paramount role.

Some miminum friction is required to pull the slab into the roll gap:

$$\mu \geqslant \tan \alpha \tag{3}$$

and the maximum absolute reduction is

$$(h_1 - h_2)_{max} = \mu^2 R \tag{4}$$

Therefore, high friction and/or large roll diameter is necessary for heavy reductions.

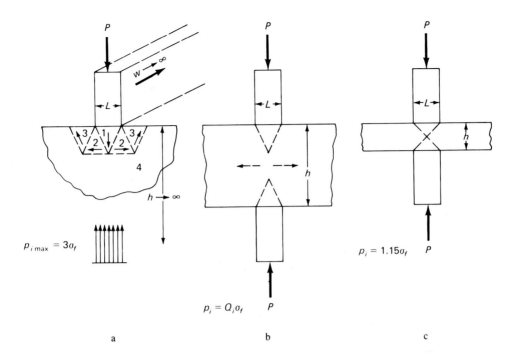

FIGURE 5. Deformation modes and interface pressures in indenting (a) a semiinfinite body, (b) a thick workpiece (h/L > 1), and (c) a workpiece with h/L = 1. (From **Schey, J. A.,** *Introduction to Manufacturing Processes,* McGraw-Hill, New York, 1977. With permission.)

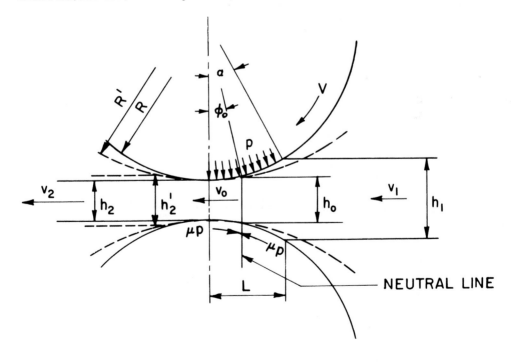

FIGURE 6. Geometry of roll pass and associated velocities. (From **Schey, J. A., Ed.,** *Metal Deformation Processes: Friction and Lubrication,* Marcel Dekker, New York, 1970. With permission.)

Since the same material volume must enter and leave the rolls in unit time, the product of velocities and slab thicknesses must be the same everywhere in the absence of spread. The slab moves with the rolls at the neutral point (Figure 6) but moves slower (backward slip) towards the entry and faster (forward slip) towards the exit. Since the position of the neutral point is governed by friction in the roll gap, forward slip s_f is a very sensitive measure of the efficiency of lubricants. With decreasing friction, forward slip diminishes:

$$s_f = \frac{v_2 - v_0}{v_0} = \frac{1}{2}\phi_0^2 \left(\frac{2R}{h_2} - 1 \right)$$

(5)

where the neutral angle depends on μ in the following relation:

$$\phi_0 = \frac{h_1 - h_2}{4R} - \frac{1}{\mu}\frac{h_1 - h_2}{4R}$$

(6)

Interface pressures reach a maximum at the neutral point, and the friction hill is similar to that observed in the compression of slabs (Figure 3). When $\mu p = \sigma_f$ sticking sets in and the friction hill becomes rounded.

Rolling of very thin strips presents difficulties because roll flattening (broken lines in Figure 6) becomes commensurate with strip thickness. Interface pressures can be reduced with small-diameter rolls and application of front and back tensions to the strip; even so, good lubrication is indispensible. Some minimum friction is still needed, otherwise the strip may skid in the rolls and lateral strip movement becomes difficult to control.

Wire, Bar, and Tube Drawing

As the product is pulled through, deformation is attained by compressive stresses at the stationary die (Figure 7). Most drawing operations are performed on round wire (Figure 7a), although shaped wire and bar are also drawn. All sliding is unidirectional, and the draw force is opposed by frictional stresses at the interface. The draw stress is

$$\sigma_{exit} = \sigma_f \left(1 + \frac{\mu}{\alpha} \right) \ln \frac{A_0}{A_1}$$

(7)

If the strength of the drawn product is insufficient to carry the draw force, the product will be torn off. This sets a limit of 30% or less reduction in cross-sectional area per pass, even with a good lubricant. Interface pressures are always below the compressive flow strength of the material.

As in forging, a large h/L ratio increases inhomogeneity of deformation. Since this ratio increases and friction forces decrease with increasing die angles, an optimum angle exists at which draw force is a minimum. Inhomogeneous deformation results in greater elongation of the surface layers, thus putting the center of the wire in tension. These secondary tensile stresses, combined with the drawing stresses, can lead to internal fracture (centerburst or arrowhead defect) in materials of limited ductility. Friction promotes this defect by increasing the drawing stresses.

In tube drawing without an internal die (Figure 7b), frictional conditions are the same as in wire drawing. More frequently, however, a short plug (British) or mandrel (U.S.) controls the internal diameter (Figure 7c). Friction on this plug increases drawing stresses, thus reducing the maximum attainable reduction. However, when the internal die element is a long mandrel (British) or bar (U.S.), this die element moves together with the drawn product (Figure 7d) and some of the drawing stresses are transmitted to it by interface friction. Higher friction on the internal surface actually increases the attainable reduction.

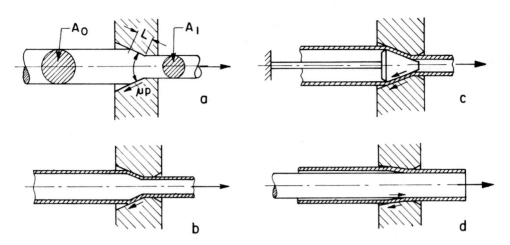

FIGURE 7. Basic bar and tube drawing operations. (From **Schey, J. A., Ed.,** *Metal Deformation Processes: Friction and Lubrication,* Marcel Dekker, New York, 1970. With permission.)

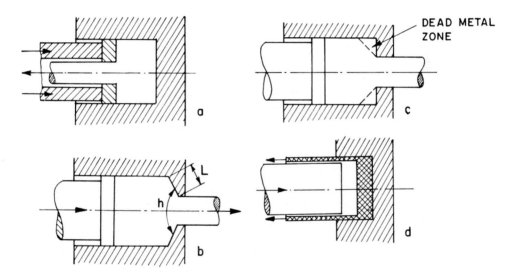

FIGURE 8. Basic forms of extrusion. (From **Schey, J. A., Ed.,** *Metal Deformation Processes: Friction and Lubrication,* Marcel Dekker, New York, 1970. With permission.)

Extrusion

Since the billet is pushed through the die (Figure 8), strength of the extruded product is immaterial and attainable reduction is limited only by the strength of the container and punch.

In reverse or indirect extrusion the billet remains stationary in the container (Figure 8a), and the magnitude of friction plays no role; die pressure p_e is a function of extrusion ratio $R = A_o/A_1$

$$p_e = \sigma_f (a + b \ln R) \tag{8}$$

where a and b are constants for a given die geometry.

In forward extrusion (Figure 8b), the stresses necessary to overcome friction add to the die pressure, often limiting the length of billet that can be extruded.

When a lubricant is used, the die entry must be tapered to facilitate material flow along the die face. Alternatively, one can extrude without any lubricant whatsoever; the die has a flat face and the material, in seeking a minimum-energy path, shears at some angle determined by the extrusion ratio. Material behind the shear surface forms a dead-metal zone (Figure 8c).

To minimize friction in reverse extrusion of tubes, the container is kept as short as possible and a short land is formed on the punch (Figure 8d). Lubricant between the workpiece and punch end face must be gradually metered out to protect the freshly formed, highly extended surfaces. The extrusion ratio and die pressures diminish as the punch diameter decreases. With a small diameter, however, the process changes to indentation and punch pressures can never be less than $p = 3\sigma_f$. As with all inhomogeneous deformation, lubrication is relatively ineffective in reducing punch pressure; nevertheless it is still desirable to prevent metal pickup and punch wear.

A special case is hydrostatic extrusion in which the extrusion pressure is supplied by a high-pressure fluid; container friction is eliminated and die friction reduced but the cycle time is long.

As in wire drawing, a high h/L ratio can lead to centerburst defects. In contrast to drawing, friction increases the pressure in the deformation zone, reduces secondary tensile stresses, and delays the onset of the defect.

Sheet Metalworking

Sheet metalworking is always a secondary process on previously rolled flat products such as sheet, strip and plate. The first such operation is usually shearing (or slitting, blanking, or punching). Separation of adjacent metal parts occurs through highly localized plastic deformation followed by shear failure (Figure 9a) and seems unaffected by friction. Nevertheless, lubrication is necessary to protect against rapid wear and die pickup. Many parts are formed by bending, and bending forces are affected by friction when the workpiece slides over some die element (Figure 9b).

Friction becomes extremely important when shapes of three-dimensional geometry are formed through stretching, deep drawing, or their combination. In pure stretching the sheet is firmly clamped (Figures 10a and b). In the absence of friction, thinning is most severe and fracture occurs at the apex of the stretched part. With increasing friction, free thinning over the punch nose is hindered and the fracture point moves further down the side of the part.

In deep drawing (Figure 11), blanks of large diameter-to-sheet thickness ratio would buckle (wrinkle) and must be kept flat by applying pressure through a blankholder. Frictional stresses increase the force required for deformation, and when the force exceeds the strength of the partially drawn product, fracture occurs. Friction must be minimized to reach the maximum possible draw.

Cups of large depth-to-diameter ratio must be produced with several redrawing steps. When the cup wall is to be reduced substantially, the drawn cup is pushed through an ironing die, and wall thickness reduction takes place under high normal pressures and severe sliding, calling for a much heavier-duty lubricant. Friction on the punch surface is again beneficial.

Many sheet metal parts in the automotive and appliance industries are of complex shapes produced by simultaneous stretching and drawing. Lubrication is essential to prevent die pickup and surface damage. To restrict free drawing-in of the sheet, a draw bead is incorporated (Figure 10b); this imposes severe conditions on the lubricant.

LUBRICATING MECHANISMS

Lubricants in deformation processes have to survive under an extremely wide range of conditions. Interface pressures range from a fraction of flow stress σ_f up to multiples of σ_f (up to 4 GPa or 500 kpsi). Sliding velocities range from zero (at sticking friction) to 50 m/s (10,000 fpm) sometimes combined with approach velocities of up to 20 m/s (66 ft/sec).

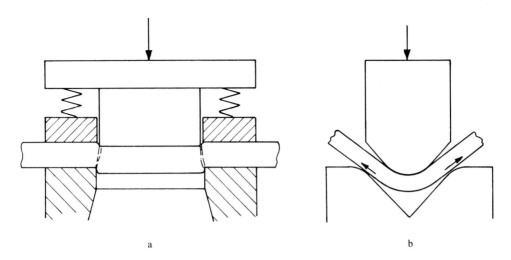

FIGURE 9. The process of (a) shearing or blanking and (b) bending.

CLAMP

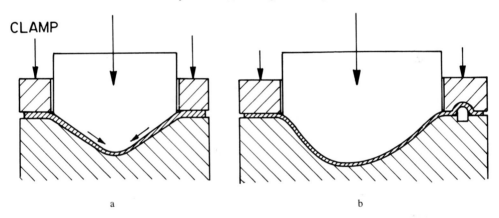

FIGURE 10. Stretching processes. (a) Stretch forming and (b) drawing with a draw bead.

to change as a result of both plastic and elastic deformations, and surface roughness of the tool and workpiece create highly variable conditions. Lubrication of the surfaces freshly exposed by physical expansion is most important. If the workpiece material shows high adhesion to the die materials, cold welding at asperities may result in pickup (in the extreme, total seizure) unless an effective lubricant is interposed.

Films of Low-Shear Strength

In principle, any continuous film of a material that has a shear strength τ_s lower than that of the workpiece material τ_f will serve as a lubricant and prevent metal-to-metal contact. Soft metals may themselves act as lubricants; in addition they may make lubrication easier by virtue of their composition and/or surface characteristics. Because their shear strength τ_s is largely independent of pressure (Figure 12), they become particularly useful at high working pressures. Oxides grown on the workpiece material may not only reduce adhesion but also behave as lubricants if their shear yield strength is low.

Plastics (polymers) deposited as thin films fulfill the same function as thin metal films. However, their shear strength is usually directly proportional to normal pressure, resulting in a constant coefficient of friction that seldom drops below 0.05 (Figure 12). They lack spreadability and cannot prevent pickup when the film is discontinuous (either from poor deposition or surface expansion).

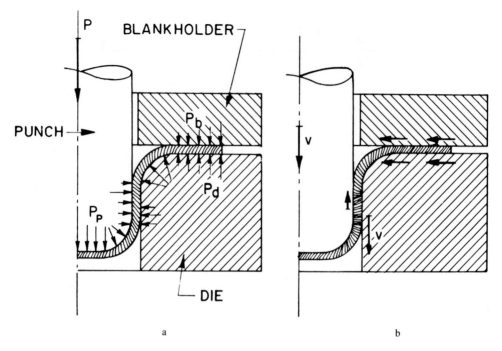

FIGURE 11.   Deep drawing with blankholder. (a) Variation of interface pressures and (b) relative sliding velocities. (From **Schey, J. A., Ed.,** *Metal Deformation Processes: Friction and Lubrication,* Marcel Dekker, New York, 1970. With permission.)

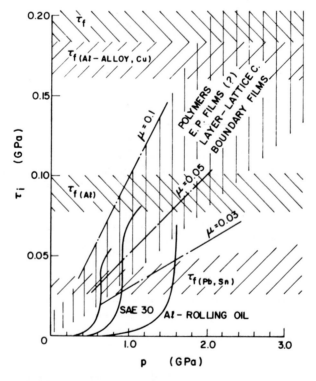

FIGURE 12.   The variation of interface shear strength as a function of interface (normal) pressure. (From **Schey, J. A.,** in *Metal Forming Plasticity,* Lippmann, H., Ed., Springer-Verlag, Berlin, 1979. With permission.)

Layer-lattice compounds such as MoS_2 and graphite are effective if deposited in a continous film. Their shear strength is also pressure-sensitive. In common with other solid films, they cannot protect surfaces if the lubricant film breaks down.

Boundary Films

Very thin films formed on the die or workpiece may prevent adhesion and also reduce friction. Extreme-pressure (EP) compounds rely on reactions that take place only if sufficient time is available at temperature and if substrate composition is favorable. Contact time under metalworking conditions is usually too brief, but if contact is repeated (as in cold heading, wire drawing, sheet metal working, and sometimes rolling), reaction is possible with the die.

Boundary films form almost instantaneously and are among the most important lubricants for reactive metals, particularly aluminum, copper, and to a lesser extent, steel. Their shear strength is pressure- and temperature-dependent. Breakdown at elevated temperatures limits them to cold working.

Full-Fluid Film Lubrication

In this regime, tool and workpiece surfaces are separated by a liquid film of sufficient thickness to avoid asperity interaction. Plastohydrodynamic theory can account for the effects of process geometry, sliding speed, and lubricant properties in maintaining such a film. The pressure and temperature sensitivity of viscosity must be considered, together with the possibility of the lubricant becoming a polymer-like solid at higher pressures (Figure 12).

Mixed Lubrication

In most practical situations, only some portion of the total contact area is separated by a thick lubricant film. Other parts of the contact area are in boundary contact, making boundary or EP additives a necessity in almost all metalworking fluids. Depending on process conditions and lubricant viscosity, the deformed surface may be smoother than before deformation. It may also be roughened by the formation of entrapped lubricant pockets.

Surface Roughness Effects

Although asperities piercing through the lubricant film can create adhesion problems, tool and workpiece surfaces need not always be very smooth. To avoid skidding in rolling, the roll surface is kept somewhat rough. A moderately coarse, nondirectional (such as bead-blasted) surface finish is desirable in maintaining graphite or MoS_2 supply in hot forging. A smooth, polished die surface is, however, desirable for liquid lubricants; any remaining roughness is preferably oriented in the direction of material flow.

Moderate roughness of the workpiece surface helps carry liquid lubricants into the interface, especially if the roughness is perpendicular to the direction of feeding. If sliding is multidirectional, a random (bead-blasted) finish is preferable (as on automotive body sheets).

Lubricating Regimes

The range for various lubricating mechanisms is summarized in Figure 13. For an interface pressure of $p = \sigma_f$, friction cannot attain a value higher than that corresponding to sticking. The presence of a solid film (other than metal) reduces the coefficient of friction to about 0.05 to 0.1, apparently irrespective of the nature of solid film. With a liquid, mixed lubrication is attained once velocity and viscosity combine to sustain the pressures required for plastic deformation. With increasing velocity and/or viscosity, the proportion of surface area lubricated by the fluid film increases and the coefficient of friction drops to typically 0.03 to 0.05. True plastohydrodynamic lubrication is rare.

With increasing interface pressures, the apparent coefficient of friction drops even for

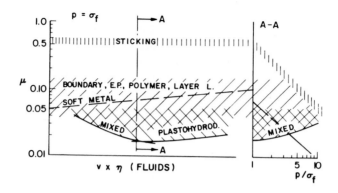

FIGURE 13.   Lubrication regimes in metal deformation processes. (From **Schey, J. A.,** in *Metal Forming Plasticity,* Lippmann, H., Ed., Springer-Verlag, Berlin, 1979. With permission.)

sticking (right-hand side of Figure 13), whereas it remains constant for solid films. Therefore, sticking can be attained even on a lubricated surface. In the presence of fluids, increasing pressure increases the proportion of boundary-lubricated areas in the regime of mixed lubrication.

WEAR

With the prevalence of boundary and mixed lubrication in metalworking processes, wear can become a serious problem.

Adhesive wear is predominant when there is great affinity between die and workpiece material. Adhesion leads to pressure welding (die pickup) at asperities, followed by shearing and wear particle formation. Most wear is concentrated on the workpiece. In rolling, however, large amounts of debris enter the lubricant and the rolls have to be refinished at regular intervals. The same is true of dies in wire and bar drawing. High die hardness is generally helpful, as can be surface or diffusion coatings, although there may be interference with reactions on which the efficiency of some lubricants depends.

The presence of hard particles, such as oxides or wear debris, accelerate three-body wear. Thus, removal of scale particles (by high-pressure water or mechanical descaling of the workpiece) becomes important in forging and hot rolling of steel, and filtration is a must in all recirculating lubricant/coolant systems.

Adhesive and abrasive wear generally follows the same laws as in machine elements, however, the rate of wear is often accelerated by the generation of virgin surfaces during plastic deformation. High workpiece temperatures subject colder dies to severe thermal shock and thermal fatigue then contributes to wear and die failure. Ceramic die coatings or die inserts can give useful improvements.

LUBRICANT SELECTION

Lubricant Testing

Lubricant properties such as composition, viscosity, stability, and resistance to bacterial attack may be measured by standard techniques. Only two measurements are unique to metalworking processes: testing for friction under deformation conditions, and for staining propensity during annealing.

Friction Testing

The extremely broad range of process conditions coupled with surface extension during deformation makes the formulation of a single test technique virtually impossible. Standard laboratory bench tests (such as the four-ball test) occasionally give successful correlation with metalworking practice; these cases are, however, exceptions and are perhaps fortuitous. Most attractive are tests in which friction balance leads to a readily measurable change in deformation. The ring compression test (Figure 4) is one which measures lubricant behavior under normal contact to simulate forging. Forward slip (Equation 5), which directly reflects friction balance in the roll gap gives similarly obvious results under rolling conditions.

A second group of test measures interface shear strength. Thus, a sheet drawing test (Figure 14a) reveals the magnitude of friction under moderate pressures for light-duty sheet-metalworking lubricants. Higher interface pressures and testing for lubricant durability are possible using a hollow specimen with an annular end face pressed and rotated against a flat anvil (Figure 14b); reasonable correlation is found with severe cold working such as extrusion.

In a third group of tests, the magnitude of friction is judged from the force required to perform deformation such as upsetting of a cylinder, extrusion, or wire drawing. Lubricant ranking is made simply by comparing the magnitudes of forces.

No single test provides information for all conceivable metalworking conditions. For a broader evaluation at least three tests have to be performed, such as ring compression for normal approach, plane-strain compression for deformation with extensive sliding, and twist compression for lubricant starvation situations. Die and workpiece material, surface preparation and roughness, interface velocity, and entry zone geometry should be the same or as closely scaled as possible to the actual process.

Testing for Staining

Lubricants are sometimes left on the deformed workpiece surface for corrosion protection, and testing in typical industrial atmospheres is necessary. In other instances, the workpiece is subjected to subsequent operations, such as annealing, joining, etc. Testing for staining propensity can be done if air access is controlled to reproduce that typical of annealing in coils.

Hot-Working Lubricants

Typical hot-working temperatures are in excess of 400°C for aluminum and magnesium alloys, over 600°C for copper alloys, and over 900°C for steels and nickel-base alloys. Lubricants are limited to those resisting the workpiece (or interface) temperature, such as recommended in Table 1.

Oxides formed during heating the workpiece can fulfill a useful parting function, provided they are ductile at the interface temperature. This condition is partially satisfied only by iron oxides, cuprous oxide, and refractory metal oxides. Other harder oxides (such as ZnO) are effective only when they can break up into a powdery form. Yet others (such as aluminum and titanium oxides) are hard, brittle, cannot follow surface extension at all, and do not protect once broken up.

Glasses of proper viscosity (typically 20 Pa·sec at the mean of the die and workpiece surface temperatures) can act as true hydrodynamic lubricants. If the process geometry is favorable (such as in extrusion), a thick glass mat may gradually melt off to provide a continuous coating on the deformed product. Glass may be applied either as glass fiber or powder to the die or hot workpiece, or in the form of a slurry with a polymeric bonding agent to the workpiece prior to heating. In the latter case it may also protect from oxidation and other reactions during the heating period.

Graphite is effective in forging steel or nickel-base alloys, if uniformly deposited on the die surfaces from an aqueous or sometimes oily base. Wetting a hot surface is difficult but special formulations (sometimes with polymeric binders) have been developed. Application

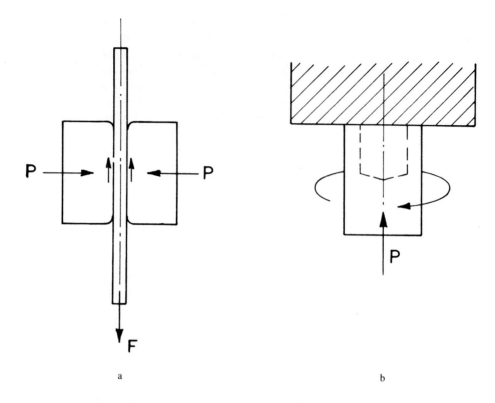

FIGURE 14.   Schematic of (a) sheet-drawing and (b) twist-compression tests.

Table 1
TYPICAL LUBRICANTS USED IN HOT WORKING

Process	Steel	Stainless steel Ni-base	Ti	Cu, brass	Al, Mg
Rolling	None (GR suspension) (MO-FA-EM)	As steel		MO-FA-EM	MO-FA-EM
Extrusion	GL (GR)	GL	GL	None (GR) (GL)	None
Forging	None GR	GR	GL MoS$_2$	GR	GR MoS$_2$

Note: Less frequent usage is shown in parentheses. Hyphenation indicates that several components are used in lubricant. EM = emulsion (the listed lubricants are emulsified and 1 to 5% is dispersed in water); FA = fatty acids, alcohols, amines, and esters; GL = glass (sometimes in conjunction with GR on the die); GR = graphite, and MO = mineral oil.

to the workpiece prior to heating is less effective because condensable vapors necessary to assure easy shearing of the graphite are driven off. Molybdenum disulfide and other layer-lattice compounds are also effective, particularly if reaction with the substrate material is possible and if temperatures are not over 500°C. They must, however, be avoided for high-Ni alloys because a low-melting eutectic forms and causes hot shortness. Oily lubricants are seldom effective, but water emulsions of oils, fatty oils, and EP compounds can be applied in a flood. The practice is universal in aluminum and magnesium rolling, widespread

Table 2
TYPICAL LUBRICANTS USED IN COLD WORKING

Process	Steel	Stainless steel[a] Ni-base[c]	Ti[b]	Cu[c] brass	Al, Mg[d]
Rolling	FO FO-EM (FO-MO)	CL-MO CL-FO-EM	FO-MO MO on oxidized surface SP	FO-MO(10—50) FO-MO-EM	1—5% FA-MO (5—20) (or synthetic MO)
Extrusion Light	EP-MO SP on PH	CL-MO SP on oxalate	SP or GR grease on fluoride-PH	FO-MO GR-FO GR-grease	Lanolin Zn stearate SP on PH
Severe	MoS₂+SP on PH				
Forging Light	EP-MO SP on PH	EP-MO CL-MO SP on oxalate	As extr.	FO SP	FO Lanolin
Severe					
Wire drawing Light	SP-FO-EM SP on lime or borax	CL-MO SP on oxalate PC, or CL-MO	As extr.	SP-FO-EM FO-MO(20—80) SP-FO-EM	FO-MO(20—40)
Heavy	EP-FO-MO on lime or PH Grease or GR-grease	CL-EP-MO	As extr., or PC	FO-MO SP	FO-MO(100—400)
Bar drawing	EP-FO-MO SP on PH	SP on oxalate PC	As bar	As bar	SP FO-MO(50—400)
Tube drawing	EP-FO-MO SP on PH	SP on oxalate PC	As bar	As bar	As bar
Light pressing	MO; SP-EM	EP-MO EP-MO-EM	MoS₂-MO	MO-EM	FA-MO
Heavy sheet drawing	FO; FO-EP-MO; SP; SP on PH; pigmented FO-SP	SP CL-MO	GR-grease[e]	FO-MO Pigmented FO-SP	FO

Table 2 (continued)
TYPICAL LUBRICANTS USED IN COLD WORKING

Process	Steel	Stainless steel[a] Ni-base[c]	Ti[b]	Cu[c] brass	Al, Mg[d]
Heavy ironing	EP-Gr-grease SP on PH	SP on oxalate	GR-grease[c] GL+GR	FO SP	Lanolin

Note: Hyphenation indicates that several components are used in lubricant. CL = chlorinated paraffin; EM = emulsion (the listed lubricants are emulsified and 1 to 20% dispersed in water); EP = "extreme pressure" compounds (containing S, Cl, and/or P); FA = fatty acids, alcohols, amines, and esters; FO = fats and fatty oils, e.g., palm oil, synthetic palm oil; GR = graphite; MO = mineral oil (viscosity in units of centistoke [=mm²/sec] at 40°C); PH = phosphate surface conversion; PC = polymer coating; and SP = soap (powder, or dried aqueous solution, or as a component of an EM).

a   Chlorine is the most effective EP agent on stainless steel.
b   Chlorine is avoided for Ti.
c   Sulfur is avoided for Ni because of reaction and for Cu because of staining.
d   Magnesium alloys are usually worked warm (above 200°C).
e   Usually conducted hot.

for copper-base alloys, and is gaining some acceptance for steel rolling. Chemical solutions (of rust preventatives, etc.) are often adequate when a protective oxide is present. The water fulfills the important function of heat extraction.

Cold-Working Lubricants

The relatively low temperatures attained during cold deformation permit use of a wide range of lubricants such as indicated in Table 2.

Solid films are of particular value for severe deformation. Although soft metals have declined in importance, tin on mild steel sheet is employed in the production of drawn and ironed tin cans. Polymeric films, interposed as a separate film or deposited on the workpiece surface, find growing but still limited application. Of greatest importance are surface conversion coatings such as produced by phosphating of steel. They present a strongly adhering film of sufficient porosity or surface detail to provide a mechanical key for the superimposed lubricant layer, typically a soap. Film attachment is further enhanced by reacting the soap with the phosphate film. Layer-lattice compounds are mostly used as additives to other lubricants to provide a last defense in case of lubricant breakdown.

Oil-based lubricants represent a large proportion of the total used. The viscosity of natural or synthetic oils is chosen to give mixed-film or occasionally almost full-fluid film lubrication, but not so high to induce excessive surface roughening or to drop friction below an acceptable limit. Because of the impossiblity of avoiding all asperity contact, lubricants always contain boundary additives: typically fatty acids, alcohols, esters, or natural fatty oils. When contact is repetitive, EP additives may also be useful, particularly for metals (stainless steel, titanium) on which fatty additives are ineffective. When conditions are unfavorable to developing hydrodynamic films, grease may be used. Whenever cooling is important, the oil is applied in a recirculating system. A flood of aqueous emulsions or dispersions is even more effective, but staining may be a problem (e.g., on Al or Mg). Removal of wear particles by filtration is an essential requirement in all recirculating systems.

REFERENCES

1. **Bastian, E. L. H.,** *Metalworking Lubricants,* McGraw-Hill, New York, 1951.
2. **Schey, J. A., Ed.,** *Metal Deformation Processes: Friction and Lubrication,* Marcel Dekker, New York, 1970.
3. *Tribology in Iron and Steel Works,* Publ. No. 125, Iron and Steel Institute, London, 1969.
4. **Rowe, G. W.,** *Mech. Mach. Electr.,* 266, 20, 1972.
5. **Schey, J. A.,** in *Proc. Tribology Workshop Atlanta,* F. F. Ling, Ed., National Science Foundation, Washington, D.C., 1974, 428.
6. **Schey, J. A.,** *Introduction to Manufacturing Processes,* McGraw-Hill, New York, 1977.
7. **Wilson, W. R. D.,** in *Mechanics of Sheet Metal Forming,* Plenum Press, New York, 1978, 157.
8. *Proc. 1st Int. Conf. Lubr. Challenges in Metalworking and Processing,* IIT Research Institute, Chicago, 1978.
9. *Proc. 2nd Inst. Conf. Lubr. Challenges in Metalworking and Processing,* IIT Research Institute, Chicago, 1979.
10. **Schey, J. A.,** in *Metal Forming Plasticity,* Lippmann, H., Ed., Springer-Verlag, Berlin, 1979, 336.
11. **Wilson, W. R. D.,** *J. Appl. Metalworking,* 1, 7, 1979.
12. **Schey, J. A.,** in *Proc. 4th Int. Conf. Prod. Eng.,* Japan Society of Precision Engineering, Tokyo, 1980, 102.
13. **Kalpakjian, S. and Jain, S. C., Eds.,** *Metalworking Lubrication,* American Society of Mechanical Engineers, New York, 1980.

METAL REMOVAL

Milton C. Shaw

INTRODUCTION

A good deal of effort in the manufacture of hard goods is concerned with providing a desired shape and accuracy to machine parts and the removal of material represents one important way of doing this. The entire area of material removal may be divided into metal cutting in which relatively large chips of uniform geometry are formed and abrasive processing in which relatively small chips having a wide dispersion of geometries are produced.

It has been estimated that about 10% of the gross national product are spent in material removal operations in the U.S. It is, therefore, important that such processes be understood and efficiently performed if productivity (effective use of labor and capital) is to be achieved. Since friction, wear, and lubrication play important roles in material removal operations, it is pertinent to consider these operations here.

There are a wide variety of removal operations which use tools of different geometry and kinematic relationship between tool and work. Some of the more important operations include the following:

1. Turning to produce cylindrical surfaces
2. Milling to produce flat surfaces and surfaces of complex geometry
3. Drilling, boring, and reaming to produce round holes
4. A wide variety of grinding operations

In addition to these, there are a host of more specialized operations designed to do a particular job more effectively or which are better suited to mass production. However, all removal operations involve tools that penetrate the work to peel off unwanted material. What goes on at the tip of these tools is essentially the same regardless of geometry or kinematics.

CUTTING MECHANICS

Orthogonal Machining

Figure 1 is a photomicrograph of a partially formed chip.[1] This was produced by moving a workpiece against a stationary two-dimensional tool, abruptly stopping the operation in midcut and then sectioning and metallographically polishing the "chip root". The magnified view of the etched surface reveals a great deal concerning the action of a metal cutting tool when removing a chip. As the material approaches line AB there is essentially no plastic flow until AB is reached. At this point a sudden concentrated shear occurs and then the material proceeds upward along the tool face with essentially no further plastic deformation. In the case of Figure 1, the cutting edge was stationary and straight, extending perpendicular to the plane of the paper. This is called orthogonal cutting since the cutting edge is perpendicular to resultant velocity vector (V) which is in the horizontal direction in Figure 1. Merchant[2] and Piispanen[3] first discussed orthogonal cutting in fundamental terms.

Figure 2 is a diagrammatic representation of Figure 1. Line AB is the trace of the surface on which the concentrated shearing action occurs and is called the shear plane. The angle the shear plane makes with velocity vector (V) is the shear angle (ϕ) while the angle between the face of the tool (rake face) and the normal to the velocity vector is called the rake angle (α). Also shown in Figure 2 is the clearance angle (γ). The thickness of the layer removed is the undeformed chip thickness (t) and the width of cut perpendicular to the paper will be

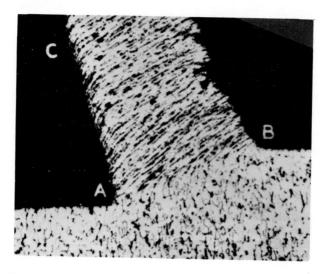

FIGURE 1. Photomicrograph of partially formed chip. AISI 1015 steel cut at 24 fpm (7.3 m/min) using 0.1% $NaNO_2$ solution. Vertical distance AB, undeformed chip thickness = 0.005 in. (0.125 mm).

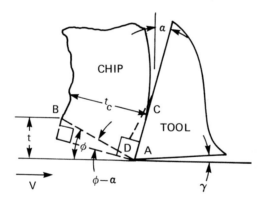

FIGURE 2. Diagrammatic representation of Figure 1.

designated (b). In general, the chip will curl away from the rake face of the tool, giving rise to a contact length AC.

If one looks closely at Figure 1 there is evidence of some plastic flow in the chip adjacent to the tool face. This is called secondary shear and results from the material in the chip being so highly loaded that it is on the verge of flowing plastically. The additional shear stress on the face of the chip is enough to cause subsurface shear in the chip in a direction essentially perpendicular to the primary shear that occurs on the shear plane.

The Built-Up Edge

The situation at the tip of a cutting tool is not always as well behaved as that shown in Figure 1. Sometimes, a ductile shear fracture instability occurs to produce a built-up edge (BUE). The photomicrograph[1] of Figure 3 shows a large BUE while Figure 4 is a diagrammatic representation. Since the dominant shear process occurs along AB (Figures 1 and 4) we should naturally expect this to be the plane of maximum shear stress. Since angle BAC (Figure 2) is usually not 90°, direction AC will not be the second direction of maximum shear stress and strain but direction CD will be. This is one reason the chip sometimes

FIGURE 3. Photomicrograph showing large BUE and portions of BUE along finished surface and along face of chip. AISI 1020 steel cut dry at 90 fpm (27.4 m/min). Undeformed chip thickness = 0.005 in. (0.125 mm).

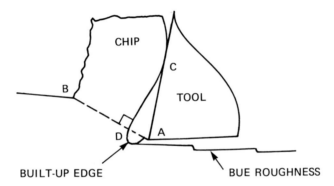

FIGURE 4. Diagrammatic representation of Figure 3 showing BUE-induced roughness on finished surface.

fractures internally in shear leaving behind the stationary body of metal attached to the tip of the tool (BUE). Another reason is associated with the fact that metals (notably steel) exhibit minimum ductility (strain at fracture) somewhat above room temperature. This is called "blue brittleness", since the minimum strain-at-fracture temperature for steel corresponds to that where the thickness of surface oxide produced gives rise to a blue interference color. At relatively low-cutting speeds (low-tool face temperature), the temperature along CD will be closer to the blue brittle temperature than along AC and then the strain at fracture along CD will be less than along AC.

The inherent instability of a large BUE is very troublesome relative to surface finish. As cutting proceeds, BUE tends to grow slowly until it reaches a critical size and then it leaves abruptly with the chip. Since the BUE grows downward as well as outward (point D below A in Figure 4), the surface produced by the periodic change in size of the BUE is as shown in Figure 4. This is one of the sources of surface roughness when cutting at relatively low speed. An unstable BUE can also decrease tool life due to abrasive action of pieces of BUE on the tool face and wear land; on the other hand, a small, stable BUE can be beneficial in protecting the tip of the tool from wear.

The most important way of avoiding a BUE is to increase cutting speed. Above a certain

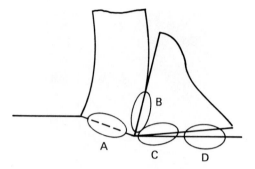

FIGURE 5.   Regions of importance in metal cutting.

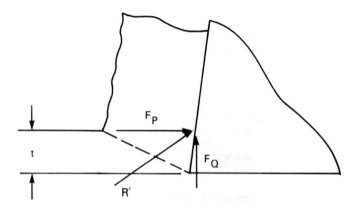

FIGURE 6.   Forces acting on tool for orthogonal cut.

cutting speed ($\sim$ 1m/sec or 200 ft/min) a large unstable BUE will not tend to form. Another way of decreasing BUE formation is to increase the rake angle of the tool. This causes angle DCA in Figure 2 to approach 0° and hence CD to become parallel to AC in Figure 4. On the other hand, making the effective rake angle more negative will tend to promote the formation of a BUE.

Regions of Importance

Figure 5 identifies four regions of importance in cutting operations. Region A concerns the shear process associated with chip formation by concentrated shear. Region B involves friction-like behavior along the tool face. Region C concerns action on the clearance face of the tool while region D concerns the characteristics of the finished surface.

Cutting Forces

By mounting the workpiece on a dynamometer, it is possible to measure cutting forces in the direction of the velocity vector (F_P) and in two orthogonal directions to $V - F_Q$ in the plane of the paper and F_R perpendicular to the plane of the paper (Figure 6). In orthogonal cutting $F_R = 0$. Figure 7 is a free body diagram of an orthogonal chip produced under steady-state conditions as a free body. Only two forces act on this free body — R (between chip and tool) and R' (between chip and work). These must be equal in magnitude and oppositely directed. This very important result explains the strong interaction observed between what happens on the tool face and what happens on the shear plane. Both the processes on the shear plane and tool face must be studied at the same time; it is very difficult to identify cause and effect in interpreting metal cutting data.

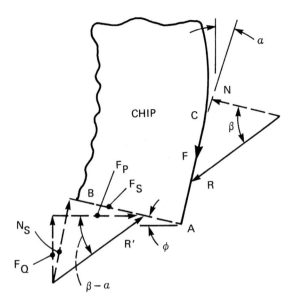

FIGURE 7.   Free-body diagram of chip showing forces acting on
tool face (AC) and shear plane (AB).

Instead of considering forces R and R', it is convenient to resolve these forces parallel
and normal to the tool face and shear plane respectively as shown by the dotted lines in
Figure 7. Adopting a simple strength-of-materials approach, we can assume a uniform
distribution of stress along AB and AC and obtain the following equations:

Along AB:

$$\tau = \frac{F_s}{A_s} = \frac{F_p \cos \phi - F_Q \sin \phi}{bt/\sin \phi} \tag{1}$$

$$\sigma = \frac{N_s}{A_s} = \frac{F_Q \cos \Phi + F_p \sin \phi}{bt/\sin \phi} \tag{2}$$

Along AC:

$$\mu = \tan \beta = \frac{F}{N} = \frac{F_Q + F_P \tan \alpha}{F_P - F_Q \tan \alpha} \tag{3}$$

where τ = mean shear stress on shear plane AB, σ = mean normal stress on shear plane
AB, A_s = area of shear plane, t = undeformed chip thickness, b = chip width, μ =
coefficient of tool face friction, α = rake angle, and ϕ = shear angle.

Cutting Ratio

While the rake angle may be directly measured on the tool, the shear angle (ϕ) is most
conveniently determined from the cutting ratio (r) where (Figure 8)

$$r = \frac{t}{t_c} = \frac{AB \sin \phi}{AB \cos (\phi - \alpha)} \tag{4}$$

or solving for ϕ:

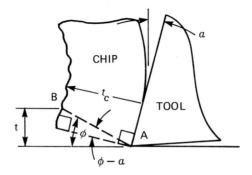

FIGURE 8. Construction for use in deriving expression connecting shear angle (ϕ) and cutting ratio (r).

$$\tan \phi = \frac{r \cos \alpha}{1 - (r \sin \alpha)} \tag{5}$$

The value of mean chip thickness in Equation 4 is not easily measured directly because most chips are relatively rough on the free surface. A more convenient method of obtaining r is from the chip length ratio (ℓ_c/ℓ). When any material is deformed plastically there is a negligible change in volume. Hence,

$$\ell \, b \, t = \ell_c \, b_c \, t_c \tag{6}$$

where the quantities ℓ, b, and t without subscripts are the length, width, and depth of the undeformed chip while subscript (c) indicates corresponding quantities in the deformed chip. From Equation 6

$$r = \frac{t}{t_c} = \frac{\ell_c \, b_c}{\ell b} \tag{7}$$

Since in most cases b $\simeq$ (10 t), chips are formed under essentially plane strain conditions which means there will be negligible side flow and

$$b \cong b_c \tag{8}$$

Cutting ratio (r) and shear angle (ϕ) play a very important role in metal cutting. In general, as shear angle (ϕ) increases, the area of the shear plane (A_S) decreases and hence the energy consumed on the shear plane decreases. Thus, the cutting ratio is a convenient measure of the efficiency — the larger the cutting ratio (r) the more efficiently the cutting operation is being performed from the point of view of the energy expended. An increase in cutting ratio may be accomplished by a change in workpiece chemistry or structure, or by a decrease in friction on the tool face.

Shear Strain in Chip Formation

Figure 9 shows an idealized model of the cutting process involving the previously described concentrated shear mechanism. Active shear planes which contain one or more imperfections are relatively far apart (compared with atomic spacing) and, therefore, the basic chip forming process may be approximated by a deck of cards sliding over each other as they cross the shear plane (Figure 9a). Figure 9b shows how the shear strain in cutting may be approximated as follows:

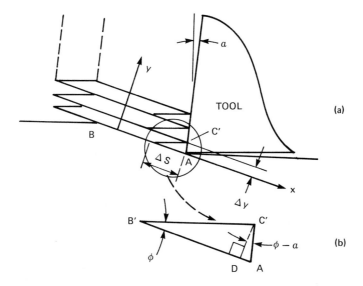

FIGURE 9. Construction for use in deriving expression for shear strain (γ) in cutting.

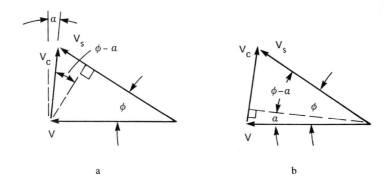

FIGURE 10. Velocity diagram for metal cutting.

$$\gamma = \frac{\Delta S}{\Delta Y} = \frac{AB'}{C'D'} = \frac{AD'}{C'D'} + \frac{D'B'}{C'D'} = \tan(\phi - \alpha) + \cot \phi \qquad (9)$$

It may also be readily shown that

$$\gamma = \frac{\cos \alpha}{\sin \phi \cos (\phi - \alpha)} \qquad (10)$$

Kinematics

There are three velocities of importance in the cutting process (Figure 10).

1. The cutting velocity (V) which is the velocity of the tool relative to the work.
2. The velocity of the chip relative to the tool (V_c) which is directed along the tool face.
3. The shear velocity (V_s) which is the velocity of the chip relative to the work and directed along the shear plane.

In accordance with the principles of kinematics these three velocities must form a closed diagram as shown in Figure 10, from which it follows that

$$\frac{V_c}{V} = \frac{\sin \phi}{\cos (\phi - \alpha)} = r \quad \text{(Figure 10a)} \tag{11}$$

$$\frac{V_S}{V} = \frac{\cos \alpha}{\cos (\phi - \alpha)} = \gamma \sin \phi \quad \text{(Figure 10b)} \tag{12}$$

As already mentioned slip does not occur on every atomic plane but only on bands of planes in the vicinity of a defect. This results in the strain in a chip being very inhomogeneous, being small within a slip band and large between. It also results in the strain rate of chip formation being very large since

$$\dot{\gamma} = \frac{d\gamma}{dt} = \frac{V_s}{\Delta y} \tag{13}$$

where Δy is the width of a slip band which is normally about 10^{-4} in. (25×10^{-4} mm). For example, for a moderate cutting speed of 100 fpm (0.5 m/sec), a rake angle (α) of $0°$, a shear angle (ϕ) of $20°$, and $\Delta y = 10^{-4}$ in. (2.5μm), the strain rate is found to be approximately 0.2×10^6/sec. This is recognized as a very high value when it is realized that the strain rate in an ordinary tensile test is about 10^{-3}/sec.

The strain in cutting not only takes place at a high strain rate but also at high temperature. However, these two items have opposite effects on flow stress and it is generally considered that the flow stress of material being cut at high speed is not very different than that for a specimen tested at an ordinary strain rate and room temperature.

Specific Cutting Energy

The concept of energy per unit volume (specific energy) is a useful one here as it is in other areas of mechanics. The total specific energy (u) is

$$u = \frac{F_p V}{(Vbt)} = \frac{F_p}{bt} \tag{14}$$

where (Vbt) is the volume rate of material removal. The specific cutting energy (u) has only two significant components — u_s, specific shear energy, and u_F, specific friction energy, where

$$u_s = \frac{F_s V_s}{Vbt} = \tau\gamma \tag{15}$$

$$u_F = \frac{F V_c}{Vbt} = \frac{Fr}{bt} = \left(\frac{F_p \sin \alpha + F_Q \cos \alpha}{bt} \right) r \tag{16}$$

Since specific surface energy and specific momentum energy are negligible, in all metal removal operations including grinding

$$u = u_s + u_F \tag{17}$$

In metal cutting operations, the main variables influencing specific energy (u) are work-piece chemistry and hardness. Values of u given in Table 1 are useful in estimating the power component of cutting force (F_p) as well as the power required for a specific cutting operation. The component of the feed force may be estimated to a first approximation by assuming it to be half the value of F_p.

Table 1
APPROXIMATE VALUES OF SPECIFIC
CUTTING ENERGY (u) FOR MEDIUM
SIZE CUTS (b = 0.1 in., t = 0.010 in./r)
AND A RAKE ANGLE OF 0°

Material	u (in.lb/in.³)	u (MPa)
Pure aluminum	100,000	690
Pure copper	200,000	1,380
Gray cast iron	200,000	1,380
Mild steel	300,000	2,070
Alloy steel	400,000	2,760
Stainless steel, titanium alloys and high-temperature alloys	500,000	3,450

$$F_Q \simeq F_p/2 \tag{18}$$

As an example consider a straight turning operation as follows:

Workpiece material	= AISI 1020 steel
Cutting speed (V)	= 200 fpm (60 m/min)
Feed (t)	= 0.010 in./r (0.25 mm/r)
Depth of cut (b)	= 0.100 in. (2.5 mm)

From Equation 14: F_p = ubt = (300,000) (0.1) (0.01) = 300 lb (1330 N); horsepower = F_p V/33,000 = 300 (200)/33000 = 1.82 hp (1.36 kW); and cutting force from Equation 18: $F_Q \simeq$ 300/2 = 150 lb (670 N).

These calculations are of value in providing a first approximation to cutting forces and horsepower required.

In general, cutting speed and depth of cut have a small influence on specific cutting energy, but rake angle and undeformed chip thickness have an appreciable influence. The specific cutting energy decreases about 1% per degree increase in rake angle (α), while u varies approximately as follows with undeformed chip thickness (t) in metal cutting

$$u \sim 1/t^{0.2} \tag{19}$$

Tool Face Temperature

Practically all of the energy involved in cutting ends up as heat most of which is convected away by the chip in the case of high-speed cutting. As a first approximation to tool face temperature (θ) we assume that all of the specific energy is converted to thermal energy which is uniformly distributed in the chip and passes off with the chip. Then

$$\theta \sim \frac{u}{J\,(\rho c)} \tag{20}$$

where J = mechanical equivalent of heat and (ρc) = volume specific heat.

Representative Example

To provide an appreciation for the magnitudes of the quantities discussed in this section, a typical example is presented in Table 2. An examination of this table reveals that metal cutting involves unusually high values of stress, strain, strain rate, and temperature. The

Table 2
REPRESENTATIVE VALUES

Job description
  Operation: turning
  Tool material: tungsten carbide
  Rake angle (α): 50
  Cutting edge direction: radial (0° side cutting edge angle)
  Depth of cut (b): 0.25 in. (6.35 mm)
  Feed (t): 0.010 ipr (0.25 mm)
  Cutting speed (V): 550 fpm (168 m/min)
  Cutting fluid: dry (air)
  Work material: AISI 1045 steel ($H_B = 180$)
Observed
  Cutting ratio (r): 0.40
  Cutting force (F_p): 790 lb (3514 N)
  Cutting force (F_Q): 395 lb (1757 N)

Calculated	Equation no.
Shear angle (ϕ): 22.4°	5
Shear strain (γ): 2.74	10
Coefficient of friction (μ): 0.61	3
Mean shear stress (τ): 88,450 psi (610 MPa)	1
Mean normal stress (σ): 101,700 psi (701 MPa)	2
Chip velocity (V_c): 220 fpm (67 m/min)	11
Shear plane velocity (V_s): 574 fpm (175 m/min)	12
Total specific energy (u): 316,000 in.lb/in.^{3} (2,179 × 10^6 Nm/m^3)	14
Specific shear energy (u_s): 242,000 in.lb/in.^{3} (1,669 × 10^6 Nm/m^3)	15
Specific friction energy (u_F): 74,000 in.lb/in.^{3} (510 × 10^6 Nm/m^3)	16
u_s/u: 0.77	
Horsepower (F_pV/33,000): 13.2hp (9.85 kW)	

total energy involved in metal cutting is distributed approximately as follows: 3/4 on shear plane and 1/4 on tool face.

Values of strain that are several times larger than those involved in ordinary materials tests are possible without gross fracture of the chip due to the fact that a very large normal stress is present on the shear plane. However, it is believed that microcracks of limited extent do form on the shear plane when cutting with a continuous chip and that the formation and rewelding of these microcracks plays an important role in the mechanism of concentrated plastic shear strain that occurs on the shear plan.

Three-Dimensional Cutting

The foregoing discussion is concerned with orthogonal cutting in which a straight cutting edge that extends normal to the cutting speed vector produces the chip. If the cutting edge has an angle to V other than 90°, the deviation from 90° is referred to as the inclination angle. The main influence an inclination has is to cause the chip to flow up the tool face at an angle other than 90° to the cutting edge. This has the effect of cutting with a larger positive rake angle without a corresponding reduction in the included angle at the tool tip with its attendant weakening effect. The result of the larger effective rake angle is to decrease cutting forces and energy and to reduce the size of the BUE. In all other respects the performance of three-dimensional tools is the same as two-dimensional tools. An example of a three-dimensional cutting tool is the face milling cutter shown in Figure 11. In this case the inclination angle is the helix angle of the cutting edge.

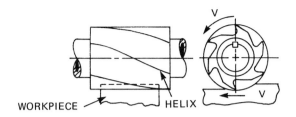

FIGURE 11. Helical milling cutter.

TOOL WEAR AND TOOL LIFE

Types of Cutting Tool Wear
Tool wear may be classified as follows:

1. Adhesive wear (attritious)
2. Abrasive wear
3. Corrosive wear (including tool decomposition and diffusion)
4. Fatigue

All of these are generally present in cutting situations *in combination*, the predominant wear mechanism depending upon cutting conditions.

In addition to the above sources of tool failure, the following also pertain: microchipping, gross fracture, and plastic deformation. These, however, are usually readily identified and corrected. The fracture modes result from subjecting the tool to too high a cutting force, operating with too large a BUE, or using a tool material that is too brittle. Plastic flow of the tool tip arises when the temperature is too high relative to the softening point of the tool material. When plastic flow occurs at the tool tip, tool clearance is lost, the temperature rises abruptly, and total tool failure occurs rather rapidly. The obvious solution to the latter difficulty is to use a lower cutting speed or a tool material that is more refractory. While cemented tungsten carbide is a very refractory material, even this material will suffer from plastic flow at the tool tip if the temperature is too high.[4]

Cutting tools wear in different ways, depending on cutting conditions (principally cutting speed, V and undeformed chip thickness, t). Figure 12 due to Opitz,[5] shows the principal types of tool wear together with the approximate ranges of values of V and t where each type of wear is predominant. The results of Figure 12 reflect primarily that the temperature on either the wear land or tool face varies as the product ($Vt^{0.6}$). By dimensional reasoning[6] the temperature rise above ambient at tool face or flank (θ) is given by

$$\theta^2 \sim Vt^{0.6} \tag{20a}$$

This means that the classification of wear types suggested by Opitz[5] is of thermal origin.

At low values of the product ($Vt^{0.6}$) tool wear consists predominantly of a rounding of the cutting point and a loss of tool sharpness. This type of wear predominates in broaching where both V and t are small. The most important types of tool wear for single point turning tools are wear land formation (Figure 12c) and crater formation (Figure 12d). Crater formation is apt to be more important than wear land formation in situations where cutting temperatures are high, which leads to decompositon of tool material and removal of the decomposition products by diffusion into the chip.

Tool Materials
Tool materials in common use include the following:

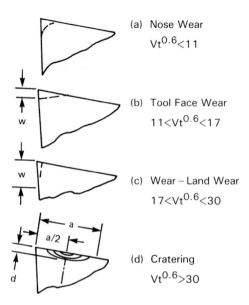

(a) Nose Wear

$$Vt^{0.6} < 11$$

(b) Tool Face Wear

$$11 < Vt^{0.6} < 17$$

(c) Wear – Land Wear

$$17 < Vt^{0.6} < 30$$

(d) Cratering

$$Vt^{0.6} > 30$$

FIGURE 12.   Types of tool wear depending on value of product ($Vt^{0.6}$).

1.   High speed steels (HSS)
2.   Sintered tungsten carbide (WC)
3.   Sintered titanium carbide (TiC)
4.   Sintered aluminum oxide (ceramic)
5.   Sintered cubic boron carbide (CBN)
6.   Sintered and single-crystal diamond tools (D)

As we proceed down the list, these materials are harder but more brittle. In general, if chipping of the cutting edge is a problem (which normally occurs as the tool finishes a cut rather than when it is beginning a cut), it is necessary to move upward in the list. This is done reluctantly, however, since the resulting tool has a lower resistance to attritious and abrasive wear.

There are several varieties of HSS and WC tools that have varying degrees of wear and shock resistance. As the wear resistance increases, however, shock resistance generally decreases. As vanadium and cobalt content increases in HSS tools, they become more wear resistant but more difficult to grind. The shock resistance of WC tools is increased by increasing the cobalt content while crater resistance is increased by substituting TiC, TaC, or CbC for some of the WC or by use of a thin (0.0002 in. or 5 μm) coating of TiC, TaC, or Al_2O_3 on a carbide substrate.

Certain tool materials are chemically incompatible with certain work materials. For example, poor results are obtained if ceramic (Al_2O_3) tools are used to machine titanium or aluminum alloys. This is because strong bonds tend to form between the work material (chip) and the oxide tool. Similarly, poor tool life results are obtained when a diamond tool is used to machine steel. The high temperature iron on the face of the tool tends to dissolve carbon from the diamond tool faces in this case. CBN gives much better results for ferrous work materials. Diamond tools are used primarily to machine nonferrous materials such as aluminum and copper base alloys.

Taylor Tool Life Equation

Tool life (T) measured in minutes is generally found to vary with cutting speed (V) in

Table 3
REPRESENTATIVE
VALUES OF TAYLOR
EXPONENT (n)

Tool material	n
High-speed steel	0.1
Tungsten carbide	0.2
Ceramic (Al$_2$O$_3$)	0.4

accordance with the following equation initially suggested by Taylor[7]

$$VT^n = C \tag{21}$$

where n and C are empirical constants depending upon tool and work materials and the undeformed chip thickness (t) which are held constant in Equation 21. Representative values for n are given in Table 3. Thus, for a high-speed steel tool, tool life $T \sim V^{10}$ and from Equation 20a $T \sim \theta^{20}$. This suggests that the major cause for the strong dependance of tool life (T) on cutting speed (V) is due to tool temperature (θ).

By use of Equation 20a, the Taylor equation may be generalized as follows for cases where both V and t vary

$$V^{1/n_1} t^{1/n_2} T = C \tag{22}$$

where $n_2 > n_1$. This equation may in turn be generalized to include a variable depth of cut (b) as follows:

$$V^{1/n_1} t^{1/n_2} b^{1/n_3} T = C \tag{23}$$

where $n_1 < n_2 < n_3$. Typical values for n_1, n_2, and n_3 for a high-speed steel single-point cutting tool are 0.10, 0.17, and 0.25, respectively.

Wear Land Tool Wear

Wear land tool wear (Figure 12c) is probably the most important type of wear for single-point tools that are well matched to the material being machined. When the extent of the wear land is plotted vs. cutting time at a constant rate of metal removal, curves such as those shown in Figure 13 are obtained. Wear rate is high at the beginning and at the end of the test.[8]

Crater Wear

The volume of wear that may be tolerated on the wear land is very much less than on the tool face. This is fortunate since the temperature on the tool face is very much higher and hence the volume rate of wear is much higher on the tool face. The maximum temperature on the tool face which occurs at about the midpoint of the contact length between chip and tool is usually sufficient to give thermal softening and a relatively large real area of contact for the tool face. The crater that forms has its maximum depth at the point of maximum temperature and, hence, the crater develops as shown in Figure 14, with its maximum point at approximately a constant distance from the cutting edge. The rate of crater formation depends on the stability of the constituents of the tool, the rate of diffusion of the products of decomposition, and their influence on the strength of the work material.

When a tungsten carbide tool is used to cut a low-carbon steel, the rate of crater formation

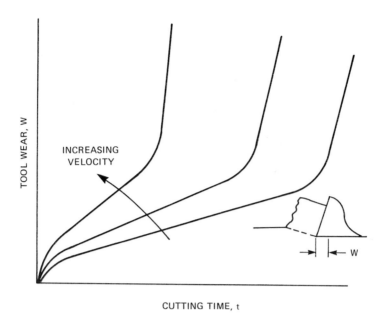

FIGURE 13.   Variation of wear land with cutting time for different values of cutting speed.

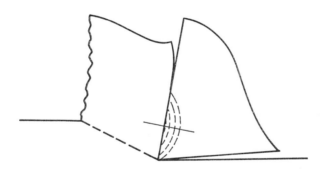

FIGURE 14.   Development of crater on rake face of cutting tool. Maximum depth remains at approximately constant distance from cutting edge.

is very high. This is because the iron on the surface of the chip has such a high affinity for carbon it causes the WC particles to decompose resulting in a high wear rate. This is not the case when machining gray cast iron since the presence of excess carbon in the form of graphite prevents WC from being decomposed. Thus, it was not possible to machine low-alloy steels with straight tungsten carbide tools for a whole decade after their introduction in the late 1920s. McKenna[9] finally found that a steel cutting grade of carbide could be produced by substituting TiC and TaC for some of the WC present in the original tools that had only been useful for machining gray cast iron and nonferrous metals. The so-called steel cutting grades of cemented tungsten carbide revolutionized machining and made it necessary to design and manufacture higher speed, more rigid machine tools of greater power.

Special Deoxidized Steels

An important development pioneered in Germany[10] and later extended in Japan[11] and the U.S.[12] is the use of special deoxidation methods to produce steel that has a lesser tendency to cause carbide tools to crater in high-speed machining. This technique involves the use of calcium and ferrosilicon as deoxidizing materials. A resulting low-melting ternary inclusion then spreads over the tool face in high-speed machining and acts as a diffusion barrier.

Coated Tools

In the late 1960s crater formation was found to be retarded by the vapor phase deposition of a very thin (0.0002 in., 5 μm) coating of TiC on a steel cutting grade carbide tool. It was later found that such coatings also tended to reduce the wear-land wear. The TiC coating provides a material in contact with the iron surface of the chip that is less likely to give up its carbon than WC. Use of such a coating avoids the tool-weakening effect associated with relatively large additions of TiC or TaC to the tool in bulk.

Tools have also been coated with Al_2O_3 and this material is thought to function as a diffusion barrier which is one of the mechanisms believed responsible for the success of the specially deoxidized steels.

In addition to TiC, other coating materials is use include TiN, HfC and HfN. All of these appear to provide a more stable material relative to decomposition in the presence of hot iron and at the same time act as diffusion barriers. A TiN coating gives lower tool-face friction than TiC when cutting a low-alloy steel. Although TiC is more stable than WC there will be some decomposition of TiC. The carbon that is released will be absorbed by the low-carbon steel, thus strengthening it and causing an increase in tool-face friction. TiN does not appear to be as good a diffusion barrier as TiC, but when it decomposes the nitrogen released does not have as great a strengthening action on the steel as does carbon. This results in lower tool-face friction for a TiN coating than for a TiC coating.[13]

Tools have also been coated with Al_2O_3 which also acts as a diffusion barrier. The subject of tool coating is a rapidly developing and further important developments are expected.

Cutting Fluids

Cutting fluids have a dual role — cooling and lubrication. In high-speed cutting operations where the tool remains buried in the cut most of the time, such as in turning, the function of the fluid is primarily one of cooling. In low-speed operations involving intermittent cutting, such as in broaching or tapping, lubrication is important. Both water-base and oil base lubricants are used, the former are generally better coolants while the latter are better lubricants. There are many secondary considerations associated with cutting fluids such as chip disposal, corrosion prevention, health and safety considerations, etc. A discussion of these aspects of cutting fluid technology is to be found in Reference 14 and in the next chapter of this handbook.

It is unlikely that anything but a vapor can penetrate the interface between chip and tool during a continuous cutting operation. This is due to the near perfect contact between chip and tool and the normally high-speed motion of the chip counter to fluid penetration. From the latter point of view it is unlikely that any penetration will be from the side of the tool (i.e., parallel to the cutting edge).

CHIP FORMATION AND CHIP CONTROL

Figure 15 shows several types of chips formed under different operating conditions. Discontinuous chip formation is shown at (a) and cutting with a large BUE is shown at (b). Cutting with the continuous ribbon-like chip previously discussed is illustrated at (c). When the work material is very soft and extensive, strain-hardening occurs during cutting or when the friction between chip and tool is high, the narrow shear zone shown in (c) assumes a fan shape as shown at (d).

Cooling the back of the chip will frequently cause the chip to curl away from the tool (e) resulting in a decrease in the contact length between chip and tool. There is an optimum

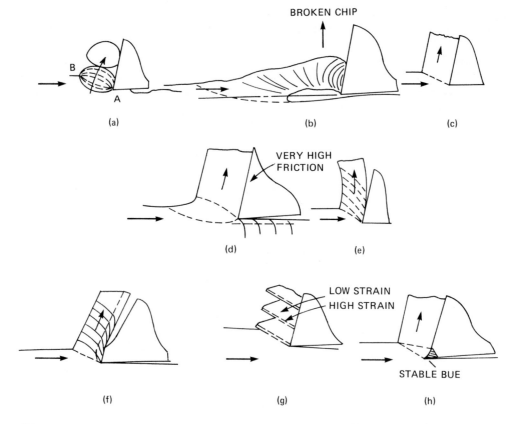

FIGURE 15.   Different types of chip formation: (a) discontinuous chip; (b) very large BUE; (c) continuous ribbon-like chip; (d) fan-shaped shear zone obtained when cutting very soft material with high tool-face friction and large undeformed chip thickness; (e) tightly curling chip; (f) controlled contact tool; (g) inhomogeneous chip; and (h) honed cutting edge with small protective BUE.

tool-chip contact length. In general, the length of shear plane decreases with chip curl but the resultant cutting force is also moved closer to the tool tip. While the former is beneficial to tool life the latter may decrease tool life. It is therefore possible to overdo the reduction in tool contact length and this is sometimes the case when a water-base cutting fluid strongly cools the back of a *thin* steel chip. Figure 15f illustrates a tool design sometimes used to control tool-chip contact length.

Certain materials (notably titanium alloys) exhibit very inhomogeneous strain in chip formation (Figure 15). Strain begins at a particularly strong defect and continues for a relatively long time due to thermal softening on the shear plane being greater than the degree of strain hardening. This type of chip formation, which can lead to chatter and poor finish, is most apt to occur when cutting materials having a low coefficient of thermal conductivity and a low-volume specific heat. Titanium alloys, stainless steels, and certain high-temperature alloys have such thermal properties and hence tend to exhibit chips as shown at (g). The stable (BUE) induced by providing a small negative rake angle at the tool tip is illustrated by Figure 15h.

The most important aspect of chip control involves the use of chip curlers and chip breakers (Figure 16) to prevent the long continuous chips that are dangerous to both machine and operator and frequently difficult to remove from the machine. This subject is discussed in detail in References 15 to 17.

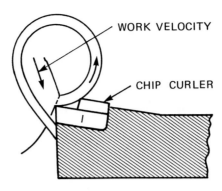

FIGURE 16. Carbide chip curler used to periodically break chips to simplify chip disposal.
I = tungsten carbide insert.

MACHINING ECONOMICS

In a single-tool turning operation, the machining cost per part will depend upon the rate of metal removal ($\dot{Z}$, in.^{3}/min) or (mm^3/min).

$$\dot{Z} = 12 \, V \, b \, t \tag{24}$$

where V = cutting speed, fpm (m/mm); b = depth of cut, in. (m); t = feed rate, in./r (mm/v) = undeformed chip thickness. The depth of cut will normally be fixed by the amount of material to be removed, but the operator may select the values of speed (V) and feed (t). The cost to make a cut of ℓ in.(m) axial length on a bar of D in.(mm) diameter will be

$$\overset{\text{I}}{} \qquad \overset{\text{II}}{} \qquad \overset{\text{III}}{}$$

$$\phi = x \, T_c + (x \, T_d + y) \, \frac{T_c}{T} + x \, T_w \tag{25}$$

where x = value of machine, operator and overhead (¢/min), T_c = cutting time (min), T_d = down time to change the tool (min), y = mean value of cutting edge (¢), T = tool life (min), and T_w = work changing time (min). Item I = machine and labor cost per part, item II = tool and tool changing cost per part, and item III = work changing cost per part.

Certain values of V and t will give a minimum cost per part and since an increase in t will cause a smaller decrease in tool life (T) than will an increase in V, it is advantageous to adjust t relative to machining cost before adjusting V. However, the cost optimum value of t usually lies beyond the practical range in turning. That is, the cost per part will decrease with increase in t, but before the cost optimum value can be reached, (t*), a constraint will be encountered such as finish, power, force on tool (breakage), chatter, surface integrity, etc. On the other hand, the cost optimum value of V (V*) usually lies within the practical range.

For a constant value of t, the Taylor Equation 21 relates tool life (T) in min and the cutting speed (V) in fpm $VT^n = C$, where n and C are constants. The value of cutting time (T_c) will be

$$T_c = \left(\frac{\ell}{t}\right) \frac{\pi D}{(12V)} \, , \text{min} = \frac{C_1}{V} \tag{26}$$

where ℓ = axial length of cut, in. (m), D = diameter of work, in. (mm), and C_1 = a constant for a constant value of t.

Substituting Equations 21 and 26 into Equation 25:

$$\cent = \frac{C_1}{V} \quad x + \frac{xT_d + V}{(C/V)^{1/n}} \quad + xT_w \tag{27}$$

The quantity (xT_w) may be considered independent of V. The cost per part $(\cent)$ will be a minimum when $\partial\cent/\partial V = 0$. This occurs when the cost optimum value of tool life is

$$T^* = \left(\frac{xT_d + y}{x}\right)\left(\frac{1}{n} - 1\right) = (R)\left(\frac{1}{n} - 1\right) \tag{28}$$

For a carbide tool $n \simeq 0.2$ and, therefore, $T^* = 4R$ if $x = 10$ ¢/min, $T_d = 10$ min, and $y = 100$¢. Then, $R = 20$ min^{-1} and $T^* = 80$ min. For a HSS tool, $n \simeq 0.1$ while for a ceramic tool $n \simeq 0.4$ and the corresponding values of T^* in the above example would be 180 min (HSS) and 30 min (ceramic).

When more than one tool is used, the value of R to be used is approximately the sum of the values of R for the individual tools. This results in the cost optimum tool life (T^*) being greater the greater the number of tools used at one time.

Manual Adaptive Control[18] is a technique for adaptively controlling a machine tool with practically no investment in capital equipment beyond that required for ordinary machining. In this case, the operator acts as the group of sensors required in adaptive control and also serves as the interface between the machine tool and the low cost ($\sim$ \$250) computer. After each tool change, he enters the time elapsed from the last tool change and the number of parts produced. The programed computer then tells the operator to increase or decrease the production rate in order to approach minimum cost per part.

GRINDING

Grinding is one of the most versatile methods of removing material from machine parts to provide precise geometry. However, the process is very complex and difficult to study because of the small size of the individual chips produced by hard abrasive particles having a wide range of shape, spacing, and relative elevation.

Grinding operations may conveniently be classified in terms of whether the wheel is dressed or whether the wear of the wheel is sufficiently high that it is self-dressing. In form-and-finishing grinding (FFG), individual chips are relatively small and wear flats develop on the active grains in the surface. Periodically the dull grains are removed or "sharpened" by dressing the wheel with a diamond tool. Typical FFG operations are horizontal spindle surface grinding, internal grinding, cylindrical grinding, and centerless grinding. In stock removal grinding (SRG) wheel wear is relatively high and the wheel is self-dressing. Examples of SRG are abrasive cut off processes, conditioning of slabs and billets in a steel mill and vertical spindle surface grinding.

The rate of wear of a grinding wheel is usually important to the economics and performance of the process. In the case of SRG wear is usually expressed in terms of a grinding ratio (G).

$$G = \frac{\text{volume of work removed}}{\text{volume of wheel consumed}}$$

In such cases, the rate of change of wheel diameter is relatively large and G may easily be measured.

In the case of FFG there is usually a negligible change in wheel volume during grinding and essentially all of the wheel consumption is associated with dressing. In such cases, the

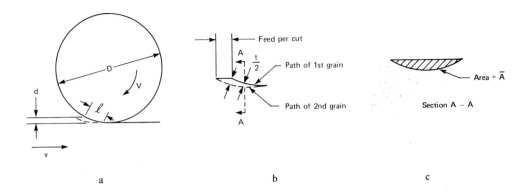

FIGURE 17.   Undeformed chip shape in grinding.

grinding ratio is not a convenient way of measuring wheel life. Instead, some means of accurately measuring the change in wheel radius, or alternatively some means of measuring the development of wear flats on the active abrasive grains must be devised. Such measurements are very difficult to make on a complete grinding wheel.

Grinding Mechanics

The mechanics of grinding is discussed in detail in Reference 19 and, therefore, only a few essentials will be reviewed here. Figure 17a shows a plunge surface grinding operation. The term plunge infers that there is no cross feed in the direction of the wheel axis. The wheel width is greater than the work width (b) and the wheel and work speeds are V and v, respectively. The wheel depth of cut is d. The mean undeformed shape of the chip is shown in Figure 17b, while Figure 17c shows the cross section of an undeformed chip at its midpoint. The maximum undeformed chip thickness (t) may be found as follows:

$$\text{Volume of single chip} = \frac{\text{volume cut/time}}{\text{cuts/time}}$$

$$= \frac{vbd}{CbV} = \frac{vd}{CV} \qquad (29)$$

where C is the effective number of cutting points per square inch. The volume of a single chip can also be found by assuming the shape of the chip to be a long slender triangle, in which case

$$\text{Volume of single chip} = \tfrac{1}{2}\,\bar{A}\,\ell \qquad (30)$$

where $\bar{A}$ is the mean cross-section of the chip ($= b't/2$) and b' is the effective chip width. It is convenient to define b' in terms of the ratio (r) of chip width (b') to chip thickness (t), thus

$$r = \frac{b'}{t} \qquad (31)$$

Substituting Equations 31 into 30 and equating the chip volumes from Equations 29 and 30:

$$t = \left[\frac{4v\,d}{VCr\ell} \right]^{1/2} \qquad (32)$$

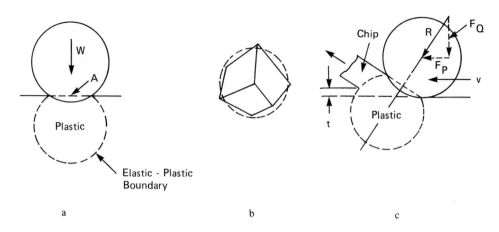

FIGURE 18. Chip-forming mechanism in fine grinding and its relation to hardness indentation.

The quantity t has been shown to be the most important quantity relative to grinding performance[20] and can be readily estimated if C, r′, and ℓ are known. The difficulty in estimating C and some of the techniques used can be found in References 21 to 23. The length of scratch (ℓ, Figure 17a) may be estimated as follows, if local wheel and work deflection are negligible

$$\ell = \sqrt{Dd} \tag{33}$$

The value of t in FFG is generally about 100 μin. For such small cuts a very small part of the abrasive grain is active and the removal mechanism differs substantially from that of the concentrated shear mechanism that pertains in cutting (as shown in Figure 2).

In fine grinding, the removal mechanism is more akin to that of an indentation hardness test.[24] Figure 18a shows the plastic zone that develops beneath a spherical indenter.[25] Figure 18b shows a typical blocky abrasive grain. Only a small percentage of such a grain produces a chip in fine grinding and the effective rake angle will be even more negative than that corresponding to the inclined faces at the lowest point on the grain of Figure 18b. Actually, the very small radius at the active point on the grain will be responsible for the effective rake angle as shown in Figure 18c. Here the action is likened to that of a spherical indenter subjected to an inclined load. The effective radius of this indenter will of course be much smaller than that of the dotted circle shown in Figure 18b. Material to the front of the indenter of Figure 18c will be plastic but unsupported and will flow upward to generate the chip.

The consequences of the chip-forming mechanism shown in Figure 18c are the following:

1. Deformation of a much larger volume of material than escapes as a chip.
2. Considerably greater subsurface flow than with the cutting mechanism shown in Figure 2, leading to a greater tendency for subsurface cracking and for residual stresses in grinding than in cutting.
3. A greater proportion of the total energy in fine grinding will end up in the workpiece surface than in metal cutting. In high speed cutting, 90% or more of the energy consumed ends up in the chip, while in fine grinding most of the energy ends up in the work (~80%).

The fact that the specific energy in fine grinding is 50 times or more greater than that for cutting the same material is in agreement with the extrusion-like mechanism shown in Figure

18 as are other experimental observations. Also adding to the specific grinding energy is the fact that metal is pushed from side to side several times before leaving as a chip.

As the undeformed chip thickness (t) increases, a greater percentage of the abrasive particle is active and the effective rake angle increases. For SRG where chips are quite large, the removal mechanism approaches that of Figure 2 and there is much less danger of overheating the work since more energy consumed ends up in the chip rather than in the workpiece. In the abrasive cut-off operation, chips are relatively large and the specific grinding energy will be only about 1×10^6 in.lb/in.^{3} (6.9×10^6 kPa) instead of about 30×10^6 in.lb/ in.^{3} (207×10^6 kPa) for horizontal surface grinding and as much as 100×10^6 in.lb/in.^{3} (690×10^6 kPa) for internal grinding.

SURFACE INTEGRITY

In many cases the quality of the finished surface is an item of major concern. This involves items such as

1. Surface finish
2. Residual surface stresses
3. Thermally induced damaged: oxidation and burning, overtempering, surface cracks
4. Mechanically induced surface cracks and highly strained areas (BUE)

From the standpoint of brittle fracture and fatigue, any residual surface stresses should be compressive. When surface integrity is a problem, the use of sharp tools and cutting conditions to avoid BUE formation or overheating the surface are principal items for consideration. Since the cutting temperature varies as $(V)^{1/2}$, high temperature problems are more apt to occur in grinding since V will normally be 6000 fpm (30 m/sec) or greater. When excess grinding temperatures are a problem, a shift to a lower wheel speed ($\sim$ 2000 fpm or 10 m/sec), use of a sharp wheel (frequent dressing), lower removal rates and an active oil-base lubricant will usually be helpful.

In addition to the references cited above, References 26 to 35 provide further background concerning material removal operations.

REFERENCES

1. **Shaw, M. C.,** *Metal Cutting Principles,* MIT Press, Cambridge, 1954.
2. **Merchant, M. E.,** Mechanics of the cutting process, *J. Appl. Phys.,* 16, 267, 1945.
3. **Piispanen, V.,** Lastumuodostumisen teoriaa, *Tek. Aikak.,* 27, 315, 1937.
4. **Trent, E. M.,** Advances in machine tool design and research, *Proc. 8th Machine Tool Design Res. Conf.,* Tobias S. A. and Koenigsberger, F. M., Eds., Pergamon Press, Oxford, 1967, 629.
5. **Opitz, H.,** Der Heutige Stand der Zerspannungsforschung, *Werkstatttechnik Maschinenbau,* 46, 210, 1956.
6. **Shaw, M. C.,** Wear mechanisms in metal processing, *Proc. Int. Tribology Conf.,* MIT Press, Cambridge, 1979.
7. **Taylor, F. W.,** On the art of metal cutting, *Trans. ASME,* 28, 31, 1907.
8. **Vilenski, D. and Shaw, M. C.,** The importance of workpiece softening on machinability, *Ann. CIRP,* 18, 623, 1969.
9. **McKenna, P. W.,** U.S. Patent 2,113,353, 1938.
10. **Opitz, H. and Koenig, W.,** *Ind. Anzieger,* 87, 46 (Part I), 26, 1965; *Ind. Anzieger,* 87, 845 (Part II); 43, 1965; *Ind. Anzeiger,* 87, 1033 (Part III), 51, 1965.
11. **Sata, T., Chairman,** *Working Group on Machinability in Japan,* Bull. 3(1), Japan Society of Precision Engineers, Tokyo, 1969.

12. **Tipnis, V. A. and Joseph, R. A.,** *J. Eng. Ind., Trans. ASME,* 93, 571, 1971.
13. **Rao, S. B., Kumar, K. V., and Shaw, M. C.,** Friction characteristics of coated tungsten carbide cutting tools, *Wear,* 49, 353, 1978.
14. **Shaw, M. C.,** Grinding fluids, *Manuf. Eng. Trans.,* 1, 1972.
15. **Henriksen, E. K.,** Balanced design will fit the chip breaker to the job, *Am. Machinist,* April 26, 1954.
16. **Nakayama, K.,** A study on chip breaker, *Bull. Jpn. Soc. Mech. Eng.,* 5(17), 142, 1962.
17. **Subramanian, K. L. and Bhattacharya, A.,** Mechanics of chip breakers, *Int. J. Prod. Res.,* 4(1), 37, 1965.
18. **Shaw, M. C. and Komanduri, R.,** Manual adaptive control, *Am. Mach.,* 121, 205, 1977.
19. **Shaw, M. C.,** Fundamentals of grinding, keynote paper, 1st Int. Grinding Conf., Carnegie Mellon Univ., April 1972, in *New Developments in Grinding,* Carnegie Press, Pittsburgh, 1972.
20. **Snoeys, R.,** The significance of chip thickness in grinding, *Ann. CIRP,* 23(2), 227, 1974.
21. **Brecker, J. N. and Shaw, M. C.,** Measurement of the effective number of cutting points in the surface of a grinding wheel, Proc. Int. Conf. Prod. Eng. Tokyo, 1974, 740.
22. **Peklenik, J.,** Ermittlung von geometrischen und physikalischen Kenngrossen fur die Grundlagen des Schleifens, Dissertation T.H. Aachen, 1957.
23. **Nakayama, K. and Shaw, M. C.,** Study of the finish produced in surface grinding. II. Analytical, *Proc. Inst. Mech. Eng. (London),* 182(3K), 182, 1967-1968.
24. **Shaw, M. C.,** *A New Theory of Grinding,* Institution of Engineers, Australia, 1972, 73.
25. **Shaw, M. C. and DeSalvo, G. J.,** The role of elasticity in hardness testing, *Metals Eng. Quart.,* 12, 1, 1972.
26. **Amarego, E. J. A., and Brown, R. H.,** *The Machining of Metals,* Prentice-Hall, Englewood Cliffs, N. J., 1966.
27. **Boothroyd, G.,** *Fundamentals of Metal Machining,* McGraw-Hill, New York, 1975.
28. **Ernst, H. et al.,** *Machine Theory and Practice,* American Society of Metals, Metals Park, Ohio, 1950.
29. **Kronenberg, M.,** *Machining Science and Applications,* Pergamon Press, Oxford, 1966.
30. **Trent, E. M.,** *Metal Cutting,* Butterworths, London, 1977.
31. **Shaw, M. C., Ed.,** *International Research in Production Engineering,* American Society of Mechanical Engineers, New York, 1963.
32. **Zorev, N. N.,** *Metal Cutting Mechanics,* Pergamon Press, Oxford, 1966.
33. **CIRP,** Yearly Proc. Int. Inst., Prod. Eng. Res., Paris, *Ann. CIRP,* 1952 to Present.
34. Machine Tool Design and Research - Yearly Proceedings of Conference held in Great Britain UMIST, Manchester, 1962 to Present.
35. Proceedings of Yearly North American Metal Working Research Conferences (NAMRC) Society of Manufacturing Engineers, Dearborn, Mich., 1972 to Present.

CUTTING FLUIDS

Ralph Kelly and Gregory Foltz

INTRODUCTION

The primary function of any cutting fluid is to control heat.[1] A cutting tool generates temperatures of 375 to 750°C and the resulting chip as it slides up the tool face creates tremendous pressures (up to 1,379,000 kPa). About 75% of the heat is generated by deformation of the metal, the other 25% by friction between the chip and the tool. By controlling the temperature generated in the cut zone, tool wear can be controlled and tool life increased.[2]

As the cutting tool cuts (or the grinding wheel grinds), metal deforms by shear or plastic flow along a shear plane extending from the top of the tool to the surface of the metal (Figure 1). Below the shear plane is undisturbed metal; above it, the deformed metal forms a chip. Reduced friction at the chip-tool interface increases the shear angle, produces a thin chip, and deforms less metal.

Where a tool face is examined under a microscope, rough peaks and valleys can be seen (Figure 2). These tiny projections collide with the chip as it slides up the tool face and weld to the chip under the conditions of very high heat and pressure. Continuous shearing of these welds results in tool wear, the tip of the tool becomes cratered, and heat concentrates at this point. Small pieces of sheared-off metal form a built-up edge on the face of the tool, a major cause of poor surface finish.

When a cutting fluid is introduced between the tool and chip, friction is reduced, the shear angle increases, the chip becomes thinner, the power requirement is reduced, and less heat is generated. Also, the built-up edge disappears, the finish smooths out, and size control improves. The nascent metal exposed under the high temperature and pressure conditions reacts with chemicals in the cutting fluid to form a low shear strength solid between the chip and the tool. The chip slides freely up the tool face, tools last much longer, speeds and feeds can be stepped up, and more work can be done with each tool. The cooling and lubricating mechanisms are dependent on the job: slow-speed, slow-feed operations need more lubrication while high-speed, high-feed operations need more cooling.

TYPES OF CUTTING FLUIDS[3]

Cutting Oils

A cutting oil contains mineral oil, fatty oil, or a combination of these. Mineral oils are petroleum derivatives; fatty oils are derived from animal or vegetable sources.

Extreme pressure (EP) sulfur, chlorine, or phosphorus additives are employed to improve antiweld properties for heavy-duty applications. Sulfur forms a better lubricant, but chlorine is more reactive than sulfur and breaks down to form the EP lubricant at lower temperatures. Phosphorus is not as effective as either sulfur or chlorine and its use is less common. Cutting oils are often classified as active or inactive; an inactive oil will not darken a copper strip immersed in it for 3 hr at 100°C while an active oil will. Inactive oils are straight mineral oils, containing sulfurized fatty oils. Active oils are sulfurized or sulfochlorinated mineral or fatty oils.

Straight mineral oil — Used for light-duty machining of ferrous or nonferrous metals, its major function is as the base fluid for the blends and additive oils listed below.

Straight fatty oils — Very limited use because of their expense and frequent odor problems. They find their greatest application in blends with mineral oils. Palm oil, lard oil, and coconut oil are the most popular.

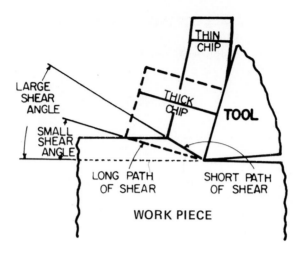

FIGURE 1.   The amount of deformation of the metal is controlled by the size of the shear angle.

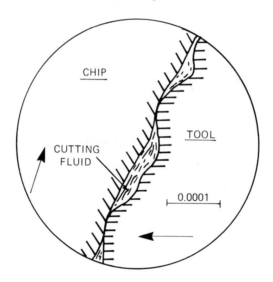

FIGURE 2.   Minute capillaries between chip and tool draw in the cutting fluid; then chemical reaction with the chip metal produces a solid film that prevents metal-to-metal contact.

Mineral fatty oil blend — Combinations of one or more fatty oils, blended into mineral oil. The fatty oils act as wetting agents and improve the lubrication of the mineral oil. These products are nonstaining to ferrous and nonferrous metals and are used where high surface finish and precision are required, especially automatic screw machines.

Sulfurized fatty mineral oil blend — These oils, containing both fatty oil and sulfur additives, have excellent lubricity. They stain less than sulfurized mineral oil since the sulfur is added as a sulfurized fat in which strong chemical bonding keeps the sulfur from being readily released until the temperature reaches 265°C. They can be used on both ferrous and nonferrous metals. Chlorine may also be added to increase the antiweld properties at low temperatures and pressures, thus producing a heavy-duty fluid for a wide range of operations.

Sulfurized mineral oil — Sulfur is dissolved in the mineral oil and forms an iron sulfide

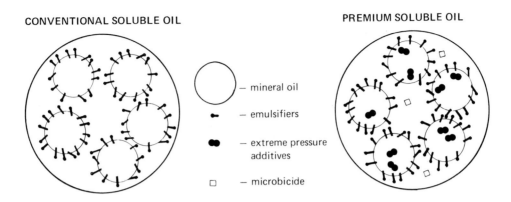

CONVENTIONAL SOLUBLE OIL

PREMIUM SOLUBLE OIL

— mineral oil

— emulsifiers

— extreme pressure additives

— microbicide

FIGURE 3.   Emulsified oils (soluble oils).

film in the machining process. This reduces friction and built-up edge, and provides antiweld properties. These oils are useful for machining tough, ductile metals. Reactivity of the sulfur makes them unsuitable for copper or copper alloys.

Sulfo-chlorinated mineral oil — Combination of sulfur and chlorine additives produces products with exceptional antiweld properties over a wide temperature range. They are used for machining (especially threading) tough, low-carbon steels. Fatty oils added to this type of product produce a cutting oil for a wide range of heavy-duty and slow-speed operations.

Cutting oils generally provide the excellent lubrication needed in low-clearance, low-speed operations; especially where a high quality surface finish is required. They have good rust control. Sump life is long since rancidity-causing bacteria do not grow in pure oil unless it is contaminated with water. The straight-oil cutting fluids do allow buildup of excessive heat, since oil dissipates heat only half as fast as water and because it is much more viscous than water-based fluids. These oils are also somewhat of a safety hazard in that they smoke and burn. In addition, their high misting properties cause the parts and surrounding area to become slippery and dirty.

Emulsified Oils (Soluble Oils)

These mixtures of mineral oil and emulsifiers (Figure 3) are supplied as concentrates which are added to water at the ratio of 1 part concentrate to 5 to 20 parts water. The oil is made soluble by emulsifying agents, primarily sulfonates. The emulsified particles range in size from 200 to 80 μm, large enough to reflect light and create a milky, opaque appearance when mixed with water. Premium grades may contain bactericides and corrosion inhibitors. Addition of fatty oils, fatty acids, or esters produces a superfatted emulsion for heavy-duty use on both ferrous and nonferrous metals. Sulfur, chlorine, or phosphorous, in addition to the fat, form an extreme pressure emulsion for very heavy-duty operations, including replacement of straight cutting oil in some applications.

Used as general-purpose products, soluble or emulsified oils offer lubrication because they contain oil, and the water aids in dissipating heat. Speeds and feeds can be stepped up and better size control obtained.

Soluble oils have several disadvantages. When mixed with hard water, some soluble oils form a precipitate which is deposited on parts and machines, and which can interfere with filtration. In extreme cases, the emulsion may be broken. At strong concentrations, mist from a soluble oil can leave machines and work areas in a messy, slippery condition. Depending on the amount of rust preventives added, rust can be a problem. The water can support bacterial growth, leading to rancid odors and short sump life if proper bactericides are not present.

CHEMICAL EMULSION

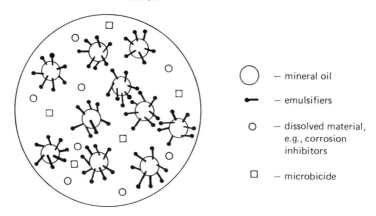

○ — mineral oil

● — emulsifiers

○ — dissolved material,
    e.g., corrosion
    inhibitors

□ — microbicide

FIGURE 4.   Chemical emulsions (semisynthetics).

Chemical Emulsions (Semisynthetics)

These (Figure 4) have a much lower mineral oil content than soluble oils (anywhere from 5 to 30% in the concentrate). They have a high emulsifier content and smaller oil globule formation, which results in a mix that is translucent or transparent. Chemical emulsions provide wetting and lubrication properties, corrosion control, and microbial control. These fluids may also contain fatty acids, sulfur, chlorine, and phosphorous to enhance EP lubrication.

They generally have enough lubricity for moderate- to heavy-duty applications. With better wetting properties than soluble oils, chemical emulsions make possible higher speeds and feeds. Chemical emulsions usually have better settling and cleaning properties than soluble oils, which contribute to a long and trouble-free sump life. Many chemical emulsions contain wetting and cleaning agents to keep the machines clean. Because they contain very little oil, chemical emulsions do not smoke. This reduction in oil also contributes to less mist.

If hard water is used to make the mixture, these products, like soluble oils, may form a hard-water scum. The cleaning action may cause some chemical emulsions to foam. While their initial purchase price is generally higher, these products are usually less costly to use than soluble oils.

Chemical Solutions (Synthetic Fluids)

These products (Figure 5) are true solutions and are completely clear. The simple type consists of organic and inorganic salts dissolved in water. The more complex moderate- or heavy-duty synthetics contain wetting agents which allow the fluid to spread more completely over the metal surface, increasing the lubrication properties of the fluid. These products usually have excellent microbial control.

The simple types are mainly used as grinding fluids since they offer rust protection and good heat removal. The more complex types are good general-purpose products, offering both good lubrication and fast heat removal. Use dilutions range from 5 to 0.5%, depending on the type of solution and the operation. They keep the wheels open and free grinding, they enable operators to see the work, and they produce considerably less mist than the other types of cutting fluids. Of water-based fluids, solutions are the least troubled by rancidity and their superior settling and cleaning properties help extend the fluid life. Their excellent cooling capability makes possible high speeds and feeds, high production rates, and good size control. Chemical solutions are stable, even in hard water. Although they contain no oil, the chemical lubrication afforded through wetting agents does supply sufficient

CHEMICAL SOLUTION
SIMPLE

CHEMICAL SOLUTION
COMPLEX

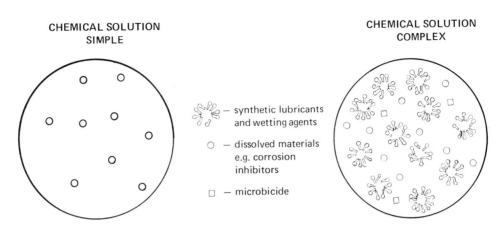

— synthetic lubricants
and wetting agents

○ — dissolved materials
e.g. corrosion
inhibitors

□ — microbicide

FIGURE 5.   Chemical solutions (synthetic fluids).

lubricity for many moderate- to heavy-duty cutting and grinding operations where machine parts have separate slideway lubrication systems or where the machine has no sliding members. Some chemical solutions may foam under conditions of moderate to high agitation.

Gases and Solids

Compressed air can be directed into the cut zone, both for heat removal and for removing chips. Inert gases and those boiling below room temperature, such as carbon dioxide, may also be used. High cost generally outweighs any added benefits. Low-boiling point liquids like Freon® (1,1,2 trichloro- and 1,2,2 trifluoroethane) are occasionally used where extremely clean parts are needed.

Solids are used on a very limited basis since they must be dispersed in some carrier. Some sulfur compounds are used and form a film on the workpiece in the same manner as extreme pressure additives.

CUTTING FLUID SELECTION

Both needs of the job and performance properties must be considered in selecting a cutting fluid.[4]

Type of Shop

A general-purpose cutting fluid is the logical choice for job shops with their wide range of parts, materials, and operations. High-production shops, with many machines doing the same operation, can use a cutting fluid designed to meet specific requirements.

Type of Machine

Many machines built before 1949 required cutting oil to serve both as the cutting fluid and as the slideway lubricant. Modern machines, however, are designed for use of water-based fluids. They have good lubrication systems, effective seals around the spindles, and wipers on the slideways. With the exception of some six-spindle automatics and some gear cutting equipment, most machines built since 1949 use either type of fluid.

Type of Operation[3,5]

Threading, reaming, tapping, and broaching are perhaps the most difficult machining operations, because of chip crowding, and generally require a cutting fluid that reduces friction. Turning, drilling, boring, and similar less difficult operations can be performed with products having low to moderate friction-reducing properties. Some grinding can be performed with cutting fluids that provide little friction reduction. Others, such as form grinding, require a heavy-duty fluid. If performed regularly, the most critical operation in a shop may dictate the cutting fluid selection.

Severity of the Operation[3,5]

Although stock removal rates, feeds, speeds, and finish requirements must be considered, operations can be divided into four general categories: (1) metal removal in which adhesion between the chip and tool interface is minimal is considered *light duty*. Examples would be turning 1112 steel or surface grinding cast iron; (2) *moderate-duty* operations are those in which adhesion between the chip and tool is noticeable and contributes to poor finishes and poor size control. Examples would be key seat milling and internal race grinding; (3) a *heavy-duty* operation involves severe adhesion between the chip and tool. Examples would be sawing of large parts, deep slotting, or centerless grinding heavy parts; and (4) *extremely heavy-duty* operations involve extremely severe adhesion between the chip and tool and small tool clearances, thus creating severe rubbing conditions between the tool and workpiece. Examples are thread chasing and tapping 4140 steel, deep hole drilling, or form and thread grinding. As severity increases, it may also be necessary to increase the lubricity of the fluid.

Materials[5]

A product providing excellent ferrous corrosion protection may stain aluminum alloys. Glass, molybdenum, and nodular iron require a fluid with excellent grit-settling characteristics. Magnesium alloys are generally machined or ground with straight oil products because of the danger of hydrogen gas evolution in water-based products. Good control of chip removal has allowed some use of soluble oils for magnesium machining.

Machinability is one means of differentiating between metals. A machinable metal allows tooling to last longer, can be machined easily at high speeds, or exhibits in some other terms its ability to be machined. Variations in microstructure, amounts of alloy material, heat treatment, and many other factors can affect machinability. Taking SAE 1112 steel as 100% machinable, other metals are divided in Table 1 into six machinability classes.[3] These machinability classes have been used in the fluid selection chart of Table 2. Because various components or additives in a cutting fluid may stain some alloys of a given metal, corrosion tests should be run to insure compatibility.

Type of System[6]

Premix tanks, proportioners, periodic machine cleaning, and frequent concentration checks are very helpful in maintaining the fluid in a good working condition. Care is especially important in machines with no filtration other than simple settling.

In nonrecirculating systems, the fluid is applied and then discharged in a one-time use. This is practical with many portable tools and in mist applications. Immediate performance of the product is the important aspect. Such things as rancidity control and dirt settling properties are not required.

In central filtration,[7] settling tank and weir systems require a cutting fluid with good settling properties. A cutting fluid with controllable foaming properties is used in a flotation filter. Magnetic separators are compatible with all product types. Centrifuges are generally used in conjunction with positive filters to remove extraneous oil and small fines, but they may cause some coarse or weak emulsion products to break. Coalescers and skimmers can also be used to remove tramp oil. Cyclones separate most particles from the fluid but should be used with low-foaming products.

With positive filters, a product with low-foaming tendency is needed which will not form

Table 1
MACHINABILITY GROUPS

Ferrous (>70%)	Ferrous (50—70%)	Ferrous (≤40—50%)	Ferrous (<40%)	Nonferrous (>100%)	Nonferrous (≤100%)
A-4023	A-2317	A-1320	317	Aluminum	Titanium
A-4024	A-31__	A-1330	418		
A-5120	A-4027	A-1340	440	108	Ti
B-1111	A-4028	A-2330	A-2515	A132	Ti-5Al-2.5Sn
B-1112	A-4032	A-2340	E-3310	195	Ti-8Mn
B-1113	A-4037	A-2515	E-50100	214	Ti-6Al-4V
C-1016	A-4042	A-3240	E-51100	355	Ti-8Al-1Mo-1V
C-1018	A-4047	A-4340	E-52100	A356	Ti-6Al-6V-2Sn
C-1019	A-4130	A-4815	E-9315	A357	Ti-13V-11Cr-3Al
C-1021	A-4145	A-4820	Tool steel	750	
C-1022	A-43__	A-6120		2011	Nickel
C-1026	A-46__	A-6150		2014	Copper
C-1030	A-51__	A-8645		2024	Brass
C-1109	A-86__	A-8650		3003	
C-1116	A-87__	A-8750		3004	
C-1117	A-94__	A-9445		5052	
C-1118	A-98__	C-1008		5056	
C-1119	C-1020	C-1010		6061	
C-1120	C-1035	C-1015		7075	
C-1137	C-1040	C-1050			
C-1141	C-1045	C-1070		Leaded brass	
C-1144	C-1060	E-2512		Magnesium alloys	
Cast steel	C-1070	E-3310		wrought + cast	
Malleable iron	E-41__	Wrought iron		Zinc	
Stainless iron	Cast iron	Stainless 18-8		Silicon bronze	

Note: These ratings were based upon a comparison with B1112 steel and given in AISI listings.

hard-water soaps to plug the filter media. Products that contain some emulsified oil agglomerate the fines and create a better, more open cake on the filter media. Solution-type fluids often allow dense chip packing on the filter media. If diatomaceous earth or other material is used as a precoat for the filter, a solution-type product is usually needed since precoat materials may filter out certain ingredients in emulsion-type products. The capacity of settling systems should allow a minimum of 10 min retention time (system volume divided by flow rate) to permit the fines to settle out. Insufficient system volume is the single greatest problem encountered with cutting fluid systems.

Water Quality[8]

Lake water and river water can be hard or soft depending upon proximity of mineral deposits and weather conditions. Well water is usually harder than surface water. Total hardness is due mainly to calcium, magnesium, and ferrous cations. These can react with soaps, wetting agents, and emulsifiers to form insoluble compounds, deplete rust inhibitors and microbicides, coat pipes, clog filters, break emulsions, and gum up a machine tool with sticky residues. High-cationic content (over 200 ppm) in the water source will be most detrimental to soluble oils and then emulsions. Excessive negative ions, sulfate, chloride, carbonate, and bicarbonate, can deteriorate mix stability, decrease product life, cause skin and residue problems, promote pitting and staining of materials, and reduce overall product performance.[9]

As water evaporates from a cutting fluid mix and makeup is added, anions and cations

Table 2
CUTTING FLUID RECOMMENDATION BY MACHINABILITY GROUPS

Type of operation	Ferrous (>70%)	Ferrous (50—70%)	Ferrous (>40—50%)	Ferrous (<40%)	Nonferrous (>100%)	Nonferrous (≤100%)
Broaching						
Internal	4,7	4,7	4,5,7	4,1,7	1,7	6,7
Surface	4,7,8	4,7	4,7	4,1,7	1,7,8	6,7,8
Tapping	4,1,7	4,1,7	4,1,7	4,1,7	3,1,7	3,1,7
Threading	4,7,8	4,7,8	4,7	4,7	3,7	2,7
Gear shaving	6	6	6	4	3	3
Reaming	6,7,8	6,7,8	4,7,8	4,7,8	3,7,8	3,7,8
Gear cutting	6	6	6	4	3	3
Drilling	10,8,7,4	10,8,7,4	10,8,7,4	10,8,7,4	10,8,7,3	10,8,7,3
Deep hole drilling	4,6,7,8	4,6,7,8	4,6,7,8	4,6,7,8	3,7,8	3,7,8
Milling	8,7,6	8,7,6	8,7,6	8,7,6	8,7,3	8,7,3
Boring	4,7	4,7	4,7	4,7	3,7	3,7
Planing	7,8,10	7,8,10	7,8,10	7,8,10	7,8,10	7,8,10
Turning	10,7,8	10,7,8	10,7,8	10,8,7	10,8,7	10,8,7
Sawing	4,7,8	4,7,8	4,7,8	4,7,8	3,7,8	3,7,8
Surface grinding	10,9,8	10,9	10,9	10,9	9,10	9,10
Thread grinding	10,8,7	10,8	4,5,8	4,7,8	1,7	1,7,8
Form grinding	10,8,7	10,8	4,5,8	4,7,8	1,7	1,7,8

Note:
1. Straight mineral oil
2. Straight fatty oil
3. Blend of mineral and fatty oil
4. Sulfurized mineral fatty oil blend
5. Sulfurized mineral oil
6. Sulfo-chlorinated mineral oil
7. Soluble oil
8. Chemical emulsion
9. Simple chemical solution
10. Complex chemical solution

Water-based fluids should not be used on magnesium grinding and great care must be used in fluid selection for magnesium machining.

may concentrate as much as 5-fold in a 30-day period. If these levels are too high, pretreatment of the water by softening, deionization, or reverse osmosis is beneficial. Ion-free treated water of zero hardness can lead to foam problems.

Amount of Hand Contract[10]

If machines are fed parts automatically and the operator contacts the fluid only when he gauges parts, a less mild product can be used. The potential of most cutting fluids to cause skin irritation increases as the concentration and contamination increases.

Amount of Contamination[11]

Lubrication oils, way oils, heat treat solutions, floor cleaners, cigarette butts, and candy wrappers are among items often found in a cutting fluid system. Some contaminants contribute to growth of bacteria, rust, dermatitis, concentration control difficulties, residue, and foam. Water-based products may either reject tramp oil contamination or emulsify up to 20% oil. Cleaner and rust preventive solutions can increase alkalinity to a point where skin problems result. Dirt (metal fines, abrasive grit, or other particulate matter) can be a problem if deposited on tool holders and fixtures.

Ease of Use[12-14]

A product should be stable and not separate during normal storage. Cutting fluid concentrates should be stored in steel or plastic containers; copper, brass, aluminum, or their alloys should not be used because of potential chemical attack. For products that are mixed, the concentrate should be added to water to avoid an inverted emulsion.

Ease of Disposal

Prolonging cutting fluid life is the first step of any waste management program, but waste treatment at some time will be inevitable. Most prevalent waste treatment[15,16] is acidification followed by the addition of alum, ferric chloride or ferric sulfate, polyelectrolytes, or polymer-inorganic combinations. Physical methods are based on separation of water from the pollutants by evaporation, distillation, filtration, ultrafiltration,[17] reverse osmosis, centrifugation, or activated carbon. Many plants that generate insufficient waste to justify their own treatment have it hauled away by a commercial disposal company.

Freedom from Undesirable Side Effects[18]

Cutting fluids should not leave an objectionable residue on either the parts or the machines; and they should be nontoxic and nonirritating to eyes, nose, and skin. Rancidity,[19,20] accompanied by its unpleasant odors, should ideally be controlled by the product or by microbicides added to the mix if needed. Rancidity can be controlled by good sanitation practices, correct concentration, aeration, and the use of good water. A cutting fluid should provide rust protection on parts for at least 72 hr under favorable conditions. Cutting fluid mixes should not smoke, burn, or damage paint on machine tools. Since World War II, most machine tool manufacturers have changed to resistant acrylic, epoxy, and polyurethane coatings.

Performance vs. Cost[21]

All cost factors should be considered — tool life, cleaning the machines, resharpening tools, part cleaning, recharging frequency, and a host of others. Performance factors in cutting fluid selection include tool life, fluid life, operator satisfaction, shop cleanliness, freedom from corrosion and rancidity, adaptability to several jobs, and elimination of health and safety hazards.

CUTTING FLUID CONTROLS

Control of the system, which includes maintenance of the mechanical components as well as the cutting fluid, is equally important as cutting fluid selection in prolonging the life of the fluid. The problems that beset cutting fluids in central system applications are the same as those in individual machines, only the magnitude is greater. A program to accomplish this control should include the following basic steps to obtain long fluid life and avoid problems.[12-14]

1. Assign the responsibility of control to one individual
2. Clean the system thoroughly before charging with a fresh mix
3. Maintain the cutting fluid mix concentration at the recommended dilution
4. Keep the cutting fluid free of chips and grit
5. Use water that has a low dissolved solids content
6. Aerate the cutting fluid
7. Provide good chip flushing at the machines and in the trenches
8. Employ good housekeeping practices
9. Remove extraneous or tramp oil

CUTTING FLUID APPLICATION

Improved tool life, better surface finish, lower power consumption, greater accuracy, and efficient chip flushing can be achieved when clean fluid floods the cutting zone. Generally, a large volume of cutting fluid at low pressure is most effective.[22] Higher fluid pressures are required for grinding because of air currents generated by the grinding wheel. A good rule to follow on flow rate is

1. General-purpose machining and grinding, m^3/sec = machining kW/120
2. High-production machining and grinding, m^3/sec = machining kW/60 to 30

An exception to the rule of low pressure is in gun drilling. Here the cutting fluid is fed, under pressure, through the tool shank. High pressure also helps where chip packing is a problem in vertical milling, reaming, etc.

The reservoir capacity must allow sufficient retention time to settle fines and to cool the fluid. For general-purpose machining and grinding operations:

1. Grinding — tank volume = flow/min × 10
2. Machining cast iron and aluminum — tank volume = flow/min × 7
3. Machining steel — tank volume = flow/min × 5

For high stock removal, tank sizes obey the same formulas since flow rate has already been increased in relation to machine horsepower.

Manual Application

Manual application consists of brushing, dripping, or squirting cutting fluid on the cutting area. This method is seldom recommended except in conjunction with flood application systems. For instance, tapping compound is often manually applied on a tapping or threading operation where extra friction-reducing chemicals are needed to provide the tool life and finish required.

Mist Application

Cutting fluid is sometimes applied as a mist generated by pumping the cutting fluid through a special nozzle where it mixes with air. Mist application has found its greatest use where, because of part size or configuration, cutting fluid could not be rechanneled to the reservoir. Mist is used in operations such as high-speed sawing of extruded aluminum window and door frames where the machine is not equipped with a cutting fluid recirculating system. Mist application also makes the use of fluids practical with portable tools.

Since the limit for oil mist in air that machine operators can breathe is only 5 mg/m^3 according to Occupational Safety and Health Act (OSHA) regulations, only preformed chemical emulsions and chemical solutions are recommended. If soluble oil or straight oil products are used, or if machines leak excessive hydraulic, way, and other lubricating oils, local ventilation should be provided or mist collectors installed.

Flood Application

Flooding is the most common application method.[12] In turning and facing, the cutting fluid is directed to the area where the chip is formed using two nozzles — one above and one below the tool. In slab milling the cutting fluid is directed to both sides of the cutter, again using two nozzles. One nozzle insures that the cutting fluid reaches the cutting zone; the other washes out the chips. In the case of face milling, a ring-type distributor can direct as many streams of fluid as needed to flood each tooth of the cutter. For drilling, reaming,

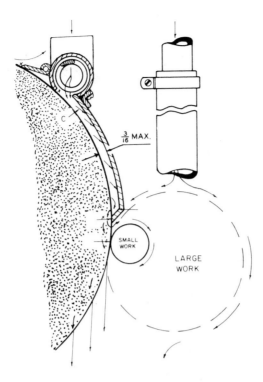

FIGURE 6.   High-velocity grinding nozzle.

or tapping through-holes with hollow-shank tools, a cutting-fluid retainer assures that all cutting edges will be flooded and that chips will be flushed from the hole. For thread chasing, direct the cutting fluid to the cutting edges of the tool; and when using a self-opening die, direct the cutting fluid to each chaser in the die head. A ring-type distributor is most effective in internal broaching.

High Pressure Application

Application of cutting fluid to grinding operations requires the part to be cooled, the grinding wheel kept clean, and friction reduced as the chip is formed. The heat produced in grinding is caused by friction; at each grain interface temperatures normally reach 950 to 1400°C. Grinding dry, or otherwise failing to reduce heat, can affect wheel wear, grinding accuracy, and the physical characteristics of a part. To control grinding heat, the cutting fluid must penetrate directly to the cutting zone through the high-pressure air bubble surrounding the wheel.[23] One method of getting fluids into this air pocket is to force them either through the bore or sides of the wheel. The fluid cannot be passed through the wheel, however, unless it has been filtered to 3 μm to avoid the wheel quickly filling with swarf. Use of a special high-velocity nozzle avoids additional filtering problems and fog can usually be reduced to acceptable limits (Figure 6). If the nozzle is functioning properly, the spark stream will virtually disappear.

There is no need for the fog to become objectionable. Fluid flow should be decreased until the part temperature or out-of-roundness just begins to increase; the flow should then be increased slightly to give a reasonable operating margin. If fog is still a problem, exhaust systems or extra guarding may be necessary.

Internal grinding requires the cutting fluid to remove grinding chips and grit from the hole being ground. Position the nozzle so that wheel rotation carries the fluid between the grinding wheel and the work.

Only when a large volume of clean fluid bathes the tool and reaches the cutting zone are full benefits obtained from cutting fluids. Relatively simple, inexpensive modifications to existing machine tools will often increase tool life as much as 100%, improve surface finish, give greater accuracy, reduce or eliminate steam, smoke, and operator complaints.

SAFETY AND HEALTH FACTORS

Cutting fluids can be tested using procedures specified in the Federal Hazardous Substances Act for such factors as acute inhalation and oral toxicity, primary skin and eye irritation, and acute dermal toxicity. The Environmental Protection Agency (EPA) administers the Federal Insecticide, Fungicide and Rodenticide Act (FIFRA), which requires registration of bactericides and fungicides, many of which are used as additives for cutting fluids. In order to get EPA registration of a biocide, the manufacturer must run many tests which prove that the biocide is effective but not extremely harmful to aquatic life, wildfowl, and man.

The Department of Transportation (DOT) administers a law which governs the shipment of hazardous materials, including chemicals and cutting fluids. The law specifies appropriate labeling and training of all employees involved. If one reships between plant locations, or repackages, these regulations must be followed.

The Occupational Safety and Health Act (OSHA), enacted by Congress in 1970, sets 5 mg/m^3 as the maximum allowable level of oil in air that operators breathe. This has resulted in installation of new ventilation or exhaust equipment in many shops. Other manufacturers switched to metalworking fluids which contained little or no oil.

Under the Clean Water Act of 1977, EPA will put severe restrictions on toxic pollutants in industrial effluents. Two alternatives exist: either eliminate products which contain materials listed or install wastewater treatment equipment to remove the pollutants from the effluent.

Although there are no government regulations in the U.S. regarding nitrosamines in cutting fluids, the National Institute for Occupational Safety and Health (NIOSH) issued a 1976 Intelligence Bulletin, announcing the discovery of a nitrosamine, diethanolnitrosamine, in cutting fluids. NIOSH pointed out that, "Although nitrosamines are suspected to be human carcinogens, their carcinogenic potential in man has not been proved."[24] Diethanolnitrosamine forms from a reaction between secondary or tertiary amines (or both) and an oxide of nitrogen. Many manufacturers have removed the nitrite from their cutting fluids to eliminate this problem. Tests are being conducted by NIOSH; in the meantime OSHA is considering guidelines that probably will recommend housekeeping and hygiene measures.

On May 5, 1980, the EPA released the Resource Conservation and Recovery Act (RCRA), which deals with all hazardous wastes and affects all generators, transporters, and disposers. RCRA deals with the identification of wastes as hazardous based on the following characteristics; ignitability, corrosivity, reactivity, and toxicity. It also provides standards for facilities which store, treat, or dispose of hazardous wastes. A cutting fluid may be classified as hazardous depending on its composition or through contamination from use. Applicability of RCRA regulations is dependent on individual situations.[25]

Many additional state and local standards exist, covering all facets of the metalworking industry. Responsibility for compliance lies with both the manufacturers and the users.

REFERENCES

1. **Cookson, J. O.**, An introduction to cutting fluids, *Tribol. Int.*, 5, February 1977.
2. **Merchant, M. E.**, Fundamentals of cutting fluid action, *Lubr. Eng.*, 163, August 1950.
3. **Dwyer, J. J.**, Cutting fluids, Special Rep. 548, *Am. Mach.*, 105, March 1964.
4. **Kelly, R.**, *Selection and Maintenance of Cutting Fluids*, Tech. Pap. MR73-110, American Society of Mechanical Engineers, New York, 1973.
5. **Springborn, R. K.**, *Cutting and Grinding Fluids: Selection and Application*, American Society of Tool and Manufacturing Engineers, Dearborn, Mich., 1967.
6. **Joseph, J. J.**, Cleaning metalworking fluids, *Am. Mach.*, 75, March 1971.
7. **Brandt, R. H.**, *Fluid Longevity and Central Clarification Systems*, Tech. Pap. MR74-171, American Society of Mechanical Engineers, New York, 1974.
8. **Bennett, E. O.**, Water quality and coolant life, *J. ASLE*, 30, 549, 1974.
9. **Bennett, E. O.**, *The Effect of Water Hardness on the Deterioration of Cutting Fluids*, Tech. Pap. MR72-226, American Society of Mechanical Engineers, New York, 1972.
10. **Suskind, R.**, Occupational skin problems, *J. Occup. Med.*, No. 1, 39, 1959; No. 2, 119, 1959; No. 4, 230, 1959.
11. **Ciesko, R.**, *The Effects of Water Soluble Cutting Fluids on Operating Conditions in Machining*, Tech. Pap. EM75-380, American Society of Mechanical Engineers, New York, 1975.
12. **Russ, G. A.**, Coolant control of a large central system, *J. ASLE*, Preprint No. 79-AM-2A-1, 1979.
13. **Tomko, J.**, *Cutting Fluid Maintenance*, Tech. Pap. MR71-804, American Society of Mechanical Engineers, New York, 1971.
14. **McCoy, J. S.**, A practical approach to central system control, *J. ASLE*, preprint No. 77-AM-1E-1, 1979.
15. **Kulowiec, J. J.**, Techniques for removing oil and grease from industrial wastewater, *Pollut. Eng.*, 49, February 1979.
16. **Nemerow, N. L.**, *Liquid Waste of Industry, Theories, Practices, and Treatment*, Addison-Wesley, Reading, Mass., 1971.
17. **Priest, W.**, Treatment of waste oil emulsions by ultrafiltration, *Water Waste Treat.*, 21, 42, 1978.
18. **Vaughn, R. L. and Miller, H. B.**, *Cutting Fluids and Environmental Compatibility*, Tech. Pap. MR70-714, American Society of Mechanical Engineers, New York, 1970.
19. **Bennett, E. O.**, The biology of metalworking fluids, *Lubr. Eng.*, 28, 237, 1972.
20. **Rossmore, H. W.**, *Microbiological Causes of Cutting Fluid Deterioration*, Tech. Pap. MR74-169, American Society of Mechanical Engineers, New York, 1974.
21. **Mason, J. W.**, *Cost Savings Through Cutting Fluid Selection*, Tech. Pap. MR69-259, American Society of Mechanical Engineers, New York, 1969.
22. **ASME**, *Machine Design Considerations for Improving Metalworking Fluid Performance*, Tech. Pap. MR76-252, American Society of Mechanical Engineers, New York, 1976.
23. **Gettelman, K. and Fisher, R. C.**, The not-so-fine art of precision grinding, *Mod. Mach. Shop*, 78, April 1975.
24. **Finklea, J. F.**, Current intelligence bulletin: nitrosamines in cutting fluids, U.S. Department of Health, Education and Welfare, Washington, D.C., 1976.
25. **Anon.**, Hazardous waste and consolidated permit regulations, *Fed. Reg.*, 45(98), 33063, May 1980.

CUTTING FLUIDS — MICROBIAL ACTION

E. O. Bennett

INTRODUCTION

A cutting-fluid manufacturer has the responsibility of producing a product which performs the desired metal-working operation in the most efficient way possible, has no adverse effect upon human health, is as resistant as is possible to spoilage, and can be disposed of via some currently available method. The producer must also be able to provide technical information pertaining to usage and appropriate preservatives that may be employed in the product.

The user has the responsibility to employ the product properly. This includes minimizing its contamination with different coolants, metals, tramp oils, dirt, food, and other materials; keeping the proper coolant dilution, employing satisfactory sanitation, minimizing human contact with the coolant, adding a preservative periodically to prevent rancidity, and not discarding it in such a way as to produce an environmental problem.

Cutting-fluid users often purchase these increasingly expensive lubricants and use them under conditions which cannot possibly provide satisfactory coolant life or employee well being. Evidence for this practice is indicated in numerous NIOSH Health Hazard Evaluation Reports regarding inspections of machining operations in a number of plants. All of these reports note infractions of standards for proper usage as the cause of the problems noted; none cite the cutting-fluid manufacturer nor do any as yet recommend substitution of another cutting fluid for the one being used.

In order to avoid problems in this area, greater attention must be given to choosing coolants that are least deleterious to human health, selecting machines which minimize worker contact and rancidity problems, minimizing oil mist and noise levels, ensuring better housekeeping practices, using proper preservatives, and keeping coolants in a sanitary condition. If this is not done, the industry will undoubtedly be subjected to increasing supervision and restrictions imposed by state and federal governments.

TESTING METHODS

In order to minimize microbial problems and to have safe sanitary working coolants, a complete knowledge of microbiology is not required and outside microbiological laboratories need not be employed to analyze coolants. Highly complicated tests are not usually necessary in order to keep a working coolant under microbial control. Mastering one simple procedure can provide much of the information required to solve most spoilage problems.

Several simple test kits are available which can be used to determine the microbial content of a working coolant and these kits can be used by people untrained in microbiology.[1-5] Several of the kits involve first placing a dipstick or paper coated with nutrients into the coolant and then placing it in a container which excludes further contamination. A second type of test kit involves nothing more than pipetting a small amount of coolant through five tubes of nutrient material. The kits can be stored in a desk drawer for one or two days and then their appearance is compared with charts furnished with the kits. The kits contain full instructions for use and coolant manufacturers or preservative suppliers usually can provide information as to where they may be obtained.

Those who have the backgrounds or financial resources may use other tests which require instruments or additional expertise.[2,5-10] Even charts for keeping the information obtained from these and other tests have been described.[1] Any one of these procedures will provide the means to accomplish a coolant control program for any company.

QUALITY CONTROL OF CONCENTRATES

One of the first concerns should be the quality of the cutting-fluid concentrates purchased for use within the plant. A few concentrates can contain large numbers of microorganisms which may already be capable of attacking the product and may be resistant to the coolant. The lubrication engineer should be aware of the problem and be able to use the technique described previously to determine the quality of concentrates as they arrive. Products with counts higher than 10,000 organisms per milliliter should be avoided.

SOURCES OF ORGANISMS IN COOLANTS

Unprotected cutting fluids quickly develop large populations of microorganisms after they are placed in service. The primary source of contamination of a fresh coolant comes from the residue of biomass remaining in the system from the previous charge. While cleaning the system between charges is a good practice because it reduces the amount of biomass, it does not normally remove all of this material.[11,12] The microbial content of a fresh coolant can increase from about 470,000 to 4 million organisms per milliliter after only one cycle through the system.[7] If the coolant is not protected, 2 to 3 days later it may contain from 40 to 100 million organisms per milliliter as additional contamination from the concentrate, the water used to dilute the coolant, the air, workers, parts, dirt, and from other sources that enter the coolant.[12]

TYPES OF ORGANISMS FOUND IN COOLANTS

The types of organisms that can attack these lubricants are quite varied and include bacteria, yeasts, and molds. Engineers need not burden themselves with a thorough knowledge of the different organisms encountered in order to accomplish rancidity control; however, there are reports available concerning this matter for those who require this information.[13-19] Those interested in the mechanism of attack may wish to study several papers on this subject.[9,20-22]

PRESERVATIVE USE

Due to their susceptibility to microbial attack, cutting fluids must be treated with preservatives in order to provide maximum life. The convenient method of treatment involves the coolant manufacturer adding a preservative to the concentrate. The effective and economical way is for the user to add it to a working coolant. Most manufacturers offer products containing preservatives to satisfy user demand; however, a study of randomly selected "preserved" products (Table 1) shows that most exhibit no antimicrobial properties and the rest offer inadequate protection.

The reasons for the poor antimicrobial properties of concentrates are related to manufacturing problems. It is a violation of federal law to use a preservative in excess of the maximum permissible level allowed in the registration of the antimicrobial agent. Since coolant manufacturers normally do not know the ultimate use-dilution level for their products, they protect themselves by adding reduced levels of preservatives to their concentrates. In addition, it may not be possible to add adequate levels of preservatives to concentrates due to instability problems. Most cutting-fluid preservatives quickly lose activity in cutting-fluid concentrates. It should be noted though, that a few cutting-fluid manufacturers have been successful in developing concentrates which are compatible with certain preservatives. Even when this occurs, there is an additional factor which makes concentrate preservation of limited value. Even if a preservative can be added to the concentrate, it will not provide rancidity control

Table 1
THE BIOLOGICAL
STABILITY OF WATER-BASE
CUTTING FLUIDS

No. of coolants	Days of microbial control[a]
31	Less than 7
10	1—14
2	21—28
8	35—49
3	70—100
2	100—140
3	140+

Note: Each coolant concentrate diluted 1:40.

[a] Control is defined as the capacity of preservative to keep microbial count under 1 million organisms/mℓ.

for the life of the coolant. As soon as a lubricant is placed in use, the preservative level begins to decline as the chemical reacts with microorganisms. The preservative usually is the first component to be consumed and this occurs long before the life of the lubricant has been exhausted. Even the most active preservatives will normally last no longer than 2 to 4 months in a working coolant. It is necessary to add a preservative to a working system at periodic intervals in order to provide maximum life.

The next step involves acquiring knowledge concerning the different cutting-fluid preservatives. Several publications provide information pertaining to the sources of these products, their chemical composition, toxicity, and other useful information.[11,23,24] Those interested in the environmental effects of these compounds may wish to review reports on this topic.[25,26]

Due to the extremely diverse composition of cutting fluids, no individual preservative is effective in all coolants.[23] Even the most commonly used cutting fluid preservative is effective in less than one half of all products. For this reason, each individual coolant must be studied in order to determine which preservative is most effective in the product. There are several reports demonstrating this point in the literature.[27-30]

HOW TO FIND AN EFFECTIVE PRESERVATIVE FOR A COOLANT

A number of tests are available which can establish the effectiveness of a preservative in a coolant.[3,27,31-35] Some of these tests are easy to do and require no more equipment than glass jars.[2] Test counts of less than 100,000 organisms per milliliter for 60 days or more under laboratory conditions demonstrate the effective preservative for the product and its potential use under industrial conditions. Coolant suppliers or preservative producers usually will answer questions and provide help in setting up one of these procedures.

Users sometimes perform engineering tests on cutting fluids in order to determine the best product for use within a plant. Several fluids may be found to be about equally effective in this regard. A preservative-coolant test can be used to provide additional information. Table 2 shows the results of such a test done on candidate coolants that passed engineering tests conducted by a major user. It may be noted that condidates 1 and 2 constitute cutting fluids in which none of the preservatives were effective. The selection of these products for use within the plant could result in considerable problems. Product 5 can be adequately preserved by three antimicrobial agents and the selection of this product for use provides several options in regards to microbial control.

Table 2
EASE OF PRESERVATION OF
CUTTING FLUIDS

Coolants	Preservative (days of inhibition)			
	A	**B**	**C**	**D**
1	28	7	28	42
2	14	7	14	56
3	126	84	35	91
4	112	91	77	63
5	35	126	140[a]	140[a]

Note: 1:40 oil to water ratio.

[a] Still inhibitory when taken off test.

MICROBIAL CONTROL OF WORKING COOLANTS

Microbial counts should be done on working coolants at periodic intervals. The procedure which should be followed for taking the samples may be found in the literature.[2] At first, counts should be made at weekly intervals until an understanding of the system is attained and the frequency of addition of a preservative is established. Once this has been accomplished and all working conditions remain relatively constant, then counts can be made at less frequent intervals. Individual machine sump systems do not create a major problem. Usually machines that perform similar functions and use the same coolant will be about equal in their preservative requirements. One or two machines within the group can be selected for testing and the information may be applied to the others.

Anytime there is a change in operating procedure such as changing the coolant, changing the oil-water ratio, adding an extra shift, changing the type of metal being worked, or using a different hydraulic fluid in the machines, then counts should be done each week once again until the new pattern of preservative addition is established.

Proper control of a working coolant involves adding a preservative as soon as the microbial content exceeds the maximum desired level and making additional treatments any time tests shown the microbial content again exceeds this level. Companies that already use this technique have set different maximum levels ranging from 1 to 20 million organisms per milliliter. Keeping the level under 10 million organisms at all times can extend coolant life from a few months to one or more years.

The method of adding the preservative must be established for a system. An antimicrobial agent may be routinely added in low concentrations at frequent intervals such as daily, once each week in the makeup, or it may be added at the maximum recommended concentration at less frequent intervals. The most effective and economical procedure normally involves adding a maximum dose whenever the microbial count exceeds the desired level.

Of course, the frequency of addition is related to working conditions. In clean, well-engineered plants it is possible to treat a system once every 4 to 8 weeks. In dirty plants and in certain metal-working operations it may be advisable to treat on a weekly basis or sometimes even on a daily basis.

After coolant control practices have been initiated, records should be kept to determine if coolant costs are really being reduced. Some cutting fluids cannot be controlled by any preservative, and the treatment of these coolants is nothing but an added expense. When using cheaper coolants, it is often more economical to use the coolant without a preservative and replace it when it has spoiled rather than periodically treating it with preservatives.

All preservatives do not function equally well in a particular product. Sometimes a more expensive preservative is more economical than a cheaper one because the interval between treatments is greater and the amount of material required is less. Sometimes it is more economical to purchase a more expensive coolant and treat it with a cheaper preservative than buying an expensive preservative and using it in a cheap coolant. The purpose of microbiological control is to reduce costs, but companies have practiced coolant control without initiating studies to determine if costs are actually reduced. Once an individual has learned to determine the microbial load of a working system and has selected the proper preservative, there are a number of things which should be kept in mind.

FACTORS THAT INFLUENCE DETERIORATION CONTROL

It is impossible to entirely prevent the introduction of organisms into working fluids;[11] however, it is possible to minimize contamination of a system. Construction of a new plant offers an opportunity to take advantage of a number of factors which may influence coolant life. Plants should never be placed downwind of sewage treatment plants, flour mills, bakeries, feed mills, fertilizer plants or cooling towers which may produce either airborne nutrients or contaminants.

The internal construction of the plant should be engineered so that it can be cleaned properly with a minimum of effort. Consideration should be given to the design of the circulation system, floor elevation, dragout recovery, coolant storage area, pipe work, pump rates, reclamation equipment and other factors.[37,38] Machines should be selected which minimize coolant contact with workers and the environment. Newly acquired reconditioned machines can be sources of major contamination problems and should be thoroughly steam cleaned before being put into operation.

Many companies design their systems so that there is too much agitation of the coolant. This practice is usually done in order to move chips or to prevent or reduce coolant odors. Few individuals appear to recognize that the greater the agitation of a coolant, usually the greater the microbial attack.[39] Circulation should be adequate for desired performance but it should not be overdone.

Petroleum-base fluids normally are subject to bacterial deterioration while synthetic and semisynthetic products are more likely to encounter mold (slime) problems. Formaldehyde-releasing preservatives are somewhat weaker against molds than against bacteria.[28,29,40] Extra care must be taken when using this type of preservative in synthetic and semisynthetic coolants because if the system is improperly treated, slime problems can occur.

A circulation system should never be underdosed[41] with a preservative. Most antimicrobial agents will markedly stimulate growth when employed in low concentrations.[42] The use of a small amount of preservative in a system may produce more growth than if nothing is done.

Preservatives should never be mixed indiscriminately. Practically all combinations of the readily available preservatives are incompatible with each other and mixing them can result in less control than using one product alone.

Coolant control in one system cannot normally be applied to other systems.[41] Each system has characteristics which are unique for that system alone. The character of each system must be learned through tests and observations.

Machines, floors, circulation systems, and parts should be kept as clean as is possible. Stock stored in outside yards may be dirty and serve as a source of contamination of the coolant. Tote baskets holding parts should be designed so that the stock can be subjected to high-pressure water jets prior to being moved to the machines.

It is sometimes necessary to shut down machines or systems for extended periods of time. If the system is to be inactive for only a few days, it is best to treat with a biocide and continue to circulate the coolant. If the shutdown period is going to be prolonged, and if it is possible and practical, it is advisable to drain and clean the circulation system and leave it dry until it is operated again. If possible, the drained coolant may be used as makeup for working systems.

There is a misconception that when a coolant develops odor or slime it has spoiled and is beyond recovery. If other criteria are used, such as rust protection, tool life, emulsion stability or finish, they may indicate that the coolant is beyond salvage. On the other hand, even though slime or odor has developed, if the engineering qualities are still satisfactory it may be possible to save the system via effective preservative treatment.

When a system which contains a great deal of slime or odor is under treatment, precautions should be taken. The breakdown of millions of microbes killed by the preservative can produce considerable amounts of organic matter which produces frothing. When this occurs foaming will start 24 to 48 hr after the addition of the preservative. This is a temporary phenomenon which will last only a few hours. Unfortunately the coolant is often discarded after the foam appears when the addition of a small amount of antifoam agent would have eliminated the problem.

Pumps and filtration equipment should be watched when biocide treatment is underway. A system that has accumulated large amounts of slime can give trouble when this material is dislodged in large masses. This release can temporarily increase the viscosity of the coolant, placing an increased load upon the pump motors and it may plug lines and pumps. These obstructing masses should be removed as soon as they are detected.

Additional factors can influence rancidity control. As confidence is gained in protecting working coolants, the engineer may wish to achieve even better coolant life by understanding these factors.

General discussions of these factors have been published;[2,17,38,43] however, those interested in more detained treatments can study reports which deal with the effects of coolant temperature,[11,44] water hardness,[45,46] water quality,[17,47,48] urine,[17] metals,[38] dragout,[36] hydraulic fluids,[36,38] oil-water ratios,[49] differences in the sensitivities of different systems to preservatives,[50] and chelating agents.[30] Those interested in medical problems[51-53] or disposal problems[54,55] may wish to read these communications.

Coolant control is not difficult to accomplish. Doing nothing more than determining the microbial content of a working coolant at periodic intervals and adding a preservative when the count reaches a certain level can produce a significant increase in coolant life. Where coolants have lasted a few weeks, it is possible to experience a doubling of coolant life. Where coolants have functioned properly for several months, a 60% improvement in coolant life is not unusual. Some users have already undertaken quality control of their fluids and two publications have appeared concerning their success in this area.[10,56]

REFERENCES

1. **Kane, E. L.,** *A Chart for Recording and Analyzing Factors Influencing Coolant Life,* ASLE Preprint No. 73AM-4C-2, American Society of Lubrication Engineers, Park Ridge, Ill., 1973.
2. **Bennett, E. O.,** The biological testing of cutting fluids, *Lubr. Eng.,* 30, 128, 1974.
3. **Hill, E. C., Gibbon, O., and Davies, P.,** Biocides for use in oil emulsions, *Tribology,* 121, June 1976.
4. **Hill, E. C.,** Some aspects of microbial degradation of aluminium rolling coolants, *Proc. 3rd Int. Biodegrad. Symp.,* 3, 243, 1976.
5. **Holdon, R. S.,** Microbial spoilage of engineering materials. VI. Improving monitoring and control, *Tribology,* 10, 273, 1977.

6. **Yanis, R. J. and Wolfe, G. F.**, Test procedures for the evaluation of cutting fluids, *Lubr. Eng.*, 164, April 1960.
7. **Hill, E. C.**, Microbiological examination of petroleum products, *Tribology*, 5, February 1969.
8. **Rossmoore, H. W.**, Methylene blue reduction for rapid inplant detection of coolant breakdown, *Int. Biodetn. Bull.*, 7, 147, 1971.
9. **Rossmoore, H. W., Holtzman, G. H., and Kondek, L.**, Microbial ecology with a cutting edge, *Proc. 3rd Int. Biodegrad. Symp.*, 3, 221, 1976.
10. **McCoy, J. S.**, A practical approach to central system control, *Lubr. Eng.*, 34, 180, 1978.
11. **Kane, E. L. and Pfuhl, W.**, Preservation and preservatives in the aluminum hotrolling and beverage can processing industry, *Lubr. Eng.*, 32, 249, 1976.
12. **Bennett, E. O.**, The deterioration of metal cutting fluids, *Prog. Ind. Microbiol.*, 13, 121, 1974.
13. **Tant, C. O. and Bennett, E. O.**, The isolation of pathogenic bacteria from used emulsion oils, *Appl. Microbiol.*, 4, 332, 1956.
14. **Tant, C. O. and Bennett, E. O.**, The growth of aerobic bacteria in metal-cutting fluids, *Appl. Microbiol.*, 6, 388, 1958.
15. **Bennett, E. O.**, The role of sulfate-reducing bacteria in the deterioration of cutting emulsions, *Lubr. Eng.*, 13, 215, 1957.
16. **Kitzke, E. D. and McGray, R. G.**, The occurrence of moulds in modern industrial cutting fluids, paper presented at the 17th ASLE Meet., St. Louis, Preprint No. 62 AM 4B-3, 1962.
17. **Bennett, E. O.**, The biology of metalworking fluids, *Lubr. Eng.*, 28, 237, 1972.
18. **Rossmoore, H. W. and Holtzman, G. H.**, Growth of fungi in cutting fluids, *Dev. Ind. Microbiol.*, 15, 273, 1974.
19. **Wort, M. D., Lloyd, G. I., and Schofield, J.**, Microbiological examination of six industrial soluble oil emulsion samples, *Tribology*, 35, 1976.
20. **Guynes, G. J. and Bennett, E. O.**, Bacterial deterioration of emulsion oils. I. Relationship between aerobes and sulfate-reducing bacteria in deterioration, *Appl. Microbiol.*, 7, 117, 1959.
21. **Isenberg, D. L. and Bennett, E. O.**, Bacterial deterioration of emulsion oils. II. Nature of the relationship between aerobes and sulfate-reducing bacteria, *Appl. Microbiol.*, 7, 121, 1959.
22. **Vamos, E. and Csop, A.**, The microbiological corrosive action of metal machining oils, *Corros. Week*, 41, 1029, 1970.
23. **Smith, T. F. H.**, Toxicological and microbiological aspects of cutting fluid preservatives, *Lubr. Eng.*, 25, 313, 1969.
24. **Paulus, W.**, Problems encountered with formaldehyde-releasing compounds used as preservatives in aqueous systems, especially lubricoolants — possible solutions to the problems, *Proc. 3rd Int. Biodegrad. Symp.*, 3, 1075, 1976.
25. **Pauli, O. and Franke, G.**, Behavior and degradation of technical preservatives in the biological purification of sewage, *Biodeterioration of Materials*, Vol. 2, Halsted Press, New York, 1972, 52.
26. **Voets, J. P., Pipyn, P., Van Lancker, P., and Verstraete, W.**, Degradation of microbiocides under different environmental conditions, *J. Appl. Bacteriol.*, 40, 67, 1976.
27. **Rossmoore, H. W. and Williams, B. W.**, An evaluation of a laboratory and plant procedure for preservation of cutting fluids, *Biodetn. Bull.*, 7, 55, 1971.
28. **DeMare, J., Rossmoore, H. W., and Smith, T. H.**, Comparative study of triazine biocides, *Dev. Ind. Microbiol.*, 13, 341, 1972.
29. **Bennett, E. O.**, Formaldehyde preservatives for cutting fluids, *Int. Biodetn. Bull.*, 9, 95, 1973.
30. **Izzat, I. N. and Bennett, E. O.**, *The Potentiation of the Antimicrobial Activities of Cutting Fluid Preservatives by EDTA*, Preprint No. 78-AM-5C-1, American Society of Lubrication Engineers, Park Ridge, Ill., 1978, 1.
31. **Pivnick, H. and Fabian, F. W.**, Methods for testing the germicidal value of chemical compounds for disinfecting soluble oil emulsions, *Appl. Microbiol.*, 1, 204, 1953.
32. **Kitzke, E. D. and McGray, R. J.**, Coolant microbiology: the role of industrial research, paper No. 59AM 3A-3, 14th ASLE Natl. Meet., Buffalo, 1959.
33. **Brandeberry, L. J. and Myers, H. V.**, Test procedures for compounds used as preservatives in industrial coolants, *Lubr. Eng.*, 16, 161, 1960.
34. **Himmelfarb, P. and Scott, A.**, Simple circulating tank test for evaluation of germicides in cutting fluid emulsions, *Appl. Microbiol.*, 16, 1437, 1968.
35. **Rogers, M. R., Kaplan, A. M., and Baumont, E.**, A laboratory inplant analysis of a test procedure for biocides in metalworking fluids, *Lubr. Eng.*, 31, 301, 1975.
36. **Bennett, E. O.**, *Effect of Dragout and Hydraulic Fluid Contamination on Rancidity Control in Cutting Fluids*, Preprint No. 76-AM-18-1, American Society of Lubrication Engineers, Park Ridge, Ill., 1976, 1.
37. **Smith, M. D. and West, C. H.**, *How Plant Practices Affect Employee Health in the Presence of Metalworking Fluids*, American Society of Lubrication Engineers, Park Ridge, Ill., August 1969, 321.

38. **Bennett, E. O.,** *Microbiological Aspects of Metalworking Fluids,* Tech. Pap. No. MR73-826, American Society of Mechanical Engineers, New York, 1973, 1.

39. **Rossmooore, H. W., Sceszny, P., and Rossmoore, L. A.,** *Evaluation of Source of Bacterial Inoculum in Development of a Cutting Fluid Test Procedure,* No. 76-AM-1B-2, American Society of Lubrication Engineers, Park Ridge, Ill., 1976, 1.

40. **Rossmoore, H. W., De Mare, J., and Smith, T. H. F.,** Anti- and pro-microbial activity of hexahydro-1,3,5-tris(2-hydroxyethyl-s-triazine in cutting fluid emulsions, in *Biodeterioration of Materials,* Vol. 2, Halsted Press, New York, 1972, 286.

41. **Bennett, E. O.,** Factors involved in the preservation of metal cutting fluids, *Dev. Ind. Microbiol.,* 3, 273, 1961.

42. **Bauerle, R. H. and Bennett, E. O.,** The effects of 2,4-dinitrophenol on the oxidation of fatty acids by *Pseudomonas aeruginosa, Ant. van Leeuwenhoek J.,* 26, 225, 1960.

43. **Hill, E. C.,** *Biodeterioration of Metal Working Fluids and Its Significance,* Publ. No. MR72-214, American Society of Mechanical Engineers, New York, 1972, 1.

44. **Hill, E. C.,** The significance and control of microorganisms in rolling mill oils and emulsions, *Met. Mater.,* 294, September 1967.

45. **Feisal, E. V. and Bennett, E. O.,** The effect of water hardness on the growth of *Pseudomonas aeruginosa* in metal cutting fluids, *J. Appl. Bacteriol.,* 24, 125, 1961.

46. **Bennett, E. O.,** *The Effect of Water Hardness on the Deterioration of Cutting Fluids,* Tech. Pap. No. MR72-226, Society of Mechanical Engineers, New York, 1972, 1.

47. **Humnicky, S.,** Pure water improves coolant mix, *Tooling Prod.,* 48, February 1971.

48. **Bennett, E.O.,** Water quality and coolant life, *Lubr. Eng.,* 30, 549, 1974.

49. **Carlson, V. and Bennett, E. O.,** The relationship between the oil-water ratio and the effectiveness of inhibitors in oil soluble emulsions, *Lubr. Eng.,* 16, 572, 1960.

50. **Bennett, E. O., Adamson, C. F., and Feisal, V. E.,** Factors involved in the control of microbial deterioration. I. Variation in sensitivity of different strains of the same species, *Appl. Microbiol.,* 7, 368, 1959.

51. **Bennett, E. O. and Wheeler, H. O.,** Survival of bacteria in cutting oil, *Appl. Microbiol.,* 2, 368, 1954.

52. **Rossmoore, H. W. and Williams, B. W.,** Survival of coagulase-positive staphylococci in soluble cutting oils, *Health Lab. Sci.,* 4, 160, 1967.

53. **Holdom, R. S.,** Microbial spoilage of engineering materials. Are infected oil emulsions a health hazard to workers and to the public? *Tribology,* 9, 271, 1976.

54. **Bennett, E. O.,** The disposal of metal cutting fluids, *Lubr. Eng.,* 29, 300, 1973.

55. **Bennett, E. O.,** *The Disposal of Metal Cutting Fluids,* Publ. No. 73AM-4C-EB, American Society of Lubrication Engineers, Park Ridge, Ill., 1973, 1.

56. **Vermooten, C. A. L.,** Microbiological destruction of soluble oil emulsion in steel plant hydraulic systems, paper read at South African Soc. Plant Pathol. Microbiol. Meet., 1975.

LUBRICANT APPLICATION METHODS

Edward J. Gesdorf

INTRODUCTION

Modern lubrication standards for industrial equipment in mass production industries such as automotive, steel, mining, rubber, etc. usually start with the following goals: safety of personnel, uninterrupted production, extended machinery life, and good housekeeping. A fifth item could easily be added — a reduction in operating costs. Since the days of low-cost labor and lubricants are gone, it now becomes extremely important to select the most efficient method of applying lubricants.

TRADITIONAL LUBRICANT APPLICATION DEVICES

Some of the older, simple lubricant application devices include:

1. Oil squirt can
2. Screw-type grease gun
3. Grease gun
4. Drop oiler
5. Vibrating pin bottle oiler
6. Thermal oiler
7. Wick-pad and waste-feed oilers
8. Splash-lubrication system
9. Ring, chain, and collar oilers
10. Mechanical positive-feed

Design, selection, and maintenance of this equipment is covered in reference material.[1,2] During the early days of the industrial revolution the only devices available for applying oil or grease to a bearing were oil and grease cups as shown in Figure 1. To eliminate their feast or famine nature, automatic pressure feeding grease cups were introduced, as illustrated in Figure 2.

In this device, the spring on top of the large piston exerts a constant pressure on the lubricant in the cup. This pressure forces the lubricant around the screw thread on the reservoir pin, which is closely fitted to the outlet bore. As grease is discharged to the bearing, the lowering compression of the spring is compensated by the screw thread of the resistance pin passing out of the discharge bore, lowering the resistance to flow. By this method a constant feed of grease is provided to the bearing.

The constant pressure supplied by the spring, combined with a restricted orifice at the outlet, caused many greases to bleed or separate oil from the soap. The cup would then load up with a cake of hard soap preventing further delivery of lubricant. Special grease was therefore required to ensure proper operation of the device, whether the grease was suitable for the bearing or not. While many of these grease cups performed an outstanding job, they suffered limitations for universal application.

During this period, mechanical force-feed lubricators (Figure 3) came into use. These devices were originally designed for engine room lubrication where they still serve better than any other type of lubricator. Today many forced-feed lubricator applications have been incorporated in centralized systems. This enables up to several hundred bearings to be served without the necessity of running a bundle of pipes or tubing from one central box.

FIGURE 1.   Typical oil and grease cups.

The first really modern lubricant dispensing equipment was the pressure gun and grease fitting introduced for motor car chassis lubrication. This equipment provided a quick convenient method for delivery of clean grease to a bearing under high pressure, and at the same time flush out most of the old lubricant. Savings in lubricant, cleanliness, and better lubrication with the pressure gun and fitting eventually led to their widespread use in practically all manufacturing plants.

The next step consisted of grouping the fittings or nipples at central stations accessible from the floor, and running individual tubes from each fitting to a bearing. Flexible hose connections were used to connect to movable points of application, as illustrated in Figure 4.

The greaser or oiler could connect to some 70 or 80 grease fittings on this press, and lubricate all bearings without having to shut down or climb over the machine. Because the bearings were not visible to the oiler when making the gun connection to the several fittings, however, some bearings received too much lubricant, some too little, and some were missed entirely. This lack of control led to the development of the centralized pressure system of lubrication. The first such system was installed on a motor car chassis in the early 1920s.

OIL AND GREASE CENTRALIZED SYSTEMS — OPERATING PRINCIPLES

An understanding of the characteristics and basic operating principles of the six available types of centralized lubricating systems is beneficial in selecting the most suitable for a specific application. Important features of each type are compared in Table 1. Over and above all other considerations, dependability of operation is most important.

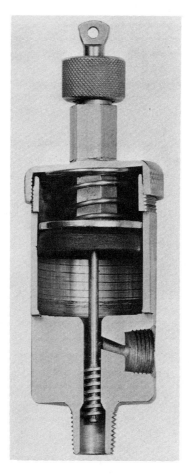

FIGURE 2.   Automatic pressure-feed-
ing grease cup.

Single Line System — Spring Return

Figure 5 illustrates a single-line system, employing spring return measuring valves. Some modern systems have included indication and adjustment at each measuring valve. The main system elements consist of a reservoir, pump, three-way valve, and a series of spring-operated measuring valves connected to a common supply line. If the pump is operated with the three-way valve in the position shown, a pressure is developed in the main supply line which becomes effective on the top of the main piston in the measuring valve, causing it to move downward and forcing the oil or grease ahead of the piston into the bearing. Pumping is continued until all measuring valves have discharged, which is noted by the operation of a poppet type indicator stem, or pressure gauge.

To recycle the system, the hand lever of the three-way valve is turned 90° to connect the main supply line to the relief line to exhaust pressure in the system. When the main line pressure drops below the force of the compression spring in the measuring valve, the spring returns the valve piston to its original position. Reloading of the measuring valve takes place when the valve piston reaches the enlarged section of the piston bore. In this type system, the pistons are moved hydraulically in one direction and by spring force in the opposite direction.

Progressive Nonreversing System

Figure 6 illustrates a system which divides delivery from the pump into the several bearing

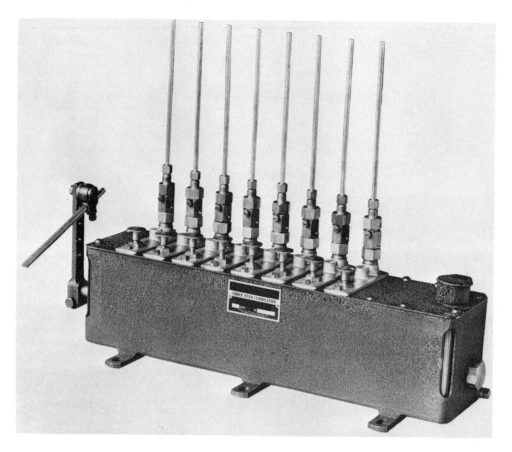

FIGURE 3.   Mechanical force-feed lubricator.

outlets. The measuring valve is progressive and nonadjustable with indication usually at the pump. The main system elements consist of a reservoir, pump, main supply line, and the measuring valve manifold. If the pump is put in operation, lubricant is taken from the reservoir and passed through the main supply line to the measuring valves. In position 1 of the diagram, the measuring valve sections are connected together with a common central pressure port. The center section of the piston of the bottom measuring valve permits a flow of pressure to the left end of the piston in the center measuring valve to hold the piston in the right end of its bore. Likewise, the center section of the middle measuring valve piston permits flow of pressure to the left end of the piston in the top measuring valve to hold the piston to the right. The center section of the piston in the top measuring valve permits pressure to flow through the valve porting to the extreme right end of the piston in the lower measuring valve, causing the piston to move to the left, pushing the oil or grease in the left end of the bore through valve porting out to a bearing at outlet 1, shown in position 2 of the diagram.

As the bottom piston moved from right to left, the center section of the piston changed the porting to permit pressure to flow to the right end of the center piston. This causes the center piston to move to the left, displacing lubricant in the left end of the bore through valve porting out to a bearing at outlet 2, as shown in position 3 of the diagram.

In moving from right to left, the center piston changes the porting through its center section to permit pressure to flow to the right end of the top piston. This causes the top piston to move to the left, pushing the lubricant in the left end of the bore through valve porting out to a bearing at outlet 3, as shown in position 4 of the diagram. The top piston,

FIGURE 4. Multiple tube system.

in moving from right to left, changes the porting at its center section to permit a flow of pressure to the left end of the lower piston to move to the right, displacing the lubricant in the right end of the bore to a bearing.

In following the foregoing sequence, a continuous cycling mechanism operates as long as there is flow from the pump.

Progressive Reversing System

Figure 7 illustrates a loop system which operates on the principle of reversible flow in

Table 1
COMPARISON OF SIX SYSTEM PRINCIPLES

Features	System identification					
	Single-line spring return (Figure 5)	Progressive nonreversing (Figure 6)	Progressive reversible flow (Figure 7)	Dual line (Figure 8)	Orifice metering oil (Figure 9)	Orifice metering oil mist (Figure 10)
Adjustable measuring valves	Yes	No	No	Yes	No	No
Measuring valves operate	Independently	In series	In series	Independently	Independently	Independently
Measuring valve actuation	Hydraulic spring return	Hydraulic	Hydraulic	Hydraulic	No moving parts	No moving parts
Metering principle	Positive displacement	Positive displacement	Positive displacement	Positive displacement	Orifice metering	Orifice metering
Measuring valve indicators	Yes	Yes	No	Yes	No	No
Measuring valve piston sealing	Packing washers	Metal-to-metal	Metal-to-metal	Metal-to-metal	No pistons	No pistons
System will handle grease	Yes	Yes	Yes	Yes	No	No
System will handle oil	Yes	Yes	Yes	Yes	Yes	Yes
Can add or subtract lube points economically	Yes	No	Yes	Yes	Yes	Yes
Economical monitoring — main supply lines	Yes	No	Yes	Yes	Yes	Yes
Measuring valve monitoring	Yes	Yes	Yes	Yes	No	No
Currently popular	Yes	Yes	No	Yes	Yes	Yes

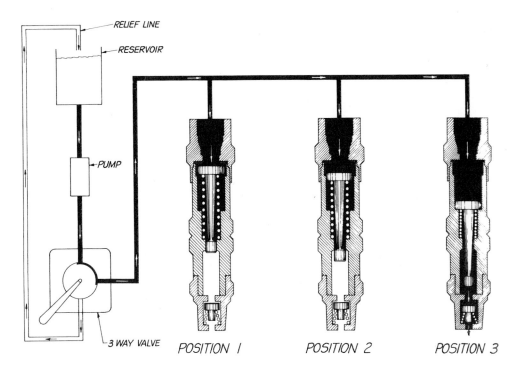

FIGURE 5.   Single-line system, spring-actuated valve.

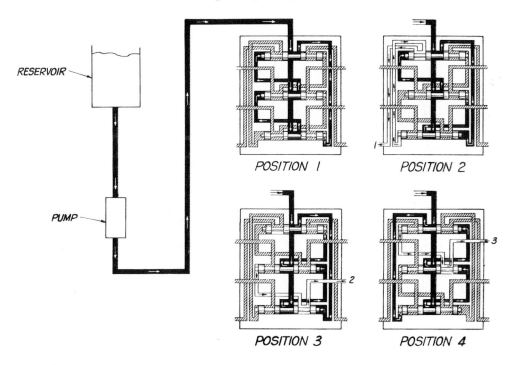

FIGURE 6.   Progressive system — nonreversing.

the main supply line. The measuring valves are progressive and nonadjustable, with indication at the end of the loop.

The main system elements consist of a reservoir, pump, four-way valve, main supply

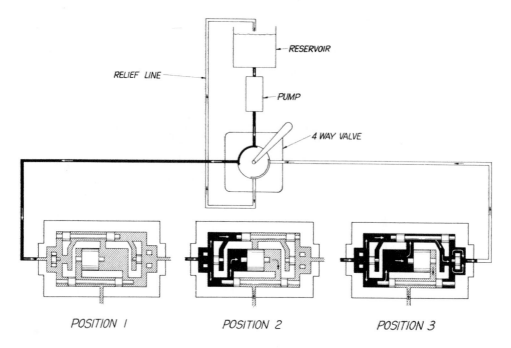

FIGURE 7.   Progressive system — reversible flow.

line, and a series of measuring valves inserted in the main supply line. If the pump is operated with the four-way valve in the position shown, lubricant is taken from the reservoir and directed into the main line loop discharging from the four-way valve at the left and entering the first measuring valve, as shown in position 1. After moving a check piston and two pilot pistons, lubricant is permitted to flow into the main cylinder, forcing the measuring piston to the right, and displacing the lubricant in the right end of the cylinder to a bearing. This is shown in position 2. At the end of its travel, the measuring valve piston uncovers a by-pass port to allow flow of lubricant from the pump to pass around the measuring valve piston and through an outlet check valve as shown in position 3. The lubricant then moves on to the next measuring valve in the circuit, where the valve operation sequence is repeated. Pumping is continued until all measuring valves have operated and until a flow of lubricant reaches the central pumping unit to operate an indicator, thus completing one lubricating cycle.

To recycle the system, the hand lever on the four-way valve is shifted 90° to change the porting of the four-way valve to permit flow of lubricant from the pump to be directed into the main supply line loop, discharging from the four-way valve at the right and entering the first measuring valve on the right end of the circuit. In effect, shifting the four-way valve reverses the flow of lubricant in the main supply loop. The measuring valves are again operated and the sequence of valve operation is as described in the foregoing except that the main line flow is in the opposite direction.

Dual Line System

Figure 8 illustrates a system which uses alternating pressure applications in two main supply lines. The measuring valves operate independently and offer indication and adjustment for each bearing. The main system elements consist of a reservoir, pump, four-way valve, and dual main line piping to which all measuring valves are attached.

If the pump is operated with the four-way valve in the position shown, lubricant is taken from the reservoir and directed into main supply line 1. As the pressure increases, it becomes

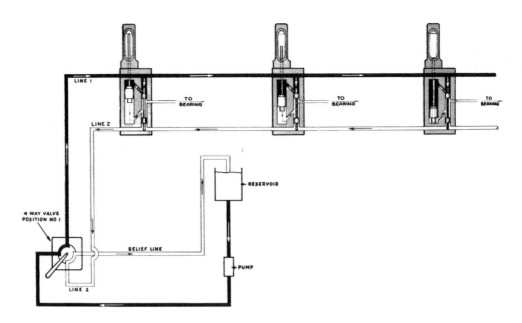

FIGURE 8. Dual line system.

effective on the top side of the pilot piston in the measuring valve, causing it to move downward. After a certain amount of travel, an angle port is opened permitting lubricant to flow into the main measuring cylinder, forcing the main piston downward. As the main piston moves downward, lubricant in the lower portion of the cylinder is forced through a second angle port and out the discharge port to a bearing. Pumping is continued until all measuring valves have operated.

To recycle the system, the hand lever of the four-way valve is shifted 90° which relieves the pressure in line 1 back to the reservoir and ports the pump to line 2. As pressure is developed in line 2, the measuring valve operation sequence described in the foregoing is repeated but with the valve pistons moving in the opposite direction. Indicator stems attached to the main pistons of the measuring valves provide a means for periodic inspection of valve operation. Valve discharge adjustment is accomplished with two flat adjusting screws in the packing gland which control main piston travel.

Orifice Oil System

Figure 9 illustrates an orifice metering system for oil which operates on the principle of pressurizing a common main supply line rapidly and bleeding the pressure off through various-sized orifices. The metering orifices operate independently of each other and are nonadjustable. The main system elements consist of a reservoir, pump, main supply line, and orifice meter assemblies.

The lubricator is of the spring discharge type and is operated by pushing the lever down which raises the piston and compresses a spring. By releasing the lever a fixed volume of oil is discharged into the supply line which is then dissipated through the orifice meters at the bearings.

Orifice Oil Mist System

Figure 10 illustrates an orifice metering system for oil mist. Oil is broken up into fine particles and dispersed in air for conveying through pipeline to the point of application. The main system elements are a filter and water separator, solenoid-operated air valve, air pressure regulator, misting head, oil reservoir, mist distribution manifold, and reclassifying fittings at the bearings.

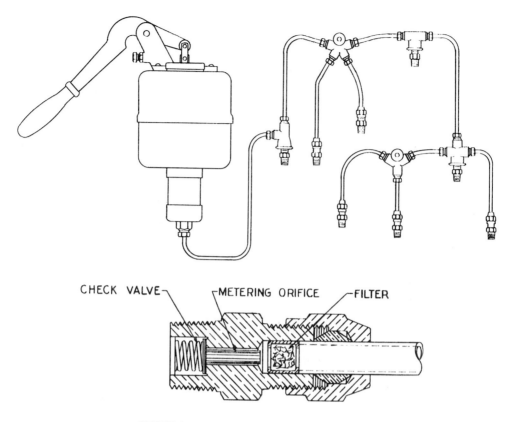

FIGURE 9.   Orifice metering system — for oil only.

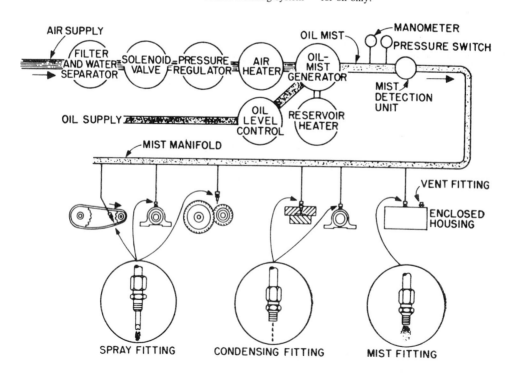

FIGURE 10.   Orifice metering system — for oil mist.

Table 2
COMPARISON OF MIST SYSTEM AND AIR LINE LUBRICATOR

Comparison	Mist system	Air line lubricator
Particle size	1—3 μm (0.00004—00012 in.)	50—100 μm (0.002—0.004 in.)
Conveying distance	Up to 91.46 m (300 ft)	25.4—30.48 cm (10—12 in.)
Air function	Conveying	Power
Air pressure	49.91 g/cm^2 (0.71 lb/in.^{2})	4,218—7,030 g/cm^2 (60—100 lb/in.^{2})
Air velocity	4.26—5.48 m/sec (14—18 ft/sec	74 m/sec (244 ft/sec)

Clean dry air is supplied to the mist generator at a controlled pressure. By using the Venturi or Vortex principle the mist unit generates a mixture of small oil particles dispersed in air. This "dry mist" can be conveyed a considerable distance in a pipeline for use at bearings, slides, chains, gears, etc. However, before the dry mist can be used to lubricate, it must be converted to wet mist with reclassifier or condensing fittings at the lubrication points. The reclassifier orifices cause the small oil particles in the dry mist to join and become heavier and wet, allowing the reclassifiers to discharge a wet spray, a wet mist, or oil droplets, all of which depends on the design or the reclassifier fittings.

While early applications of oil mist lubrication were limited to oils with viscosities below 190 cSt at 40°C, air heating has extended this limit to 900 cSt at 40°C. Polymer oil additives with molecular weights in the 50,000 to 150,000 range have greatly reduced the volume of stray oil escaping into the atmosphere.[3] From experience to date, 90% of general purpose machinery in petrochemical plants are candidates for mist lubrication, with potentially improved reliability, improved efficiency and reduced contamination problems.[4]

Aerosol Lubricators

Since a comprehensive discussion on aerosol lubricators has already been written by Faust[5] and orifice oil mist systems have been explained in the foregoing section, only a comparison of the two is required (see Table 2). Air line lubricators as shown in Figure 11 provide a mixture of oil and air. The mixture is usually generated by the Venturi principle and since the oil particles in the mixture are of a large size they can be conveyed only a short distance.

SYSTEM ACCESSORIES FOR CENTRALIZED SYSTEMS

Accessories of one type of another — alarms, lights, signals, counters, timers, and interlocking controls are popular for protection against failure.

In general, indicators on measuring valves are of two types — one indicating measuring valve performance, the other indicating measuring valve failure. During lubricating system inspections, which are essential at periodic intervals to determine *performance* and to insure that all lines are intact, the measuring valve indicators should be checked for operation, regardless of type used.

Many favor interlocking automatic lubricating systems with equipment startup and shutdown, together with use of low-level alarms on lubricant reservoirs. Lubricating systems then require only periodic inspection and reservoir filling.

SYSTEM PLANNING AND INSTALLATION

Planning of a centralized lubrication system revolves around the question "How much lubricant does a bearing require per unit of time"? The basic premise for sleeve bearing lubrication by a centralized system is the need for replacing 1/3 of the bearing clearance volume every 4 hr for grease, or every 2 hr for oil. This is given by the formula:

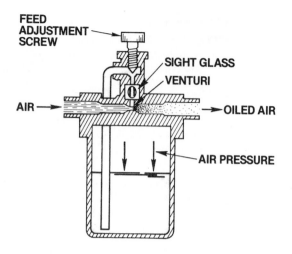

FIGURE 11.   Typical air line lubricator.

1/3 Bearing makeup volume = 3.1416 × diameter × length × radial clearance/3

To eliminate the extensive amount of arithmetic that would be required for a variety of bearing sizes on a large machine, a bearing makeup volume chart for plain bearings is shown in Figure 12. This chart is also adapted for use with rolling element bearings by following Table 3 and using bearing inside diameter as shaft diameter.

Many factors and operating conditions influence quantity of lubricant required by a bearing. These include speed (r/min), temperature, water, dirt, dust, scale, and type of lubricant. The makeup volume formula is usually modified by a service factor to increase or decrease the amount of lubricant in accordance with operating conditions. System manufacturers provide on request an abundance of information and assistance on system planning.

Installation should be carried out with skilled and qualified craftsmen to assure that the finished job not only operates correctly, but also has the greatest aesthetic value. Equipment should be located in protected areas wherever possible, to avoid damage. High-temperature zones and attachment to vibrating supports should be avoided. At the same time, consideration must be given to accessibility from the standpoint of inspection and servicing of equipment. All tubing and pipelines should be securely clamped to protect from possible damage.

LUBRICANTS DISPENSED BY SYSTEMS

Manufacturers of centralized lubricating systems seldom make recommendations on lubricants as they feel this is the responsibility of lubricant suppliers, machine designers, and plant operators. A centralized system supplier can give guidance, however, on consistency, pumpability, and stability of lubricants that affect lubricating system performance. Lubricants should always be selected first to satisfy bearing requirements; satisfying system requirements should be a secondary consideration. Greases having consistency ratings in the range of NLGI #00 to NLGI #2 are being handled successfully in centralized systems. Oils with viscosities as high as 50 cSt at 100°C (250 SSU at 210°F) have been handled but heavy oils must be maintained at temperatures at least 10°C (20°F) above the pour point. Light and medium weight oils are readily handled in centralized systems.

CENTRALIZED OPEN GEAR SPRAY SYSTEMS

Centralized spray lubricating systems, as shown in Figure 13, were first introduced in

SYSTEM PLANNING

HOW TO FIND MAKE-UP LUBE VOLUME "V" IN CUBIC INCHES*

> **PLAIN BEARINGS**
>
> $V = V_i$ from graph below times bearing length
>
> Example: How much oil is needed for a plain bearing 5″ long by 4″ diameter. Speed = 100 R.P.M.
>
> V_i from curve = .011
> $V = 5 \times .011 = .055$ cu. in. every 2 hours

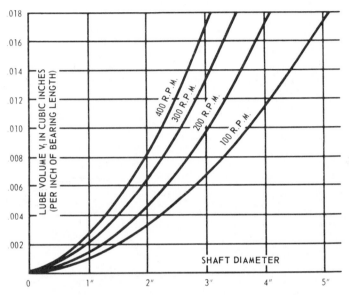

*Lube to be applied at intervals of 2 hours for oil or 4 hours for grease.

FIGURE 12.   Bearing makeup volume chart for plain bearings.

Table 3
MAKEUP VOLUME FOR ROLLING ELEMENT BEARINGS

Type bearing	Apply bearing length
Single-row ball	1 in. (2.5 cm)
Double-row ball	2 in. (5.0 cm)
Roller (square rollers)	1 in. (2.5 cm)
Roller (long cylinderical or tapered)	1/2 Bearing length

1950 for applying heavy residual gear lubricants to the pressure side of gear teeth in open gear installations. While these systems have become extremely popular for gear and pinion sets on ball mills, rod mills, and kilns, they are also being used for lubricating chains, wire rope, and slide surfaces.

Air and measured quantities of lubricant come together in a nozzle and are released in the form of a fine spray to coat gear teeth. Since the systems can be pumped manually or automatically and are intermittent in their operation, they are used as lubricant film maintaining devices. Controlled amounts of lubricant can be applied to open gear installations with considerable efficiency over the manual, drip, pour, swab, and brush methods.

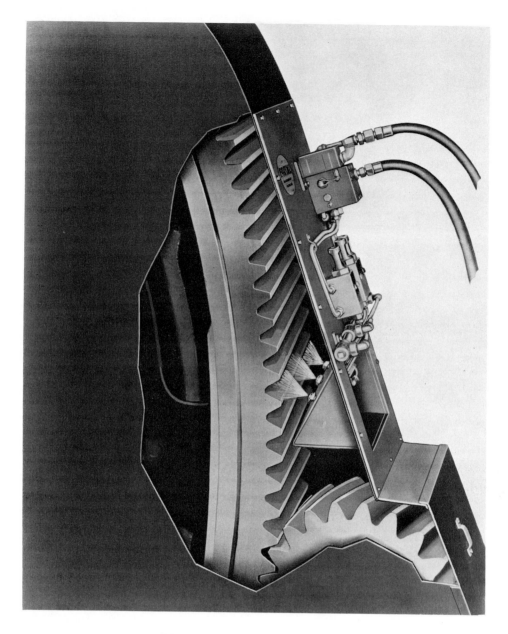

FIGURE 13.　Typical gear spray system.

BULK GREASE HANDLING SYSTEMS

Following broad acceptance of centralized systems, three important developments occurred in the lubrication field. One was the introduction of multipurpose grease; the second was the acceptance of the standardized 181.436-kg (400-lb) container as a replacement for a variety of drum sizes that were in use; and finally, bulk grease systems were introduced in the late 1950s for large grease consuming plants in the steel, cement, rubber, and automotive industries.

Grease could now be delivered by special transport trucks to the users plant and pumped directly into stationary bulk tanks that hold as much as 11,400 kg (25,000 lb) of grease.

The bulk tanks are usually centrally located in the plant so that grease can be pumped through 5.08- or 7.62-cm (2- or 3-in.) pipe distribution lines to the consuming areas. Some bulk grease systems are designed around 1814.36-kg (4000-lb) grease bulk bins that are transportable between the suppliers bulk grease plant and the consumer.

It is not unusual to keep the bulk distribution lines under a constant pressure of 175,767 to 211,110 g/km^2 (2500 to 3000 psi). Grease under pressure is therefore instantly available to operate tote hoses, hose reels, fill centralized system reservoirs, operate area control panels, etc. Several tangible and intangible benefits brought about by bulk grease handling systems including elimination of residual grease waste, drum cost and drum handling, improvement of plant safety and housekeeping, and most important — elimination of grease contamination.

REFERENCES

1. **Clower, J. I.,** *Lubricants and Lubrication,* 1st ed., McGraw-Hill, New York, 1939, chap. 9.
2. **O'Connor, J. J. and Boyd, J.,** *Standard Handbook of Lubrication Engineering,* McGraw-Hill, New York, 1968, chap. 25.
3. **Gulker, E. and Huttenwerke, H.,** New oil mist lubrication concepts for higher efficiency and better environment, in *Tribology in Metal Working: New Developments,* Institute of Mechanical Engineers, London, 1980, 43.
4. **Bloch, H.P.,** Application of pure oil mist lubrication, *Mech. Eng.,* 103(5), 30, 1981.
5. **Faust, D.G.,** *Standard Handbook of Lubrication Engineering,* McGraw-Hill, New York, 1968, 25.

CIRCULATING OIL SYSTEMS

A. J. Twidale and D. C. J. Williams

INTRODUCTION

Use of a circulating system is commonly dictated by requirements of machine elements for a constant and copious supply of clean oil delivered at the correct temperature and pressure. An essential feature of the design is to ensure that the oil does not become too hot and is cleaned of its contaminants as a regular procedure.

All circulating systems need certain basic equipment: a reservoir, a pump, and pipework. Filtration and control of temperature, pressure and flow are also necessary; and isolating valves have to be provided to facilitate the maintenance of these controls and the machine being lubricated. These in turn call for suitable safeguards such as relief valves, pressure gauges, pressure and flow signaling devices, a standby pump, etc. When ingress of water or dirt is expected, a second reservoir may be necessary. Although variations are numerous, Figure 1 shows a typical arrangement. Basic capacities and components are listed in Table 1.

RESERVOIRS

The reservoir provides storage for the oil, allows sufficient dwell time for separation of air entrained as foam, for water and solid impurities to settle by gravity, and provides a reserve against losses. The reservoir should be located beneath the equipment on a 75-mm high subbase so that the oil can return by gravity through drain lines. Where this is not possible, a smaller collecting tank and sump pump must be provided.

Capacity

The minimum total reservoir capacity is determined by the following:

Working capacity — This consists of the total machine requirement at maximum anticipated flow multiplied by dwell time in minutes — see Table 1.

Drain back capacity — The total amount of oil in circulation will normally return to the reservoir when the pumps are stopped in no longer than 3 min. This volume may be calculated by multiplying the flow rate by a factor of 1 to 3. Add to this the capacity of any header, pressure, or collector tanks.

Air space — Allowance for thermal expansion, foam, and venting normally comprises approximately 1 min of flow rate or a minimum of 100-mm height.

Allowance for drain back capacity and air space must exceed the volume of oil retained in the system filter, cooler, delivery pipework and machine sumps that does not drain back when the pumps are stopped. This is to ensure adequate reservoir size to receive the oil required to fill the system initially and during maintenance.

Proportions

Head room and available space are important factors when deciding reservoir proportions. The largest oil surface area possible is desirable where entrained air is anticipated. The most suitable porportions frequently are width = height = length/2.

Connections

A sloping bottom of 1 in 30 with the pump suction at the high end and drain connections at the lowest point allows water and other impurities to be drained off. A floating suction

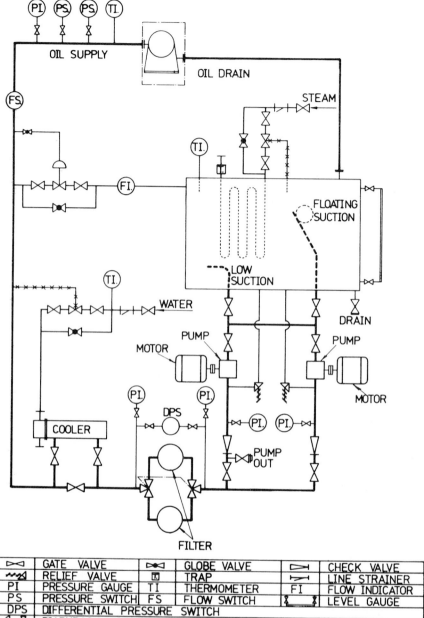

FIGURE 1. Typical system arrangement.

is necessary where heavy contamination is possible, as in large rolling mill systems where mill scale, water, and other impurities may occur. This is arranged to take oil appreciably above the water level but sufficiently below the surface to avoid entraining surface debris or foam. To minimize splashing and foam generation, the returning oil should discharge at or just below the operating level. In some cases, drain piping may desirably be positioned so that return oil can be inspected and a return filter provided (see Figure 2).

Suction and drain connections may be separated by baffles. By promoting a tortuous path

Table 1
CHARACTERISTICS OF TYPICAL OIL CIRCULATING SYSTEMS

| Application | Duty | Machine requirement | | | | | | Pump | | | Filtration | | Cooler | | Reservoir |
| | | Oil viscosity at 40°C | | gal/min | ℓ/min | lb/in.² | Bar | Type | Main driver | Standby driver | Type | μm | Type | Temperature reduction (°C) | Dwell time (min) |
		SSU	cSt												
Small electrical machinery	Bearings	350	75	1	4	15	1	Gear	Motor		Line strainer or mechanically cleaning	150	Shell and tube	45—40	5
Paper mill dryer section	Bearings and gears	700	150	20	75	45	3	Gear	Motor	Motor	Dual basket and centrifuge	120	Shell and tube	60—40	40 (Twin reservoirs to permit settling)
Steel mill	Bearings	1500	300	170	650	60	4	Gear	Motor	Motor	Dual basket and centrifuge	120	Shell and tube	50—40	40 (Twin reservoirs to permit settling)
Steel mill	Gears	1620	350	240	900	50	3.5	Gear	Motor	Motor	Mechanically cleaning (motorized)	150	Plate	50—40	20
General	Bearings	1000	200	10	40	15	1	Gear	Machine shaft or motor	Motor	Dual basket	100	Shell and tube	50—40	12
Grinding mills Petroleum gas compressors	Shaft jacking Compressor bearings and seals	2000 150	450 32	5 50	20 200	3000 150	200	Hydraulic Screw	Motor Steam turbine	— Motor	Disposable Disposable	20 10	— Radiator or shell and tube	65—50	12 8
Petroleum steam turbines and generators	Bearings and controls	150	32	1700	6500	90	6	Centrifugal	Steam turbine	Motor plus DC motor-driven emergency pump	Disposable	10	Radiator	60—45	8

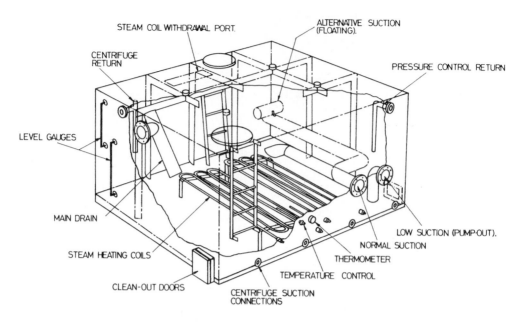

FIGURE 2.   Typical reservoir layout.

between inlet and outlet, baffles enable the best use of the reservoir volume, eliminating dead pockets and assisting in deaeration and settling contaminants. A single-vertical baffle can form a weir to localize contamination and turbulence. Oil should flow across the weir as smoothly as possible to avoid cascading and frothing.

Due to fluctuation of oil levels and to permit the escape of moisture and vapor, the reservoir must be adequately vented. In dirty environments, these vents may incorporate filters to minimize intake of foreign matter. Access is necessary for inspection and cleaning.

Finish

After fabrication the internal surfaces must be thoroughly cleaned, either mechanically or by grit blasting, and a coat of oil-resistant paint applied to all surfaces excluding welds. After a 24-hr hydrostatic test, the painting should be completed.

Reservoir Heating

Heaters should raise the oil from ambient to the specified operating temperature, normally 40°C, within 4 hr. A further controlled increase in temperature will aid release of contaminants. Permanent heating arrangements are usually electric or steam.

Steam heaters are normally in the form of a continuous coil of 3/4 in. (20 mm) nominal bore seamless steel pipe. For every 4000 ℓ reservoir capacity, the coils should be 12-m long and will consume 30 kg/hr of steam at 3.5 bar. Pipe joints must be welded and pressure tested; no mechanical connections should be permitted within the reservoir.

Electric heaters fitted through the side of the reservoir should be provided with sheaths containing replaceable elements. This facilitates removal of a faulty element without having to empty the tank. The heaters should be controlled by thermostats and a low-level switch should interrupt the electrical supply should the oil level drop exposing the heaters. They should be 1 kW for every 400 ℓ reservoir capacity.

Whichever type of heater is used, temperature of the heat transfer surface must be low enough to avoid decomposition of the oil. For steam, the pressure should not exceed 3.5 bar which is equivalent to 150°C. For electric heaters, surface temperature is controlled by limiting the watts rating per surface area which varies depending on viscosity.

Viscosity		Watts rating	
SSU	cSt	in.²	cm²
300	65	8	1.25
300—1160	65—250	6	1.0
1160 and above	250 and above	4	0.75

PUMPS

The following system requirements governing pump size may also influence the choice of pump type.

Rate of flow — Some excess capacity must be provided as a safety margin to offset changes in system demand, a worn pump, wear in bearings, seals, etc. Basic pump sizing is between 110 and 125% of the equipment requirement for large systems; actual selection may exceed this because of standard pump sizes available.

Viscosity — Operating temperature range and the resultant viscosity variance affects the pump size and its driving power requirements.

Suction conditions — The type of pump and/or actual positioning of the pump is governed by suction conditions. If installed above the oil level, the pump must be self priming and capable of producing the necessary suction lift.

Supply pressure — Design pressure at the pump is the sum of (1) required pressure at point of application, (2) static delivery head, (3) friction losses in piping and equipment, (4) filter drop when cleaning is necessary, and (5) an allowance for pump protection as follows.

Protection — Relief valves for protection of positive displacement pumps and drivers should be sized to pass full pump flow at a pressure not more than 25% above the valve "set-pressure". To determine "set-pressure" add 0.7 bar (10 psi) to the maximum pumping pressure in 0 to 7 bar range or add 10% for pump operating above 7 bar. Select the valve to be used, then ascertain the actual pressure when passing full flow at maximum viscosity. When the valve is integral with the pump, the manufacturer will have taken this into account.

Choice of Pump Type

Positive displacement rotary pumps are generally employed for lubrication purposes because their rate-of-delivery is substantially unaffected by changes in delivery pressure and viscosity. Centrifugal pumps can also be used; since rate of delivery is then a function of pressure, related factors such as flow-demand, temperature and viscosity must be accurately controlled. The pump supplier can confirm the choice of pump and determine maximum absorbed power so that the driver can be selected. Pump types frequently used are as follows.

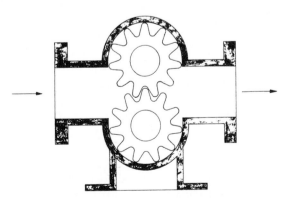

Gear pump — Positive displacement, relatively cheap, compact, simple in design. Where quieter operation is necessary, helical or double helical pattern may be used. Both types capable of handling dirty oil. Available to deliver up to 150 ℓ/min and 70 bar (1000 psi). Self-priming with 7.5 m maximum suction lift.

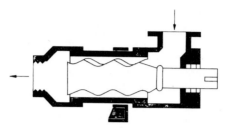

Screw pump — Positive displacement, quiet running, pulseless flow, ideal for pumping low-viscosity oils, can operate continuously at high speeds over long periods, lower power consumption. Available to delivery up to 5000 ℓ/min and 70 bar. Self priming with 7.5 m maximum suction lift.

Centrifugal pump — High rate of delivery at moderate pressure; can operate with greatly restricted output, but protection against overheating necessary with no flow condition by permitting continuous bleed through restrictor in bypass line. Will handle dirty oil. Limited to oil viscosities below 250 cSt. Available to deliver up to 9000 ℓ/min and 12 bar. Flooded suction required to prime.

PURIFICATION AND FILTRATION

The degree of cleanliness chosen will depend on machine requirements. For most industrial applications, general practice uses filtration of 50 to 150 μm. The most common contaminants are water, dirt and dust, or carbon and sludge which can develop from oxidation of the oil. Purification schemes frequently used are as follows.

Settling

This is frequently the first resort in oil reconditioning and separates water and solid impurities from the oil when the system is at rest. Effectiveness of settling is indicated by the following settlement rates for particles in still, 30 cSt oil.

Particle diameter (μm)	Order of size	Time to settle
1	Fine dust	6 months
10	Silt	4 days
100	Fine sand	30 min
1,000	Coarse sand	15 sec

The table shows that only large particles will settle within practical time limits. Water settling is hastened by raising the bulk temperature to 70°C after which it can be drained from the bottom of the reservoir.

Filtration

This is the most universal method and many filter materials and designs are available. Filters should be selected so that under clean conditions and maximum working viscosity the pressure loss does not exceed 0.3 bar (5 psi). Cleaning is normally recommended when pressure loss increases by 1 bar. Filters usually fall within the following types.

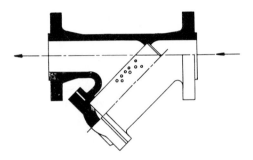

Line strainer — Woven gauze or perforated metal element; easily removed for hand cleaning when system stopped. Normally 150 μm and above.

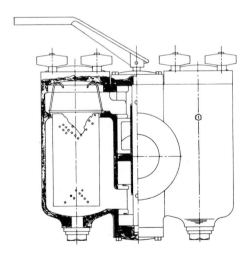

Dual basket filter — Woven gauze or perforated metal element with changeover valve; elements easily removed for hand cleaning when system in operation. Magnets can be incorporated. Normally 50 μm and above.

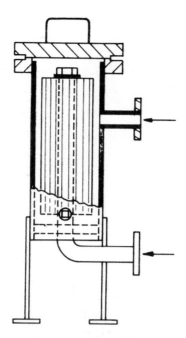

Disposable element filter — Element of materials such as treated paper, felt, and nylon easily replaced when system stopped. Dual versions with changeover valve permit element replacement when system operating. Normally below 50 μm.

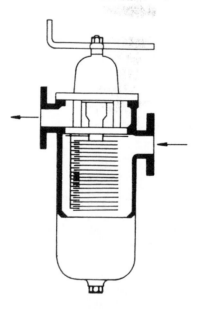

Mechanically cleaning filter — Interleaved radial plates plough the dirt from the gaps between metal discs when the filter pack is rotated, which can be while the system is operating. May be motorized. Periodically drain contaminant from sump when system is stopped. Normally 150 μm and above.

Various other designs of mechanically cleaning filter, such as wire wound and back flushing, are available.

Centrifuging

Installed in a by-pass circuit. Removal of sediment and water can be carried out while the oil system is in operation. For maximum efficiency, the oil should be centrifuged at 70°C with provision of an inline oil heater.

COOLERS

Constant oil temperature desirably enables constant flow and pressure control as these are both affected by changes in viscosity. Most machine designers recommend that the working temperature of the oil be 40°C. High-ambient temperatures, heat generation from bearings and gears, and machine and oil pump inlet power all transfer heat into the oil. The amount of heat is normally specified by the machine designer or based on experience with similar units. This commonly amounts to an oil temperature increase through the machine in the region of 10°C which needs to be removed by a cooler. Prolonged working temperatures above 60°C shorten oil life.

Water and air are the common cooling mediums. When considering water, its cleanliness, corrosion characteristics, hardness, and pressure will affect selection of cooler materials. Temperature, quality, and quantity of cooling medium available are important in obtaining the most efficient cooler. Pressure losses for oil or water through the cooler should not exceed 0.7 bar.

Cooler types frequently used are as follows:

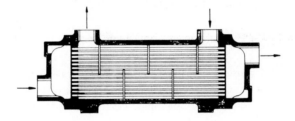

Shell and tube — Oil through shell, water through tubes. Requires space for tube removal.

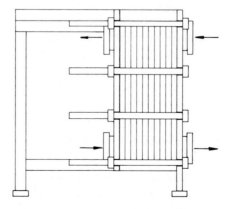

Plate — Oil and water between alternate plates. Compact and readily separated for cleaning.

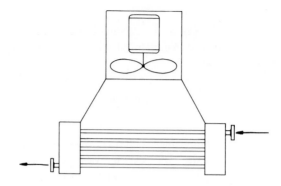

Radiator — Oil through tubes, air over tubes motivated by fan.

TEMPERATURE CONTROL

Along with heaters and coolers, controls are required to cope with inevitable temperature fluctuations.

Reservoir Heater

The control thermostat or thermostatic valve probe, Figure 3, should be positioned near the pump suction and at about middepth of the oil in the reservoir to sense the average temperature. Ensure that the thermostat is never exposed or placed near any localized hot spot such as a heater.

Oil Cooler

One control method is to regulate the water/air flow, the other to regulate the flow of oil to be cooled. Water flow can be governed by a hand control valve as its temperature usually only fluctuates on a seasonal basis. Automatic control of air is essential as its temperature fluctuates daily.

Automatic control utilizes a direct-acting modulating valve in the cooling water supply line which is controlled by a sensing element in the cooling oil outlet. The effect on oil temperature is not instantaneous but is generally acceptable for industrial systems. Cooling water pressure should be reasonably stable, otherwise a pressure regulating valve will be required.

Alternatively, where instantaneous response to control is vital and/or where air is the cooling medium, the cooler should be provided with a bypass line and a control valve to divert flow into the bypass. The valve is of a three-way type with overlapping ports, Figure 4. Mixing of the oil streams within the valve produces an average temperature and a thermostatic element detects any deviation in temperature and corrects the valve position.

PRESSURE CONTROL

A pressure control valve is necessary to spill-off surplus oil from the pump, to regulate any flow variation due to temperature fluctuations, and to accommodate any changes in demand from the machine being lubricated. The following are typical methods of control.

Spring-loaded relief valve — Provides coarse control and is sensitive to viscosity changes. Generally used on smaller, simple systems.

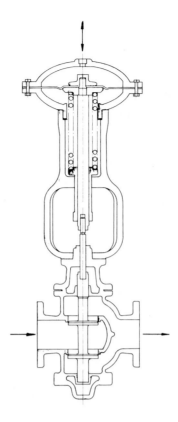

Direct-operated diaphragm valve — The diaphragm chamber is connected so that system pressure is transmitted to the diaphragm. The spring counter-balancing the diaphragm load is adjustable and determines the system pressure. Any change in demand will tend to vary the system pressure and the diaphragm, sensing this, will reposition the valve to adjust the spill-off rate. This valve will maintain the pressure within acceptable limits provided the viscosity remains reasonably stable.

Pneumatically controlled diaphragm valve — The diaphragm is air actuated via a control instrument. This valve is normally selected when very accurate control is necessary or if the system operating pressure is too great for the direct-acting valve diaphragm.

Header tank — While space requirements often preclude this simple form of control, the procedure is to place a tank at the required height. Filled directly with a line teed off the main pump supply, the tank is fitted with an overflow connection. This method ensures the continual change of tank contents to maintain the oil at system temperature. As an added advantage, if the system pumps fail the tank will discharge its contents via the fill connection to the equipment being lubricated. Pump check valves will prevent oil returning directly to the main reservoir.

Providing different pressures within the same system involves the use of pressure-reducing valves. These normally comprise a restricting orifice, or valve opening, which is controlled by imposing the outlet pressure on the valve control diaphragm.

EMERGENCY EQUIPMENT AND SYSTEM CONTROL

If a machine must continue to run after a failure in the lubrication system main pump, it is imperative to arrange fully automatic starting of a second, and maybe a third pump. This can be done by use of flow or pressure-operated switches, or both. Control panel lights

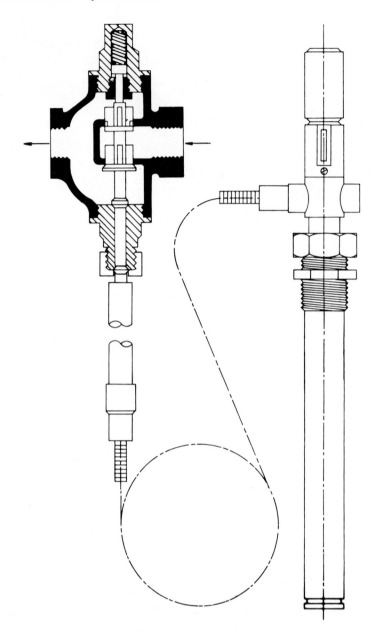

FIGURE 3.   Thermostatic valve.

should indicate that normal running has been interrupted. A continued fall in pump output, due to failure of the emergency action, will operate switches which give an alarm signal and, if desired, cause the machine to stop.

Should it be impractical to stop the machine immediately upon failure of the lubrication system, a header or pressure tank may be incorporated downstream of the cooler to provide a diminishing supply of oil for up to 5 min (see Figure 5). The pressure tank is designed so that air trapped in the top at system pressure will force oil from the bottom when pump pressure is lost. To calculate total volume of the pressure tank, first decide the quantity of oil, Q, required for emergency purposes. Using Boyle's law, P1V1 = P2V2 where P1V1 = pressure (absolute) and volume of air, oil system operating; and P2V2 = pressure

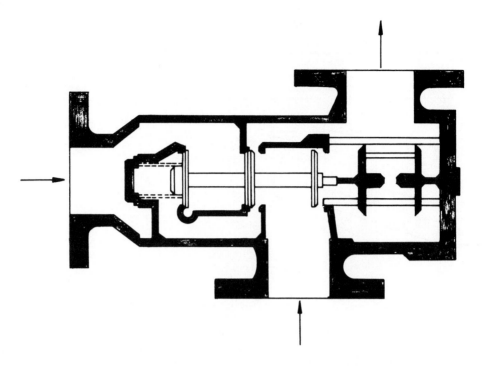

FIGURE 4.   Three-way mixing valve.

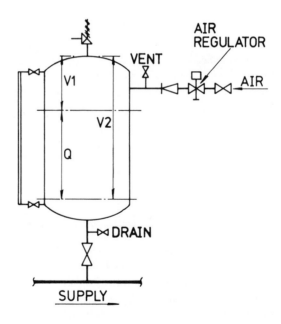

FIGURE 5.   Pressure tank.

(absolute) and volume of air, oil system at rest; since V2 = V1 + Q; therefore V1 = P2(V1 + Q)/P1. Total volume of tank = V1 + Q + volume of bottom dished end. When the system is operating, the tank should be adjusted until the required quantity of oil is visible in the tank level glass, Q + volume of bottom dished end (normally two-thirds full).

To ensure that this balance between the air/oil volumes is not upset when the system is stopped, the tank should not be fully discharged. For this reason, the volume of bottom

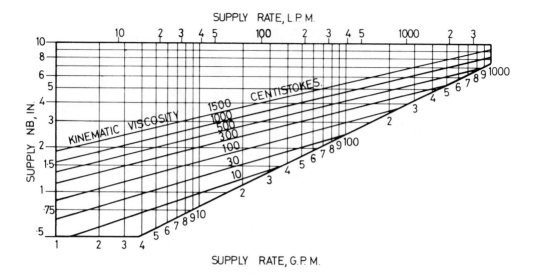

FIGURE 6.   Supply line sizing.

dished end is included in the tank sizing calculations. As some air will be absorbed in the oil, it will be necessary from time to time to add air. A check valve should be fitted in the air supply line to prevent oil from entering the air main and an air regulator installed to avoid accidental overpressurizing.

SYSTEM PIPING

Sizes of interconnecting pipes should be considered in relation to oil viscosity, velocity, and resultant friction losses. Pipes should be large enough to prevent cavitation in pump suction lines, to avoid undue pressure drop in pump supply lines (minimizing pump drive power), and to avoid backup in drain lines.

Suction
Pipe runs should be short. Right angle bends and tee pieces should be kept to a minimum. Nominal bore of pipe to be one size larger than supply.

Supply
Friction loss due to viscosity frequently outweighs velocity considerations, particularly with heavier oils. Figure 6 is based on a friction loss of 0.1-m head per m of pipe and restricted to velocities under an acceptable 2 m/ sec. The viscosity at specified operating temperature should be used to determine pipe nominal bore.

To determine the friction loss in pipework, multiply the length by 0.1 m head. Friction loss caused by fittings and valves can be determined by converting them to equivalent pipe lengths in Table 2. These, together with pressure losses through the filter and cooler, will determine the system losses.

Drain
Drain pipes should be sized to run not more than half full so as to encourage escape of entrained air, provide space for any foam and give a margin of safety. Drain pipes should be vented. Flow rate, slope, and viscosity govern their size. A minimum slope of 1 in 40 is essential but use should be made of all available drop.

The pipe nominal bore may be determined from Figure 7. At startup, pipework and any

Table 2
EQUIVALENT LENGTHS OF
FITTINGS

Fitting/valve	Equivalent lengths in pipe diameters
45° Elbow	15
90° Elbow	25
Tee	60
Gate valve	6
Globe valve	300

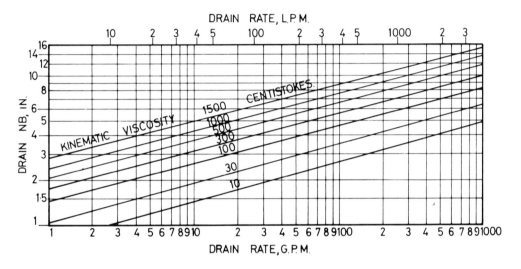

FIGURE 7.   Drain line sizing.

oil retained in the machine sumps will be cold, and the oil viscosity at ambient temperature should be used to determine pipe nominal bore.

Cleanliness

Many troubles and faults experienced with lubrication systems arise from internal bores not being clean at final installation. Steel pipe must be internally shotblasted or pickled and phosphated before fabrication. All components and pipe fittings should be inspected for internal cleanliness. During fabrication, care must be taken to ensure that bores are kept clean. All welds should be mechanically cleaned, and completed pipe fabrications may also be chemically cleaned. During shop assembly or prior to dispatch of pipes to site, all bores should be internally oiled to prevent corrosion.

Following site installation, chemical cleaning may be carried out while bypassing bearings, gearboxes, etc. The system should then be flushed for 48 hr, again bypassing bearings, gearboxes, etc. The pipework should be hammered and oil temperature varied between ambient and 70°C at 4-hr intervals by adjustment to reservoir heating. For the first 24 hr the filter elements should be removed. After 48 hr or when filter elements remain clean, the system should be drained, reservoir cleaned, and machine piping connected.

Design Principles

JOURNAL AND THRUST BEARINGS

A.A. Raimondi and A.Z. Szeri

INTRODUCTION

This chapter applies hydrodynamic lubrication theory to the analysis and design of self-acting fluid film journal and thrust bearings, in contrast to earlier chapters which emphasize lubrication theory and solution techniques. Most of the material has been summarized in design charts and tables for estimating performance of a variety of applications.

It is not within the scope of this chapter to recommend bearing proportions, allowable temperature rise, etc. These are left to the designer to decide on the basis of experience and test. The charts provided here will serve for performance calculations on many representative bearings; similar information is available in the literature for a variety of designs. Computer programs are also available for studying design parameters for specific applications.

LUBRICANT PROPERTIES IN BEARING DESIGN

The lubricant property of greatest concern in fluid film bearings is the absolute viscosity, or just viscosity, μ. Its SI unit is Pa $\cdot$ sec (Pascal second), and in English units it is usually expressed in $lb_F \cdot sec/in.^2$ (reyn). The ratio of absolute viscosity to density (ρ) is termed the kinematic viscosity, $\nu = \mu/\rho$. It is measured in m^2/sec in SI units and commonly in $in.^2/sec$ in English units. Table 1 contains conversion factors for commonly used viscosity units.

Increasing temperature lowers the viscosity of lubricating oils as shown in Figure 1 for typical industrial petroleum lubricants in the various ISO viscosity grades. The viscosity of a number of other fluids is given in Figure 2.

Average Viscosity

In numerous applications, the temperature rise in the bearing film remains relatively small. However, in estimating bearing performance on the basis of classical (isothermal) theory, the calculations should employ an effective viscosity compatible with the mean bearing temperature rise.[1,19] This calculation might be based on the assumptions that:

1. All heat, H, generated in the film by viscous action is carried out by the lubricant.
2. The lubricant which leaves the bearing by its sides has a uniform average temperature $T_s = (T_i + \Delta T/2)$, where $\Delta T = T_o - T_i$ is the mean temperature rise across the bearing.

This mean temperature rise, ΔT, can be calculated from a simple energy balance which gives:

$$\Delta T = H/[\rho c(Q - Q_s/2)] \qquad (1)$$

For typical petroleum oils, $\rho c = 112$ $lb_F/in.^2 F$ (139 $N/cm^2 C$) and for water $\rho c = 327$ $lb_F/in.^2 F$ (406 $N/cm^2 C$).

Rather than assuming a uniform effective viscosity at a mean bearing temperature, using the actual variation in viscosity resulting from temperature changes as the lubricant flows through the bearing will result in considerably improved accuracy in calculating performance. However, this increases the complexity of the solution, usually requires special computer

Table 1
VISCOSITY CONVERSION FACTORS

To convert from	To	Multiply by
Absolute viscosity		
Centipoise (0.01 g/cm·sec)	Pa·sec	10^{-3}
reyn (lb$_F$sec/in.^{2})	Pa·sec	6.895×10^3
lb/in.sec	Pa·sec	1.786×10
Centipoise	reyn	1.45×10^{-7}
Kinematic viscosity		
Centistoke (0.01 cm^2/sec)	m^2/sec	10^{-6}
in.^{2}/sec	m^2/sec	6.452×10^{-4}
Centistoke	in.^{2}/sec	1.550×10^{-3}

Note: To convert Saybolt seconds universal (SSU) to centistokes (cSt), use ν (cSt) = 0.22 SSU − 180/SSU.

programs, and precludes the recasting of results as generalized design charts. Theory that incorporates pointwise variation of lubricant temperature and viscosity is called thermohydrodynamic (THD) theory of lubrication; some results of this will be given in later sections of the chapter.

Density values of lubricants are required in performance calculations involving turbulence and inertia. Typical density for many petroleum oils is 0.83 kg/cm^3 (0.030 lb/in.^{3} or 7.77 $\times$ 10^{-5}lb-sec^2/in.^{4}).

Heat capacity of petroleum lubricants is typically 1675 J/kgK (0.40 Btu/lbm F, 1.44 $\times$ 10^{6}in.^{2}/sec^{2}F). Detailed property information is supplied in other chapters and in literature from lubricant suppliers.

FLUID FILM PHENOMENA

Laminar Flow

For low-surface velocities, small film thicknesses, and large viscosities, the lubricant flow is an orderly laminar flow. The equations of motion and continuity combine into Reynolds equation for lubricant pressure:[2]

$$\frac{\partial}{\partial x}\left(\frac{h^3}{\mu}\frac{\partial p}{\partial x}\right) + \frac{\partial}{\partial z}\left(\frac{h^3}{\mu}\frac{\partial p}{\partial z}\right) = 6\,U_o\frac{\partial h}{\partial x} + 12\,V_o \qquad (2)$$

When applying Equation 2 to journal bearings $U_o = U_1 + U_2$, where U_1 and U_2 are the tangential velocities of the bearing and shaft, respectively. In thrust bearings only translation is involved and $U_o = U_1 - U_2$, where U_1 is the velocity of the runner and U_2 is the velocity of the bearing. Pressure generation thus depends on the sum of the tangential velocities in journal bearings but on the difference of tangential velocities in thrust bearings. In either case, V_o is the normal velocity of separation of the surfaces.

In order to generate a load-carrying (positive) pressure, the right hand side of Equation 2 must be negative; that is, $U_o\,(\partial h/\partial x) < 0$ and/or $V_o < 0$. The first condition specifies a film that is convergent in space (in the direction of U_o), and the second condition specifies a film that is convergent in time.

Boundary Conditions

In the purely convergent lubricant film shape of a plane thrust bearing (see Figure 8), lubricant pressure relative to ambient is positive everywhere, reducing to zero only at the

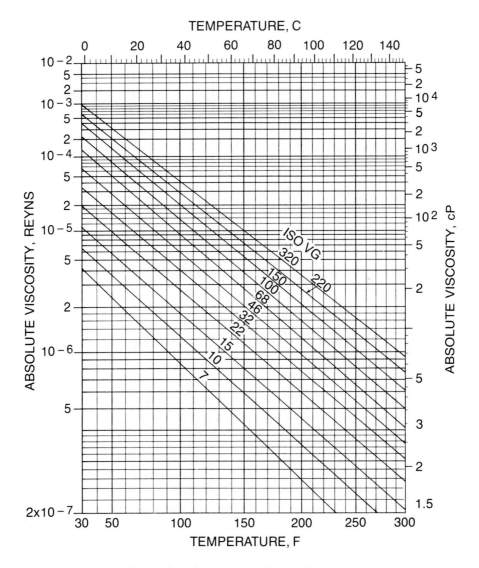

FIGURE 1. Viscosity-temperature curves for typical petroleum oils in ISO viscosity grades.

edges of the film. Consequently, the film is continuous. In the convergent-divergent geometry typical of journal bearings (Figure 22) and crowned thrust pads (Figure 20), theory predicts below-ambient pressures in diverging portions of the film. Two possibilities arise:[2]

1. If ambient pressure is high enough that $p \geqslant p_{cav}$ everywhere in the bearing (here p represents absolute pressure of the lubricant and p_{cav} is the cavitation pressure, usually assumed to be 14.7 psi) a continuous film results (Type 1 boundary condition).
2. If the ambient pressure is not high enough to prevent the absolute pressure from falling below p_{cav}, the film will cavitate (rupture) at the position where the local absolute pressure reaches cavitation pressure (Type 2 boundary condition).

Unless criteria 1 for the Type 1 boundary condition is met, it is customary to assume $p_{cav} = p_{atm}$ and the Type 2 boundary condition with ruptured film.

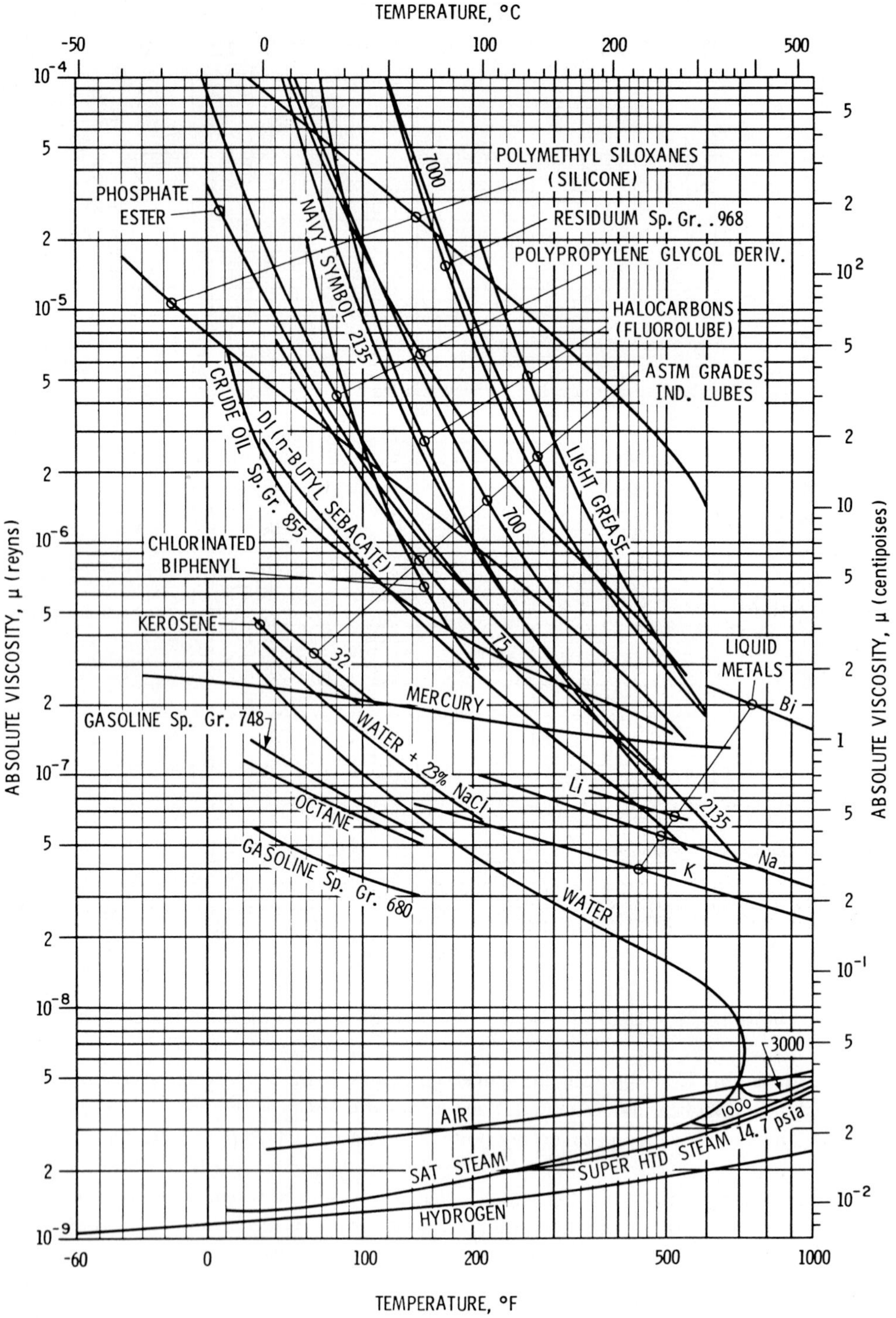

FIGURE 2.   Viscosity-temperature curves for various fluids.

Flow Transition

Two basic modes of flow occur in nature: laminar and turbulent. Flow transition from laminar to turbulent in bearings is preceded by flow instability in one of two basic forms: (1) centifugal instability in flows with curved streamlines, or (2) parallel flow instability characterized by propagating waves in the boundary layer.

Instability between *concentric cylinders* was studied by Taylor.[3] He found that when the Taylor number (Ta = Re^2C/R = $R\omega^2C^3/\nu$) reaches its critical value of 1707.8, laminar flow becomes unstable. The equivalent critical *reduced* Reynolds number is $\sqrt{C/R}$ Re = 41.3. The instability manifests itself in cellular, toroidal vortices that are equally spaced along the axis.

As the Taylor number is increased above its critical value, the axisymmetric Taylor vortices become unstable to produce nonaxisymmetric disturbances, and turbulence eventually makes its appearance.[4] If the Reynolds number reaches 2000 before the Taylor number achieves its critical value, turbulence is introduced rapidly[5] without appearance of a secondary laminar flow.

Eccentricity plays a role in defining critical conditions as covered in the chapter on Hydrodynamic Lubrication (Volume II). A positive radial *temperature gradient* in the clearance space, such as found in journal bearings at the position of minimum film thickness, is also destabilizing,[6] as is heat generation by viscous dissipation.[7] These statements draw support from experimental journal bearing data.[8] The local critical Reynolds number R_h = $R\omega h/\nu$ seems to be in the 400 to 900 range.[2] Accepting R_h = 900 for onset of turbulence, the critical value of global Reynolds number is approximately Re = 900 $\bar{\nu}/(1 - \epsilon)$. Here, $\bar{\nu}$ is the ratio of the lowest value of the kinematic viscosity in the film to its value at the leading edge. Thus, for a fourfold decrease in viscosity and an 0.8 eccentricity ratio, the critical global Reynolds number is Re = 1125. A global value of 1000 has been used in later examples as a criterion for onset of turbulence.

In *thrust bearings,* it was found[9] turbulent transition takes place within the range 580 < Re < 800, where Re = $U_a h_a/\nu$ is calculated on average conditions. Reference 10 reports agreement, but after replacing the average film thickness with the minimum film thickness in Re.

Turbulence

Turbulence is an irregular fluid motion in which properties such as velocity and pressure show random variation with time and with position. Once a relationship is established between the mean flow and Reynolds stresses, averaged equations of motion and continuity can again be combined to yield an equation in the (stochastic) average pressure $\bar{p}$:

$$\frac{\partial}{\partial x}\left(\frac{h^3}{\mu k_x}\frac{\partial \bar{p}}{\partial x}\right) + \frac{\partial}{\partial z}\left(\frac{h^3}{\mu k_z}\frac{\partial \bar{p}}{\partial z}\right) = \frac{U_o}{2}\frac{\partial h}{\partial x} + V_o \qquad (3)$$

Calculations in later sections of this chapter make use of a linearized theory[11] for turbulence functions k_x and k_z. The main contribution of isothermal turbulence to bearing performance is a significant increase in both load-carrying capacity and power loss.

Thermal Effects

If the bearing is large or if loading conditions are severe, pointwise variation of viscosity in the lubricant film is significant. Assuming negligible temperature variation in the axial direction, thermohydrodynamic (THD) journal bearing lubrication is represented by the following equations of pressure and temperature:[12]

$$\frac{\partial}{\partial x}\left(\frac{h^3}{\mu_i k_x}\frac{\partial \bar{p}}{\partial x}\right) + \frac{\partial}{\partial z}\left(\frac{h^3}{\mu_i k_z}\frac{\partial \bar{p}}{\partial z}\right) = U_o\frac{\partial}{\partial x}(hF) + V_o \qquad (4)$$

Table 2
PARAMETERS FOR THD
SAMPLE SOLUTION

C/R	= 0.002	β	= 80°
L/D	= 0.75	k/k_b	= 0.00292
Λ	= 0.0206	T_i	= 50°C
Pr	= 239	Oil:	ISO VG 32

$$U \frac{\partial \overline{T}}{\partial x} + V \frac{\partial \overline{T}}{\partial y} = \alpha \left(\frac{\partial^2 \overline{T}}{\partial x^2} + \frac{\partial^2 \overline{T}}{\partial y^2} \right) - \frac{\partial (\overline{V't'})}{\partial y} + \frac{\nu}{c} \overline{\phi} \tag{5}$$

Here, $\overline{\phi}$ the mean dissipation and the turbulent functions k_x, k_z, and F, as well as the velocity temperature correlation $\overline{V't'}$, depend on the turbulence model used. A minimum list of nondimensional parameters that characterize bearing performance according to THD theory must include

$$S = \frac{\mu_i N}{P} \left(\frac{R}{C} \right)^2$$

$$Re = \frac{R\omega C}{\nu_i}$$

$$\Lambda = \frac{\mu_i \omega}{\rho c T_i} \left(\frac{R}{C} \right)^2$$

$$Pe = \frac{\rho \omega c C^2}{k} \left(\frac{C}{R} \right) \tag{6}$$

Consideration here is limited to varying S, Re, and Pe while keeping all other parameters constant at the values given in Table 2 in an example which covers transition from laminar to turbulent operation.[12,13]

A significant result of THD theory is the strong effect of turbulence on bearing temperature illustrated in Figure 3. Transition to turbulence (occurring at D ~ 25 cm) is beneficial for limiting bearing temperatures, especially at low loads. Coefficient of friction is strongly dependent on bearing specific load in the laminar regime (Figure 4), but this dependence lessens as diameter is increased.

Inertia Effects

Lubricant inertia can have a significant effect on bearing performance if Re (C/R) ≥ 1.[2] While the equations of motion are nonlinear when convective inertia is retained, the problem becomes tractable as pointwise lubricant inertia is replaced by its average value, obtained via integration across the film.[14,15] The averaged equations of motion and continuity combine in a single equation in lubricant pressure. For journal bearings:

$$\frac{\partial}{\partial \psi} \left(\overline{h} \frac{\partial \overline{p}}{\partial \psi} \right) + \left(\frac{D}{L} \right)^2 \frac{\partial}{\partial \psi} \left(h \frac{\partial \overline{p}}{\partial \gamma} \right) = \frac{2\pi k_z}{\overline{h}^2} \left(\frac{\partial \overline{h}}{\partial t} + \frac{1}{2} \frac{\partial \overline{h}}{\partial \psi} \right)$$

$$+ \pi Re \left(\frac{R}{C} \right) \left[2 \frac{\partial^2 \overline{h}}{\partial \psi \partial \overline{t}} + \frac{1}{2} \frac{\partial^2 \overline{h}}{\partial \psi^2} + \frac{\partial^2 \overline{h}}{\partial \overline{t}^2} - \frac{4}{\overline{h}} \left(\frac{\partial \overline{h}}{\partial \overline{t}} + \frac{1}{2} \frac{\partial \overline{h}^2}{\partial \psi} \right) \right] \tag{7}$$

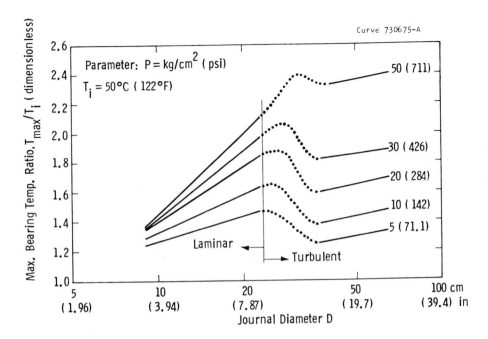

FIGURE 3.   Bearing maximum temperature pattern in the laminar-to-turbulent transition. (From Suganawi, T. and Szeri, A. Z., *Trans. ASME*, 92, 473-481, 1970. With permission.)

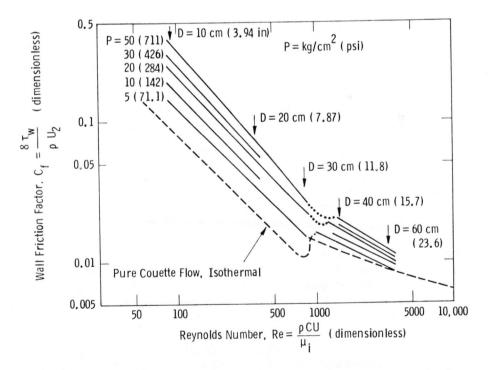

FIGURE 4.   Dependence of bearing friction on unit load P decreases as larger diameter introduces turbulence. (From Suganawi, T. and Szeri, A. Z., *Trans. ASME*, 92, 473-481, 1970. With permission.)

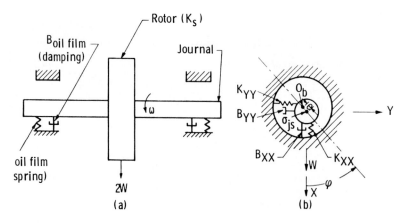

FIGURE 5.   Dynamical elements of rotor-shaft configuration.

Equation 7 was made nondimensional through:

$$h = C\bar{h}$$

$$\bar{t} = t\omega$$

$$x = R\psi$$

$$z = \gamma/2$$

$$p = \mu N(R/C)^2 \bar{p}$$

The entries of Table 3 were calculated from Equation 7 for a 160° partial arc journal bearing (L/D = 1.0).

Dynamic Properties of Lubricant Films

Figure 5 represents an idealized configuration where rotor weight (2W) is supported on two bearings. Under steady load W (Figure 5b), the journal center O_{js} is displaced from the bearing center to the steady operating position shown.

Rotor response to a small excitation, say imbalance, assuming the bearings to be rigid supports, will be as shown in Figure 6. (The same curve applies with rolling contact bearings.) Such rotors cannot be operated at the critical speed and can become "hung" on the critical when attempting to drive through. When hydrodynamic bearings are used, the lubricant film adds another spring (in addition to the shaft spring in bending) and, importantly, considerable damping. Two effects can be noticed in the rotor response curve: (1) critical speed is lowered below that calculated for rigid supports, and (2) vibration amplitude is reduced.

In this example, the excitation is imbalance and occurs at running speed. In practice, exciting frequencies can be different from the shaft speed: magnetic pulls, gear impacts, out-of-round shaft, steam, or aerodynamic forces[16] on turbine or compressor blades, etc. The latter has been known to cause large self-excited vibrations. In addition, lubricant films themselves can originate destructive self-excited vibrations. A classical case is oil whip at slightly less than one-half running speed.[17] Another oil film phenomenon is a self-excited vibration at exactly one-half (or other exact submultiple) of the running speed known as subharmonic resonance.[18]

The rotor-shaft configuration of Figure 5 is reduced to a simple dynamical system of springs and dashpots in Figure 7. A mass W/g (one half the rotor weight) can be imagined to be concentrated at O_{js}, the steady running position of the journal. If some excitation F

Table 3
JOURNAL BEARING PERFORMANCE INCLUDING INERTIA AND TURBULENCE (L/D = 1.0, β = 160°)

ε	Re(C/R)	S	φ°	$\bar{K}_{xx}$ [a]	$\bar{K}_{xy}$	$\bar{K}_{yx}$	$\bar{K}_{yy}$	$\bar{B}_{xx}$	$\bar{B}_{xy}$	$\bar{B}_{yx}$	$\bar{B}_{yy}$	$\bar{D}_{xx}$	$\bar{D}_{xy}$	$\bar{D}_{yx}$	$\bar{D}_{yy}$
0.05	10^{-3}	3.33	83.90	1.52	20.24	−5.50	1.17	40.43	1.91	1.18	11.05	0.00	0.00	0.00	0.00
	1	2.92	88.23	0.93	20.16	−5.43	0.73	40.23	4.32	−0.06	10.97	2.91	0.01	0.01	0.79
	10	1.45	90.70	−2.45	20.41	−5.58	0.07	40.61	16.13	−2.81	11.33	14.47	0.23	0.23	3.95
	100	0.260	92.24	−5.42	20.76	−5.72	−0.91	41.42	27.25	−6.23	11.55	26.00	0.31	0.31	7.09
	1000	0.031	92.80	−6.62	20.94	−5.79	−1.26	41.81	31.88	−7.53	11.67	30.68	0.36	0.36	8.37
0.1	10^{-3}	1.64	78.03	1.54	10.47	−2.82	1.18	20.86	2.00	1.21	5.74	0.00	0.00	0.00	0.00
	1	1.438	80.54	1.13	10.48	−2.81	1.04	20.81	3.29	0.76	5.76	1.45	0.04	0.03	0.40
	10	0.710	90.17	−0.50	10.44	−2.77	0.48	20.56	8.54	−1.01	5.77	7.09	0.14	0.14	1.94
	100	0.126	91.54	−2.09	10.68	−2.92	0.02	21.13	13.91	−2.55	6.05	12.62	0.31	0.31	3.45
	1000	0.015	91.83	−2.77	10.81	−3.03	−0.07	21.30	16.29	−3.00	6.33	14.85	0.49	0.49	4.08
0.3	10^{-3}	0.459	60.81	1.80	4.58	−1.15	1.25	8.92	2.30	1.46	2.61	0.00	0.00	0.00	0.00
	1	0.406	62.81	1.63	4.58	−1.12	1.19	8.82	2.61	1.26	2.60	0.46	0.03	0.03	0.13
	10	0.207	71.00	0.92	4.56	−1.09	0.95	8.55	3.99	0.55	2.64	2.27	0.17	0.17	0.63
	100	0.036	79.93	0.23	4.41	−1.08	0.72	8.25	5.29	−0.05	2.64	3.87	0.29	0.29	1.08
	1000	0.0042	82.59	0.00	4.43	−1.09	0.66	8.21	5.84	−0.22	2.68	4.45	0.34	0.34	1.25
0.5	10^{-3}	0.190	51.29	2.37	3.99	−0.82	1.36	7.38	2.81	1.76	2.19	0.00	0.00	0.00	0.00
	1	0.172	52.16	2.28	3.97	−0.80	1.29	7.38	2.98	1.70	2.26	0.23	0.02	0.02	0.06
	10	0.093	58.09	1.68	3.90	−0.75	1.18	6.96	3.43	1.20	2.31	1.22	0.15	0.15	0.34
	100	0.017	65.39	1.07	3.72	−0.73	0.99	6.54	3.86	0.66	2.23	2.18	0.28	0.28	0.63
	1000	0.0020	67.80	0.90	3.70	−0.74	0.95	6.49	4.11	0.52	2.25	2.52	0.33	0.33	0.73
0.7	10^{-3}	0.070	42.40	4.36	4.41	−0.32	1.22	7.50	2.98	1.67	1.53	0.00	0.00	0.00	0.00
	1	0.065	43.29	4.15	4.41	−0.30	1.25	7.46	3.08	1.70	1.62	0.10	0.01	0.01	0.02
	10	0.039	47.42	3.33	4.25	−0.28	1.26	6.98	3.19	1.49	1.77	0.63	0.06	0.06	0.14
	100	0.008	53.59	2.29	3.94	−0.37	1.21	6.44	3.38	1.20	1.95	1.32	0.20	0.20	0.34
	1000	0.00099	55.80	2.01	3.87	−0.40	1.20	6.33	3.51	1.10	2.02	1.57	0.26	0.26	0.43

Table 3 (continued)
JOURNAL BEARING PERFORMANCE INCLUDING INERTIA AND TURBULENCE (L/D = 1.0, β = 160°)

ε	Re(C/R)	S	φ°	$\bar{K}_{xx}$[a]	$\bar{K}_{xy}$	$\bar{K}_{yx}$	$\bar{K}_{yy}$	$\bar{B}_{xx}$	$\bar{B}_{xy}$	$\bar{B}_{yx}$	$\bar{B}_{yy}$	$\bar{D}_{xx}$	$\bar{D}_{xy}$	$\bar{D}_{yx}$	$\bar{D}_{yy}$
0.9	10^{-3}	0.012	31.29	13.18	8.19	0.52	1.42	10.05	4.06	1.88	1.14	0.00	0.00	0.00	0.00
	1	0.011	31.61	12.95	8.11	0.57	1.43	10.69	3.96	1.82	1.13	0.02	0.00	0.00	0.00
	10	0.0025	33.60	11.11	7.72	0.58	1.54	10.35	4.11	2.01	1.37	0.19	0.00	0.00	0.03
	100	0.0026	38.62	7.58	6.41	0.64	1.61	8.36	3.48	1.66	1.42	0.59	0.04	0.04	0.09
	1000	0.00035	40.90	6.20	5.80	0.46	1.55	7.87	3.37	1.50	1.50	0.80	0.09	0.09	0.14

$\bar{K} = (C/W)K, B = (C/W)\omega\bar{B}, D = (C/W)\omega^2\bar{D}$

[a] $\bar{K} = (C/W)K, B = (C/W)\omega B, D = (C/W)\omega^2\bar{D}$

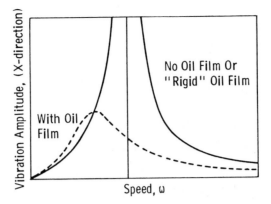

FIGURE 6.   Effect of oil film on response.

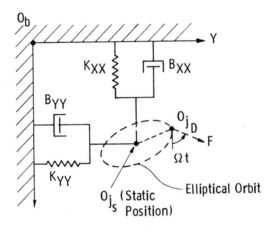

FIGURE 7.   Representation of oil film as a simple dynamical system of springs and dampers (cross-film springs K_{xy}, K_{yx} and dampers B_{xy}, B_{yx} are not shown).

occurs at frequency Ω the mass center will respond by orbiting about O_{j_s}, its instantaneous orbital position denoted by O_{j_D}. This portrayal implies that dynamic displacement (O_{j_s} − O_{j_D}) will be very small in relation to the steady running displacement. When dynamic displacement is ''large'', lubricant film behavior is highly nonlinear and the system of Figure 7 becomes very approximate.

Displacements (Figure 7) in the (x,y) directions caused by excitation F are opposed by the dynamic oil film force whose components are

$$-F_x = K_{xx}x + K_{xy}y + B_{xx}\dot{x} + B_{xy}\dot{y}$$

$$-F_y = K_{yy}y + K_{yx}x + B_{yy}\dot{y} + B_{yx}\dot{x} \qquad (8)$$

The four spring coefficients (K_{xx}, K_{xy}, K_{yx}, K_{yy}) and the four damping coefficients (B_{xx},....B_{yy}) enable linear representation of the oil film force. For example, if a vibrating system consisted solely of a mass m running on a hydrodynamic oil film and excited by a force F at frequency Ω, the system equations of motion become:

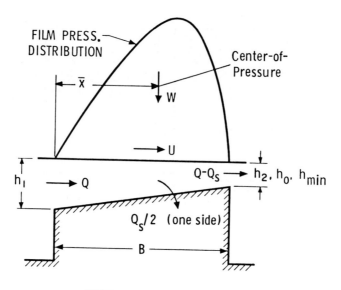

FIGURE 8.   Fixed-pad slider bearing.

$$m\ddot{x} + B_{xx}\dot{x} + B_{xy}\dot{y} + K_{xx}x + K_{xy}y = F \cos \Omega t$$

$$m\ddot{y} + B_{yy}\dot{y} + B_{yx}\dot{x} + K_{yy}y + K_{yx}x = F \sin \Omega t \qquad (9)$$

Steady state response (x,y, phase angle) to excitation F can then be obtained by solving the above linear equations. Equally important, existence of any self-excited vibrations can be investigated by performing a conventional stability analysis[18] on the homogeneous equations obtained by setting the right hand side of Equation 9 equal to zero. When the oil film is only one element in a more complicated linear dynamical chain, it can be incorporated in the basic equations of motion for the vibrating system, as shown. All eight oil film coefficients are required in order to make accurate dynamical analyses of rotor-shaft configurations.

In some instances, lubricant inertia should also be accounted for in linear representation of the oil film.[15] This effect becomes important when Re (C/R) $\geq$1, and oil film force Equation 8 should be amended as follows to include acceleration (inertia) coefficients:

$$-F_x = K_{xx}x + K_{xy}y + B_{xx}\dot{x} + B_{xy}\dot{y} + D_{xx}\ddot{x} + D_{xy}\ddot{y}$$

$$-F_y = K_{yy}y + K_{yx}x + B_{yy}\dot{y} + B_{yx}\dot{x} + D_{yy}\ddot{y} + D_{yx}\ddot{x} \qquad (10)$$

Table 3 gives approximate D inertia coefficients for a 160° partial-arc bearing.

THRUST BEARINGS

Fixed-Type Thrust Bearings

The fixed-pad slider bearing in Figure 8 is the most basic configuration. In practice, this type is commonly constructed as shown in Figure 9a with a flat area following a machined taper to support the sliding surface when it is at rest. Basic design charts for these configurations can be found in Reference 19.

If the pads are arranged in an annular configuration with radial oil distribution grooves, a complete thrust bearing (Figure 10) is achieved. Approximate performance calculations of this bearing can be made by relating the rectangular slider bearing (width B, length L)

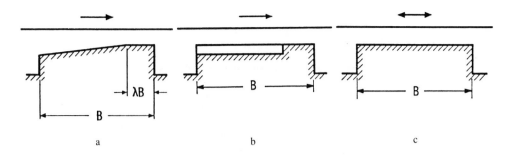

FIGURE 9. Practical variations of fixed-pad bearing; (a) taper-flat, $\lambda = 0.2$, (b) step pad, and (c) parallel surface.

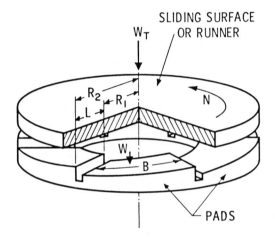

FIGURE 10. Fixed-pad thrust bearing.

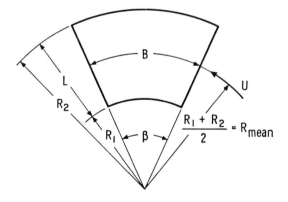

FIGURE 11. Sector for fixed-land (or pivoted-pad) thrust bearing.

to the sector configuration of Figure 11. Total load capacity W_T (Figure 10) is the sum of the individual load capacities, W.

The step-pad bearing, a shrouded step-pad bearing with side walls on the lower entrance step (Figure 9b), an initial pocket followed by a pad, and the parallel surface bearing (Figure 9c) represent other types of fixed pad bearings.[20,21] The parallel surface type is used primarily for positioning since its load capacity is limited (P = W/BL = 70 to 140 kPa, 10 to 20 psi).[22] The spiral groove bearing of Figure 12 is widely used in gas bearing applications.[23]

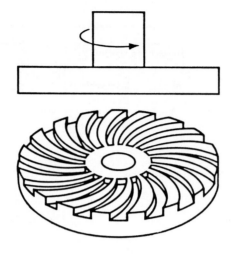

FIGURE 12.   Spiral groove bearing.

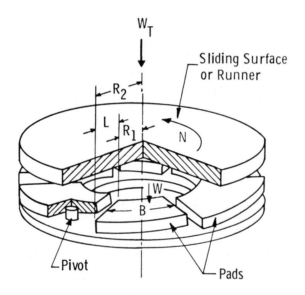

FIGURE 13.   Pivoted-pad thrust bearing.

Pivoted-Pad Thrust Bearings

This type (Figure 13) employs a supporting pivot at the center of film pressure ($\bar{x}$ in Figure 8) developed on the surface of a slider bearing. While performance is theoretically identical to a fixed-pad bearing designed with the same slope, the pivoted type has the advantages of (a) self-aligning capability, (b) automatically adjusting pad inclination to optimally match the needs of varying speed and load, and (c) operation in either direction of rotation. Theoretically, the pivoted pad can be optimized for all speeds and loads by judicious pivot positioning, whereas the fixed-pad bearing can be designed for optimum performance only for one operating condition. Although pivoted pad bearings involve somewhat greater complexity, standard designs are readily available for medium to large size machines.

When the thrust bearing is divided into a set of pie-shaped segments, circumferential length of each sector at its mean radium (R_{mean}) is commonly set approximately equal to the

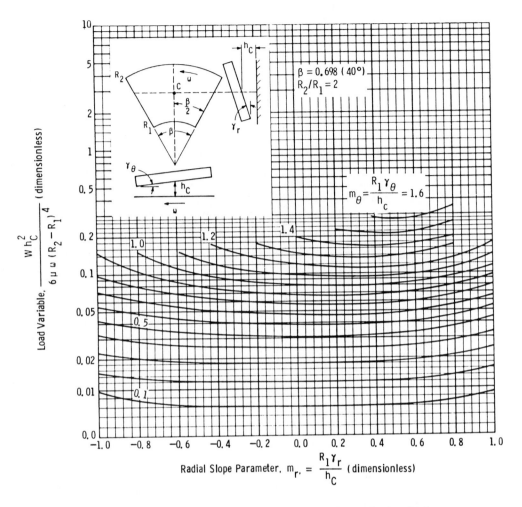

FIGURE 14.   Load capacity chart for fixed- or tilting pad sector (laminar flow).

radial pad width to give approximately "square" pads (L = B, Figure 11). This in turn sets the number of pad sectors to be employed.

Performance Charts

Thrust bearing performance charts (Figures 14 to 19) are conveniently entered on a trial basis with an assumed tangential slope parameter m_θ and assumed radial slope parameter such as $m_r = 0.0$. Load capacity, minimum film thickness, power loss, flow, and pivot location (if a pivoted-pad type) are then determined and the procedure repeated on a trial basis to find an optimum design. Figures 14 to 17 are applicable to fixed-sector bearings (Figure 10) as well as the pivoted type. Figures 18 and 19 provide pivot locations for tilting pad sectors.

These thrust bearing charts are for a ratio of outside radius to inside radius of 2 and a sector angular length of 40°. While this angle corresponds to seven sectors for a full thrust bearing, the results should generally give a preliminary indication of performance for other sector geometries with the same surface area and mean radius. Although English engineering units are used in the examples which follow, the charts are generally dimensionless for ready use of SI units.

Example: calculate thrust pad sector performance when given the following:

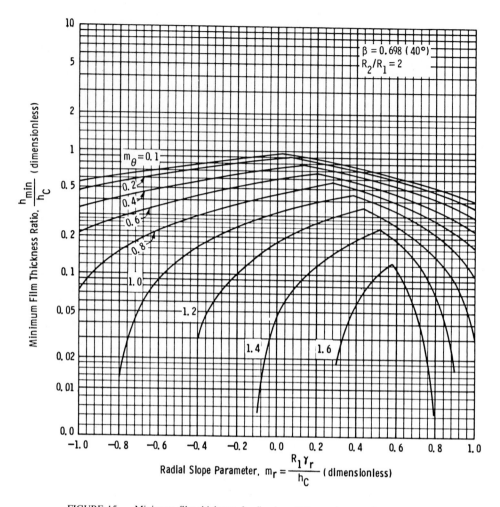

FIGURE 15.   Minimum film thickness for fixed- or tilting pad sector (laminar flow).

β = 40° (0.70 rad)

R_2 = 5.50 in.

R_1 = 2.75 in.

h_c = 0.002 in. (at pad center)

N = 50 r/sec ($\omega = 2\pi \cdot 50 = 314$ rad/sec)

γ_r = 0 (no radial tilt)

γ_θ = 5.82 × 10^{-4} rad (0.0333°)

μ = 2 × 10^{-6} lb sec/in.^{2} (ISO VG 32 oil at T_s = 135 F, Figure 1)

Calculating first

$$m_\theta = (R_1/h_c)\gamma_\theta = (2.75/0.002)\,5.82 \times 10^{-4} = 0.80$$

Enter Figures 14 to 17 with $m_\theta = 0.80$ and $m_r = 0.0$:

$$Wh_c^2/[6\,\mu\,\omega\,(R_2 - R_1)^4] = 0.058$$

Load capacity per sector is then calculated as

$$W = 0.058 \times 6 \times 2 \times 10^{-6} \times 314(5.50 - 2.75)^4/(0.002)^2 = 3125 \text{ lb}$$

$$h_{min}/h_c = 0.45;\ h_{min} = 0.45 \times 0.002 = 0.00090 \text{ in.}$$

$$Hh_c/[\mu\omega^2(R_2 - R_1)^4] = 3.04$$

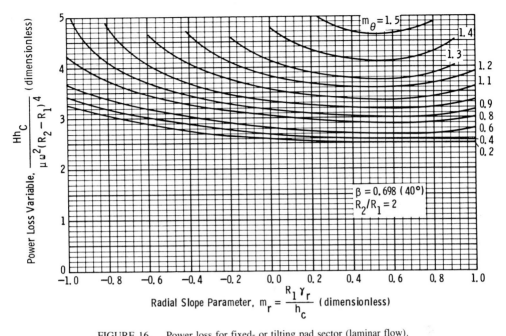

FIGURE 16.   Power loss for fixed- or tilting pad sector (laminar flow).

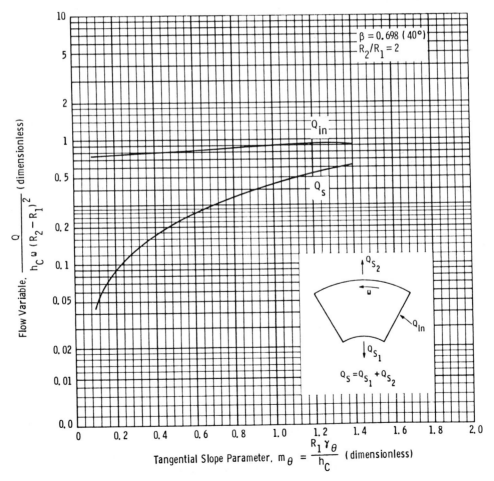

FIGURE 17.   Flow chart for fixed- or tilting pad sector (laminar flow).

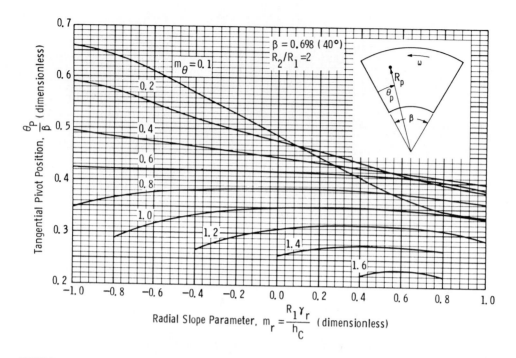

FIGURE 18.   Tangential location of center-of-pressure (fixed-pad sector), or pivot position (tilting pad sector) (laminar flow).

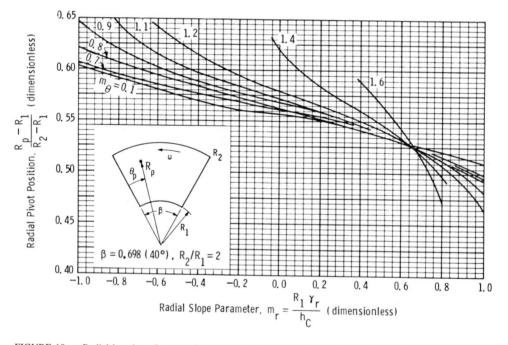

FIGURE 19.   Radial location of center-of-pressure (fixed-pad sector), or pivot position (tilting pad sector) (laminar flow).

From which the power loss per sector is

$$H = 3.04 \times 2 \times 10^{-6} \times (314)^2 (2.75)^4 / 0.002 = 1.70 \times 10^4 \text{ lb in./sec (2.58 hp)}$$

From Figure 17, sector flow rates are

$$Q_{in}/[h_c \omega (R_2 - R_1)^2] = 0.86$$

$$Q_s/[h_c \omega (R_2 - R_1)^2] = 0.35$$

$$Q_{in} = 4.08 \text{ in.}^3/\text{sec} (1.06 \text{ gal/min})$$

$$Q_s = 1.66 \text{ in.}^3/\text{sec} (0.43 \text{ gal/min})$$

Temperature rise can then be calculated from Equation 1 with $\rho c = 112$ as typical for petroleum oils.

$$\Delta T = 1.71 \times 10^4/[112(4.08 - 1.66/2)] = 47.0 \text{ F}$$

This temperature rise implies an inlet temperature T_i of

$$T_i = T_s - \Delta T/2 = 135 - 47.0/2 = 111.5 \text{ F}$$

If this did not match the oil inlet temperature, a new effective oil viscosity would be assumed and the steps repeated.

From Figures 18 and 19, the pivot must be placed at $\theta_p/\beta = 0.39$ and $(r_p - R_1)/(R_2 - R_1) = 0.56$ to achieve the above calculated performance. This performance is also achieved with a fixed-type bearing by machining slope angles $\gamma_\theta = 0.0333°$ and $\gamma_r = 0$.

Centrally Pivoted Pads

If the sector pad surface is truly flat, then basic theory (isothermal) requires that the tangential pivot location be offset toward the pad trailing edge for appreciable load capacity. For example, the optimum pivot location for the basic slider bearing (Figure 8) is $\bar{x} = 0.58B$ (for a "square", $L = B$ pad). However, flat pads are frequently constructed with a central pivot ($\bar{x} = 0.5B$) for design simplicity and to apply them for either direction of rotation. The following phenomena commonly enable nearly optimum load capacity for centrally pivoted flat pads when operating with lubricating oil or other relatively high viscosity fluid: (1) viscosity varies as the lubricant passes through the film to alter the pressure distribution, (2) film pressure elastically deforms (crowns) the pad, and (3) film heating thermally deforms the pad.[24,25]

In low-viscosity fluid applications involving water, liquid metals, and gases, it is usually necessary to deliberately crown the pad spherically or cylindrically by precise manufacturing techniques to ensure high-load capacity.[25] Figure 20 indicates that the optimum spherical crown is very small, about 0.6 times the minimum film thickness, h_o, and will result in a load capacity nearly equivalent to that of a flat pad with its pivot optimally placed.

Cylindrical and spherically crowned pads can be designed by employing Tables 4 and 5 and Figure 20. These data should be used in conjunction with the following performance equations:

$$W = C_{W,L} \, C_{W,T} \mu \, U \, B^2 L/h_p^2 \qquad (11)$$

$$h_o = C_{h_o,L} \, C_{h_o,T} \, h_p \qquad (12)$$

$$H = C_{H,L} \, C_{H,T} \mu \, U^2 B \, L/h_p \qquad (13)$$

$$Q_{in} = C_{Q_{in},L} \, C_{Q_{in},T} \, U \, h_p L \qquad (14)$$

$$Q_s = C_{Q_s,L} \, C_{Q_s,T} \, Q_{in} \qquad (15)$$

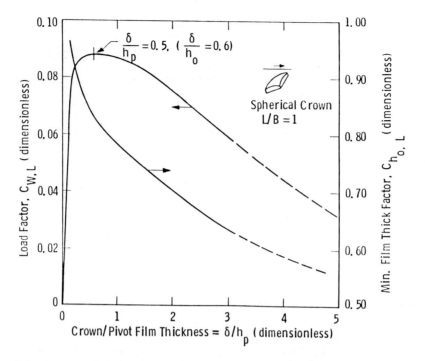

FIGURE 20.   Design chart for spherically crowned rectangular pad with central pivot ($\overline{x}$ = 0.5 B). (From Raimondi, A.A., *ASLE Trans.*, 3, 265, 1960. With permission.)

Table 4
LAMINAR FLOW PERFORMANCE FACTORS FOR RECTANGULAR TILTING PAD BEARINGS

Performance factor	Laminar flow performance value					
	Flat pad ($\overline{x}/B$ = 0.58)			Cylindrically crowned pad ($\overline{x}/B$ = 0.50)		
Aspect ratio (L/B) =	0.5	1.0	1.5	0.5	1.0	1.5
$C_{W,L}$	0.0456	0.136	0.202	0.071	0.132	0.158
$C_{h_o,L}$	0.761	0.708	0.684	0.70	0.77	0.75
$C_{H,L}$	0.992	1.03	1.06	1.00	1.04	1.05
$C_{Q_{in},L}$	0.624	0.597	0.563	1.10	0.735	0.729
$C_{Q_s,L}$	0.344	0.308	0.256	0.53	0.40	0.32

From Malinowsky, S. B., *Mach. Design*, 45, 100, 1973. With permission.

To correct the results for operation at high speed, the turbulence correction factors given in Figure 21[26] should be used, as illustrated in the following example.

Example: calculate the load capacity, minimum film thickness, and cylindrical crown δ required when given the following.

$$\mu = 4.5 \times 10^{-8} \text{ lb sec/in.}^2 \text{ (200 F water, Figure 2)}$$
$$\rho = 9.07 \times 10^{-5} \text{ lb sec}^2/\text{in.}^4$$
$$\nu = \mu/\rho = 4.96 \times 10^{-4} \text{ in.}^2/\text{sec}$$
$$U = 1296 \text{ in./sec}$$
$$B = L = 2.75 \text{ in.}$$
$$\overline{x}/B = 0.50$$
$$h_p = 0.0016 \text{ in.}$$

Table 5
CROWN-PIVOT FILM
THICKNESS RATION (δ/h_p)
FOR OPTIMUM LOAD
CAPACITY, CYLINDRICALLY
CROWNED PADS

	Aspect ratio (L/B)		
$Re = Uh_p/\nu$	**0.5**	**1**	**1.5**
Laminar	1.0	0.6	0.73
2,000	1.6	1.1	1.0
5,000	1.6	1.1	1.0
10,000	1.8	1.2	1.1
30,000	2.0	1.2	1.2

From Malinowsky, S. B., *Mach. Design*, 45, 100, 1973. With permission.

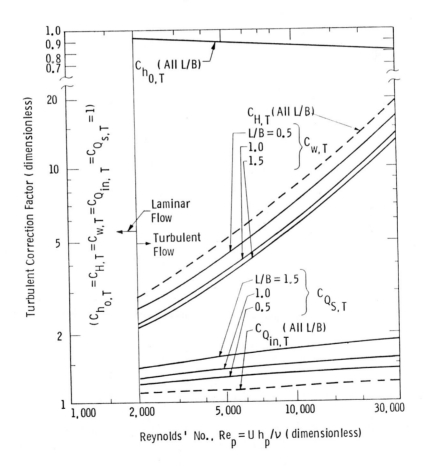

FIGURE 21. Turbulence correction factors for pivoted thrust pads. (From Malinowski, S.B., *Mach. Design*, 45, 100, 1973. With permission.)

Laminar flow solution: from Table 4, load factor $C_{W,L} = 0.132$; from Equation 11, using $C_{W,T} = 1$ for laminar flow:

$$W = 0.132 \times 1 \times 4.5 \times 10^{-8} \times 1296 \times (2.75)^2 \times 2.75/(0.0016)^2 = 62.5 \text{ lb}$$

for load capacity ($P = W/BL = 8.26$ psi). Similarly from Equation 13, using C_{h_o}, $L = 0.77$:

$$h_o = 0.77 \times 1.0 \times 0.0016 = 0.0012 \text{ in.}$$

Power loss, H, and flow requirements (Q_{in}, Q_s) can be calculated in a similar manner from Table 4 and Equations 14, 15, and 16 to yield

$$
\begin{aligned}
H &= 372 \text{ in. lb/sec } (0.056 \text{ hp}) \\
Q_{in} &= 4.19 \text{ in.}^3/\text{sec } (1.09 \text{ gpm}) \\
Q_s &= 1.68 \text{ in.}^3/\text{sec } (0.44 \text{ gpm})
\end{aligned}
$$

The cylindrical crown required to attain maximum load capacity is given in Table 5 as

$$\delta/h_p = 0.6$$

$$\delta = 0.6 \times 0.016 = 0.00096 \text{ in.}$$

For *turbulent flow* correction, the Reynolds number calculation gives

$$Re_p = Uh_p/\nu = 1296 \times 0.0016/4.96 \times 10^{-4} = 4180$$

As Re_p is above 2000, entering Figure 21 gives as turbulent correction factors

$$C_{W,T} = 3.2$$

$$C_{h_o,T} = 0.9$$

$$C_{H,T} = 4.3$$

$$C_{Q_{in},T} = 1.1$$

$$C_{Q_s,T} = 1.4$$

Using Equations 12 to 16 gives turbulent performance as:

$$
\begin{aligned}
W &= 62.5 \times 3.2 = 200 \text{ lb. } (P = 26 \text{ psi}) \\
h_o &= 0.0012 \times 0.9 = 0.0011 \text{ in.} \\
H &= 372 \times 4.3 = 1600 \text{ in. lb/sec } (0.24 \text{ hp}) \\
Q_{in} &= 4.19 \times 1.1 = 4.61 \text{ in.}^3/\text{sec } (1.20 \text{ gpm}) \\
Q_s &= 1.68 \times 1.4 = 2.35 \text{ in.}^3/\text{sec } (0.61 \text{ gpm})
\end{aligned}
$$

Interpolating for $Re_p = 4180$ in Table 5 indicates that the cylindrical crown should be changed so that $\delta/h_p = 1.1$. The required optimum cylindrical crown then becomes

$$\delta/h_p = 1.1$$

$$\delta = 1.1 \times 0.0016 = 0.0018 \text{ in.}$$

Flexible-Type Thrust Bearings

While a bearing pad is not as free to pivot with spring, rubber, or other elastic support

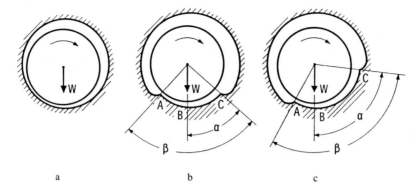

FIGURE 22.   Fixed-type journal bearings: (a) full 360° bearing, (b) centrally loaded partial bearing, and (c) offset loaded partial bearing.

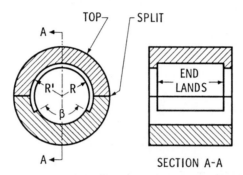

FIGURE 23.   Partial bearing with relief and end lands in top.

as in a pivoted-pad bearing, it can partially respond to changes in speed, load, viscosity, and temperature. Advantages include (1) better load distribution than with fixed pads, and (2) less space requirement and simpler design than pivoted-pad bearings.

Flexible-type bearings are difficult to analyze since the elastic and thermal deflections in the system must be considered.[27] Some data relative to film thickness and power loss are given in Reference 28.

FIXED-TYPE JOURNAL BEARINGS

The full and the partial bearings are the two basic configurations of fixed-type hydrodynamic journal bearings (Figure 22). Most other designs found in practice (Figure 23, for example) can be considered as variations of these two.

The active (load-carrying) bearing arc extends entirely around the journal for the full journal bearing; in the partial bearing, the active arc only partially surrounds the journal. The partial bearing is "centrally loaded" if the load direction bisects the active arc, and "offset" or eccentrically loaded" if it does not. Design charts are presented here for the most commonly used bearings: the full journal bearing and the centrally loaded partial bearing.

Centrally Loaded Partial Bearings

Figures 24 to 30 are design charts for a 160° centrally loaded partial bearing with a length-to-diameter (L/D) ratio equal to 1 (see Reference 29 for offset loading). These were derived from numerical solutions of basic Equation 2 with Type 2 boundary condition (film rupture

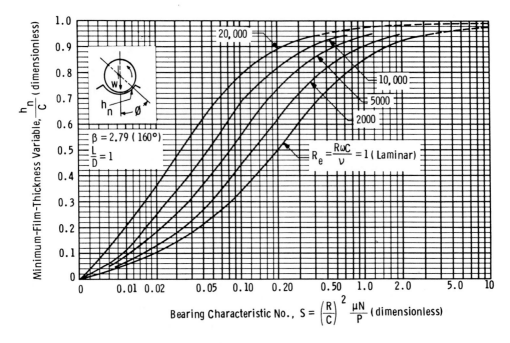

FIGURE 24. Minimum film thickness for partial arc journal bearing.

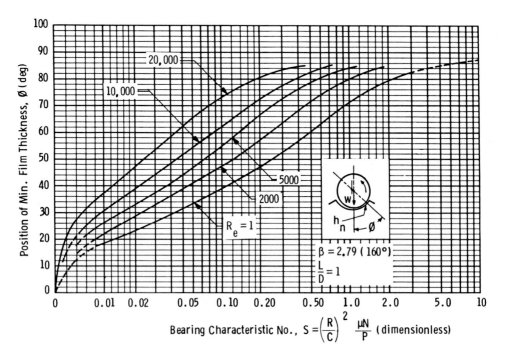

FIGURE 25. Position of minimum film thickness in partial arc journal bearing.

in the divergent region). Additional but more approximate data based on a short bearing type solution in Table 3 contain acceleration coefficients D_{xx}, D_{xy}, D_{yx}, D_{yy}. Use of the charts is illustrated by the following examples.

Example: calculate the performance of a centrally loaded partial arc journal bearing given the following:

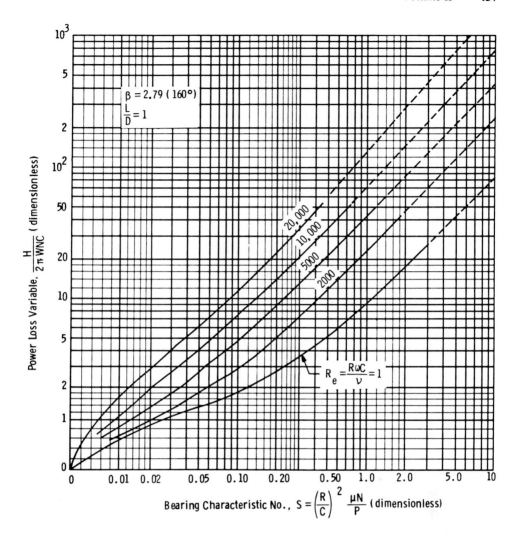

FIGURE 26.   Power loss for partial arc journal bearing.

β = 2.79 rad (160°)   N = 40 r/sec (ω = $2\pi N$ = 251 rad/sec)
D = L = 20 in. (L/D = 1) μ = 1.8 × 10^{-6} lb sec/in.^{2} (ISO VG 32
C = 0.020 in.   oil at 140 F, Figure 1)
W = 80,000 lb (P = W/DL = 200 psi) ρ = 7.77 × 10^{-5} lb sec^2/in.^{4}
                                            ν = μ/ρ = 2.32 × 10^{-2} in.^{2}/sec

Calculating the Reynolds and bearing characteristic numbers:

$$Re = R\omega C/\nu = 10 \times 251 \times 0.020/2.32 \times 10^{-2} = 2160 \text{ (turbulent)}$$

$$S = \mu N (R/C)^2/P = 1.8 \times 10^{-6} \times 40 \times (10/0.020)^2/200 = 0.090$$

Minimum film thickness — Entering Figure 24 with S = 0.090 and Re = 2160:

$$h_n/C = 0.44$$
$$h_n = 0.44 \times .020 = 0.0088 \text{ in.}$$

Position of minimum film thickness — Similarly, from Figure 25, attitude angle ϕ = 48°.

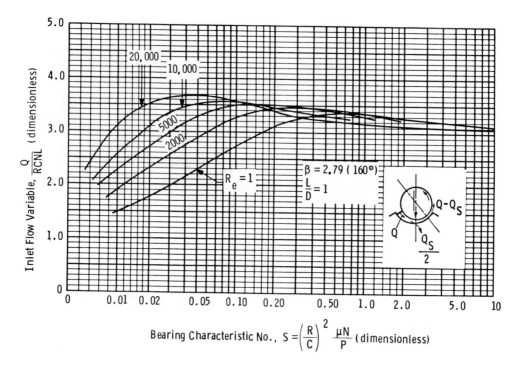

FIGURE 27.   Inlet flow for partial arc journal bearing.

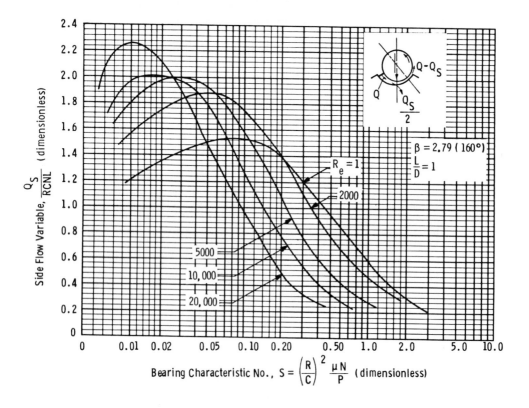

FIGURE 28.   Side flow for partial arc journal bearing.

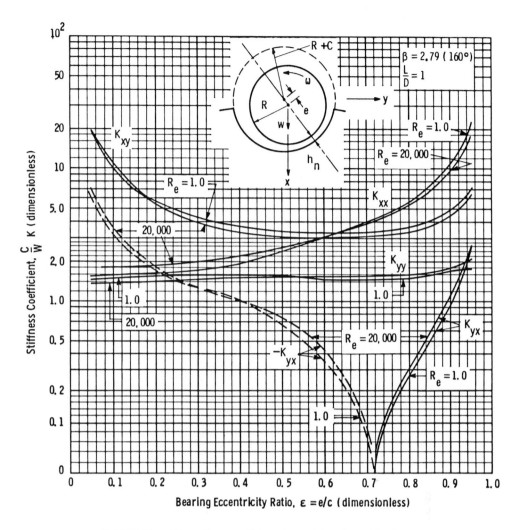

FIGURE 29.   Film stiffness coefficients for partial arc journal bearing.

Power loss — Figure 26 gives H/(2πWNC) = 2.8, from which

$$H = 2.8 \times 2\pi \times 80,000 \times 40 \times 0.020 = 1.13 \times 10^6 \text{ lb in./sec (171 hp)}$$

Lubricant flow — Figure 27 gives Q/(RCNL) = 3.3, from which Q = 528 in.^{3}/sec (137 gpm). This inlet flow is drawn into the leading edge of the arc by journal rotation. Of this amount, Q_s which escapes laterally from both sides of the bearing arc is found from Figure 28 which gives Q_s/(RCNL) = 1.8; then Q_s = 288 in.^{3}/sec (74.8 gpm). If this bearing is operating submerged in fluid, ample lubricant will always be available at the leading edge of the bearing arc. If, however, ample lubricant is to be supplied to the leading edge by external means, the feed rate must be at least (1) equal Q if there is no carryover by the shaft back to the leading edge or (2) equal Q_s with carryover. Insufficient external supply and starved operation does not imply that a bearing will necessarily fail, but its performance will be altered.[30] On the other hand, if lubricant is supplied at a pressure greater than ambient (for example, into the top in Figure 23), part of the lubricant may by-pass the active arc, considerably increase total lubricant flow, and again alter the performance characteristics. Certain types of grooving will also affect flow and performance.

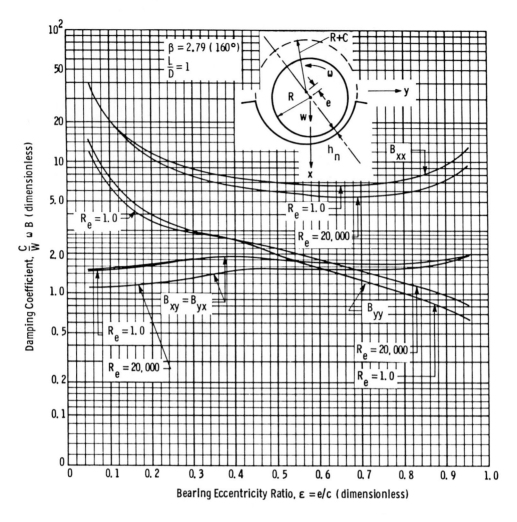

FIGURE 30.   Film damping coefficients for partial arc journal bearing.

Temperature rise — If none of the lubricant leaving the trailing edge is carried over to the leading edge by the journal, Equation 1 gives for the above example

$$\Delta T = 1.13 \times 10^6 / (112 \, (528 - 288/2))$$

from which the temperature rise $\Delta T = 26.2F$. This corresponds to an inlet temperature T_i to the bearing arc of

$$T_i = T_s - \Delta T/2 = 140 - 26.2/2 = 126.9 \, F$$

Dynamic coefficients — Stiffness and damping coefficients are obtained from Figures 29 and 30 with the following eccentricity ratio:

$$\epsilon = 1 - h_n/C = 1 - 0.44 = 0.56$$

Figure 29 gives $(C/W)K_{xx} = 2.8$, from which $K_{xx} = 2.8 \times 80,000/0.020 = 11.2 \times 10^6$ lb/in. Similarly, $(C/W)K_{yy} = 1.55$, $(C/W)K_{xy} = 3.0$, and $(C/W)K_{yx} = -0.50$ which yield,

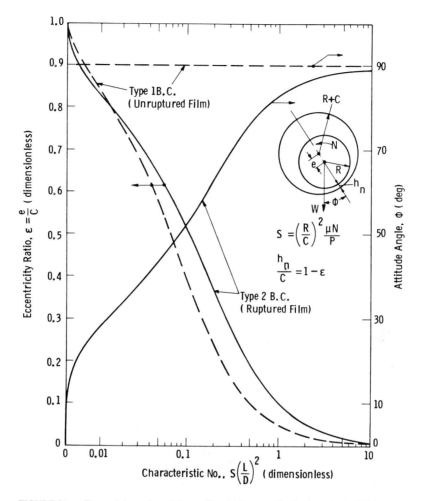

FIGURE 31. Eccentricity ratio, minimum film thickness and attitude angle for full-journal bearings (laminar flow, Re < 1000).

respectively, $K_{yy} = 6.2 \times 10^6$, $K_{xy} = 12.0 \times 10^6$, and $K_{yx} = -2.00 \times 10^6$ lb/in. Figure 30 gives (C/W) $\omega B_{xx} = 6.8$, from which $B_{xx} = 6.8 \times 80,000/0.020 \times (1/251) = 0.11 \times 10^6$ lb sec/in. In an identical manner, we find $B_{yy} = 0.029 \times 10^6$ and $B_{xy} = B_{yx} = 0.028 \times 10^6$ lb sec/in.

Full Journal Bearing

Performance of the full journal bearing can be estimated from the design charts presented in Figures 31 to 34. These data are based on a short journal bearing approximate solution[31] and are most accurate for small L/D ratios (L/D ≤1). Curves are presented for both Type 1 (continuous film) and Type 2 (ruptured film) boundary conditions (BC). While the Type 1 BC is especially useful for pump bearings completely submerged in a high-ambient pressure, Type 2 BC is commonly found in most other applications.

Use of Figures 31 to 34 is similar to that shown by the examples for a partial arc bearing. Flow rates Q and Q_s (Type 2 BC only) can be calculated from the following equations with eccentricity ratio ϵ taken from Figure 31:

$$Q = \pi R N L C (1 + \epsilon)$$

$$Q_s = 2 \pi R N L C \epsilon$$

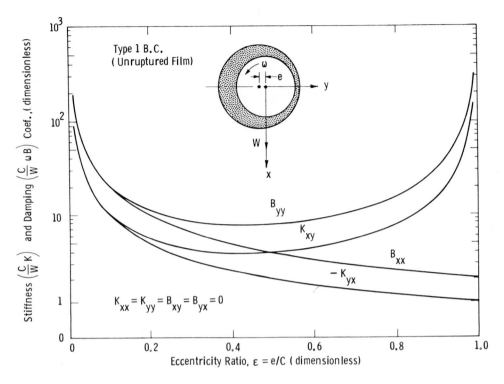

FIGURE 32.   Unruptured film stiffness and damping for full-journal bearing (laminar flow, Re < 1000).

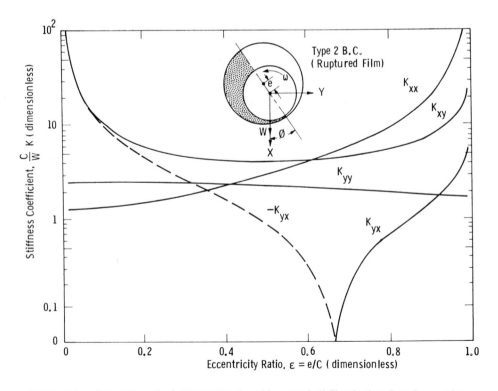

FIGURE 33.   Film stiffness for full journal bearing with ruptured oil film (laminar flow, Re < 1000).

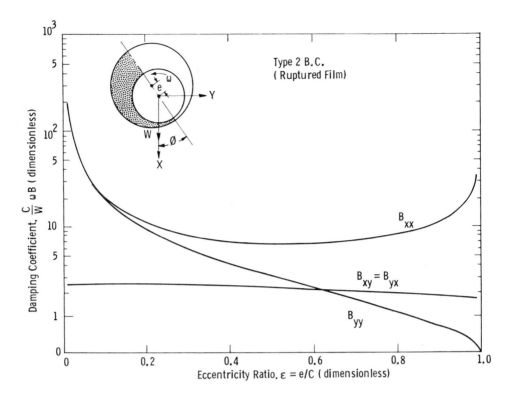

FIGURE 34.   Film damping for full journal bearing with ruptured oil film (laminar flow, Re < 1000).

Stability of Full Journal Bearings

The stability characteristics of a rotor carried on full journal bearings (ruptured film, Type 2 BC) supported on a rigid foundation can be estimated from Figure 35. The full journal bearing (unruptured continuous film, Type 1 BC) is inherently unstable over its entire operating range. External damping in the system must be present to yield stable operation. More stable fixed-type bearing and the influence of bearing support flexibility can be found in References 32 and 33.

Example: determine the oil film-rotor stability given the following for a horizontal rotor:

D = 2R = 5 in.	N = 90 r/sec (5400 rpm)
L = 2.5 in.	K_s = 5 × 10⁶ lb/in. (rotor stiffness)
C = 0.005 in.	Rotor weight = 5000 lb (M = 5000/386 = 13.0 lb sec²/in.)
	μ = 2 × 10⁻⁶ lb sec/in.²

From Figure 35: P = W/(DL) = 1/2 × 5000/(5 × 2.5) = 200 psi; S = 0.225 with S(L/D)² = .0563, (C/W)K_s = 10, and (C/W)Mω^2 = (0.005/2500) × 13.0 × (2 π × 90)² = 8.28. Since coordinate point S(L/D)² = 0.0563, (C/W)Mω^2 = 8.28 lies below the curve for (C/W)K_s = 10, the rotor is free of oil whip instability.

PIVOTED-PAD JOURNAL BEARINGS

Tilting-pad journal bearings consist of a number of individually pivoted pads or shoes. This pivoting capability enables relatively high loading in applications where shaft deflection or misalignment is a factor. Another advantage is their inherent stability since the load component from each pad passes through the journal center. A further advantage is that clearance can be closely controlled by making the pivots adjustable radially, thus enabling

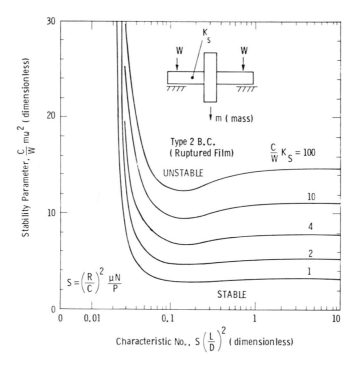

FIGURE 35.   Stability of single-mass rotor on full-journal bearings (ruptured film, laminar flow).

(1) operation with smaller clearances than considered proper for a plain journal bearing, or (2) preloading the journal to achieve relatively high stiffness (important with vertical rotor).

Load Orientation

While not a necessity, usual practice is to construct all pads alike, to space them uniformly around the circumference, and to use an aspect ratio (L/B') approximately equal to one. If the number of pads is large, it makes little difference whether the load line of action passes through the center of one pad or lies midway between two pads. If the number of pads is small, however, load-between-pads orientation is preferred because it gives (1) greater load capacity for a given minimum film thickness, (2) lower pad temperature rise by distributing the load more uniformly, and (3) greater lateral stiffness and damping for horizontal rotor applications.

Pivot Position

Pivots are usually positioned at the pad center (at $\beta/2$) to obtain identical performance independent of the direction of journal rotation. With oil as a lubricant, load capacity is not unduly sensitive to the pivot location. However, when using low-viscosity fluids such as water, liquid metals, and particularly gases, load capacity is sensitive to pivot locations and offset (toward the trailing edge) pivots are preferable.[34]

Pad Contour (Preloading)

Figure 36 shows a pad machined to radius R + C (position 1). Assuming the pad does not tilt, the film thickness is uniform (equal to C) and develops no hydrodynamic force. If the pad is moved to position 2 by displacing the pivot *radially inward* (C − C'), film thickness is no longer uniform and a hydrodynamic force ''preloads'' the journal.

Bearings in vertical machines often undergo little if any radial load (magnetic pull,

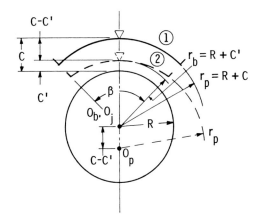

FIGURE 36.   Preloading of a pad. (1) As machined and (2) preloaded; O_b = bearing (pivot circle) center, O_j = journal center, O_p = pad center, r_b = bearing (pivot circle) radius, r_p = pad radius, and R = journal radius.

imbalance, misalignment, etc.) and thus operate nearly concentric ($\epsilon_o \sim 0$) with the journal. This is often undesirable because of the low-radial stiffness and possible spragging. Therefore, vertical machines often resort to preloaded guide bearings.

Preload is measured by preload coefficient $m = (C - C')/C$ which represents the fraction of the radial clearance C "used up" by moving the pad radially. A value of $m = 1$ means that the pad has been moved inward a distance C to touch the shaft ($C' = 0$); $m = 0$ represents a pad which is radially at its machined position ($C' = C$, no preload). While each pad pivot is usually displaced inward the same amount, only the top (normally unloaded) pads are sometimes preloaded to prevent spragging in horizontal machines. If the pad pivot were moved *radially outward,* the pad will have negative preload, an undesirable situation. Fitted pads machined to the journal contour ($C = 0$) result in (infinite) negative preload when assembled with a bearing clearance C'. Their use is suspect.

Spragging

Spragging usually causes no difficulty in ordinary applications. With high speeds and low-viscosity fluids like water, liquid metals, and gases, however, an unloaded or lightly loaded pad may become tipped forward with its leading edge pulled in toward the shaft.[34] A divergent film is momentarily created which sucks the loading edge of the pad tighter against the journal. With a very lightly loaded journal, this decrease in pressure at one pad will change the load on the other pads and cause the journal to shift its position in the bearing. This sometimes results in a persistent spragging of each pad in succession as the journal migrates through the clearance space. Spragging can be controlled by:

1.   Stops or springs to prevent improper tipping of the pads.
2.   A clearance ratio which permits the pads to realign automatically (preloading).
3.   Relieving the leading edge of the pads.

Pad Inertia

A pad supported on a rigid pivot (Figure 37) will track journal vibration by rocking (pitching) about the pivot. The influence of pad inertia on dynamic spring and damping coefficients is negligible except when approaching pad resonance when the phase angle between a zero-inertia pad and a pad with a finite inertia becomes 90°, implying that journal

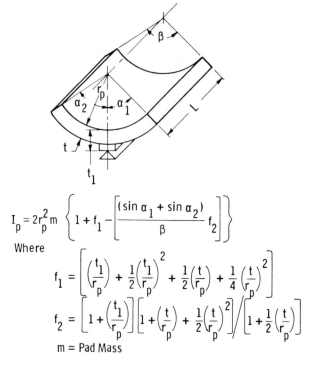

$$I_p = 2r_p^2 m \left\{ 1 + f_1 - \left[\frac{(\sin\alpha_1 + \sin\alpha_2)}{\beta} f_2 \right] \right\}$$

Where

$$f_1 = \left[\left(\frac{t_1}{r_p}\right) + \frac{1}{2}\left(\frac{t_1}{r_p}\right)^2 + \frac{1}{2}\left(\frac{t}{r_p}\right) + \frac{1}{4}\left(\frac{t}{r_p}\right)^2 \right]$$

$$f_2 = \left[1 + \left(\frac{t_1}{r_p}\right) \right]\left[1 + \left(\frac{t}{r_p}\right) + \frac{1}{2}\left(\frac{t}{r_p}\right)^2 \right] \Big/ \left[1 + \frac{1}{2}\left(\frac{t}{r_p}\right) \right]$$

m = Pad Mass

FIGURE 37.   Mass moment of inertia of pad around axial axis.

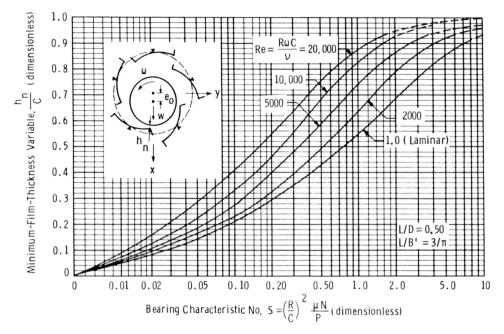

FIGURE 38.   Minimum film thickness [five 60° tilting pads, centrally pivoted, no preload (C′ = C), L/D = 0.5].

motion and pad motion are 90° out of phase.[35] Onset of pad resonance can be determined from the ''critical pad mass parameter'' and requires calculation of the pad pitch inertia I_p. Design data given in this section are based on a pivot fictitiously located on the pad surface above the actual pivot. For this case, I_p can be calculated from the general expression given

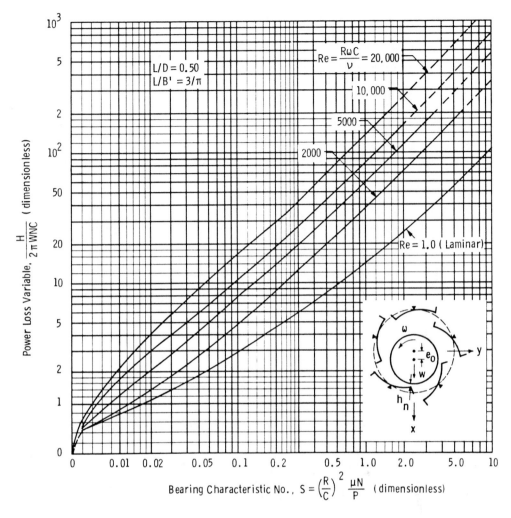

FIGURE 39.   Total power loss (five 60° tilting pads, centrally pivoted, no preload, L/D = 0.5).

in Figure 37 by taking $t_1/r_p = t/r_p = 0$. Design data accounting for pad inertia near resonance can be found in Reference 35.

Excitation Frequency

Unlike fixed-arc bearings, dynamic spring and damping coefficients of tilting pad journal bearings are dependent upon the frequency Ω of the excitation force. These coefficients are usually presented for the common case of unbalance excitation ($\Omega/\omega = 1.0$). Following presentation of a variety of performance data for five-pad bearings in Figures 38 to 44, the effect of excitation frequency is provided in Figures 45 to 48 for a five-pad bearing.

Influence of preload on the stability of vertical (essentially unloaded) guide bearings employing four, five, six, and eight pads is given in Figures 49 to 52.

Example: (horizontal rotor): find the performance of a five-pad bearing given:

β = 1.05 rad (60°)	W = 2500 lb
D = 5 in.	μ = 2 × 10⁻⁶ lb sec/in.² (ISO VG 32 oil
L = 2.5 in. (L/D = 0.5)	at 135 F avg. temp., T_s, Figure 1)
C = C' = 0.005 in. (no preload)	ρ = 7.77 × 10⁻⁵ lb sec²/in.⁴
N = 60 r/sec	$\nu = \mu/\rho$ = 2.57 × 10⁻² in.²/sec

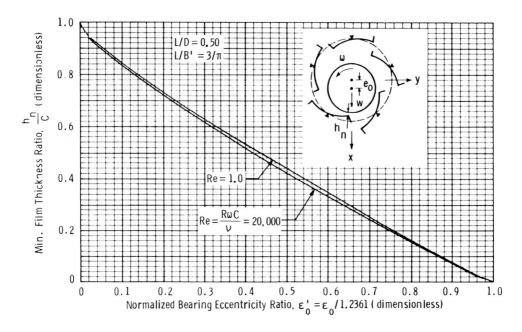

FIGURE 40.   Normalized bearing eccentricity ratio (five 60° tilting pads, centrally pivoted, no preload, L/D = 0.5).

Calculating the Reynolds and bearing characteristic numbers:

$$Re = R\omega C/\nu = (5/2) \times 2\pi \times 60 \times 0.005/2.57 \times 10^{-2} = 183$$

$$S = \left(\frac{R}{C}\right)^2 \frac{\mu N}{P} = \left(\frac{5/2}{0.005}\right)^2 \frac{2 \times 10^{-6} \times 60}{2500/(5 \times 2.5)} = 0.15$$

Minimum film thickness — Because Re is below 1000, flow is laminar and Re = 1.0, curve in Figure 38 gives:

$$h_n/C = 0.26$$

$$h_n = 0.26 \times 0.005 = 0.0013 \text{ in.}$$

Power loss — From Figure 39, $H/(2\pi WNC) = 3.9$ from which $H = 1.84 \times 10^4$ lb-in./sec (2.78 hp).

Normalized bearing eccentricity ratio — Entering Figure 40 with $h_n/C = 0.26$ gives $\epsilon'_o = \epsilon_o/1.2361 = 0.67$. This value will be used to enter the charts to obtain the dynamic stiffness and damping coefficients. As a matter of interest, the journal displacement (eccentricity) is

$$e_o = \epsilon_o C' = 0.67 \times 1.2361 \times 0.005 = 0.0041 \text{ in.}$$

Dynamic performance — ($\Omega/\omega = 1$, unbalance excitation). Entering Figures 41 to 43 with $\epsilon'_o = 0.67$ gives $(C/W)K_{xx} = 4.9$ from which

$$K_{xx} = 4.9(W/C) = 4.9(2500/0.005) = 2.5 \times 10^6 \text{ lb/in.}$$

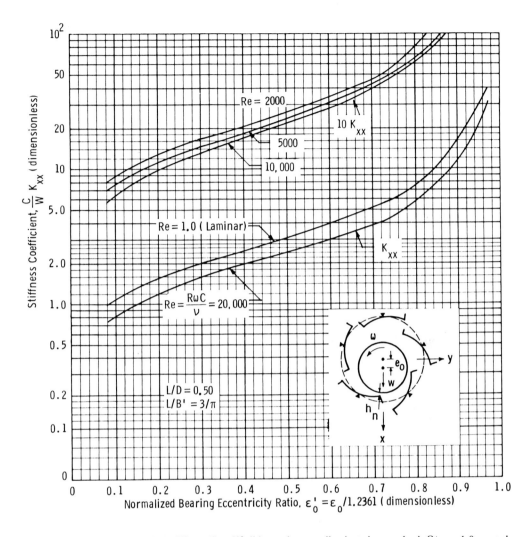

FIGURE 41.   Bearing vertical stiffness (five 60° tilting pads, centrally pivoted, no preload, $\Omega/\omega = 1.0$, no pad inertia, $L/D = 0.5$).

Similarly,

$$(C/W)K_{yy} = 2.5$$

$$(C/W)\omega B_{xx} = 3.6$$

$$(C/W)\omega B_{yy} = 2.0$$

from which

$$K_{yy} = 1.2 \times 10^6 \text{ lb/in.}$$

$$B_{xx} = 4.8 \times 10^3 \text{ lb sec/in.}$$

$$B_{yy} = 2.7 \times 10^3 \text{ lb sec/in.}$$

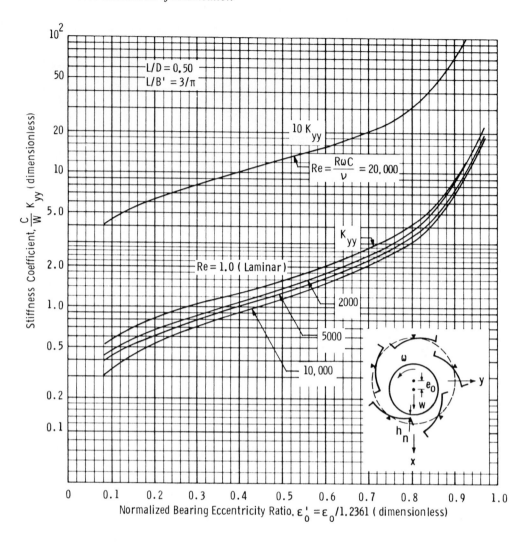

FIGURE 42. Bearing horizontal stiffness (five 60° tilting pads, centrally pivoted, no preload, ($\Omega/\omega = 1.0$, no pad inertia, L/D = 0.5).

If the excitation frequency ratio were different from $\Omega/\omega = 1.0$, say 2.0, the stiffness and damping coefficients could be obtained directly by entering Figures 45 to 48 with $\epsilon_o' = 0.67$, as in the above example, and $\Omega/\omega = 2.0$. Figures 45 to 48, although valid only for laminar flow, can also be used for turbulent flow (Re $\geq$ 1000) to approximate the stiffness and damping coefficients since they are not strongly dependent on Reynolds number. To do this, ϵ_o' should first be obtained as shown in the above example through Figures 38 and 40 using the appropriate curve for the actual value of Re.

Critical mass — From Figure 44, the critical pad mass parameter is

$$CWM_{crit}/[\mu DL(R/C)^2]^2 = 0.45$$

$$M_{crit} = 0.45 \times [2 \times 10^{-6} \times 5 \times 2.5 \times (500)^2]^2/[0.005 \times 2500]$$

$$= 1.41 \text{ lb sec}^2/\text{in.}$$

$$(I_p)_{crit} = M_{crit} \, r_p^2 = 1.41 \times (5/2)^2 = 8.81 \text{ lb sec}^2 \text{ in.}$$

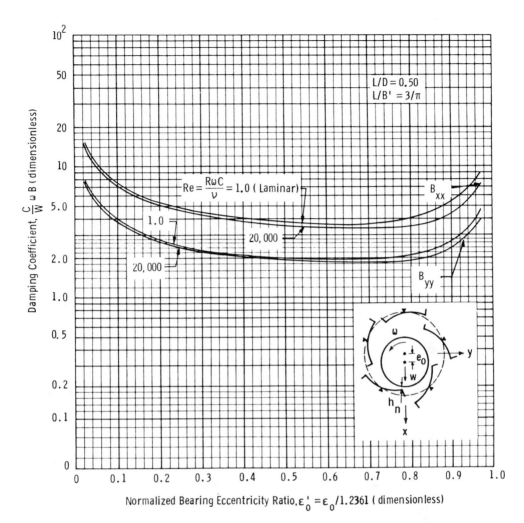

FIGURE 43.   Bearing Vertical and Horizontal Damping (five 60° tilting pads, centrally pivoted, no preload, $\Omega/\omega = 1.0$, no pad inertia, L/D = 0.5).

From Figure 37 with $\alpha_1 = \alpha_2 = \beta/2 = 30°$, $t_1 = t = 0$

$$(I_p)_{actual} = 2\,r_p^2 M[1 - 2\sin 30°/1.05] = 0.595\,M$$

If actual pad mass, M, is less than $8.81/0.595 = 14.8$ lb sec^2 in., there is no danger of resonance. This corresponds to a pad weight of $386 \times 14.8 = 5710$ lb.

Example: (vertical guide bearing): find the stiffness and damping coefficients for a tilting pad guide bearing using the same oil as in the previous sample example and given the following:

n = 6	N = 30 r/sec
β = 0.89 rad (51°)	C = 0.015 in.
D = 15 in.	C' = 0.0075 in.
L = 6.75 (L/D = 0.45)	ρ = 7.77 × 10^{-5} lb sec^2/in.^{4}
μ = 2 × 10^{-6} lb sec/in.^{2}	ν = 2.57 × 10^{-2} in.^{2}/sec

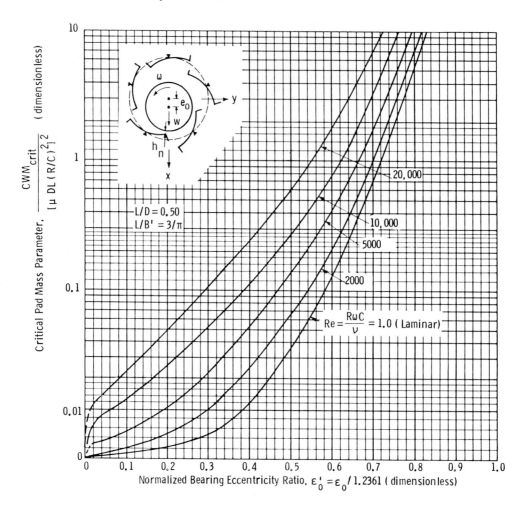

FIGURE 44. Pad critical mass (five 60° tilting pads, centrally pivoted, no preload, $\Omega/\omega = 1.0$, L/D = 0.5).

The following indicates laminar flow because Re < 1000.

$$Re = R\omega C/\nu = (15/2) \times 2\pi \times 30 \times 0.015/2.57 \times 10^{-2} = 824$$

$$\xi = 2(R/C)^3 \mu NL = 2(7.5/0.015)^3 \times 2 \times 10^{-6} \times 30 \times 6.75 = 1.01 \times 10^5 \text{ lb/in.}$$

$$m = 1 - C'/C = 1 - 0.0075/0.015 = 0.50$$

Entering Figure 51 with m = 0.50 gives

$$K/\xi = 9.5$$

$$\omega B/\xi = 11.3$$

$$\xi/(W/C) = 14.6$$

$$H/(2\pi NC^2 \xi) = 33$$

$$10^2 M_{crit}/(\xi/N^2) = 1.0$$

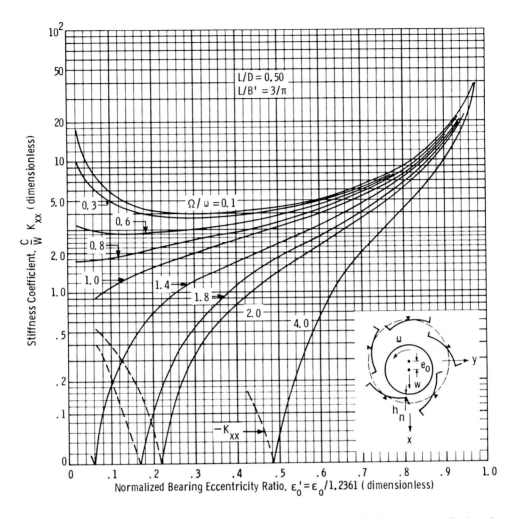

FIGURE 45.   Effect of excitation frequency on bearing vertical stiffness (five 60° tilting pads, centrally pivoted, no preload, no pad inertia, laminar flow).

Hence

$$K = 9.5 \times 1.01 \times 10^5 = 9.6 \times 10^5 \text{ lb/in. stiffness}$$

$$B = 11.3 \times 1.01 \times 10^5/(2\pi \times 30) = 6.05 \times 10^3 \text{ lb sec/in. damping}$$

These are the dynamic stiffness and damping coefficients for calculating rotor response to an unbalance excitation.

A load $W = \xi C/14.6 = 104$ lb is required to produce a very small static displacement ($e_o = \epsilon_o C' = 0.01 \times 0.0075 = 75 \times 10^{-6}$ in.). The likelihood of pad resonance can be calculated from the critical pad mass $M_{crit} = 1.12$ lb sec^2/in. as shown in the previous example. The power loss is found to be $H = 1.45 \times 10^5$ lb in./sec (21.4 hp).

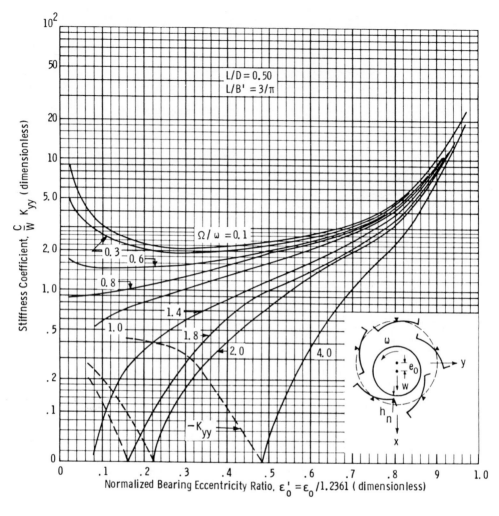

FIGURE 46.   Effect of excitation frequency on bearing horizontal stiffness (five 60° tilting pads, centrally pivoted, no preload, no pad inertia, laminar flow).

NOMENCLATURE

B	=	Slider bearing width (in direction of motion), in.
B, B_{xx}, B_{xy}, B_{yx}, B_{yy}	=	Lubricant film damping coefficient, lb-sec/in.
B'	=	$r_p\beta$, Pad arc length (tilting pad journal bearing), in.
C	=	$r_p - R$ = Pad or partial arc radial clearance, in.
C'	=	$r_b - R$ = Tilting pad journal bearing (pivot circle) radial clearance, in.
$C_{w,L}$ $C_{H,L}$; $C_{Qin,L}$; $C_{Qs,L}$; $C_{ho,L}$	=	Laminar flow performance factors for load, power loss, inlet flow, side flow, and minimum film thickness, respectively, dimensionless
$C_{w,T}$; $C_{H,T}$; $C_{Qin,T}$; $C_{Qs,T}$; $C_{ho,T}$	=	Turbulent flow correction factors for load, power loss, inlet flow, side flow, and minimum film thickness, respectively, dimensionless
D	=	2R = Journal diameter, in.

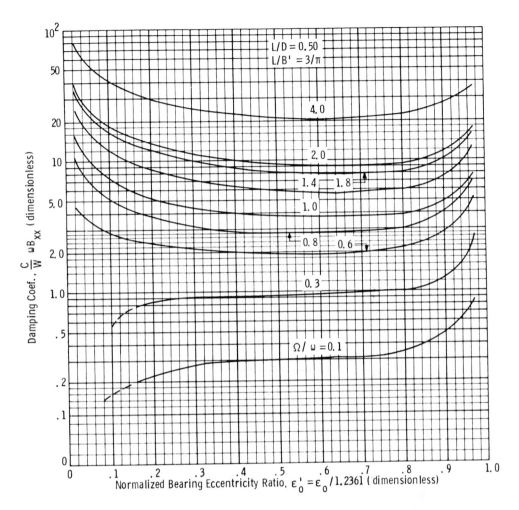

FIGURE 47. Effect of excitation frequency on bearing vertical damping (five 60° tilting pads, centrally pivoted, no preload, no pad inertia, laminar flow).

D_{xx}, D_{xy}, D_{yx}, D_{yy}	=	Lubricant film acceleration coefficients, lb-sec^2/in.
F	=	Friction force or excitation force, lb
F	=	Thermohydrodynamic (THD) turbulence function, dimensionless
F_x, F_y	=	Dynamic lubricant film force components, lb
H	=	Power loss, lb-in./sec
I_p	=	Mass moment of inertia of pad around axial axis (Figure 37), lb-in.-sec^2
K, K_{xx}, K_{xy}, K_{yx}, K_{yy}	=	Lubricant film stiffness coefficient, lb/in.
K_S	=	Rotor stiffness (Figure 35), lb/in.
L	=	Length (perpendicular to motion), in.
M	=	I_p/r_p^2 = Equivalent pad mass, lb-sec^2/in.
M_{crit}	=	Value of M giving resonance, lb-sec^2/in.
N	=	Speed, rev/sec
P	=	Unit load, W/DL (journal bearing), = W/BL (slider bearing), lb/in.^{2}
Pe	=	Peclet number = $\rho c \omega C^2/k$, dimensionless

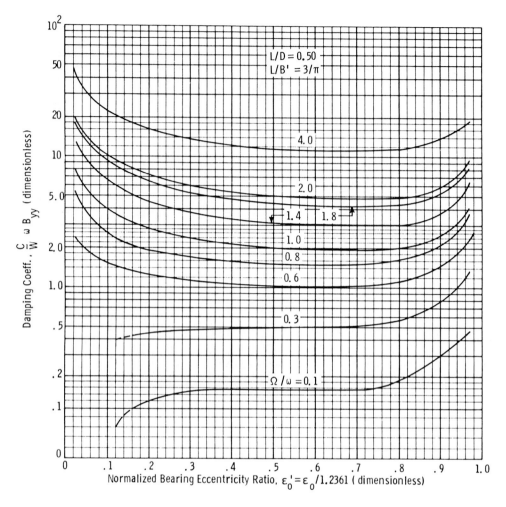

FIGURE 48.   Effect of excitation frequency on bearing horizontal damping (five 60° tilting pads, centrally pivoted, no preload, no pad inertia, laminar flow).

Pr	=	Prandtl number = ν/α, dimensionless
Q,Q_{in}	=	Flow rate into bearing, in.³/sec
Q_s,Q_{s1}, Q_{s2}	=	Side flow flow, in.³/sec
R	=	Journal radius, in.
R_1,R_2	=	Sector pad inner radius, outer radius, in.
R_{mean}	=	(R_1 + R_2)/2, in.
R_p	=	Pivot or center-of-pressure radial location, in.
R_e	=	Global Reynolds number = $R\omega C/\nu$, dimensionless
R_h	=	Local Reynolds number = $R\omega h/\nu$, dimensionless
Re_p	=	Slider bearing Reynolds number = Uh_p/ν, dimensionless
S	=	Bearing characteristic number = $(R/C)^2(\mu N/P)$, dimensionless
T	=	Temperature, °F
$\overline{T}$	=	Mean (turbulent) value of T,°F
T_i,T_o,T_s	=	Temperature at inlet, outlet, side, °F
T_{max}	=	Maximum bearing temperature, °F

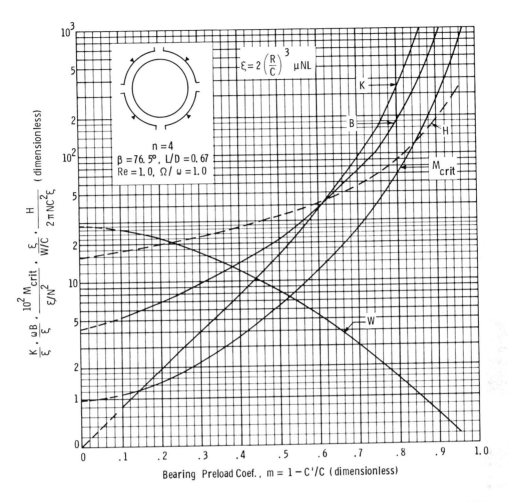

FIGURE 49. Effect of preload on four-pad bearing (vertical rotor with slight radial load giving $\epsilon_o = 0.01$, laminar flow, no pad inertia).

ΔT	$=$	Temperature rise, °F
Ta	$=$	Taylor number $= (C/R)(R\omega C/\nu^2)$, dimensionless
U, U_a	$=$	Linear velocity, average value, in./sec
U_1, U_2	$=$	Tangential velocities, in./sec
V_o	$=$	Normal velocity, in./sec
W	$=$	Load, lb
c_f	$=$	Coefficient of wall stress $= 8\tau_w/\rho U_2^2$, dimensionless
c	$=$	Specific heat, in.²/sec² °F
e	$=$	Eccentricity or displacement of journal with respect to pad or partial arc, in.
e_o	$=$	Eccentricity (displacement) of journal with respect to bearing $(\overline{O_b O_j})$, in.
$e_{o_{max}}$	$=$	Maximum possible eccentricity, in.
f	$=$	Coefficient of friction $= F/W$, dimensionless
h, h_a	$=$	Film thickness, average value, in.
$\overline{h}$	$=$	Dimensionless film thickness $= h/C$

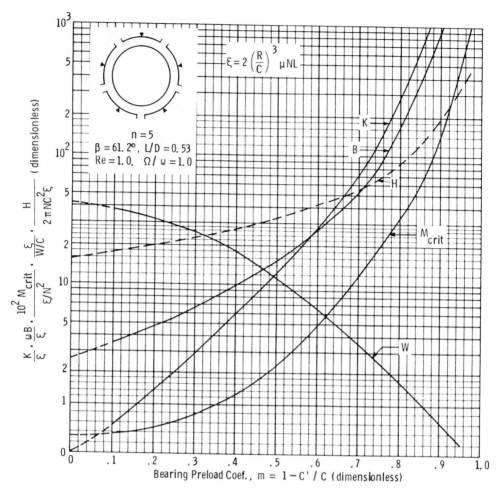

FIGURE 50.   Effect of preload on five-pad bearing (vertical rotor with slight radial load giving $\epsilon_o = 0.01$, laminar flow, no pad inertia).

h_c	=	Film thickness at geometric center of sector pad, in.
h_p	=	Film thickness at pivot location, in.
h_o	=	Minimum film thickness (crowned pad), in.
h_{min}	=	Minimum film thickness (sector pad), in.
h_2	=	Outlet film thickness (crowned pad), minimum film thickness (slider bearing, Figure 8), in.
h_n	=	Minimum film thickness (journal bearing), in.
k, k_b	=	Heat conductivity of oil, bearing, lb/sec °F
k_x, k_z	=	Turbulence functions, dimensionless
m	=	Mass, lb-sec²/in.
m	=	$(C - C')/C$ = Preload coefficient, dimensionless
m_r	=	Radial slope parameter = $R_1 \gamma_r/h_c$, dimensionless
m_θ	=	Tangential slope parameter = $R_1 \gamma_\theta/h_c$, dimensionless
n	=	Number of pads, dimensionless
p	=	Lubricant film pressure, lb/in.²
p_{cav}, p_{atm}	=	Cavitation, atmospheric pressure, lb/in.²

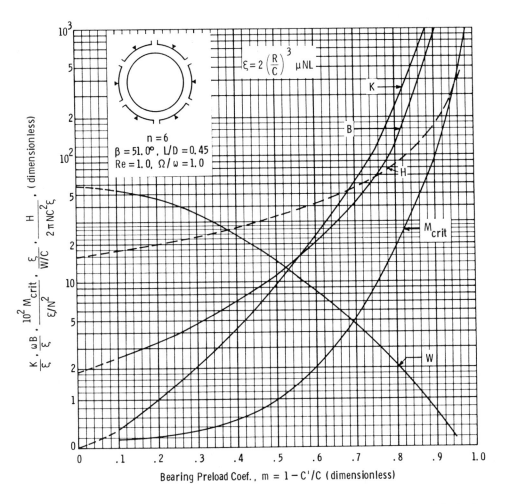

FIGURE 51. Effect of preload on six-pad bearing (vertical rotor with slight radial load giving $\epsilon_o = 0.01$, laminar flow, no pad inertia).

$\overline{p}$	=	Mean (turbulent) pressure, lb/in.^{2}
r_b	=	Tilting pad journal bearing (pivot circle) radius, in.
r_p	=	Pad or partial arc radius, in.
t	=	Time, sec
$\overline{t}$	=	Dimensionless time $= t\omega$
t'	=	Fluctuating component of T,°F
$\overline{x},y,z$	=	Rectangular Cartesian coordinates, in.
x	=	Pivot or center-of-pressure location, measured from leading edge, in.
Ω	=	Excitation speed, rad/sec
Λ	=	Dissipation number $= \mu\omega(R/C)^2/\rho cT$, dimensionless
α	=	Diffusivity $= k/\rho c$, in.^{2}/sec
β	=	Angular extent of pad, sector, or partial-arc, rad
ϵ	=	e/c = Pad, or partial-arc eccentricity ratio, dimensionless

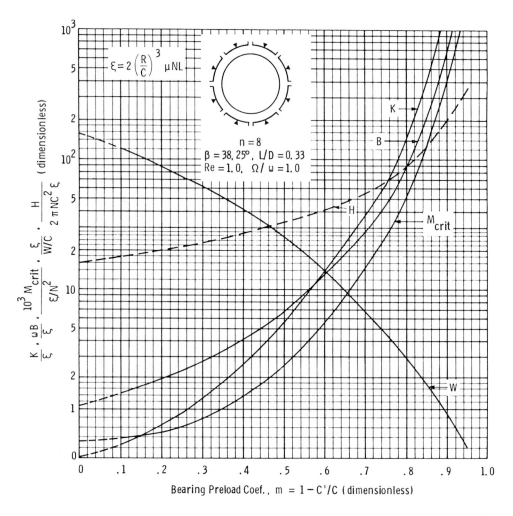

FIGURE 52.   Effect of preload on eight-pad bearing (vertical rotor with slight radial load giving $\epsilon_o = 0.01$, laminar flow, no pad inertia).

$\epsilon_{o_{max}}$	=	$e_{o_{max}}/C' = 1.2361$ (For five-pad bearing, Figure 40), dimensionless
ϵ_o'	=	$e_o/e_{o_{max}} = \epsilon_o/\epsilon_{o_{max}} = \epsilon_o/1.2361$ = Normalized bearing eccentricity ratio (for five-pad tilting pad journal bearing only), dimensionless
ϕ	=	Attitude angle, deg
μ	=	Absolute viscosity, lb-sec/in.^{2}
ν	=	Kinematic viscosity, in.^{2}/sec
ρ	=	Density, lb-sec^2/in.^{4}
τ_w	=	Wall stress, lb/in.^{2}
θ_p	=	Pivot or center-of-pressure location, measured from trailing edge, rad
δ	=	Crown, in.
γ_r, γ_θ	=	Radial, tangential slope of pad, rad
ω	=	$2\pi N$ = Rotation speed, rad/sec
ξ	=	$2(R/C)^3\mu NL$, lb/in.
$(\ )_i$	=	Quantity evaluated at inlet

REFERENCES

1. **Kaufman, H. N., Szeri, A. Z., and Raimondi, A. A.,** Performance of a centrifugal disk-lubricated bearing, *Trans. ASLE,* 21, 315, 1978.
2. **Szeri, A. Z., Ed.,***Tribology: Friction, Lubrication and Wear,* Hemisphere Publishing, Washington, D.C., 1980.
3. **Taylor, G. I.,** Stability of a viscous liquid contained between two rotating cylinders, *Phil. Trans. R. Soc. Ser. A,* 223, 289, 1923.
4. **Coles, D.,** Transition in circular couette flow, *J. Fluid Mech.,* 21, 385, 1965.
5. **DiPrima, R. C.,** A note on the stability of flow in loaded journal bearings, *Trans. ASLE,* 6, 249, 1963.
6. **Li, C. H.,** The effect of thermal diffusion on flow stability between two rotating cylinders, *Trans. ASME Ser. F,* 99, 318, 1977.
7. **Li, C. H.,** The influence of variable density and viscosity on flow transition between two concentric rotating cylinders, *Trans. ASME Ser. F,* 100, 260, 1978.
8. **Gardner, W. W. and Ulschmid, J. G.,** Turbulence effects in two journal bearing applications, *Trans. ASME Ser. F,* 96, 15, 1974.
9. **Abramovitz, S.,** Turbulence in a tilting-pad thrust bearing, *Trans. ASME,* 78, 7, 1956.
10. **Gregory, R. S.,** Performance of thrust bearings at high operating speeds, *Trans. ASME Ser. F,* 96, 7, 1974.
11. **Ng, C. W. and Pan, C. H. T.,** A linearized turbulent lubrication theory, *Trans. ASME Ser. D,* 87, 675, 1965.
12. **Suganami, T. and Szeri, A. Z.,** A thermohydrodynamic analysis of journal bearings, *Trans. ASME Ser. F,* 101, 21, 1979.
13. **Suganami, T. and Szeri, A. Z.,** A parametric study of journal bearing performance: the 80 degree partial arc bearing, *Trans. ASME Ser. F,* 486, 1979.
14. **Constantinescu, V. N.,** On the influence of inertia forces in turbulent and laminar self-acting films, *Trans. ASME Ser. F,* 92, 473, 1970.
15. **Szeri, A. Z., Raimondi, A. A., and Giron, A.,** Linear force coefficients for squeeze-film damper, *Trans. ASME Ser. F,* in press.
16. **Alford, J. S.,** Protecting turbomachinery from self-excited rotor whirl, *ASME J. Eng. Power Ser. A,* 87, 333, 1965.
17. **Hagg, A. C.,** Influence of oil-film journal bearings on the stability of rotating machines, J. Appl. Mech., *Trans. ASME,* 68, A211, 1946.
18. **DenHartog, J. P.,** *Mechanical Vibrations,* 4th ed., McGraw-Hill, New York, 1956.
19. **Raimondi, A. A. and Boyd, J.,** Applying bearing theory to the analysis and design of pad-type bearings, *Trans. ASME,* 77, 287, 1955.
20. **Johnston, R. C. R. and Kettleborough, C.F.,** An experimental investigation into stepped thrust bearings, *Proc. Inst. Mech. Eng.,* 170, 511, 1956.
21. **Wilcock, D. F.,** The hydrodynamic pocket bearing, *Trans. ASME,* 77, 311, 1955.
22. **Raimondi, A. A.,** Adiabatic solution for the finite slider bearing, *ASLE Trans.,* 9, 283, 1966.
23. **Gross, W. A., Matsch, L. A., Castelli, V., Eshel, A., Vohr, J. H., and Wildmann, M.,** *Fluid Film Lubrication,* John Wiley & Sons, New York, 1980.
24. **Baudry, R. A., Kuhn, E. C., and Wise, W. W.,** Influence of load and thermal distortion on the design of large thrust bearings, *Trans. ASME,* 80, 807, 1958.
25. **Raimondi, A. A.,** The influence of longitudinal and transverse profile on the load capacity of pivoted pad bearings, *ASLE Trans.,* 3, 265, 1960.
26. **Malinowski, S. B.,** Rerate tilting-pad thrust bearings, *Mach. Design,* 45, 100, 1973.
27. **Vohr, J. H.,** Prediction of the operating temperature of thrust bearings, *Trans. ASME J. Lubr. Technol.,* 103, 97, 1981.
28. **Wilcock, D. F. and Booser, E. R.,** *Bearing Design and Application,* McGraw-Hill, New York, 1957.
29. **Raimondi, A. A.,** A theoretical study of the effect of offset loads on the performance of a 120° partial journal bearing, *ASLE Trans.,* 2, 147, 1959.
30. **Raimondi, A. A., Boyd, J., and Kaufman, H. N.,** Analysis and design of sliding bearings, in *Standard Handbook of Lubrication Engineering,* McGraw-Hill, New York, 1968, chap 5.
31. **DuBois, G. B. and Ocvirk, F. W.,** Analytical Derivation and Experimental Evaluation of Short-Bearing Approximation for Full Journal Bearings, NASA TR1157 and TN2808, National Aeronautics and Space Administration, Washington, D.C., 1952.
32. **Allaire, P. E.,** Design of journal bearings for high speed rotating machinery, in *Fundamentals of the Design of Fluid Film Bearings,* American Society of Mechanical Engineers, New York, 1979, 45.
33. **Warner, R. E. and Soler, A. I.,** Stability of rotor-bearing systems with generalized support flexibility and damping and aerodynamic cross-coupling, *ASME J. Lubr. Technol.,* 7F, 461, 1975.

34. **Boyd, J. and Raimondi, A. A.,** Clearance considerations in pivoted pad journal bearings, *ASLE Trans.*, 5, 418, 1962.

35. **Lund, J. W.,** Spring and damping coefficients for the tilting-pad journal bearing, *ASLE Trans.*, 7, 342, 1964.

SLIDING BEARING MATERIALS

A. O. DeHart

BEARING MATERIAL PROPERTIES

Selection of materials for sliding surface bearings is a multifunctional optimizational problem: no one material is best for all applications. Commonly considered material properties include score resistance, conformability, embedability, compressive strength, fatigue, corrosion, thermal properties, wear resistance, and cost. Unfortunately, a selection based upon the best value for one of these properties may be improper when all factors are considered.

In spite of the difficulty, bearing materials are selected for many different applications every day. If we consider each of the required properties separately, a rational basis should develop to aid in making the best compromise in finding the right material for the job at hand.[1-3]

Score Resistance

Score resistance, also termed antiweld and antigalling, is the vital ability of the bearing material to resist welding to the journal under what can be highly distressful conditions. Many engineering tests have been run to assess the ability of bearing materials to resist welding to the steel or cast iron commonly employed for journals. Roach et al.[4] showed that the only elemental metals that have satisfactory score resistance against steel are in the B subgroup of the periodic table and are either insoluble with iron or form weak intermetallic compounds. Relative score resistance for various elements is given in Table 1. Although bearing alloys and mixtures are much more complicated, the performance of elements can be used as a guideline. For example, adding more of a good material (e.g., lead) to a bronze will generally improve score resistance, while the addition of a poor metal (e.g., zinc) will degrade the score resistance.

Strength

Several bearing properties have a relationship to material strength — compressive strength, fatigue strength, embedability, and conformability. Compressive strength, a basic requirement for support of the applied load without cracking or extruding, is closely related to normally reported physical properties. But the effect of temperature should be reecognized when choosing a particular babbitt (Figure 1). Ultimate strength for typical babbitt compositions is given in Figure 2.

One method of improving the effective compressive strength of weaker materials is by using a thin layer on a strong substrate such as steel. Providing the bond is adequate, a thin layer of soft bearing material tends to adopt the stiffness and strength of the substrate.

Fatigue Strength

Fatigue strength is important in bearings subjected to load reversals such as are encountered with connecting rod and main engine bearings. Not only is the fatigue problem due to the dynamic nature of the load, but also to the attendant flexing of the support structure. While fatigue strength varies with temperature and application, Table 2 gives an approximate guide for various materials. It is clear that fatigue ratings are opposite of conformability ratings. Fatigue strength is enhanced by bonding a thin layer of bearing material to a steel back to form the bimetal bearing, particularly when the bearing material thickness is less than 0.1 mm (Figure 3).

Table 1
SCORE RESISTANCE OF ELEMENTS AGAINST 1045 STEEL

Good	Fair	Poor	Very poor	
Germanium	Carbon	Magnesium	Beryllium	Molybdenum
Silver	Copper	Aluminum	Silicon	Rhodium
Cadmium	Selenium	Copper	Calcium	Palladium
Indium	Cadmium	Zinc	Titanium	Cerium
Tin	Tellurium	Barium	Chromium	Tantalum
Antimony		Tungsten	Iron	Iridium
Thallium			Cobalt	Platinum
Lead			Nickel	Gold
Bismuth			Zirconium	Thorium
			Columbium	Uranium

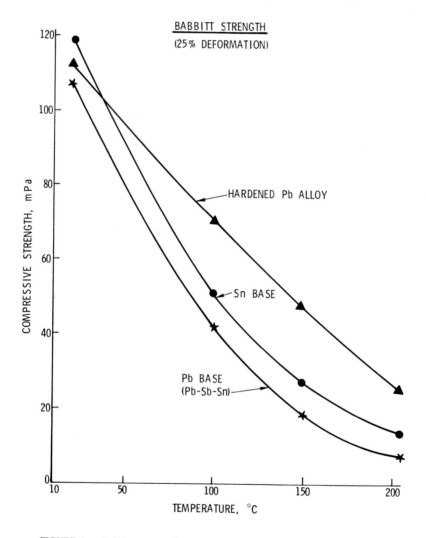

FIGURE 1.   Babbitt compressive strength for 25% deformation vs. temperature.

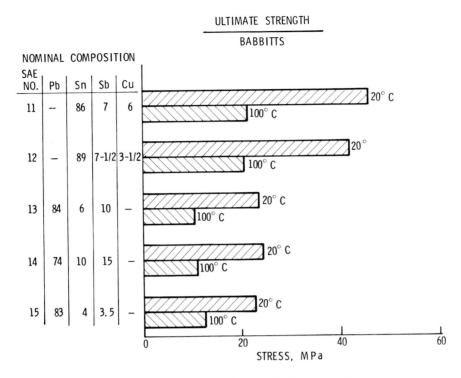

FIGURE 2.   Ultimate strength for typical babbitt compositions.

Table 2
FATIGUE STRENGTH FOR
TYPICAL BEARING METALS

Material	Approximate maximum dynamic load capacity (MPa)
Bronze	100
Silver	80
Copper	80
Copper lead with tin	70
Aluminum alloys	25—60
Copper lead	20—50
Thin babbitt	10—25
Thick babbitt	5—10

While bonding to a steel back increases the fatigue limit of a bearing material, the multilayered bimetal bearing is subjected to a peculiar type of fatigue damage. The fatigue cracks propagate near the bond layer and parallel to the steel back. Whenever vertical cracks form in the material, platelets of the metal become loosened and may pass through the bearing.

Embedability
Undesirable solid contaminants in the form of machining chips and grinding debris are among the enemies of sliding surface bearings. In general, softer materials such as babbitt permit a certain amount of particles to embed with minimum wear damage to the journal. In harder bearing materials such as bronze, the abrasive particles are only partially embedded

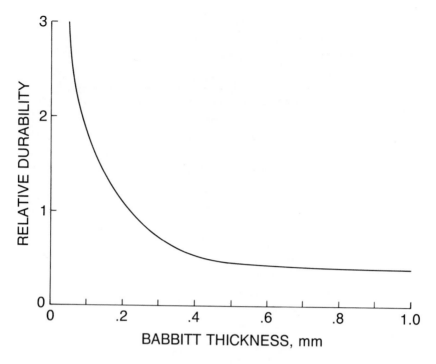

FIGURE 3.   Effect of thickness of a bonded babbitt on fatigue life. (From Crankshaw, E., *Sleeve Bearing Materials,* American Society for Metals, Metals Park, Ohio, 1949, 150. With permission.)

and tend to act as abrasive cutting tools which can cause destructive journal wear. Where high load dictates use of harder and stronger bearing materials, electroplating onto the surface a thin film of babbitt-like material composed of lead, tin, and sometimes copper enhances embedability. The thicker babbitt bimetal bearings have the best embedability properties.

Experience shows that embedability is highly rate-dependent. The bearing can tolerate much larger quantities of abrasive particles if they are injected over a long period of time rather than the same quantity in a short time interval.

Corrosion

Historically, corrosion of bearings was a significant problem. As oils oxidize, they formed complex organic acids that attacked many bearing materials. Unprotected copper and lead were particularly vulnerable. The resulting corrosion either weakened the structure so that the bearing failed by fatigue, or there would be sufficient corrosive wear that the bearing would have excessive clearance. This corrosion problem has been largely overcome by compounding oils to resist oxidation degradation — provided the lubricant is changed at appropriate intervals. Additional protection is conferred by using materials of higher corrosion resistance such as aluminum and tin which are usually not attacked; if at least 3% tin is added to a lead babbitt the corrosion resistance is usually adequate for most applications. Antimony also reduces the tendency for corrosion.

Thermal Properties

Frictional heat is removed from the bearing by the lubricant and by conduction through both the journal and the bearing. Conduction through the bearing can be a major consideration, particularly for high-speed bearings where, in general, materials of higher thermal conductivity would be better employed. Even silver has been used successfully for high-speed bearings.

Matching *thermal expansion* of the bearing to that of the journal is important to maintain the correct clearance. This clearance is readily controlled with bimetal or trimetal bearings with steel back construction. But solid wall bearings, where one material is used for the entire bearing, can cause problems. Poor heat transfer and high expansion rates are major concerns that have prevented the adoption of plastic bearings in high-speed applications. Even solid bronze bearings must be used carefully to prevent loss of clearance or retention. Such bearings must be mechanically located in their housings, because press fits cannot be relied upon for retention. As noted previously, material strength, and therefore load capacity, generally decrease with increasing temperature.

Wear

Bearing wear is many faceted. It can take the form of adhesion, abrasion, corrosion, and fatigue as well as any combination of these. Adhesive wear is associated with score resistance. If the bearing material welds to the journal surface, the material with the weaker bond is torn away and wear results. Abrasive wear is a more mechanical process where the harder material abrades or machines a softer material. Corrosive wear is related, of course, to the corrosion resistance. Shearing of oil and journal sliding action tend to remove any passivating films that may form on the bearing surface, so that materials deficient in corrosion resistance can suffer very rapid wear in corrosive environments. Fatigue wear of bearings can occur in the large scale already described or on a micro basis at small surface peaks or stress sites.

Score resistance, conformability, compressive strength, fatigue, corrosion, thermal properties, wear resistance, and cost: in general, these performance requirements cannot be met with a single material. A modern-day, high-performance automotive bearing might have several layers of mixtures and alloys — each engineered to meet a particular set of performance parameters.

METALLIC BEARING MATERIALS

Basically, modern materials fall into six classes: babbitts, copper-based bearings, aluminum-based bearings, silver bearings, and porous metal bearings. Representative properties of the various classes are given in Table 3.[2]

Babbitt

Babbitt, named after Issac Babbitt who obtained the first American patent on a special bearing material in 1839, is used today to describe a number of soft lead- and tin-based bearing materials bonded to a harder and stronger shell. While bronzes have been widely used for the backing material, current practice commonly makes use of steel. Where babbitts have sufficient strength, they are good materials for most applications. They have superior embedability and conformability and excellent antiscore qualities. Unfortunately, their strength is limited by temperature.

Effective strength of babbitts can be improved by reducing thickness. The highest strength babbitts are obtained by electrodeposition of lead and tin or lead, tin, and copper onto a bearing substrate. On the other hand, in large electrical machinery or in some marine applications, babbitts as heavy as 10-mm thick are cast onto steel or cast iron supporting structures and are often mechanically keyed into place. Typically, these bearing systems are designed for very low unit loads (on the order of 1.4 mPa) with life expectancies of over 20 years. Table 4 gives nominal compositions for tin- and lead-based babbitts.

Tin Babbitt

Tin-base babbitts are the material of choice for corrosive conditions where the increased cost can be justified. Tin-base babbitts are composed of up to 90% tin with copper and

Table 3

PHYSICAL PROPERTIES OF SLIDING BEARING MATERIALS

	Hardness, Brinell	Specific gravity	Tensile strength (MN/m²)[a]	Modulus of elasticity (GN/m²)[a]	Thermal conductivity [W/(m·K)]	Coefficient of expansion (10⁻⁶/°C)
Metals						
Lead babbitt	21	10.1	69	29	24	25
Tin babbitt	25	7.4	79	52	55	23
Copper lead	25	9.0	55	52	290	20
Lead bronze	60	8.9	230	97	47	18
Tin bronze	70	8.8	310	110	50	18
Aluminum alloy	45	2.9	150	71	210	24
Cadmium	35	8.6		55	92	30
Silver	25	10.5	160	76	410	20
Steel	150	7.8	520	210	50	12
Cast iron	180	7.2	240	160	52	10
Porous metals						
Bronze	40	6.4	120		29	19
Iron	50	6.1	170		28	12
Aluminum	H55[b]	2.3	100			
Plastics						
Nylon	M79[b]	1.14	79	2.8	0.24	170
Acetal	M94	1.42	69	2.8	0.22	80
PTFE	D60[c]	2.17	21	0.4	0.24	170
Phenolic	M100	1.36	69	6.9	0.28	28
Polyester	D78[c]	1.45	17	7.1	0.59	52
Polyimide	E52[b]	1.43	73	3.2	0.43	50
Other nonmetallics						
Carbon graphite	75[d]	1.7	14	14	17	3.1
Wood		0.68	8	12	0.19	5
Rubber		1.2			0.16	77
Tungsten carbide	A91[b]	14.2	900	560	70	6
Al₂O₃	A85	3.9	210	340	2.8	15

[a] To convert N/m² to lb/in.² divide by 6.895.
[b] Rockwell.
[c] Shore durometer.
[d] Shore scleroscope.

Table 4

NOMINAL PERCENTAGE COMPOSITION OF TIN- AND LEAD-BASE BABBITTS

SAE No.	Tin	Antimony	Lead	Copper
		Tin Base		
11	86.0	7.0	0.5	6.0
12	88.0	7.5	0.5	3.5
		Lead Base		
13	6.0	10.0	83	0.5
14	10.0	15.0	74	0.5
15	1.0	15.0	82	0.5
16	4.0	3.5	92	0.1

antimony additions to increase hardness and strength. Of course, these additions generally decrease embedability and conformability. SAE Specifications 11 and 12 define two common tin-base babbitts. These can be cast onto steel, bronze, or brass backs or directly into a bearing housing. The materials are soft and corrosion resistant with moderate fatigue resistance. They are used for main and connecting rod bearings and for motor bearings in corrosive situations. This material operates well on either hard or soft journals.

Lead-Based Babbitt

While tin babbitts had always been the bearing materials of choice over lead, the World War II shortage of tin forced the general use of lead babbitts. Many users were surprised to find that lead babbitts worked as well as, or, in some cases, better than tin. While their relative merits were debated, there seemed to be agreement that lead babbitts were at least equivalent of tin in thin linings, but in thicker linings the tin may be superior. Additions of tin and antimony have been found generally to correct the inadequate corrosion resistance with some of the wartime lead babbitts.

Today, most high-performance bearings use some type of plated lead babbitt of nominally 10% tin, with about 3% copper ofteen used to confer additional hardness. SAE Specifications 19 and 190 cover these overplated materials. SAE Specifications 13 through 16 cover four reepresentative bearing alloys containing enough tin or antimony to confer adequate corrosion resistance, athough SAE 15 should not be used where there are heavy concentrations of organic acids. SAE 15 is normally cast onto a steel back while the other alloys are cast onto steel, bronze, brass, or directly into the bearing housing. SAE 16 is normally cast into and on a porous-sintered matrix — usually copper-nickel bonded to steel. These babbitts are soft and moderately fatigue-resistant materials that are widely used in main bearing and connecting rod applications. While they will operate with hard or soft journal surfaces, they perform best when the journal surfaces are smooth — below 0.3 mm R_a.

Copper-Base Bearings

Copper-base bearing materials are widely used both as solid wall bearings where the bearing material is also the wall, and as multilayered materials where the copper is either cast or sintered onto a steel backing. The solid wall bearing may be either cast or sintered, with the porous bearing being a special case where the sintering is done to maintain an open structure for lubricant. The solid, cast copper materials are usually classified as tin, lead, or aluminum bronzess.

Tin Bronzes

Tin bronzes are made up of copper with tin being the major alloy constituent. Zinc is often added to improve castability with some penalty to score resistance. These materials are generally strong and hard with reduced score resistance and embedability compared to most other materials. Although tin itself has good score resitance, it does not improve that of copper until over 60% has been added (Figure 4). These tin bronzes must rely upon the lead content to obtain good score resistance at lower tin concentrations. Consequently, they are best used for high-load situations where low speeds are maintained, with ample supply of clean lubricant, and with increased journal hardness. SAE Alloys 791 and 795 in Table 5 are typical solid bronze bearing materials that have high hardness and good fatigue resistance.

Leaded Bronzes

Leaded bronzes, which are softer than the tin bronzes, will tolerate more misalignment due to their improved conformability. The increased lead content also confers additional embedability and eases their lubricant requirement. In fact, increased lead content improves most bearing characteristics except those concerned with strength, such as fatigue resistance.

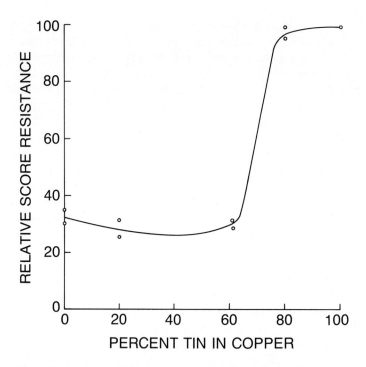

FIGURE 4.   Influence of tin content on score resistance of bronze. (From Roach, A. E. and Goodzeit, C. L., *Gen. Motors Eng. J.*, 2(5), 25, 1955. With permission.)

Table 5
NOMINAL PERCENTAGE COMPOSITION OF
COPPER-BASE BEARING ALLOYS

SAE No.	Copper	Lead	Tin	Zinc	Silver
Typical Bronzes					
791	89	4	4	3	—
792 or 797	77 min	10	10	0.5	—
793 or 798	83 min	8	4	4 max	—
794 or 799	74	23	3.5	3 max	—
795	90	—	0.5	9	—
Copper-Lead Alloys					
48	71	28	0.3	0.1	1.5
480	64	35	0.5	—	1.5
481	60	40	0.3	0.1	1.5
482	67	28	5	—	—
484	55	42	3	—	—
485	46	51	3	—	—
49	75	24	0.3	—	—

Leaded bronzes are generally cast or sintered onto a steel back. SAE 792, 793, and 794 are cast materials while 797, 798, and 799 are sintered. They all have sufficient tin for corrosion resistance in most situations and are generally good bearing materials. SAE 792 and 797, with the balanced composition of 10% tin and 10% lead, are excellent bearing materials for a wide range of applications. These have been replaced to a large degree by the CA932 alloy (7Sn, 7Pb, 3Zn) for easier casting of solid wall bearings.

Copper-Lead Alloys

Containing higher percentages of lead than the leaded bronzes, these have the best score resistance and embedability of the copper-based bearing materials. Lead content of this group may go as high as 51% (SAE 485). Copper-lead alloys SAE 48, 49, 480, and 481 have minimum amounts of tin and are somewhat subject to oil corrosion. SAE 482, 484, and 485 have 3 to 5% tin. The steel-backed, copper-lead bearing was developed about 1923 by the Allison Engineering Company and the U.S. Army Air Corp. The first bearings were constructed for the liberty engine connecting rods.

All copper-lead alloys are either cast or sintered onto a steel back. They are moderately hard with good to fairly good fatigue resistance. They are used for main bearing and connecting rod bearings in a wide variety of engines. They may be used with or without plated overlays but recent experience has shown that unplated copper-lead bearings tend to suffer from corrosion damage in engines operated with unleaded fuel.

In copper-lead-tin bearing alloys the copper is used to support the load, and lead confers score resistance and embedability to the copper. Since lead is readily attacked by organic acids, tin is mixed with the lead to prevent corrosion, which it does very well in most situations as long as tin content is over 3%. After extended periods of time at elevated temperature, the tin is depleted from the lead by diffusion into the copper where copper-tin intermetallic compounds are formed that may degrade the score resistance of the basic bearing material. At the same time, corrosion resistance of the lead is reduced.

Aluminum Bronzes

Aluminum bronzes are popular for manufacturing plant equipment, where great strength is required without too much concern about score resistance and embedability. These materials will support high loads at temperatures as high as 260°C (500°F), but they will not tolerate misalignment or inadequate lubrication. They are widely used as wear-resistant bearing plates and bushings in machine tools.

Aluminum

Aluminum-based bearings have superior corrosion resistance and good thermal conductivity. Aluminum has long been used as a solid wall bearing in heavy and medium duty diesel engines as an aluminum, tin, silicon, and copper composition (SAE 780) or aluminum, tin, copper, and nickel (SAE 770). Positive bearing location by means such as pinning is used to keep the bearing in the desired location because the high thermal expansion causes the bearing to lose retention capability due to yielding of the solid aluminum.

Aluminum materials have fair to excellent antiscore properties. Superior antiscore properties result when 4% silicon and 1/2% cadmium are added to the basic aluminum while SAE 782 uses 3% cadmium to obtain the desired score resistance. Both SAE 781 and 782 bearing alloys are typically bonded to a steel back and have a thin babbitt electroplated overlay for superior antiscore and embedability characteristics.

The venerable 6% tin-aluminum bearing has been used as a solid wall bearing material as well as steel-backed and overplated for engine applications. The composition is nominally 6% tin with copper and nickel about 1%. SAE 770 and 780 are typical compositions for the 6% tin-aluminum. SAE 780 is bonded to a steel back with $1^1/_2$% silicon addition as well.

In the early 1940s, it was discovered that added silicon considerably improved the score resistance of aluminum bearings. Cadmium additions from 1 to 3% were also found to improve bearing properties. Examples of this class of materials are SAE 781 and 782. These alloys are bonded to steel backs and typically overplated with a thin babbitt to provide bearings for heavy duty automotive and diesel application; 11% silicon aluminum bearing alloys are used in Europe for highly loaded diesel bearings.

Special techniques were found to increase the tin content of tin aluminum. A 20% tin-aluminum material was developed in Great Britain in 1958. The composition is nominally 20% tin and 1% copper, with special cold-working and heat-treating to provide a reticular tin-aluminum which is bonded to a steel back without any overplate. It has become the dominant bearing material in European automotive use.

An alternative to tin-aluminum was developed in this country in 1968. This bimetal aluminum bearing featured a lead babbitt that was mixed into the aluminum bearing materials by a novel casting process to develop a lead gradient across the thickness of the aluminum alloy. The nominal composition of the bearing material of the surface was 10% lead, $1\frac{1}{2}$% tin, 2% silicon, 1% cadmium, 1% copper, $\frac{1}{2}$% magnesium, with the balance aluminum. This alloy, bonded to a steel back, was developed to fill the gap between the babbitts and the high performance, more expensive trimetal bearings. Both the high tin and high lead aluminum materials operate satisfactorily on soft crank journals in a wide variety of connecting rod and main bearings in modern automotive engines.

More recently, a sintered lead-aluminum material has been developed for automotive applications. Nominal composition is $8\frac{1}{2}$% lead, 4% silicon, $1\frac{1}{2}$% tin, $\frac{1}{2}$% copper, and the balance aluminum. These lead-containing aluminums have excellent antiscore properties, good embedability and fatigue strength, and excellent corrosion resistance. The tin addition confers adequate corrosion resistance to the lead contained in the aluminum structure, while the copper improves fatigue strength.

Zinc-aluminum materials are of interest in Germany and Japan. A 5% zinc-aluminum alloy with 1% addition of nickel, lead, magnesium, and silicon are used as solid wall automotive main bearings. This bearing is overplated with a lead-copper alloy to improve compatability. Aluminum alloys of higher zinc content have been publicized in the U.S. but little comparative performance data are available.

Silver

Silver was widely used in high-performance aircraft engine bearings in World War II. Their use surged when it was found that a plated, lead-tin or lead-indium overlay (0.1- to 0.025-mm thick) greatly improved the reliability of this bearing. About 0.3-mm silver (99.9% pure) was electroplated onto a carbon-steel back. These bearings are used today in specialized applications in high-performance diesel engines and turbochargers. These bearings are readily made where there is access to a conventional machine shop and to plating facilities. When properly manufactured, they provide very high load capacity and tend to have a forgiving mode of failure. After a momentary overload has caused high temperature and resulting shut down, these bearings can often be restarted to perform satisfactorily. This self-healing characteristic can be a big recommendation for use in experimental mechanisms when only a few bearings are needed. Particular attention should be paid to proper lubricant choice and maintenance to prevent scoring and corrosion damage.

POROUS MATERIALS

Porous metal bushings are widely used all around us for fractional horsepower electric motors, fans, electric can openers, and the like. They can be premachined or bored in place, with many bushings being made with simple, self-aligning capability. Since these porous metal bushings are made by sintering metal powders, they are available in a large number of material combinations; however, most porous metal bearings are made of bronze or iron-based materials. By far the most common sintered bushing is 90% copper and 10% tin in which the pores have been filled with oil. Since they have from 10 to 35% interconnected porosity, the oil flows freely from the oil reservoir to the bearing surface. Other materials used are leaded bronzes, iron bronzes, iron with copper, and leaded iron.

Table 6
LIMITING SERVICE FACTORS FOR NONMETALLIC AND
POROUS METAL BEARINGS

Bearing material	PV limit		Max temp (°C)	Load capacity (MPa)	Max speed (m/sec)
	MPa·m/sec	psi × ft/min			
Polyimides	4	110,000	260	—	—
Porous bronze	1.8	50,000	125	14	6
Porous iron	1.1	30,000	125	21	2
PTFE fabric	0.9	25,000	250	400	0.8
Filled PTE	0.5	15,000	250	17	5
Carbon-graphite	0.5	15,000	400	4	13
Wood	0.4	12,000	70	14	10
Phenolics	0.18	5,000	120	41	13
Acetal	0.10	3,000	100	14	3
Nylon	0.09	3,000	90	14	3
PTFE	0.04	1,000	250	3	0.3
Polycarbonate	0.03	1,000	105	7	5

PLASTIC AND OTHER NONMETALLIC BEARING MATERIALS

Plastic-based materials are often used both dry and lubricated in applications where speeds, loads, and temperatures are low. They are inexpensive and generally compatible with steel surfaces. The least expensive bushing for minor bearing requirements is molded of nylon or polyacetal. While these materials can operate without lubricant, their durability is much increased with some type of lubrication.

Service applications of plastic bearings are limited by thermal conductivity, thermal expansion problems, and thermal degradation. In fact, an empirical thermal relationship has been developed that limits the product of the applied load and speed for unlubricated applications. The generally accepted service limit for these materials is given by a limiting PV product where P is load in pascals and V is surface speed in meters per second. Typical values are given in Table 6 for a range of materials.[1,2] Most experimenters develop PV limits such that the wear will not exceed a given value in a certain number of test hours. Since the tests vary, there is often disagreement between limiting values. Maximum PV values can often be increased by using a steel back with a thin sheet of the plastic. Even higher bearing loads are possible when the plastic is bonded directly to a sintered metal matrix which is in turn bonded to a steel back to improve thermal conductivity and control thermal expansion.

When high-temperature plastics such as PTFE or polyimides are used, superior properties are obtained but the bearings are no longer inexpensive. They do have the advantage of being able to operate at high temperature without lubricant in specialized applications. Relatively low-friction coefficients are possible with unlubricated PTFE-based bearings at high load and at low speed.

Another method of utilizing plastics combines a thermosetting resin such as phenolic or polyester with a woven fabric such as cotton, linen, or asbestos. They are strong — being suitable for high-impact applications such as rolling-mill bearings. They are also useful in marine applications and respond well to water lubrication. Due to thermal instability, they are not useful above about 107°C.

Rubber bushings also perform well with water lubrication. These bearings are usually fluted or dimpled and are cooled by axial water flow through the bearing. When combined with a hard, rust-resistant shaft, they resist dirt damage.

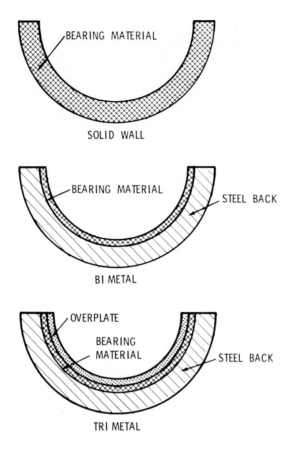

FIGURE 5.   Sleeve-type half bearings.

Carbon graphites are worthy of consideration for specialized application. Their self-lubricating property enables applications where contamination from a conventional lubricant would be undesirable — such as in textile or food processing machinery. They are also useful for temperatures up to 750°F and are resistant to attack from most chemicals and solvents. Since graphite depends upon absorbed water vapor for its excellent frictional characteristics, care must be exercised in high-temperature, high-speed, or high-altitude applications which would tend to remove the valuable moisture. Special proprietary processes are available to transfer the surface to silicon for extreme resistance to wear. Care must be exercised to protect these materials from mechanical shock, because they are brittle. This problem can be reduced by using a metal-filled carbon graphite which also provides improved material compatibility. Silver, copper, and lead are popular metal additives to carbon graphite.

BEARING CONSTRUCTION

Solid Wall Bearings

Solid wall bearings (Figure 5) can be made from most bearing materials. Two areas of concern are retention in the receiving hole and maintenance of the required press fit throughout the operating range. In general, the common solid wall materials (bronze, aluminum alloys, and plastics) have higher coefficients of thermal expansion than do iron or steel housings. This causes problems when the bearing and housing are exposed to a wide range of operating temperatures. If sufficient press is used to retain the bearing at low temperature, the material may yield when subjected to higher temperature — thereby losing the press at low temperature. Consequently, good practice dictates that retaining means other than the simple press be relied upon for solid wall bearings.

Bi-Metal

Bi-metal bearings involve a steel backing (or occasionally bronze) lined with a bearing material. The steel backing tends to strengthen the bearing material, increases its fatigue and compressive strength, and eases retaining the bearing within the receiving hole over a wide range of service conditions. Substantial press fits may be used to aid in the retention of a bushing or bearing. Rolled bushings — with and without locking lug construction — are used in many applications. Where sleeve half-bearings are used, as in automotive engine applications, bearing crush is used to provide the required press fit for adequate retention. The backing steel may be stressed to the yield point in applications requiring high crush.

The main complication that arises with the bi-metal bearing is failure of the bearing material bond to the back. Various techniques are used to ensure a good bond in commercial bearings. Some manufacturers of aluminum bearings use a thin layer of pure aluminum between the steel back and the aluminum bearing alloy. Other manufacturers employ elaborate chemical and mechanical cleaning operations immediately prior to the bonding operation. Copper-lead bearings can achieve satisfactory bond by simply casting the material directly onto the steel back. Heating of sintered bearing material in a controlled-atmosphere furnace automatically sinters the material and simultaneously effects the bond. Some bi-metal bearings involving plastics are manufactured by first sintering a porous metal matrix layer onto the steel back and then causing a plastic material to permeate the matrix and leave a thin coating over top. A similar technique is used with lead-base babbitt to permeate a sintered copper-nickel matrix.

Tri-Metal

In these bearings an additional layer is added, most often composed of a thin electroplated surface of lead and tin, or lead, tin, and copper. This thin layer can substantially increase the score resistance, embedability, and conformability of the basic bi-metal construction. Addition of copper reportedly increases strength of the overplate and reduces the wear rate. Many overplates use a nickel barrier layer of approximately 1-μm thickness under the overplate which may range from 10- to 25-μm thick. This barrier layer reduces diffusion of tin from the overplate to avoid leaving the lead in a tin-depleted state where it is subject to corrosion by organic acids in the oil. This is particularly useful for copper-lead bearings, but the nickel barrier is also used with some aluminum bearing constructions. After extended operation at elevated temperature the tin is found to diffuse into the nickel layer to form nickel-tin intermetallic compounds, some of which are detrimental to score resistance.

In the elegant gridded-silver bearing a pure silver layer is bonded to a steel back and then gridded and processed so that about 50% of the surface is comprised of about 50% lead-tin babbitt. Such bearings were used in high-performance aircraft engines in World War II. These bearings have superior fatigue strength, outstanding antiscore characteristics, and excellent embedability. They resist corrosion damage in most engine oils. A somewhat less expensive bearing having almost as good performance can be made by substituting oxygen-free, high-conductivity copper for the silver.

REFERENCES

1. **Anon.,** Sliding-Bearing Materials, *Mach. Design,* 53(14), 148, 1981.
2. **Booser, E. R.,** Bearing materials, in *Encyclopedia of Chemical Technology,* Vol. 3, 3rd ed., John Wiley & Sons, New York, 1978, 670.
3. **Pratt, G. C.,** Materials for plain bearings, *Int. Metall. Rev.,* 18, 62, 1973.
4. **Roach, A. E., Goodzeit, C. L., and Hunnicutt, R. P.,** Scoring characteristics of thirty-eight different elemental metals on high-speed sliding contact with steel, *Trans. ASME,* 78, 1659, 1956.
5. **Crankshaw, E.,** Mechanical features of steel backed bearings, in *Sleeve Bearing Materials,* American Society for Metals, Metals Park, Ohio, 1949, 150.
6. **Roach, A. E., and Goodzeit, C. L.,** Why bearings seize, *Gen. Motors Eng. J.,* 2(5), 25, 1955.

SLIDING BEARING DAMAGE

H. N. Kaufman

INTRODUCTION

Bearings in service are damaged by various causes, which in some cases can be determined if the damage is discovered before it becomes catastrophic. Excessive vibration or mis-alignment of parts can cause damage by imposing much greater loads than were anticipated in the design. Poor bonding of the bearing material to its backing, too little or too much bearing clearance, contaminated environments, too high or too low temperatures, and high humidity can initiate trouble. Bearing damage is often discovered during normal inspection of machinery. In other cases, damage is indicated by machinery instrumentation such as temperature and/or vibration sensors. Damage in journal and thrust bearings can usually be classified into the following categories: fatigue, wear (mechanical, electrical, and thermal), wiping, corrosion, and erosion.

FATIGUE

Fatigue occurs when a bearing is subjected to loads which impose cyclic stresses in the metal. These stresses can be produced (1) by cyclic loads as in automotive bearings, (2) from vibratory conditions existing in the machine, or (3) as a result of flexing of the bearing structure due to an inadequate bearing and/or support design. Improper bonding of the bearing material to its backing can also result in fatigue by flexing.

Fatigue has been observed in most bearing materials, but it occurs more often in the babbitt materials which have low fatigue strengths. Fatigue strength can be increased, as in automotive bearings which are subject to high cyclic loads, by using thin babbitt linings as indicated in Figure 1.

Fatigue results in initiation of cracks at the surface of the bearing which then propagate into the bearing material. In some cases these cracks intersect within the thickness of the bearing material and in other cases the cracks extend to the bond, turn, and propagate above the bond but parallel to it (Figure 2). In either case, intersecting cracks result in pieces of babbitt which are interlocked similar to a jigsaw puzzle. Eventually, pieces of babbitt become loose (Figure 3) and small fragments are carried into the bearing clearance by the journal. Wiping damage (surface melting and smearing) is usually observed in conjunction with fatigue damage.

Babbitt fatigue of a 178-mm (7-in.) diameter journal bearing is shown in Figure 4. Fatigue damage has extended to the bond in some areas, as indicated by the machined grooves in the steel backing. The babbitt in a turbine bearing can be as thick as 13 mm (0.5 in.). Similar fatigue can also occur in thin babbitt in automotive bearings subjected to severe cyclic loads. Both thrust and journal bearings are subject to fatigue damage.

WEAR IN BEARINGS

Wear is the removal of material from the bearing and/or the journal surfaces. Mechanical, electrical, or thermal phenomena, or combinations of these phenomena can cause wear.

Mechanical Wear

Mechanical wear results from the abrasive action of debris which enters the bearing with the lubricant. This debris often results from the processes used in the manufacture of the

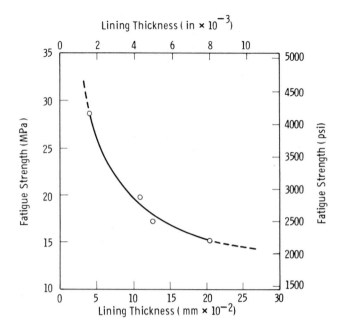

FIGURE 1.   The effect of lining thickness on babbitt fatigue strength. (From Szeri, A. Z., Ed., *Tribology: Friction, Lubrication, and Wear,* Hemisphere Publishing, Washington, D.C., 1980. With permission.)

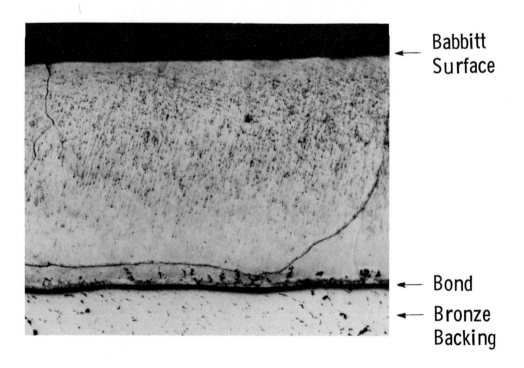

FIGURE 2.   Section through babbitt thickness showing advanced stage of fatigue cracking in babbitt with cracks extending to bond and intersecting. (Magnification × 75.) (From Burgess, P. B., *Lubr. Eng.*, 9(6), 309, 1953. With permission.)

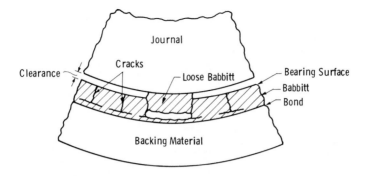

FIGURE 3.   Sketch showing babbitt fatigue initiating wiping damage.

FIGURE 4.   Babbitt fatigue in 178-mm (7-in.) diameter turbine bearing
with thick babbitt layer.[1,2]

machinery in which the bearings are used and is in the form of weld spatter, grinding wheel abrasive, and foundry sand. Other debris can enter the bearing from the environment.[3] Regardless of the source, the debris can cause scoring and tracking of the bearing surface and embedding of the debris in the surface. Journal scoring can also occur, but its severity is dependent upon the relative hardness of the debris and journal materials. In some cases the debris becomes completely embedded in the bearing material and results in only superficial wear damage with no significant effect on bearing performance. In other cases, severe wear damages both the bearing and journal. Increased wear and ultimate failure of the bearing occurs. Figure 5 shows circumferential scoring, tracking, and embedded debris in a 76-mm (3-in.) diameter tin-base babbitt bearing with an embedded debris particle at the end of a score mark.

Self-propagating mechanical wear by debris can also occur. One such type of wear has

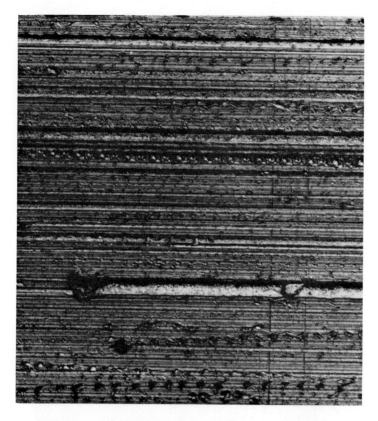

FIGURE 5. Scoring, tracking, and embedded debris in a tin-base babbitt bearing. (Magnification × 15.)[1,2,7]

FIGURE 6. Machining-type wear of a high-chromium (12%) steel journal.[1,2]

FIGURE 7. Damaged babbitt bearing used with journal of Figure 6.[1,2]

been termed wire-wooling or machining-type wear because the debris generated resembles wire wool or metal turnings by a machine tool.[4,5] In this type of wear the journal is usually more severely worn than the bearing. Severity of the damage is dependent upon the chromium content of the journal steel. High chromium, e.g., 12% chromium, is particularly susceptible to this type of wear damage as seen in Figure 6. Figure 7 is the mating bearing which was used. The damage is initiated by debris embedded in the bearing material which generates additional journal steel debris and forms a hard steel scab in the babbitt surface. This scab acts as a tool which further propagates the wear and often results in catastrophic damage. Simple scoring damage on a journal can sometimes be incorrectly identified as wire-wooling or machining-type wear, but in many cases is not self-propagating.

Another type of mechanical wear results from self-loading or radial binding of a bearing and its journal. Jamming of debris in the bearing clearance, too tight a radial fit, or dimensional interference from differential thermal expansion[6] can cause high wear of both bearing and journal. Figures 8 and 9 show a graphite bearing and its journal which were worn by self-loading. Both axial and circumferential cracks occurred in the graphite bearing.

Electrical Damage

Wear is sometimes experienced in rotating machinery as a result of the passage of current between the bearing and its mating surface, journal, or thrust runner.[7,8] Sparking between the surfaces causes pitting damage to both surfaces. Pits on the harder journal or runner surfaces are usually considerably smaller than those on the bearing. Multiple pits, closely spaced, produce a frosted appearance of the surfaces; and the removal of fused metal particles cause the surfaces to be rough. The rough surfaces produce further wear by mechanical abrasion. An additional consequence of sparking is deterioration of the lubricant and possible contamination of the lubricant and the lubricating system by spark debris. In extreme cases, the passage of current can cause an increase in the temperature of the parts which may damage the bearing or the lubricant.

FIGURE 8.   Wear and cracking due to self-loading of a graphite bearing.[1,2]

FIGURE 9.   Damage on journal used with bearing on Figure 8.[1,2]

The extent and size of electrical pits depend on many variables including the voltage, current, film thickness, type of material, etc.[8] Figure 10 is a portion of a tin-base babbitt sleeve bearing in which a high current (30 A) caused pits ranging in diameter from 0.20 mm (0.008 in.) to about 0.25 mm (0.010 in.). Figure 11 is a portion of a tin-base babbitt thrust bearing in which a low current (2.5 mA) produced microscopic pits. In many cases pitting occurs at potentials much less than 1 V.

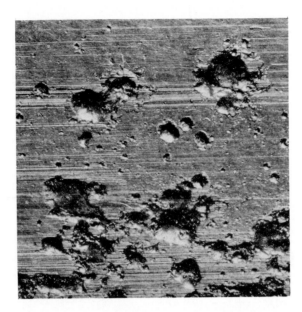

FIGURE 10.   Electrical pitting in tin-base babbitt produced by high current. (Magnification × 15.) (From Boyd, J. and Kaufman, H. N., *Lubr. Eng.*, 15(1), 28, 1959. With permission.)

FIGURE 11.   Electrical pitting in tin-base babbitt produced by low current. (Magnification × 15.) (From Boyd, J. and Kaufman, H. N., *Lubr. Eng.*, 5(1), 28, 1959. With permission.)

Regardless of their size, electrical pits have a characteristic appearance. The bottoms are rounded and have a smooth, shiny, melted appearance (Figure 10). Usually the periphery of the pit at the bearing surface has a ridge of melted metal. In some cases this ridge is worn away by contact with the journal. For cases in which wiping damage has been superimposed, electrical pitting can be identified as the cause by examining the harder journal surface which operated against the bearing.

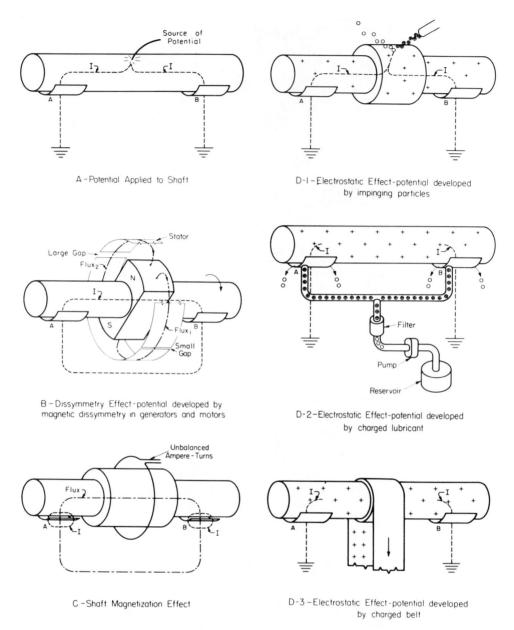

FIGURE 12.   Principal sources of bearing current in rotating machinery. (From Boyd, J. and Kaufman, H. N., *Lubr. Eng.*, 15(1), 28, 1959. With permission.)

Figure 12 shows schematically some sources of bearing currents in rotating machinery. Table 1 reviews these sources and methods of eliminating or reducing the bearing currents. Because bearing currents may be produced by a variety of conditions, no single method of measuring potentials is suitable in all cases. Table 1 indicates the best location for taking measurements. In the case of the dissymmetry effect, measuring the potential between the shaft and one bearing may be unreliable. This is because both bearings are in series with the generated emf and a temporary large resistance in the one bearing may make the potential across the other bearing negligible. The most reliable method is to measure the potential between the extremities of the shaft. Using this method on ordinary electrical apparatus with

Table 1
PRINCIPAL SOURCES OF BEARING CURRENT AND METHODS OF CURRENT CONTROL

Type[a]	Source of bearing current	Location of potential differences	Type of current	Method(s) of current control
A	*Potential applied to shaft* Intentional or accidental application of potential to shaft	Between shaft and each bearing	AC or DC depending upon source	Insulate all bearings or Ground shaft
B	*Dissymmetry effect* Variation of flux with angular position of shaft due to magnetic dissymmetry in machine	Between extremities of shaft Between each bearing and the shaft; (checking for the possibility of this type of bearing currents by measuring potential difference between the shaft and one bearing not always reliable, see text; better to check between extremities of shaft)	AC, Frequency depending upon construction and speed of machine	Insulate one bearing; if one bearing of a machine generating bearing currents is insulated and if the shaft on that end of the machine is metallically connected to the shafts of other machines, all bearings of the other machines should be insulated. or Ground *both* ends of shaft
C	*Shaft magnetization effect* (also referred to as Homopolar effect) Unbalanced DC or AC ampere-turns encircling shaft causing axial magnetization of shaft	Between points on the shaft located at opposite ends of each bearing Between each end of each bearing and its journal	DC or AC depending upon frequency in ampere turns	Balance ampere turns by addition of neutralizing coil or Increase reluctance of magnetic circuit by substitutung non-magnetic material for magnetic material in bearing support or Ground *both* ends of *each* journal
D	*Electrostatic effect* Potential developed by impinging particles	Between shaft and each bearing	DC, Measurements apt to be erratic (see text)	Insulate all bearings for high voltage or Ground shaft
	Potential developed by charged lubricant	Between shaft and each bearing	DC, Measurements apt to be erratic (see text)	Ground shaft and all bearings

Table 1 (continued)
PRINCIPAL SOURCES OF BEARING CURRENT AND METHODS OF CURRENT CONTROL

Type[a]	Source of bearing current	Location of potential differences	Type of current	Method(s) of current control
D	*Electrostatic effect*			
	Potential developed by impinging particles	Between shaft and each bearing	DC, Measurements apt to be erratic (see text)	Insulate all bearings for high voltage or Ground shaft
	Potential developed by charged belt	Between shaft and each bearing	DC, Measurements apt to be erratic (see text)	Insulate all bearings for high voltage or Ground shaft

[a] See Figure 12.

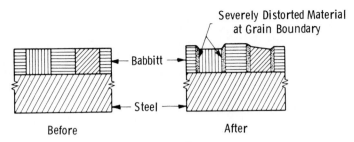

FIGURE 13.   Sketch showing effect of thermal cycling on tin-base babbitt grains.[1,2]

FIGURE 14.   Surface mottling on 178-mm (7-in.) diameter bearing due to anisotropy of tin-base babbitt.[1,2]

journal bearings, it was found that potentials less than 300 mV cause no significant electrical damage. The corresponding value for machines with ball and roller bearings is 100 mV.

Since electrostatic effects are greatly influenced by humidity and by surface conditions, the measurement of electrostatic potential is apt to be extremely erratic. Absence of electrostatic potential during a set of measurements does not necessarily mean that such potentials are not present under other conditions. Generally speaking, electrostatic potentials ordinarily do not produce sustained currents of large magnitude. The intermittent charging and discharging, however, can eventually produce enough bearing damage to cause failure.

The erratic nature of electrostatic potentials makes it difficult to set practical limits for satisfactory operation. It is known that peak voltages of 20 V or more can produce bearing damage. Reducing the voltage to the order of 1 V by some form of grounding device ordinarily eliminates the trouble.

The main methods of eliminating or reducing damage due to bearing currents include: (1) eliminating the source, (2) insulating the machine parts, (3) grounding the shaft, and (4) modifying the machine design.

Damage from Thermal Effects

The physical properties of some materials, such as the tin-base babbitts can differ along different axes of the grains making up their structure. Such anisotropic properties, coupled with differences in orientation of grain axes, can result in grain distortion when thermal cycling is imposed. This effect is shown schematically in Figure 13. If this grain distortion occurs in a babbitt bearing, the journal can contact the distorted or raised grains and result in slight wear or burnishing. This produces a mottled appearance of the bearing surface as shown in Figure 14. Mottling is usually not detrimental to bearing performance. However,

FIGURE 15. Wiping damage on a 178-mm (7-in.) diameter bearing.

FIGURE 16. Developed view of babbitt smeared by wiping.[1,2]

in some cases of severe grain distortion, cracks can occur in the babbitt surface along the grain boundaries as fatigue from thermal cycling.

Another thermal phenomenon results from the reduced strength properties of babbitt with increasing temperature. At elevated temperatures babbitt will undergo creep with rippling of the surface and subsequent wiping.[9]

WIPING

Wiping is the smearing or removal of bearing material from one point and the redeposition at another point on two surfaces in sliding contact. Superficial wiping in which bearing performance is not significantly affected can occur from either a temporary overload or temporary loss of lubricant. If either the overload or loss of lubricant is of long duration, severe damage frequently results. Bearing misalignment often results in wiping damage. Wiping damage is shown in Figure 15 on a 178-mm (7-in.). diameter bearing. Babbitt smeared by wiping is shown in the developed view of Figure 16.

FIGURE 17. Section through babbitted steel showing babbitt blister formation at locations above inclusions in steel.

Wiping sometimes is the indirect result of blistering at the interface of babbitt metal bonded to steel. This is a rare occurrence caused by hydrogen gas, trapped in the steel during manufacture, later diffusing to the interface where sufficient pressure is developed to cause the babbitt to blister and to be wiped by the journal or runner. Figure 17 is a section through babbitted steel which shows the inclusions in steel through which the gas can migrate to cause a blister. Figure 18 is a section through a blister showing the separation at the bond.

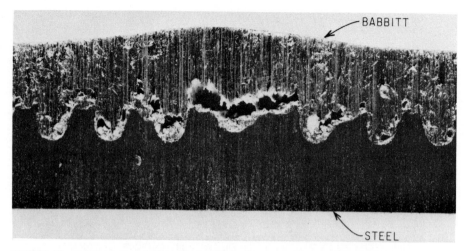

FIGURE 18. Section through babbitt showing blister due to gas diffusion from steel backing. (Magnification × 2.)[1,2]

FIGURE 19. Lead corrosion on surface of a 127-mm (5-in.) diameter lead-base babbitt bearing. (From Boyd, J. and Kaufman, H. N., *Lubr. Eng.*, 15(1), 28, 1959. With permission.)

CORROSION

Chemical

Chemical attack of the bearing material can result in either the removal of some component of the material or the buildup of a deposit due to chemical reaction on the bearing surface.[10] The corrosive agents may be compounds formed by oil deterioration or certain organic acids, or they may be contaminants in the environment. Lead corrosion can occur in lead-base babbitts and leaded bronze bearing metals; copper corrosion can occur in copper-lead bearing metals.[11]

Figure 19 shows lead corrosion on the surface of a lead-base babbitt bearing. Figure 20

FIGURE 20. Corrosion and fatigue of a leaded-bronze railroad diesel bearing.[1,2]

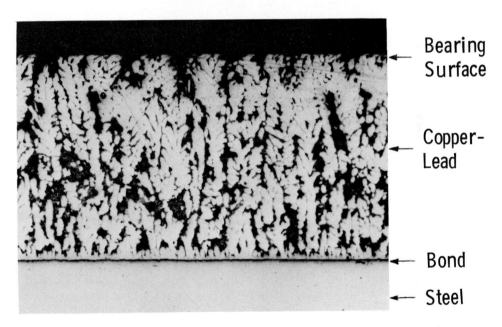

FIGURE 21. Section through a copper-lead bearing showing lead corrosion. (Magnification × 150.) (From Burgess, P. B., *Lubr. Eng.*, 9(6), 309, 1953. With permission.)

shows leaded bronze corrosion combined with fatigue on a 229-mm (9-in.) diameter railroad diesel bearing. Figure 21 shows a metallographic section through a copper-lead bearing in which lead corrosion occurred. The black voids at the bearing surface were pockets of lead removed by the corrosive attack. Figure 22 is a metallographic section showing copper corrosion in a copper-lead bearing. The white copper grains at the bearing surface have been chemically attacked.

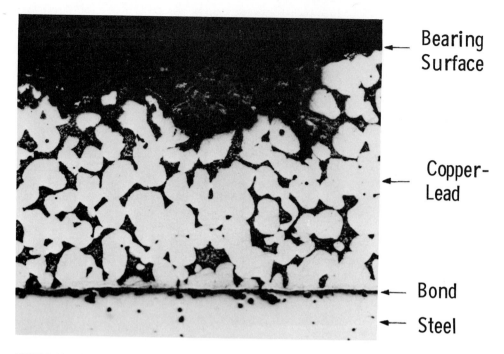

FIGURE 22.   Section through a copper-lead bearing showing copper corrosion. (Magnification × 150.) (From Burgess, P. B., *Lubr. Eng.*, 9(6), 309, 1953. With permission.)

FIGURE 23.   Fretting corrosion of 25-mm (1-in.) diameter bronze bushing and steel pin.[1,2]

Fretting

Fretting corrosion is produced on materials in intimate contact which are subjected to the combined action of chemical attack and small oscillatory motions. The 25-mm (1-in.) diameter bronze bushing and steel pin in Figure 23 show damage due to fretting corrosion. Figure 24 shows an enlargement of the fretting damage in the bushing. Fretting is less severe on the harder steel pin.

FIGURE 24. Enlargement of fretting corrosion of bronze bushing of Figure 23. (Magnification × 15.)[1,2]

FIGURE 25. Cavitation erosion of a 44-mm (1.75-in.) diameter aluminum bearing.[1,2]

EROSION

Erosion is the removal of material from the bearing surface by fluid action which results in the formation of voids or pits in the surface. It can be produced by changes in the direction of flow of high-velocity fluid streams or by the abrasive action of debris in the fluid stream as it impinges on the bearing material.

Cavitation erosion is a type of erosion in which the formation and collapse of gas bubbles in the lubricant produces high localized pressures which result in fatigue pitting of the bearing surface. Figure 25 shows erosion damage on a 44-mm (1.15 in.) diameter aluminum bearing from a high-speed gas compressor. Figure 26 is an enlargement of the damage.

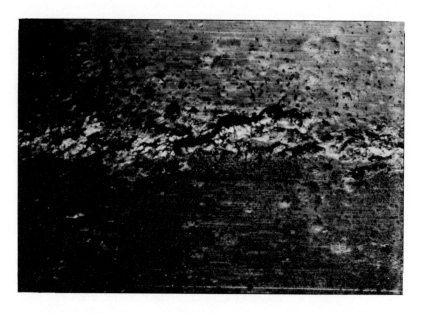

FIGURE 26. Enlargement of cavitation erosion of bearing of Figure 25.

REFERENCES

1. **Szeri, A. Z., Ed.,** *Tribology: Friction, Lubrication, and Wear,* Hemisphere Publishing, Washington, D.C., 1980.
2. **O'Connor, J. J. and Boyd, J., Eds.,** *Standard Handbook of Lubrication Engineering,* McGraw-Hill, New York, 1968.
3. **Elwell, R. C.,** Foreign object damage in journal bearings, *Lubr. Eng.* 34, 187, 1978.
4. **Kaufman, H. N.,** An evaluation of various journal materials with regard to machining type wear, *Lubr. Eng.,* 24(5), 204, 1968.
5. **Dawson, P. H. and Fidler, F.,** Wire-wool type bearing failures. I. The formation of the wire wool, *Proc. Inst. Mech. Eng.,* 180(21), 1965—1966.
6. **Conway-Jones, J. M. and Leopard, A. J.,** Plain bearing damage, Proc. 4th Turbomachinery Symp., Institute of Mechanical Engineers, 1975.
7. **Boyd, J. and Kaufman, H. N.,** The causes and control of electrical currents in bearings, *Lubr. Eng.,* 15(1), 28, 1959.
8. **Kaufman, H. N. and Boyd, J.,** The conduction of current in bearings, *ASLE Trans.,* 2(1), 67, 1959.
9. **Booser, E. R., Ryan, F. D., and Linkinhoker, C. L.,** Maximum temperatures for hydrodynamic bearings under steady load, *Lubr. Eng.,* 26, 226, 1970.
10. **Fowle, T. I.,** Problems in the lubrication systems of turbomachinery, *Proc. Inst. Mech. Eng.,* 186(60/72), 705, 1972.
11. **Burgess, P. B.,** Mechanisms of sleeve bearing failure, *Lubr. Eng.,* 9(6), 309, 1953.

ROLLING ELEMENT BEARINGS

W. J. Derner and E. E. Pfaffenberger

ROLLING BEARING TYPES

Rolling contact bearings are generally categorized by the type of rolling element and the manner in which it is used. The most obvious divisions are between ball and roller bearings and between radial and thrust bearings, while some angular contact bearings are utilized for both radial and thrust loads.

Individual sections in this chapter will cover (1) types of rolling element bearings and their selection criteria, (2) dimensional standards, (3) characteristics of materials employed, (4) rolling bearing theory, (5) load, speed, and related application limits, (6) lubrication, and (7) failure analysis. Load rating for individual bearings, their application ranges, and many other details are available in catalogs and related literature from bearing suppliers.

Ball Bearings (Figure 1)

Radial and Angular Contact

The most common design of radial ball bearing is the Conrad type where, in general, five to nine balls are inserted between the inner and outer rings. Where a greater radial capacity is desired, filler type rings utilize notches on one shoulder so that more balls can be inserted. An optimum capacity is achieved through the use of a split inner or outer ring wherein a maximum number of balls can be inserted with the retainers. This latter design, however, requires external means for holding the ring halves together so that load can be divided between contacts on both halves.

Self-aligning ball bearings are available in double row varieties in which the outer ring raceway is of a larger radius than the ball and the inner has two raceways ground in it, one for each row of balls.

Thrust Ball Bearings

These are available in designs for single direction as well as double direction thrust and normally are found with 90° contact angles.

Roller Bearings

A variety of rollers have developed for use in bearings including the early spring-wound cylindrical, solid and hollow cylindrical, tapered, and spherical rollers. Retainers (cages) used to space the rolling elements may be land riding, roller riding, or supported on a raceway. Retainers may be (1) machined out of solid, (2) stamped and formed, (3) fabricated and fastened by cold heading or riveting, or (4) molded of one or two elements. Since retainer contact with the roller or ring involves some sliding, a lubricant should be chosen which is compatible with the nature of the contact as well as the material of the retainer.

Radial and Angular Contact (Figure 2)

Tapered roller bearings — Supported between two cones of different angles, the tapered roller centers itself between them and recognizes a certain axial force which maintains its contact with a lip or rib generally on the inner ring. Its contact with that rib involves sliding which must be lubricated to prevent wear and to dissipate the heat generated.

Cylindrical roller bearings — This type generally runs cooler than other roller bearings because of the narrow and uniform shape of the Hertzian contact with no more roller end contact with the ribs or flanges than is required to provide guidance or location. In high-

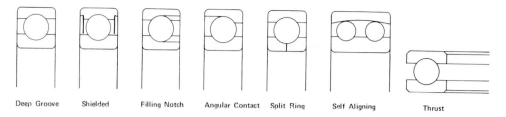

FIGURE 1.   Common types of ball bearings.

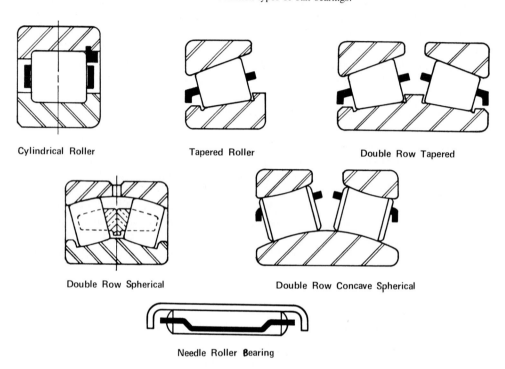

FIGURE 2.   Common types of radial roller bearings.

speed application, positive guidance is essential to reduce parasitic losses due to the working of the roller back and forth between the ribs.

Spherical roller bearings — These are produced in both asymmetrical and symmetrical roller designs which have their unique requirements for lubrication and cooling. The design with asymmetrical rollers has a flange contact similar to that in tapered roller bearings and requires careful lubrication. The symmetrical roller design is supplied either with a retainer guided roller or with stabilizing rings which utilize the retainer as a spacer or separator only. These latter two spherical bearings are comparable in their performance and generally are utilized in applications which require greater load carrying capacity and/or run at higher speeds. Concave roller designs are generally produced with stamped steel retainers and are sensitive to loading variations which may produce undesirable skewing of the rollers. Because of the reliance on formed steel retainers for guidance, these bearings generally run warmer than those with convex rollers.

Thrust Bearings (Figure 3)

Tapered roller bearings — Tapered roller bearings with contact angles of 45° and more are used for heavy duty thrust applications that include slow speed, high shock load appli-

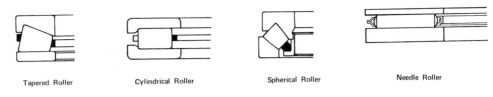

Tapered Roller   Cylindrical Roller   Spherical Roller   Needle Roller

FIGURE 3.   Roller thrust bearings.

cations similar to those in screw-down bearings in rolling mills. Since many of these run at low speeds or are even subjected to only momentary, limited rotation (and/or oscillation), lubrication requirements are highly specialized and require careful evaluation.

Cylindrical roller bearings — Conventional roller thrust bearings with flat plate surfaces are utilized in low-speed applications where high thrust capacity is essential. Tandem cylindrical roller thrust bearings are used in many applications wherein single-helical gears are run at low speeds. Because of the relatively high slip in these applications, as well as the generally severe contact conditions in all slow speed and/or oscillating operation, special attention to cooling and extreme pressure (EP) additives in the lubricant are required.

Spherical roller bearings — Spherical thrust bearings require the same attention as do most tapers, but absorb misalignments which would not be acceptable to cylindrical and tapered varieties. Spherical thrust bearings are provided with both symmetrical and asymmetrical roller designs. The latter are somewhat limited in having a moderately loaded rib or locating flange as a sliding contact which must be lubricated as with the tapered roller bearings. In lower-speed applications where hydrodynamic films are not readily generated at the rib roller end contact, wear becomes a significant factor limiting the life or function of these bearings. In all cases of angular contact, spherical or tapered roller thrust bearings, the relation of thrust to radial load must be carefully controlled to ensure that the bearing does not come apart. Manufacturers' recommendations must be carefully adhered to.

Mounted-Bearing Units

A significant proportion of rolling element bearings are supplied in integral housings with seals which offer advantages in that they do not require a large, continuous machined housing. While a great majority are grease lubricated, for adverse environmental conditions some are provided with complete lubricating systems to include cooling and filtration. For severe contamination, flushable seals (Figure 4) have lube fittings separated from the main lubrication system of the bearing. Some applications are so severe that frequent and heavy relubrication is relied on to purge the system of contaminants and to exclude water vapors due to "breathing" where intermittent operation is encountered.

ROLLING BEARING STANDARDS

Boundary Dimensions

The great majority of the world rolling element bearings are in compliance with boundary dimension plans adopted by the International Standards Organization (ISO). Domestic standards originated with the Anti-Friction Bearing Manufacturer's Association (AFBMA) have been taken over by American National Standards Institute (ANSI) which, as a participating body in ISO work, offers the most complete and authoritative set of standards for use in this country.

A great majority of the so-called inch series of bearings have been superceded by metric series which fit a number of well-established boundary plans.[1] Of particular note, a new series of tapered roller bearings has achieved a reduction in the multiplicity and complication of sizes with worldwide acceptance in an ISO standard.[2]

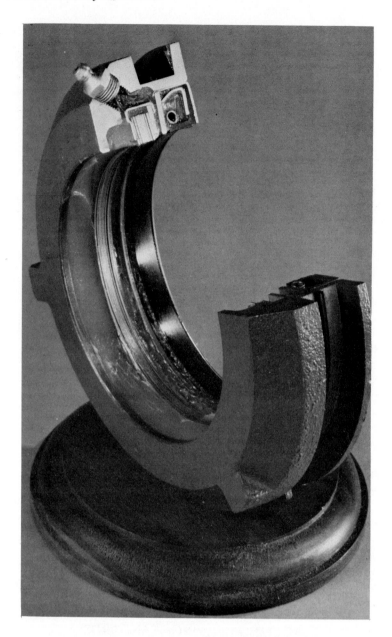

FIGURE 4. Flushable seals with lubricant fittings for severe contamination.

As will be seen in Figure 5, the basic metric series cover bore, OD, and width and fully describe the annuli for each increment of bore size. There are some exceptions, but over the range of 15 to 130 mm most series have bores with 5-mm increments. Above 130 mm increments are normally 10 mm or larger.

Most bearings sold in this country offer dimensions which are inch equivalents of even millimeters. For a full description of all standard bearings which are currently manufactured, it is recommended that ANSI be contacted for the current standards and then refer back to manufacturer's catalogs for availability of bearings fitting these boundary dimension series.

While it is possible to justify the design, development, manufacturing, and application of bearings which do not meet the standard boundary dimension plans, application of special

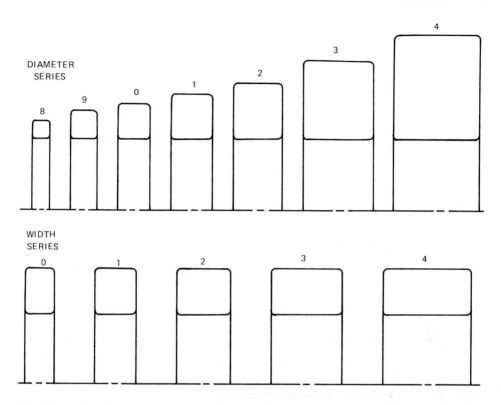

FIGURE 5. Graphical representation of the diameter and width series for radial bearings. (From *Metric Ball and Roller Bearings Conforming to Basic Boundary Plans,* ANSI/AFBMA Standard 20, Anti-Friction Bearing Manufacturers Association, Arlington, Va., 1977. With permission.)

designs requires careful consideration and prior evaluation. The useful life of most machinery is often extended by overhaul, at which time replacement of bearings may be essential. At overhaul the availability of new bearings may be a real problem and any selection of a bearing which does not fit standard published boundary dimensions must include a plan for resupply. In some industries the replacement bearings are procured at the time of original manufacture and then are stocked either by the user or the supplier. Machines manufactured in higher volumes quite often will employ special bearings designed or selected by the manufacturer and stocked by his dealers around the country. Typical would be the automotive service parts groups.

Because the wide variety of boundary plans, it is suggested that special bearings be considered only when weight or size limitations are a problem.

Internal Clearances

Also available through ANSI are standards for the clearance ranges for all sizes of bearings as shown in Tables 1 to 3. While the standard clearance range is normally designated as symbol 0, some manufacturers have their own nomenclature. Reference to bearing manufacturers' catalogs and sales literature will relate exact clearance ranges to requirements of applications for each type of bearing. It is desirable to coordinate quality control procedures with the supplier to ensure a minimum of confusion and delay in qualifying bearings for internal clearance.

BEARING MATERIALS

The four essential properties of bearing material—strength, fatigue resistance, wear resistance, and dimensional stability will be considered separately.

Table 1
RADIAL INTERNAL CLEARANCE, SINGLE-ROW, RADIAL CONTACT, BALL BEARINGS

Tolerance Limits in 0.0001 In.

Basic bore diameter (mm)		Symbol 2[a]				Symbol O (normal)[a]				Symbol 3[a]				Symbol 4[a]			
		Manufacturing limits		Acceptance limits		Manufacturing limits		Acceptance limits		Manufacturing limits		Acceptance limits		Manufacturing limits		Acceptance limits	
Over	Incl.	Low	High	Low	High	Low	High	Low	High	Low	High	Low	High	Low	High	Low	High
2.5	10	—	2.5	1	3.0	1.5	4.5	3	5	4	8	3	9	8	12	7	13
10	18	—	3.0	1	3.5	2	6	4	7	5	9	4	10	9	13	8	14
18	24	—	3.5	2	4.0	3	7	5	8	6	10	5	11	10	15	9	16
24	30	—	4	2	4.5	3	7	5	8	6	10	5	11	10	15	9	16
30	40	—	4	2	4.5	3	8	6	9	7	12	6	13	12	17	11	18
40	50	—	4	2	4.5	4	10	7	11	8	13	7	14	13	19	12	20
50	65	—	5	3	6	5	11	9	12	10	16	9	17	16	23	15	24
65	80	—	5	4	6	6	13	10	14	11	19	10	20	19	27	18	28
80	100	—	6	5	7	7	15	12	16	13	22	12	23	22	32	21	33
100	120	—	7	6	8	8	18	14	19	15	25	14	26	25	37	24	38
120	140	—	8	7	9	8	20	16	21	18	30	16	32	30	43	28	45
140	160	—	8	7	9	9	23	18	24	20	34	18	36	34	49	32	51
160	180	—	9	8	10	10	25	21	28	23	38	21	40	38	55	36	58
180	200	—	11	10	12	11	27	25	32	27	44	25	46	44	62	42	64

Note: Tolerance limits for radial internal clearance of single-row, radial contact ball bearings under no load (applicable to bearings of ABEC 1, ABEC 5, ABEC 7, and ABEC 9 Tolerance Classes).

[a] These symbols relate to AFBMA Standard Section 5 — Identification Code.

Table 2
RADIAL INTERNAL CLEARANCE, CYLINDRICAL ROLLER BEARINGS

Tolerance Limits in 0.0001 in.

Basic bore diameter (mm) Over	Incl.	Symbol 2[a] Low (Interch.)	Matched[c] Low	Matched[c] High	High (Interch.)	Symbol O[a] (normal) Low (Interch.)	Matched[c] Low	Matched[c] High	High (Interch.)	Symbol 3[a] Low (Interch.)	Matched[c] Low	Matched[c] High	High (Interch.)	Symbol 4[a] Low (Interch.)	Matched[c] Low	Matched[c] High	High (Interch.)
—	10	0	4	8	12	4	8	12	16	10	14	18	22	14	18	22	26
10	18	0	4	8	12	4	8	12	16	10	14	18	22	14	18	22	26
18	24	0	4	8	12	4	8	12	16	10	14	18	22	14	18	22	26
24	30	0	4	10	12	4	10	14	18	12	16	20	26	16	20	24	28
30	40	0	5	10	14	6	10	16	20	14	18	22	28	18	22	28	32
40	50	2	6	12	16	8	12	18	22	16	20	26	30	22	26	32	35
50	65	2	6	14	18	8	14	20	26	18	22	30	35	26	30	35	41
65	80	2	8	16	22	10	16	24	30	22	28	35	41	30	35	43	49
80	100	4	10	18	24	12	18	28	32	26	32	41	45	35	41	49	55
100	120	4	10	20	26	14	20	32	35	32	37	47	53	41	47	57	63
120	140	4	12	24	30	16	24	35	41	35	41	53	61	45	53	63	71
140	160	6	14	26	32	20	26	39	45	39	45	59	65	51	59	71	77
160	180	8	14	30	34	24	30	43	49	43	49	65	69	59	65	79	85
180	200	10	16	32	37	26	32	47	53	49	55	71	77	65	71	87	92
200	225	12	18	35	41	30	—	—	59	55	—	—	85	71	—	—	100
225	250	16	20	39	45	35	—	—	65	61	—	—	90	81	—	—	110
250	280	18	22	43	49	39	—	—	71	69	—	—	100	90	—	—	122
280	315	20	24	47	52	43	—	—	77	77	—	—	110	100	—	—	134
315	355	22	26	53	57	49	—	—	85	85	—	—	120	110	—	—	146
355	400	26	30	59	63	55	—	—	93	96	—	—	134	126	—	—	163

Interchangeable[b] refers to the outer Low and High limit columns of each symbol group.

Table 2 (continued)
RADIAL INTERNAL CLEARANCE, CYLINDRICAL ROLLER BEARINGS

Tolerance Limits in 0.0001 in.

Basic bore diameter (mm)		Symbol 2[a]		Symbol 0[a] (normal)				Symbol 3[a]				Symbol 4[a]		
		Matched[c]		Matched[c]		Interchangeable[b]		Matched[c]		Interchangeable[b]		Matched[c]		
Over	Incl.	Low	High	Low	High	Low	High	Low	High	Low	High	Low	High	High
400	450	28	—	61	75	106	108	—	—	140	153	—	—	179
450	500	33	—	71	81	118	118	—	—	155	165	—	—	202

Note: Tolerance limits for radial internal clearance of cylindrical roller bearings under no load. This table applies to bearings of bore, O.D. and width dimensions conforming to AFBMA Standard Section 2, Table 1 and chamfer dimensions conforming to AFBMA Standard Section 3, Tables 5.1 and 5.2

[a]   These symbols relate to the AFBMA Standard Section 5 — identification code (edition later than December 1962).

[b]   The term "interchangeable" refers to such assembly of rings and rollers that the separable ring *can* be replaced by any other ring of the same design and manufacture.

[c]   The term "matched" refers to such assembly of rings and rollers that the separable ring *cannot* be replaced by any other ring.

Table 3

**RADIAL INTERNAL CLEARANCE SELF-ALIGNING ROLLER
BEARINGS, METRIC**

Tolerance Limits in 0.0001 In.

Basic bore diameter (mm)		Bearings with cylindrical bore							
		Symbol 2[a]		Symbol O[a] (normal)		Symbol 3[a]		Symbol 4[a]	
Over	Incl.	Low	High	Low	High	Low	High	Low	High
14	24	4	8	8	14	14	18	18	24
24	30	6	10	10	16	16	22	22	28
30	40	6	12	12	18	18	24	24	32
40	50	8	14	14	22	22	30	30	39
50	65	10	17	17	26	26	36	36	47
65	80	12	20	20	32	32	44	44	57
80	100	14	25	25	39	39	53	53	71
100	120	17	31	31	48	48	64	64	83
120	140	20	38	38	57	57	75	75	95
140	160	24	43	43	65	65	87	87	110
160	180	26	47	47	71	71	95	95	122
180	200	28	51	51	79	79	103	103	133
200	225	32	55	55	87	87	114	114	149
225	250	36	59	59	95	95	126	126	165
250	280	39	67	67	103	103	138	138	180
280	315	44	75	75	110	110	145	145	197
315	355	47	79	79	122	122	161	161	216
355	400	51	87	87	134	134	177	177	236
400	450	55	95	95	145	145	197	197	260
450	500	55	103	103	162	162	216	216	284
500	560	59	110	110	173	173	236	236	307
560	630	67	122	122	189	189	256	256	335
360	710	75	138	138	209	209	276	276	362
710	800	83	153	153	228	228	303	303	397
800	900	91	169	169	256	256	338	338	440
900	1000	103	188	188	280	280	365	365	480

Note: Tolerance limits for radial internal clearance in self-aligning roller bearings under no
load. This table applies to bearings conforming to the Basic Plan for Boundary
Dimensions of Metric Radial Bearings, AFBMA Standards, Section 2, Table 1.

[a]   These symbols relate to the AFBMA Standard Section 5 — identification code.

Strength

Cyclic contract stresses may vary from 700 to 2800 MPa (100,000 to over 400,000 psi)
in operation and shock loads may range to 4100 MPa (600,000 psi). Fatigue life is inversely
proportional to the 7th to 10th power of the stress under the most heavily loaded rolling
element. Maximum hardnesses are limited by material sensitivity to cracking and metallurgical stability.

Due to the high-contact stress and subsurface shear stress in rolling element bearings,
surface or near-surface elements must be hardened to a minimum of Rockwell C59. To
support the high-subsurface stresses, the supporting core must be of equal hardness or provide
a hardness gradient that insures structural integrity and fatigue life. Strength and hardness

Table 4
APPROXIMATE TEMPERATURE LIMITS
FOR ROLLING BEARING STEELS[b]

Through-hardening steel	Carburizing Steel	Temperature	
		°C	°F
TBS-9	4118	1018	
Stroloy 503A	5120	1026	
52100[a]	8620	150	300
1070M	4620		
	4720		
	4320		
52100 Type 1	4820		
52100 Type 2	9310	175	350
440C	3310		
TBS-600	CBS-600	205	400
52CB		260	500
MHT			
14-4 (Mod. 440C)			
Halmo		315	600
M-50			
TBS-1000	CBS-1000	370	700
M-10		430	800
M-1		480	900
M-2			
WB-49		540	1000

[a]   Has been used successfully at 205°C (400°F) when heat stabilized, with increased internal clearance in bearing, and very light load.

[b]   Based on hot-hardness and temper resistance.

at operating temperature must be considered since commercial bearing materials are required to operate at 120°C (250°F) with no derating. At higher temperatures, either increased scatter in early life or reduced average life must be anticipated. Over 175°C (350°F), materials with high hot-hardness such as the tool steels are required. Table 4 gives commonly used materials and their temperature limits.

Where static loads are encountered with bearings not rotating, resistance to plastic indentation becomes important. Static capacities are normally chosen to insure no significant indentation of the raceways. Some bearings, however, can operate subsequently at lower loads with slight brinelling with no significant effect on system behavior. When carburized materials with higher core strength are chosen, these static loads should not exceed the crush strength of the case and it is advisable to determine these limits by laboratory tests.

Fatigue Resistance

Whereas high carbon chromium bearing alloys were once felt essential to adequate fatigue life, most materials, properly hardened, and with sufficient core strength to support the shear stress can insure satisfactory life and life dispersion. Fatigue resistance is also a function of residual stresses remaining after bearing manufacture. Tensile stresses are to be avoided and in many cases it is advisable to specially process inner rings in a manner to limit these. In those applications where abrasion, overheating, lubricant depletion, or abusive assembly stresses in the inner and outer rings are anticipated, provisions must be made to limit the stress or else reduced bearing life may be anticipated.

Wear

Fortunately, the more commonplace materials used today, namely 52100, TBS9 (Timken®), and the various carburizing steels have excellent resistance to wear under normal operating conditions with full film lubrication. Any abrasion is detrimental due to the damaging effects of increased osculation between the rolling elements and inner rings. Wear between the rolling elements and retainer can be controlled for the most part by proper selection of the retainer materials and by optimizing the geometry of the contact of the rolling element and retainer.

Dimensional Stability

In case of thru-hardening steels, operating temperatures must be limited to 120°C (250°F) or provisions made to accommodate potential distortion/growth of the inner rings. For higher operating temperatures, parts must be "stabilized" by additional heat treatment or tempering and/or deep freezing to insure transformation of all retained austenite. In the case of carburized parts where the volume of retained austenite is small, transformation is not significant. In the transformation of retained austenite in the finished part made of conventional thru-hardening materials, shrinking is encountered after relatively few hours of operation. This is followed by growth during extended operation or exposure to higher temperature. AFBMA standards are available as a guide in selection of parts for stability.

For corrosive conditions, it is advisable to consider means for protecting the bearing other than resorting to exotic materials. Some stainless steels are used occasionally in rolling element bearing for specialized applications and must be chosen with care. Fatigue data and relative life of these materials are not well substantiated in many cases.

Retainer Materials

Commercial bearings in past years generally used a leaded bronze or an aluminum bronze for retainers. More recently, ferrous materials have been used either as strip steel or castings. Recently the use of polymeric retainers has developed, primarily heat-stabilized nylon 6-6 both with and without glass reinforcement. In ball bearings polymers are of great value since they are better able to endure the adverse operating conditions induced by moment loads.

For high-temperature applications, particularly with marginal lubrication, coated retainers generally made of various ferrous materials have been selected. The aerospace industry has utilized an iron-silicon bronze, both plated and unplated, for adequate strength at higher temperatures. Oiling systems in early gas turbine engines could supply no additional lubricant to a bearing for long periods at startup. Under these conditions, break-in or sacrificial films were essential and the bronze retainers used were generally silver plated or silver plated with a lead-tin overlay.

Nodular-graphite iron retainers have good wear properties when properly prepared and resist some gases which may be found in lubricants. In ammonia compressors, they endure the environment detrimental to most normally used bronzes. Nylon polymers are sensitive to materials with free phosphorus on the surface, to a limited number of diluents and solvents, and to some lubricants.

Breakdown products must be anticipated from operating lubricants and their additive packages at high temperatures. Material selection must anticipate attack from EP additives at high temperature. Many boundary lubricants are particularly detrimental. Moisture is doubly harmful in that it can aid breakdown products to become corrosive and attack bearing surfaces.

ROLLING BEARING THEORY

Contact of Elastic Bodies

Two basic theoretical types of contact occur between elastic solids, namely point contact

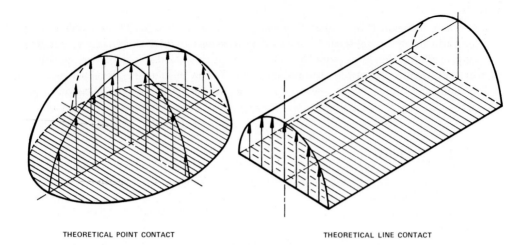

THEORETICAL POINT CONTACT   THEORETICAL LINE CONTACT

FIGURE 6.   Stress distributions over elliptical "point" contact area and rectangular "line" contact (no end effects).

and line contact. Point contact refers to the conjunction of two elastic bodies such that the initial contact, under no load, is a single point. As load is applied, the bodies deform elastically and the contact spreads out into a finite area of elliptical shape. Assuming that the pressure is entirely normal to the contact surface (i.e., there are no tangential or frictional forces) the contact pressure assumes a semiellipsoidal distribution as shown in Figure 6.

As defined schematically in Figure 7, typical osculation values for ball bearings are 51 to 52% for the inner race and 52 to 53% for the outer race. For spherical roller bearings, osculation is normally given as the ratio of the radius of curvature of the rolling element to the radius of curvature of the ring raceway. Typical values range from 94 to 99%. Osculation clearance is the clearance at the end of the effective length of contact between a roller and a raceway.

Theoretical line contact is that which occurs between two cylinders with parallel axes, or a cylinder and a flat plate of infinite length in the cylinder axial direction. Under load, the bodies deform elastically and the contact area becomes a rectangle. The distribution of pressure over this contact surface is uniform in the axial direction and elliptical in the direction perpendicular to the axis, as illustrated in Figure 6. Since actual bearing contacts are not infinitely long, true line contact is only approached under certain conditions in real bearings.

Roller Crowning

If a cylinder of some finite length is loaded against another cylinder or flat, a stress concentration develops at the end of contact due to the discontinuity. For lightly loaded cylindrical or tapered roller bearings, this stress concentration is not serious. With high load or with misalignment, however, it can drastically reduce bearing life. Therefore most (but by no means all) cylindrical and tapered roller bearings utilize crowned rollers such as illustrated in Figure 8. The full crown is a circular arc that results in a contact ellipse with a length to width ratio on the order of 50. To overcome the rather short length of contact with a full crown under light load, the partial crown provides a straight length of about 50 to 60% of the roller length, with only the ends relieved. This type of crown has two disadvantages: (1) stress concentrations under heavy loads at the junction between the crown and the straight length, and (2) less tolerance of misalignment than the full crown. At least one manufacturer provides a modified full crown: flatter in the center of the roller with increasing curvature toward the ends. Typical values of roller crown radius range from 1 to

BALL BEARING OSCULATION

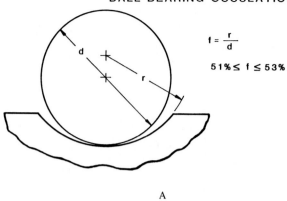

A

SPHERICAL ROLLER BEARING OSCULATION

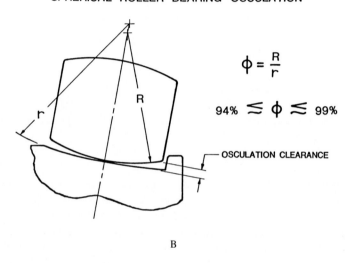

B

FIGURE 7. Osculation for (A) ball and (B) spherical roller bearings.

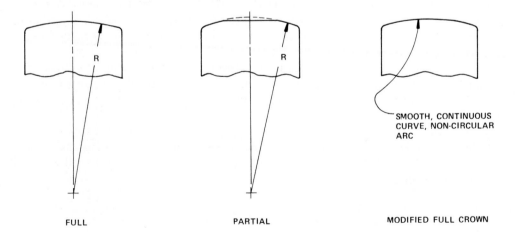

FIGURE 8. Three types of roller crowns.

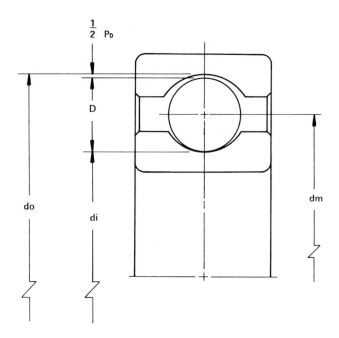

FIGURE 9. Radial ball bearing nonmenclature.

25 m (50 to 1000 in.). Osculation clearances are normally in the range of 5 to 50 μm (0.0002 to 0.002 in.).

Contact Stresses

The maximum principle stress, which is often used as a measure of load severity, is the compressive stress normal to the contacting surface at the center of the contact ellipse. Normally, however, one of the sheer stresses at some depth below the contact surface is decisive with regard to the load-carrying ability of the contact. Typically, maximum compressive stress for the most heavily loaded rolling element for normally loaded rolling bearings is in the range of 700 to 1400 MPa (100,000 to 200,000 psi), with roller bearings tending toward the lower end of that range, and ball bearings the higher. A high stress in a very heavily loaded rolling bearing would be 2500 MPa (360,000 psi). In a nonrotating bearing, stresses up to 4000 MPa (580,000 psi) and higher sometimes can be tolerated without significant impairment of subsequent operation. Reference 3 gives the most general method of calculating stresses in a concentrated contact, while References 4 to 8 give procedures particularly adapted to rolling element bearings. Hartnett[7] provides a means of estimating potential stress concentrations with crowned rollers. Hamrock[8] has developed a very simple method which may be carried out on a hand calculator.

Nomenclature

Definitions required for discussion of loads and motions within rolling bearings are listed below and are illustrated in Figures 9 to 12 for ball, cylindrical roller, spherical roller, and tapered roller bearings.

C	=	Basic load rating
D	=	Rolling element diameter
d_m	=	Pitch diameter of the rolling element set
F_a, F_r	=	Bearing applied axial load, radial load
f_c	=	A factor used to determine basic load rating (Equations 9 to 12); depends on geometry and material

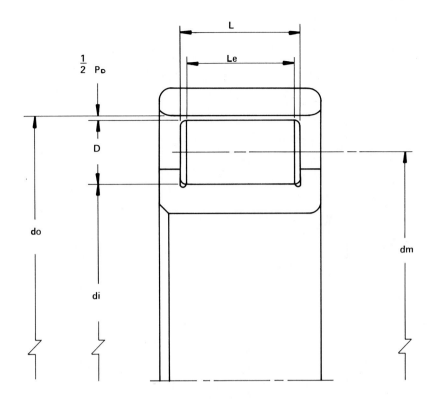

FIGURE 10.   Cylindrical roller bearing nomenclature.

f_i, F_o	=	Inner and outer race osculations for ball bearing (see Figure 7)
i	=	Number of rows of rolling elements
K	=	Coefficient in the load-deflection equation; depends upon material, elastic properties, and osculation
l	=	Roller length
l_{eff}	=	Effective length of contact between roller and ring raceway
N	=	Number of stress cycles
N_1, N_o, N_c	=	Rotational speed of inner ring, outer ring, and cage
N_r	=	Rolling element rotational speed about its own axis (i.e., relative to the cage)
Q	=	Rolling element load
S	=	Probability of survival
t	=	Exponent in the load-deflection equation; depends upon osculation
V	=	Stressed volume
y	=	D cos α/d_m
Z	=	Number of rolling elements per row
z_o	=	Depth of maximum orthogonal shear stress
$\propto$	=	Contact angle (also 1/2 included cup angle for tapered roller bearings)
β	=	1/2 Included cone angle (tapered roller bearings)
γ	=	1/2 Included roller centerline angle (taper roller bearings)
δ	=	Deflection
λ	=	Life reduction factor for stress concentrations
ν	=	1/2 Included roller angle (tapered roller bearings)
τ_o	=	Maximum orthogonal shear stress

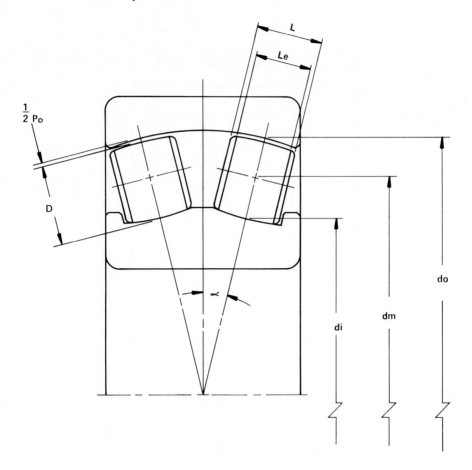

FIGURE 11. Spherical roller bearing nomenclature.

Contact Deformations and Load Distribution

To determine the magnitude of the contact stresses occurring in a rolling bearing, one must first determine the magnitude of the load on each rolling element. Figure 13 shows a free-body diagram of a radially loaded bearing. The equations of static equilibrium are not sufficient to determine the distribution of load among the rolling elements; load-deflection relationships at the rolling element-raceway contacts are also required. In addition, it is necessary to assume that the ring is round before loading and remains round after loading, and the only deflections are those at the Hertzian contacts. In general, load-deflection relationship for a Hertzian contact can be expressed by:

$$Q = K\delta^t \tag{1}$$

Typical values for coefficient K and exponent t are given in Table 5. They do not vary greatly for rolling element contacts for bearings with the usual range of power transmission applications, say 25- to 300-mm (1- to 12-in.) shaft size. Heavily loaded ball bearings, where the contact ellipse is of significant size in relation to the ball diameter, may be somewhat stiffer than indicated by the foregoing relationship; light to normally loaded spherical roller bearings, as well as tapered and cylindrical roller bearings with crowned rollers, seldom operate with rectangular contact areas and are usually somewhat less rigid than indicated by the line contact exponent. For normal engineering purposes, however, values given in Table 5 are adequate.

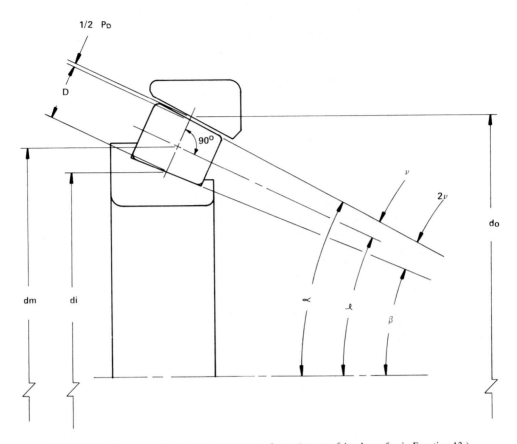

FIGURE 12.   Tapered roller thrust bearing nomenclature (use cos ℓ in place of α in Equation 13.)

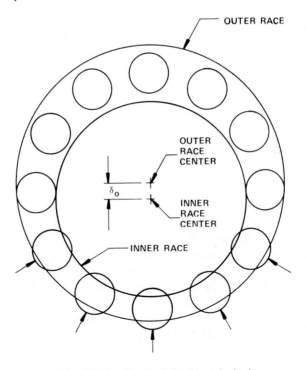

FIGURE 13.   Bearing deflection under load.

Table 5
ELASTIC CONTACTS
ORDER OF MAGNITUDE

Type of contacts	Exponent (t)	Coefficient (K)
Point	1.5	18×10^6
Line	1.11	7×10^6

Note: Load in pounds; deflection in inches.

Consider the case of an inner raceway displaced a distance δ_o with respect to the outer race as shown in Figure 13. Given the diameters of the raceways and the magnitude and direction of displacement, mutual approach of the two raceways can be calculated at any roller location. Deflection across a roller at that position is then equal to the approach less any initial clearance in the bearing. The roller load can then be calculated from Equation 1. The vector sum of all roller loads will yield the applied radial load F_r required to produce displacement δ_o. If it is assumed that the bearing has zero radial clearance and that the rings are circumferentially rigid as previously described, for any practical number of rolling elements for bearings with a load-deflection exponent of 1.5

$$Q_{max} = \frac{4.37 \, F_r}{iZ\cos\alpha} \tag{2}$$

and

$$Q_{max} = \frac{4.08 \, F_r}{iZ\cos\alpha} \tag{3}$$

for bearings with a load deflection exponent of 1.11. Since few bearings operate with zero clearance, the radial load on the most heavily loaded ball or roller is commonly calculated from the following equation:

$$Q_{max} = \frac{5 \, F_r}{iZ\cos\alpha} \tag{4}$$

For thrust and angular contact ball and roller bearings

$$Q_{max} = \frac{F_a}{Z\sin\alpha} \tag{5}$$

Rolling Bearing Kinematics

Rotational speed of a rolling element or the cage (the orbital speed of the rolling element) can be determined by analyzing the bearing as an epicyclic gear train, if no slip is assumed at the rolling-element/raceway contacts. Significant slip is possible, however, with bearings operating under extremely light load at very high speed. For the general case where both inner and outer rings rotate, rotational speeds of the cage and rolling elements are given by the following equations:

$$N_c = 1/2[N_i(1 - y) + N_o(1 + y)] \tag{6}$$

$$N_r = \frac{d_m}{2D} (1 - y) (1 + y) (N_o - N_i) \tag{7}$$

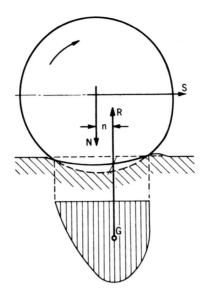

FIGURE 14. Forces on rolling element with bulge created by tangential force.

Table 6
AVERAGE BEARING COEFFICIENTS OF FRICTION MEASURED AT SHAFT SURFACE

	Coefficient of friction					
	Ball bearings		Spherical roller bearings		Cylindrical roller bearings	
Type of load	Starting	Running	Starting	Running	Starting	Running
Radial	0.0025	0.0015	0.0030	0.0018	0.0020	0.0011
Thrust	0.0060	0.0040	0.0120	0.0080	—	—

Note: Bearing torque = coefficient of friction × load × shaft radius.

For the more typical case where only one ring rotates, the speed of the stationary ring is zero in the above equations. These relationships are useful for determining centrifugal and gyroscopic forces in high-speed bearings, relating the number of contact stress cycles to inner and/or outer ring rotation, and determining the degree of slip by comparing the calculated cage speed to a measured cage speed.

Friction in Rolling Bearings

One source of resistance to rolling is the internal friction (hysteresis) due to cyclic stressing of rolling contact surfaces. Another source of friction in rolling contacts is illustrated in Figure 14. The tangential force required for rolling results in a slight bulge in front of the rolling element and a depression behind. Due to these small elastic deformations, slippage on a microscopic scale (microslip) occurs within the contact areas.

For many bearing types (such as ball bearings, spherical roller bearings, and most thrust roller bearings), sliding due simply to the lack of theoretical true rolling is another source of friction. In addition, pure sliding contacts in rolling bearings include roller-end/flange contacts, rolling-elements/cage contacts, and cage/land contacts. References 4 to 6 give methods for estimating friction torque for rolling bearings. For rough estimates of torque, Table 6 may be used.

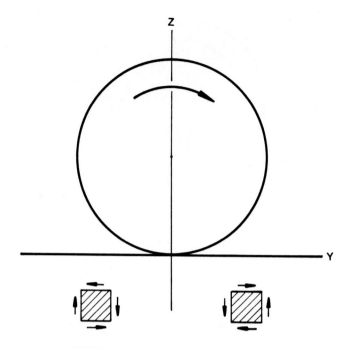

FIGURE 15.   Illustration of orthogonal shear stress.

Static Capacity

Occasionally rolling element bearings are subjected to very heavy loads while stationary. To determine whether or not such loading has any effect on subsequent rotational operation, the concept of a static load rating has been developed. Permanent flat spots, or "brinelling", usually manifest themselves as noise or vibration. They do not seem to significantly effect torque or fatigue life of the bearing unless they are of extreme magnitude. Departure from elastic behavior cannot be detected at a stress level significantly less than 4000 MPa (580,000 psi). Formulas for calculation of static load ratings involve not only elastic material properties but also details of the internal bearing geometry.[10,11]

Rolling Bearing Endurance

The mathematical model generally used for rolling contact fatigue calculations was originally developed by Lundberg and Palmgren.[12,13] It may be expressed mathematically as follows:

$$\log_e \left(\frac{1}{S}\right) \propto \frac{\tau_o^c N^e V}{z_o^h} \tag{8}$$

The maximum orthogonal shear stress, τ_o, occurs on planes parallel and perpendicular to the rolling direction as illustrated in Figure 15. This stress has its maximum value at some depth below the surface and at some distance ahead of the center of the contact in the rolling direction. At that same depth and at the same distance behind the center of the contact, τ_o reaches the same magnitude but in the opposite direction. Its total amplitude determines the severity of loading for a rolling contact with regard to fatigue. Determination of τ_o involves material properties and contact geometry along with load distribution among the rolling elements. Depth z_o, also a function of these variables, is in the denominator of Equation 8 because the deeper below the contact surface that the crack initiates, the longer for it to propagate to the surface and result in a spall. The number of stress cycles, N, occurring at any point in the rolling contact surfaces, can be related to the number of inner ring revolutions through the kinematic relationships described earlier. The volume of stressed material, V, is included on the theory that resistance to fatigue crack initiation varies throughout the material. The greater the volume of material subjected to cyclic stress, the greater the probability of a crack initiating.

By making simplifying assumptions, Equation 8 can be rewritten in terms of bearing geometry factors. For arbitrary conditions of 90% reliability (S = 0.9) for 1 million inner ring revolutions, the following equations result.

For radial ball bearings:

$$C = f_c (i \cos \alpha)^{0.7} Z^{2/3} D^{1.8} \tag{9}$$

Note: when D > 1 in., or 25.4 mm, use exponent 1.4 instead of 1.8, and where

$$f_c = 3050 \lambda \left\{ 1 + \left[1.04 \left(\frac{1-y}{1+y} \right)^{1.72} \left(\frac{f_i}{f_o} \cdot \frac{2f_o - 1}{2f_i - 1} \right)^{0.41} \right]^{10/3} \right\}^{0.3} y^{0.3} \left[\frac{(1-y)^{1.39}}{(1+y)^{1/3}} \right] \left(\frac{2f_i}{2f_o - 1} \right)^{0.41} \tag{10}$$

for C in pounds and D in inches. (To obtain C in N when D is in millimeters, replace 3050 with 40.1 for D ≤ 25.4 mm and with 146.2 for D > 25.4 mm.)

For radial roller bearings:

$$C = f_c (i \cos \alpha L_{eff})^{7/9} Z^{3/4} D^{29/27} \tag{11}$$

$$f_c = 18673 \lambda \nu \left\{ 1 + \left[1.04 \left(\frac{1-y}{1+y} \right)^{143/108} \right]^{9/2} \right\}^{-2/9} y^{2/9} \left[\frac{(1-y)^{29/27}}{(1+y)^{1/4}} \right] \tag{12}$$

for C in pounds and l_{eff} and D in inches. To obtain C in Newtons with l_{eff} and D in millimeters, use 207.9 in place of 18673. Reduction factor λ accounts for stress concentrations within contacts and is covered in the next section.

ROLLING BEARING APPLICATION CONSIDERATIONS

In this section, an attempt will be made to point out application and operating conditions that differ significantly from the standard assumptions, and how they might affect bearing performance.

Load

Basic load rating C discussed earlier is itself an extremely high load; a load nearly as high as the C rating would not normally be applied in practice. The expected range of relative loading is indicated in Table 7, which is taken from AFBMA Standard 7. To this might be added two additional load categories: extremely light load, less than 1 to 2% of C, and extremely heavy load, greater than 25 to 30% of C.

Providing the fitting practices recommended in AFBMA Standard 7 are followed for light to heavy loading, no problems should be expected. In the extremely light load range however, especially in combination with very high-speed and high-viscosity lubricants, rolling elements sometimes slide rather than roll. This problem is not uncommon in gas turbine engine bearings. Solutions have included preloaded hollow rollers and out-of-round ring raceways.

At the other end of the scale are extremely heavy loads, usually accompanied by very low speed, such as might occur in a crawler tractor final drive. Operation is usually in the boundary lubrication regime and surface origin failures result. Also, under extremely heavy loads (i.e., above 25 to 30% of C) it is often difficult to prevent movement of a press fit

Table 7
DEFINITION OF LOAD
RANGES FOR ROLLING
ELEMENT BEARINGS

Load	Ball bearings	Roller bearings
Light	Up to 7% C	Up to 8% C
Normal	From 7% C	From 8% C
	Up to 15% C	Up to 18% C
Heavy	Over 15% C	Over 18% C

inner ring on its shaft, even using the fits recommended for a heavy load. The usual solution to this problem is a threaded shaft and locknut to clamp the press fit inner ring against a shaft shoulder. These characteristics of boundary lubrication, high load, and heavy press fits combine to produce a very highly stressed component. Under such conditions it is important that the bearing and mating components (e.g., shaft, housing) are free from flaws such as nicks and dents. While rolling element bearings can be quite forgiving of such flaws under normal conditions, under extremely heavy load they can lead to premature failure and bulk cracking either as a primary or, more often, a secondary failure mode.

Load direction can often determine how well a rolling bearing performs. On angular contact bearings, axial load is simply combined with the radial load to get an equivalent radial load to estimate life. For radial cylindrical roller bearings, however, axial load is resisted only by the roller ends and ring flanges; it is not considered in the fatigue life calculation. Opposing flanges on inner and outer rings are intended to provide only axial location and not to support applied thrust loads such as occur in helical gear drives. Axial load capacity of these bearings can be improved by careful attention to such characteristics as roughness, rounding of sharp corners, roller end to race squareness, and rib to ring race squareness.

Some types of bearings and bearing arrangements (e.g., "back-to-back" angular contact bearings) are designed specifically to resist moment loading. Spherical roller bearings, on the other hand, cannot support any moment load. For other roller bearings, application of a moment load usually results in edge loading of the rollers and early failures. For ball bearings, a certain amount of moment load can be tolerated without causing the balls to ride too high in the groove. Such loading, however, is very detrimental to the typical riveted or welded ribbon-type metallic cage, and can result in early fatigue failure of the cage. Such failures are common in agricultural tillage implements subject to distortions as they bump and bounce across rugged terrain. Polymeric cages (e.g., injection-molded nylon) have largely overcome this problem.

Speed

Operating speed of a rolling bearing has ramifications other than simply determining how long it takes to reach its normal fatigue life. Very low speed can result in surface distress damage to the rolling contact due to inadequate EHD lubricant film thickness. Very high speeds can result in skidding damage, less than expected EHD film thickness due to starvation, abnormal distribution of loading between the inner and outer raceways due to centrifugal forces on rolling elements, strain on the cage due to its own centrifugal expansion and to resisting gyroscopic moments of rollers, centrifugal expansion of the gyroscopic moments of rollers, centrifugal expansion of the inner race causing a reduction or complete loss of a press fit, and centrifugal force throwing lubricant away from the inner race and resulting in ineffective lubrication. These problems generally occur only at very high speeds such as in aircraft gas turbine mainshaft bearings and solutions are reported in the literature.

Table 8
APPROXIMATE NORMAL SPEED LIMITS FOR ROLLING ELEMENT BEARINGS

	Normal speed limit (DN)[a]		
Bearing type	Grease or oil bath	Circulating oil	Highest DN attained under ideal conditions
Radial or angular contact groove ball bearings	300K[b]	500K	3500K
Cylindrical roller bearings	300K	500K	3500K
Tapered roller bearings	150K	300K	3500K
Spherical roller bearings	150K	300K	1000K

[a] DN = bore (D), mm × speed (N), rpm.
[b] K indicates × 1000 (e.g., 300K = 300,000).

The most common problem associated with high speed, however, is simply overheating. For this reason, most bearing catalogs give a speed limit for each type of bearing. Typical values are shown in Table 8. Not absolute limits, these are simply speeds below which ordinary lubricants and lubrication methods will normally give satisfactory performance. Particular attention to bearing components, the lubricants used, and the lubricant application method enables satisfactory performance at speeds well beyond typical catalog limits. The highest speeds are also indicated for various types of bearings under laboratory conditions.

Temperature
The range of temperatures over which rolling element bearing can perform successfully depends upon the materials of the rolling bearing components, materials of auxiliary components such as seals, and lubricants.

Two aspects of material hardness-temperature relations are to be considered: short-term hot-hardness and long-term temperature resistance. The former refers to inadequate hardness of bearing steels at abnormally high operating temperature even though they recover full hardness when reduced to room temperature as illustrated in Figure 16. Temper resistance, on the other hand, refers to the ability of a bearing steel to recover its original room temperature hardness after being exposed to high temperature for some period of time as illustrated in Figure 17. Based on this phenomenon and a minimum acceptable hardness of 59 Rockwell C at operating temperatures, approximate temperature limits for bearing steels are given in Table 4.

Typical temperature limits for common rolling bearing seal materials are given in Table 9. Temperature limits for lubricants can vary considerably depending upon the additives they might contain. For example, a hypoid gear lubricant containing highly active EP additives may be limited to a significantly lower temperature than that shown in Table 10. Maximum temperatures for some of the more common nonmetallic cage materials are shown in Table 11. Metallic cages rarely limit the allowable operating temperature.

Misalignment and Contact Geometry
Even though the underlying load rating relations are based on perfect geometric shapes with smooth surfaces in perfect alignment, Lundberg and Palmgren[12,13] introduced reduction factor λ in Equations 10 and 12 to account for stress concentrations within contacts. Stress concentrations can result from, for example, edge loading due to heavy loading or improper crown, inadequate crown blend, misalignment, imperfect geometry, and rough surfaces. For ball bearings only the latter three are significant; consequently, values of λ are usually much

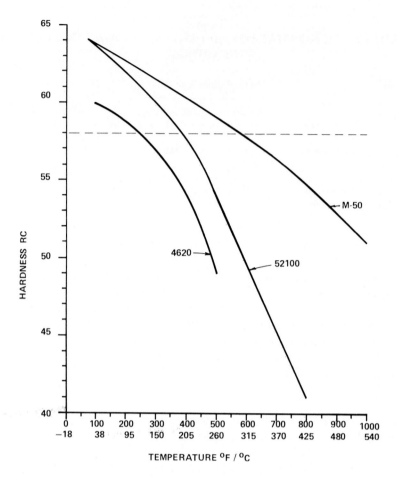

FIGURE 16.   Hot hardness of representative rolling element bearing steels.

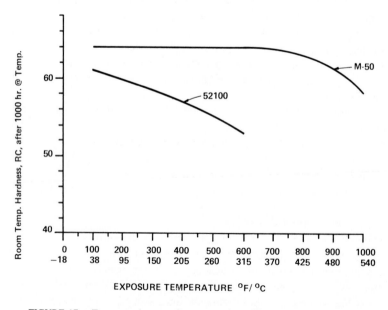

FIGURE 17.   Temper resistance of representative rolling element bearing steels.

Table 9
TEMPERATURE
LIMITS FOR SEAL
MATERIALS

Material	Temperature limit	
	°C	°F
Felt	95	200
Buna N	110	225
Polyacrylic	150	300
Viton	260	500

Table 10
APPROXIMATE TEMPERATURE LIMITS FOR
ROLLING BEARING LUBRICANTS

Lubricant	Temperature	
	°C	°F
Standard petroleum greases	110	225
Special petroleum greases, mineral oils	150	300
Synthetic greases and oils	230	450
Solid film	Bearing material limit	

Table 11
TEMPERATURE LIMITS FOR
NONMETALLIC CAGE MATERIALS

Material	Temperature Limit	
	°C	°F
Heat-stabilized nylon 66 phenolic	120	250
Fiberglass-reinforced nylon 66	150	300
Polyphenylene-sulfide	230	450

higher. Values of λ in Table 12 may be used in Equations 10 and 12 and are used in the load rating standards. They are applicable to bearings made to standard tolerances associated with ABEC/RBEC Class 1 Precision. For roller bearings, this value also assumes rollers crowned so as to achieve optimized contact.

In an attempt to define good and poor alignment, Table 13 is provided. The lower value of misalignment in each range may be considered as the limit of "good" alignment. The λ factor has already provided compensation for misalignment up to this value. At misalignments greater than the higher value in the range, serious reduction in life should be expected. Between these two values, performance will depend strongly upon the specific design as well as load, speed, and other operating conditions. Some bearings may perform satisfactorily, others may suffer premature failure.

Clearance and Ring Support

Internal bearing clearance is usually defined as the total possible radial play of one ring

Table 12
LIFE REDUCTION FACTOR
FOR STRESS
CONCENTRATIONS

Bearing type	λ
Ball bearings	
Single-row radial	
Double row radial	0.95
Angular contact	
Ball bearings	
Double row radial	0.90
Contact	
Ball bearings	
Self-aligning	1.0
Roller bearings	
Cylindrical, tapered, and spheri-	0.61[a]
cal, crowned rollers	

[a] Use in conjunction with $\nu = 1.36$ so that $\lambda \nu = 0.83$.

Table 13
ALLOWABLE MISALIGNMENT
WITH NO REDUCTION IN LIFE FOR
ROLLING BEARINGS UNDER
NORMAL LOAD

Bearing type	Misalignment
Cylindrical and tapered roller bearings	0.0005—0.001 in/in.
Ball bearings	0.003—0.005 in/in.
Spherical roller bearings	0.008—0.025 in/in.

with respect to the other with no external load. For some bearing types, axial clearance, or end play, is more significant. Radial clearance can be difficult to measure and the definition is not especially meaningful for some single row, angular contact bearing types (such as tapered roller bearings) where the radial play depends upon the relative axial location of the two rings. A more rigorous and general definition can be expressed as follows in terms of component dimensions:

$$P_D = (d_o - d_i)/\cos\alpha - 2D \qquad (13)$$

If the calculated clearance P_D is positive, it is called clearance. If negative, i.e., if the roller is larger than the available space, it is called preload. Some bearing types, principally tapered roller bearings, are adjustable: the user sets the clearance or preload when mounting the bearing. For such bearings clearance is usually determined by measuring end play, while preload is determined by measuring bearing rotational torque. Some angular contact ball bearings, while theoretically adjustable, are actually made to be mounted in a manner that precisely determines the internal clearance. For the many nonadjustable bearing types, such as single row deep groove ball bearings, cylindrical roller bearings, and spherical roller bearings, the initial unmounted clearance is built into the bearing by selective assembly by the bearing manufacturer. These are the clearances that are standardized for various bearing types and sizes.

When a bearing is installed, at least one ring (sometimes both) is installed with an interference fit. This results in a change in the ring diameter of from 50 to 80% of more of the amount of interference. This reduced clearance may be called clearance. Once a bearing has reached its equilibrium operating temperature, clearance may be further changed by thermal expansion or contraction. For most applications operating in a typical ambient temperature with no external heat sources, temperature effects may be safely neglected.

Once a bearing is mounted and subjected to radial load, one ring is displaced eccentrically with respect to the other and clearance is described by angular extent of the arc wherein the rolling elements are loaded. A radial bearing with precisely zero operating clearance will have a "load zone" of 180°. For the normal case of some nominal clearance, the load zone can vary significantly as radial load varies from light to heavy. For zero clearance or a heavy preload, the load zone may not appreciably change from either 180° or 360°, respectively, for a fairly wide range of loads. For the case of a light preload and heavy applied radial load, the load zone may vary significantly between limits of 180 and 360°. Finally, the load zone can vary greatly for angular contact bearings with varying ratio of radial to axial load.

The theoretical derivations used by Lundberg and Palmgren[12,13] assumed that radially loaded bearings have precisely zero operating clearance. However, the formulas contain several empirical coefficients and exponents that were determined from fatigue life tests of actual bearings that contained some clearance. Consequently the methods given are applicable to rolling element bearings containing some nominal clearance.

If the bearing is assembled in accordance with the manufacturer's standard unmounted clearance specification, is mounted with shaft and housing fits corresponding to normal loading, and operates within a temperature range acceptable for standard bearing components, one need not be concerned about internal clearance. Usually when a bearing is assembled to some nonstandard unmounted clearance range, it is because of an unusual loading condition or expected unusual temperature condition and not because some nonstandard operating clearance is sought.

Because of the greater resiliency of the contact, ball bearings are commonly assembled with much less clearance than roller bearings. A slight preload under the extremes of normal clearance and fitting practice need not be cause for concern: a light preload can yield greater than "standard" life due to the better distribution of load. Roller bearings, on the other hand, are usually much stiffer and require more clearance; preload may not be tolerated. The bearing manufacturer should be consulted for operation outside the following limits:

> Ball bearings: Preload greater than 5 to 10 μm
> A load zone much less than about 160°
> Roller bearings: A load zone greater than 180°
> A load zone much less than about 120°

These upper limits apply of course only to radial contact bearings, or both rows of double row angular contact bearings under radial load. Single-row angular contact bearings, including thrust bearings, operate quite normally with load zones of 360°.

The preceding discussion of clearance assumes that both inner and outer rings are adequately supported. This is a satisfactory assumption for the great majority of commercial bearing applications. Some notable exceptions however are planet gear bearings, such as illustrated in Figure 18, and metalworking backup roll bearings. Calculation of bearing life for such applications is beyond the scope of this chapter.

Mounting

Most bearing manufacturers' catalogs or service instructions give the information necessary to accomplish a successful mounting. In addition, AFBMA publishes standardized shaft and housing fits, along with a method for selecting the appropriate fits for most applications.

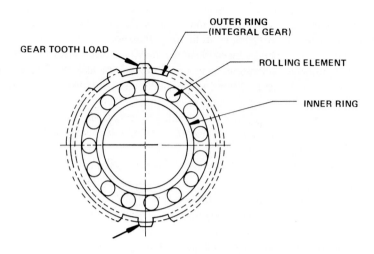

FIGURE 18.   Planet gear bearing showing gear tooth loading.

The mounting arrangement must be stiff enough to support all the static and dynamic, radial and axial loads. In addition, the mounting often must provide resistance to axial movement of the rings with respect to the shaft or housing. Finally, if excessive wear is to be avoided, circumferential motion of the rings with respect to the shaft or housing must be prevented. As a general rule, the ring that rotates with respect to the load must be secured against circumferential movement relative to its supporting member. Excessive clearance at the shaft or housing bearing interface can result in either ring changing shape under load to concentrate the load on too few rolling elements. Likewise, abrupt changes in housing shape, or webs for stiffening, can result in "hard" spots in outer ring support.

Normally the most reliable means of mounting an inner ring on a shaft is with an interference fit. Specific dimensions of the fit should be established in accordance with AFBMA standards and the fit is usually accomplished either by pushing the inner ring squarely onto the shaft or by heating the inner ring so that it may be slipped onto the shaft and then allowed to shrink tight. When the shrink-fit method is used, the inner ring should be heated in an oil bath or a hot air oven (far removed from the heating elements in either case) to a temperature not greater than about 120°C (250°F) to avoid metallurgical transformations. Other possible causes of unsatisfactory interference fits include a low shaft shoulder, out-of-roundness, rough surfaces, concentricity or angularity errors in stepped shafts, and excessive shaft deflection.

Another popular inner ring mounting is the split sleeve adapter with a tapered OD, locknut, and lockwasher, used in conjunction with a tapered bore inner ring. The locknut on the sleeve is tightened until a measured amount of internal clearance is removed from the bearing. It is a reliable mounting style and allows the use of much looser shaft tolerance. Cost of manual labor in tightening and adjusting clearance usually more than offsets savings in cost of the shaft, however, and tightening to a precise axial position is difficult.

The inner ring of many small size (up to 100 mm shaft) bearings are locked to the shaft with either setscrews or a cam locking collar. Setscrews must be adequately tightened for successful performance and must either be at least 15 points Rockwell C harder than the shaft, or the shaft must be spot drilled or flatted. The eccentric locking collar, locked in the direction of shaft rotation, relies upon eccentric pinch for its considerable holding power. All of these mounting types are relatively inexpensive and easy to install with no precise adjustment. They do require, however, a very close clearance fit between the inner ring bore and shaft with rather close shaft tolerances. They also are limited in their load carrying ability. The capacity of a slip fit mounting is given by Price and Galambus.[15]

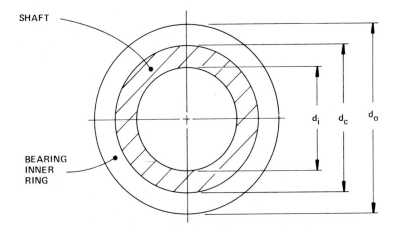

FIGURE 19. Inner ring/shaft nomenclature. Symbols: P_c = constant pressure at interface, δ = effective interference, E_i = modulus of elasticity (inner membrane), E_o = modulus of elasticity (outer membrane) ν_i = Poisson's ratio (inner membrane), ν_o = Poisson's ratio (outer membrane), and Δd = change in diameter. Any consistent system of units may be used.

Housing, or outer ring mountings, encounter many of the same problems. Housings do perform some other functions such as providing a lubricant reservoir and providing seals to prevent the egress of lubricant and ingress of contaminants. Housings are often purchased from the bearing manufacturer, may be either solid or split, and are usually sized for a loose fit with the bearing OD. Where the load is stationary with respect to the inner ring and rotates with respect to the outer ring, however, the outer ring is usually mounted with an interference fit. An example would be a vibrating screen.

Both shaft and housing mounting are susceptible to fretting. This physical, chemical wear phenomenon results from relative motion of very small amplitude between surfaces. Fretting is especially troublesome in the slip fit mounting, and determines the limiting load given above.[15] It also occurs in other mounting types if the interface pressure at mounting surfaces is insufficient to support the shear stresses developed by application loads.

For many types of angular contact bearings, such as tapered roller bearings, the mounting arrangement must provide for adjustment of clearance within the bearings. The change in diameter of a press fit ring can be estimated from the following equations (see Figure 19 for definitions of terms):

$$P_c = \frac{\delta}{\dfrac{d_c^2 + d_i^2}{E_i(d_c^2 - d_i^2)} + \dfrac{d_o^2 + d_c^2}{E_o(d_o^2 - d_c^2)} - \dfrac{\nu_i}{E_i} + \dfrac{\nu_o}{E_o}} \tag{14}$$

$$\Delta d_o = P_c \left(\frac{d_o}{E_o}\right)\left(\frac{d_c^2}{d_o^2 - d_c^2}\right)$$

$$\Delta d_i = -P_c \left(\frac{d_i}{E_i}\right)\left(\frac{d_c^2}{d_c^2 - d_i^2}\right) \tag{15}$$

Where both members are steel and $d_i = 0$, e.g., an inner ring on a solid shaft, the above equations reduce to

$$\Delta d_o = \left(\frac{d_o}{d_c}\right)\delta \tag{16}$$

Materials and Processing

For cylindrical and needle roller bearings, one or both rings are often omitted with the rollers running directly on the shaft or in the housing. To achieve "catalog" life, shaft or housing material must meet the same hardness and cleanliness standards as the omitted ring. Where a lower hardness than R_c59 must be used, the reduction in life can be estimated from the following:

$$\frac{L_{10} \text{ @ hardness H}}{L_{10} \text{ @ } 59R_c} = e^{0.1(H-69)} \tag{17}$$

Cleanliness for bearing steels is usually specified in accordance with ASTM Specification A295 for through hardening steels or A534 for carburizing steels. Both of these specifications require vacuum degassing to minimize nonmetallic inclusions. Premium vacuum remelted steels with extraordinarily few and small nonmetallics are used for special high-reliability bearings such as for aircraft gas turbine engines with three to five times standard life.

ROLLING BEARING LUBRICATION

Basic Principles

The principle function of a lubricant in a rolling element bearing is to minimize friction and wear. Other important functions may include (1) protection from corrosion, (2) dissipation of heat, (3) exclusion of contaminants, and (4) flushing away of wear products. It has long been recognized that bearings can operate for extended periods under certain conditions with no evidence of wear. Presence of the original grinding, honing, or other finishing marks suggested that an oil film separated the rolling surfaces. It wasn't until Grubin and Vinogradova[16] combined elasticity and pressure viscosity relationships with hydrodynamic theory, however, that an acceptable explanation was obtained for the mechanism of forming an oil film. Dowson and Higginson[17] subsequently developed solution techniques that do not require the prior assumption of a Hertzian pressure profile. Details of elastohydrodynamic theory and related calculations are provided in an earlier chapter.

By applying EHL calculations for the conditions shown schematically in Figure 20, film thicknesses are readily determined to be of the same order of magnitude as surface roughnesses. These roughness values for rolling bearing components commonly range from a low of 1 to 2 μin. AA for balls or rollers to a high of 16 μin. AA or more for as-ground raceways.[18] It is reasonable to define the following three regimes of lubrication.

Full film — In this regime, bearing geometry, application conditions, and lubricant properties combine to yield a lubricant film so thick that there is no contact of even the highest peaks through the film. Unusually long bearing lives can be obtained.

Boundary lubrication — Here the lubricant film is practically nonexistent and the load is supported by intimate contact of the rolling component surfaces. Under these conditions bearings life is not readily predictable and often is rather short. Physical and chemical properties of the system components (e.g., bearing, lubricant, environment) all play a role in determining overall performance.[19-22]

Partial film — In this transition regime the load is supported partially by an EHL film and partially by intimate contact of the rolling surfaces. If the film is so thin as as to allow significant interaction of asperities, performance may be more nearly like boundary lubrication. On the other hand, if the film is thicker with only occasional interaction of the highest peaks, then the contact may behave more nearly as full film. It is within this regime that most bearings operate and can be expected to achieve "catalog" life.

Specific film thickness Λ, the ratio of EHD film thickness to composite surface roughness, has come into use to define the state of lubrication. Composite surface roughness, in turn,

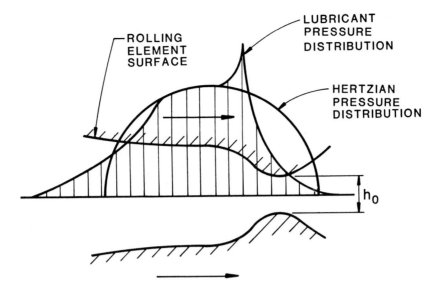

FIGURE 20.   Pressure distribution in EHD contact with Newtonian fluid.

may be taken as the sum of the AA (CLA) roughnesses of the two rolling contact surfaces. For Λ greater than about four, full EHL film conditions are obtained; for Λ values less than about one, boundary lubrication conditions apply; and for Λ between one and four, partial film conditions exist. Correlations of the Λ value with a fatigue life adjustment factor (see previous discussion on loading ratings) are shown in Figures 21 and 22.

A summary of EHL formulas applicable to rolling element bearings is given in Table 14. Two notes of caution are in order. First, operation at a lambda value less than one may result in unpredictable performance. Figure 22 for example, is strictly applicable only to the specific combination of bearing materials, lubricants, and environments used to derive the graph. A second note of warning has to do with high-speed operation. Even though a bearing may be supplied with copious amounts of lubricant, the lubricant can be pushed out of the rolling element path and not have sufficient time to flow back before the next rolling element passes that point.[26] Under such conditions, the lubricant film thickness will be less than expected. Although there is no general rule of thumb, this could occur at speed as low as 200,000 to 300,00 DN. For grease lubrication, film thickness may only be on the order of 50 to 75% of that estimated from base oil properties.

Mineral Oil Lubricants

Greases

Grease is a combination of petroleum or synthetic oil and 3 to 30% or more of a suitable thickener. Grease consistency or stiffness is determined primarily by the thickener percentage and the base oil viscosity. A grease of given consistency may be compounded in many ways by varying thickener percentage and oil viscosity. Because of this, greases of equal stiffness cannot be assumed equal in performance.

In the past, the most widely used thickeners were either sodium or calcium soaps; modern greases, however, contain many types of thickeners. Each grease has certain characteristics, but categorizing them for usage according to thickener type can be misleading. For general industrial use, sodium, calcium, lithium, aluminum, and mixed soap types, and various synthetic and nonsoap-base greases are widely used. Considerable effort has been directed to development of a multipurpose lubricant, but no grease yet satisfies all requirements.

Grease will deteriorate as a result of time, temperature, shear (mechanical working in the bearing), and contamination. These characteristics are therefore measured by standard tests

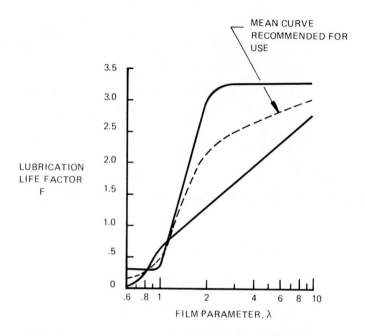

FIGURE 21.   Lubrication life adjustment factor for ball and roller bearings. (From *Life Adjustment Factors for Ball and Roller Bearings, An Engineering Design Guide,* American Society of Mechanical Engineers, New York, 1971. With permission.)

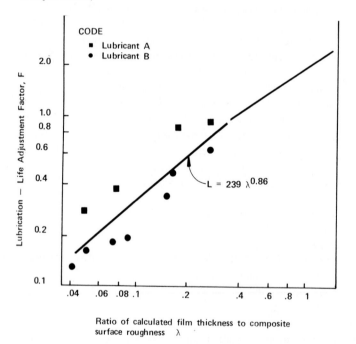

FIGURE 22.   Lubrication life adjustment factor for tapered roller bearing. (From Danner, C. H., *ASLE Trans.,* 13, 241, 1970. With permission.)

of oxidation resistance, water resistance, mechanical stability, oil separation, dropping or melting point, and evaporation. Stiffness is measured by a hardness or penetration test. A change in this property during operation may have various causes: excessive working or

Table 14
EHD FILM THICKNESS FORMULAS

For Point Contacts[25]

$$H = 3.63 \, U^{0.68} \, G^{0.49} \, W_p^{-0.073} \, [1 - e^{-0.68k}]$$

For Line Contacts[17]

$$H = 1.6 \, U^{0.7} \, G^{0.6} \, W_l^{-0.13}$$

Where

$$H = \frac{h_o}{R}$$

h_o = Minimum film thickness

$$R = \frac{1}{1/R_1 + 1/R_2}$$

R_1, R_2 = Radii of curvature of body 1 and body 2, respectively, in plane parallel to rolling direction

$$U = \frac{\eta_o \, u}{E' \, R}$$

η_o = Absolute viscosity at contact entry conditions

u = Mean surface velocity = $(u_1 + u_2)/2$

$$E' = \frac{2}{\dfrac{1 - \nu_1^2}{E_1} + \dfrac{1 - \nu_2^2}{E_2}}$$

ν_1, ν_2 = Poisson's ratio for body 1 and body 2, respectively

E_1, E_2 = Young's modulus for body 1 and 2, respectively

$G = \alpha E'$

α = Lubricant pressure-viscosity coefficient

$$W_p = \frac{F}{E' \, R^2}$$

F = Applied load

$$W_l = \frac{p}{E' \, R}$$

p = Load per unit length of contact

Note: Any consistent system of units may be used.

churning, oil separation or vaporization, change in oil viscosity due to oxidation, etc. In some cases greases have been found excellent in laboratory evaluation and completely unsuitable in field performance. In other instances the reverse has been true. For this reason, field testing and field development of lubrication requirements for a particular equipment installation are often necessary.

Bearings and bearing units are designed for service ranging from nonregreasable (lubricated-for-life) to almost continuous relubrication by means of automatic systems. Advantages of grease over oil lubrication include the ease of sealing it within the bearing, the ability of grease to seal out contaminants, and its ability to coat parts and provide good corrosion protection. Disadvantages of grease include its inability to remove heat or flush away wear products, the possibility of accumulating dirt or other abrasive contamination, and a potential incompatibility problem if thickeners of different types are mixed.

Oils

Oil can be pumped, circulated, filtered, cleaned, heated, cooled, and atomized. Its advantages over grease include its ability to remove heat, flush away wear products and contaminants, and to be recycled. It is more versatile than grease and is suitable for many severe applications involving extreme speeds and high temperatures. On the other hand, it is more difficult to seal or retain in bearings and housings. Oil level or oil flow in high-speed bearings is critical and must be properly controlled.

Selection of proper oil viscosity is essential and is based primarily on expected operating temperature, speed, and bearing geometry. Excessive oil viscosity many cause skidding of rolling elements and undue lubricant friction with severe overheating and raceway damage. Insufficient oil viscosity may result in metal contact and possible premature failure. Other oil properties such as viscosity index, flash point, pour point, neutralization number, carbon residue, and corrosion protection are of varying significance in specific installations.

Synthetic Lubricants

Development of synthetic lubricants was initially prompted largely by the extreme environmental demands of military and aerospace activities. Currently the following classes of synthetic oils are available as bearing lubricants: (1) synthetic hydrocarbons, such as alkylated aromatics and olefin oligomers, (2) organic esters, such as dibasic acid esters, polyol esters and polyesters, (3) others, such as halogenated hydrocarbons, phosphate esters, polyglycol ethers, polyphenyl ethers, silicate esters, and silicones, and (4) blends, which would include mixtures of any of the above.

Use of a synthetic lubricant in a commercial application may be dictated by extreme operating conditions, for fire resistance, to meet a specification or code requirement, or to conserve petroleum-based lubricants. Although synthetic lubricants usually permit a much broader operating temperature range, temperature limits for synthetics are often misunderstood. For example, in various aircraft and space applications operation at extremely high temperature is essential but life requirements may be very short. Since industrial requirements are usually for much longer periods of operation, temperature limits for a given synthetic in industry can be much lower. Some synthetic lubricants may also have other limiting characteristics such as in load-carrying ability and high-speed operation.

Dry Lubricants

Dry, or solid lubricants are usually used under conditions of high temperature or where boundary lubrication prevails. For example, notable success has been achieved by solid lubrication of kiln car wheels, conveyor wheels, and furnace roll bearings. These high-temperature applications involve extremely low speed where ample torque is available to rotate the bearing at a relatively high coefficient of friction.

Solid lubricants may simply be dusted as a dry powder on parts to be lubricated, or it may be placed in a liquid carrier. The liquid may either be a fluid intended to evaporate or it may itself be a lubricating liquid or grease. Solid films are also applied as a bonded

Table 15
TYPICAL OPERATING
TEMPERATURE LIMITS FOR
GREASES

Thickener	Temperature limit	
	°C	°F
Calcium soap	65	150
Calcium complex		
Sodium soap	120	250
Sodium-calcium		
Lithium soap		
Aluminum complex	150	300
Synthetics	205	400

coating. Some of the more common materials used are graphites, molybdenum disulfide, cadmium iodide, and fluorinated polyethylenes. Typical bonding agents are resins, silicone, ceramics, and sodium silicate. Another method incorporates the lubricant into one or more of the bearing components, typically a bearing retainer. Soft metals such as silver and tin could be used for this process. In such cases the dry lubricant is transferred from the cage to the bearing raceways by the rolling elements rubbing against the cage. Bearing life is governed by the wear-out life or depletion of the lubricant. Since these special bearings are usually quite expensive, practical industrial practice is to design equipment for use of conventionally lubricated bearings.

Lubricant Temperature Limits

Temperature is the major factor affecting life of a rolling bearing lubricant. Lubricant temperature is influenced primarily by bearing speed, bearing load, ambient temperature, and lubricant system design. With two different greases used on identical applications, base oil type and viscosity, thickeners, and chemical structure can all contribute to different operating temperatures. Some greases will churn in high-speed bearings and cause over-heating, whereas a channeling type grease may function satisfactorily at a much reduced temperature.

Extremely low temperatures must also be considered. The lubricant must permit an ac-ceptable starting torque and must not freeze or become too stiff. While the lubricant must permit equipment turnover at the lowest temperature, it must also have adequate viscosity at the higher operating temperatures to provide sufficient oil film strength. For example, a petroleum type lubricant with very low viscosity oil considered for startup at $-40°C$ and operation at $40°C$ may be unsuitable for operation at $80°C$. In such cases, a synthetic oil or grease may be required to function satisfactorily at both the high and low limits.

Tables 15 and 16 give approximate operating temperature limits for greases and oils. As mentioned previously, however, performance can vary widely depending upon the specific details of a given application. Additives can also affect the suitable operating temperature limits. They can, for example, be somewhat extended by oxidation-inhibiting additives or they may be somewhat reduced by EP or antiwear additives. Earlier chapters of this hand-book, along with References 27 through 29, provide more detailed information on various lubricant factors. Consultation with a reputable lubricant supplier is highly recommended.

Lubricant Selection

Table 17 illustrates "critical" ranges of extreme load, speed, or temperature where special

Table 16
APPROXIMATE OPERATING TEMPERATURE LIMITS FOR OILS

Lubricant	Long term[a]		Short term[b]	
	°C	°F	°C	°F
Petroleum oils	95—120	200—250	135—150	275—300
Superrefined petroleum oils	175—230	350—450	315—345	600—650
Synthetic hydrocarbons	175—230	350—450	315—345	600—650
Organic esters	175—190	350—375	220—230	425—450
Phosphate esters	95—175	200—350	135—230	275—450
Polyglycols	160—175	325—350	205—220	400—425
Polyphenyl ethers	315—370	600—700	425—480	800—900
Silicate esters	190—220	375—425	260—290	500—550
Silicones	220—275	425—525	315—345	600—650
EP oils	60—80	140—175	65—85	150—185

[a] Long term ≈ hundreds of hours.
[b] Short term ≈ few hours.

From Szeri, A. Z., *Tribology — Friction, Lubrication, and Wear*, Hemisphere Publ., Washington, D.C., 1980. With permission.

lubrication provisions may be required. Whether an application is "normal" or critical", optimum performance is obtained when lubricant viscosity is selected for full film separation of the metal surfaces. Rotational speed is by far the most significant factor. While the required viscosity may be determined by calculations involving all pertinent variables according to the equations summarized in Table 14, a good approximation for standard bearings and a normal range of conditions may be found from the following:

$$\text{viscosity at operating temp., SUS} = 70 \left(\frac{\text{maximum catalog speed}}{\text{actual speed}} \right)^{0.7} \quad (18)$$

In this approximation, the recommended viscosity for all bearings at their maximum normal speed limit and lubricated with a naphthenic type oil is a constant 70 SUS at operating temperature. This relationship is considered valid up to the normal speed limit for a given bearing except for some adjustments in the heavy-load or low-speed range as indicated in Table 17. Some modificiation of the viscosity calculation may be necessary with synthetic lubricants due to differences in the pressure-viscosity coefficient.

Table 17 provides a general guide to lubricant selection for bearings operating in the "critical" range. In this guide many other environmental factors and maintenance procedures are presumed to be normal. Moisture, dirt, radiation, chemicals, vacuum, and oxidizing atmospheres may also influence lubricant selection. Consultation with the bearing manufacturer or a lubricant supplier is recommended when unusual conditions exist.

Lubricating Systems

A large variety of manual lubricating devices are available, including drip feed from cups. These devices permit longer intervals between manual servicing, but still require trained personnel and rigid observance of servicing schedules. A lubrication system can improve reliability. There are three basic types of centralized lubrication systems: centralized greasing, centralized circulating oil, and oil mist or fog systems. These systems are often custom-designed for individual application; their installation should be considered in the initial planning of any machine or machinery complex.

Grease lubrication systems generally may be adjusted for amount and frequency of lu-

Table 17
CRITICAL OPERATING RANGES FOR ROLLING ELEMENT BEARING LUBRICATION

Critical range	Oils	Greases	Synthetics (fluid or grease)
Load—high Over 18% of bearing "C" rating or shock and vibration	Increase calculated viscosity by factory of 1.5 to 2.0 EP types desirable	Extreme pressure types recommended; viscosity of oil in grease in chosen on same basis as for oil lubrication	Use with caution as some products may not provide adequate film strength; consultation recommended
Speed—high Over 75% of maximum cataloged speed	Oils may be subject to foaming, and high temperatures, and should be selected carefully; circulation or frequent replacement may be required; oil levels must be controlled	Grease not normally recommended; some products suitable depending on other operating conditions; consultation advisable	Use with caution; consultation recommended
Speed — low Up to 10% of maximum cataloged speed	Use EP type of highest viscosity possible; computed viscosity may be increased by factor of 2.0	Use EP types having highest viscosity oil content; prefer softer grade than used on medium-speed applications	High viscosity and EP characteristics desirable; consultation recommended
Temperature — high Over 95°C (200°F)	Deterioration rate increases rapidly above 95°C (200°F); circulation or frequent replacement indicated; select high-quality oils with low-carbonizing characteristics	Deterioration rate increases rapidly above 95°C (200°F); regreasing frequency must be increased accordingly and must often be determined experimentally	Superior in heat resistance and operating temperature range; prefer medium speeds and medium or light loads
Temperature — low Under −5°C (20°F)	Requires oil viscosity to permit cold start and correct viscosity at subsequent operating temperature; many conventional oils suitable to −34°C (−30°F)	Requires grease to permit cold start and not thin excessively at subsequent operating temperature; many conventional greases suitable to −29°C (−20°F)	Usually required at temperatures lower than −29°C (−20°F); excellent temperature range characteristics

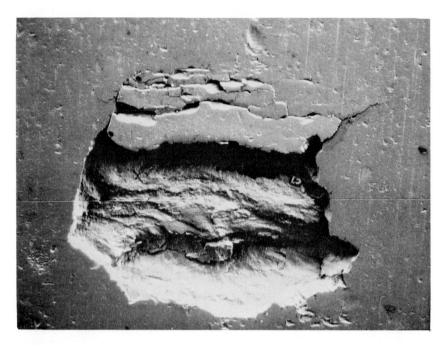

FIGURE 23.   Subsurface nucleated spall on cylindrical bearing inner ring raceway. (Magnification × 50.)

brication. Visual gages are usually provided to facilitate checking for a continuous lubricant supply to all bearings in the system. In cases where separate bearings operate under different conditions of temperature, speed, and load, use of more than one system may be necessary to meet the correct lubrication needs of the individual bearings.

Circulating oil lubrication systems are most beneficial when bearings must be cooled continuously and when abrasive materials must be flushed away to assure safe operation. Circulating oil lubrication systems nearly always have filter and heat exchanger elements in addition to their oil reservoir and pump. They may also have a centrifuge or a sump for separating and removing foreign material, remote controls, warning devices, automatic cut-off switches, etc. These are particularly useful in meeting the special requirements of paper mills, lumber mills, steel mills, coal processing plants, and similar applications.

Oil mist lubrication systems use an air stream to provide oil to the bearings. The air pressure maintains a positive pressure within the bearing chamber which effectively prevents foreign matter from entering. The air flow can be regulated to produce minimum lubricant friction and the concomitant lubrication friction temperature effect. The air flow will not, however, provide significant cooling.

Air flowing out of a mist-lubricated bearing may discharge a fine oil vapor. This vapor may be objectionable, especially in the food and textile industries. In such cases, it is necessary to vent to other areas or provide air cleaning systems. Drainage of bearing reservoirs, provision for proper oil levels during bearing start-up, and timing of the mist flow must meet precise specifications. For this reason the system manufacturer should be relied upon to adjust the system for correct operation. Detailed information on lubricating systems is given in other chapters.

FAILURE ANALYSIS

Selection, application, and installation of rolling element bearings is based on subsurface nucleated fatigue. In the field, however, only 5 to 10% of the bearings removed from service are found to have developed this type of failure such as illustrated in Figure 23.

Table 18
FAILURE MODES THAT LIMIT PERFORMANCE

Failure modes	Cause	Results
Brinelling	Dents in race or balls caused by impact or improper loading	Rough-running bearing surface origin fatigue
Cracked rings	Improper fits—abuse—cracked mounting	Catastrophic failure
Electric discharge	Flow of electric current through bearing	Rough-running bearing (fluting)
Fretting	Small amplitude motion of contact surfaces under load — causes indentation in race or localized material removal from mounting surfaces	Rough-running bearing — loss of bearing retention on shaft or in housing — stress concentration cracking
Galvanic action	Exposure of bearing to electrolyte (salt water in oil) and metal higher in galvanic series (copper, aluminum)	Pitting, leading to surface origin fatigue
Grease hardening/increased viscosity oil	Overheating of bearing — oil bleed loss — excessive bearing cavity ventilation — high ambient temperature	Excessive friction — wear — cage failure — surface origin fatigue
Lock-up	Temperature gradient due to shaft heating, causing loss of clearance	When steady state; preloading resulting in inner ring turning on shaft and severe shaft damage
		When intermittent: premature fatigue failure of rollers and rings
Lubricant decomposition	Attack by lubricant decomposition products	Pitting, surface origin fatigue
Peeling-frosting	Low-viscosity lubricant, insufficient EHD film	Surface origin fatigue — increased clearance — rough operation
Race deformation	Overload and/or overheating	Destruction of geometry — increased clearance — cage failure — jamming
Rib wear and denting	Heavy impact loads — excessive thrust loads — abrasive contaminants — loss of lubricant or sparse lubricant	Loss of roller control where rib-guided — jamming of bearing — cage failure
Roller end wear	Skewing of cylindrical rollers under bearing misalignment — poor lubrication — abrasive contaminants	Slow deterioration leading to roller jamming — increased clearance
Rolling element banding	Excessive thrust load or overheating	Increase in internal clearance — rough-running after bearing stops and is started again
Rolling element scoring	Trapping of hard particles between retainer pocket and roller, causing deep cuts or scratches	Surface origin fatigue
Rusting	Exposure to humid or salt environment with insufficient protection by lubricant	Rough-running — surface origin fatigue
Smearing	Material transfer during skidding — loss of traction in load zone — very light loads — rapid acceleration	Rough-running bearing — surface origin fatigue
	Smearing of roller ends and guide flange	Overheating — loss of guidance — increase in axial float

Table 18 (continued)
FAILURE MODES THAT LIMIT PERFORMANCE

Failure modes	Cause	Results
Wear of cages	Misalignment with nonaligning bearings — loss of lubricant or sparse lubricant — abrasive contaminants	Cage breakage — cage drop and jamming of bearing
Wear of rolling elements	Loss of or sparse lubricant — abrasive contamination	Jamming of bearing by debris — excessive internal clearance — retainer breakage

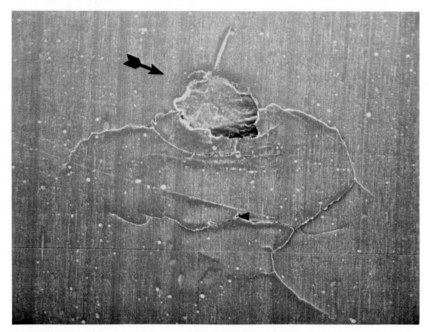

FIGURE 24.   Scanning electron micrograph of surface nucleated spall.

Fatigue can often be induced by maldistribution of load in bearings due to varying stiffness of the mounting or support surfaces, housings, or shafts. Recognizing the sensitivity of rolling element bearing life to the variations in stress under the most heavily loaded rolling element (ball bearing life $\sim (1/\text{Stress})^{8\text{-}10}$, roller bearing life $\sim (1/\text{Stress})^{7\text{-}9}$), the designer must carefully consider the mounting, its stiffness, and the influence of mutual deflections of all components in the system. Distortions due to temperature distributions are equally important and transient conditions must be properly accounted for.

Damage commonly results from imposed loads which differ considerably from those anticipated in a machine design. Misalignment or fitting errors in mounting a bearing, misalignment or coupling faults between two machines, differential thermal expansion in a frame and shaft system, and rotor unbalance are among such factors. Simple visual or low power microscopic analysis of the ball paths in a ball bearings will frequently enable a useful evaluation of the magnitude and nature of these operating conditions.[30]

Table 18 lists failure modes that limit the performance of rolling element bearings. Several bearing companies have published similar lists and several volumes have been written on the subject. Of particular note is Reference 31. Detailed failure analysis should be correlated with the bearing company involved since their laboratory, background, and experience enable them to draw conclusions and make recommendations.

FIGURE 25. SEM of ground surface after running. (Magnifications × 100 and 500.)

Bearings which have been grease-lubricated require special attention since they generally show surface effects which are the combination of many operating regimes. Grease lubrication in many bearings is a variable which depends on frequency of relubrication or on cyclic temperature variations to which the bearing is exposed. Many greased bearings operate with depleted films and significant wear will obliterate many original evidences of loading. Caution must be exercised against relying on the obvious conclusions while paying insufficient attention to the minor findings evidenced on careful examination.

Failure analysis has been greatly aided by utilization of the scanning electron microscope.[31,32] Small differences in the surface can indicate either the immediate condition of the bearing or some condition which has resulted from its previous operation. Figures 24 to 27 show scanning electron micrographs of bearing components which had been removed from service. Equally important is analysis of lubricants which can indicate the extent of contamination and deterioration. More recently, "ferrographic" analysis of filtrants or tailings from lubricants has become a valuable tool in monitoring transient bearing condition.

Since catastrophic failures have reduced value in aiding the troubleshooter, every effort must be made to look at units which have not failed completely, preferably a number of them with different periods of operation in order to detect and trace the incipient failure mode. Of particular importance is the observation of changes in surfaces, the lubricant, housing, and shaft as well as the bearing to correct outside influences that can cause early bearing failure.

In many instances, misalignment of sufficient magnitude to cause moment loading to run the rolling elements off the raceway induces a violent premature fatigue failure. Obviously, severe misalignment of tapered and cylindrical bearings must be corrected initially. Practical limits for misalignment are shown in Table 13.

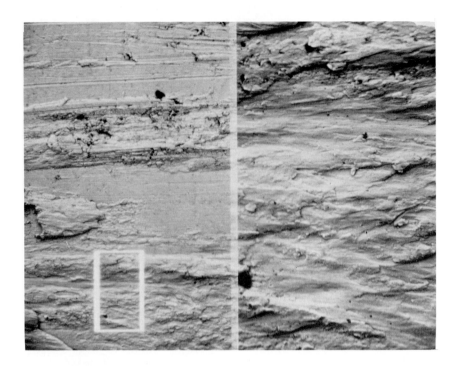

FIGURE 26.   SEM of low-velocity galled roller. (Magnifications × 100 and 500.)

FIGURE 27.   SEM of fretting wear in bore. (Magnification × 100.)

REFERENCES

1. *Metric Ball and Roller Bearings Conforming to Basic Boundary Plans,* ANSI/AFBMA Standard 20, Anti-Friction Bearing Manufacturers Association, Arlington, Va., 1977.
2. *Tapered Roller Bearings — Radial Inch Design,* ANSI/AFBMA Standard 19, Anti-Friction Bearing Manufacturers Association, Arlington, Va., 1974.
3. **Boresi, A. P., Sidebottom, O. M., Seely, F. B., and Smith, J. O.,** *Advanced Mechanics of Materials,* 3rd ed. John Wiley & Sons, New York, 1978, 581.
4. **Harris, T. A.,** *Rolling Bearing Analysis,* John Wiley & Sons, New York, 1966.
5. **Palmgren, A.,** *Ball and Roller Bearing Engineering,* SKF Industries, Philadelphia, 1959.
6. **Eshmann, Hasbargen, and Weigand,** *Ball and Roller Bearings, Their Theory, Design, and Application,* K. G. Heyden & Co., London, 1958.
7. **Hartnett, M. J.,** The analysis of contact stresses in rolling element bearings, *ASME Trans. J. Lubr. Technol.* 101(1), 105, 1979.
8. **Hamrock, B. J.,** Stresses and Deformations in Elliptical Contacts, Tech. Memo. 81535, National Aeronautics and Space Administration, Washington, D.C., 1981.
9. **Lundberg, G.,** *Cylinder Compressed Between Two Plane Bodies,* Aktiebolaget, Svenska Kullagerfabriken, Goteborg, 1949.
10. *Load Ratings and Fatigue Life for Ball Bearings,* ANSI/AFBMA Standard, Anti-Friction Bearing Manufacturers Association, Arligton, Va., 1978.
11. *Load Ratings and Fatigue Life of Roller Bearings,* ANSI/AFBMA Standard II, Anti-Friction Bearing Manufacturers Association, Arlington, Va., 1978.
12. **Lundberg, G. and Palmgren, A.,** Dynamic capacity of rolling bearings, *Acta Polytech. Mech. Eng. Ser.,* 1 (3), 1952.
13. **Lundberg, G. and Palmgren, A.,** Dynamic capacity of roller bearings, *Acta Polytech. Mech. Eng. Ser.,* 2(4), 1952.
14. **Moyer, C. A. and McKelvey, R. E.,** A rating formula for tapered roller bearings, *SAE Trans.,* 71, 490, 1963.
15. **Price, C. E. and Galambus, M.,** *Bearing Application for Material Conveying Equipment,* Paper No. 80-3011, American Society of Agricultural Engineers, St. Joseph, Mich., 1980.
16. **Grubin, A. N. and Vinogradova, I. E.,** Investigation of the contact of machine components, *TsNIITMASh,* book No. 30, Department of Scientific and Industrial Research, London, 1949.
17. **Dowson, D. and Higginson, G. R.,** A numberical solution to the elastohydrodynamic problem, *J. Mech Eng. Sci.,* 1(1), 6, 1959.
18. **Puckett, S. J. and Pflaffenberger, E. E.,** *Rolling Contact Bearing Surfaces — The Current Relationship Between Requirements and Processing,* Paper No. 1073, Society of Manufacturing Engineers, Dearborn, Mich., 1973.
19. **Tallin, T. E.,** On competing failure modes in rolling contact, *ASLE Trans.,* 11, 418, 1967.
20. **Littmann, W. E., Widner, R. L., Wolfe, J. O., and Stover, J. D.,** The role of lubrication in propagation of contact fatigue cracks, *ASME Trans. J. Lubr. Technol. Ser. F,* 90(1), 89, 1968.
21. **Rounds, F. G.,** Some effects of additives on rolling contact fatigue, *ASLE Trans.,* 10, 243, 1967.
22. **Bock, F. C., Bhattacharyya, S., and Howes, M. A. H.,** Equations relating contact fatigue life to some material, lubricant, and operating variables, *ASLE Trans.,* 22(1), 1, 1979.
23. *Life Adjustment Factors for Ball and Roller Bearings, An Engineering Design Guide,* American Society of Mechanical Engineers, New York, 1971.
24. **Danner, C. H.** Fatigue life of tapered roller bearings under minimal lubricant films, *ASLE Trans.,* 13, 241, 1970.
25. **Hamrock, B. J. and Dowson, D.,** Isothermal elastohydrodynamic lubrication of point contacts. III. Fully flooded results, *ASME Trans. J. Lubr. Technol.,* 99(2), 264, 1977.
26. **Coy, J. J. and Zaretsky, E. V.,** Some limitations in applying classical EHD film thickness formulae to a high speed bearing, *ASME J. Lubr. Technol.,* 103(2), 295, 1981.
27. **Neale, M. J., Ed.,** *Tribology Handbook,* John Wiley & Sons, New York, 1973.
28. **Szeri, A. Z., Ed.,** *Tribology — Friction, Lubrication, and Wear,* Hemisphere Publishing, Washington, D.C., 1980.
29. **Hatton, R. E.,** Synthetic oils, in *Interdisciplinary Approach to Liquid Lubricant Technology,* NASA SP-318, National Aeronautics and Space Administration, Washington, D.C., 1973.
30. ASLE Manual, *Interpreting Service Damage in Rolling Type Bearings,* American Society of Lubrication Engineers, Park Ridge, Ill., 1953.
31. **Tallian, T. E., Baile, G. H., Dalal, H., and Gustafson, O. G.,** *Rolling Bearing Damage Atlas,* SKF Industries, King of Prussia, Pa., 1974.
32. **Derner, W. J.,** *The Use of the Scanning Electron Microscope in Analyzing Rolling Contact Surfaces,* Paper 790851, Society of Automotive Engineers, Warrendale, Pa., 1979.

GEARS

J. L. Radovich

INTRODUCTION

Many studies have been made in recent years to understand more fully the lubrication requirements of gears. The ideal situation would be a theoretical solution which would predict the optimum lubricant for a specific set of gears and operating conditions based on easily measured system parameters. To date, gear lubrication has not been reduced to this pure science. Consequently, experience is still one of the most valuable tools for proper lubricant selection.

Lubrication provides the vital function of separating the contacting surfaces of the gear teeth by an easily sheared film which reduces friction, improves efficiency, and extends the useful life. In addition, lubrication may also provide cooling and flushing of the gear tooth surfaces, corrosion protection, and chemical modification of the surface material. Although proper lubrication is a necessity for successful operation of a set of gears, it is not a cure for inadequate design, manufacture, or improper operation.

GEAR TYPES AND TERMINOLOGY

Figure 1 shows a spur gear and pinion in mesh and displays several terms used in gearing. A central element is the pitch diameter which is calculated as follows:

$$D = \frac{2 \times C \times T}{T + t} \quad \text{or} \quad D = \frac{2 \times C}{1 + 1/R}$$

$$d = \frac{2 \times C \times t}{T + t} \quad \text{or} \quad d = \frac{2 \times C}{R + 1}$$

where D and d are the pitch diameter of the gear and pinion, respectively; C is the operating center distance; T and t are the number of teeth in the gear and pinion, respectively; and R is the ratio of the gear set $(R = T/t)$.

Pitch line velocity is the peripheral speed of the pitch diameter in meters per second or feet per minute.

$$V = \frac{\pi}{60} d N$$

where v = pitch line velocity in meters per second; d = pinion pitch diameter in meters; and N = pinion speed in RPM; or

$$V = \frac{\pi}{12} d N$$

where v = pitch line velocity in feet per minute; d = pinion pitch diameter in inches; and N = pinion speed in RPM.

There are several types of gear configurations (Figure 2). Each gear type has different design advantages and some have special lubrication requirements.

Spur — Gear shafts are parallel and gear teeth are cut in line with the shaft centerline (Figure 2a). For spur gears with a transverse contact ratio (contact length divided by base pitch) of less than two, the tangential load is carried by two teeth at the beginning of the

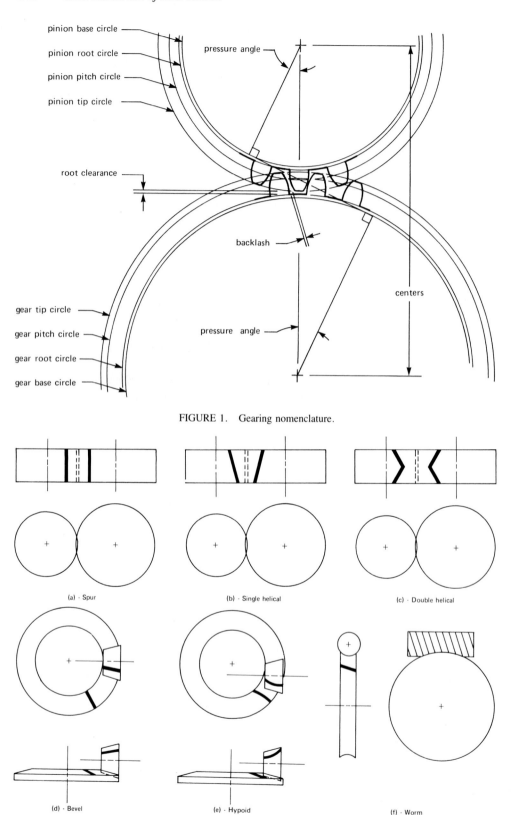

FIGURE 1.   Gearing nomenclature.

(a) - Spur

(b) - Single helical

(c) - Double helical

(d) - Bevel

(e) - Hypoid

(f) - Worm

FIGURE 2.   Gear types.

contact cycle. The load is then carried by one tooth only as one of the teeth leaves mesh and then by two again as the next tooth comes into mesh.

Helical — Gear shafts are usually parallel, but may be at any angle to each other. Gear teeth are cut at an angle to the shaft certerline (Figure 2b). The transfer of load from one tooth to the next is more uniform than spur because several teeth are always in contact along some portion of the tooth face at the same time. Because of the helix angle this gear type generates a thrust load along the axis of the gear shaft.

Double helical — Gear shafts are parallel. The gear face is split into two sections, each with helical teeth. The two helical sections have equal helix angles, but opposite hands (Figure 2c). Contact conditions are the same as single helical, but since the thrust load from each helix is equal in magnitude and opposite in direction, no net thrust load is imposed on the gear shaft. However, one of the elements must be free to move axially with respect to the other in order to equalize the tooth loads on each helix. If this is not done, single helix loading will occur.

Bevel — Shaft centerlines are orthogonal and intersecting. Bevel gears can be straight or spiral (Figure 2d).

Hypoid — Basically the same as bevel gears except that shaft centerlines do not intersect (Figure 2e). Because of this offset, relative sliding velocity between contacting surfaces is higher than for bevels. Because of this sliding and the high contact stresses, an extreme pressure lubricant compounded with friction modifying additives is required.

Worm — Shaft centerlines are orthogonal and nonintersecting. The worm resembles a screw thread and drives the worm gear. Both elements are in the same plane (Figure 2f). Since the worm rotates like a screw, high-sliding velocity is developed between contacting surfaces on the worm and wheel. As a result, a lubricant containing friction modifiers is necessary to reduce friction and improve efficiency.

DESIGN CONSIDERATIONS AND GEAR MATERIALS

In considering a gear application, the power to be transmitted and input speed and gear ratio are usually specified. Orientation of the input shaft to the output shaft may also be indicated. Standard formulas for determining the allowable power which can be transmitted by a gear set have been developed by the American Gear Manufacturers Association (AGMA). Using these formulas in conjunction with the information specified, the designer then has to balance the following variables.

Gear type — If the input and output shafts are required to operate at right angles, or some condition other than parallel, bevel, hypoid, helical, or worm gears must be used. If the shafts are parallel to one another, spur, helical, or double helical gears can be used. The gear type will also influence the type of bearings and the housing design required to support the gear forces.

Center distance — As the linear distance between the centerlines of two mating gears is increased, for the same transmitted power, the tangential tooth load decreases since the torque is generated with a longer moment arm. The pitch diameters of the gear and pinion would increase and, consequently, the pitch line velocity would increase also. This increase in center distance would allow a narrower face width or softer material and less stringent lubrication requirements. The disadvantages are that the larger gears take more space and tend to cost more.

Face width — By increasing the width of the gear face, the contact area is lengthened and unit loading is reduced. This would allow the use of softer gear material or a reduction of the center distance. The disadvantage is that as the face width increases, the shafting must be made more rigid so that dynamic deflection of the gear shaft will not reduce the effective contact of mating teeth.

Helix angle — Increasing the angle that is made by the centerline of the tooth with the centerline of the shaft will increase the face contact ratio. This means that more teeth are in mesh in the contact zone, which distributes the load more uniformly. Adversely, an increase in helix angle causes an increase in the thrust load generated by single-helical gears. This increases the loads on bearings and other structural components.

Tooth size — A larger tooth with a greater cross-sectional area will support a larger load before stresses in the root section break the tooth away from the gear. As tooth size increases, the number of teeth in the same pitch diameter decreases. This decreases the transverse contact ratio, with fewer teeth in contact in the load zone, which hinders the uniform transfer of load from one tooth to the next.

Material — The most important consideration is generally hardness, which is a measure of load-carrying capacity of the material. Other factors can be material cost, availability, machinability, wear resistance, compatibility with the operating environment, etc. The hardness of two mating elements should be proportioned with the pinion being harder than the gear. Each tooth on the pinion undergoes a number of load cycles equal to the gear ratio for every load cycle of a tooth on the gear. Proportioning the hardnesses will tend to equalize the wear.

The most common material used for commercial gearing is steel. Carbon content determines what hardness levels can be achieved, while alloying elements such as nickel, chromium, molybdenum, manganese, etc. are used to improve the hardenability, strength, and toughness of plain carbon steels. Hardenability, the ability to increase hardness below the surface of the material, is important in gears which are to have teeth cut after hardening. Since the material does not have to be heat treated after machining, distortion is minimal. Through hardened steels are used in the range of 180 to 440 BHN.

Steel can be further hardened at the gear tooth surface by carburizing, nitriding or flame or induction hardening. When such case hardening is employed, the case hardness must prevail for an appropriate distance below the tooth surface. The minimum case depth should be approximately one sixth of the tooth thickness at the pitch diameter. If case hardness depth is insufficient, loads and deflections applied to tooth surfaces may cause subsurface cracks. Through repeated load cycles, these cracks could propagate toward the surface causing portions of the hardened tooth surface to break away. Grinding of the tooth flanks may be required after surface hardening to correct for distortion caused by the process.

Containing additional carbon in the form of free graphite, cast iron has less ductility, lower tensile strength, and a lower modulus of elasticity than steel. Cast iron is generally used in applications where intensity of gear tooth loading is less severe. Since the free graphite gives a sound deadening quality, cast iron may be preferred over steel in low-sound level applications. The graphite may also act as a lubricant and help gears survive under conditions of insufficient lubrication.

Of nonferrous gear materials, bronze is the most common. Typical bronze alloys for gearing are 86 to 90% copper, 9 to 12% tin, and 3% or less of lead, zinc, and phosphorus. This material does not rust, is nonmagnetic, and offers a good balance of strength and hardness. It is frequently used for worm wheels which, when run with hardened, ground steel worms, create a system of dissimilar metals that does not seize and score under the loads and high-sliding contact associated with worm drives. Selection of other nonferrous materials is usually based on weight reduction or suitability for manufacturing processes such as die casting. Some lubricant additives which work well on ferrous materials may have little or no affect on nonferrous materials.

Nonmetallic gears are made from a variety of plastics, laminations of impregnated fabrics, etc. Because of the low-material modulus, gear tooth stiffness is reduced, giving a reduction in dynamic tooth loading and increased surface endurance. Nonmetallic gears are most effective when meshing with hardened steel gears. They offer quiet operation, abrasion

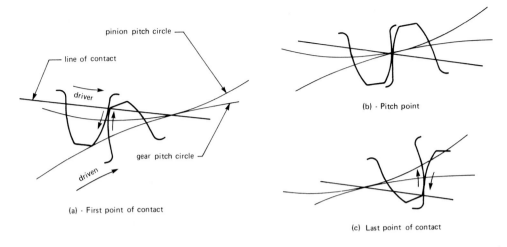

FIGURE 3.   Relative motion of meshing gear tooth surfaces.

resistance, and resistance to impact loads because of the material resilience. Suitability of lubricants depends on compatibility with the gear material. Some nonmetallic gears will provide adequate service with no lubrication.

The type and grade of lubricant to be used may be an early design consideration. Most often, though, design and material parameters are first balanced to suit the specified application and then a lubricant is selected to meet the gearing requirements.

GEAR LUBRICATION

Gear teeth may operate in three conditions of lubrication: boundary, mixed, and full film. Boundary lubrication occurs when gear sets start or stop. Here the chemical properties of the lubricant are most important to prevent scoring of the surfaces due to metal-to-metal contact. If gear sets were operated under conditions of boundary lubrication for extended periods of time, wear would be rapid and severe.

With increased relative motion, the gearing moves into mixed lubrication. Here, tooth surface asperities are close enough to influence the coefficient of friction. In this regime, wear would occur at a slower rate than with boundary lubrication, but could still be too rapid for reasonable service life of the gear set.

Optimum lubrication is full-film where gear tooth surfaces are completely separated by an elastohydrodynamic (EHD) oil film at least two to three times as thick as the composite surface roughness. Since the lubricant viscosity is the most important characteristic in full-film lubrication, the proper lubricant grade selection is important.

Contacting conditions of two gear teeth in mesh as shown in Figure 3 are typical of spur, helical, and bevel gears. The contact starts as high-relative sliding and some rolling. The sliding decelerates toward the pitch line. At the pitch line, the only motion is rolling. After the pitch line, sliding again takes place and accelerates until the teeth leave mesh. The radii of curvature of the gear tooth flanks also changes constantly. It is least at the root of the tooth and greatest at the tip.

To evaluate the role of lubricants in the operation of gear sets, it is necessary to understand the contacting conditions of gear teeth in mesh. In the past 2 decades, much work has been done to show that tooth surfaces are not perfectly rigid, but deflect elastically in the contact zone due to the very high-contact pressure as shown in Figure 4.

Viscosity does not remain constant through the meshing cycle, but increases rapidly with pressure. When the elastic deflection is considered along with the increase in viscosity in

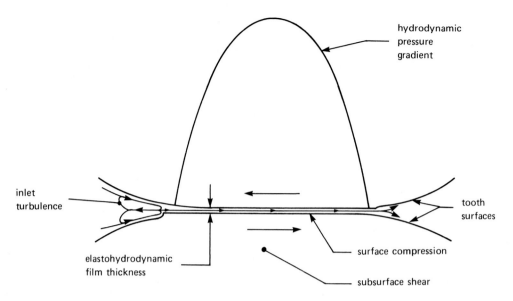

FIGURE 4. Deflection of gear tooth surfaces in contact.

the contact zone, film thicknesses are predicted which corresponds with laboratory results and practical experience.

Elastohydrodynamic theory has shown that a film approximately 1-μm thick can separate gear teeth and that, under the high-contact pressures, this film may become more rigid than the gear tooth surfaces.[1] The formulas for calculating film thicknesses[2] provide values that are difficult to measure under actual operating conditions.[3] However, they do point out trends. Increases in relative velocity of gear tooth surfaces, in lubricant viscosity at ambient conditions, and in the radius of curvature at the point of contact will increase the oil film thickness more dramatically than proportionate reductions in the unit tooth load, or elastic modulus of the gear material.[4]

For proper selection of the type, grade, and method of lubricant application, it is necessary to evaluate the following factors.

Type of gearing — Spur, worm, hypoid, etc.

Size — Pitch diameter, whole depth and face width as an indication of the surface area to be lubricated.

Mounting and enclosure — Type of housing surrounding the gears. Whether this housing will keep splashed oil and oil vapors in and keep dirt, water vapor, and other contaminants out must be considered as well as whether bearings are lubricated by the same system.

Speed — Pitch line velocity of the gear set.

Load characteristics — Consider whether the transmitted power is continuous, steady, cyclic; are shock loads possible; is high vibration present; does the unit start and stop frequently; etc.

Temperature — Range over which the lubricant must perform. This would include the lowest ambient temperature anticipated at startup to the highest operating temperature.

TYPES OF LUBRICANTS

The most common lubricants used in industry today are those blended from petroleum base stocks. These fall into the following five major categories.

Inhibited Oils

These are mineral oils containing rust and oxidation (R&O) inhibitors and perhaps antifoam or antiwear agents. Rust inhibitors protect ferrous surfaces against rust. Oxidation inhibitors retard the formation of acidic contaminants, carbonaceous material, and increased viscosity. Inhibited oils are generally suitable for spur, helical, and bevel gears transmitting light to moderate loads. They perform well over a wide range of sizes and speeds in the temperature range of approximately -20 to $120°C$. This type of lubricant is ideal for bearings if both bearings and gears must be lubricated from the same system. Constant relubrication of gear teeth is preferred since the oil does not adhere to tooth surfaces. This type of oil can be used effectively to cool the gear mesh and flush the tooth surfaces of wear particles or debris. The lubricant can be easily conditioned with filters and heat exchangers for consistent temperature and cleanliness.

Extreme Pressure (EP) Oils

These are basically inhibited oils with added extreme pressure additives. The EP agent controls wear in the boundary lubrication phase: stopping, starting, shock loads, etc. There are two basic types of EP agents. The first is chemically active, such as sulfur phosphorous, which reacts with gear tooth surface material under high temperature to form a thin film of easily sheared material. If high points of mating surfaces come in contact, the contacting material will shear rather than fuse and cause scoring. Therefore, controlled wear is exchanged for destructive wear. The second type of EP agent is a solid lubricant in suspension. The solid particles (such as graphite, molybdenum disulfide, borate, etc.) get between tooth surfaces and prevent metal-to-metal contact. EP oils are used for spur, helical, and bevel gears where loads are too heavy for non-EP mineral oils. EP oils are also used for worm and hypoid gears. The useful temperature range is -20 to $120°C$.

If bearings are to be lubricated from the same system, some caution is required. Active EP agents must be noncorrosive to bronze if any is present in the bearings. Solid lubricants may reduce internal clearances in low clearance, precision bearings causing high temperature, and probable failure. Constant relubrication of gearing with EP oils is preferred since the oil will not adhere to tooth surfaces. Extremely fine filtration will remove solid lubricant additives, however, conditioning through cooling and filtration is easily accomplished and this type of lubricant can be used to reduce gear mesh temperatures.

Compounded Oils

These are usually steam cylinder stocks compounded with acidless fat or tallow as lubricity additives to reduce friction. These oils are primarily used in worm gear drives where the high-sliding action of the gear teeth requires a friction reducing agent to reduce heat and improve efficiency. The useful temperature range is approximately 5 to $120°C$. Bearings can be lubricated with this type of lubricant without difficulty. Constant relubrication of gear teeth is recommended since this type of oil does not cling to gear teeth and will be wiped off the gear teeth in mesh. This type of lubricant can be conditioned through cooling and filtration and can be used to reduce operating gear mesh temperatures.

Open Gear Compounds

These are heavy-bodied lubricants for large, slow-speed, heavily loaded gear sets. These lubricants contain tackiness additives to adhere to gear teeth and resist being thrown off or squeezed out of mesh. Some of these compounds are so heavy they require solvents to make them soft enough to apply. The solvents evaporate and leave the thick protective film on the gearing. Since this type of lubricant adheres to gear teeth, constant application is not necessary. The useful temperature range is approximately 5 to $120°C$. Bearings should be evaluated to see if they can be properly lubricated with this type of lubricant. Being so viscous and adherent to gear teeth, this type of lubricant does not offer the advantage either of cooling or flushing the gear mesh.

Table 1
TYPES OF LUBRICANT USED WITH VARIOUS GEAR APPLICATIONS

	Gear types				
Lubricant	Spur	Helical	Worm	Bevel	Hypoid
R & O oil (non-EP)	Normal loads	Normal loads	Light loads Slow speeds only	Normal loads	**Not** recommended
EP oil	Heavy or shock loading	Heavy or shock loading	Satisfactory for most applications	Heavy or shock loading	Required for most applications
Compounded oil (ca. 5% tallow)	Not normally used	Not normally used	Preferred by most gear manufacturers	Not normally used	For light loading only
Heavy-bodied open gear oils	Slow-speed Open gearing	Slow-speed Open gearing	Slow-speeds **Only** EP additive desirable	Slow-speeds Open gearing	Slow-speeds **Only** EP additive required
Grease	Slow-speed Open gearing	Slow-speed Open gearing	Slow-speeds **Only** EP additive desirable	Slow-speeds Open gearing	**Not** recommended

From Root, D. C., *Lubr. Eng.*, 32, 8, 1976. With permission.

Greases

These are liquid lubricants thickened with soaps to a gelatinous consistency. The soap holds the liquid portion and releases it as necessary, analogous to oil in a sponge. The liquid portion does the lubricating with the advantages and limitations of this base oil for load and operating temperature. The advantage is that lubricant does not have to be added continuously; extended intervals between relubrication are possible. Components which would be difficult to lubricate, or where lubrication maintenance is undesirable, can be "packed for life". Examples are gears and bearings in small power tools or household appliances. Grease can act as a seal to keep out dirt and moisture. The disadvantage is that, unlike a continuous flow of oil, grease will not carry away heat or wear particles from the gear mesh until new grease is introduced and the old grease is displaced.

Grease may be used in spur, helical, worm, and bevel gears. It is generally restricted to slow-speeds, or very small, lightly loaded components with intermittent service. Higher speeds will cause the grease to "channel" or be displaced by the gear teeth, preventing lubricant from reaching areas where it is needed. The selection of a proper National Lubricating Grease Institute (NLGI) consistency is very important. Useful operating temperatures are limited by both the base oil and the soap. For multipurpose greases, the useful range is approximately −30 to 120°C.

Application of these various types of lubricants to various gear types is summarized in Table 1.

In addition to the petroleum based lubricants, a wide variety of synthetic lubricants are available. The cost of synthetic lubricants is higher than petroleum based products, but they have definite advantages for special applications. Since they contain no waxes to solidify at low temperatures and no carbon to form deposits at high temperatures, synthetic lubricants will function from approximately −73 to 260°C. Since many synthetics have much higher viscosity indices than petroleum based lubricants, their viscosity does not change as much when the operating temperature increases or decreases. Synthetic lubricants may also be used where fire resistance is required. When use of a synthetic lubricant is considered, lubricity and load-carrying capacity should be evaluated as well as compatibility with paints, plastics, seals, and gasket materials.

Another class of lubricating materials consist of solid lubricants such as graphite, molybdenum disulfide, fluorocarbon polymers, etc. Some solid lubricants can be used for very high-temperature applications of 540°C and greater.[5] Solid lubricants can be bonded, applied dry, or used with a liquid carrier which evaporates or decomposes, leaving the lubricating material behind. Solid lubricants can be used where access for relubrication is difficult, or where lubricant leakage could contaminate surrounding components.

LUBRICANT GRADE

After selecting the type of lubricant to be used, the proper viscosity grade must be determined. A lubricant which is too heavy will cause excessive heat to be generated with excessive power losses and inefficiency. A lubricant which is too light will cause rapid wear of components, resulting in reduced service life. Therefore, the lightest lubricant which will keep wear rates within acceptable limits is the most desirable.

If bearings are to be lubricated by the same system, gearing requirements are generally more severe and will determine the lubricant grade required. In reduction units consisting of several gear sets, the lowest speed gearing is usually the most critical, since speeds are slower and torque is greater, and the lubricant grade would be selected to meet the requirements of this set. If a multiple reduction unit employed worm gears for one reduction, this would represent the most severe lubricant criteria. In general, each component in complex systems must have any special lubrication requirements evaluated. Final lubricant selection must be suitable for all of the components.

A theoretical approach to the selection of the proper lubricant grade is very involved. The best method would be to build a prototype, run it under the expected range of operating conditions, and measure the wear and horsepower losses for various lubricant grades. If this approach is not practical and past experience is not available, it becomes necessary to use published empirical data.

A good source for this data, as it applies to industrial gearing, is information published by the AGMA. Tables 2 through 8, taken from the AGMA Lubrication Standards, show suggested viscosity grades for gears operating under normal loads over a range of speeds and ambient temperatures.

Increased ambient or operating temperatures require heavier oils, since increased oil temperature lowers the operating viscosity and reduces the oil film thickness. Heavier oils may be required if the unit is subjected to shock loads or high levels of vibration. These load pulsations could cause higher temperatures in the mesh, which would lower the operating viscosity and reduce the oil film thickness. Oils required to perform over a wide range of temperatures should be selected with high-viscosity indices to reduce the effects of temperature on the oil viscosity.

METHODS OF LUBRICATION

Lubricant can be applied to gear teeth in a variety of ways. Easily flowing liquid lubricants such as inhibited, EP, and compounded oils are usually applied to gear teeth by means of a splash system or a more complex force-fed system.

In the splash system shown in Figure 5, lubricant is applied by allowing the gear to run partially submerged in the oil. Oil picked up by the gear is then carried into the gear mesh where it is needed. Oil tends to be thrown off the gear set due to rotation; and this oil, as well as oil mist from churning and windage will wet the inside of the gear case. As this oil

Table 2
VISCOSITY RANGES FOR AGMA LUBRICANTS

R & O inhibited gear oils (AGMA lubricant no.)	Viscosity range [cSt (mm²/sec) at 40 °C][a]	Equivalent ISO grade[b]	EP gear lubricants (AGMA lubricant no.)[c]	Viscosities of AGMA former system (SSU at 100 °F)[d]
1	41.4—50.6	46		193—235
2	61.2—74.8	68	2 EP	284—347
3	90—110	100	3 EP	417—510
4	135—165	150	4 EP	626—765
5	198—242	220	5 EP	918—1122
6	288—352	320	6 EP	1335—1632
7 Comp[e]	414—506	460	7 EP	1919—2346
8 Comp[e]	612—748	680	8 EP	2837—3467
8A Comp[e]	900—1100	1000	8A EP	4171—5098

Note: Viscosity ranges for AGMA lubricant numbers will henceforth be identical to those of the ASTM System (footnote a).

[a] "Viscosity System for Industrial Fluid Lubricants," ASTM 2422; also British Standards Institute, B.S. 4231.

[b] "Industrial Liquid Lubricants — ISO Viscosity Classification," International Standard, ISO 3448.

[c] Extreme pressure lubricants should be used **only** when recommended by the gear drive manufacturer.

[d] AGMA 250.03, May 1972 and AGMA 251.02, November 1974.

[e] Oils marked Comp are compounded with 3—10% fatty or synthetic fatty oils.

From Standard AGMA 250.04, *Lubrication of Industrial Enclosed Gear Drives*, American Gear Manufacturers Association, Arlington, Va., 1974. With permission.

Table 3
AGMA LUBRICANT NUMBER RECOMMENDATIONS FOR ENCLOSED HELICAL, HERRINGBONE, STRAIGHT BEVEL, SPIRAL BEVEL, AND SPUR GEAR DRIVES

Low-speed center distance		AGMA Lubricant no.[b,c]	
		Ambient temp[d]	
Type of unit[a]	(Size of unit)	−10—+10°C (15—50 °F)[e]	10—50 °C (50—125 °F)
Parallel shaft	(Single		
Up to 200 mm	reduction)	2—3	3—4
	(to 8 in.)[f]		
Over 200 mm, up to 500 mm	(8 to 20 in.)[f]	2—3	4—5
Over 500 mm	(Over 20 in.)	3—4	4—5
Parallel shaft	(Double		
	reduction)		
Up to 200 mm	(To 8 in.)[f]	2—3	3—4
Over 200 mm	(Over 8 in.)	3—4	4—5
Parallel shaft	(Triple		
	reduction)		
Up to 200 mm	(To 8 in.)	2—3	3—4
Over 200 mm, up to 500 mm	(8 to 20 in.)[f]	3—4	4—5
Over 500 mm	(Over 20 in.)[f]	4—5	5—6
Planetary gear units	(Housing		
	diameter)		
Up to 400 mm	(To 16 in.) OD	2—3	3—4
Over 400 mm	(Over 16 in.) OD	3—4	4—5
Straight or spiral bevel gear units			
Cone distance up to 300 mm	(To 12 in.)[f]	2—3	4—5
Cone distance over 300 mm	(Over 12 in.)[f]	3—4	5—6
Gearmotors and shaft-mounted units		2—3	4—5
High-speed units[g]		1	2

[a] Drives incorporating overrunning clutches as backstopping devices should be referred to the gear drive manufacturer as certain types of lubricants may adversely affect clutch performance.

[b] Ranges are provided to allow for variations in operating conditions such as surface finish, temperature rise, loading, speed, etc.

[c] AGMA viscosity number recommendations listed above refer to R & O gear oils shown in Table 2, EP gear lubricants in the corresponding viscosity grades may be substituted where deemed necessary by the gear drive manufacturer.

[d] For ambient temperatures outside the ranges shown, consult the gear manufacturer. Some synthetic oils have been used successfully for high- or low-temperature applications.

[e] Pour point of lubricant selected should be at least 5°C lower than the expected minimum ambient starting temperature. If the ambient starting temperature approaches lubricant pour point, oil sump heaters may be required to facilitate starting and insure proper lubrication.

[f] Inch unit as shown are approximations.

[g] High-speed units are those operating at speeds above 3600 rpm or pitch line velocities above 25 m/sec (5000 fpm) or both. Refer to Standard AGMA 421, *Practice for High Speed Helical and Herringbone Gear Units,* for detailed lubrication recommendations.

From Standard AGMA 250.04, *Lubrication of Industrial Enclosed Gear Drives,* American Gear Manufacturers Association, Arlington, Va., 1974. With permission.

Table 4
AGMA LUBRICANT NUMBER RECOMMENDATIONS FOR ENCLOSED CYLINDRICAL AND DOUBLE-ENVELOPING WORM GEAR DRIVES

Type (worm gear drive)	Worm speed up to (rpm)	AGMA Lubricant no.[a] Ambient temp.[b]		Worm speed above (rpm)[d]	AGMA Lubricant no.[a] Ambient temp.[b]	
		−10—+10 °C (15—50 °F)[c]	10—50 °C (50—125 °F)[c]		−10—+10 °C (15—50 °F)[c]	10—50 °C (50—125 °F)[c]
Cylindrical worm[e]						
Up to 150 mm (to 6 in.)	700	7 Comp, 7 EP	8 Comp, 8 EP	700	7 Comp, 7 EP	8 Comp, 8 EP
Over 150 mm, to 300 mm (6 to 12 in.)	450	7 Comp, 7 EP	8 Comp, 8 EP	450	7 Comp, 7 EP	7 Comp, 7 EP
Over 300 mm, to 450 mm (12 to 18 in.)	300	7 Comp, 7 EP	8 Comp, 8 EP	300	7 Comp, 7 EP	7 Comp, 7 EP
Over 450 mm, to 600 mm (18 to 24 in.)	250	7 Comp, 7 EP	8 Comp, 8 EP	250	7 Comp, 7 EP	7 Comp, 7 EP
Over 600 mm (over 24 in.)	200	7 Comp, 7 EP	8 Comp, 8 EP	200	7 Comp, 7 EP	7 Comp, 7 EP
Double-enveloping worm[e]						
Up to 150 mm (to 6 in.)	700	8 Comp	8A Comp	700	8 Comp	8 Comp
Over 150 mm, to 300 mm (6 to 12 in.)	450	8 Comp	8A Comp	450	8 Comp	8 Comp
Over 300 mm, to 450 mm (12 to 18 in.)	300	8 Comp	8A Comp	300	8 Comp	8 Comp
Over 450 mm, to 600 mm (18 to 24 in.)	250	8 Comp	8A Comp	250	8 Comp	8 Comp
Over 600 mm (over 24 in.)	200	8 Comp	8A Comp	200	8 Comp	8 Comp

[a]   Both EP and compounded oils are considered suitable for cylindrical worm gear service. Equivalent grades of both are listed in the table. For double-enveloping worm gearing, EP oils in the corresponding viscosity grades may be substituted only where deemed necessary by the worm gear manufacturer.

[b]   Pour point of the oil used should be less than the minimum ambient temperature expected. Consult gear manufacturer on lube recommendations for ambient temperatures below −10°C (approximately 15°F).

[c]   Center distances in inches and temperature ranges in degrees Fahrenheit are approximations of millimeters and degree Celsius shown.

d   Worm gears of either type operating at speeds above 2400 rpm or 10 m/sec (2000 fpm) rubbing speed may require force-feed lubrication. In general, a lubricant of lower viscosity than recommended in the above table shall be used with a force-feed system.

e   Worm gear drives may also operate satisfactorily using other types of oils. Such oils should be used, however, only upon approval by the manufacturer.

From Standard AGMA 250.04, *Lubrication of Industrial Enclosed Gear Drives*, American Gear Manufacturers Association, Arlington, Va., 1974. With permission.

Table 5
VISCOSITY RANGES FOR AGMA OPEN GEAR LUBRICANTS

R & O gear oils (AGMA lubricant no.)	Viscosity ranges [SSU at 100 °F (cSt at 37.8 °C)]	EP gear oils (AGMA lubricant no.)	Residual compounds (AGMA lubricant no.)[a]	Viscosity Ranges [SSU at 210 °F (cSt at 98.9 °C)][a]
4	626—765 (140—170)	4 EP	14R	2,000—4,000 (428.5—856.0)
5	918—1,122 (200—250)	5 EP	15R	4,000—8,000 (857.0—1714.0)
6	1,335—1,632 (300—360)	6 EP		
7	1,919—2,346 (420—500)	7 EP		
8	2,837—3,467 (650—800)	8 EP		
9	6,260—7,650 (1400—1700)	9 EP		
10	13,350—16,320 (3000—3600)	10 EP		
11	19,190—23,460 (4200—5200)	11 EP		
12	28,370—34,670 (6300—7700)	12 EP		
13	850—1,000 (190—220) at 210 °F (at 98.9 °C)[b]	13 EP		

a   Residual compounds-diluent type, commonly known as solvent cutbacks, are heavy-bodied oils containing a volatile, nonflammable diluent for ease of application. The diluent evaporates leaving a thick film of lubricant on the gear teeth. Viscosities listed are for the base compound without diluent. **Caution** — these lubricants may require special handling and storage procedures. Diluents can be toxic or irritating to the skin. Consult lubricant supplier's instructions.

b   Viscosities of AGMA lubricant numbers, 13 and above are specified at 210 °F (98.9 °C) as measurement of Saybolt viscosities of these heavy lubricants at 100 °F (37.9 °C) would not be practical.

From Standard AGMA 251.02, *Lubrication of Industrial Open Gearing*, American Gear Manufacturers Association, Arlington, Va., November 1974.

Table 6
RECOMMENDED AGMA LUBRICANTS (FOR CONTINUOUS METHODS OF APPLICATION)

Ambient temp in degrees Fahrenheit (Celsius)[a]	Character of operation	Pitch line velocity				
		Pressure lubrication		Splash lubrication		Idler immersion
		Under 1000 ft/min (5m/sec)	Over 100 ft/min (5 m/sec)	Under 1000 ft/min (5 m/sec)	1000—2000 ft/min (10 m/sec)	Up to 300 ft/min (1.5 m/sec)
15—60[b] (−9—16)	Continuous	5 or 5 EP	4 or 4 EP	5 or 5 EP	4 or 4 EP	8—9 / 8 EP—9 EP
	Reversing or frequent "start-stop"	5 or 5 EP	4 or 4 EP	7 or 7 EP	6 or 6 EP	8—9 / 8 EP—9 EP
50—125[b] (10—52)	Continuous	7 or 7 EP	6 or 6 EP	7 or 7 EP / 9—10[c]	6 or 6 EP / 8—9[d]	11 or 11 EP
	Reversing or frequent "start-stop"	7 or 7 EP	6 or 6 EP	9 EP—10 EP	8 EP—9 EP	11 or 11 EP

Note: AGMA Viscosity number recommendations listed above refer to gear lubricants shown in Table 5. Although both R & O and EP oils are listed, the EP is preferred.

a Temperature in vicinity of the operating gears.

b When ambient temperatures approach the lower end of the given range, lubrication systems must be equipped with suitable heating units for proper circulation of lubricant and prevention of channeling. Check with lubricant and pump suppliers.

c When ambient temperature remains between 90 and 125°F (32 and 52°C) at all times use 10 or 10 EP.

d When ambient temperature remains between 90 and 125°F (32 and 52°C) at all times use 9 or 9 EP.

From Standard AGMA 251.02, *Lubrication of Industrial Open Gearing*, American Gear Manufacturers Association, Arlington, Va., November 1974. With permission.

Table 7
RECOMMENDED AGMA LUBRICANTS (FOR INTERMITTENT METHODS OF APPLICATION LIMITED TO 1500 FT/MIN (8 M/SEC) PITCH LINE VELOCITY)[a]

| Ambient temp in degrees Fahrenheit (Celsius)[b] | Mechanical spray systems[c] | | Gravity feed or forced drop method using EP lubricant |
	EP lubricant	Residual compound[d]	
15—60 (−9—16)	—	14R	—
40—100 (4—38)	12 EP	15R	12 EP
70—125 (21—52)	13 EP	15R	13 EP

Note: AGMA Viscosity number recommendations listed above refer to gear oils shown in Table 5.

[a] Feeder must be capable of handling lubricant selected.
[b] Ambient temperature is temperature in vicinity of the gears.
[c] Greases are sometimes used in mechanical spray systems to lubricate open gearing. A general purpose EP grease of number 1 consistency (NGLI) is preferred. Consult gear manufacturer and spray system manufacturer before proceeding.
[d] Diluents must be used to facilitate flow through applicators.

From Standard AGMA 251.02, *Lubrication of Industrial Open Gearing*, American Gear Manufacturers Association, Arlington, Va., November 1974. With permission.

runs off the gear case, it can be collected in troughs and directed to bearings. For very slow gearing using heavy oil, it may be necessary to use scrapers to remove oil from the sides of the gear rim for use as a bearing lubricant. If for other reasons the sump oil level is higher than required for gearing lubrication, an oil pan can be used to limit the amount of oil that the gear contacts. The flow of oil into the pan is controlled by small holes below the oil level to eliminate excessive churning and improve efficiency. Oil pans are recommended for gears with pitch line velocities over 13 m/sec (2500 ft/min) and work well for slow- and moderate speed gearing with pitch line velocities up to 18 m/sec (3500 ft/min). Higher speed gearing will tend to centrifugally throw off most of the oil before it gets into mesh. The lighter oil generally used with high-speed gearing tends to compound the problem. Higher power losses caused by churning also tend to make splash lubrication impractical for high-speed gearing.

Another common means for applying easy flowing liquid lubricants is a force-fed system (Figure 6). In this system oil is taken from the gear case, pumped through a filter, heat exchanger, and pressure relief valve, and delivered back to the unit under pressure. The amount of oil delivered to the bearings can be set by an orifice or other flow-control device. Oil is applied to the gear mesh by spray nozzles in a manifold.

Frictional heat generation may range, typically, from 0.5 to 1% of the transmitted horsepower per mesh for spur or helical gears. Sump capacity and oil flow rate can be established for a desired sump temperature (95°C maximum is recommended for petroleum based lubricants). Inlet oil temperature and flow rate should be selected to maintain a suitable operating oil viscosity in the mesh. For enclosed, industrial gears, inlet oil temperatures of 38 to 54°C and temperature rises in the mesh of 17 to 28°C are typical for circulating systems.

The amount of oil is controlled by the size of the nozzles and oil pressure. Oil velocity must be sufficient to get the oil down into the tooth space before windage and tooth rotation

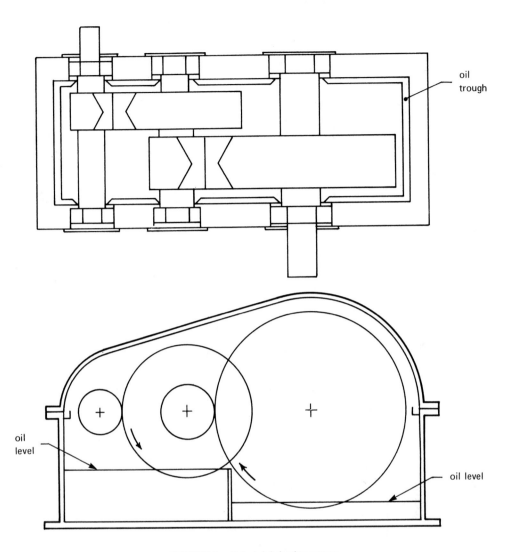

FIGURE 5. Splash lubrication system.

deflect and block the spray. Spray nozzle pressures of 15 to 50 psi are usually adequate for industrial gearing. Advantages of this system are that the oil delivered is of controlled quantity, cleanliness, and temperature. The oil spray can flush wear particles and debris from the gear mesh. Heat will also be carried away with the excess lubricant. In high-speed gearing where excessive lubricant in the mesh is not desirable, high-flow sprays should be directed at the gear tooth surfaces as they leave mesh for maximum cooling. Low-flow sprays should be directed at the teeth just before mesh for maximum lubrication. Screening is sometimes installed in high-speed gear cases above the sump oil level to minimize the windage effects on oil foaming. There are no practical limiting speeds for force-fed systems.

Heavy open gear compounds are generally applied by paddle or brush, slush pan or automatic lubricator. The manual paddle or brush method is crude, with the distribution of lubricant being erratic.

The slush pan method utilizes a pan filled with lubricant into which the gear dips. The lubricant is then carried into mesh where excess material is squeezed out and drops back into the pan. The method provides for continuous lubrication and applies lubricant evenly across the entire tooth. It can be used with gearing with pitch line velocities up to 10 m/sec (2000 ft/min).

Table 8
RECOMMENDED QUANTITIES OF LUBRICANT
(FOR INTERMITTENT METHODS OF APPLICATION WHERE PITCH LINE VELOCITY DOES NOT EXCEED 1500 FT/MIN (8 M/SEC) FOR AUTOMATIC, SEMIAUTOMATIC, HAND SPRAY, GRAVITY FEED, OR FORCED DRIP SYSTEMS)

Ounces (cm²) per application at intervals of:[a]

Gear diameter in feet (meters)	1/4 Hr[b] Face width in inches (cm)					1 Hr Face width in inches (cm)					4 Hr[b] Face width in inches (cm)				
	8 (20)	16 (40)	24 (60)	32 (80)	40 (100)	8 (20)	16 (40)	24 (60)	32 (80)	40 (100)	8 (20)	16 (40)	24 (60)	32 (60)	40 (100)
10 (3.0)	0.2 (6)	0.3 (9)	0.4 (12)	0.5 (15)	0.6 (18)	0.8 (24)	1.2 (35)	1.6 (47)	2.0 (59)	2.4 (71)	5.0 (148)	6.0 (177)	8.0 (237)	10.0 (296)	12.0 (355)
12 (3.7)	0.3 (9)	0.3 (9)	0.4 (12)	0.5 (15)	0.6 (18)	1.2 (35)	1.4 (41)	1.8 (53)	2.2 (65)	2.6 (77)	6.0 (177)	7.0 (207)	9.0 (266)	11.0 (325)	13.0 (384)
14 (4.3)	0.3 (9)	0.4 (12)	0.5 (15)	0.6 (18)	0.7 (21)	1.4 (41)	1.6 (47)	2.0 (59)	2.4 (71)	2.8 (83)	7.0 (207)	8.0 (237)	10.0 (296)	12.0 (355)	14.0 (414)
16 (4.9)	0.4 (12)	0.5 (15)	0.6 (18)	0.7 (21)	0.8 (24)	1.6 (47)	2.0 (59)	2.4 (71)	2.8 (83)	3.2 (95)	8.0 (237)	10.0 (296)	12.0 (355)	14.0 (414)	16.0 (473)
18 (5.5)	0.5 (15)	0.6 (18)	0.7 (21)	0.8 (24)	0.9 (27)	2.0 (59)	2.4 (71)	2.8 (83)	3.2 (95)	3.6 (106)	10.0 (296)	12.0 (355)	14.0 (414)	16.0 (473)	18.0 (532)
20 (6.1)	0.6 (18)	0.7 (21)	0.8 (24)	0.9 (27)	1.0 (30)	2.4 (71)	2.8 (83)	3.2 (95)	3.6 (106)	4.4 (130)	12.0 (355)	14.0 (414)	16.0 (473)	18.0 (532)	20.0 (591)
22 (6.7)	0.7 (21)	0.8 (24)	0.9 (27)	1.0 (30)	1.1 (33)	2.8 (83)	3.2 (95)	3.6 (106)	4.0 (118)	4.8 (142)	14.0 (414)	16.0 (473)	18.0 (532)	20.0 (591)	22.0 (651)
24 (7.3)	0.8 (24)	0.9 (27)	1.0 (30)	1.1 (33)	1.2 (35)	3.2 (95)	3.6 (106)	4.0 (118)	4.4 (130)	5.2 (154)	16.0 (473)	18.0 (532)	20.0 (591)	22.0 (651)	24.0 (710)
26 (7.9)	0.9 (27)	1.0 (30)	1.1 (33)	1.2 (35)	1.3 (38)	3.6 (106)	4.0 (118)	4.4 (130)	4.8 (142)	5.6 (166)	18.0 (532)	20.0 (591)	22.0 (651)	24.0 (710)	26.0 (769)
28 (8.5)	1.0 (30)	1.1 (33)	1.2 (35)	1.3 (38)	1.4 (41)	4.0 (118)	4.4 (130)	4.8 (142)	5.2 (154)	6.0 (177)	20.0 (591)	22.0 (651)	24.0 (710)	26.0 (769)	28.0 (828)

a   The spraying time should equal the time for 1 and preferably 2 revolutions of the gear to ensure complete coverage. Periodic inspections should be made to ensure that sufficient lubricant is being applied to give proper protection.

b   Four hours is the maximum interval permitted between applications of lubricant. More frequent application of smaller quantities is preferred. However, where diluents are used to thin lubricants for spraying intervals must not be so short as to prevent diluent evaporation.

From Standard AGMA 251.02, *Lubrication of Industrial Open Gearing*, American Gear Manufacturers Association, Arlington, Va., November 1974. With permission.

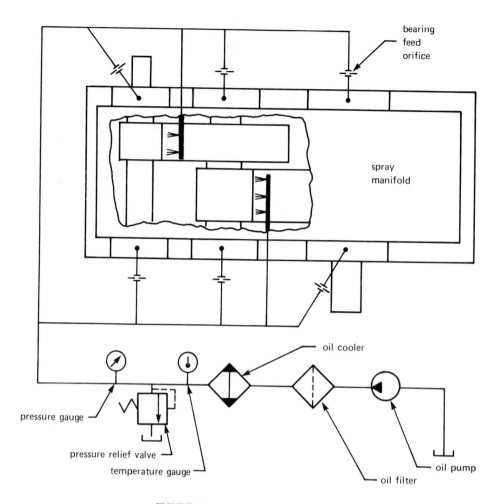

FIGURE 6. Force-fed lubrication system.

An automatic lubricator may be either a drip or spray type. The drip type applies a small amount of lubricant to the gear teeth at regular intervals. When the material gets into mesh it is spread across the teeth. The spray type deposits a thin uniform film of lubricant over the entire gear tooth surface. Automatic lubricators usually require that the lubricant be thinned with a volatile solvent which will evaporate, leaving a heavy film on the teeth. This method is effective for pitch line velocities up to 8 m/sec (1500 ft/min). Suggested lubricant quantities are given in Table 8.

Grease can be applied to gear teeth using the slush pan arrangement described above. Alternately, the gearing can be surrounded by a close fitting enclosure which will keep the grease in contact with the gear tooth surfaces. The enclosure also prevents the grease from being thrown off centrifugally, allowing the teeth to run dry.

GEAR WEAR

Light wear may be beneficial in gear sets. During the break-in period, initial wear can improve the finish of contacting surfaces by wearing down high points or asperities. Since one tooth on one element contacts many teeth on the other element before coming into contact with the original tooth again, there tends to be an averaging and overall reduction of profile errors. If dynamic alignment of two mating elements is off slightly, initial wear will remove some material from the area of hardest contact and spread the load over more of the face-width. These minor modifications from light, initial wear improve the gear tooth conformity and make it possible for the gears to operate under full-load with less chance of damage. Initial wear can be intentionally induced by using a lighter lubricant for a break-in period. How light a lubricant depends on the initial load and the desired rate of wear. Extreme caution should be exercised and tooth surfaces should be examined frequently to prevent rapid, destructive wear. Under normal operating conditions, initial wear may occur and stop by itself when sufficient conformity has been achieved to support the load.

Moderate wear would result from increased rate of surface material removal due to more significant surface irregularities, gear tooth misalignment, dynamic load pulsations, insufficient lubricant viscosity, or any conditions that would cause the gearing to operate under conditions of boundary of mixed film lubrication. It could also be caused by abrasive material in the lubricant. This wear would probably not stop by itself but would continue, slowly, over a long period of time. Depending on the anticipated life of the gearing, this type of wear may, or may not be acceptable.

Heavy wear would involve rapid removal of surface material, destroying the tooth form, and hindering the smooth operation of the gear set. This can be caused, for example, by operating the gears without any lubricant or under conditions of heavy overload or severe misalignment of contacting tooth surfaces. This destruction of the tooth form will lead to a very short-service life for the gear set if the causes are not found and corrected. Excessive loading is the most common cause of rapid wear, although lubrication receives considerable attention since it is easier to change.

Examining a gear set to determine the cause of excessive wear or failure is not always an easy task. The final mode of failure may be the result of previous wear mechanisms of an entirely different nature. A discussion of several common forms of wear may help to establish the probable sequence of events leading to excessive wear or failure in a gear set.

Breakage (Figure 7) — Catastrophic tooth failure is caused by a gear tooth being subjected to loads and, consequently, bending stresses in excess of the endurance limit of the gear material. This will cause a tooth or portion of a tooth to break away. The breakage usually starts as a crack which propagates with repeated load cycles showing a typical fatigue failure. Occasionally, a tooth may fail from a single-load cycle. This would not show a fatigue type failure. Breakage can be caused by shock loads, loads induced by high vibration, large pieces of debris passing through the gear mesh, or by misalignment causing the tooth load to be carried by a small portion of the face-width. Obviously, lubrication is not a factor and the mechanical defects must be located and corrected.

Pitting (Figure 8) — This common form of surface fatigue occurs when small localized high spots on the tooth surface are overstressed due to high-unit loading. When this occurs, small subsurface cracks are formed which propagate to the surface after repeated cycles. When an area of material is no longer supported, it leaves the surface of the tooth creating a small depression or pit. Small pits may form in a new gear set as high spots from machining are removed. Once enough load bearing surface is in contact, pitting may stop and the surface will begin to ''polish over''. If this takes place, the pitting is considered ''initial'' and is not harmful. If the pitting continues to spread, leaving increasingly less surface to carry the load, it is considered progressive. If the condition is not corrected, material will continue to be removed from the tooth until, eventually, tooth breakage may occur. This is not a lubrication failure. The condition can be caused by misalignment of the gear teeth, causing a relatively small area to carry the load with resultant high stresses. Gear material too soft for the application or operating loads greater than those for which the gearing was designed can also cause pitting. The use of a heavier oil may help spread the load over a larger area, but lubricant adjustments will generally give little, if any, help. The use of some EP lubricants or additives may extend the life of the gearing but will not correct the problem.

FIGURE 7. Tooth breakage. (From Standard AGMA 110.4, *Nomenclature of Gear Tooth Failure Modes,* American Gear Manufacturers Association, Arlington, Va., August 1980. With permission.)

FIGURE 8. Pitting. (From Standard AGMA 110.4, *Nomenclature of Gear Tooth Failure Modes,* American Gear Manufacturers Association, Arlington, Va., August 1980. With permission.)

FIGURE 9. Spalling. (From Standard AGMA 110.4, *Nomenclature of Gear Tooth Failure Modes,* American Gear Manufacturers Association, Arlington, Va., August 1980. With permission.)

Spalling (Figure 9) — This mechanism is the same as for pitting. Large flakes or chips may be removed from through hardened or case hardened gears from subsurface flaws or stresses caused by improper heat treatment. The joining of pits as the metal between them is removed is also a form of spalling. Again, this is not a lubrication problem. It is a result of material defects, excessive load, or other application problems.

Plastic flow (Figure 10) — This type of wear represents gear tooth surface deformation caused by heavy loads stressing the surface material beyond its elastic limit. Usually occurring in softer metals, the surface material may be extruded out along the ends of the teeth and along the tip causing fins to form. Prominent ridges at the pitch line or depressions in the dedendum may also be indications of this type of wear. If the wear is caused by high vibration or shock load, a heavier lubricant may cushion the load somewhat. However, this type of wear is a material failure and lubricant changes will not correct it.

Scratching (Figure 11) — This is a type of abrasive wear. When hard particles, which are larger than the oil film thickness separating the gear teeth, pass through the gear mesh, they scratch the gear tooth surfaces in the direction of sliding. These particles can be dirt, sand, casting scale, welding slag, gear or bearing material, or any debris which finds its way into the lubrication system. The material can be airborne and enter through openings in poorly fitting enclosures or uncovered inspection ports. It may be the result of poorly cleaned housings or rotating elements prior to assembly. The debris may also be from wear within the gear unit. Laboratory analysis can generally indicate the type of particle material. Increasing the lubricant viscosity will increase the film thickness and may ease the problem, but not cure it. A better solution is to remove the abrasive particles through finer lubricant filtration, improved maintenance, etc. Once the problem is corrected, gear surface deterioration will cease.

Scoring (Figure 12) — When small local high spots or asperities on mating gear tooth surfaces break through the oil film and cause metal-to-metal contact, they may weld together

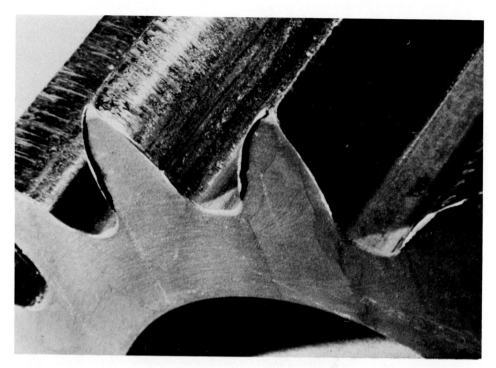

FIGURE 10.   Plastic flow. (From Standard AGMA 110.4, *Nomenclature of Gear Tooth Failure Modes,* American Gear Manufacturers Association, Arlington, Va., August 1980. With permission.)

FIGURE 11.   Scratching. (From Standard AGMA 110.4, *Nomenclature of Gear Tooth Failure Modes,* American Gear Manufacturers Association, Arlington, Va., August 1980. With permission.)

FIGURE 12.   Scoring. (From Standard AGMA 110.4, *Nomenclature of Gear Tooth Failure Modes,* American Gear Manufacturers Association, Arlington, Va., August 1980. With permission.)

and then tear out a portion of the softer material. This material, fused to one tooth, will gouge the mating surface and leave furrows in the direction of sliding. This can lead to rapid removal of material and reduce the smoothness of action of the gearing. This type of wear represents a lubrication failure: the oil film is insufficient to keep the gear tooth surfaces separated. Increasing the viscosity will increase the film thickness. EP additives will smooth the gear tooth surfaces by preventing the "weld and break" scoring mechanism. If the surfaces are badly scored, careful stoning or other smoothing means should hasten the healing process after corrective lubrication measures have been taken.

Corrosive wear — Light pitting or rust on gear tooth surfaces or other exposed, unpainted metal surfaces indicates corrosion and corrosive wear. Corrosion can be caused by water in the oil from condensation, leaks in water cooled heat exchangers, etc. It can also be caused by acids or corrosive additives in lubricating oil. The corroded material is removed by the meshing action of the gear, exposing fresh surfaces to be attacked. Some lubricant additives inhibit corrosion by protecting surfaces against rust. Others inhibit the formation of acids from oil oxidation. If the cause of the problem is known to be external, it should be corrected at the source.

Burning — If a gear tooth is subjected to intense heat, material hardness will be lost. This softened material will be rapidly removed since it cannot adequately carry the load. This phenomena is characterized by temperature discoloration of the gear tooth material. If the heat sources are external, they should be corrected. If friction is the problem, the type and method of lubrication should be reevaluated.

REFERENCES

1. **Fowle, T. I.,** Gear lubrication: relating theory to practice, *Lubr. Eng.,* 32, 17, 1976.
2. **Fowles, P. E.,** EHL film thickness — practical significance and simple computation, *Lubr. Eng.,* 32, 166, 1975.
3. **Ku, P. M.,** Gear failure modes — importance of lubrication and mechanics, *ASLE Trans.,* T19, 239, 1976.
4. **Wedeven, L. D.,** What is EHD?, *Lubr. Eng.,* 31, 291, 1975.
5. **Lipp, L. C.,** Solid lubricants — their advantages and limitations, *Lubr. Eng.,* 32, 574, 1976.
6. **Hersey, M. D.,** Gear lubrications, in *Theory and Research in Lubrication,* John Wiley & Sons, New York, 1966, chap. 11.
7. **Root, D. C.,** Selecting the right gear oil, *Lubr. Eng.,* 32, 8, 1976.
8. Standard AGMA 250.04, *Lubrication of Industrial Enclosed Gear Drives,* American Gear Manufacturers Association, Arlington, Va., 1974.
9. Standard AGMA 251.02, *Lubrication of Industrial Open Gearing,* American Gear Manufacturers Association, Arlington, Va., November 1974.
10. Standard AGMA 110.4, *Nomenclature of Gear Tooth Failure Modes,* American Gear Manufacturers Association, Arlington, Va., August 1980.

MECHANICAL SHAFT COUPLINGS

Michael M. Calistrat

INTRODUCTION

A mechanical shaft coupling is that part of a machine which transmits torque from one shaft to another. There are significant differences between different types of mechanical shaft couplings, and for a better understanding of their working principles we should first categorize them. First, shaft couplings can be divided into rigid and flexible couplings.

Rigid couplings — usually in the form of a long collar or a couple of bolted flanges, perform the torque transmission in the most efficient way. Unfortunately they can be used only when the two connected shafts are perfectly aligned. Rigid couplings do not require lubrication.

Flexible couplings — transmit torque without slip, and accommodate misalignment between the driving and driven shafts.[1] They too can be divided into two categories depending on the means used to accommodate the misalignment: couplings using the flexing of one or more of their components, and couplings that use sliding of two or more of their components. Some couplings use both of these methods in their design. The couplings that accommodate misalignment only through flexing do not require lubrication and fall outside the scope of this manual. Couplings that use sliding for accommodating misalignment must be lubricated in order to minimize wear.

Nonlubricated, or dry, couplings use either an elastomer or one or more thin metal disks which flex in order to accommodate misalignment. Since useful life of these flexing elements is limited by such factors as fatigue, fretting corrosion, or elastomer aging, they have to be replaced periodically. Besides that, elastomer couplings are usually larger and heavier than lubricated couplings with similar ratings. Hence, the need for lubrication is compensated in many instances by less maintenance and/or smaller size. Dry couplings are used mainly in fractional and low-horsepower applications (< 200 kW or 300 hp); lubricated couplings are particularly popular in large-horsepower applications.

DESCRIPTION

Although there is an endless variety of lubricated shaft couplings, three designs are most frequently found: the gear, chain, and the steel grid couplings, shown in Figures 1 to 3, respectively.

Gear Couplings

The gear coupling (Figure 1) has five major components: two hubs, two sleeves, and in some cases a spacer (not shown). It also has a number of bolts, nuts, lockwashers, and two or more seals. The spacer is omitted when the separation between the connected shafts is small. The hubs have a row of external teeth with involute profile; the sleeves have matching internal teeth. Each gear mesh acts as a spline; to accommodate misalignment, the external teeth are slightly thinner than the space between the internal teeth. The space thus generated between the teeth is called the backlash and it allows the hub teeth to assume an angular position, as shown in Figure 4.

The need for lubrication can be understood by considering the sliding motion of the hub teeth on the sleeve teeth. Figure 5 illustrates a section through half of a gear coupling, i.e., a hub and a sleeve. Because the coupling is misaligned, the centerline of the hub teeth, BB, does not coincide with the centerline of the sleeve teeth, AA. Hence, the lower hub tooth

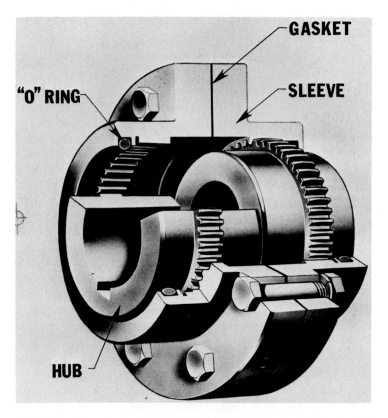

FIGURE 1.   Gear type coupling. (Courtesy of Koppers Co., Inc., Pittsburgh, Pa.)

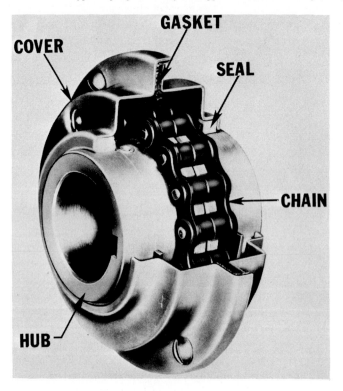

FIGURE 2.   Chain type coupling. (Courtesy of FMC Corporation, Chicago, Ill.)

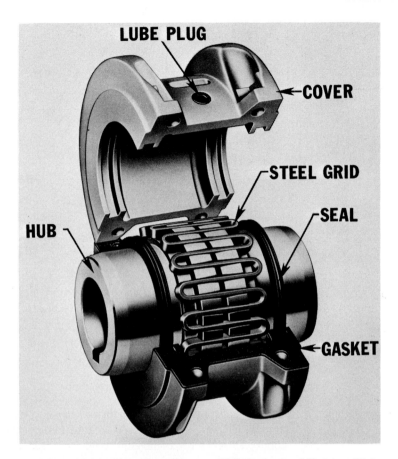

FIGURE 3. Steel grid coupling. (Courtesy of Falk Corporation, Milwaukee, Wis.)

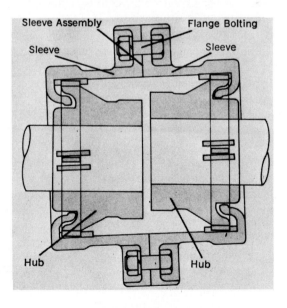

FIGURE 4. Teeth engagement in gear type couplings. (Courtesy of Koppers Co., Inc., Pittsburgh, Pa.)

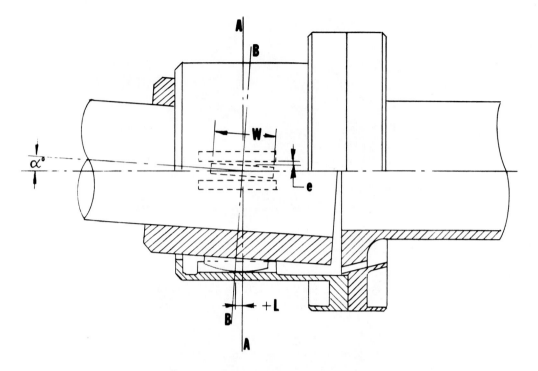

FIGURE 5.   Section through a misaligned coupling.

is at the left of the line AA, but after half a revolution the tooth will move to the right of the line AA. Although each hub tooth remains always engaged with the same sleeve tooth, it slides over it in the axial direction, back and forth, completing a full cycle each revolution. The total travel of a hub tooth depends on the angle of misalignment and on the pitch diameter of the gear mesh. One can expect the travel to be in the order 0.6 mm (0.025 in.); this small motion and the relatively high frequency of oscillation create poor lubrication conditions.

Fortunately, these detrimental sliding conditions are more than offset by the beneficial effect on the lubricant of centrifugal forces, whose magnitude is given in the graph of Figure 6. It can be seen that even in a coupling operating at a motor speed of 3600 rpm, the centrifugal forces can exceed 500 G's. The relative acceleration, G, as related to earth gravity, can be calculated as

$$g = 14.2 \ \overline{PD} \ \ \overline{rpm}^2 \ \ 10^{-6}$$

where $\overline{PD}$ is the pitch diameter in inches (cm/2.54).

Because centrifugal force is a function of the square of rotational speed while frequency of the hub tooth oscillatory motion only increases in direct proportion with speed, coupling lubrication improves at high speeds. The reverse is true when a coupling operates at very low speeds. Centrifugal forces decrease rapidly with lower speed, and below a given level they can no longer force the lubricant between the coupling teeth. Rapid wear commonly results.

Chain Couplings

The chain coupling has only three components: two sprockets and a short length of double-row roller chain. Chain couplings are generally used at low speeds, but they can be used at higher speeds with the addition of a metal or plastic cover to retain the lubricant. This type

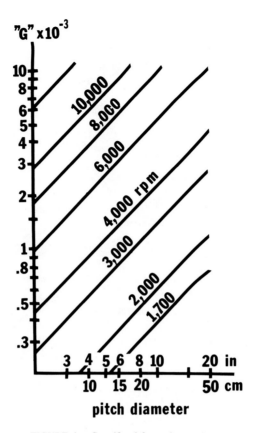

FIGURE 6.   Centrifugal forces in couplings.

of coupling is used in applications having short shaft separation. The principle of operation is similar to the gear couplings when considering the sprockets as hubs and the double-row chain as the two sleeves. Only when used with a cover can the chain couplings benefit from the effect of centrifugal forces in the lubricant. From a torsional point of view, the chain couplings are less stiff than gear couplings.

Steel Grid Couplings

Steel grid couplings are even more flexible, torsionally, than chain couplings. They also operate similarly to a gear coupling, having two toothed hubs and a sleeve in a form of a convoluted spring steel band. Because of the special profile of the teeth, the steel grid flexes under torque, as shown in Figure 7.

To accommodate misalignment the hub teeth slide over the steel grid just as in a gear coupling. A split cover is always provided in order to retain both the steel grid and the lubricant within the coupling. Similar to chain couplings, the steel grid couplings can be used only when the gap between the shafts is small.

DESIGN AND MATERIALS

The Gear Coupling

The three major types of gear couplings are

1.   Standard — medium speeds and medium-torque couplings
2.   Spindle — low speeds and high-torque couplings
3.   High performance — high speeds and medium-torque couplings

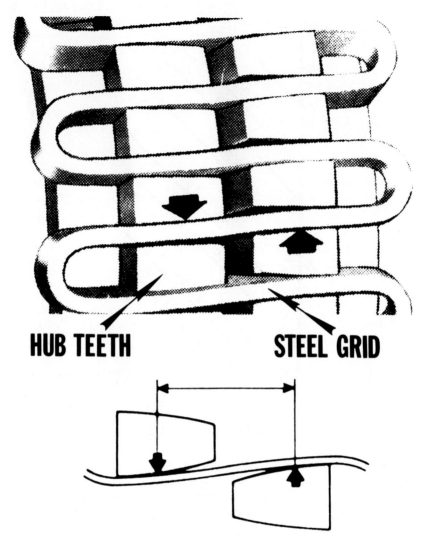

FIGURE 7. Steel grid flexing under torque. (Courtesy of Falk Corporation, Milwaukee, Wis.)

Although all gear couplings use the same basic components, the designs and materials vary significantly.

Standard Couplings

These are usually made of carbon steel and are not heat-treated. The gear mesh has roughly 60 teeth, which have an involute profile. Most manufacturers use a 20° pressure angle, but there are exceptions. The standard hub bore diameter is roughly equal to the pitch radius; and depending on the design, the hub can be overbored to various degrees.

In order to accommodate misalignment, gear coupling teeth have a built-in backlash. To increase even farther the amount of misalignment under which gear couplings can operate, some manufacturers make the teeth flanks barreled. The smaller the radius of curvature of the tooth flank, however, the larger the contact pressure between the teeth as shown in Figure 8.

Standard couplings use two basic types of seal to retain the lubricant: the all-metal seal

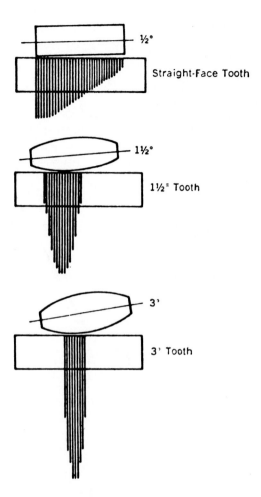

FIGURE 8. Typical force patterns in gear couplings.

(Figure 9) and the elastomer seal (Figure 10). The main difference between the two types of couplings in Figures 9 and 10 is the lubricant capacity; for couplings with identical flanges, the metal seal coupling has a larger lubricant reservoir than the elastomer seal coupling. Although physically possible, couplings with different lubricant capacities should not be bolted together because one half can become starved of lubricant and wear rapidly.

Spindle Couplings

Spindle couplings are used in special applications that require high-torques and high-misalignment capability, but at relatively low speeds. A typical example is the rolling mill in steel mills. To be able to transmit high torques, spindle couplings are made of high-strength alloy steels, and the teeth have a hardened surface obtained usually through nitriding or carburizing. To further increase the strength of the teeth, spindle couplings have fewer teeth than standard couplings, usually 40 teeth. To accommodate high misalignments, all spindle couplings use barreled teeth with a relatively small radius of curvature. Another design characteristic of spindle couplings is a quite large distance between the driving and driven shafts, and the couplings use a "floating shaft" to bridge this distance.

High-Performance Couplings

In many cases these couplings (Figure 11) operate at speeds in excess of 10,000 rpm and

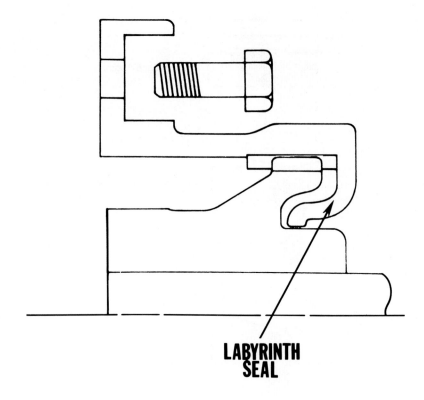

FIGURE 9. All-metal labyrinth seal.

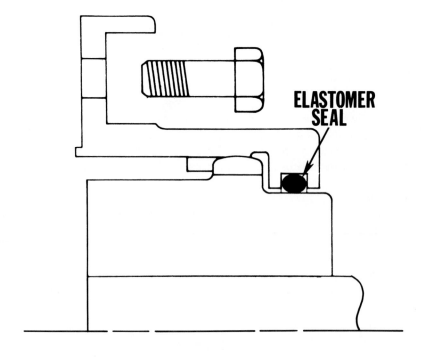

FIGURE 10. Elastomer seal.

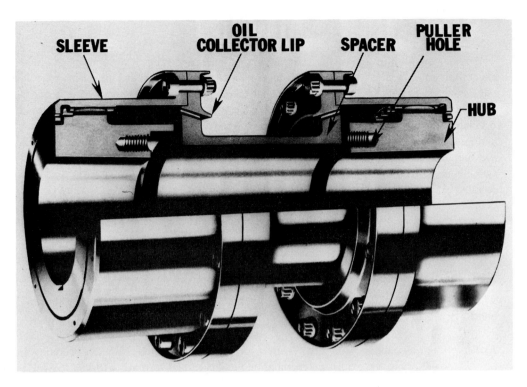

FIGURE 11.   High-speed coupling. (Courtesy of Koppers Co., Inc., Pittsburgh, Pa.)

are characterized by their relatively light weight. Although they do not transmit the high torques of spindle couplings, they are also made of high-strength, heat-treated alloy steels in order to minimize their weight. Because this type of coupling operates under poor lubrication conditions, it also has hardened teeth surfaces, usually obtained through nitriding. High-performance couplings are dynamically balanced because residual unbalance induces high-amplitude vibration at high speed.

Chain Coupling
The sprockets are made of medium carbon steel, with the teeth hardened in order to minimize wear. The sprockets have approximately 18 teeth, with the hub extending to one side only. The chain is usually a double-strand roller construction; some manufacturers use a specially constructed divided roller chain. All the chains are equipped with one connecting link which enables installation and removal. The housings (covers) are split either in a plane perpendicular to the axis, or in an axial plane. The first type is usually stamped steel, the latter is molded plastic. The housings are always provided with one or two lube plugs, two seals, and a gasket at the split plane. They are connected by four or more bolts.

Steel Grid Couplings
The hubs are made of steel and the grid is made of hardened chrome vanadium alloy steel. Under normal conditions the grid material is stressed below its endurance limit to prevent fatigue failure. As with chain couplings, the cover is split either in an axial plane, in which case it is made of aluminum, or in a plane perpendicular to the axis, in which case it is made of stamped steel. Each half cover is provided with a seal and a lubrication hole. A gasket is located between the two half covers.

COUPLING LUBRICATION

Whether the coupling is of the gear, chain, or steel grid type, it accommodates misalignment through a sliding motion between the hub (or sprocket) teeth and the mating element (sleeve, chain, or steel grid). The torque that the coupling transmits generates a force between these mating surfaces, and the rotation combined with the misalignment generate a rapid oscillatory motion. The forces and the motion tend to wipe off the lubricant from the contact surfaces. As the hub tooth slides in one direction, however, the lubricant has a chance to wet the area left uncovered behind the tooth. To do that, the lubricant has to work against two elements: its own viscosity and the very short time available during one cycle. The teeth would operate dry if it was not for the centrifugal forces created by the rotation of the coupling. The centrifugal forces generate a pressure in the lubricant annulus high enough to force the lubricant in the spaces between the teeth. The pressure in the lubricant can be determined by the formula:

$$p = \rho\, \omega^2\, (R_1^2 - R_2^2) / 2$$

where ρ = specific mass, ω = angular velocity (rad/sec), R_1 = major radius of the annulus, and R_2 = minor radius of the annulus.

Example: For a coupling of Figure 4, with a 152-mm (6 in.-) pitch diameter, operating at 3600 rpm, the pressure generated by the centrifugal forces is 410 kPa (60 psi).

Methods of Lubrication

The two basic methods of lubrication are batch and continuous oil flow. Only gear couplings use the latter. Whether one or the other method should be used depends primarily on the capability of the coupling to dissipate the heat generated by friction. When a coupling operates under a horseshoe-type cover, the circulation of air induced by the rotational speed provides all the cooling the coupling needs. When the couplings operate in a sealed enclosure, as is the case with most high-performance couplings, special cooling must be provided, usually by a continuous flow of oil.

A batch-lubricated coupling can be filled with either oil or grease. The main problem with oil-filled couplings is that the seals tend to allow the oil to drip away *under static conditions* and the coupling would then operate dry when it is restarted. Only one type of gear coupling is designed with a labyrinth seal so that it can operate with oil. Grease is much easier to seal in a coupling, and it is the most frequently recommended lubricant. It is interesting that when the coupling is operating, centrifugal forces prevent the lubricant from escaping; seals are required both to prevent dirt from penetrating into the coupling and to retain the lubricant when the coupling *is not operating*.

Grease as a Coupling Lubricant

Because the gear, chain, and grid couplings are similar in the manner through which they accommodate misalignment and transmit torque, a grease that is satisfactory for one type generally gives good results for the others. Nevertheless, various manufacturers unfortunately have quite different recommendations for lubricants and a common denominator is difficult to find.

The most important conclusion of one study is that the wear rate of a type gear coupling is greatly influenced by the viscosity of the base oil of the grease: the higher the viscosity, the lower the wear rate.[2] This phenomenon might explain why many coupling manufacturers prefer to recommend a NLGI No. 2 grease over a No. 1 or 0. However, the penetration of a grease seldom reflects its base oil viscosity, which generally should exceed 198 cSt at 40°C for optimum performance.

Especially demanding conditions are the high centrifugal forces imposed on a grease in a coupling. Even under normal gravity, greases tend to bleed some oil; under the high G level in couplings, the stratification of additives, soap, and oil takes place rapidly and in many cases can be total.[3,4] A grease which is separated into oil and soap is no longer suitable for coupling lubrication. First, the oil will then leak out of the coupling, particularly when the coupling is at a standstill. Second, the centrifugal force will force the denser soap and fillers away from the axis of rotation and eventually coat the inside surfaces of the coupling housings (or sleeves). In Figures 1, 2, and 3 one can see that the hub teeth are located close to the housings, which is done to reduce the tangential forces on the teeth. The soap separated from the grease under the influence of centrifugal forces will surround the torque transmitting elements, whether they are gear teeth, chains or steel grids, and *will prevent the oil from lubricating the surfaces that are under sliding contact.* Under these conditions, a coupling can wear rapidly although it is full of grease. Hence, the resistance of a grease to separation under centrifugal forces is an important factor in selecting a good lubricant for couplings. A third factor which should be considered is the volumetric percentage of oil in the grease. Under high-centrifugal forces, practically all the greases will eventually separate at a rate which depends on various factors. Some of the soap will start flowing toward the torque-carrying elements right after the coupling is started. The more soap a grease contains, the higher are the chances that the hub teeth will be flooded by soap. It is better, then, to use a grease that has a small percentage of soap or fillers.

Oil as a Coupling Lubricant

Oil lubrication is used almost exclusively in high-speed gear couplings. As mentioned previously, the flow of oil is necessary primarily to remove the heat generated by friction.[5] The efficiency of a good-quality, high-performance gear coupling is excellent, usually better than 99.9%. Since high-performance couplings can transmit high power, however, the amount of power loss can be considerable. For instance, a coupling transmitting 20,000 hp at 99.9% efficiency will absorb 15 kW, which it will transform into heat. If this heat is not dissipated it will damage the coupling. The oil used for the lubrication and cooling of couplings is taken from the lubrication system of the machines connected by the coupling. In most cases this oil is a light turbine oil which is not a good coupling lubricant. To offset this marginal lubrication, many high-performance couplings have nitrided teeth.

The high centrifugal forces present in a coupling do not affect the oil as long as it is a homogeneous fluid; however, the oil not only has components which are added for the purpose of enhancing its properties, but also has unwanted impurities. Under centrifugal forces, some of the additives and most of the impurities are separated and retained within the coupling.

When the quantity of sludge in a coupling becomes excessive, it will (1) impair axial movement of one shaft with respect to the other, (2) corrode the teeth and accelerate their wear rate, and (3) reduce or completely stop the flow of oil, with high temperatures and high wear rates as a consequence.

Sludge can be of two types: a moist sludge with the appearance and consistency of a NLGI No. 3 grease, or a dry sludge which is hard and crumbles like clay. Moist sludge is caused by an accumulation of oil additives in the coupling, in particular silicone antifoam compounds. Moist sludge is a lubricant and its only drawback is that it prevents oil from circulating through the coupling. Dry sludge is caused by an accumulation of impurities from the oil such as sand in the desert, or fertilizer near a chemical plant.

Whether sludge is moist or dry, the worst conditions are caused by water or other corrosive liquids that might enter the coupling. Besides causing rapid wear and shortening coupling life, corrosion can reduce the fatigue resistance of the coupling sleeves, which in turn can

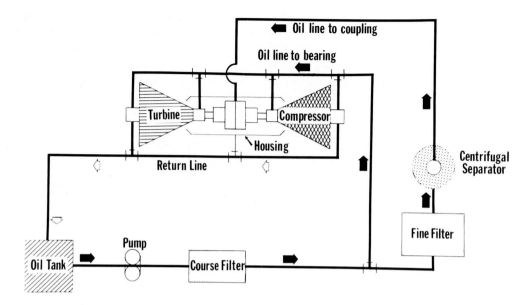

FIGURE 12.   Oil circuit for continuous lubrication.

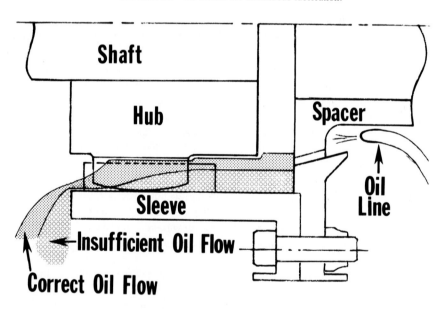

FIGURE 13.   Oil flow in damless couplings.

result in catastrophic failures. The designers of lubrication systems are generally aware of the problems that impurities in the oil can cause, and take necessary precautions to prevent sludge accumulation in the couplings. Figure 12 illustrates an oil circuit that provides special filtration for couplings. (API 671 recommends a 10-μm filter or less.)

Figure 13 illustrates a type of high-performance gear coupling which was developed so that it will not retain sludge (known as *damless* or *antisludge* coupling). A large flow of oil is required in order to maintain a sufficient level of lubricant in the coupling. It can be seen that if the oil flow is insufficient, much of the tooth surface remains nonlubricated. Following the manufacturers' recommendations for the oil flow is very important for these types of

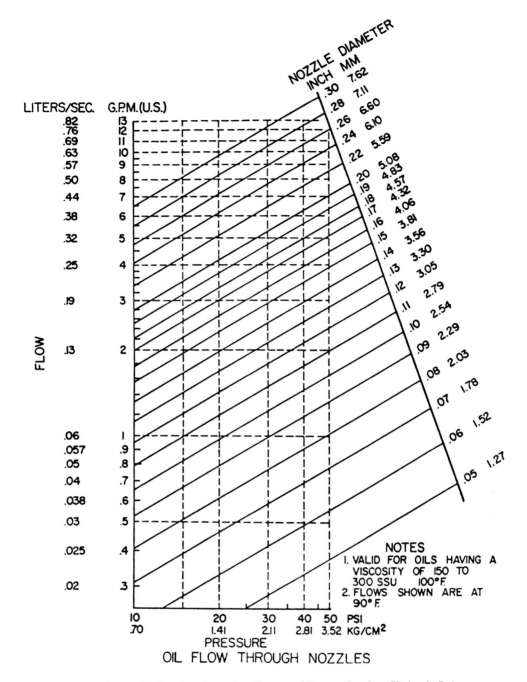

FIGURE 14. Oil flow through nozzles. (Courtesy of Koppers Co., Inc., Pittsburgh, Pa.)

couplings. While in a dam-type coupling the lubrication is ensured by the oil annulus and the oil flow is determined by the cooling requirements, the damless coupling needs a large oil flow to be properly lubricated. In order to determine the nozzle size as a function of the required flow and available pressure, the graph from Figure 14 can be used.

Lubricant Selection

Lubricant selection depends on the type of coupling and particularly on the application.

Even the best lubricant for one application can cause rapid wear if used in a different application. This fact cannot be sufficiently emphasized. An industrial plant that lubricates all couplings with whatever grease is available cannot expect good performance from the couplings. Substantial savings in maintenance costs can be realized through the use of proper lubricants.

To start with, the best approach is to read and follow coupling manufacturers' recommendations. If they are available, one must verify that they are up to date, particularly because lubricant manufacturers have changed many of their oils and greases in the last few years. If lubricant recommendations are not available, the following can be used as a guide:

1. Steel grid coupling manufacturers recommend an NLGI No. 2 grease.
2. Chain coupling manufacturers recommend an NLGI No. 1 grease when the coupling is used with a housing, and a No. 2 grease if the chain is exposed.
3. Gear coupling greases must be selected to match the application:

   a. *Normal applications* are those where the centrifugal forces do not exceed 220 G's (see Figure 6). If the pitch diameter is not known, one can use the formula

   $$\text{rpm} \leqslant 2800 / \sqrt{d}$$

   where d is the shaft diameter in inches (cm/2.54). Also, normal applications are those where misalignment at each hub is less than 3/4°, and where the peak torque is less than two times the continuous torque. For these applications one should use an NLGI No. 1 or 2 grease with a high-viscosity base oil (preferably higher than 198 cSt at 40°C), and not more than 8% soap content.

   b. *Low-speed applications* are those where centrifugal forces are lower than 10 G's. If pitch diameter is not known, one can use the formula

   $$\text{rpm} \leqslant 200 / \sqrt{d}$$

   Same limits are valid for misalignment and shock torque as for normal applications. An NLGI No. 0 or 1 grease should be used, with a high-viscosity base oil (preferably higher than 198 cSt at 40°C), and not more than 8% soap content.

   c. *Fluctuating torque applications* are those where the centrifugal forces do not exceed 220 G's, misalignment at each hub is less than 3/4°, but the torque fluctuates widely or even reverses directions Examples of such applications are reciprocating compressors and internal combustion engines running frequently at idling speeds. Because the centrifugal effect cannot force a grease in the backlash that switches rapidly from one side of the teeth to the other, these applications will cause a coupling to chatter. For these applications one should use a highly viscous *oil*, usually recommended for the lubrication of open gears, chains, and wire ropes. These oils are compounded from residual mineral oils and are characterized by their high viscosity (over 6000 cSt at 38°C), adhesiveness, and dark black color. For insertion in the coupling, they should be heated or pumped with a grease gun. One should avoid using oils that are thinned with a solvent, because as the solvent evaporates it leaves only a little oil in the coupling.

   d. *High-speed applications* can be defined as those where the centrifugal forces are higher than 220 G's, misalignment at each hub is less than 1/2°, and the torque transmitted is fairly uniform. For these applications, the lubricant must have a very good resistance to centrifugal separation. Data on the separation characteristics under centrifugal forces of greases (or oils) are available from either the lubricant or coupling manufacturer.

e. *High-torque, high-misalignment applications* are those where the centrifugal forces are lower than 220 G's, misalignment is larger than 3/4° (usually between 1.5 and 3°) and the shock loads exceed 2.5 times the continuous torque. Many such applications also have high-ambient temperatures, and only a few greases can perform satisfactorily. Besides the characteristics of a grease for "normal applications", the grease should also have antifriction and antiwear additives, extreme pressure (EP) additives, a Timken® OK load greater than 40 lb, and a minimum dropping point of 150°C.

4. Gear coupling oils should always be of high viscosity grade, (no less than 150 SSU at 100°C). Although the viscosity cannot be *too* high for satisfactory coupling operation, oils with viscosities higher than 1000 SSU at 100°C should not be used since they cannot practically be poured into a coupling.

   Continuous oil flow lubrication uses the oil from the system, which is seldom a high viscosity oil. To increase its viscosity, the oil should be cooled before it enters the coupling.

Relubrication Procedure

If manufacturers' recommendations are not available, oil-filled couplings should be relubricated every six months and grease-filled couplings once a year. Coupling guards should be observed periodically for evidence of lubricant escaping from the couplings. The causes for this malfunction (improper sealing) should be found and corrected, and the coupling refilled with lubricant before restarting. Unless the grease used has no oil separation under the centrifugal forces present in the coupling, it is advisable to open and clean the coupling before relubrication. Without cleaning, additional *soap* is introduced in the coupling; and as indicated previously, too much soap is detrimental to coupling performance.

The quantity of lubricant that should be used depends on the internal volume of the coupling, which varies not only with the size of the coupling, but also with the coupling type and make. The lubricant volume of every coupling can be found in the catalog or instruction manual. If lubricant volume is not available, one can use the following method: the two halves of the coupling should be so assembled that the lube plugs of the halves are diametrically opposite; the couplings should be rotated until the lube plugs are at 45° to the vertical plane; both lube plugs should be removed; grease should be pumped through the lower hole until it flows out the upper hole. This method may cause some overfilling, in which case some lubricant escapes past the seals on start-up. Excessive overfilling should be avoided because it generates high-thrust forces on the equipment bearings.

REFERENCES

1. **AGMA,** *Nomenclature for Flexible Couplings,* Standard No. 510.01, American Gear Manufacturers Association, Arlington, Va., 1965.
2. **Calistrat, M. M.,** Wear and lubrication of gear couplings, *Mech. Eng.,* 28, October 1975.
3. **Clapp, A. M.,** Fundamentals of lubrication relating to operation and maintenance of turbomachinery, *2nd Turbomachinery Symp.,* Texas A & M University, October 1973.
4. **Calistrat, M. M.,** Grease Separation Under Centrifugal Forces, ASME Paper 75 PTG-3, American Society of Mechanical Engineers, New York, 1975.
5. **Filepp, L.,** Lubricant as a coolant in high speed gear couplings, *J. Lubr. Technol. Trans. ASME,* 178, January 1970.

DYNAMIC SEALS

W. K. Stair

CLASSIFICATION

Fluid seals are divided into two main classes — static seals and dynamic seals. Static seals are gaskets, O-ring joints, packed joints, welded joints, and similar devices used to seal static connections or openings with little or no relative motion between mating parts. A dynamic seal is any device used to restrict flow of fluid through an aperture closed by relatively moving surfaces. Some dynamic seals also include static sealing elements in their design.

Seals are also frequently classified as contact seals or clearance seals. Some seal elements may operate as clearance seals under certain conditions and as contact seals under others. The term seal may refer to a system rather than a single device. A sealing system may require a mechanical seal, a viscoseal, and a labyrinth seal in order to produce the desired end result.

Table 1 shows the dynamic seal elements which make up the bulk of industrial, commercial, utility, and transportation sealing applications. Discussion follows of the selection and design factors involved with each of these types.

POSITIVE RUBBING CONTACT SEALS

Mechanical Face Seal

The mechanical face seal, or end face seal, in Figure 1 is a device for sealing the annular space between a rotating shaft and a housing. A rotary and a stationary face are forced towards rubbing contact by mechanical means and by fluid pressure acting on the rear of one of the sealing faces. The two contacting faces are usually compatible materials capable of operation with boundary lubrication. Low coefficients of friction and high thermal conductivities are generally desirable.

The face seal mating surfaces are lapped flat to within 0.5 to 1.5 μm (20 to 60 μin.). Excessive roughness causes high friction, accelerated wear, and short life. However, extremely smooth and flat surfaces lack the ability to generate hydrodynamic pressure in the fluid film which also leads to high friction, rapid wear, and short life.

Stator and Rotor Arrangement

The stationary seal face or the rotating face may be flexibly mounted, usually with one or more springs to keep the faces in contact. This seal face is referred to as the seal head and the opposing ring as the seal seat. The seal head may be internally or externally mounted (Figure 2). With pressure acting on the outer diameter of the internal seal (Figures 2a, 2e, and 2f), the seal rings are in compression. This affords a wider choice of materials, many of which are hard and brittle and should not be subjected to tensile stress.

Rotary seal heads (Figures 2a, 2b, and 2c) are convenient to install on the shaft which generally is made from acceptable materials to acceptable tolerances. The stuffing box housing requires minimal machining and the stationary seat can employ a wide range of designs and materials. Dynamic balance of the rotating assembly is more difficult and rotary seal heads are usually employed at seal face speeds below 25 to 30 m/sec (5000 to 6000 ft/min).

Stationary seal heads (Figures 2d, 2e, and 2f) avoid rotation of the spring assembly and are therefore often preferred for higher speeds. Better tolerances and finishes are required

Table 1
CATEGORIES OF DYNAMIC SEALS

Operational principle for dynamic seal element	Normal motion	Extent of use	Energy loss	Leakage	Life
Positive contact					
Face	Ro Os	H	L	L	M—H
Rings	Ro Os Re	H	H	L	L—M
Lip	Ro Os Re	H	L	L	L—M
Packings	Ro Os Re	H	L—H	L—H	L
Diaphragms	Os Re	L	L	L	H
Controlled clearance					
Hydrodynamic	Ro	L	L	L—M	M—H
Hydrostatic	Ro Os	L	L	L—M	M—H
Floating bushing	Ro Os Re	M	M	M—H	H
Fixed-geometry clearance					
Labyrinth	Ro Os Re	H	H	H	H
Bushing	Ro Os Re	M	H	H	M—H
Special control of fluid					
Freeze	Ro Os	L	L	M	L—M
Magnetic fluid	Ro Os Re	L	L	L	M
Centrifugal	Ro	L	M	L	H
Screw	Ro	L	M	L	H
Magnetic	Ro Os	L	L	M	M

Note: Ro = rotary, Os = oscillatory, Re = reciprocating, H = high, M = moderate, and L = low.

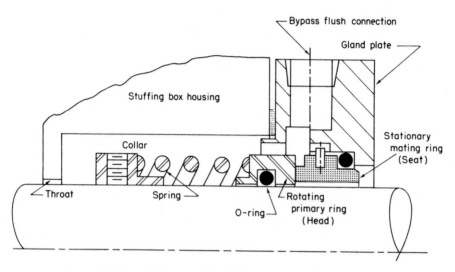

FIGURE 1. Single, unbalanced, inside mechanical face seal.

for the stuffing box housing and gland plate. Also, some stationary designs require special liners or sleeves if the seal is to be pressure loaded, Figure 2e.

Springs and other parts of the internal seal head assembly are usually in contact with the sealed fluid which, if corrosive, requires more care in the choice of materials. The internal seal may require considerable stuffing box space, requires accurate positioning of the seal head assembly, and may be more difficult to inspect, disassemble, and adjust.

External seal operation in Figures 2b, 2c, and 2d is observable and installation, removal,

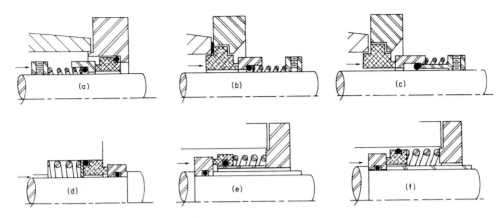

FIGURE 2. (a) Internal rotating head. Pressure on OD of faces; seats in compression; pressure loaded; springs exposed. (b) External rotating head. Pressure on ID of faces; seats in tension; pressure unloaded; springs not exposed. (c) External rotating head. Pressure on ID of faces; seats in tension; pressure loading depends on balance; springs not exposed. (d) External stationary head. Pressure on ID of faces; seats in tension; pressure loaded; springs exposed. (e) Internal stationary head. Pressure on OD of faces; seats in compression; pressure loaded; springs exposed. (f) Internal stationary head. Pressure on OD of faces; seats in compression; pressure unloaded; springs not exposed.

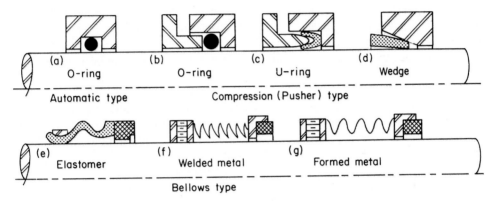

FIGURE 3. Secondary seal configurations for shafts.

or adjustment is easier. This seal can be designed to function as a relief valve and is useful when stuffing box space is limited. Face loading can be adjusted and seal seat wear can be monitored. The external seal is widely used in chemical and other applications with corrosive liquids since the springs are not exposed to the sealed fluid. The external seal is more vulnerable to impact damage and is usually pressure unloaded. Pressure limits are lower for both unbalanced and balanced designs. The pressure acts inside the seal rings and subjects them to undesirable tensile stress. Also, this seal contains blind cavities which act as centrifuges and collect debris from the fluid.

Secondary Seals

The semistatic seal between a rotating seal head and the shaft or between a stationary seal head and the housing is termed the secondary seal. Various types are typified in Figure 3. The automatic secondary seal, Figures 1 and 3a, is energized by the sealed pressure and does not require mechanical compression. In addition to the O-ring shown in Figure 3a, U-, V-, and piston rings can be employed.

The compression seal utilizes a mechanical load provided by a metal follower loaded by the spring which loads the primary seal faces (O-, V-, U-, wedge, and square packing rings

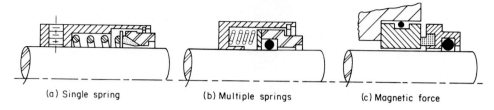

(a) Single spring   (b) Multiple springs   (c) Magnetic force

FIGURE 4.   Face loading devices.

may be used). As the primary seal face wears, the secondary sealing element in automatic and compression seals is pushed forward along the shaft. In a bellows secondary seal, wear is taken up by extension of the bellows made from molded elastomers, formed metal, or welded metal discs. The pusher type seal is more susceptible to dirt, which may increase sliding friction, than the bellows type. The bellows seal is pressure limited since high pressures may deflect the bellows radially enough to alter the effective bellows diameter. Formed and welded metal bellows seals may employ loading springs in addition to the spring provided by the bellows.

Seal Loading Devices

Lapped sealing faces are held in contact by sufficient preloading at assembly to keep the seal closed before hydraulic loading is developed, to withstand pressure reversals, and to overcome secondary seal friction. Preload should be just sufficient to keep the seal closed at the maximum expected axial excursion. Unnecessary preload tends to increase face load and shorten seal life.

The single preload spring, Figure 4a, has the advantage of simplicity and the relatively large wire cross section provides greater resistance to deterioration by corrosion. The multiple spring seal, Figure 4b, requires less axial space, gives more uniform seal face load, has better resistance to centrifugal forces, can have face preload adjusted by using a different number of springs, and a large number of seal sizes can be fitted with the same springs. Wave springs, finger springs, Belleville springs, slotted washers, and curved washers may be employed in seals requiring minimum axial space. These seals must be carefully designed and installed to obtain desired preload since they have a high-spring rate.

The magnetic seal, Figure 4c, eliminates the need for springs and permits a compact design. A disadvantage is the attraction of magnetic debris to the seal faces. A bellows, Figure 3f, may be used as a combined secondary seal and face loading device or in combination with a single spring.

Seal Balance

Rate of energy dissipation between seal faces can be expressed as:

$$E = \eta F_f V = \eta P_f A_f V \qquad (1)$$

where F_f = seal face normal force, η = coefficient of friction, V = mean seal face velocity, P_f = average seal face pressure, and A_f = projected seal face area.

The $P_f V$ term represents the energy dissipation per unit of projected seal face area for a unity friction coefficient. For effective sealing and an acceptable wear rate, design factors which determine $P_f V$ must be controlled.

Assuming zero discharge pressure, P_l, the forces in Figure 5 tending to close the seal are hydrostatic force, F_p, arising from the pressure being sealed, and spring force, F_s, necessary to maintain contact between the faces at start-up and shutdown. Hydraulic force, F_o, which acts to separate the seal faces, is controlled by the characteristics of the interface flow process

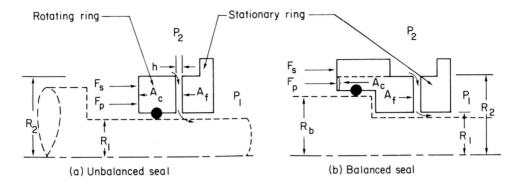

FIGURE 5. Forces acting on face seal.

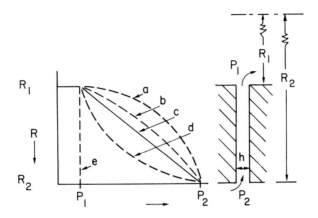

FIGURE 6. (a) Clearance (h) decreasing in flow direction. (b) Parallel faces, $(R_2 - R_1)$ R_2 not small. (c) Parallel faces, $R_2 - R_1$ $\ll R_2$, R_1. (d) Clearance increasing in flow direction. (e) Clearance (h) = flow = 0.

as illustrated in Figure 6. A pressure profile factor, β, multiplied by the differential pressure and interface area represents the opening force due to the interface pressure distribution. Referring to Figure 5b,

$$F_{closing} = F_c = P_2 A_c + P_1 (A_f - A_c) + F_s \tag{2}$$

$$F_{opening} = F_o = \left(P_1 + \beta(P_2 - P_1) \right) A_f \tag{3}$$

$$F_{net} = F_f = F_c - F_o = P_2 (A_c - \beta A_f) - P_1 (A_c - \beta A_f) .+ F_s \tag{4}$$

Let $b = A_c/A_f$ = balance factor, then

$$F_f = (P_2 - P_1)(b - \beta) A_f + F_s \tag{5}$$

$$P_f = (P_2 - P_1)(b - \beta) + P_s \tag{6}$$

β is about 0.5 for a linear pressure distribution across the seal face, $\beta > 0.5$ for a converging flow path, and $\beta < 0.5$ for diverging flow.

For the unbalanced seal, $b \geqslant 1$ and contact pressure P_f can be greater than the sealed pressure. In Figure 5b a step in the shaft or a stepped sleeve permits (A_o/A_f) to be made

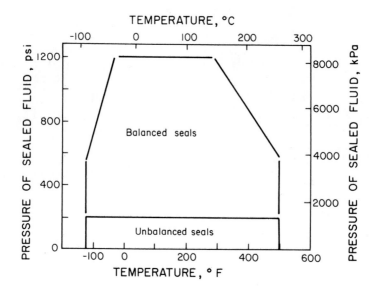

FIGURE 7. Approximate envelope of manufacturers recommended operating limits for inside seals. (From Bernd, L. H., *Lubr. Eng.*, 24(10), 479, 1968. With permission.)

less than the unity, which allows contact pressure P_f and the energy dissipation to be reduced. Note that if b equals 0.5 and one assumed a linear pressure profile, $(b - \beta)$ becomes zero and only spring pressure keeps the seal faces closed. The balance ratio chosen in practice is usually in the range of 0.58 to 0.75.

Ordinary Pressure-Temperature-Speed Limits

High-quality, general purpose mechanical face seals will meet a large majority of ordinary sealing requirements. These involve sealing of clean, abrasive-free, safe, and only slightly corrosive fluids which provide good seal face lubrication under the following conditions:

1. Seal cavity pressure between 2.8 MPa (400 psi) and 1.3 Pa (0.01 torr).
2. Seal cavity temperature between 200°C (400°F) and −40°C (−40°F).
3. Seal face speeds less than 23 m/sec (4500 ft/min).

The approximate envelope of operating conditions for ordinary inside seals (Figure 1) is shown in Figure 7. Pressure limits for unbalanced external seals, Figure 2b, are about 20% of those in Figure 7, while balanced external seals, Figure 2c, have pressure limits about 40% of those of Figure 7. Some inside balanced seals designed specifically for high pressure have been used at pressures in excess of 17 MPa (2500 psi) at shaft speeds of 23 to 33 m/sec (4500 to 6500 ft/min). Stuffing box pressure limits for unbalanced seals have been set rather arbitrarily irrespective of service conditions at 0.7 to 1.4 MPa by some manufacturers, more conservatively by some users in Table 2.

The upper PV limit for unbalanced seals is frequently taken to be 0.7 MPa·m/sec. (200,000 psi·ft/min). The limiting PV is useful in expressing the relative merit of various face material combinations. Some definitions used for PV factor follow (refer to Figure 5):

$$PV = (P_2 - P_1)V = \Delta P \cdot V \tag{7}$$

$$PV = \Delta P \cdot b \cdot V \tag{8} \text{ (Reference 3)}$$

$$PV = P_f \cdot V = (\Delta P (b - \beta) + P_s)V \tag{9} \text{ (from Equation 6)}$$

Table 2
PRESSURE LIMITS FOR UNBALANCED
SEALS

Seal inside diameter		Shaft speed (rpm)	Sealed pressure	
mm	in.		kPa	psi
13—50	0.5—2	to 1800	690	100
		1801 to 3600	345	50
Over 50 to 100	Over 2 to 4	to 1800	345	50
		1801 to 3600	172	25

From **API**, *Centrifugal Pumps for General Refining Services*,
Standard 610, American Petroleum Institute, Washington,
D.C., 1971. With permission.

For an unbalanced seal with b = 1, Equation 9 becomes

$$PV = \left(\Delta P \left(1 - \beta\right) + P_s\right)V \tag{10}$$

Equation 9 is the most logical even though values for β are not precisely known. Other manufacturers use Equation 7 and establish PV ratings for specific fluids, seal types, and face material combinations which will give a seal life of approximately 15,000 hr. For example, a 4-in., unbalanced, inside seal using ceramic vs. carbon-graphite seal faces at a speed of 1750 rpm may have a recommended PV rating of 0.5 MPa·m/sec (150,000 psi·ft/min) when operating in 65°C (150°F) water, and 0.7 MPa·m/sec (205,000 psi·ft/min) with light hydrocarbon oils. The user is cautioned to ascertain the PV definition when comparing face seal performances and to use PV limits simply as guides. In some applications, sealing hydrocarbons with specific gravity < 0.65 for example, a balanced seal is recommended regardless of pressure, speed, or seal size.[4]

Stuffing box temperature is very significant in seal design and selection. Temperature at seal faces can be several hundred degrees above ambient due to friction: once a seal design is established, seal cavity cooling should be set so as not to exceed the temperature limits of materials involved. Properties of most liquids change with increasing temperature to make sealing more difficult. Corrosion rate approximately doubles with each 17°C (30°F) temperature increase, and for most oils viscosity is halved. Similar viscosity reductions are noted for water, light hydrocarbons, and acids for temperature increases of about 40°C (70°F). Stuffing box temperature should be kept 15 to 30°C (25 to 50°F) below the product boiling point at the stuffing box pressure.

Leakage

For "average" commercial seals in good condition, leakage ranges from 0 to about 10 cc/min. Approximate water leakage rate for a mechanical face seal at 690 kPa (100 psi) varies from 1×10^{-4} to 10^{-1} mℓ/hr/m of periphery/kPa (2×10^{-5} to 2×10^{-2} mℓ/hr/in. of periphery/psi), where 1 m$\ell \approx 10$ to 20 drops.[5]

Life

For very high speed and poor lubrication in a cryogenic liquid seal in a turbopump, life may be a matter of minutes. In water pump service, a carefully designed and maintained seal has a life of about two years. In refinery service, typical seal life is 3 to 12 months.

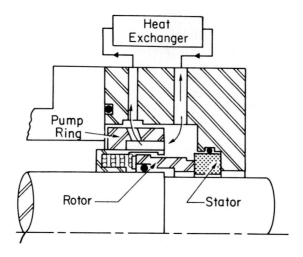

FIGURE 8.   Internal pumping ring.

Many seal manufacturers base their design and maximum PV recommendations on an estimated life of about 15,000 hr.

Temperature Extremes

Seals for temperatures above 200°C or below −40°C often use metal bellows. Elastomers become unserviceable much beyond these limits. The usual temperature range for metal bellows seals is −240 to 650°C (−400 to 1200°F), but Inconel X-750, Rene 41, or refractory alloys have been suggested for temperatures to about 1100°C (2000°F).[6]

Temperature control, either cooling or heating, can be obtained by bypass or circulating ring flushing, a water or steam jacketed stuffing box, or a quenching connection in the gland plate as shown in Figure 1. Clean fluid from the pump discharge can be cooled, injected to cool seal parts, and then directed through the restricted stuffing box throat back to pump suction. An alternate arrangement uses a small pumping ring to circulate a small quantity of clean fluid through a small external heat exchanger (Figure 8).[7] Occasionally, the material being pumped solidifies at ambient temperature. In such cases, the seal region must be heated, for example, by using a steam-heated gland.

Pressure Extremes

Pressures greater than about 2.8 MPa (400 psi) may distort seal faces and other components. Conversely, high vacuum causes elastomers to outgas and destroy the vacuum. The outgas problem can be solved by using metal bellows seals. High pressures require seal balance and careful design of ring geometries. Cross-sectional twisting especially must be reduced, and the gland plate must ensure flatness and accurate alignment. Some manufacturers insist that the gland plate be provided as part of the high-pressure seal assembly. Seal cavity pressure can be borne by two seals in tandem to accommodate high system pressures. A clean process fluid stream or buffer fluid at pressure P_b is circulated through the outer seal chamber, usually set so ($P_{sys} - P_b$) is approximately the same as ($P_b - P_{atm}$). The outer seal is considered a backup in the event the inner seal fails.

High Speed

At seal speeds over 23 m/sec (4500 ft/min), the mechanical seal requires matched springs and careful assembly to avoid unbalance of the rotating assembly. At speeds over 33 m/sec (6500 ft/min), the seal head is usually made stationary and special designs are used for speeds up to 64 m/sec (12,500 ft/min) and above.

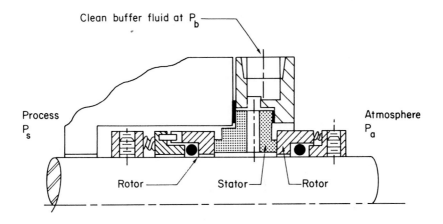

Clean buffer fluid at P_b

Process P_s

Atmosphere P_a

Rotor —— Stator —— Rotor

FIGURE 9. Internal-external seal. $P_b > P_s$ for double seal; $P_b < P_s$ for tandem seal.

Abrasive, Corrosive, and Hazardous Fluids

Design for abrasive and corrosive fluids can follow either of two avenues: (1) fabrication of seal components from exotic abrasion and corrosion resistant materials, or (2) creation of a compatible environment to isolate the seal. Hazardous fluids may not be hostile to seal components, but safety considerations usually dictate seal environment control.

Abrasives in the sealed process fluid may be due to (1) the inherent nature of slurries, or liquids containing foreign matter such as sand, dirt, or oxides, (2) crystalline particles which result from evaporation or from contact with atmosphere, and (3) crystalline particles which result from heating or cooling. A clean, process-compatible liquid is injected through the flush connection to cool and isolate seal parts from abrasive particles in the process fluid. The flush liquid flows through a close-clearance bushing at the throat back to the process. Amount of liquid injected can be controlled by the supply pressure and the restriction: a plain bushing, a floating bushing, or a lip type bushing. The clean injected fluid may be pump discharge from which abrasive particles have been removed by centrifugal separation or a settling tank.

For process fluids which crystallize upon contact with air, an auxiliary connection can be used to inject low-pressure water to wash away the seal leakage and prevent abrasives from forming at seal faces. Such an arrangement can also be used to dilute and drain away dangerous fluid leakage. Where dilution of the process fluid by flow through the throat bushing cannot be permitted, a double seal is employed with clean buffer fluid circulated by auxiliary means between the two seal elements to provide an almost complete isolation from the process fluid. Where seal housing space is limited, an internal-external multiple seal may be arranged as in Figure 9. With the buffer fluid pressure P_b greater than process pressure P_s, both seals are lubricated by the buffer field. The arrangement can serve as a tandem seal when $P_b < P_s$.

Materials

Seal components and gland ring parts for noncorrosive fluids such as gasoline, hydrocarbons, and oils are usually made from ferritic stainless steels such as 502 or 430. For moderate corrosion resistance in environments such as water, sea water, dilute acids, fatty acids and alkalis, austenitic stainless steels such as 302, 304, and 316 are widely used. For highly corrosive environments such as strong mineral acids and strong alkalis, nickel-copper base materials such as Monel or nickel-molybdenum alloys such as Hastelloy B or Hastelloy C are frequently employed. Temperature range for these materials is -100 to $400°C$ (-150 to $750°F$). Table 3 presents seal face material combinations for various environments. Tables

Table 3
FACE SEAL MATERIAL COMBINATIONS[4,8]

Common end face seal and material combinations used in a variety of fluids

Seal nose material **vs.** Seal seat material

Seal nose material	Seal seat material
Hard faced stainless	Carbon graphite
Stellite	Carbon graphite
Chromium boride	Carbon graphite
Ceramic	Carbon graphite
Hastelloy A.B.C.	Carbon graphite
Carbon-filled Teflon®	Carbon graphite
Glass-filled Teflon®	Carbon graphite
Cast iron	Carbon graphite
Ni-Resist	Carbon graphite
Nitralloy	Carbon graphite
Stellite-faced stainless	Carbon graphite
400 Series stainless steel	Carbon graphite
Bronze	Carbon graphite
Tungsten carbide	Carbon graphite
Sintered iron or bronze	Carbon graphite
Tool steel, hardened	Carbon graphite
Monel	Carbon graphite
Nickel cast iron	Carbon graphite
Chrome plate	Carbon graphite
Ceramic faced stainless	Carbon graphite
Chrome carbide	Carbon graphite
Ceramic	Carbon
Tungsten carbide	Carbon
Laminated plastic	Bronze
Tungsten carbide	Bronze
Tungsten	Tungsten
Tungsten carbide	Tungsten carbide
Stellite	Ceramic
Glass-filled PFTE	Ceramic
PFTE	Ceramic
Ceramic	Ceramic
Ceramic faced stainless	Ceramic
Phosphor bronze	Carbon babbitt
Aluminum bronze	Carbon babbitt
Aluminum bronze	Stellite on stainless
Boron carbide	Boron carbide

Environment

Environment	Hard faced stainless / CG	Stellite / CG	Chromium boride / CG	Ceramic / CG	Hastelloy A.B.C. / CG	Carbon-filled Teflon / CG	Glass-filled Teflon / CG	Cast iron / CG	Ni-Resist / CG	Nitralloy / CG	Stellite-faced stainless / CG	400 Series stainless / CG	Bronze / CG	Tungsten carbide / CG	Sintered iron or bronze / CG	Tool steel, hardened / CG	Monel / CG	Nickel cast iron / CG	Chrome plate / CG	Ceramic faced stainless / CG	Chrome carbide / CG	Ceramic / Carbon	Tungsten carbide / Carbon	Laminated plastic / Bronze	Tungsten carbide / Bronze	Tungsten / Tungsten	Tungsten carbide / TC	Stellite / Ceramic	Glass-filled PFTE / Ceramic	PFTE / Ceramic	Ceramic / Ceramic	Ceramic faced stainless / Ceramic	Phosphor bronze / Carbon babbitt	Aluminum bronze / Carbon babbitt	Aluminum bronze / Stellite on stainless	Boron carbide / Boron carbide
Acids	A	A	A	A	1																							7	6	A						
Caustics	2					A																														
Gasoline				A	A	A	A	A	A		3																									
Gas (O_2, N_2, H_2, He, CO_2)				A		5		A	A	A	A	A	A	A	A				A		A															
Heat transfer				A		5		A			A	A							A	A	A				5											
Oil				A		5	A	A	A	A	A	A	A	A											5											
Oxidizing fluids				A													A																			
Salt solution		A										A																			A					
Salt water	A											3	A	A		A									A	A					A		A	A		
Slurry													A	A											A											
Water	A			A		A	A		A				A	4	A	A						A	A		A	A	A				A		A	A	A	A

Note: 1. For nonoxidizing acids. 2. Nonmetallic carbon graphite. 3. Metallic carbon graphite. 4. For constant operation only. 5. For high temperature (approx. 700 °F). 6. For oxidizing acids. 7. Attacked by many mineral acids. A = Acceptable.

Table 4
RECOMMENDED
TEMPERATURE LIMITS FOR
FACE SEAL MATERIALS[a]

Material	Maximum temperature	
	°F	°C
Tungsten carbide	750	400
Stainless steel	600	316
Carbon-graphite	525	275
Stellite	450	232
Nickel-cast iron	350	177
Leaded bronze	350	177
Alumina[b]	350	177
Glass-filled TFE	350	177

[a] Product temperature; maximum working temperature is higher.
[b] Subject to thermal shock fracture.

From *Guide to Modern Mechanical Sealing*, Durametallic Corporation, Kalamazoo, Mich., 1971. With permission.

Table 5
RECOMMENDED TEMPERATURE LIMITS FOR
SECONDARY SEAL MATERIALS[a]

Material	Minimum temperature		Maximum temperature	
	°F	°C	°F	°C
Nitrile-low	− 40	− 40	176	80
Nitrile-medium	− 22	− 30	194	90
Nitrile-high	− 4	− 20	212	100
Neoprene	− 58	− 50	212	100
Butyl	− 40	− 40	194	90
Silicone, fluorosilicone	− 76	− 60	392	200
Fluorocarbon	− 58	− 50	437	225
TFE	−100	− 73	350	177
Glass-filled TFE	−175	−115	450	232
Graphite	−450	−268	750	400

[a] Product temperature.

From *Guide to Modern Mechanical Sealing*, Durametallic Corporation, Kalamazoo, Mich., 1971. With permission.

4, 5, and 6 show recommended temperature limits for seal faces, secondary seal materials, and springs.

Ring Seals

Split or segmented rings of metallic or nonmetallic material are used as piston ring, rod, and circumferential seals, Figure 10. The piston ring (expanding ring) and rod seals (contracting ring) are used principally in reciprocating applications, circumferential seals mainly

Table 6
RECOMMENDED TEMPERATURE
LIMIT FOR SEAL SPRING MATERIAL

Material	Maximum temperature	
	°F	°C
Carbon steel (music wire)	248	120
Carbon steel (hard drawn)	392	200
Monel	392	200
K-Monel	446	230
18-8 stainless steel	527	275
Nimonic 90	752	400
Inconel X	1022	550

From Lymer, A. and Greenshield, A. L., *Pumps,* 24(7), 209, 1968. With permission.

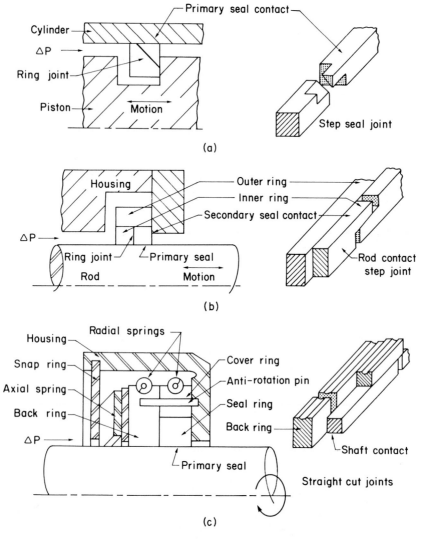

FIGURE 10.   Ring seal designs. (a) Piston ring seal (expanding split ring, single-ring design); (b) rod seal (contracting split ring, two-ring design); and (c) circumferential seal (three-ring design).

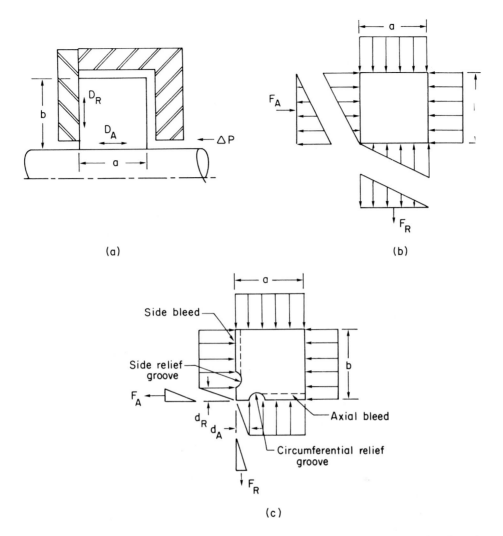

FIGURE 11.   Pressure-induced forces on a seal ring. (a) Drag forces on rod seal ring; a = face dimension; b = wall dimension; D_R = radial drag = $n_R F_A$; D_A = axial drag = $n_A F_R$; n = coefficient of friction. (b) Contact forces on ring; F_A = axial unbalance; F_R = radial unbalance. (c) Contact forces on pressure-relieved ring.

as rotary seals. A single-stage ring seal may employ one, two, or three split or segmented rings in a single groove or housing for rubbing contact with either the shaft or bore. The pressure of the sealed medium forces the ring into axial and radial contact and the initial or static contact load is provided by the elastic properties of the ring (expanding or contracting seals) and/or by auxiliary springs (circumferential seals). Ring seals have three potential leakage paths: (1) between ring and bore or ring and shaft, (2) between ring and side wall of groove or housing, and (3) the ring gap.

Contact loads and drag forces caused by fluid pressure on a ring are depicted in Figures 11a and 11b. These loads increase with increasing pressure with an accompanying increase in wear. High pressure may also prevent the ring from following dynamic excursions of the shaft, piston, or rod. Improved dynamic response and wear reduction in one-directional seal rings can be obtained by pressure relief grooves (Figure 11c) for a "balanced" ring. The grooving tends to increase leakage by reducing the leak path length across the seal dam. Good design requires a compromise between wear, frictional heating, and leakage. Ring

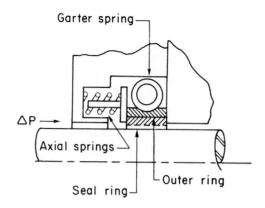

FIGURE 12.   Two-piece rod seal.

and sealing dam dimensions are usually selected such that axial and radial forces are about the same. Auxiliary springs are required in low-pressure applications to maintain contact in both axial and radial directions.

Split Ring Seals

Expanding split rings (piston rings) and contracting split rings (rod seals) are used as piston head, rod, rotary, butterfly valve, and static seals to control leakage of hot combustion gases or fluids. The most common expanding ring application is to seal between the reciprocating piston and cylinder wall in internal combustion engines and reciprocating compressors. The contracting ring seal is used in hydraulic cylinders where high-pressure, high-temperatures, thermal fatigue, and reliability requirements make elastomeric packings undesirable.

Split rings may be used singly or in series. A second step joint ring will reduce leakage by approximately 15%. Although a third ring will provide little additional leakage improvement, it may extend the overhaul period by coming into operation when the first and second rings are worn.

Expanding rings are manufactured with free-ring dimensions to produce uniform radial pressures of about 70 to 550 kPa (10 to 80 psi) when installed. Auxiliary springs are normally unnecessary. The contracting ring, however, can provide only limited tension and frequently employs auxiliary springs to insure conformation to the rod surface as shown in Figure 12. Expanding split rings (Figure 13) are frequently used as rotary seals on hydraulic transmissions and clutches, torque converters, hydrostatic transmissions, crackcase seals on large engines, and turbosuperchargers. Step seal rings have a face dimension greater than the wall dimension. Relative motion and wear take place at the side contact area; this prevents wear grooves in the housing which would prevent removal of the shaft from the housing during overhaul.

Lubrication is necessary when using metallic ring seals. In oxygen compressors, oil-free air compressors, food processing plants, and certain chemical processes where lubricants cannot be tolerated, nonmetallic materials may be employed for seal rings. Table 7 gives recommended temperature limits for commonly used split ring materials. Typical performance ranges of split ring seals are shown in Table 8.

Circumferential Seals

Adaptation of split and segmented rings to prevent leakage of high-temperature air and combustion gases into the bearing cavities of aircraft gas turbine engines led to development of high-performance, high-speed, elevated temperature circumferential seals, Figure 10c. The basic arrangement and balance considerations in Figures 10 and 11 also apply to circumferential seals.[12]

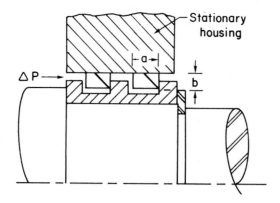

FIGURE 13.   Rotating shaft seal, 1.1 < a/b < 1.5.

Table 7
TEMPERATURE LIMITS FOR RING SEAL
MATERIALS[10,11]

Material	Temperature	
	°F	°C
Glass-filled nylon	350—400	177—204
Filled tetrafluoroethylene	−450—500	−268—260
Bronze	600	316
Low-alloy grey iron	650	343
Malleable iron	700	371
Ductile iron	700	371
Polyimide	750	399
Ni-Resist	800	427
Tool steel, R_c 62—65	900	482
410 Stainless steel	900	482
17—4 PH stainless steel	900	482
Carbon (high temperature)	950	510
S-Monel	950	510
Ductile Ni-Resist	1000	538
Stellite #31	1200	649
Inconel X	1200	649
Surface treatments		
Chromium plate	500	260
Silver plate	600	316
Tin plate	700	371
Cadmium-nickel plate	1000	538
Flame plate	1000—1600	538—871

Circumferential seals, in comparison with mechanical face seals, are low cost, compact, light in weight — 0.03 to 0.06 N/mm of diameter (0.17 to 0.34 lb/in.), permit large axial shaft excursions, and often facilitate assembly. The general configurations, Figure 14, consist of one or more relatively stationary, flexible split or segmented rings mating with a rotating shaft or bore. The rings are loosely keyed to the stationary member by antirotation pins or locks. As rings wear, they become arch bound and function as minimum clearance floating bushing seals with increased leakage, but they are rarely subject to catastrophic failure. The seal elements are retained and held in contact by axial and radial springs which produce contact loadings of about 0.05 to 0.15 N/mm (0.28 to 0.86 lb/in.) of seal circumference.

Table 8
TYPICAL PERFORMANCE RANGES FOR SPLIT RING
SEALS[10,11]

Diameter, mm (in.)	Pressure, MPa (lb/in.²)	Temperature, °C (°F)	Speed, m/sec (ft/sec)	Leakage, cm³/min
Expanding Ring Seals (Piston Head)				
12—3000 (11/2—120)	0.1—27.6 (15—4000)	15.6—482 (60—900)		2—200
Contracting Ring Seals (Rod Seals)				
22—76 (7/8—3)	20.7—27.6 (3000—4000)	204—399 (400—750)		0.1—10
Rotary Seals				
19—1220 (3/4—48)	0—10.3 (0—1500)	37.8—260 (100—500)	1.8—102 (6—336)	0—270

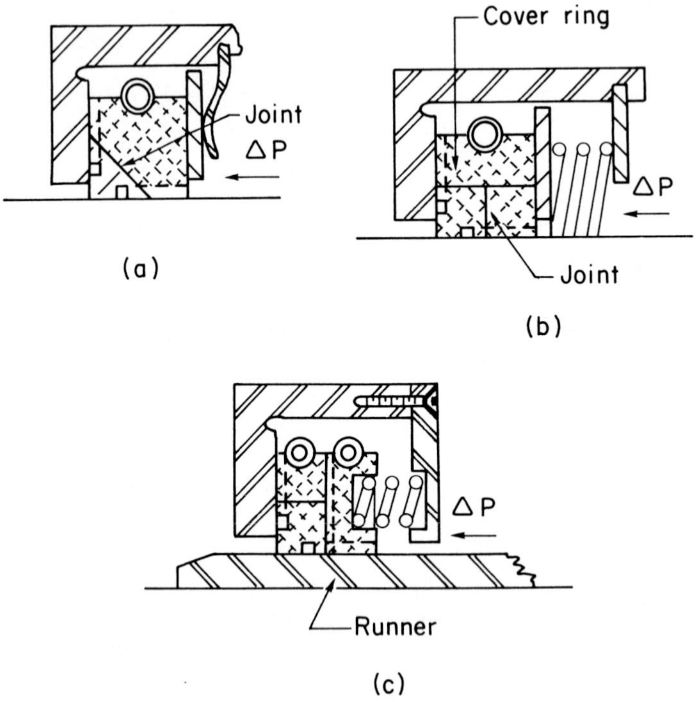

FIGURE 14.   Typical circumferential seals. (a) Single-split ring design: finger spring for axial loading; seal step joint (triangle or rectangle) split ring; requires very precise fabrication; antirotation lock location is critical; small number of parts. (b) Double split ring design: single-coil spring for axial loading; step joint inner split ring; square-cut cover ring seals inner joint; antirotation locks must be carefully located. (c) Triple segmented ring design: multiple coil spring axial loading; most widely used design; square-cut segmented rings; gap geometry easily controlled; simple antirotation pin arrangement; large number of parts.

The small seal mass provides tolerance to a significant degree of shaft whirl. The shaft should be round within 12.7 μm (0.0005 in.) for shaft diameters up to 12.7 cm (5 in.), 1 μm/cm (0.0001 in./in.) for larger diameters. Surface finish should be 0.20 μm (8 μin.)

RMS, minimum shaft or runner hardness 55 Rockwell C, and shaft radial runout 25 μm (0.001 in.) TIR or less. While greater runout can be tolerated by increasing radial spring load, this will cause greater wear. Seal housing side wall surfaces should have similar roughness and hardness values as the shaft. Flatness of the seal secondary surface and the corresponding housing side wall should be within 4 to 6 helium bands (1.17 to 1.75 μm), and the shaft or runner should be free of taper.

The segmented ring arrangement shown in Figure 14c is the most widely used type and is a highly effective gas seal. Viscous liquid in the primary seal interface, however, hydrodynamically lifts the seal rings to cause unacceptable leakage except for low-pressure differentials of 68.96 kPa (10 psi) or less. Circumferential seals can be employed as liquid seals by applying a slight buffer gas overpressure of about 27.6 kPa (4 psi) between two seal elements. Gas leakage keeps the ring segments free of liquid.

Leakage for circumferential seals is generally about an order of magnitude less than with labyrinth seals. Wear life (which depends on speed, pressure differential, temperature, material and design) ranges from about 100 hr to over 10,000 hr. To avoid shaft wear, most high-performance seals employ a clamped-on shaft sleeve or runner which can be replaced or refinished during overhaul. Typical characteristics of a pressure-relieved three-ring segmented circumferential seal are about as follows: diameter 165 mm (6.5 in.), speed 61 m/sec (200 ft/sec), pressure differential between 316°C (600°F) air and 121°C (250°F) oil, 448.2 kPa (65 psi), wear life in excess of 600 hr with leakage of 3.5×10^{-4} m³/sec (0.75 standard ft³/min), and radial unbalance force of 306 N/m (1.75 lb/in.) of seal circumference.

Frictional drag force of approximately 12 N (2.7 lb) without cooling would generate wear track temperatures of 550 to 650°C (1020 to 1200°F). To avoid severe damage to the carbon-graphite seals, oil cooling jets at 9 to 12 m/sec (30 to 40 ft/sec) are directed on the runner near the wear track or under a cantilevered runner.[12] Operation in ambient temperatures of 540°C (1000°F) has been reported.[10] Friction and temperature rise can be minimized through use of pressure-relieved seals, Figure 15. Commonly used axial dam width, E, is about 1 to 1.25 mm (0.039 to 0.049 in.), which produces a radial force of 0.50 to 0.63 N/m/kPa (0.020 to 0.025 lb/in./psi). A narrow axial sealing dam reduces frictional heating, but the wear rate and fragile nature of normal seal rings must be considered. Radial dam width, D, is usually 1 to 1.4 mm (0.039 to 0.055 in.).

Materials employed for circumferential seals are included in Table 7. While sealing rings have been made of bronze, ceramic, filled PTFE, and filled polyimide, the most common material is carbon-graphite.[13] Carbon provides a reasonably low coefficient of friction with most metals and ceramics and has excellent wear properties. Standard grades of carbon-graphite tend to oxidize at temperatures of 320°C (600°F) or above, but some impregnated carbons are satisfactory up to 510°C (950°F). Seal runners are made of low-alloy steels, series 300 and 400 stainless steels, tool steel, 4140 and 4340 steels, Stellite 31, Hastelloy C and F, and Inconel X. Runners are often hard chrome plated or flame plated with aluminum oxide or tungsten carbide. A preferred substrate for hard surfacing, such as low-alloy carbon steel, has high thermal conductivity and is hardenable to Rockwell C 34 to 36.[12]

Radial Lip Seals

Elastomeric radial lip seals are used principally to retain liquids in and to exclude contaminants from equipment having rotating shafts. Over 200 lip seal styles have been developed for an enormous variety of applications involving shaft sizes from 5 to 1525 mm (0.20 to 60 in.). Retained fluids are commonly lubricating oils or liquids having some lubricating qualities, sump temperatures vary from −60 to 200°C (−76 to 390°F), and peripheral speeds range up to 20 m/sec (4000 ft/min). Sealed pressures are moderate, in the range of 20 to 100 kPa (2.9 to 14.5 psi), but special seal designs have been used up to 3450 kPa at 15 m/sec (500 psi at 3000 ft/min).[14]

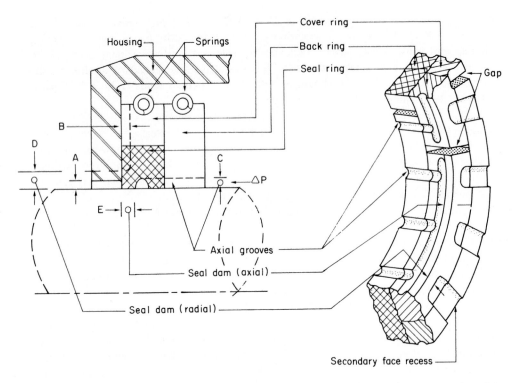

FIGURE 15.   Pressure relieved circumferential seal; three-segmented rings. A = housing clearance, B = face recess depth, C = axial groove depth, D = radial dam width, E = axial dam width, D = 2A + C, and C ≃ wear limit.

Radial lip seals have advantages of small space requirements, low cost, ease of installation, high sealing effectiveness, tolerance for modest shaft runout and misalignment, ability to seal a wide variety of fluids, and availability in numerous standard types and sizes. Disadvantages include limited useful pressure range, limited operating life, stick-slip phenomenon which causes leakage, and their need to be lubricated.

A typical dual lip seal is shown in Figure 16. Sealing is due to surface tension effects in the thin fluid film at the air-side edge of the seal lip-shaft contact. Radial load depends upon lip interference, fluid pressure, spring loading, and elastomer properties. Under stationary conditions, the lip contacts the shaft and forms a leakage barrier about 0.125 to 1.25 mm (0.005 to 0.050 in.) wide. Under dynamic conditions, the lip rides on a thin fluid film about 0.5 to 2.5 μm (20 to 100 μin.) thick. If the film is too thick, the seal leaks; if too thin, the seal may fail due to frictional heating and elastomer deterioration. Changes such as oil swell, stress relaxation, thermal expansion, wear, and hardening influence the contact pressure which is the most important performance factor. Figure 17 shows examples of available lip seal designs.

Life and Leakage

Experimental studies have produced seal life in excess of 10,000 hr.[18,19] As a general guide for elastomeric lip seals in continuous operating, general purpose applications, long life is in excess of 1000 hr, medium life is 400 to 800 hr, and short life is 100 hr or less of operating time before observable leakage. Short life usually results from poor seal selection, poor shaft workmanship, careless installation, or severe working conditions.

Approximate oil leakage rates for radial lip seals at 35 kPa (5 psi) vary from 1×10^{-4} to 1×10^{-2} mℓ/hr/m of periphery per kPa ΔP (2×10^{-5} to 3×10^{-3} mℓ/hr/in. of periphery

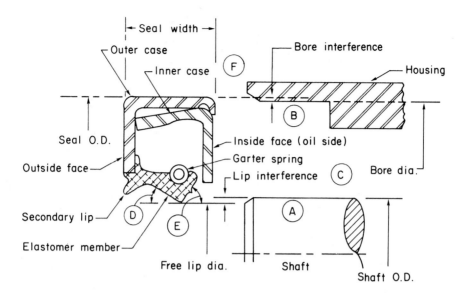

FIGURE 16. Typical bonded, dual lip elastomeric lip seal. [A] Alloy or stainless steel hardened to a minimum Rockwell C-30, C-45 recommended, finished to 0.25 to 0.50 μm (10 to 20 μin.). Finish must *not have machine lead*. Chamfer 15 to 30°, 1.75 to 3 mm (0.060 to 0.118 in.) wide with 0.8 μm (32 μin.) finish. [B] Maximum inside corner radius of 1.2 mm (3/64 in.). Bore depth equals seal width plus 0.4 mm (1/64 in.). Bore finish of 2.54 μm (100 μin.) or better. Chamfer 15 to 30°, 1.75 to 2.25 mm (0.060 to 0.090 in.) wide. [C] Interference varies from 1 mm (0.040 in.) for small shaft diameters, 25 mm (1 in.), to about 3 mm (0.120 in.) for shaft diameters of 127 mm (5 in.) or larger. [D] Barrel angle, usually in range of 20 to 35°. [E] Oil-side angle, generally between 40 to 70°. [F] Optional inner case is sometimes used for additional strength for pressfits and to protect seal lip.

per psi ΔP).[5] Approximately 80% of high quality, carefully installed lip seals, fabricated from suitable elastomers, will leak about 0.002 g/hr or about 1 drop per 8-hr shift in continuous operation. About 20% of such seals will leak 0.002 to 0.1 g/hr or about 1 to 50 drops per 8 hr.

Lip contact load — This most important performance parameter commonly ranges from 0.05 to 0.12 N/mm (0.3 to 0.7 lb/in.) of circumference. Lip load should be as low as possible without leaking. Hydrodynamic type lip seals can operate with lower lip contact loads.

Temperature — Excessive lip temperature is a prime cause of seal failure. Underlip temperature rise between the lip and ambient fluid is a function of the viscosity of the sealed fluid, the product $D^{1/3} \omega^{2/3}$ (D = shaft diameter and ω = rotational speed), and heat transfer from the contact zone.[20] Underlip temperature rise for an unwetted seal is about five to six times greater than that for operation fully submerged in the sealed fluid. Typical sump temperatures for many applications range from 70 to 130°C (158 to 266°F). Conventional lip seals, operating about 50% submerged, experience underlip temperature rises on the order of 10 to 36°C (18 to 65°F). Some newer hydrodynamic seals have underlip temperature rises about 15 to 30% less than conventional seals. High-lip temperatures may degrade the elastomer, increase chemical reaction between the elastomer and sealed fluid, and thermally degrade the sealed fluid (sludge deposits and carbonized abrasive particles). At low temperatures, -30°C (-22°F) or below, some elastomers become hard, brittle, and unable to follow shaft excursions. Leakage results. The higher modulus of elasticity also increases lip loading which causes wear. Extreme temperature problems can generally be solved by heating or cooling the sump, selecting suitable lubricants, and giving special attention to elastomer selection (see Tables 9 and 10).

Pressure — Increasing fluid pressures cause higher lip loads, thinner fluid films, higher underlip temperatures and shorter life. General purpose seals should be selected according

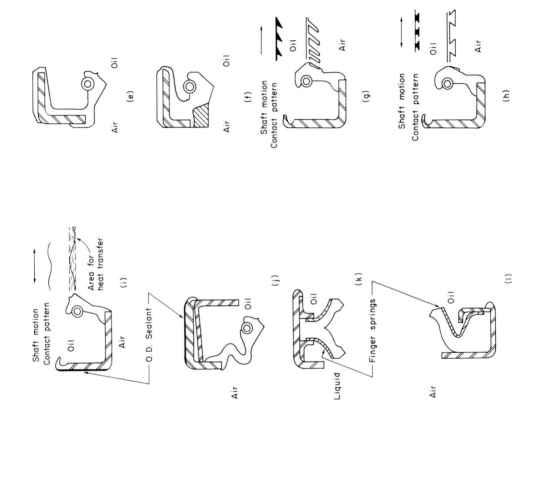

FIGURE 17. Radial lip seal design variations. (a) *Springless, single-lip seal* — economical; used to retain highly viscous materials or to exclude dust in lower speed applications, e.g., 10.2 m/sec (2000 ft/min) or lower. When used as dirt excluder, the seal is installed with the primary lip facing the contaminant. (b) *Single-lip, spring-loaded seal* — this design is the most widely used; economical general purpose seal used to retain lower viscosity fluids at speeds up to about 20 m/sec (3950 ft/min) in numerous industrial and automotive applications. The optional inner case may be incorporated to give additional strength to case and protect lip. (c) *Dual lip, spring-loaded seal with shotgun case* — the nonlubricated secondary seal is used as a dirt excluder; secondary lip is not spring loaded and has a very light contact pressure. Not generally used above 5 m/sec (1000 ft/min) because of frictional heating of sealing element. The heavy shotgun case outer flange positions the seal accurately, prevents seal cocking, and also reduces secondary leakage past the seal-to-bore interface. (d) *Assembled case, Teflon® spring-loaded seal* — the assembled case seal, in contrast to the bonded designs above, incorporates leather or an elastomer which has poor bonding properties, in a multipart assembled seal; more expensive than bonded seals; useful procedure when making small number of special seals. The Teflon® seal is a premium seal for use with sparse lubrication, high speeds, and chemically active fluids; Teflon® may be used as a springless seal. (e) *Single-lip, spring-loaded seal with rubber-covered case* — covered cases are used to prevent corrosion of the metal cases; they provide maximum protection against bore leakage due to deficiencies in the bore surface or to compensate for differential thermal expansion. This design is difficult to fabricate from high viscosity elastomers. (f) *Single-lip, spring-loaded high-pressure seal* — These seals are made with heavier cross-sections and thicker flex sections; some use antiextrusion rings behind the primary lip; such seals can withstand pressures up to 10,500 kPa (1500 psi) at low speeds.[14] (g) *Helix seal* — hydrodynamic lip seal utilizing ribs, set at an angle, molded on the air side of the continuous seal lip. This viscoseal concept applied to the lip seal is in fact a pump and tends to pump leakage back into the sump; design can also pump liquids and solid contaminants into the sump and is unidirectional and has high-pumping efficiency. Radial lip load can be lower than in conventional lip seal design; should not be used with a secondary lip; more expensive than conventional seals.[15] (h) *Triangle seal* — one of a family of bidirectional hydrodynamic seal designs which pump leakage back into sump irrespective of direction of shaft rotation; not as efficient a pump as helix seal but can be used in many applications instead of conventional lip seal; has lower radial lip load; is more expensive than conventional seal.[16] (i) *Wave seal* — a bidirectional hydrodynamic seal having a smooth normal width lip molded in a wave like pattern; amplitude and number of waves are fixed to provide adequate pumping of leakage back into sump. The static contact path is about 0.125- to 0.25-mm (0.005- to 0.010-in.) wide while the swept area for heat transfer is two to four times as wide; low lip pressures, low under-lip temperature rise, and longer life is obtained in comparison with conventional seal.[17] (j) *High-shaft eccentricity seal* — this seal has a special elastomer element design giving added radial freedom of motion. The maximum total eccentricity for this seal design can be as much as 1.5 mm (0.059 in.) and a greater tolerance of shaft misalignment is provided (see Table 12 for normal MTE tolerances); OD sealant is shown on I and J cases. (k) *Dual lip seal with both lips spring-loaded* — this arrangement, shown here with finger springs rather than garter springs, is employed to exclude liquid contaminants from the sealed system. A similar arrangement can be used to keep two different system fluids from mixing. (l) *External seal* — this type of seal is often used when the shaft is stationary and the bore rotates as in oil lubricated automotive wheel applications. The bore surface finish must be as specified for the shaft (see Figure 16) or a wear ring is inserted in the bore.

Table 9
CHEMICAL RESISTANCE OF LIP SEAL ELEMENTS

Fluid medium	Nitrile (BF,BG,BK,CH)[a]	Polyacrylate (DF,DH)	Silicone (FC,FE,GE)	Fluoroelastomer			Fluoroplastic PTFE
				Fluorosilicone (FK)	Fluorocarbon (HK)		
Engine oil	Good	Good	Good	Good	Good		Good
ATF-A	Fair	Good	Good	Good	Good		Good
Grease	Good	Fair	Fair	Fair	Good		Good
EP Lube	Fair-poor	Good	Poor	Fair	Good		Poor
SAE90 (nonadditive)	Good	Good	Good	Good	Good		Good
MIL-L-7808	Fair	Poor	Good	Good	Good		Good
MIL-L-23699	Fair	Poor	Good	Good	Good		Good
MIL-L-6082-A	Good	Good	Good	Good	Good		Good
MIL-L-5606	Good	Good	Poor	Good	Good		Good
MIL-L-2105	Fair	Good	Poor	Fair	Fair		Good
MIL-G-10924	Good	Good	Poor	Good	Good		Good
Fresh or salt water	Good	Poor	Good	Good	Good		Good
Acetic acid	Poor	Poor	Poor	Poor	Fair		Good
Ammonium gas	Good	Poor	Fair	Poor	Poor		Good
Brake fluid	Good	Good	Fair	Good	Good		Good
Butane	Good	Good	Fair	Fair	Good		Good
Freon 12	Good	Poor	Poor	Poor	Fair		Good
Fuel oil	Good	Fair	Poor	Poor	Fair		Good
Kerosene	Good	Fair	Poor	Poor	Fair		Good
Gasoline	Good	Fair	Poor	Poor	Fair		Good
Ketones (MEK)	Poor	Poor	Poor	Poor	Poor		Good
Methyl chloride	Poor	Poor	Poor	Poor	Good		Good
Molybdenum disulfide	Good	Good	Good	Good	Good		Good
Oxygen	Good	Good	Fair	Fair	Good		Good
Perchlorethylene	Fair	Poor	Poor	Poor	Good		Good
Petroleum base hydraulic oil	Good	Good	Good	Good	Good		Good
Phosphate ester	Poor	Poor	Good	Good	Good		Good
Trichlorethylene	Poor	Poor	Poor	Poor	Good		Good

[a] ASTM D2000/SAE J200 type and class designations.

From Dreger, D. R., Ed., *Mach. Design*, 52(8), 1980. With permission.

Table 10

SEAL ELEMENT SELECTION GUIDE[21,22]

Material	Cost factor	Sump temp range °C (°F)	Advantages	Disadvantages
Nitrile (BF, BG, BK, CH)[a]	1	−46 to 107°C (−50 to 225°F)	Low cost; low swell; easily processed; good oil resistance; good low-temperature properties; good abrasion resistance	Poor resistance to EP additives; poor high-temperature resistance
Polyacrylate (DF, DH)	1.2	−40 to 135°C (−40 to 275°F)	Good oil resistance including EP lubricants; low swell; good high-temperature resistance; High-oxidation resistance	Poor water resistance; fair wear resistance; Poor low-temperature characteristics; poor compression set; poor abrasion resistance
Silicone (FC, FE, GE)	1.3	−62 to 149°C (−80 to 300°F)	Wide temperature range; very flexible; easily molded; high-lubricant absorbency; good water resistance	Easily torn or cut; high swell; poor resistance to oxidized oil; poor abrasion resistance; poor dry running properties
Fluorosilicone (FK)	2.0	−62 to 149°C (−80 to 300°F)	Good oil and chemical resistance; good low-temperature properties; wide temperature range; low-compression set	Poor abrasion resistance; poor dry running; poor tear strength; expensive; difficult to process; poor wear resistance
Fluorocarbon (elastomer) (HK)	2.0	−40 to 177°C (−40 to 350°F)	Very good oil and chemical resistance; excellent heat resistance; wide temperature range low swell; good wear properties	Difficult to process; expensive; becomes stiff at low temperatures; poor followability at low temperatures
Fluorocarbon thermoplastic PTFE	3	−96 to 204°C (−140 to 400°F)	Excellent temperature range; excellent oil and chemical resistance; low friction; no swell; good dry running	Difficult to process; limited design options; high cost; easily damaged; nonelastic; becomes stiff at low temperatures; poor followability at low temperatures

a   ASTM D2000/SAE J 200 Type and Class Designations

Table 11
LIP SEAL OPERATING PRESSURE LIMITS[23,24]

Shaft surface speed		Maximum pressure			
ft/min	m/sec	psi		kPa	
0—1000	0—5.1	15	7	103.4	48.3
1000—2000	5.1—10.2	10	5	69	34.5
2000—3600	10.2—18.3	5	3	34.5	20.7

to Reference 23 limits in Table 11, premium seals to Reference 24 limits, and applications involving higher pressures require custom designs. Sumps for operation at ambient pressure must have clear vents to prevent pressure buildup.

Shaft speed — Since underlip temperature rise strongly depends upon shaft surface speed, the speed will often determine other design parameters such as pressure, shaft to bore misalignment, dynamic runout,[24] shaft tolerance,[6] shaft finish, lubricants, and seal design. The upper speed limit is about 20 m/sec (4000 ft/min) but special seal designs have been operated at 30 m/sec (6000 ft/min). High-speed radial lip seals may operate under full hydrodynamic conditions.[19] In high-speed applications, the lower pressure limits in Table 11 Reference 23, should be used. The ideal surface finish is 0.38 μm (15 μ in.) and should be no finer than 0.25 μm (10 μin.) since too smooth finishes inhibit fluid film formation. Shaft out-of-roundness should be less than 5 μm (200 μin.) and the number of lobes or geometric perturbations should be minimized.

Lubrication — The sealed fluid must be thermally and chemically compatible with the sealing material, Tables 9 and 10. Dry running will cause high lip temperatures, rapid wear, and in some cases, stick-slip failure. A more desirable thermal environment is provided with the seal submerged in the lubricant. Applications subject to start-stop cycling with relatively long idle periods should have an oil storage reservoir in contact with the seal to prevent dry start-up.

Installation — Careless installation is a major cause of seal failure. Proper installation involves care in handling the seal, care to be sure the shaft and seal bore are free from nicks, burrs, scratches, and foreign particles, and the seal should be firmly seated with appropriate installation tools. The shaft and bore should be provided with chamfers (Figure 16) and thimbles or sleeves used to protect the sealing lip as it passes over splines, keyways, and shaft apertures. Seal should be square with shaft centerline within 0.25 mm (0.010 in.) TIR.

Materials

Cases are made of cold-rolled low-carbon steel, stainless steel, brass or aluminum. Carbon steel cases may be plated for corrosion protection or coated to provide better elastomer bonding. Aluminum and brass are sometimes used to minimize differential expansion between seal cases and nonferrous housing bores. Finger springs are made from stainless steel. Garter springs are normally made from hard-drawn carbon steel wire. Stainless steel garter springs are recommended for elevated temperatures or corrosive fluids.

The elastomer element must be compatible with the sealed fluid, withstand the high temperatures with minimum degradation, remain elastic at the lowest temperature anticipated, be low in cost, and easy to process into a bonded or assembled seal. Early radial lip seals utilized leather or felt but now employ primarily the five families of materials summarized in Tables 9 and 10. Nitrile, a copolymer of 18 to 50% acrylonitrile and butadiene, is the most widely used. Low nitriles exhibit better low-temperature flexibility and high nitriles have lower swell in petroleum oils and fuels. Carboxylated nitriles are more expensive and

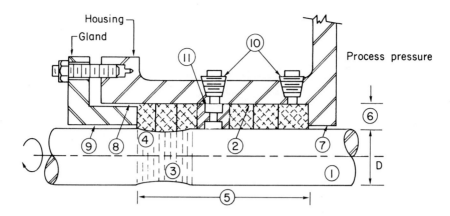

FIGURE 18. Typical pump stuffing box with compression packing. [1] Shaft finish = 0.25 to 0.50 μm (10 to 20 μin.) CLA; shaft hardness = Rockwell C-50; shaft runout should not exceed 0.025 mm (0.001 in.) TIR. [2] Bore finish = 1 to 1.5 μm (40 to 60 μin.) CLA. [3] Rings nearest gland are deformed most; approximately 70% of wear under first 30% of packing. [4] Harder end rings are sometimes used at gland and at throat. [5] Packing length ≈ 1.5 D. [6] Packing radial thickness ≈ 0.15 to 0.3 D. [7] Throat clearance 0.2 to 0.4 mm (0.008 to 0.015 in.); 0.8 mm maximum. [8] Gland-to-bore clearance 0.125 to 0.25 mm (0.005 to 0.010 in.). [9] Gland-to-shaft clearance 0.4 to 0.8 mm (0.015 to 0.030 in.). [10] Tap locations for lantern gland inlet. [11] Lantern ring.

have better high temperature wear resistance with a sacrifice in low temperature flexibility. PTFE, a thermoplastic rather than an elastomer, has a wide temperature range and is resistant to almost all fluids. It is difficult to process and is usually employed as assembled seals. Butyl, epichlorhydrin, an ethylene-propylene terpolymer (EPDM) are used in special purpose seals.

Packing Seals

Mechanical shaft packings include compression packing, automatic or lip packing, and squeeze packing. Compression packings are a pliable material compressed between the throat and gland of a stuffing box for reciprocating, oscillating, and rotating applications. Leakage in dynamic applications is usually on the order of 50 to 500 mℓ/hr, but may be essentially zero in semistatic valve stem applications. Automatic packings utilize a flexible lip energized by the contained fluid pressure. Employed primarily for reciprocating applications, heat dissipation problems restrict rare rotating applications to speeds below 1 m/sec (200 ft/min).

Squeeze packings utilize precision-molded elastomer rings, such as the O-ring, installed in precisely machined grooves (glands) on cylinders, pistons, or rods in hydraulic or pneumatic devices.[25,26] Squeeze packings are most frequently used in reciprocating service or in low-speed oscillating applications such as valve stems. Rotary applications are recommended only under well-lubricated low speed conditions, 1.75 to 4 m/sec (350 to 800 ft/min). None of these packing devices are bearings. Side loads due to out-of-round parts, warped shafts, or poor bearing supports will cause rapid wear and inadequate sealing.

Compression Packing

The soft packing, jamb packing, or compression packing, Figure 18, is the most common fluid seal. It consists of a number of deformable packing rings or a long rope-like material spiral wrapped around the shaft or rod, compressed by the gland to seal against the housing bore and shaft. Leakage on the order of 0.01 mℓ/hr/m/kPa (0.0018 mℓ/hr-in.-psi) is necessary to lubricate and cool the packing. Leakage from a compression packing will be approximately 5 to 100 times that from a mechanical face seal under the same service conditions and friction loss will be about three times greater. Compression packing has the advantage of being replaceable without disassembly of equipment and a gradual leakage increase usually

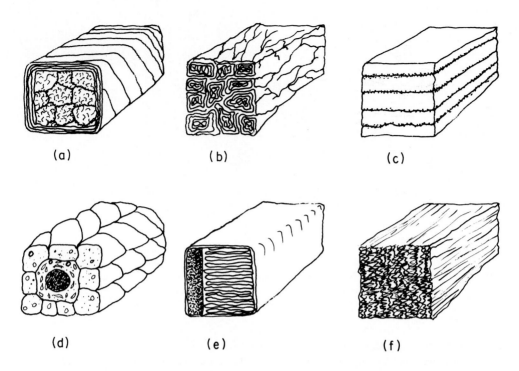

FIGURE 19. Typical soft packing and commonly used materials: (a) spiral-wrapped metal foil over reinforced braided asbestos core; (b) crumpled metal foil, graphited; (c) cotton duck laminated with synthetic rubber; (d) lead wire reinforced flax braid over synthetic rubber core; (e) folded and wrapped asbestos fabric, soft rubber core at housing bore; and (f) graphite foil wound around shaft and then compressed.

COMMONLY USED MATERIALS

Fibers and fabrics	Metal wire and foil	Fluid lubricants	Dry lubricants
Asbestos	Aluminum	Beeswax	Graphite
Cotton	Copper	Castor oil	Mica
Flax	Inconel	Fluorolubes	Molybdenum disulfide
Graphite	Iron	Glycerol	PTFE
Jute	Lead	Grease	Talc
Metal	Monel	Linseed oil	Tungsten disulfide
Plastic	Nickel	Oil	
Synthetic rubber	Stainless steel	Paraffin	
	Zinc	Silicone	
		Soap	
		Waxes	

provides adequate warning of impending failure. While initial cost of compression packings is lower, their periodic maintenance and adjustment for wear and loss of packing volume frequently swing total cost in favor of mechanical seals.

Compression packings are used extensively in rotary applications such as pumps up to about 15 m/sec for pressures up to 1000 kPa (145 psi) and valve stems under semistatic conditions up to 34,500 kPa (5000 psi). Compression packing are sometimes used for sealing reciprocating shafts but they have the disadvantage of high friction.

Figure 19 shows representative designs and the most frequently used materials. Representative packings, lubricants, temperature limits, and applications are shown in Table 12. Soft packing, usually square cross section rings or long continuous pieces which can be

Table 12
REPRESENTATIVE COMPRESSION PACKINGS AND APPLICATIONS

Base material	Lubricant	Approximate upper temperature limit °C	°F	Cold water or sludge	Hot water	Low-pressure steam	High-pressure steam	Cold gasoline and oil	Hot gasoline and oil	Air and gas	Acids and alkali
Cotton, jute, flax, plaited or braided	Graphite, mica, or talc + oil or grease (35—55%)	66	150	X							
Laminated cotton duck and elastomer (tucks)	Graphite and/or oil	121	250	X	X						
PTFE braided	PTFE dispersion	232	450	X	X	X		X		X	X
Asbestos, braided	PTFE fibers and dispersion	260	500	X	X	X	X	X	X	X	X
Polyamide, braided	PTFE dispersion	288	550	X	X					X	
Asbestos, braided	Graphite with oil and/or grease (30—50%)	302	575	X	X	X	X	X	X	X	X
Graphite yarn or foil	None	(316)[a] 538	(600)[a] 1000	X	X	X	X	X	X	X	X
Asbestos, braided with wire reinforcement	Graphite (3—7%)	788	1450			X				X	X
Metal foil, lead, spiral wound aluminum	Graphite and oil	177 427	350 800				X	X	X	X	

[a] Oxidizing environment.

field fabricated, is generally composed of fiber roving, yarn, or filament braids. Impregnating solid and/or fluid lubricants may constitute 3 to 60% by weight of packing. Metal wire or elastomeric filaments are common reinforcements. Metal wire mesh, spiral wrapped metal foil, or crumpled metal foil calendered into square cross sections are frequently used for severe service with oil and graphite lubrication. Pure nickel foil, for example, is one of the few packings which can effectively seal caustic alkalis. Zinc is primarily used as a sacrificial anode to prevent galvanic corrosion when using graphite packing with some stainless steels. Flexible graphite in filament, yarn, fabric, or foil form provides high thermal conductivity for friction heat dissipation. This permits cool operation even at reduced leakage rates and graphite packings frequently exhibit leakage performance similar to that with mechanical seals.

About 75% of shaft packing problems arise from improper packing selection and poor installation. The greatest packing deformation and most severe shaft wear occur near the outboard end of the stuffing box. Localized wear can be minimized by cutting the packing rings on a mandrel the same diameter as the shaft, tamping each ring fully as it is installed, and spacing ring joints at 120°. Allow ample initial leakage and compress the packing in a careful step-wise procedure over the break-in period of 8 to 100 hr. During break-in, some 15 to 30% of the lubricant content of the packing may be lost.

For compression packings, it is best to use die-formed rings which may be purchased as a set or prefabricated by the user in a mold of correct dimensions. These rings minimize gland take-up during break-in, enhance extrusion resistance, reduce the break-in period, tend to exclude abrasives, and allow sealing at higher pressures. The ring OD may be slightly oversize to provide good housing bore fit. A typical packing set may use very dense "anti-extrusion" rings at the throat and gland with intermediate rings graded from soft near the throat to hard near the gland.[27]

A lantern ring, Figure 18, is frequently used in compression packings for rotary applications, especially at high pressures and temperatures. The lantern ring has an H cross section and is made of rigid material such as brass, aluminum, stainless steel, or PTFE. The ring is adjacent to openings in the stuffing box wall for injecting coolants or lubricants, and a discharge can be provided on the opposite side of the housing. The lantern ring can also be used to (1) introduce fluid from pump discharge when pump suction is subatmospheric to prevent air leaking in, and (2) introduce a clean external buffer liquid to seal against abrasives, slurries, toxic liquids, and gases. The buffer fluid pressure should be about 20 to 70 kPa (3 to 10 psi) above the pump suction. The lantern ring is usually located about midway in the packing set but its exact location may be dictated by suction pressure, lubricant viscosity, or buffer fluid pressure.

Automatic Packing

Pressure-energized lip-type automatic packings, the most widely used seal in the high pressure hydraulic and pneumatic field, are generally installed with a very small interference. Contact force and area increase with fluid pressure, improving the seal. Used almost exclusively for reciprocating applications, contact force, area, and friction on an unpressurized return stroke are lower than on the pressure stroke and produce a "breathing" action that helps lubricate the seals. The friction of automatic packing is approximately proportional to pressure up to about 7000 kPa (1015 psi). Above this, the rate of friction increase with pressure decreases and becomes quite small at about 14,000 kPa (2030 psi).[28] Automatic packings are depicted in Figure 20 in order of increasing pressure limits. They are available in a wide variety of homogeneous elastomers or fabric-reinforced compositions.

Cup and flange packing — These are the simplest designs, require a minimum of space, and are easily installed (Figure 21). The flange packing OD and cup packing ID are sealed by mechanical compression, which limits maximum operating pressure to approximately 3500 kPa (500 psi). Excessive tightening of the inside follower tends to crush and extrude the cup packing against the cylinder wall, which causes high friction, wear, and reduced sealing effectiveness. Similar crushing of the flange packing may result from gland over-tightening. Cup and flange packings are less effective seals than U- or V-rings but are frequently used because of space limitations. Leather continues to be much used for flange packing along with various synthetic rubbers, PTFE, nylon, and other plastics. Fabric-reinforced elastomers greatly reduce problems with mechanical clamping.

U-ring packing — These low-friction packings of leather, elastomer, or fabric-reinforced elastomer are used singly in continuous (nonsplit) rings. They are infrequently used in tandem. U-rings are chiefly employed as piston seals but can be arranged in glands. In double-acting piston seals, the U-ring must be used heel-to-heel. A lip-to-lip arrangement will create a pressure trap and cause rapid seal wear and failure. Homogeneous U-rings in Shore A hardness of 70 can be used up to about 10,000 kPa (1450 psi) in precision machined parts. Maximum radial clearance should be about 0.075 mm (0.003 in.). For higher pressures or for applications with excess clearance, harder U-rings up to Shore A of 90 and/or fabric-reinforced rings should be used. U-rings with metal-reinforced bases have been used up to 35,000 kPa (5100 psi). Some proprietary U-ring designs having long thick-walled static sealing lips can be installed with enough interference to make pedestal rings unnecessary.

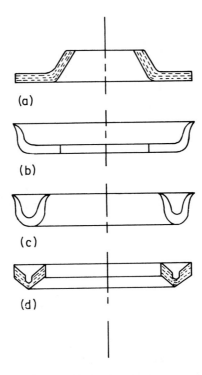

FIGURE 20. Principal automatic packing types. (a) Flange or ''hat'' packing — unbalanced seal for rod seals up to about 3500 kPa (500 psi). flange compression can cause reduced sealing performance; seals on rod OD. (b) Cup packing — for large-displacement hydraulic and pneumatic service up to about 3500 kPa (500 psi); unbalanced seal affected by bottom compression; seals on cylinder ID. (c) U-ring packing — balanced seal, used singly up to about 10,000 kPa (1450 psi); mainly used on pistons but can be used in glands; not used in split ring form. (c) V-ring packing — balanced seal used in stacked sets with number of rings depending on pressure; can be used as split rings; usually installed in glands but can be used in pistons; pressure limit is approximately 32,000 kPa (4600 psi) per ring.

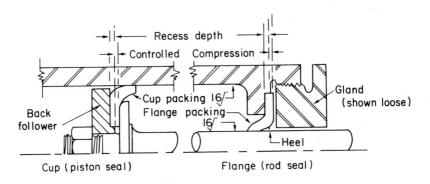

FIGURE 21. Cup and flange packing.

V-ring packing — These are the most widely used automatic packing since they cover a wide pressure range. Rings are usually assembled in a stacked set between male and female support rings. The stack is held together by a spring on the pressure side, Figure 22, or by shims between the outer support ring and the gland follower. Springs, when used, generally provide a load of about 0.8 to 2.5 N/mm (4.6 to 14.3 lb/in.) of mean V-ring circumference. Since V-rings are generally installed in multiple, split rings may be used with a diagonal

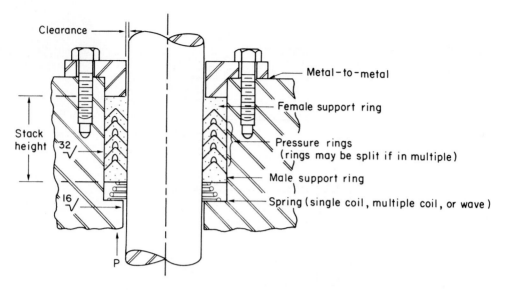

FIGURE 22. V-ring automatic gland seal.

cut, the joints spaced at 120°, to simplify replacement without machine disassembly. V-rings are available in leather, homogeneous elastomers, fabric-reinforced elastomers, and PTFE. Split rings are usually fabric reinforced. Homogeneous rings are used up to about 20,000 kPa (2900 psi). At pressures around 35,000 kPa (5100 psi), homogeneous rings can be mixed with leather or PTFE rings, or a combination of different hardness rings can be used with softer, more leak-tight rings placed nearest the high pressure. At pressures above 45,000 kPa (6500 psi), endless fabric-reinforced elastomer or PTFE rings are common, and thin metal separators frequently support each pressure ring. V-rings can be used as piston seals but are more commonly used in rod seal glands. V-rings can be designed to withstand almost 45,000 kPa (6500 psi) per ring, but this practice results in poor seal life. Three rings are usually the fewest employed even at modest pressures. At 35,000 kPa (5100 psi), a typical packing set would have five or six rings. The male and female support rings are usually made from the same material as the pressure rings when used at low pressures, less than 20,000 kPa (2900 psi). For higher pressures, support rings are available in PTFE, rockhard duck and rubber, metal and phenolic.

Installation — Industry standardization is greater for automatic packing than for any other seal type. Many failures result from a disregard of design and dimensional information provided by the packing manufacturer. A problem common to lip-type automatic packings is extrusion due to high pressure and excess clearance. Metal surfaces in sliding contact with automatic packing should be finished to 0.2 to 0.4 μm (8 to 16 μin.). Finish should not be smoother than about 0.13 μm (5 μin.) because slight roughness helps retain lubricant. The static surface in contact with the packing should be finished to 0.8 μm (32 μin.).

Squeeze Packing

Squeeze packings are made in several shapes, in a large number of standardized sizes,[25] and from over a dozen elastomers with hardness ranging from 10 to 100 Shore A.[21] These seals, Figure 23, are low in cost, require minimum space, are easy to install, require no adjustment, seal in both directions, have low friction, can be used as piston or gland seals, can be selected for compatibility with a wide range of fluids, and are readily available for industrial, aerospace, and military applications. Squeeze rings, though simple in form, are made with closely held diametral and cross section tolerances. To ensure long life and effective sealing, recommended groove dimensions, surface finishes, and diametral clearances must be carefully followed.

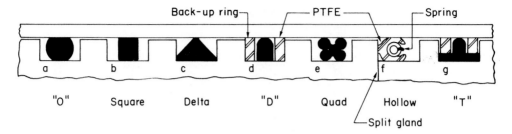

FIGURE 23.   Typical squeeze-type packing rings.

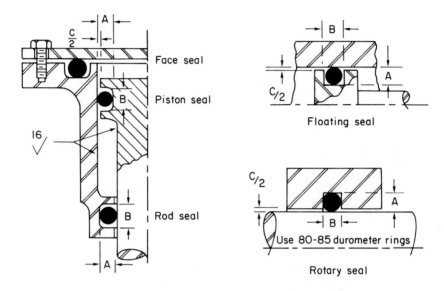

FIGURE 24.   Typical O-ring seal arrangements.

The most widely used and highly standardized squeeze packing is the O-ring. It sometimes undergoes spiral failure in reciprocating applications initiated by part of the ring rolling while the remainder slides against a cylinder or shaft. The T, delta, and D-ring are less prone to rolling, but all ring shapes are subject to torsional or spiral failure. Slight rolling of an O-ring due to pressure or direction reversal reduces break-out friction below that of a sliding ring such as the square, delta, or D-ring. These have higher friction than the O-ring, and the square ring is rarely used for dynamic applications. The quad and T-rings are usually installed with low squeeze and thus low friction. The T-ring is capable of high pressures, and quad rings can be used in rotary seals at some 30% higher speed than O-rings.

O-rings can be used in four basic dynamic seal arrangements (Figure 24) or as static face and gland seals. Groove dimensions for each arrangement are different,[26] and they also differ by industrial standards which permit 1 to 5% circumferential stretch in reciprocating applications or by military specification MIL-G-5514F which limits stretch to 2%. Gland detail is shown on Figure 25.

Gland design recommendations for O-rings in radial static, reciprocating, floating piston rings, and rotary seals are given in Reference 29. Gland depth for a static seal is about 70 to 80% of the O-ring cross-section dimension, giving 15 to 32% squeeze. For reciprocating seals, gland depth, A, is about 80 to 90% of the cross-section dimension, giving 8 to 25% squeeze. The floating reciprocating seal, usually for pneumatic applications, uses a bore

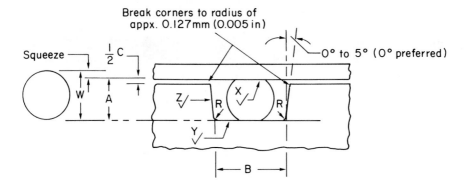

FIGURE 25.   O-ring dynamic seal gland detail. Surface finishes: X = 0.254 to 0.508 μm (10 to 20 μin.) CLA; *NOTE*: do not use less than 0.127 μm (5 μin.); Y = 0.8 μm (32 μin.) CLA; Z = 0.8 μm (32 μin.) CLA without backup rings, 1.6 μm when used with backup; and B = groove shown for no backup ring. If ring is employed use supplier's recommendation for B.

diameter slightly smaller than the O-ring OD and the groove diameter is slightly smaller than the O-ring ID. With changes in pressure and direction, a momentary leak occurs as the ring moves from one side of the groove to the other. Since this design is primarily for low-pressure pneumatic service, about 1380 kPa (200 psi), this slight leakage is generally acceptable. This arrangement can also be used in low-pressure liquid service if a few drops of leakage per cycle can be tolerated.

Dynamic O-ring seals are used primarily for well-lubricated reciprocating service. With proper design, however, they can be employed in low-speed rotary service at pressures up to about 5500 kPa (800 psi). The gland for rotary applications compresses the O-ring about 5% circumferentially. Its depth is only slightly less than the O-ring cross-section, so there is little radial squeeze. Rotary seals are not put in tension around the shaft because most elastomers if heated by friction while under tensile stress will contract. This contraction, the Gow-Joule effect, causes further contact load, increased friction and temperature, and rapid failure. O-rings and other squeeze packings are made from a large number of elastomers in hardnesses from about 55 to 90 Shore A. A standard O-ring with a hardness of 60 will seal pressures in dynamic applications to about 1750 kPa (250 psi) and about 10,500 kPa (1500 psi) with a 90 hardness. Higher pressures, up to about 20,700 kPa (3000 psi), require backup rings to prevent ring extrusion. T-ring shape can be used up to about 138,000 kPa (20,000 psi). Table 13 gives some characteristics of the most widely used elastomers.

CONTROLLED CLEARANCE SEALS

Hydrodynamic Seals

While mechanical face seals often function with separation of the sealing surfaces because of static or dynamic pressure forces,[30] controlled close clearance seals provides a definite sealing surface separation during normal operation. The hydrodynamic seal shown in Figure 26 was designed for gas, but hydrodynamic seals can also be used for liquids. Essentially, the sealing ring interface is an ordinary mechanical face seal with a fluid film bearing geometry added to give positive separation of the surfaces. The self-acting lift pads have pockets about 10 to 25 μm (0.0005 to 0.001 mℓ) deep and pocket-to-land width ratios in the circumferential direction of about 2:1. Axial and radial grooves keep pressure the same around each pad. During seat rotation, high-pressure gas is dragged into the pad and compressed as it passes over the step at the end of the pad. This creates lift forces that separate the primary seal ring and rotating seat.

Table 13
CHARACTERISTICS OF COMMON SQUEEZE SEAL MATERIALS

Elastomer	Tensile strength (MPa)	Compression set resistance	Abrasion resistance	Oxidation resistance	Resiliency — hot	Resiliency — cold	Fuels and lubricants petroleum based	Acid resistance	Alkali resistance	Water and steam resistance	
Nitrile	GE 20.7	GE	G	FG	G	FG	FE	FG	FG	FE	
Ethylene propylene	GE 20.7	FE	G	E	G	FG	P	GE	GE	E	Phosphate ester base fluids
Fluorocarbon	G 17.2	FE	FG	E	G	F	E	G	GE	FG	
Neoprene	GE 20.7	FG	G	GE	GE	G	PF	G	GE	G	High aniline oils
Silicone	P 10.3	GE	P	E	G	G	PF	FG	FG	FG	
Polyurethane	E 31.0	FG	E	G	F	F	FG	PF	PF	PF	
Buna S	GE 20.7	FG	G	FG	G	G	P	FG	G	FG	Brake fluids
Polyacrylate	F 13.8	G	FG	GE	G	P	GE	P	P	P	
Fluorosilicone	P 8.3	G	P	E	G	FG	G	P	P	G	

Note: E, G, F, and P = excellent, good, fair, and poor.

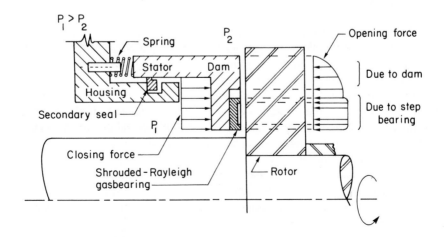

FIGURE 26. Self-acting hydrodynamic seal with Rayleigh step bearing.

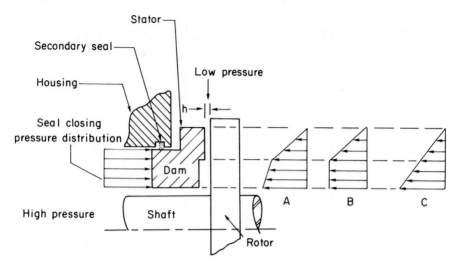

FIGURE 27.   Self-activated hydrostatic face seal. A = seal opening pressure distribution at equilibrium h, B at small h, C at large h.

The pressure drop and leakage occur across the sealing dam of the sealing ring. The fluid film bearing also contributes high film stiffness such that the seal ring can dynamically track seal seat motion. This is especially important in high-speed applications where runout could not otherwise be tolerated. A spiral groove pattern can be applied on the seal face to operate in a manner similar to the lift pads.[31] With a wide radial face, pumping action of the spiral grooves can result in zero net leakage under ideal conditions.

Hydrostatic Seals

There are two kinds of hydrostatic close clearance seals: self activated and externally pressurized. Figure 27 shows a self-activated hydrostatic seal with a shallow radial step approximately at midface. In case A (normal design separation), the hydrostatic separating force is due to one pressure drop across the recess and another across the sealing dam. This is in equilibrium with the seal closing (hydrostatic pressure) force as shown. If face separation decreases or increases, a restoring force develops due to the change in pressure profile as shown in B and C. Similar performance and stability can be achieved with a gradually converging face separation in the direction of leakage. Alternatively, a midface pocket in a flat-faced seal can be connected to the high-pressure side through an additional channel offering resistance to flow. With appropriate geometries, pressure profiles are similar to those in Figure 27. Instability problems sometimes occur with gases when operating with relatively large face separation and high leakage. Generally, these seals are used in high pressure differential applications. Rotation usually has a negligible effect in these cases (rotational speed is too low and separation is too high for significant hydrodynamic effects).

An externally pressurized hydrostatic seal is shown in Figure 28. Under all conditions of operation, the buffer pressure must be higher than the sealed pressure. The buffer fluid overpressure may be relatively low, 15 to 35 kPa (2.5 to 5 psi), and is usually dictated by the control system employed. Where abrasives are present in the sealed fluid, the buffer fluid flushes abrasives away from the sealing interface. This principle is also used for sealing toxic fluids. If the buffer fluid is not compatible with the sealed fluid, a more complex seal system is required.

Hydrodynamic and hydrostatic concepts are combined in a hybrid seal in Figure 29. At zero and low pressures, hydrodynamic pumping allows operating without face contact. Although the seal gap does increase with speed, the increase is moderate throughout a large

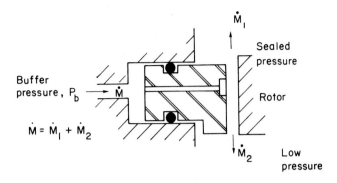

FIGURE 28.   Externally pressurized hydrostatic face seal. (From Muller, H. K., Hydrodynamic and Hydrostatic Face Seals, ASLE Seals Education Course, Houston, Tex., May 1972. With permission.)

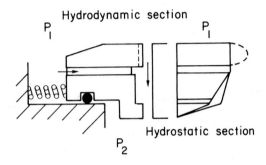

FIGURE 29.   Hybrid face seal with hydrodynamic and hydrostatic elements. (From Muller, H. K., Hydrodynamic and Hydrostatic Face Seals, ASLE Seals Education Course, Houston, Tex., May 1972. With permission.)

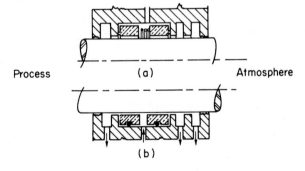

FIGURE 30.   Floating buffered bushing seal. (From Stair, W. K., Liquid buffered bushing seals for large gas circulators, Paper C5, presented at 1st Int. Conf. Fluid Sealing, BHRA, Fluid Engineering, Cranfield, Bedford, England, April 1961.)

speed range. At higher pressures, hydrostatic action stabilizes the gap. Hydrostatic and hydrodynamic face seals are not off-the-shelf units. Each requires special treatment wherein application constraints and machine characteristics must be incorporated in the overall design.[32]

Floating Bushing Seal

The floating bushing seal shown in Figure 30 can accommodate large radial shaft move-

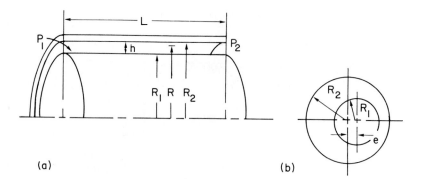

FIGURE 31. Flow model for bushing seal.

ments and still behave as a close clearance seal. Multiple short rings can be staged for better sealing and to accommodate shaft misalignment. In high-temperature applications, thermal expansion of the bushing must match that of the shaft.

The basic mass flow equations for incompressible constant area parallel flow[34] are

Laminar

$$\dot{M} = \frac{\rho h^3 W (P_1 - P_2)}{12\mu L} \tag{11}$$

Turbulent

$$\dot{M} = \frac{4.71 \rho^{4/7} h^{12/7} W (P_1 - P_2)^{4/7}}{L^{4/7} \mu^{1/7}} \tag{12}$$

The flow model for a bushing seal is shown in Figure 31. Since flow path width is W = $2\pi R$, laminar concentric annular flow between the cylindrical surfaces is

$$\dot{M} = \frac{\pi \rho R H^3 (P_1 - P_2)}{6\mu L} \tag{13}$$

For an eccentric annular film, film thickness h = $h_m (1 + \epsilon \cos \Theta)$, where Θ is reckoned from the position at which h = $h_{minimum}$, and ϵ = e/h_m. Equation 13 for laminar flow becomes:

$$\dot{M} = \frac{\pi \rho R h_m^3 (P_1 - P_2)}{6\mu L} (1 + 1.5\epsilon^2) \tag{14}$$

When the annulus is fully eccentric, ϵ = 1 and the factor $(1 + 1.5 \epsilon^2)$ becomes 2.5. Substituting $2\pi R$ for W in Equation 12 for *turbulent* concentric flow:

$$\dot{M} = \frac{9.42 \rho^{4/7} h^{12/7} \pi R (P_1 - P_2)^{4/7}}{L^{4/7} \mu^{1/7}} \tag{15}$$

The fully eccentric correction factor for full turbulence is 1.315, where $\dot{M}$ = mass velocity, L = length of flow path, W = width of flow path, h = film thickness, h_m = mean film thickness, P = pressure, R = radius, e = eccentricity, μ = absolute viscosity, and ρ = fluid density.

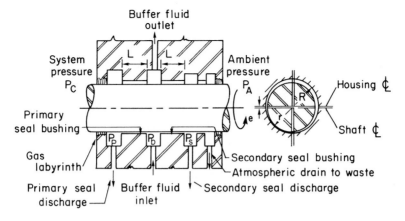

FIGURE 32.   Simple buffered bushing seal. (From Stair, W. K., Liquid buffered bushing seals for large gas circulators, Paper C5, presented at 1st Int. Conf. Fluid Sealing, BHRA, Fluid Engineering, Cranfield, Bedford, England, April 1961.)

FIXED-GEOMETRY CLEARANCE SEALS

Buffered Bushing Seal

Bushing seals depend on small clearances between relatively moving surfaces and are commonly used to limit leakage of liquids. They are frequently used as shown in Figure 32 with process fluid leakage being prevented by a reverse leak of buffer fluid. To minimize ingress of buffer fluid, the primary bushing pressure differential, $(p_b - P_p)$, should be small. On the other hand, a process gas may leak *against* a small primary bushing pressure gradient. While the buffered seal arrangement generally requires an extensive system of piping, pumps, heat exchangers, separators, and controls, the seal has much potential for large systems, particularly those containing hazardous fluids.

Labyrinth Seal

Labyrinth seals, which comprise a series of flow restrictions as shown in Figure 33, capitalize on entrance and exit losses and turbulence to minimize leakage flow. Their effectiveness is highly dependent on the annular clearance between the rotating shaft and stationary housing. The labyrinth seal has a long history and is widely used to minimize steam or gas leakage when direct contact and wear between sealing members is not feasible. Leakage rates are relatively high compared to other seal types.

Analysis of the labyrinth seal has generally considered the labyrinth as an orifice,[35] or as turbulent pipe flow. The actual process lies somewhere between. Using the former approach, Egli[36] derived the leakage equation and curves in Figure 34, where A = leakage area, α = contraction factor, ϕ = flow function, γ = carryover factor, $\dot{M}$ = mass velocity, ρ_1 = entrance fluid density, and p_1 = entrance fluid pressure.

SEALS USING SPECIALIZED CONTROL OF FLUID

Freeze Seal

Freeze seals have been used primarily by the nuclear industry as stem seals for valves handling liquid sodium, potassium, and lead (Figure 35). Basically, liquid metal solidifies in the annulus around the shaft and acts as the seal. In operation, frictional or other heat causes a thin fluid film to develop between mating parts. Properly designed, the freeze seal will have a starting torque no greater than a typical packing seal and lower running power. A typical gap is 0.76 mm (30 mil): small enough to prevent extrusion of a solid sodium

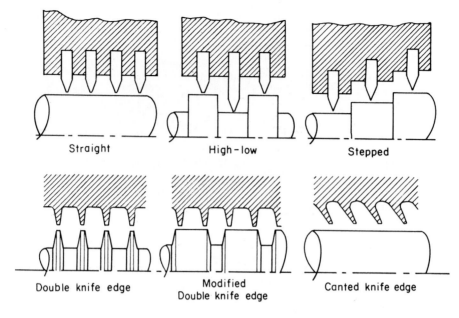

FIGURE 33. Typical labyrinth seal teeth arrangements.

plug up to a pressure differential of 103 kPa (15 psi), and large enough to prevent formation of a strong bond between the shaft and the frozen material. This type of seal has been used to temperatures of about 650°C (1200°F). Among disadvantages, the seal will leak if the auxiliary coolant system fails. High wear occurs if abrasives are present or the fluid precipitates.

Ferromagnetic Seal

A recent development of considerable interest is the ferromagnetic seal.[37] Its elements are knife edges on the shaft and magnets and pole blocks located on the stationary housing (Figure 36). A magnetic circuit through the seal holds magnetic fluid in place to prevent leakage of process fluid. The magnetic fluid is usually a colloidal suspension of ferrite particles dispersed in a carrier fluid. A number of carrier fluids can be used and their usual limitations are vapor pressure, fluid degradation at high shear rates, and magnetic saturation temperature. This type is popular as a rotary vacuum seal.

Slinger Seal

Inertia forces control fluid leakage in this seal comprised of a rotating disk enclosed in a confined housing (Figure 37). Centripetal acceleration of liquid carried around by the disk generates a pressure. Increased gas pressure on one side of the disk forces the interface up and a liquid seal results that operates analogously to a manometer. Although used in a variety of specialized applications, this seal is limited by frictional heating, cavitation erosion, a maximum pressure differential, and instability of the interface between the liquid and gas.

Viscoseal

Viscoseal operation depends on the pressure developed by viscous drag in a thin annulus or slit with grooves on a rotating shaft or plate (Figure 38). The viscoseal offers reliability, long life, and very low leakage. Under laminar conditions, seal performance can be readily predicted and the geometry optimized. Sealing performance improves markedly under turbulent conditions, although optimum laminar viscoseal geometry has been found not to be optimum for turbulent conditions.[38]

In turbulent operation of the viscoseal, ambient gas may be transported against the seal

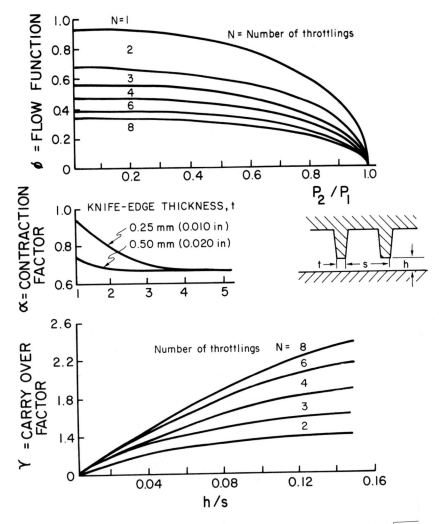

FIGURE 34.   Seal flow factors in Egli labyrinth seal leakage equation: $M = A\alpha\phi\gamma\sqrt{gp1\rho1}$.

pressure gradient and cause gas ingestion into the sealed system. Related problems with "seal breaks" and "secondary leaks" have been reported to be strongly affected by surface tension of the sealed fluid.[39] Viscoseals have been used primarily for liquids. As an example of recent applications, significant improvement in automotive sealing has been achieved by use of the viscoseal principle in modified elastomeric lip seals (Figure 17g).

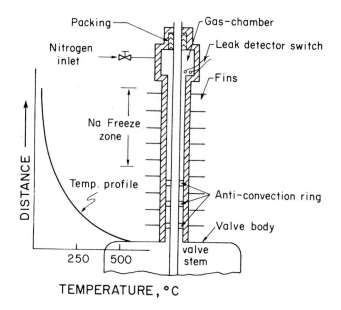

FIGURE 35.   Schematic of a freeze seal.

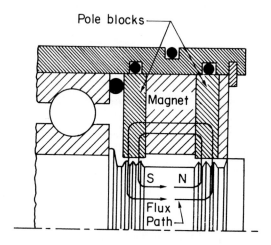

FIGURE 36.   Ferromagnetic fluid shaft seal. (From Moskowitz, R., *ASLE Trans.*, 18(2), 135, 1975. With permission.)

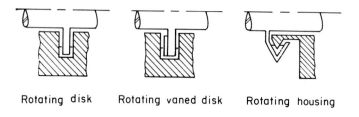

FIGURE 37.   Typical slinger seal configurations.

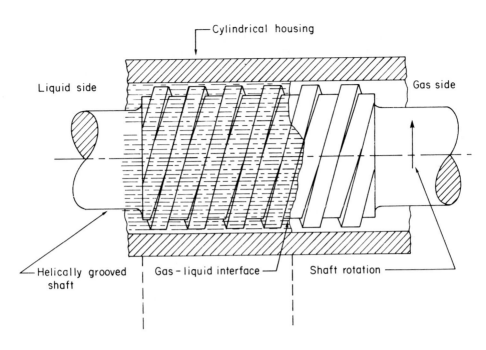

FIGURE 38. Viscoseal.

REFERENCES

1. **Bernd, L. H.,** Survey of the theory of mechanical seals. I. Characteristics of seals, *Lubr. Eng.,* 24(10), 479, 1968.
2. **API,** *Centrifugal Pumps for General Refining Services,* API Standard 610, 5th ed., American Petroleum Institute, Washington, D.C., March 1971.
3. **Ludwig, L. P. and Greiner, H. F.,** Designing mechanical face seals for improved performance. I. Basic configurations, *Mech. Eng.,* 100(11), 38, 1978.
4. **Anon.,** *Guide to Modern Mechanical Sealing,* 6th ed., Durametallic Corporation, Kalamazoo, Mich., 1971.
5. **Austin, R. M., Nau, B. S., Guy, N., and Reddy, D.,** *The Seal Users Handbook,* 2nd ed., BHRA Fluid Engineering, Cranfield, Bedford, England, 1979.
6. **Stevens, J. B.,** Face seals — metal bellows types, *Mach. Design,* 41(14), 32, 1969.
7. **Stair, W. K. and Ludwig, L. P.,** Energy conservation through sealing technology, *Lubr. Eng.,* 34(11), 618, 1978.
8. **Schoenherr, K.,** *Materials in End-Face Mechanical Seals,* No. 63-WA-254, American Society of Mechanical Engineers, New York, 1963, preprint.
9. **Lymer, A. and Greenshield, A. L.,** Thermal aspects of mechanical seals, *Pumps,* 24(7), 209, 1968.
10. **Anon.,** *Dynamic Sealing — Theory and Practice,* Koppers Company, Inc., Baltimore, Md., 1958.
11. **Anon.,** *Engineer's Handbook of Piston Rings, Seal Rings, Mechanical Shaft Seals,* 8th ed., Koppers Company, Inc., Baltimore, Md., 1968.
12. **Stein, P. C.,** Runners for circumferential seals — requirements and performance, *Lubr. Eng.,* 36(8), 475, 1980.
13. **Ruthenberg, M. L.,** Mating materials and environmental combinations for specific contact and clearance type seals, *Lubr. Eng.,* 29(2), 58, 1973.
14. **Wheelock, E. A.,** High pressure radial lip seals for rotary and recriprocating applications, *Lubr. Eng.,* 37(6), 332, 1981.
15. **Weinand, L. H.,** Helixseal — a practical hydrodynamic radial lip seal, *ASME Trans. J. Lubr. Technol.,* 90(2), 433, 1968.
16. **Taylor, E. D.,** Birotational seal designs, *Lubr. Eng.,* 29(10), 454, 1973.
17. **Horve, L. A.,** Reducing Operating Temperatures of Elastomeric Sealing Lips, SAE Int. Automotive Eng. Congr., SAE Paper No. 730050, January 8 to 12, 1973.

18. **Brink, R. V.,** The working life of a seal, *Lubr. Eng.,* 26(10), 375, 1970.
19. **Schnurle, F. and Upper, G.,** *Influence of Hydrodynamics on the Performance of Radial Lip Seals,* No. 73AM-9B-2, American Society of Lubrication Engineers, Washington, D.C., 1973, preprint.
20. **Upper, G.,** Temperature of sealing lips, Proc. 4th Int. Conf. Fluid Sealing, No. 8, May 5 to 9, 1969, preprint.
21. **Dreger, D. R., Ed.,** Materials reference issue. III and IV, *Mach. Design,* 52(8), 1980.
22. **Ostmo, O.,** How to select shaft seal materials, *Lubr. Eng.,*29(6), 240, 1973.
23. **Seneczko, M., Ed.,** Mechanical drives reference issue. III, *Mach. Design,* 52(14), 1980.
24. **Jackowski, R. A.,** Elastomeric lip seals, Proc. DOE/ASME/ASLE Seals Education Workshop, Session 9, Atlanta, Ga., October 8 to 10, 1979.
25. **SAE,** *Standard O-Ring Sizes,* Aerospace Standards AS 568, Society of Automotive Engineers, Warrendale, Pa.
26. **SAE,** *Gland Design,* Aerospace Recommended Practices ARP 1231; ARP 1232; ARP 1233; and ARP 1234, Society of Automotive Engineers, Warrendale, Pa.
27. **Hoyle, R.,** How to select and use mechanical packings, *Chem. Eng.,* 103, 1978.
28. **Anon.,** *Fluid Sealing,* 3rd ed., George Angus and Company, Ltd., Northumberland, England, 1965.
29. **Anon.,** *O-Ring Handbook,* Publ. ORD-5700, Parker Hannifin Corporation, Lexington, Ky., 1977.
30. **Findlay, J. A., Sneck, H. J., and Reilly, J. A.,** Final Rep. on Study of Dynamic and Static Seals for Liquid Rocket Engines, Contract NAS 7-434, Phase III, NASA CR 109646, General Electric Company, January 1970.
31. **Strom, T. N., Ludwig, L. P., Allen, G. P., and Johnson, R. L.,** Spiral groove face seal concepts; comparison to conventional face contact seals in sealing liquid sodium (400 to 1000°F), *ASME Trans. J. Lubr. Technol.,* 90(2), 450, 1968.
32. **Muller, H. K.,** Hydrodynamic and Hydrostatic Face Seals, ASLE Seals Education Course, Session 9, Houston, Tex., May 1972.
33. **Stair, W. K.,** Liquid buffered bushing seals for large gas circulators, Paper C5, presented at 1st Int. Conf. Fluid Sealing, BHRA Fluid Engineering, Cranfield, Bedford, England, April 1961.
34. **Stair, W. K.,** Basic theory of fluid sealing, Proc. DOE/ASME/ASLE Seals Education Workshop, Atlanta, Ga., October 8 to 10, 1979.
35. **Tao, L. H. and Donovan, W. F.,** Through-flow in concentric and eccentric annuli of fine clearance with and without relative motion of the boundaries, *ASME Trans.,* 77(11), 1291, 1955.
36. **Egli, A.,** The leakage of steam through labyrinth seals, *ASME Trans.,* 57, 115, 1935.
37. **Moskowitz, R.,** Dynamic sealing with magnetic fluids, *ASLE Trans.,* 18(2), 135, 1975.
38. **Stair, W. K. and Hale, R. H.,** The turbulent viscoseal — theory and experiment, Paper H2, presented at 3rd Int. Conf. Fluid Sealing, BHRA Fluid Engineering, Cranfield, Bedford, England, April 1967.
39. **Stair, W. K., Fisher, C. F., Jr., and Luttrull, L. H.,** Further experiments on the turbulent viscoseal, *ASLE Trans.,* 13(4), 311, 1970.

WEAR RESISTANT COATINGS AND SURFACE TREATMENTS

S. Frank Murray

INTRODUCTION

When it is necessary to upgrade the sliding characteristics and wear resistance of metal surfaces, coatings can often be used effectively without sacrificing any of the bulk property requirements of the substrate material. In addition, the use of coatings may often provide savings in both raw material and production costs. The objective of this chapter is to present an overview of current practices on the use of coatings for tribological applications.

FACTORS TO BE CONSIDERED IN SELECTING COATINGS

A wide spectrum of surface coatings or modifications are available.[1,2] These range from soft, low friction, solid lubricant films and polymers to a number of very hard coatings. Table 1 shows typical examples, classified according to the application process. When coatings are being applied by processes such as electroplating, thermal spraying, sputtering, etc., the number of possible substrate/coating combinations is very large. In contrast, chemical conversion and diffusion treatments are generally confined to specific classes of alloys.

A detailed breakdown of various lubrication, speed, load, substrate and coating factors involved in choosing a wear-resistant coating has been prepared by Czichos.[3] While some material combinations can run dry if the operating conditions are not too severe, the great majority would be much more effective with some form of lubrication, even with low-viscosity fluids such as fuel or water. Solid lubricant films can also provide satisfactory life in many applications.

An ideal bearing material combination would be two hard, smooth bearing surfaces, perfectly aligned with no edge contacts. However, cost and fabrication problems with such a precise system restrict its use to a very few premium applications. An alternative approach is to make one of the two surfaces considerably softer so that it can flow plastically under load. The following table shows the approximate order in which a few typical soft bearing alloys will conform and achieve fluid film lubrication:

Tin-base babbitt

Lead-base babbitt

Cadmium-nickel   Faster transition to
                     hydrodynamic lubrication

Copper lead

Leaded tin bronze

The key appears to be the ability to develop a better surface finish and conforming geometry in the shortest possible time.

Since most soft bearing alloys have limited structural strength and fatigue resistance, they are generally used as thin overlays on backings of steel, bronze, or aluminum alloys. As many as two or three layers of different alloys may be applied — each serving a different purpose. Application methods include casting, sintering, or electroplating of the individual layers.

Recent advances in polymer technology, particularly with the polyamide-imide, polyimide and polyphenylene sulfide plastics, have produced a number of plastic coatings with excellent

Table 1
TYPICAL PROCESSES FOR PRODUCING WEAR-RESISTANT COATINGS

Coatings Applied on the Surface

Electroplating
  Cr, Ag, Au, Rh, Rd,Pb-Sn, Pb-Sn-Cu, Sn-Ni, In
Electroless deposition
  Ni, Co, and composites of particles dispersed in nickel
  matrix
Chemical or Physical Vapor Deposition (CVD or PVD)
  Hard coatings, e.g., TiC, TiN, SiC
Sprayed coatings (oxy-acetylene, plasma or detonation)
  Various metals, carbides, and oxides, etc.
Sputtered coatings
  Various metals, ceramics and solid lubricants
Hard facings
  Stellites, Colmonoy coatings, various alloys
Organic coatings
  Solid lubricant films and polymer or elastomer coatings
Chemical conversion coatings
  Phosphating
  Anodizing
  Sulfurizing
  Thermal or chemical oxidation

Thermal treatments
  Flame hardening
  Induction hardening
  Electron beam hardening
  Laser hardening
  Chill casting
Diffusion treatments
  Carburizing
  Nitriding
  Carbonitriding
  Siliconizing
  Boriding
  Chromizing
Miscellaneous treatments
  Spark hardening
  Mechanical working
  Porous sintered layers — impregnated
  Laser alloying

chemical resistance and useful friction and wear capabilities at temperatures as high as 260°C (500°F).

Finally, fragmentary experience shows that certain hard coatings can be carefully run-in at gradually increasing loads and speeds to develop mirror finishes on the contacting surfaces. Oil-lubricated, borided 4340 steel sliding against hardened steel is one example.

ANTICIPATED MODES OF WEAR

Specific modes of wear (i.e., abrasive, adhesive, etc.), will influence the choice of a wear-resistant coating. For example, hard plasma-sprayed oxide coatings might be ideal where two surfaces contact at nominal loads and speeds. If this same coating were subjected to erosive impact by hard particles, however, it would probably disintegrate rapidly.

In applications where loose abrasive particles are trapped between sliding or rolling surfaces, the coating generally must have some minimum hardness to resist penetration by the particles. One exception: babbitt-lined sleeve or thrust bearings often tolerate dirt particles in oil by embedding the particles. Water-lubricated rubber bearings are also very tolerant of abrasives because the elastic rubber accommodates particles as they pass through.

A recent classification of wear modes by Eyre[4] ranked the severity of wear in industrial machinery as follows:

Wear modes	%
Abrasive	50
Adhesive	15
Erosive	8
Fretting	8
Chemical	5

In many cases, material selection is complicated by wear resulting from a number of mechanisms. This point is emphasized in the following discussion of major wear processes.

Abrasive Wear

Abrasive wear is caused by penetration and cutting of a surface.[5] Wear caused by sharp asperities on one surface removing material from an opposing surface is classified as *two-body abrasion*. Examples are a file shaping a metal surface, chunks of minerals sliding down a metal chute, or a rough metal surface sliding against another metal. Wear caused by foreign matter trapped between two moving surfaces is termed *three-body abrasion*. This occurs when particles are trapped in a bearing clearance or when mineral particles are being reduced by ball milling.

The amount of abrasive wear that can be tolerated varies widely. In a hydrodynamic gas bearing a single scratch might cause rapid failure. On the other hand, wear of mils per hour might be tolerable in minerals handling equipment. While abrasive wear is generally associated with sliding, a hard particle trapped between two rolling surfaces could produce a pit which would then initiate a fatigue spall.[6] One note of caution: in rolling contacts the point of maximum shear is at some finite depth below the surface. A hard coating thickness coinciding with this point of maximum shear could result in separation between the coating and the substrate.

The literature indicates that abrasive particles or asperities must have an angle of attack of about 80 to 120° to cut the surface. For this reason, two-body abrasion with fixed asperities will generally cause much more wear than the three-body mode. When loose particles are trapped between surfaces, only a small percentage actually cut metal. The rest simply plough through the surfaces or roll through the loaded contact area.

The volume of material removed by abrasive wear increases almost linearly with load and sliding distance for both two-body and three-body abrasion. Thus:

$$V \propto L \cdot D$$

where V = volume of wear, L = load, and D = distance traveled. Exceptions to this linearity are ascribed to fragmentation of the abrasive or clogging of the surface.

For pure metals and steels in the annealed condition, and for many nonmetallic hard materials such as ceramics, wear resistance is directly proportional to their penetration hardnesses. If steels are heat treated to higher hardness, their resistance to wear is increased. Chemical composition of steel could also influence the results.[7] Eyre[7] found a definite improvement in wear resistance of steel after work hardening, while Krushchov and Babichev reported no effect.[8] Figure 1 summarizes how various means of increasing hardness affect the wear resistance of metals.

For applications where abrasion, impact and shock are severe, as in mining and earth moving, a tough material is needed with high fatigue resistance.[9] Austenitic manganese steels are widely used for this type of application.[10] Although their hardness is only about 200 Bhn after they have been heat treated to improve toughness, these steels readily work-harden when they are deformed and can develop case hardnesses of 450 to as high as 550 Bhn. They can be used as solid members, replaceable wear strips or as welded overlays.

Richardson[11] showed that the hardness of surfaces must be at least half the hardness of the abrasive for any benefit in wear resistance. Hardening the surfaces more than 1.3 times the abrasive hardness gave no further improvement. Tabor[12] showed that a metal surface of indentation hardness Hs will be scratched by a point of hardness Hp if Hp is greater than or equal to 1.2 Hs. Thus, for the two-body abrasion mode, an asperity on a steel surface hardened to 60 Rc will scratch steel hardened to less than 52 Rc. Similarly, in the shop File

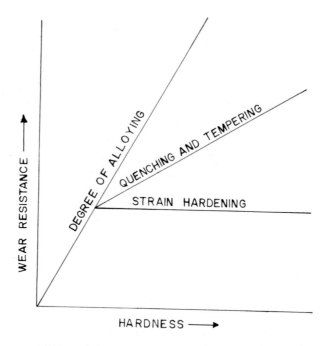

FIGURE 1.   Schematic of effect of hardness on abrasive wear resistance for various means of increasing hardness. (From Krushchov, M. M. and Babichev, M. A., *Friction and Wear in Machinery,* Vol. 19, American Society of Mechanical Engineers, New York, 1964. With permission.)

Table 2
HARDNESSES OF TYPICAL MINERAL CONTAMINANTS

Mineral	Composition	Moh hardness	Approximate Vickers hardness (Vpn)
Hematite	Fe_2O_3	5.5—6.5	680—900
Magnetite	FeO, Fe_2O_3	5.5—6.5	680—900
Monticellite	$MgCaSiO_4$	5—5.5	550—680
Diopside	$CaMg(SiO_3)_2$	5—6	550—800
Sillimanite	Al_2SiO_5	6—7	800—1000
Quartz	SiO_2	7	1000
Spinel	$MgO \cdot Al_2O_3$	8	1420
Corundum	Al_2O_3	9.4	2100

Hardness Test, a fully hardened file (Rc 66), drawn by hand, stops cutting when the hardness of a steel test piece is about 53 Rc or harder.

For case-hardened surfaces, case depth is important. Too thin a case will allow penetration or brinelling, depending on the shape and hardness of the abrasive. For mild abrasion, a case depth of at least 0.25 mm (0.01 in.) is desirable. For severe service, at least 1 to 1.5 mm is necessary. Very high metal surface hardnesses will be required to resist pure abrasion if one of the surfaces, or the abrasive particles, are minerals. Table 2 show the hardness of some typical mineral contaminants. To equal the hardness of a mineral such as Monticellite (Table 2), surface hardness of the case must be at least 5 to 5.5 Moh (52 to 58 Rc). To achieve a hardness ratio of 1.3, the case must be at least 68 Rc. Figure 2 shows the relationship between Moh hardness and indentation hardness.

Aside from higher hardness, practical steps used to reduce abrasive wear are as follows:

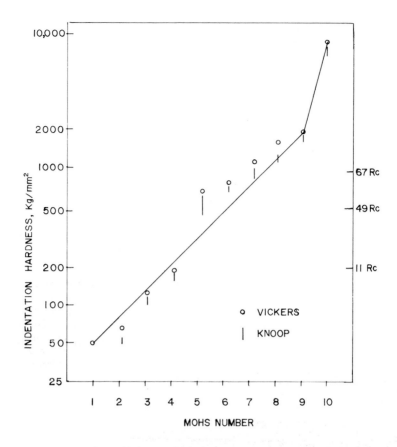

FIGURE 2. Relationship between Mohs hardness number and indentation hardness. (From Tabor, D., *Proc. Phys. Soc. (London)*, 67(3B), 249, 1957. With permission.)

Two-Body Abrasion

1. Improve the surface texture, preferably by techniques which do not produce sharp asperities.
2. Reduce the loads.
3. Consider elastomeric coatings — particularly those which can be repaired or reapplied in the field.
4. Use careful run-in at light loads to wear off asperities before applying full load.
5. If severe impact loads are also encountered, select materials for fatigue resistance (toughness), with abrasion resistance as a secondary consideration.

Three-Body Abrasion

1. Prevent entry of particles by seals.
2. Provide grooves, pockets, or soft areas in surfaces to trap particles.
3. For lubricated systems, use filtration or separators.
4. Design lubricant systems and grooves to promote flushing of debris.

Adhesive Wear

When two surfaces are brought into contact, peaks or asperities deform plastically until the real area of contact is just sufficient to support the load elastically. At these asperity contacts, strong adhesion can occur. When one surface slides over the other, further junction growth takes place until the junctions shear. Such shear may take place at the interface with little or no surface damage, or adhesion forces may be so strong that shear takes place in the bulk of the weakest member — resulting in metal transfer and wear. Under steady-state conditions, adhesive wear normally varies linearly with load and speed as long as the stresses are not too high. Thus, wear volume can be expressed as:

$$V = K \times S \times W/3$$

where S is the sliding distance, W the load, and K a constant for a given material combination.

Criteria for selecting material combinations that will slide effectively with minimal adhesion are discussed in the following paragraphs.

Hardness — An increase in hardness (yield strength) of the surfaces will reduce the real area of contact and, thus, the strength of junctions. Within a given class of materials, the harder the material the better the wear resistance, but a change in composition to obtain higher hardness will not necessarily result in lower wear. Chemical composition (which governs the type of oxide films formed), solubility, and crystal structure must also be factored into the selection process.

Nonsoluble combinations — With no tendency for two sliding surfaces to alloy or interact in any way, material transfer and welding will be minimized. Unfortunately, there are two very practical drawbacks to solubility as a major selection criterion. First, most practical bearing systems involve alloys with a heterogeneous surface composition and unpredictable solubility behavior. Secondly, this concept implies that like materials should never be run against each other. Actually, innumerable bearing combinations of like materials are being used successfully, particularly steel sliding against steel. As long as operating conditions do not exceed certain critical values of load, speed, or temperature, true area of contact is confined to a few asperities, surfaces are protected by oxide and contaminant films, and wear rates are low.[13] When both surfaces are reasonably hard and operating conditions are not too severe, solubility is rarely a strong deterent in materials selection. It becomes much more important when one or both surfaces are soft and the area of contact is large.

Hexagonal crystal structure — Materials with hexagonal crystal structures generally have low adhesion and good sliding characteristics even under high vacuum conditions.[14] Despite these findings, a serious lack of information exists on the structural characteristics of many types of coatings promoted as being wear-resistant.

Oxide film formation — In the normal air environment, metal surfaces are always covered by thin oxide films which minimize bare metal-to-metal interactions. Relative hardness of the oxide and the substrate metal is an important factor. Hard oxides on soft metal substrates are readily disrupted, while soft oxides on hard substrates are much more durable. For high-temperature (500°C +) sliding, Peterson et al.[15] have shown that particular alloy compositions can form complex, thin, adherent oxide coatings which serve as protective films. These oxides only function in relatively narrow temperature ranges under oxidizing conditions: the temperature must be high enough to regenerate the film as quickly as it is worn away, but low enough to avoid excessive oxidation. Figure 3 shows the frictional behavior of a nickel-chrome superalloy as a function of temperature.

One aspect of adhesive wear that deserves more attention is the production of loose wear debris which creates three-body abrasive wear. This debris is generally oxidized and heavily work-hardened, making it significantly harder than the parent surfaces. While a circulating oil system can employ filters to remove such debris, grease will trap it. Figure 4 shows the wear rates of two grease-lubricated bronze bearings. One bearing was removed periodically, cleaned, and put back on test. The other bearing was run continuously. Accumulation of the wear debris generated during these tests appeared to be responsible for the high wear of

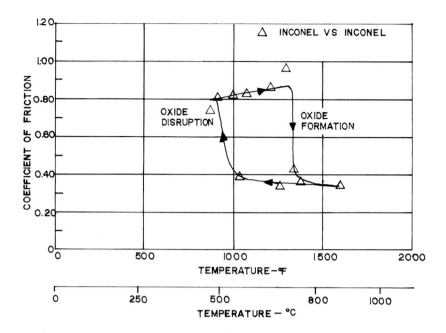

FIGURE 3.   Frictional behavior of Inconel X-750 during temperature cycling. (From Peterson, M. B., Lee, R. E., and Florek, J. J., *ASLE Trans.*, 3(1), 101, 1960. With permission.)

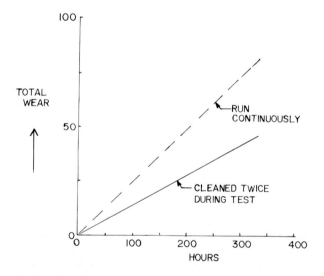

FIGURE 4. Relative wear of grease-lubricated bronze bearings run continuously or cleaned periodically.

the bearing run continuously. Proper design of grooving to trap the debris, coupled with periodic regreasing, would help to alleviate this condition.

Most bench-type wear test equipment is relatively insensitive to this abrasive wear condition. The geometry of the test specimens, e.g., a pin sliding on a disc or a block sliding on a rotating shaft, favors escape of any loose debris. While useful for evaluating adhesive wear, these tests do not reflect changes in the mode of wear from adhesion to three-body abrasion.

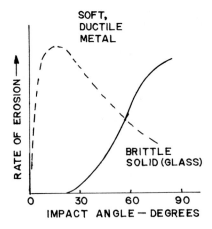

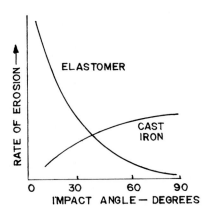

FIGURE 5.   Effect of impact angle on erosion of ductile metal and brittle solid. (From Wolfe, G. F., *Lubr. Eng.*, 19, 28, 1963. With permission.)

FIGURE 6.   Erosion as a function of impact angle. (From Eyre, T. S., *Tribol. Int.*, 11(2), 91, 1978. With permission.)

Erosion

Erosive wear can be caused by either solid particulates or liquid droplets striking the surface at high velocities.[16,17] In this chapter, emphasis is placed on erosive wear by solid particle impact. Typical problem areas include: compressor and turbine blading, helicopter rotor blades, impeller-type pumps, pressure letdown valves, etc.

Since the contact stress results from kinetic energy of the particles in the fluid stream, size and density of the particles, their velocity, and angle of impact must all be considered. Erosive weight loss is roughly proportional to the square or cube of velocity. Relative hardness and shape of the particle, its fracture characteristics, and the ductility of the solid surface also influence the resulting damage.

As illustrated in Figure 5, cutting wear predominates at low angles of impingement for ductile metals unless the particles are smooth spheres. The harder the surface, the lower the rate of material removal. At higher angles of impact, hard, brittle materials show more erosive wear as elastic properties of the surface become much more important. Annealed materials often erode less than the same alloy in the hardened state. An elastomeric coating may be a very viable solution to erosion at high angles of impingement. Figure 6 compares the relative effect of impact angle on wear rates for a metal and a rubber material.

While surface hardness becomes relatively less significant at high angles of impact, coatings of very hard materials are being used to prevent erosion damage at both high and low angles of impact. For example, Hansen et al.[18] evaluated a large variety of metals, alloys, carbides, and ceramics in a sandblast type of test at 20 and 700°C with 90 and 20° angles of impingement. As shown in Table 3, some ceramics and carbides with a low metal binder content were more erosion resistant than metals or alloys. Hard coatings that were particularly effective included: chemical vapor-deposited SiC, electrodeposited TiB_2, borided Mo, and borided WC cermets. Generally, coating thicknesses of about 50 to 80 μm (0.002 to 0.003 in.²) were necessary for adequate protection. These test results have been partially verified by evaluation of control valve components in coal gasification plants.

The above results are of fundamental interest for the following reasons:

1. Contrary to other basic studies (e.g., the results illustrated in Figure 5), some hard, brittle materials can be very effective at high-impingement angles.
2. Most metals and alloys, except molybdenum, have essentially the same erosion rates.
3. Thin, hard coatings can provide erosion protection for softer metal substrates.

Table 3
RELATIVE EROSION RESISTANCE OF VARIOUS SOLID
MATERIALS AND COATINGS AT A 90° IMPACT ANGLE

Materials	Relative erosion factor[a]	
	20°C	700°C
Metals		
Stellite 6B	1.0	1.0
Mild steel	0.75	—
Type 304 stainless	1.0	0.72
Molybdenum	0.46	—
Cemented carbides		
Tungsten carbide + 25% cobalt	1.0	1.2
Tungsten carbide + 5.8% cobalt	0.41	0.81
Tungsten carbide + <1.5% cobalt	0.11	0.12
Pure materials		
Silicon carbide	0.11	0.42
Cubic boron nitride	0[b]	0
Diamond	0	0
Hard coatings		
CVD silicon carbide	0.05	0
Electrodeposited TiB_2	0	0
Borided molybdenum	>0.1	—

[a] $$\text{Relative erosion factor} = \frac{\text{mean specimen volume loss}}{\text{mean volume loss of stellite 6B control specimens}}$$

[b] No measurable wear.

From Hansen, J. S., Kelly, J. E., and Wood, F. W., Erosion Testing of Potential Valve Materials for Coal Gasification Systems, Bureau of Mines Rep. of Investigation No. 8335, 1979, p. 26. With permission.

The criterion of hardness should not be arbitrarily set aside in choosing materials to withstand high-angle erosion. In most practical applications, impingement angles vary because of geometrical variations in the solid surface and possible turbulence in the fluid stream. From a practical viewpoint, hard coatings would be best for protecting critical surfaces, while soft, elastic coatings or annealed alloys might be more practical for the rest of the system.

Fretting Wear

Fretting wear results from small amplitude, reciprocating slip between fitted surfaces. As wear takes place, either by adhesive wear or microfatigue, debris is trapped in the contact area and tends to dam the lubricant away from this area. This wear debris is heavily work-hardened and oxidized, a condition which also results in three-body abrasion. Fretting can be very serious with large rotating shafts, flexible couplings, and highly stressed structural components. The fretting damage creates stress risers which can result in catastrophic failures by fatigue.

One solution to fretting wear is to prevent slip entirely, but this is generally impractical. Some improvements can often be realized by (1) using a surface chemical conversion coating such as phosphating, (2) choosing a more suitable lubricant, or (3) when no fluid lubricant is present, by using a bonded solid film lubricant. For small motions, the elastic properties of plastic or rubber coatings may be sufficient to take up all of the slip. Very hard coating materials, e.g., bonded carbides or borided surfaces, have also been used to prevent fretting.

Generally, rougher or more porous surfaces are less prone to lubricant depletion. The valleys or pores also serve as reservoir traps for loose debris. However, rough surfaces are also more susceptible to fatigue-type wear. Sprayed metal coatings such as molybdenum or copper have been used successfully in certain applications. These coatings tend to be porous because of oxidation of the metal during the spraying operation. Another coating system which should be promising is borided molybdenum with its surface hardness of about 3000 Knoop. The coating could be applied by spraying molybdenum on the substrate, grinding the surface smooth, and then boriding to produce a hard, porous surface. Other processes which can produce hard porous surfaces include spark hardening, selective etching, and porous chrome plating.

Chemical Wear

Exposure of fresh metal surfaces, coupled with the high pressures and flash temperatures developed at contacting asperities, create ideal conditions for chemical reactions in sliding contacts. These reaction films serve to prevent bare metal-to-metal contacts and welding or metal transfer. However, under certain conditions, an excessive amount of soft reaction product is produced which then wears away rapidly.

This corrosive wear could be attributed to a number of factors. These include:

1. Excessively high-operating temperature. This promotes lubricant oxidation to form acidic and corrosive products and also increases reaction rates.
2. Use of reactive chemical additives (EP agents). Additives containing phosphorous, sulfur, or chlorine are often used in lubricants to form protective inorganic films on heavily loaded bearing surfaces. Such compounds corrode certain bearing alloys, and also become overly reactive at high temperatures.
3. Excessive moisture in the lubricant. This problem is particularly acute in marine applications. Tin-base babbitt forms a relatively hard "scab" with seawater contamination that can abrade a steel journal. Pitting corrosion because of water contamination is a major cause of ball bearing failures on naval aircraft.[19]
4. Atmospheric corrosion. Many industrial components operate, unlubricated, in exposed locations. Rust formation of ferrous alloys and subsequent abrasion results in rapid material loss.

Changes in bearing alloy composition, electroplating, diffusion treatments, chemical conversion coatings, and organic coatings are all potential solutions to the problem. Research is also being directed toward new techniques such as ion implantation to change the characteristics of metal surfaces.

Cavitation-Erosion

Cavitation involves gas- or vapor-filled bubbles or pockets in flowing liquids as a result of the dynamic generation of low pressure. Collapse of these bubbles can generate extremely high pressures and velocities in the fluid. Adjacent solid surfaces may be rapidly pitted and eroded by this action. This type of wear is particularly serious in valves, impeller-type pumps, and propellers. Hobbs[20] correlated cavitation-erosion with ultimate resilience, expressed as follows:

$$\text{Ultimate resilience} = \frac{\frac{1}{2}(\text{tensile strength})^2}{\text{elastic modulus}}$$

Plastics and elastomers with high-tensile strength and resilience have been used successfully as protective coatings on metal substrates. Strong adhesion of the coating is essential. Metal overlays or inlays are effective in certain applications when flame- or plasma-sprayed, welded, or electroplated. In general, cavitation damage decreases with increasing hardness, particularly among materials of the same general class.

SUBSTRATE AND COATING CONSIDERATIONS

This section considers the various types of surface treatments shown in Table 1 and the processing variables involved.

Coatings Applied on the Surface
Electroplating
This process is applicable to practically any metal surface and, by suitable preparation, to plastics and many other nonconducting materials. Since it is a low-temperature process (<100°C), warpage or dimensional changes are avoided.

There are disadvantages. Hydrogen embrittlement can occur with certain alloys. Quality control and adhesion may be problems. Since electroplating is a line-of-sight process, holes, recesses, and complex shapes should be avoided.

Despite these problems, a variety of platings are used as wear-resistant coatings. These range from soft, conformable coatings such as tin- or lead-base alloys, to hard chrome. Thicknesses normally range from 2.54 (0.0001 in.) to 500 μm (0.020 in.), although platings as thick as 3180 μm (0.125 in.) are possible with some metals. Electroplated precious metals such as gold, silver, and rhodium are used for sliding electrical contacts as well as specialized bearing applications. For selective plating of worn surfaces in the field, a porous electrode impregnated with proprietary plating solutions can be used to brush-plate limited areas.[21] This technique is used to repair scratches or flaws in chrome-plated cylinders and to build up worn areas on babbitted bearings.

Electroplating can increase surface hardness, improve corrosion resistance, provide soft and conformable coatings, or create a nonsoluble material combination with lower adhesion.

Electroless Deposition
Certain metals, such as nickel, copper, and cobalt can be deposited by chemical reduction from aqueous solutions at temperatures below 100°C. Electroless nickel is most widely used. Although more expensive than electroplating, electroless deposits are uniform and protective, and complex shapes including holes and ID surfaces can be coated. Most metals, except lead, cadmium, tin, and bismuth can be plated. The deposition rate is slow; thicknesses range from 2.54 (0.0001 in.) to 180 μm (0.007 in.).

Electroless nickel plate contains about 8 to 10% phosphorous. As deposited, the hardness is about 500 Vpn (49 Rc), but can be increased to about 1000 Vpn (70 Rc) by heat treating the plated part at 400°C. Despite its hardness, practical experience with electroless nickel as a wear-resistant coating has often been disappointing. Lubrication is essential; electroless nickel is not recommended for dry sliding applications. Silver plating the opposing surface is reported to be beneficial. When considering electroless nickel for a bearing surface, evaluations should be made under conditions simulating the actual application.

Composite platings of very fine hard particles dispersed in an electroless nickel matrix appear to be much more effective than straight electroless nickel for wear resistance. The particles are suspended in the plating bath and codeposit with the nickel. Silicon carbide is widely used for the hard particles, but diamond is also commercially available. Particle size and shape are critical and the surfaces must be finished so that no sharp peaks project from the surface. These platings are used extensively for molds which must resist abrasion from glass fiber-reinforced plastic parts[22] and also for guides and rollers subjected to abrasion by textile fibers.

Vapor Deposited Coatings

Although the principles of vapor deposition have been known for over 80 years, industrial applications have been very limited. Two application techniques are being used: chemical vapor deposition (CVD) and physical vapor deposition (PVD).[23] In CVD, the coating is formed either from gaseous chemical reactants at the substrate surface, or by thermal decomposition of volatile compounds such as the carbonyls. In PVD, the coating is evaporated or sputtered from the source to the substrate. Recent interest centers around the use of thin, hard coatings on cutting tools. In CVD of titanium carbide on cemented tungsten carbide tool bits, as an example, titanium tetrachloride is vaporized, mixed with hydrogen and methane, and fed into a reaction chamber containing the tool bits. These parts are heated to 800 to 1000°C and the following reaction takes place at the surfaces:

$$TiCl_4 + CH_4 \xrightarrow{H_2} TiC + 4HCl$$

Strong bonding takes place because of some diffusion. Tool life is reportedly improved by factors of 4 to 10. Certain carbides, nitrides, borides, and oxides of metals such as titanium, silicon, tungsten, and chromium can be deposited.

CVD is also used commercially to apply hard, wear-resistant coatings of silicon carbide on carbon-graphite seal faces. Test results have shown that this material runs best against itself in displaying outstanding resistance to wear by abrasives in the fluids.[24]

This process has limitations. It is most economical when a large number of parts are treated simultaneously, but part size is limited by the size of the reaction chamber. Process temperatures are so high that many substrate alloys would be annealed. Reduction of the process temperature will retard diffusion and reduce adherence of the coating. Vapor deposition should be useful for creating hard surfaces on small parts made from stainless steels and nickel- or cobalt-base superalloys.

The lower processing temperature with PVD permits coating of high-speed steel tools without excessively softening the substrate. Adherent coatings have been obtained at temperatures below 500°C.

Sprayed Coatings

Any material that can be melted without decomposition can be sprayed as a surface coating.[25] Plasma or detonation gun coatings of ceramics, carbides, and refractory metals (Mo and W) are of particular value for upgrading wear resistance. A major disadvantage is that this is a line-of-sight process. The densest and most adherent coatings are those sprayed perpendicular to the surface. As the impact angle decreases, coating quality drops and spraying angles below 45° are definitely not recommended. The amount of heat that must be dissipated limits the ability of even specialized "mini-gun" equipment to coat bores less than about 75 mm (3 in.) in diameter and longer than 100 mm (4 in.). Flat surfaces and outside diameters are no problem.

Practically any metallic substrate can be spray coated. Size is no impediment. With reasonable care, bulk temperature of the substrate can be kept below 175°C (350°F). While steel grit blasting is normal practice, steel particles embedded in the substrate can rust and cause blisters in the coating. In critical applications, grit blasting with sharp, fresh Al_2O_3 abrasive avoids this problem. All coating should be done as soon as possible after abrasive blasting. Since it is difficult to roughen hard metal surfaces such as hardened steel by abrasive blasting, a thin-sprayed undercoat of metal such as nickel-chrome or nickel aluminide should be applied first to provide a rough base for the final coating.

For ceramic and carbide coatings, thicknesses range from about 100 (0.004 in.) to 1000 μm (0.04 in.). Heavier coatings of metals can be deposited, and plasma spraying is widely used for salvage and repair work. If the coatings are to be finish-ground for bearing or shaft surfaces, the as-sprayed coating should be at least twice as thick as the finished coating. This will ensure minimum porosity and optimum cohesion and adhesion. The finished coating thickness should be as thin as possible to minimize problems. Coating vendors should be consulted in selecting sliding combinations. Mating a sprayed oxide coating against a metal is particularly risky since the metal tends to transfer to the ceramic as islands of work-hardened material which can then severely abrade the opposing metal surface. Such combinations should be carefully evaluated before specifying them for practical applications.

Where substrate corrosion is a problem, corrosion-resistant metal undercoatings, e.g., nickel aluminide, can help. Soft nickel plate has also been used, but must be grit blasted before the final coating is applied. Care is needed to avoid exposing the substrate.

Other variations of these coatings involve spraying and fusing. The fusing step requires very high temperatures which could affect the metallurgy of the substrate.

Sputtering

In this process, atoms of material from a negatively charged target of the coating material are vaporized by bombardment with positive ions of an inert gas such as argon. These atoms are then transported, in the vapor phase, through the plasma of ionized gas and deposited on the surface to be coated.[26,27] Coatings of metals, alloys, solid lubricants, and hard materials such as oxides and carbides can be applied. Substrate heating is negligible. These coatings are characteristically uniform and very thin, ranging in thickness from 50 (500 Å) to 1000 nm (10,000 Å). The process is ideal for applying wear-resistant coatings on precision components such as gas bearings or rolling contact bearings because no subsequent finishing is required.

Reverse sputtering before coating removes any contamination from the surface and enables outstanding adherence. Even very hard coatings such as TiB_2 or Cr_2O_3 can be flexed, brinelled, or bent over a small radius without cracking. Wear life of a sputtered coating appears to compare favorably with similar coatings 1000 times thicker deposited by other processes.

Although the process requires a vacuum, it can be automated to some extent. Graded coatings can be applied, without breaking vacuum, if the equipment has provision for multiple targets.

In ion implantation, the evaporated material is ionized and accelerated to the workpiece by electrical fields. The ions actually penetrate the surface. Both wear and corrosion resistance can be affected.[28]

Hard Facings

By welding, or spraying and subsequent fusion, various wear-resistant alloys can be deposited on metal substrates.[29,30] This technique is widely used for heavy-duty industrial and construction equipment which is subject to severe wear by abrasion or impact-abrasion.

Hard facings are used on new equipment and also find wide application in building up and salvaging worn parts. The process has many advantages: wide ranges of coating materials are available, heavy deposits are feasible, repairs can be made in the field, metallurgical bonding is obtained, some coatings are corrosion resistant, and expensive materials are conserved by applying the coatings on low-cost substrates. For abrasion resistance, hard materials, e.g., metal-bonded tungsten carbide, cobalt alloys, and nickel-chrome-boron are used. For maximum impact resistance, high manganese work-hardening steels perform best. Chrome steels and low alloy or carbon steels are in many cases comparable or better in abrasion resistance than the more expensive cobalt-base alloys.

Aside from processing temperature and cost, the major drawback to hard facings is the possibility of cracking in some applications, particularly with thick deposits. The very high-surface temperatures involved may also affect the substrate.

Organic Coatings

Many plastic and elastomeric coatings have been used as wear-resistant surfaces on metal substrates. Typical examples include:

Teflon®	Acrylics
Nylon	Polyurethanes (rigid)
Vinyls	Polyamide-imides
Epoxies	Polyphenylene sulfide
Phenolics	Aromatic polyesters

In many cases, these plastics also provide corrosion protection.

By compounding powdered solid lubricants, such as MoS_2, graphite, or Teflon into a suitable resin matrix, a variety of wear-resistant solid lubricant films with outstanding frictional characteristics have been obtained. These coatings are covered in the chapter on Lubricant Types and Their Properties (Volume II).

Elastomeric coatings effectively prevent erosive or abrasive wear under certain conditions. As long as the velocity of an eroding particle is not too high, the elastomer surface deforms and recovers elastically with no damage. In abrasive-blasting booths, durable rubber gloves protect the operator's hands. Hard, tough polyurethane elastomers are used to coat steel tires on industrial equipment such as forklifts and carts which operate on rough, hard surfaces.

Some new abrasion-resistant plastic coatings have provided unique durability on the hulls of icebreakers.[31] A large number of candidates were screened and a nonsolvented polyurethane and a nonsolvented epoxy were selected for trials. After four years of service, both coatings have remained essentially intact. Similar coatings have shown promise for preventing cavitation damage on ship propellers.

These plastic and elastomeric coatings can be applied by spraying, brushing, dipping, and fluidized bed. Surface preparation is a major consideration and abrasive grit blasting has been found to be very suitable. Durable coatings usually require heat curing at temperatures up to 175°C (350°F), although some newer materials require temperatures as high as 350°C (660°F). This can be a problem with certain substrate materials, especially age-hardened aluminum alloys.

Chemical Conversion Coatings

Various processes are used to form *in situ* inorganic coatings on metals.[1,32,33] Unlike bonded solid lubricant films, these conversion coatings do not necessarily provide low friction or long life. Their primary function is to prevent bare metal-to-metal contacts and promote surface smoothing during the early stages of run-in when surface imperfections can penetrate through the lubricant film. The most common types are phosphates, sulfides, and oxides.

Phosphating to produce a complex inorganic phosphate surface is the most widely used process. For application, parts are immersed in aqueous solutions of phosphates at a temperature of about 93°C (200°F).[32] A manganese phosphate coating is generally best for wear resistance because it is relatively soft and tends to "smear" over the contact area. Zinc phosphate produces a harder coating and is used primarily as a substrate pretreatment for improving the adherence of protective polymer coatings. Coating thicknesses range from 2.54 to 38 μm (0.0001 to 0.0015 in.), depending on the application temperature and bath composition. Such coatings can be applied to cast iron, steel, zinc, and cadmium, but not to stainless or other corrosion-resistant alloys. As a rule of thumb, 50% of the coating thickness penetrates the surface and 50% appears as dimensional growth. The coatings are porous (more so in thicker layers) and this helps to retain the lubricant. Phosphating is particularly useful for applications such as gears or piston rings where initial conformity may not be ideal. Besides their beneficial effects on sliding, these coatings also provide corrosion protection.

Sulfide coatings are generally applied from molten salt baths at temperatures ranging from 190°C to about 550°C.[1,33] Both electrolytic and chemical processes are used. Coatings are characteristically less than 10-μm (0.0004-in.) thick. Because the salt baths contain cyanides or cyanates, the coating actually consists of iron nitrides as well as sulfides. Unlike the phosphates which are easily friable, sulfide coatings are very wear resistant. They can be applied to a wide variety of ferrous alloys, even low-carbon steels which normally would not respond well to nitriding.

Oxide films of significant thickness can also be produced on metal surfaces.[1,33] Anodizing aluminum to produce a hard Al_2O_3 surface is widely used to improve wear resistance. Magnesium, titanium, and beryllium can also be anodized. These anodized coatings are hard and brittle surface layers, supported on relatively soft substrates; brinnelling loads can crack the coatings, resulting in high wear. A similar result would be obtained if a hard particle were trapped between two anodized surfaces.

Since anodizing is done at temperatures below 100°C, metallurgical changes are no problem. Films as thick as 100 μm (0.004 in.) are used. By first creating a porous, hard-anodized coating and then impregnating it with Teflon® or other solid lubricant, improved sliding performance can be obtained. These anodized films have a relatively short wear life in dry sliding; however, a thin wiped film of oil or grease increases the life dramatically.

Ferrous alloys can be oxidized by various processes.[1,33] Proprietary salt baths of caustic/nitrate solutions or molten nitrate/nitrite baths are often used. Heating steel in steam at 260 to 400°C can also produce an adherent oxide coating. The latter process, part of the "Ferrox" treatment, is often used for treating piston rings. Like the phosphate treatments, these oxide coatings minimize bare metal contacts and prevent scuffing in lubricated applications such as gears, needle bearings, and piston rings.

Thermal Treatments

Steels and cast irons which contain enough carbon to be through-hardened in thin sections can be case hardened by localized surface heating to produce a hard, wear-resistant martensitic structure with a tough, ductile core. Two production techniques are being widely used: induction heating and flame hardening.[34] In addition, electron beam and laser hardening are becoming more commonplace.[35,36] Flame hardening does not lend itself to close control, but it is particularly suitable for large parts. The other three processes can be closely controlled by varying the energy input. Advantages of these thermal treatments are reduced energy consumption, ability to selectively harden surfaces which require wear resistance, high production rates, and ease of automation. Gears, cams, and shafts are among the many machinery components that can be surface hardened by these methods.

Chill casting is also used to harden critical surfaces on cast iron parts which contain about 3% carbon. Instead of allowing slow cooling with formation of graphite flakes, chills are used to cool the cast iron rapidly and cementite (Fe_3C) is formed.[37] Other carbide-forming elements such as chromium and vanadium are also added to promote surface hardness. Cast cam shafts and cam followers can be hardened by this method.

Diffusion Treatments

A variety of commercial diffusion treatments increase the wear resistance of metals, particularly steel and iron parts. Case carburizing and nitriding are the most prominent, and many modifications are available to achieve specific changes in surface chemistry and metallurgical structure. Extensive information is available in the literature[1,37,38] and from suppliers. In many cases, the difference in chemistry has significant effects on lubrication and sliding behavior.

Diffusion treatments uniquely permit selection of a steel with optimum core strength. Wear resistance is then provided by the surface diffusion process. Gears, splines and many other components subject to bending stresses particularly benefit from this approach. As an added bonus, case carburizing or nitriding improves the fatigue properties of steels. Nitriding carbon or low alloy steels also upgrades their corrosion resistance, but lowers that of stainless steel.

Most carburizing and carbonitriding processes are done above the transformation temperature of the steels. Quenching or subsequent heat treatment to achieve desired properties may induce dimensional changes or warpage. Since nitriding is done below the transformation temperature, dimensional changes are minimized. In many cases, no subsequent finishing is required.

Proprietary molten salt bath processes are used to apply very thin (10 to 100 μm), relatively soft nitride coatings on steel. These are often times very effective. Application temperatures are about 570°C (1050°F).

Table 4 compares some commercial diffusion processes. Practical limits on case depths are established by diffusion rates of the elements. The heavier the case, the higher the cost and the greater the dimensional changes. Case depth wear resistance is generally ranked as follows:

Light case	< 500 μm (0.02 in.)	Good wear resistance at low stresses
Medium case	500—1000 μm (0.02—0.04 in.)	Good wear resistance at higher stresses
Heavy case	1000—1500 μm (0.04—0.06 in.)	Sliding and abrasive wear resistance; resists crushing and fatigue
Extra heavy case	> 1500 μm (> 0.06 in.)	Wear and shock resistance

If brinelling results because of indentation-type loading, either the case depth must be increased or the substrate strength upgraded.

Other diffusion processes have been developed to upgrade wear resistance of both ferrous and nonferrous alloys. Siliconizing[39] and boriding[40] are examples.

Miscellaneous Treatments

This category includes cold-working, spark-hardening, sintered porous surface coatings impregnated with solid lubricants, and laser alloying. A variety of mechanical reduction and burnishing processes can also upgrade surface hardness to some extent, as well as provide improved surface texture and fatigue resistance.

Spark hardening is used routinely to apply thin, wear-resistant coatings such as tungsten carbide on tools, chucks, dies, etc. A positively charged electrode of the coating material is vibrated against a negatively charged substrate. Each time contact is made, current discharges from a condenser and material is deposited on the surface. The resulting surface is normally rough, but proprietary processes are available for better finishes. Wolfe[41] found that sparked silver coatings were particularly promising, possibly because the pores acted as lubricant reservoirs. This suggests that a sparked layer of silver might inhibit fretting wear.

Impregnating a porous surface layer with a solid lubricant has been the subject of a number of investigations. Best known material is probably the DU supplied by Glacier Metals, Ltd. Spherical bronze particles are sintered on a steel or bronze backing and the porous layer is then impregnated with Teflon® and lead. The material is produced as flat-strip stock which can then be machined, punched, or rolled to form washers, sleeve bearing inserts, etc.

In laser alloying, the laser creates a thin, molten layer on the metal surface. Alloying elements are then introduced into the molten skin. This technique can form a coating whose chemistry and corrosion or wear resistance is markedly different than the substrate.[36]

Table 4
COMMERCIAL DIFFUSION PROCESSES USED TO HARDEN ALLOYS, PARTICULARLY IRON AND STEEL PARTS

Treatment	Diffusing elements	Typical substrate alloys	Typical process temperature	Typical hardness	Advantages and disadvantages
Carburizing	Carbon	Low carbon steels and alloys	925°C	55—65 Rc	Case depth from 125—3800 μm (0.005—0.15 in.); growth and distortion are problems
Nitriding	Nitrogen	Steels with Al or Cr additions Stainless steel Tool steels	565°C	67—72 Rc	Long process time; steels with Al additions give hard case with limited ductility; Cr reduces hardness but increases ductility; typical case depths 125—760 μm (0.005—0.03 in.); low distortion; growth about 10% of case depth
Carbonitriding	Carbon and nitrogen	Low C steels and alloys	760—870°C	50—65 Rc	125—760 μm (0.005—0.03 in.) case depth; less distortion than carburizing
Liquid salt bath nitriding	Nitrogen and carbon	Carbon steels, low alloy and tool steels, stainless steel	510—565°C	55—75 Rc	Hard case 5 μm (0.0002 in.) depth; no subsequent finishing required; case has good ductility; hardness depends on substrate chemistry
Boriding	Boron	Low C and alloy steels. Ni and Co alloys, Mo, W, P/M[a] parts	800—1150°C	1800—3000 Vpn (depends on substrate composition)	12—100 μm (0.0005—0.004 in.)
Chromizing	Chrome	Carbon and low alloy steels Ni, Co, Mo tool steels, P/M[a] parts	950—1150°C	70 Rc (1200 Vpn)	Case depth to 125 μm (0.005 in.); not brittle; properties depend on steel chemistry; steels can later be heat treated; good oxidation resistance
Siliconizing	Silicon	Carbon and low alloy steels, iron	1000°C	—	Helps wear resistance, but main purpose is corrosion and oxidation resistance; case depth 125—2540 μm (0.005—0.1 in.); steel can be heat-treated afterward

[a] Power metal.

Table 5
SOME PRACTICAL APPLICATIONS FOR WEAR-RESISTANT COATINGS

	Modes of wear					
Type of coating process	Abrasive	Adhesive	Erosion	Fretting	Chemical	Impact sliding
Electroplating	C	A	C	C	A	—
Electroless plating	C	C	—	—	A	—
Vapor deposition	A	C	A	X	X	—
Sprayed coatings	A	A	B	B	C	—
Sputtering	—	A	—	—	—	—
Hard facings	A	C	A	C	B	A
Organic coatings	C	B	A	C	A	—
Chemical conversion	—	A	—	B	C	—
Diffusion	A	A	C	B	—	C
Thermal	A	C	B	B	—	C
Miscellaneous						
Spark hardening	—	A	—	X	—	—
Mechanical working	—	A	—	—	—	—
Impregnated porous surfaces	—	A	—	—	—	—

Note: A = major usage, B = frequent usage, C = occasional usage, X = should be applicable.

TYPICAL APPLICATIONS FOR WEAR-RESISTANT COATINGS

Most surface treatments listed in Table 1 are particularly applicable to *ferrous alloys*. Ultimate choice depends on factors such as: cost, effect of process temperature on the substrate, and the dominant modes of wear. Hard, diffusion coatings are particularly valuable for gears, cams, crankshafts, etc., where through hardening would result in a brittle material.

Stainless steels can be hardened to thin-case depths by nitriding or boriding. For thicker coatings, spraying or hard facing would be best.

Aluminum, titanium, and *magnesium* alloys can be anodized to improve wear resistance. However, where concentrated loading is encountered, substrate deformation and subsequent cracking of the coating is likely. Diffusion treatments for these metals and alloys are limited to a few proprietary processes which involve electroplating followed by thermal diffusion.[33] With the exceptions of chemical vapor deposition and hard facing, all processes listed in Table 1 for applying coatings *on* the surface (e.g., spraying, plating, sputtering, etc.) can be used with these alloys. Thin layers of tin alloys are often used to upgrade the performance of aluminum bearings.

Plasma-sprayed oxide coatings are particularly effective for hard surfacing aluminum and titanium alloys as long as solid particle erosion or impact loading is not a problem. Metal-bonded carbides would be more suitable for erosion resistance. Mismatches in thermal expansion coefficients, as on aluminum, are rarely a problem as long as the coatings are reasonably thin, on the order of 50 to 150 μm (0.002 to 0.006 in.). Like anodizing, the real problem is substrate deformation under load.

When coatings are required on *copper* alloys, electroplating or spraying techniques are applicable.

Superalloys are frequently plasma-sprayed with ceramics or carbides for improved wear resistance. Aluminum oxide or nickel-chrome bonded chrome carbide are very effective at high temperatures (to 1000°C). Boriding also produces very hard surface coatings.

Table 5 categorizes coating processes used to resolve wear problems. One obvious con-

Table 6
SUMMARY OF APPROACHES TO SELECTION OF WEAR-RESISTANT COATINGS

Purpose of coating	Basic approach	Coating process	Limitations	Use on
Improve conformability, prevent scuffing	Soft-conformable coating	Electroplated soft metal (e.g., Sn or Pb alloy)	Lubricated applications	Electroplating — most metals
	Chemical conversion coating	Phosphate, oxide, or sulfurize	Fluid film lubrication should result	Chemical conversion — generally cast iron or steel.
Reduce abrasive wear Light abrasion	Embedable material or elastomer coating	Babbitt lining or rubber bonded to metal backing		Steel, bronze, aluminum, etc.
	Increased surface hardness	Thermal or diffusion coatings	Life determined by: hardness, shape and size of abrasive and by coating thickness	Generally steels and cast iron
		Chrome plate		
		Sprayed coatings of cermets or ceramics		Most metals
Moderate abrasion	Increase surface hardness	Heavy case carburizing	Same as above, high temperature processes	Steel and cast iron
		Sprayed, fused coatings		Steel and other high melting alloys
		Hard facings		
Heavy abrasion	Increase hardness	Hard facings, e.g., high chrome steels	Same as above	
Abrasion plus oxidation or corrosion	Increase hardness and corrosion resistance	Cobalt or nickel-base hard facings	Same as above	Steel and other high-melting alloys
Abrasion plus impact	Use work hardenable fatigue resistant alloys	Facings of austenitic manganese steels	Same as above	
Reduce adhesive wear		Electroplating or electroless composite	Must be lubricated	Most metals
	Change composition of surfaces (increased hardness generally beneficial)	Chemical conversion	Limited wear life	Cast iron and steel
		Solid lubricant coatings, organic coatings	Liquid lubricants may soften coatings	Most metals
		Sprayed coatings	Quality control	Most metals
		Anodizing	Cracking because of substrate deformation	Al, Be, Mg
		Malcomizing (nitriding)	Very thin cases	Austenitic stainless steels
		Soft nitriding	Thin case	Most steels
		Sputtering	Cost, best for high-precision parts	Most metals
		Chemical vapor deposition	High-temperature process	Temperature-tolerant materials

Table 6 (continued)
SUMMARY OF APPROACHES TO SELECTION OF WEAR-RESISTANT COATINGS

Purpose of coating	Basic approach	Coating process	Limitations	Use on
Reduce erosive wear	Use hard or elastic surface coatings[a]	Diffusion	For low angle impact	Steel, cast iron
		Thermal spray and fuse	Same as above	Most metals
		Sprayed metals (annealed)	For high angle impact	Most metals
		Elastomers and plastics	For high angle impact	Most metals
Reduce fretting wear	Increase hardness and reduce adhesion	Thermal or diffusion coatings	Wear life	Steel or cast iron
	Provide lubricant reservoirs	Spark hardening		Most metals
		Porous sprayed coatings		
	If dry, provide lubricant	Bonded solid lubricants	Wear life; may be affected by liquid lubricants	Most metals
Chemical wear	Change surface composition	Electroplating		Most metals
		Nitriding		Steels
		Chemical conversion		Cast iron, steel
		Ion implantation	Wear life of coatings	Most metals
		Sprayed metals		Most metals
		Siliconizing		Steel
Cavitation-erosion	Sacrificial metal overlays	Sprayed, welded, or electroplated	Wear life of coatings and adhesion to substrate	Most metals
	Elastic surface coatings	Elastomers or plastics		Most metals

a Some very hard coatings are promising for high-angle impact.

clusion: in many applications two types of coatings with entirely different physical properties might provide equally satisfactory service. For example, a hard brittle coating or an elastomer coating might both be suitable for reducing erosive wear. Table 6 presents typical examples of coatings for resolving specific wear problems.

Currently, emphasis is being placed on variations in conventional coating techniques, such as ion nitriding, laser hardening and laser glazing or surface alloying. These offer advantages, such as better process control, shorter process times, and the ability to selectively harden surfaces. Future trends appear to be directed toward surface modifications by ion implantation, chemical vapor deposition, and selective diffusion of elements from fused salt baths.

REFERENCES

1. **Wilson, R. W.**, Surface treatments to combat wear, *First European Tribology Congress,* C278/73, Sponsored by the Tribology Group, Institute of Mechanical Engineers, Mechanical Engineering Publications, London, September 1973, 165.
2. Special feature issue, Wear resistant surfaces, *Tribol. Int.,* Vol. 11(2), 91, 1978.
3. **Czichos, H.**, *Tribology: A Systems Approach to the Science and Technology of Friction, Lubrication and Wear,* Elsevier, New York, 1978.
4. **Eyre, T. S.**, Wear characteristics of metals, *Tribol. Int.,* 9, 203, 1976.
5. **Finken, E. F.**, Abrasive wear, *Evaluation of Wear Testing,* ASTM STP 446, American Society for Testing and Materials, Philadelphia, 1969, 55.
6. **Tallian, T. E.**, Elastohydrodynamic Hertzian contacts. II, *Mech. Eng.,* 17, December 1971.
7. **Eyre, T. S.**, The mechanisms of wear, *Tribol. Int.,* 11(2), 1978.
8. **Krushchov, M. M. and Babichev, M. A.**, The effect of heat treatment and work hardening on the resistance to abrasive wear of some alloy steels, in *Friction and Wear in Machinery,* Vol. 19, American Society of Mechanical Engineers, New York, 1964, 1.
9. **Diesburg, D. E. and Borik, F.**, Optimizing abrasion resistance and toughness in steels and irons for the mining industry, in *Source Book on Wear Control Technology,* Rigny, D. A. and Glaeser, W. A., Eds., American Society for Metals, Metals Park, Ohio, 1978, 94.
10. **Avery, H. S.**, Austenitic manganese steel, in *Metals Handbook,* Vol. 1, 8th ed., American Society for Metals, Metals Park, Ohio, 1964, 834.
11. **Richardson, R. C.**, Wear of metals by relatively soft abrasives, *Wear,* 11, 245, 1968.
12. **Tabor, D.**, Moh's hardness scale — a physical interpretation, *Proc. Phys. Soc. (London),* 67(3-B), 249, 1957.
13. **Archard, J. J. and Hirst, W.**, The wear of metals under unlubricated conditions, *Proc. R. Soc. (London),* A236, 397, 1956.
14. **Buckley, D. H. and Johnson, R. L.**, The influence of crystal structure and some properties of hexagonal metals on friction and adhesion, *Wear,* 11, 405, 1968.
15. **Peterson, M. B., Lee, R. E., and Florek, J. J.**, Sliding characteristics of metals at high temperatures, *ASLE Trans.,* 3(1), 101, 1960.
16. **Preece, C.**, *Erosion,* Academic Press, New York, 1979.
17. **Finnie, I.**, Some observations on the erosion of ductile metals, *Wear,* 19, 81, 1971.
18. **Hansen, J. S., Kelly, J. E., and Wood, F. W.**, Erosion Testing of Potential Valve Materials for Coal Gasification Systems, Bureau Mines Rep. of Investigation No. 8335, 1979, 26.
19. **Cunningham, J. S. and Morgan, M. A.**, Review of aircraft bearing rejection criteria and causes, *Lubr. Eng.,* 35(8), 435, 1979.
20. **Hobbs, J. M.**, Experience with a 20-KC cavitation erosion test, in *Erosion by Cavitation or Impingement,* ASTM STP 408, American Society for Testing Materials, Philadelphia, 1967, 159.
21. **Rubenstein, M.**, Fluid power in aerospace, *Hydraul. Pneumatics,* 25(9), 202, 1973.
22. **Anon.**, Mold corrosion, abrasion checked with new silicon-carbide coating, *Mod. Plastics,* 66 and 68, July 1976.
23. **Archer, N. J.**, "Vapor deposition of wear-resistant surfaces," *Tribol. Int.,* 11(2), 135, 1978.
24. **Panel Discussion,** High performance seal materials, *Lubr. Eng.,* 35(6), 309, 1979.

25. **Committee Rep.,** Thermal spraying, in *Welding Handbook,* 6th ed., Phillips, A. L., Ed., American Welding Society, New York, 1969, chap. 29.
26. **Stupp, B. C.,** Sputtering and ion plating as industrial processes, preprint No. SAE 730547, presented at the SAE Automobile Eng. Meet., Detroit, May 1973.
27. **Spalvins, T.,** Microstructural and wear properties of sputtered carbides and silicides, in *Source Book on Wear Control Technology,* Rigny, D. A. and Glaeser, W. A., Eds., American Society for Metals, Metals Park, Ohio, 1978, 348.
28. **Dearnaley, G.,** Ion implantation of engineering components, in *Advances in Surface Coating Technology,* Welding Institute, Abington Hall, Cambridge, 1978, 111.
29. **ASM Committee**, The selection of hard facing alloys, in *Metals Handbook,* Vol. 1, 8th ed., American Society for Metals, Metals Park, Ohio, 1964, 820.
30. **Committee Rep.,** Surfacing, in *Welding Handbook,* 6th ed., Walter, S. T., Ed., American Welding Society, New York, 1969, chap. 44.
31. **Calabrese, S. J., Buxton, R., and Marsh, G.,** Frictional characteristics of materials sliding against ice, *Lubr. Eng.,* 36(5), 283, 1980.
32. **ASM Committee**, Phosphate coating, in *Metals Handbook,* Vol. 2, 8th ed., American Society for Metals, Metals Park, Ohio, 1964, 531.
33. **Gregory, J. C.,** Chemical conversion coatings of metals to resist scuffing and wear, *Tribol. Int.,* 11(2), 105, 1978.
34. **ASM Committee,** Induction hardening and tempering, Flame hardening, in *Metals Handbook,* Vol. 2, 8th ed., American Society for Metals, Metals Park, Ohio, 1964, 167.
35. **Jenkins, J. E.,** Electron beam surface hardening, *Tool Prod.,* 44(9), 76, 1978.
36. **Desforges, C. D.,** Laser heat treatment, *Tribol. Int.,* 11(2), 139, 1978.
37. **Elliot, T. L.,** Surface hardening, *Tribol. Int.,* 11(2), 121, 1978.
38. **ASM Committee** Case hardening of steel, in *Metals Handbook,* Vol. 2, 8th ed., American Society for Metals, Metals Park, Ohio, 1964, 93.
39. **Kanter, J. J.,** Siliconizing of steel, in *Metals Handbook,* 8th ed., Vol. 2, American Society for Metals, Metals Park, Ohio, 1964, 529.
40. **Fiedler, H. C. and Sieraski, R. J.,** Boriding steels for wear resistances, in *Source Book on Wear Control Technology,* Rigny, D. A. and Glaeser, W. A., Eds., American Society for Metals, Metals Park, Ohio, 1978, 364.
41. **Wolfe, G. F.,** Effect of surface coatings on the load-carrying capacity of steel, *Lubr. Eng.,* 19, 28, 1963.
42. **Tilly, G. P.,** Sand erosion of metals and plastics, *Wear,* 14, 241, 1969.

SYSTEMS ANALYSIS

Horst Czichos

INTRODUCTION

The foregoing chapters of the handbook have amply illustrated that there is a great range of technical systems to be lubricated as well as a great variety of tribological processes, i.e., contact, friction, lubrication, and wear processes, that occur in lubricated systems. Whereas complex problems of this type have been solved in the past by isolating single events and treating these in terms of simplified cause-effect relationships, today a "multi-disciplinary" or "systems" approach is needed.[1,2]

The purpose of this chapter is to present an overall systems view which may help to systematize approaches to the solution of lubrication problems, taking into account the various influencing factors, processes, and parameters. For more details the reader is referred to Reference 3, and References 4 to 7 provide examples of the general development and application of systems theory in contemporary science and technology.

THE SYSTEM CONCEPT AND ITS APPLICATION TO TRIBOLOGY

General Considerations

As a starting point for an engineering systems approach to the analysis of tribological systems, consider a typical lubricated mechanical system, namely a gearbox. The technical purpose of this system is to transform certain "inputs", i.e., torque and angular velocity, into "outputs". The transformation occurs through the contact of gears, and as a consequence of interactions of the gear teeth, friction and wear processes occur.

Lubrication represents a deliberate attempt to avoid or reduce the effect of friction and wear upon a mechanical system. A lubricant can also act, as it flows away, as a cooling agent removing heat from the location of the friction process. If the sliding or rolling surfaces are completely separated by the action of a lubricant at all times, there may be no wear process. In this event, the analysis is simplified ("no-wear model"). However, if in a lubricated state there is some contact between surfaces or between boundary lubricants on the surfaces, the interfacial tribological processes are of paramount concern. In such cases, the presence of a lubricant may complicate the analysis, partly because the reaction products present may be complex and difficult to characterize, and partly because transient conditions may be the major concern.

The first step in a systems analysis is proper identification and isolation of the problem. As shown in Figure 1 for the example of a gearbox, the two partners (or the two "systems elements") which form the tribologically interacting surfaces, i.e., gear 1 and gear 2, can be hypothetically separated from their environment by the proper choice of a "system envelope". All components of the system are then by definition within this envelope and are part of the so-called internal "structure" of the system. The structure consists of the elements (A) of the system, their relevant properties (P), and their interrelations (R), described formally by the set $S = \{A, P, R\}$.

The "external" quantities which cross this system envelope from the outer world are the "inputs" of operating variables, and the quantities which cross the system envelope from the inside are the "outputs". In other words, the inputs of the operating variables are transformed through the structure of the system into outputs which are used, the use-outputs. Simultaneously, as a consequence of interactions between the elements, loss-outputs occur, denoted in summary by the terms friction and wear losses. The way in which the inputs are transformed into outputs determines the technical function of the system.

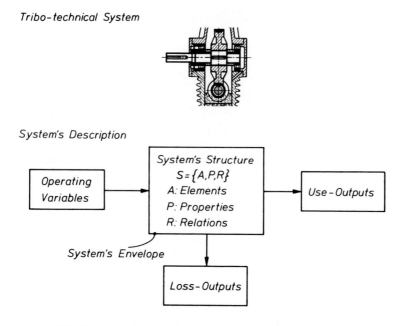

FIGURE 1.   Analysis and description of a tribo-technical system.

From the discussion of the system illustrated in Figure 1, the relevant tribological parameters of mechanical systems can be divided into the following four groups:

1.   Technical function of the tribological system
2.   Operating variables
3.   Structure of the tribological system
     a. Elements (or system components)
     b. Properties of the elements
     c. Interactions between the elements
4.   Tribological characteristics

Technical Function of Tribological Systems

Technical aims realized through moving surfaces in tribological systems may range from aerospace applications to biomechanical joints. However, from a physical point of view, four basically different groups of technical purposes can be distinguished as illustrated by the examples shown in Figure 2.

The most general technical purpose is the guidance of motion through various types of "bearings". The other basic groups are the transmission of mechanical work, the transmission of information — for instance the control of machine functions with cams — and the forming of materials. Basically the technical function of tribological systems is connected with the transmission or transformation of one or more of the basic quantities: motion, work, information, and materials. In using these or related quantities, the technical function of various tribological systems may be classified in terms of the input-output relations of these quantities.

A broad classification is given in Table 1. Invariably, motion is a characteristic of any tribological system. This motion may constitute a transfer of work, information, or materials. In some instances, the purpose of a system may be to change a rate of motion or to eliminate it altogether. It is also often desired to restrict motion, i.e., to reduce the number of degrees of freedom a machine element may possess. Mechanical devices which produce or transfer

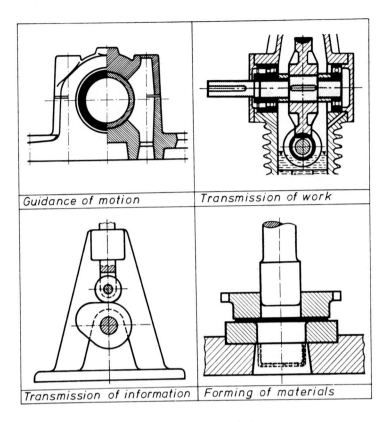

FIGURE 2.   Types of tribological systems.

Table 1
CLASSIFICATION OF TECHNICAL FUNCTIONS OF
TRIBOLOGICAL SYSTEMS

Inputs and outputs needed for technical function			
Main inputs $\{X\}$	**Main outputs** $\{Y\}$	**Primary technical function of the system**	**Examples**
	Guidance of motion		Bearings
		Coupling of motion	Clutches
		Annihilation of motion	Brakes
Motion			
	Work	Power transmission (mech., hydr., pneum.)	Gears
+			
	Information	Generation of information	Clocks; cams and followers
Work		Reproduction of information	Data transducer (audio, video; tape or record)
Motion + Materials	Materials	Transportation	Wheel/rail Pipeline
		Forming of materials	Wiredrawing

information by utilizing the motion of macroscopic bodies are steadily being replaced by devices in which there is little or no mechanical motion, for example the replacement of the mechanical clock by digital electronic clocks. In other instances materials are not merely moved but also changed in state or form.

In applying the system concept to other technical systems, e.g., electrical or electronic systems, the functional behavior of the system is often described in terms of mathematical input-output relations. However, in attempting to apply the system concept to the subject of tribology, a fundamental difference must be emphasized between the behavior and the functional description of electrical systems and mechanical systems in which friction and wear processes occur.

Compare, for example, the behavior of an electrical transformer and a mechanical gearbox. At a first glance, the functional purpose of both systems appears to be analogous, i.e., to transform certain inputs — voltage and current in the electrical system, and angular velocity and torque in the mechanical system, respectively — into outputs used for technical purposes. The function of both systems may be described formally as a transformation of the inputs into the outputs via a certain transfer function. However, the dynamic performance of both systems is accompanied by perturbations. In both systems, energy losses occur due to electromagnetic or frictional resistances. The fundamental difference between the behavior of the electrical and mechanical systems originates from their different "structure". The structure of the electrical system generally remains constant with time. In this case, the transfer function can be worked out mathematically. This has led to various applications of the powerful systems engineering method of network theory and related methods characterizing functional behavior.[8-10] In the mechanical case, however, the structure of the system generally changes with time, through friction and wear. This aspect, which is of great importance for the reliability of the system under question, is described in more detail later.

Operating Variables

The most characteristic operating variable of a tribological system is the type of relative motion between tribo-element (1) and tribo-element (2). The basic types of motion are sliding, rolling, spin, and impact. Every type of relative motion between system components can be expressed as a superposition of these four basic types of motion. In addition to characterization of the type of motion, its dependence on time should be specified, being for example: continuous, oscillating, reciprocating, or intermittent.

The other basic operating variables are the following quantities:

1. Load, F_N
2. Velocity, v
3. Temperature, T
4. Distance of motion, s
5. Operating duration, t

For some tribological systems, these physical operating variables are accompanied by material inputs, e.g., flow rate of the lubricant. Some disturbing inputs may also be present, e.g., vibration and radiation. It may also be necessary to specify derived quantities, e.g., contact pressures, temperature gradients, etc.

Structure of Tribological Systems

As described above, the structure of a tribological system is given by the system elements (the material components of the system), their relevant properties, and their interrelations, described formally by the set $S = \{A, P, R\}$.

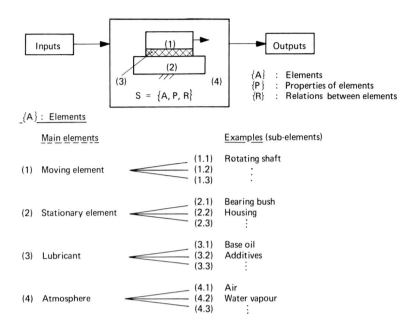

FIGURE 3.   Analysis of the structure of tribological systems.

Elements of the System, A $= \{a_i\}$

If the system envelope is located as closely as possible around the "interacting surfaces in relative motion", four different basic elements are involved in the friction and wear processes in most tribological systems. As illustrated in Figure 3 for a simple sliding system, the pair of interacting surfaces involving moving element (1) and stationary element (2). The other two basic elements are the lubricant (3) (if any) and the atmosphere (4). These main elements are linked to others or may be composed of subconstituents. For example, element (3), the lubricant, may consist of a base oil and additives. In Table 2 elementary elements or components (1), (2), (3), and (4) are listed as examples from every group of the basic tribological systems compiled in Table 1.

Properties of the Elements, P $= \{P(a_i)\}$

Behavior of any tribological system is influenced by many properties of the basic elements (1), (2), (3), and (4). Although the great variety of tribo-mechanical systems and tribological processes makes it difficult to provide a comprehensive general compilation, the following properties of the elements are of primary concern:

1.   Properties of tribo-elements (1) and (2): these can be subdivided into "volume" and "surface" properties. Volume properties: geometry, chemical composition and metallurgical structure, elastic modulus, hardness, density, thermal conductivity. Surface properties: surface roughness and surface composition.
2.   Properties of the lubricant (3): these may be classified into system-independent and system-dependent properties.
3.   Properties of the environmental atmosphere (4): primarily chemical composition and the amount and pressure of its components, especially water vapor.

Interactions Between the System Elements, R $= \{R(a_i, a_j)\}$

Tribological interactions between the elements of a mechanical system, i.e., the contact, friction, lubrication, and wear processes, are of paramount interest. Figure 4 provides sim-

Table 2
EXAMPLES OF BASIC ELEMENTS OF TRIBOLOGICAL SYSTEMS

Tribological system (or process)	Elements of the system			
	Tribo-element (1) (moving or stationary)	Tribo-element (2) (moving or stationary	Interfacial medium (3)	Surrounding medium (4)
Sliding bearing	shaft	bushing	lubricant	air
Band clutch	shaft	band	—	air
Disc brake	disc	pad	conta-minant	air
Worm gear set	worm	gear	gear oil	air
Cam and follower	cam	follower	lubricant	air
Printing unit	print-head	paper	dye	air
Audio pick-up	record	sapphire tip	—	air
Electrical contact	ring	brush	spray	cover gas
Locomotion	wheel	rail	conta-minant	air
Pipeline	fluid	pipeline	—	—
Wiredrawing	wire	die	borax	air
Hot extrusion	billet	die	glass	air
Turning	workpiece	cutting tool	cutting fluid	air

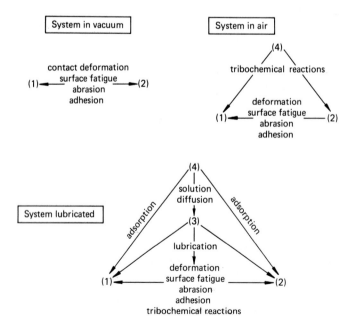

FIGURE 4.   Schematic representation of tribological interactions between system components.

plified schematic diagrams for systems of increasing complexity, i.e., increasing number of interacting elements.

In an ultrahigh vacuum, the simplest tribological system consists only of interacting partners (1) and (2). The main interactions are then covered by the terms contact deformation, surface fatigue, abrasion, and adhesion. In air, these processes are supplemented by interactions with the atmosphere (4). Finally in a lubricated system, direct (contact) interactions between moving and stationary elements are prevented or influenced through the different mechanisms of lubrication.

Also, interactions between (4) and (3) with (1) and (2) should be taken into account. For instance, the diffusion of atmospheric oxygen into the lubricant (4) → (3), followed by oxidation processes between the lubrication and the moving and stationary partners (3) → (1), (2), can distinctly influence the mechanisms of mixed and boundary lubrication.

Tribological Characteristics

Characteristics that describe the dynamic changes of a lubricated mechanical system as a consequence of friction and wear processes may be divided into the following three groups: tribo-induced changes in the system structure, tribo-induced energy losses, and tribo-induced material losses.

Depending on the processes within a lubricated mechanical system, the tribo-induced changes of a system structure (a) may concern:

1. Destruction or creation of elements, e.g., the degradation of a lubricant or, on the contrary, the creation of "frictional polymers".
2. Changes in properties of elements, for instance, changes in contact topography and surface composition.
3. Changes in interrelations between elements, for instance, changes of wear mechanisms under the action of the operating variables, or changes in the lubrication mode.

Friction-induced energy losses (b) and wear-induced materials losses (c) may be expressed formally as:

Friction losses $= f$ (operating variables; system structure)
Wear losses $= f$ (operating variables; system structure)

Consequently, friction coefficient, f, and wear rate, w, may be expressed formally as:

$$f = f(X;S) \qquad w = f(X;S)$$

Although parameter groups X and S are not independent variables since they are connected with each other through the tribological interrelations R, the above symbolic representation of friction and wear characteristics can be conveniently used as a starting point for application of the system methodology. From the above symbolic equations it follows that any systematic approach to the solution of a lubrication problem in a mechanical system must be based on the detailed knowledge of both the operating variables and the structure of the system.

Influence of Tribological Processes on Structure, Function, and Reliability of Mechanical Systems

In the upper part of Figure 5, a typical tribological system, namely a gear box, is shown schematically. As already described in Figure 1, the technical function of the system is to transform certain inputs, namely angular velocity and torque, into useful outputs via a certain transfer function.

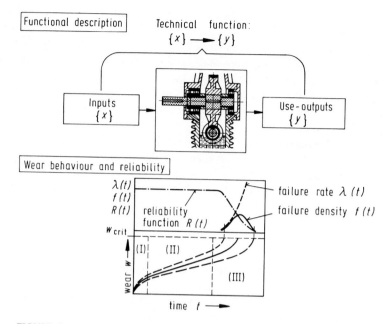

FIGURE 5.   Influence of changes of system's structure on function and reliability of a tribological system.

Dynamic performance of the system is accompanied by perturbations. For example, through the action of wear processes, the properties of the moving components may be changed and a certain material loss-output may result. As illustrated in the lower part of Figure 5, three primary characteristics may be distinguished for the loss-output:

1. Self accommodation (running-in)
2. Steady state
3. Self acceleration

These three modes may follow each other in time as indicated in Figure 5. If then the wear rate reaches a maximum admissible level, the systems structure has changed in such a way that the functional input-output relations of the system are disturbed severely. Repeated measurements show random variations in the data as indicated by the dashed lines in the wear diagram of Figure 5. From sample functions of the wear process, a distribution, a failure rate, and a corresponding reliability function results.[11,12]

Generally, the reliability of a mechanical system is expressed by a probabilistic function R(t) based on the following definitions:

$F(t)$ — Probability distribution function of the time to failure

$f(t) = \dfrac{dF(t)}{dt}$ — Density function

$\lambda(t) = \dfrac{f(t)}{1 - F(t)}$ — Failure rate
$\lambda(t)\, dt$ is a conditional probability that the system will fail during the time $t + dt$ under the condition that the system is safe until the time t

$R(t) = 1 - F(t)$ — Reliability function

MTBF — Mean time to failure
Measure reliability for repairable equipment

In some cases, the failure rate $\lambda(t)$ of a component in a system can be estimated from the point of view of the physical behavior of the material used.[13] Empirically, and sometimes theoretically, the following probabilities have been proposed:

Exponential Distribution

$$\lambda(t) \quad = \text{ constant } = C$$

$$f(t) \quad = C \cdot \exp(-Ct)$$

$$R(t) \quad = \exp(-Ct)$$

In this case, the failure rate is constant. It means physically that any failure occurs accidentally without any accumulation of fatigue-like effects during its service time. Many kinds of electronic components follow this type of failure. Components in a machine break down in this mode when the failure is brittle fracture.

Rayleigh Distribution

$$\lambda(t) = Ct$$

$$f(t) = Ct \cdot \exp(-Ct^2/2)$$

In this case, the failure rate increases with time. The constant, C, indicates the rate of deterioration of the component which depends upon the stress level applied to it.

Normal Distribution (Truncated)

$$f(t) = 1/s(2\pi)^{1/2} \exp\left\{-\tfrac{1}{2}(t - \mu/s)^2\right\}$$

Many components of machines obey this distribution, especially if the failure occurs due to wear processes. The failure rate of this distribution cannot be expressed in a simple form.

Weibull Distribution

$$\lambda(t) = \frac{C}{t_o} t^{C-1}$$

$$f(t) = \frac{C}{t_o} t^{C-1} \exp(-t^C/t_o)$$

This is a distribution with two parameters, t_o, the nominal life, and the constant C. The distribution is found to represent failure of many kinds of mechanical systems, such as fatigue in ball bearings.

Gamma Distribution

$$f(t) = C \frac{(Ct)^{x-1}}{\Gamma(x)} \exp(-Ct)$$

where $\Gamma(x)$ is a gamma function. This is also a distribution with two parameters. Theoretically, the importance of this distribution is attributed to the equation being an x-fold

Table 3
PHENOMENA OF DETERIORATION AND MODE OF FAILURE

| Physical process | Deterioration (Mode of failure) | | | | | | Probability distribution |
| | Topological change | | Geometrical | | Physical property | | |
	Fixing	Separation	Micro	Macro	Bulk	Surface	
Fracture		•					Exponential
Yielding				•			Exponential
Fatigue		•	•		•		Weibull
Creep		•		•	•		Normal
Diffusion					•		Normal
Corrosion						•	Rayleigh
Erosion						•	Rayleigh
Rusting	•					•	Rayleigh
Wear			•			•	Normal
Adhesion	•						Gamma
Staining			•			•	Exponential

convolution of the exponential function. It means physically that a component fails at the x-th shock which occurs as a Poisson statistical process.

These are representative distributions which appear in the failure process of various components and systems. As a general overview, Table 3 provides a compilation of the phenomena of deterioration and the mode of failure in connection with underlying physical processes.[14]

From the experimental determination of failure distribution curves conclusions may be drawn on the type of failure mechanism. For most tribological systems failing as a consequence of wear processes, the failure behavior is characterized by the normal distribution or the Weibull distribution. Knowledge of the failure mode and the type of failure distribution can often be used to improve the reliability of the system. For instance, this approach can be used to select the type of ball or roller bearing system to operate under a given set of operating conditions with high operation safety.[15,16] In this connection, the importance of lubrication technology on system reliability has been emphasized.[17,18]

To conclude the discussion on failure and reliability, the dependence of the failure rate on the operating duration of a system should be considered. If the failure rate is plotted as function of time, a unique "bathtub-curve" is often found, as shown in Figure 6. None of the distributions discussed above have this shape, but an approximation may be obtained by selecting an appropriate probability density function for each of the three regimes.[19] Regime (a) describes the region of the "infant death" of the system. This regime is characterized by a decrease of the failure rate with time, for example with effective running-in. Regime (b) of constant failure rate is the region of normal running. Here, failure occurs as a consequence of statistically independent factors. Regime (c) is characterized by a rising failure rate which is the normal mode of wear-induced failure of mechanical systems. Here, failure may be due to aging effects.

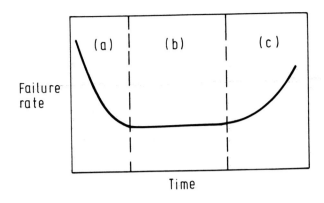

FIGURE 6.   "Bathtub" failure rate curve.

APPLICATION METHODOLOGY

The system concept described in the preceding section can be used as a convenient framework which may help to systematize approaches to the solution of lubrication problems.[3] This will be illustrated in three examples of broad areas of lubrication technology.

Comprehensive Characterization of Tribological Systems

The value of all lubrication work, whether basic or applied, would be greatly increased if experimental variables would be presented in a form suitable for subsequent independent evaluation and correlation.

With the help of a data sheet such as Table 4, relevant parameters can easily be compiled in a comprehensive manner.[20] For the typical application example in Table 4, characteristics of a journal bearing operating at the minimum of the Stribeck curve are compiled with data taken from the literature.[21] Since the procedure of compiling the data sheet is all that is really relevant here, only some main features of the example are pointed out.

1. The technical function of the system can be described as "guidance of motion".
2. Operating variables are given by the type of motion, namely "continuous sliding" of a duration of 40 min together with load F_N, velocity v, and bulk oil temperature T.
3. The structure of the system is described by the elements of the system, their relevant properties, and interrelations. In this case, tribo-element (1) is given by a bearing bushing consisting of a lead-tin-bronze and tribo-element (2) is a steel shaft. Lubricant (3) is a mineral oil of the type SAE 20W-20, and the atmosphere (4) is laboratory air. Together with the designation of the elements and materials, the data sheet contains all important properties of the elements. The data sheet also indicates the tribological interactions between the elements of the system which can be characterized as a "running-in" period leading to topographical changes which are within the initial roughness of the bearing element (2).
4. As tribological characteristics, emphasis is laid on the characterization of wear rates. Although the original wear rates are given in mass units, since the dimension of the bearing are given it is possible to estimate wear-time rates and wear-distance rates. Because the wear measurements have been performed by means of radiotracer techniques, it is possible to distinguish between the wear rates of the bearing and the shaft.

In a similar manner, characteristics of other tribological systems can be described with the help of the tribological systems data sheet. Although the data sheet may be individually

Table 4

TRIBOLOGICAL SYSTEMS DATA SHEET APPLIED TO A JOURNAL BEARING

I Technical function of the tribo-system

Journal bearing: guidance of motion

II Operating variables

Type of motion[1]: continuous sliding	Duration of operation t[ ]: 40 min

Load[λ] $F_N(t)$ [N]	Velocity[3] $v(t)$ [m/s]	Temperature[4] $T(t)$ [$^\circ$C]
F_N 17000 ... 0 20 40 t (min)	1 ... 0 20 40 t (min)	T 56 54 52 ... 0 20 40 t (min)

Other op. variables: oil flow rate: 500 cm³/min	Location lubricant

III Structure of the tribo-system

Properties of elements (initial/final)	Tribo-element (1)	Tribo-element (2)	Lubricant (3)	Atmosphere (4)
Designation of element and material	bushing bronze	shaft steel	mineral oil SAE 20 W 20	air
Geometry/Dimensions/Volume	1/d = 0.4	d = 65 mm	———	———
Chemical composition	15 Pb; 7 Sn 78 Cu	0.2 C; 0.8 Mo 0.2 Cr; 0.4 V	———	———
Phys.-mech. data: Hardness (N/mm²) Viscosity $\eta(T,p)$ other[5] thermal conductivity	H=700 45 kcal/mh$^\circ$C	H=2630 3 kcal/mh$^\circ$C	η=130 cP/20° η=17.5cP/80°C density: 0.9 g/ml	———
Topography descriptors (c.l.a.,etc) R_t (μm)	2.5 - 3.0	0.5	Other data: ratio: radial clearance / shaft radius ψ_r = 0.2 %	
Surface layer data (if different from volume)	—	—		
Contact area A[ ]	1xd = 17 cm²		Tribological interactions:	
Ratio: contact area / total wear track ε [%]	$\varepsilon^{(1)}$ 100	$\varepsilon^{(2)} \propto$ 20	running-in process	
App lubrication mode: quasi-hydrodynamic				

IV Tribological characteristics

Changes in properties of the elements[6]	Friction data (vs time t or distance s)	Wear data (vs time t or distance s)
(1): R_t < 2μm	Minimum of Stribeck curve	w 200 (μg) 100 ... 0 20 40 t (min) (2) (1)

Other characteristics (e.g.: contact resistance, vibrations, noise, etc): ———	Appearance of worn surfaces: (1): run-in (2): slightly roughened

shortened, extended, or grouped in another order, in all applications the total of the four groups of parameters compiled in Table 4 should be taken into account.

Systematic Lubricant Selection Procedure

The system concept provides a guideline which may help to systematize the lubricant selection procedure. Clearly, the systematic guideline can be only a rough skeleton which must be completed by using information from the preceding chapters. From the system point of view, in a lubricant selection procedure system-independent and system-dependent characteristics must be distinguished. Typical characteristics falling in the system-independent category are cost, availability, and physical and chemical properties such as chemical composition, density, thermal conductivity, acidity, flash and fire point, pour point, etc.

For the testing and specification of system-independent lubricant properties and characteristics, well-known tests have been worked out and standardized. (This has been done, for instance, in the U.S. by the American Society for Testing and Materials, (ASTM-D2), in the U.K. by the Institute of Petroleum, and in the Federal Republic of Germany by the Fachausschuβ Mineralöl und Brennstoffnormung in Deutschen Institut für Normung (FAM-DIN). The details of the various tests can be found in the official publications of these institutions and in other portions of this handbook.

The system-dependent characteristics of lubricants depend essentially on the specifications of the whole tribological system. Thus, all of the systems characteristics described earlier must be taken into consideration, at least in principle, to make sure that no important operational aspect or influencing parameter has been overlooked.

In contrast to the standardized tests for the system-independent physical and chemical properties of lubricants, the testing of system-dependent characteristics should be performed in connection with the technical function of the actual tribo-engineering system in which the lubricant is used. These tests assess predominantly the overall ability of a lubricant to permit rubbing surfaces to operate without scuffing, seizing, or other manifestation of material destruction. This can be broadly classified in three groups.[22]

Simplified bench tests — These tests employ simplified test geometries leading to point, line, or flat contact. Most of these tests were devised to differentiate between EP and non-EP oils, and their accuracy is sometimes not good enough to grade different levels of EP activity. Erratic results can occur if operating variables (e.g., temperature of the lubricant) are not closely controlled. Predicting the performance of lubricants on the basis of these tests alone is almost impossible. On the other hand, they are convenient for acceptance testing, for production control, and as indicators of batch variations of lubricants.

Testing with tribo-technical components — Because of the above shortcomings, a different type of lubricant testing is required to permit control of as many variables as possible while simulating actual performance requirements. A convenient way of doing this is to test lubricants in the laboratory, where operating conditions can be controlled, with the parts under test being those used in the complete tribo-engineering unit.

Full-scale tests — There is general agreement that the only satisfactory means of evaluating the performance characteristics of lubricants is by full-scale tests of their actual use in tribo-engineering systems. Since the cost of field or proving-ground tests is considerable, this type of testing is generally used only as final proof of the decisions made while developing the design of an actual tribo-engineering system.

The systematic lubricant selection procedure may follow the "flow chart" as shown in Table 5. From the technical function (A) of the mechanical system, it is often possible to make a preselection of the lubricant, i.e., to specify the "type" or "class" of the lubricant, e.g., gear oil or cutting fluid, etc.

For the further specification of the lubricant, allowable ranges should be known for the operating variables (B), such as load F_N (or pressure p), speed v, operating temperature T (including the friction-induced temperature rise ΔT), operating duration t, as well as the allowable limits of the tribological characteristic (C), such as friction coefficient, wear rate, and heat and vibration data.

The structure of the system (D) determines the other system components which interact with the lubricant. Material and surface properties of the other system components are to be considered. A crucial factor is an estimation of the tribological processes to be expected, i.e., contact conditions, interfacial friction and wear mechanisms, and the prevailing lubrication mode.

In addition to the system-dependent parameters from the groups (A) to (D), the system-

Table 5
CHARACTERISTICS TO BE CONSIDERED IN A
SYSTEMATIC LUBRICANT SELECTION
PROCEDURE

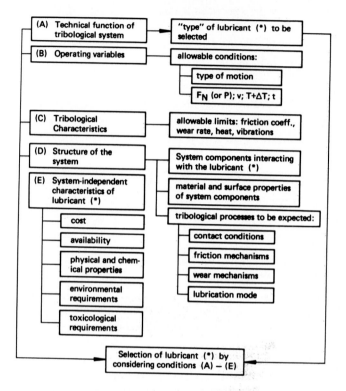

independent characteristics (E) such as the cost of the lubricant, its availability, and its physical and chemical properties are to be considered. Further, other general requirements, such as the environmental conditions (e.g., ability for recycling) or toxicological conditions should not be overlooked.

Selection of Condition Monitoring Techniques for Tribological Systems

To minimize serious failures of tribological systems, increasing attention is being given to machinery condition or health-monitoring techniques.[23,24] Emphasis is being given to on-line monitoring as a means of detecting deterioration so that remedial action can be taken before the breakdown point is reached. Clearly, on-line monitoring is especially important where deterioration leads to catastrophic damage and failure. For monitoring a given tribo-engineering system such as illustrated in Figure 1, a systematic compilation from which suited techniques may be selected is given in Table 6.

Monitoring of Operating Variables

In monitoring the operating variables, techniques for the control of the allowable load, the allowable velocity, and the allowable operating temperature of the system may be distinguished. Instruments for monitoring these quantities include, for example, force transducers, pressure gages, velocity transducers, thermocouples, or infrared (IR) instruments.

Although the use-function is connected with the above quantities, i.e., load, velocity and temperature, information on the performance of a given tribo-engineering system is often indirectly obtained by monitoring the loss-outputs (see c and d, Table 6) corresponding to

Table 6
MACHINERY CONDITION MONITORING TECHNIQUES

(a) Monitoring of operating variables
- Allowable load or contact pressure
  → force transducers, pressure gauges
- Allowable velocity
  → velocity transducers
- Allowable temperature
  → thermocouples, IR-techniques

(b) Monitoring of components and lubrication
- Misalignments, vibration of components
  →proximity transducers, accelerometers
- Surface conditions of components
  → profilometry, surface analyses
- Lubricant supply
  → sight glasses, oil flow meters
- Lubricant film thickness
  → distance meters, pressure gauges
- Lubricant effectiveness
  → viscometry, chemical oil analyses

(c) Monitoring of friction-induced energy losses
- Friction characteristics
  → force transducers, torquemeters
- Friction-induced noise
  → noise analyses, acoustic emission analyses
- Friction-induced temperatures
  → thermocouples, IR-techniques

(d) Monitoring of wear-induced material losses
- Wear debris and lubricant contaminants
  → spectrographic oil analysis procedure (SOAP)
  → magnetic chip detectors
  →ferrography
  → radioactive tracer methods

the use-input and use-output quantities. For the movement from diagnosis towards prognosis, signals from the loss-outputs indicating the ailments of machinery may be fed back by means of servo-control equipment or microprocessors to the input, thus influencing the operational inputs in order to establish a proper functioning of the whole system.

Monitoring of Components and Lubrication

Misalignment and vibration of components can be detected by proximity transducers or accelerometers. Changes in surface conditions of components or the appearance of pits or cracks are a powerful indicator of incipient failure, particularly in rolling elements and gears. However, the application of monitoring techniques is hindered by the fact that moving surfaces have to be investigated. To overcome these difficulties, attempts have been made to use lasers for *in situ* detection of surface roughness characteristics. Another possibility is to make a replica of the surface to be monitored.

A very important aspect is the monitoring of the lubrication. If the volume of lubricant to the component is inadequate or if the physical and chemical properties of the lubricant change in service, the lubrication mode may change, leading to a deterioration of the performance of the whole system. Methods of monitoring oil supply range from visually checking the oil level in the sump or oil tank using a sight glass, to the installation of oil pressure gages and oil flow meters. These detectors are often connected to an automatic alarm system. Further, the lubricant quality or effectiveness should be monitored by taking samples of the used oils at intervals and subjecting them to laboratory tests to determine whether relevant lubricant properties have changed.

Monitoring of Friction-Induced Energy Losses

An increase or decrease in the friction of system components such as bearings and gears can be an indicator of decreased performance and incipient failure. Changes in the noise spectrum can also indicate incipient failure. Two types of acoustic signals may be distinguished:[25]

1. Noise fields due to vibrations, impacts, or aerodynamic processes, emitted in a relatively low frequency range (10 to 20 kHz).
2. Impulse-like acoustic signals of low amplitude due to microstructural changes, like micro-cracking, emitted in a frequency range of about 50 kHz to 1.5 MHz (acoustic emission).

Tribo-induced heat and temperature rise may lead to thermal distortion and thermal stresses and both may adversely affect mechanical strength of machine components and properties of the lubricant, thus influencing the performance and safety of the machine.

Monitoring of Wear-Induced Material Losses

Careful examination of wear debris or lubricant contaminants can indicate their origin and allow conclusions to be drawn about their formation, and hence the conditions of inaccessible moving parts. For example, it is not possible to examine *in situ* the working parts of a jet engine, but each drop of lubricant which circulates through the moving parts carries with it evidence of its experience in passage. The wear debris contained in lubricants may be monitored by the following methods:

Spectrographic oil analysis procedure (SOAP) — Very small concentrations of metallic wear products (1 to 2 ppm) suspended in used lubricating oil can be identified by spectrographic analysis.[26] Information thus obtained often indicates which components are wearing. For example, abnormal levels of iron in the sump oil of a Diesel engine can indicate excessive cylinder bore wear.

Magnetic chip detectors — Use of a magnet in a lubrication system provides a simple and effective method for monitoring contamination. A strong magnet will attract particles in an oil stream as, for instance, metal flakes such as arise from fatigue fragmentation. The magnetic probe is replaced at regular intervals by a fresh probe while the original is retained for the assessment of the particles adhering to it.

Ferrography — Ferrography is a technique developed to separate wear debris from the lubricant and spread it according to size on a transparent substrate for examination in an optical or scanning electron microscope.[27] The analyzer consists of a pump to deliver a diluted oil sample at low rate, a magnet to provide a high-gradient magnetic field near its poles, and an inclined transparent substrate (Ferrogram slide) on which the particles are deposited. The quantity of wear particles and their size distribution can be determined by optical density measurement.

Radioactive tracer methods — The use of radioisotopes, artificially produced by neutron irradiation, offers a convenient method for following the movement of material during deformation, transfer, or the formation of wear debris. In recent years, a great reduction has been obtained in background radiation by implanting radioactive ions instead of activating the sample. A thin-layer activation technique enables differentiation between the wear of different parts of moving machine elements.[28]

From the systematic compilation given in Table 6, the most-suited condition monitoring technique may be selected for a given situation.

REFERENCES

1. **Ku, P. M., Ed.,** Interdisciplinary Approach to Friction and Wear, NASA SP-181, National Aeronautics and Space Administration, Washington, D.C., 1968.
2. **Ku, P. M., Ed.,** Interdisciplinary Approach to the Lubrication of Concentrated Contacts, NASA SP-237, National Aeronautics and Space Administration, Washington, D.C., 1970.
3. **Czichos, H.,** *Tribology — A Systems Approach to the Science and Technology of Friction, Lubrication, and Wear,* Elsevier, Amsterdam, 1978.
4. **Bertalanffy, L. von,** *General System Theory,* Penguin, London, 1971.
5. **Dixhoorn, J. J. van and Evans, F. J.,** *Physical Structure in Systems Theory,* Academic Press, London, 1974.
6. **Ropohl, G.,** *Systems Engineering — Principles and Applications,* (in German), Hanser Verlag, München, 1975.
7. **Faurre, P. and Depeyrot, M.,** *Elements of System Theory,* North-Holland, Amsterdam, 1977.
8. **Seely, S.,** *An Introduction to Engineering Systems,* Pergamon Press, Elmsford, N.Y., 1972.
9. **Thoma, J. U.,** *Introduction to Bond Graphs and Their Application,* Pergamon Press, Oxford, 1975.
10. **Karnopp, D. and Rosenberg, R.,** *Systems Dynamics: A Unified Approach,* John Wiley & Sons, New York, 1975.
11. **Thum, H.,** Reliability and wear of mechanical aggregates, (in German), *Schmierungstechnik,* 3, 139, 1972.
12. **Fleischer, G.,** Problems of reliability of machines, (in German), *Wiss. Z. Tech. Hochsch. Magdeburg,* 16, 289, 1972.
13. **Messerschmidt-Bolkow-Blohm GmbH, München, Ed.,** *Technical Reliability, (in German),* Springer-Verlag, Berlin, 1971.
14. **Yoshikawa, H.,** Fundamentals of mechanical reliability and its application to computer aided machine design, *CIRP Ann.,* 24, 297, 1975.
15. **Bergling, G.,** Reliability of ball bearings, (in German), *Kugellager-Z.,* 51, 1, 1976.
16. **Eschmann, P.,** Safety and endurance of ball bearings (in German), *Walzlagertechnik,* 13, 3, 1974.
17. **Bartz, W. J.,** Tribo-engineering as a basis for the prevention of damages of machine elements (in German), *Schmiertech. Tribol.,* 20, 50, 1973.
18. **Fleischer, G.,** Influence of lubrication technology on the reduction of wear intensity and on the maintenance of a high reliability (in German), *Standardisierung Qualität,* 21, 83, 1975.
19. **Shooman, M. L.,** *Probabilistic Reliability — An Engineering Approach,* McGraw-Hill, New York, 1968.
20. **Czichos, H.,** A systems analysis data sheet for friction and wear tests and an outline for simulative testing, *Wear,* 41, 45, 1977.
21. **Katzenmeier, G.,** Wear Behavior and Load Carrying Capacity of Journal Bearings in the Transition Region from Full Fluid Lubrication to Partial Lubrication-Investigations Utilizing Radioisotopes, (in German), *Kernforschungszentrum,* Karlsruhe, KFK, February 1972, 1569.
22. **Junemann, H.,** Mechanical tests for lubricants (in German), *Erdöl Kohle,* 29, 259, 1976.
23. **Collacott, R. A.,** *Mechanical Fault Diagnosis and Condition Monitoring,* Chapman and Hall, London, 1977.
24. **Woodley, B. J.,** Failure prediction by condition monitoring, *Mater. Eng. Appl.,* 1, 19, 1978.
25. **Ziegler, K.,** Condition monitoring of machines and components by means of acoustic measuring techniques (in German), *Schmiertech. Tribol.,* 24, 5, 1977.
26. **Beerbower, A.,** Spectrometry and other analysis tools for failure prognosis, *Lubr. Eng.,* 32, 285, 1976.
27. **Scott, D., Seifert, W. W., and Westcott, V. C.,** Ferrography — an advanced design aid for the 80's, *Wear,* 34, 251, 1975.
28. **Gerve, A.,** Applicability of radio nuclides for the investigation of the influences of design and lubrication on the wear of machines elements (in German), *VDI-Berichte,* 196, 43, 1973.

Index

INDEX

A

AA, see Arithmetic average
AAS, see Atomic absorption spectroscopy
Abrasion, 61, 171—175, 186, 190, 192, 201, 206,
    209, 216—217, 625—627, 651
  defined, 171
  three-body, 625, 627
  two-body, 625, 627
Abrasion resistance
  plastic coatings, 636
  test methods for, 171—175
Abrasive fluids, 589
Abrasives, see also specific types
  hardness of, 175
Absolute viscosity of gases, 292
Absorbed gases, 17
Absorbed water, 17
Absorption spectroscopy, 248
Accessories for centralized lubrication systems, 389
Acid esters, 528
Acoustic emission, 660
Active oils, 357
Adapters, 522
Additives, see also specific types, 54—56, 65, 210,
    219—221, 247, 258, 260—264, 301
  antiwear, 55
  base stocks and, 308
  boundary lubrication, 54, 56, 301—302
  detergent, 286, 304
  extreme-pressure, 51, 55, 57, 64, 263, 301
  functions of, 301—306
  lubricating oil, 301—315
  seal swell, 306
  tackiness, 263, 306
  wetting, 360
ADF, see Amplitude density function
Adhesion, 34, 35, 59, 61—62, 64, 169—170,
    201—202, 206, 209—210, 327, 328, 627—
    629, 651
  balance between chemical wear and, 212—216
  coefficient of, 34
Adsorption
  energy of, 210
  heats of, 210
  on surfaces, 32—33
Adsorption-desorption energy, 209
Aerosol lubricators, 389
AES, see Auger electron spectroscopy
AFBMA, see Anti-Friction Bearing Manufacturer's
    Association
AGMA, see American Gear Manufacturers
    Association
Air, see also Atmosphere
  entrainment, 241, 243
"Air pallets", 105
Aldehydes, 306
Alkali halide crystals, 29

Alkenyl polyamine succinimides, 313—314
Alkylated aromatics, 528
Alkylene oxide derivatives, 315
Alkyl hydroxyl benzyl polyamine, 314
2-Alkyl-4-mercapto-1,3,4-thiadiazole, 310
Alloys, see also specific types
  aluminum, 640
  bronze, 542
  copper, 640
  copper-lead, 471
  laser, 638
  magnesium, 640
  titanium, 350, 640
Aluminum
  alloys of, 640
  bearings of, 471—472
Aluminum bronzes, 471, 505
Aluminum grease, 259, 306
Ambient pressure, 415
American Gear Manufacturers Association
    (AGMA), 541
  Lubrication Standards of, 547
American National Standards Institute (ANSI), 497
Ammonia compressors, 505
Amonton-Coulomb law, 49
Amplitude density function (ADF), 11
Analytical surface tools, 18—25
Angle
  gear helix, 542
  shear, 357
Annular flow, 616
Annular ring, 125
Anodized coatings, 637
ANSI, see American National Standards Institute
Anticipated wear, 204
Anti-Friction Bearing Manufacturer's Association
    (AFBMA), 497, 505, 515, 521, 522
Antioxidants, 260—262
Antisludge coupling, 576
Antiwear additives, 55
AntiWear Number (AWN), 49, 66
Apparent viscosity, 265—266
Apparent wavelength of surface, 8
Application devices, 379—380
Applied load and coefficient of friction, 40—41
Area curves of bearings, 11—13
Argon ion bombardment, 169
Arithmetic average (AA) deviation, 6
Arithmetic average (AA) roughness grades, 7
Aromatic hydrocarbons, 53, 54, 306
  alkylated, 528
Arrowhead defect, 322
Asperities, 33, 56—59, 152, 187
  blunt, 190
  interlocking of, 35
  longitudinal, 155
  surface, 189
  transverse, 155

Asperity contact area, 57
Asperity load to EHL load ratio, 153
Asperity tip curvature, 51
Atmosphere, see also Air
   environmental, 649
   lubricant combinations with, 65
Atomic absorption spectroscopy (AAS), 248
Atomically clean surfaces, 61
Auger electron spectroscopy (AES), 20—23, 169
Austenite, 505
Austenitic manganese steels, 173, 625
Automatic packing, 605, 608—610
Automatic pressure feeding grease cups, 379
Automatic secondary seal, 583
Average deviation, 6
Average film thickness of EHL, 153
Average friction and EHL, 153
Average pressure, 417
Average viscosity, 413—414
AWN, see AntiWear Number
Axial clearance (end play), 520
Axial pressure gradient, 73

B

Babbitts, 489
   bearings of, 467—469
   lead-base, 490
   linings of, 477
   tin, see Tin babbitts
BAC, see Bearing area curves
Backup
   emergency lubrication, 405—408
   roller bearing, 521
Balance, 585
   dynamic, 581
   seal, 584—586, 588
Ball bearings, 146
   Conrad, 495
   radial, 508
   radial internal clearance of, 500
   self-aligning, 495
   thrust, see Thrust bearings
Bar drawing, 322—323
Barium fluoride-calcium fluoride mixture, 273, 279
Barium grease, 259
Base oils
   synthetic, 308
   viscosity of, 574
Base stocks, 308
Basic metal sulfonates, 312
Basket filters, 401
"Bathtub-curve", 654
Bead-blasted surface, 9
Beads, 324
Bearing area curves (BAC), 11—13
Bearings, see also specific types
   aluminum, 471—472
   babbitt, 467—469
   backup roller, 521

ball, see Ball bearings
bi-metal, 475
bronze, 490
capillary-restricted, 110
centrally loaded partial, 435—441
clearance of, 519—521
coefficients of friction for, 513
compliant hydrodynamic journal, 155—156
compliant hydrodynamic thrust, 155—156
connecting-rod, 133—136
Conrad ball, 495
copper-base, 469—471
copper-lead, 490, 491
current of, 485—486
cylindrical, 126—136, 495—497, 509, 524
deflection of, 511
double-acting thrust, 110—113
dry, 279, 285—286
dynamically-loaded, 121
dynamics of, 79—89, 121—138
engine, 463
fixed-land (pivoted-pad) thrust, 425—427
fixed-type journal, 435—443
fixed-type thrust, 424—425
flexible-type thrust, 434—435
frequency of, 81—83
full journal, 441—443
furnace roll, 528
hydrostatic, 105, 107, 113—116
infinite slider, 69—71
internal, 519—521
journal, see Journal bearings
lead-based babbitt, 469
leaded bronze, 490
load ratings of, 515—516
main, 136
materials for, 467—472, 499—505
metals for, 64, 467—472
mounting for, 497, 521—523
narrow, 72
natural frequency of, 81—83
needle, 524
nonrotating, 126
partial, 435—441
performance charts for, 427—431
permeable, 69
pivoted-pad journal, 443—453
planar, 121—126
planet gear, 521, 522
plastic, 473—474
radial ball, 508
response of, 83—85
retainers of, 279
roller, see Roller bearings
rolling, see Roller bearings
rubber, 624
self-aligning ball, 495
short journal, 72—73
silver, 472
single acting thrust, 109—110
solid wall, 474—475

speed ratings of, 516—517
spherical, 496, 497, 510
stability of, 85—87
tapered, 495—497, 511, 523
temperature ratings of, 517
thrust, see Thrust bearings
tin babbitt, 467—469
tri-metal, 475
wear of, 467
wheel, 266
Belleville springs, 584
Bellows seals, 588
Bench tests, 657
Bending, 324
Benzotriazole, 310
Bevel gears, 541, 543, 546, 549
Bi-metal bearings, 475
Binders, 278
BIS/GMA, 194
Bleeding, 264—265
Blistering, 489
Blunt asperities, 190
Body-centered cubic structure, 18, 19
Boiling point, 238, 291
Bombardment of argon ions, 169
Bonded coatings, 277—279
Bonding
  covalent, 26
  diffusion, 42
  solid state, 26
Bonds, 169
Bore, 522
Borided molybdenum, 632
Boroding, 638
Boron carbide film, 178
Boron nitride, 273
Boundary conditions, 414—415
Boundary dimensions, 497—499
Boundary films, 52, 64, 209, 327
  formation of, 60—61
  friction of, 59—60
  rheodynamic formation of, 60—61
  solid-like, 63
  thickness of, 60, 62
  viscosity of, 63
Boundary lubrication, 52—61, 302, 515, 524
  additives for, 54, 56, 301—302
  friction and, 49—50
  properties of, 62—65
  selection of, 65—66
  wear and, 59—60
Breakage of gear, 559
Breakaway pressure, 117
Break-in, 52
Brinelling, 514
"Broad-section" formulas, 125
Bronze, 545
  alloys of, 542
  aluminum, 471, 505
  bearings of, 490
  iron-silicon, 505

leaded, 469, 490, 505
tin, 469
Brushes, 194
BUE, see Built-up edge
Buffered bushing seals, 617
Built-up edge (BUE), 336—338, 344, 345, 350,
  355, 357
Bulk grease handling systems, 392—393
Bulk lubricants, 53—54
Bulk modulus, 240—241
Burning of gear, 563
Bushings, 472
Bushing seals
  buffered, 617
  floating, 615—616

C

Calcium fluoride-barium fluoride mixture, 273, 279
Calcium (lime) grease, 257
Calcium 12-hydroxystearate grease, 257—258
Calculated temperature, 66
Cam-follower systems and EHL, 158—160
Candellilla was, 305
Capacitance, 142
Capillary-restricted bearing, 110
Capillary tubes, 109
Capillary viscometers, 230, 234
Capillary viscosities, 235
Carbon graphites, 474
Carbonitriding, 638
Carbon monoxide, 53
Carbons, 21, 282—284
Carburizing of case, 637
Cartography of surface roughness, 9—11
Car wheels, 528
Case
  carburizing of, 637
  hardening of, 171
Casting, 637
Catastrophic wear, 56
Cavitation erosion, 179—180, 221—222, 493,
  632—633
Cavitation pressure, 415
Centerburst defect, 322
Center distance of gear, 541
Center line average (CLA) deviation, 6
Central film thickness, 146
Centralized lubrication systems, 530
  accessories for, 389
  greases for, 390
  installation of, 389—390
  oils for, 390
  planning of, 389—390
Centralized oil and grease systems, 380—389
Centralized oil mist systems, 530
Centralized open gear spray systems, 390—391
Central Limit Theorem, 12
Centrally loaded partial bearings, 435—441
Centrally pivoted pads, 431—434

Centrifugal expansion, 516
Centrifugal forces, 513, 568, 575
Centrifugal pump, 400
Centrifuging of oil, 403
Ceramics, 29, 284—285
Cermets, 284—285
Chain couplings, 568, 573
Checkmark, see Surface texture
Check valves, 117
Chemical characteristics, 260
Chemical characterization of surfaces, 24
Chemical conversion coatings, 275, 636—637
Chemical corrosion, 490—491
Chemical emulsions, 360
Chemical properties of lubricants, 229
Chemical reactivity, 212
Chemical solutions, 360—361
Chemical vapor deposition (CVD), 634
Chemical wear, 632
    balance between adhesion and, 212—216
    corrosive, 210—212
Chemisorption, 27—28
Chemistry of surfaces, 27—29
Chill casting, 637
Chip formation, 349—350
    shear strain in, 340—341
Chips
    control of, 349—350
    machining, 465
    magnetic detectors of, 660
Chlorinated polymers, 305
Christensen's stochastic theory, 153
Chromatography
    gas, 238
    gel permeation, 248
Chrome plating, 632
Chromium, 17
Circular section, 121—123
Circulating oil systems, 532
    centralized, 530
Circumferential seals, 594—597
CLA, see Center line average
Clamped surfaces, 31
Cleanliness
    oil system, 409
    standards for, 524
Clean surfaces, 27, 40, 61
Clearance
    axial (end play), 520
    bearing, 519—521
    internal, 499—503
    internal bearing, 519—521
    osculation, 506
    radial internal, 500—503
Clearance seals, 581
    controlled, 612—616
    fixed-geometry, 617
Coatings
    abrasion-resistant plastic, 636
    anodized, 637
    bonded, 277—279

conversion, 275, 333, 636—637
diffusion, 274—275, 328, 637—638
elastomeric, 630
high-temperature, 279
inorganic, 636
organic, 636
plasma-sprayed, 281
plastic, 636
selection of, 623—624
sintered porous surface, 638
soft nitride, 638
sprayed, 634—635
sulfide, 637
surface, 328
tool, 349
vapor deposited, 634
wear-resistant, 623—644
Coaxing, 43
Coefficient of adhesion, 34
Coefficient of friction, 31, 32, 35, 37—43, 48, 317
    applied load and, 40—41
    bearings, 513
    contact pressure and, 40—41
    sliding, 46—47
    sliding speed and, 39
    starting, 42
    starting rate and, 40
    static, 42, 46—47
    surface roughness and, 41
    temperature and, 39—40
    wear rate and, 41—42
Coherence, 64—65
Cohesive bonds, 169
Cohesive wear, 165
Cold-working, 331—333, 638
Collision of transverse asperities, 155
Columnwise influence coefficients, 97—98
Combustion engines, 578
Compatibility
    metallurgical, 203, 204
    sliding, 204
Compliant hydrodynamic bearings, 155—156
Composite brushes, 194
Composite platings, 633
Composites
    metal-lamellar solid, 280—282
    polymer, 280
    self-lubricating, 275, 279—285
Compounded oils, 545
Compound formation of surfaces, 28—29
Compounds
    extreme-pressure, 327
    layer-lattice, 327
    open gear, 545
Compression packing, 605—608
Compression seal, 583
Compression test, 329
Compressors
    ammonia, 505
    reciprocating, 578
Computers, 82, 145, 146

Concave rollers, 496
Concentrates, 372
Concentrations of stress, 506, 520
Concentric annular flow, 616
Concentric cylinders, 417
    rotational viscometers of, 231
Concentric flow, 616
Condition monitoring techniques for tribological
        systems, 658—660
Conduction of heat, 78
Conductivity
    electrical, 245
    thermal, 64, 244
Cone and plate viscometer, 231
Conformal surfaces, 163
Conic section, 126
Conjunction
    elliptical, 145
    temperature of, 66
Connecting-rod bearing, 133—136
Connection
    flush, 589
    quenching, 588
Conrad ball bearings, 495
Consistency of greases, 525
Constriction, 147
Construction, 144
Contact
    continuous, 49
    cylinders in, 164
    elastic bodies, 505—506, 512
    Hertzian, 164
    line, 139—144, 163
    noncontinuous, 49
    point, see Point contact
    sliding electrical, 279
    spherical, 144, 163—164
    spin roll point, 149
    starved, 146
    stress of, 508, 513
    theoretical line, 506, 508, 513
Contact area, 3, 9, 33—36, 56
    asperity, 57
    real, 56
Contact deformation, 510—512, 651
Contact fatigue, 217—221
    mathematical model for, 514
Contact film shape, 147
Contact geometry, 517—519
Contact load, 599
Contact pressures, 506, 648
    coefficient of friction and, 40—41
Contact seals, 581
    positive rubbing, 581—612
Contaminants, 41, 364
    lubricant, 660
    solid, 465
    surface, 21
Continuous contact, 49
Continuous oil flow lubrication, 579
Contour of pad, 444—445

Contracting split rings, 594
Controlled clearance seals, 612—616
Convergent film shape, 414
Conversion coatings, 333
    chemical, 275, 636—637
Conveyor wheels, 528
Coolants, 371, 375, 376
    preservatives for, 373
    sources of organisms in, 372
    types of organisms in, 372
    working, 374—375
Coolers for oil, 403—404
Copolymers, 305
Copper, 17, 27, 28
    alloys of, 471, 640
Copper-base bearings, 469—471
Copper corrosion, 490, 491
Copper-lead alloys, 471
    bearings of, 490, 491
Corrosion, 61, 178, 201, 206—207, 212, 277, 466,
        490—492
    chemical, 210—212, 490—491
    copper, 490, 491
    fluids of, 589
    fretting, 492
    gear, 563
    lead, 490
    temperature and, 632
Corrosion inhibitors, 262, 302—303
Corrugation, 168—169
Counterformal surfaces 163
Coupling lubricant, 574—579
    grease as, 574—575
    oil as, 575—577
Coupling oils, 579
Couplings
    antisludge, 576
    chain, 568, 573
    damless, 576
    flexible, 565
    gear, 565—573
    high-performance, 571—573
    mechanical shaft, 565—579
    nonlubricated (dry), 565
    rigid, 565
    spindle, 571
    standard, 570—571
    steel grid, 569, 573
Covalent bonding, 26
Cracks, 477, 488
Crater wear, 347—348
Critical speed of vibration, 420
Crowns
    full, 506
    partial, 506
    roller, 506—508
Crystalline defects, 26
Crystalline structure of surfaces, 25—26
Crystals
    alkali halide, 29
    hexagonal structure of, 628

micronized, 170
Crystal viscometers, 235
Cubic structure, 18, 19
Cup packing, 608
Current of bearings, 485—486
Curvature of asperity tip, 51
Curved washers, 584
Cushioning, 63
Cutoff, 6, 9
Cutting
  mechanics of, 335—344
  three-dimensional, 344
  wear of, 181
Cutting energy, 342—343
Cutting fluids, see also specific types, 349, 357—
     369, 371—378
  application of, 366—368
  controls of, 365
  selection of, 361—365
  types of, 357—361
Cutting forces, 338—339
Cutting oils, 357—359
Cutting ratio, 339—340
Cutting tools, 357
  life of, 345—349
  wear to, 345
CVD, see Chemical vapor deposition
Cylinders
  concentric, 417
  in contact, 164
  upsetting of, 317—319
Cylindrical coordinates, 73
Cylindrical journal beaings, 126—136
Cylindrical roller rings, 495—497, 509, 524
Cylindrical worm gear drive, 550—551

D

Damage
  electrical, 481—487
  surface, 50, 56
  thermal, 487—488
  wiping, 483
Damless coupling, 576
DBPC, 304, 311
Dead-metal zone, 324
Debris
  grinding, 465
  jamming of, 481
  loose wear, 628
  transport of, 188
  wear, 182, 660
Decomposers, 305
Deep drawing, 324
Defects
  arrowhead, 322
  centerburst, 322
  crystalline, 26
Deflection, 168
  bearing, 511

Deformation, 59, 185
  contact, 510—512, 651
  inhomogeneous, 320, 322, 324
  plastic, 25, 171
  rough surfaces and, 186—189
Degradation of polymers, 236—237
Delamination, 190
  of wear, 170—171
Delay period, 179
Density, 239—240
  amplitude, 11
  dislocation, 170
  gases, 292
  lubricants, 414
  temperature and, 240
Dental restorative materials, 194
Deoxidized steels, 348—349
Deposition
  chemical vapor, 634
  electroless, 633
  physical vapor, 634
Depressants, 305
Depth of wear, 49
Designs, 298—299
Detergents, 286, 304
Deterioration control, 375—376
Deviation
  arithmetic average, 6
  center line average, 6
  rms standard, 152
Dialkyldiphenylamine, 311
Diaphragm valve, 405
Diatomaceous earth, 363
Dibasic acid esters, 528
2,6-Di-*t*-butyl-*para*-cresol (DBPC), 304, 311
Dichalcogenides, 272
Die forging, 319—320
Dielectric materials, 194
Die pickup, 324, 328
Die wear, 320
Diffraction, 20, 27
Diffusion bonding, 42
Diffusion coatings, 274—275, 328, 637—638
Dilauryl selenide, 312
Dimensional stability, 505
Direct matrix inversion, 96—97
Direct-operated diaphragm valve, 405
Disc machines, 171, 182
Discontinuity, 506
Dislocations, 25
  density of, 170
  edge, 25, 26
  screw, 25, 26
Dispersants, 304, 306
Dispersions, 286—288
Displacement, 423
Disposable element filter, 402
Disposal, 365
Dissipation
  energy, 584
  power, 134

turbulent, 418
Dissolved nitrogen, 53
Dissolved oxygen, 53
Dissolved water, 53
Distance
    break-in sliding, 52
    gear center, 541
Distortion of grain, 487, 488
Distribution
    exponential, 653
    failure, 654
    gamma, 653—654
    Gaussian height, 11—12
    height, 11—14
    load, 510—512
    non-Gaussian height, 12—13
    normal, 653
    pressure, 140, 142—144, 147
    Rayleigh, 653
    temperature, 77
    Weibull, 653
2,4-Ditertiarybutyl-*p*-cresol (DBPC), 304, 311
Double-acting thrust bearings, 110—113
Double-enveloping worm gear drive, 550—551
Double helical gears, 541
Double seal, 589
Drain pipes, 408—409
Draw bead, 324
Drawing
    bar, 322—323
    deep, 324
    tube, 322—323
    wire, 322—323
D-rings, 611
Drip feed, 530
Drives
    gear, see Gears
    surfaces of, 31
Dropping point of greases, 264
Dry bearings, 279
    selection of materials for, 285—286
Dry (nonlubricated) couplings, 565
Dry lubricants, 528—529
Dual basket filter, 401
Dual line lubrication system, 386—387
Dual lip seal, 598
Ductile failure, 195
Dusting wear, 282
Dyes, 263—264, 306
Dynamically-loaded bearings, 121
Dynamic balance, 581
Dynamic displacement, 423
Dynamic performance, 648, 652
    of bearings, 87—88
Dynamics
    bearings, 79—89, 121—138
    films, 420—424
Dynamic seals, see also specific types, 581
    O-ring, 612

E

Eccentricity, 417
    ratio of, 73
Economics of machining, 351—352
Economy of fuel, 302
Edge
    built-up, see Built-up edge
    dislocation of, 25, 26
Effective viscosity, 413
EHL, see Elastohydrodynamic lubrication
Elastic bodies, 505—506, 512
Elastic modulus, 63
Elastic, plastic solid, 59
Elastic shear modulus, 151
Elastohydrodynamic lubrication (EHL) film, 139,
        524, 543
    average friction and, 153
    cam-follower systems and, 158—160
    friction and, 147—151, 153
    full-film, 39—151
    gears and, 157—158
    high-slip friction and, 150—151
    low-slip friction and, 147—150
    machine components and, 156—160
    partial-film, 152—155
    rolling element bearings and, 156
    temperature and, 147
    thickness of, 139—142, 153, 516, 527
Elastohydrodynamic lubrication (EHL) load, 153
Elastomeric coating, 630
Elastomers, 190—193, 588, 604, 605
    outgassing of, 588
Electrical conductivity, 245
Electrical contacts, 279
Electrical damage, 481—487
Electrical wear, 180
Electrochemical wear, 222—223
Electroless deposition, 633
Electroless nickel plate, 633
Electron beam hardening, 637
Electron diffraction, 20, 27
Electron microprobe (EM), 23
Electron microscopy, 18
    scanning, 3, 23
    transmission, 170
Electron spectroscopy, 20—23, 169
Electroplating, 13, 633
Electrostatic effects, 487
Element analysis, 99—104
Element bearings, 156
Elliptical conjunction, 145
Elliptical section, 123
EM, see Electron microprobe
Embedability, 465—466
Embrittlement, 64
    hydrogen, 633
Emergency lubrication backup, 405—408
Emissions, acoustic, 660
Emulsified oils, 359
Emulsion modifiers, 305

Emulsions
  chemical, 360
  inverted, 365
Enclosed gears, 549, 550—551
End play (axial clearance), 520
Endurance of rolling bearings, 514—515
Energy
  of adsorption, 209, 210
  dissipation of between seal faces, 584
  equation for, 77—79
  specific cutting, 342—343
  surface, 27
Energy absorption-controlling components, 31—32
Energy losses
  friction-induced, 651, 660
  tribo-induced, 651
Engine bearings, 463
Engineering systems approach, 645, 648
Engine lubricants, 307
Engine oils, 308
Engines
  gas turbine, 505
  internal combustion, 578
Entrainment, 62—64
  air, 241, 243
  film, 61
Environmental atmosphere, 649
Environmental effects, 29, 165—166
EP, see Extreme pressure
Epoxy, 194
Equilateral triangular section, 125
Erosion, see also Wear, 194, 221, 493, 630—633
  cavitation, 179—180, 221—222, 493
Errors
  of form, defined, 6
  in geometry, 51
ESCA, see X-ray photoemission spectroscopy
Esters, 306
  dibasic acid, 528
  organic, 528
  phosphate, 528
  polymethacrylic acid, 314
  polyol, 528
  silicate, 528
Etching, 13, 18
  selective, 632
Ethers, 528
Eutectic mixtures, 273
Evaporation, 265
Excitation frequency, 447—453
Exit construction, 144
Expansion
  centrifugal, 516
  linear, 175
Exponential distribution, 653
  split ring, 594
External friction, 165
Externally pressurized seals, 614
External seal operation, 582
Extreme pressure (EP), 266, 588
  functionality of, 55

Extreme pressure (EP) additives, 51, 55, 57, 64,
    263, 301
Extreme pressure (EP) compounds, 327
Extreme pressure (EP) films, 56, 64
Extreme pressure (EP) lubricant, 357
Extreme pressure (EP) oil, 50, 64, 545
Extrusion, 323—324, 329
  hydrostatic, 324
  ring, 612

F

Face-centered cubic structure, 18, 19
Face seals, 581—591, 595
Face temperature, 343
Face width, 541
Facings, 635
Failure
  analysis of, 532—535
  distribution of, 654
  grease and, 535
  mechanism of, 654
  modes of, 533—534
  spiral, 611
  torsional, 611
  wear-induced, 654
  weight viscometer, 234
"Fairing" procedures, 75
Falex tests, 267
Fatigue, 61, 170—171, 190, 192, 218, 222, 477,
    532, 534
  contact, 217—221, 514
  flow, 167—168
  resistance to, 274, 504
  surface, 201, 207, 651
  tests for, 170
  thermal, 488
Fatigue strength, 463—465, 477
Fats, 310
Fatty acids, 65
  unsaturated, 306
Fatty oils, 357
Feeding grease cups, 379
Ferrography, 181, 660
Ferromagnetic seals, 618
"Ferrox" treatment, 637
Fibers, 275
Fiddling, 43
Field ion microscope, 18
Fillers, 260—264
Film
  boron carbide, 178
  dynamic properties of, 420—424
  elastohydrodynamic, 524, 543
  entrainment of, 61
  extreme pressure, 56, 64
  friction of, 59—60
  full, 327, 333, 524
  full-fluid, 327, 333
  Hertzian inlet, 141

hydrodynamic, 497
inlet, 141, 142, 146
maximum pressure of, 134
mixed, 333
oxide, 64, 169, 628, 637
partial, 524
polymeric, 333
pressure of, 134
reaction, 273
rheodynamic boundary, 60—61
rubbed, 277
soft metal, 273—274
solid-like boundary, 63
squeeze, 69, 121—138
stiffness of, 110
surface, 51, 56
thin, 277—279
transient behavior of, 121
turbulent, 75
viscosity of, 63
Film meniscus, 142
Film shape
convergent, 414
point contact, 147
pressure distribution vs., 142—144
Film thickness, 113, 136, 144—146
average, 153
boundary, 60, 62
central, 146
chart of, 144
elastohydrodynamic, 139—142, 516, 527
formulas of, 527
load vs., 140
minimum, 134, 146
parameter of, 51
predictions of, 135
specific, 218
Filters
disposable element, 402
dual basket, 401
mechanically cleaning, 402
Filtration, 328, 362, 376
oil, 401
Finger springs, 584
Finish of surface, 32, 44
Finite difference, 93—99
Finite element analysis, 99—104
Firecracking, 176
Fit, 522
Fixed-geometry clearance seals, 617
Fixed-land (pivoted-pad) thrust bearing, 425—427
Fixed-type journal bearings, 435—443
Fixed-type thrust bearings, 424—425
Flame hardening, 637
Flange packing, 608
Flattening, 322
Flexible couplings, 565
Flexible-type thrust bearings, 434—435
Floating bushing seals, 615—616
Floating piston rings, 611
Flood application, 366—367

Flow, 106—107, 127
calculation of, 98—99
concentric, 616
continuous oil, 579
equation for, 106
fatigue of, 167—168
instability of, 417
laminar, 106, 414, 616
oil, 399, 579
plastic, 56, 59, 169, 561, 562
restrictors of, 107—109
shear stress of, 59
strength of, 56
Taylor vortex, 74
transition of, 417
turbulent, 417, 616
"Flow chart", 657
Flow rate, 122, 139
oil, 399
Fluctuating torque applications, 578
Fluids, see also specific types
abrasive, 589
corrosive, 589
cutting, see Cutting fluids
hazardous, 589
specialized control of in seals, 617—619
synthetic, 260, 360—361
Flushable seals, 497
Flush connection, 589
Foam decomposers, 305
Foaming, 243
Fog systems, 530
Force-fed lubrication, 558
mechanical, 379
Force-fed oil system, 554
Forces
centrifugal, 513, 568, 575
cutting, 338—339
gyroscopic, 513
van der Waals, 26
Force transducers, 658
Force transmitting components, 31
Forging, 317—320
impression die, 319—320
Form errors, 6
Forming of materials, 646
Formulation of lubricant, 306—309
Forward slip, 329
Four-ball methods, 266—267
Free path, 292, 299
Free time, 165
Freeze seals, 617—618
Frequency
bearing natural, 81—83
unbalance excitation, 447—453
vibration, 168
Fresh surface, 61
Fretting, 176—178, 523, 536, 631—632
Fretting corrosion, 492
Friability, 64—65
Friction

average, 153
boundary film, 59—60
boundary lubrication and, 49—50
coefficient of, see Coefficient of friction
elastohydrodynamic lubrication and, 147—151
energy losses induced by, 651, 660
external, 165
high-slip, 150—151
importance of, 50
internal, 165
kinetic, 42—43
laws of, 35—36
low-slip, 147—150
mathematical representation of, 317
measurement of, 37—39
modifiers of, 301, 302
requirements for, 65
rolling, 43
rolling bearings and, 513
severity of, 50—51
sliding, 35—36, 46—47
static, 42—43, 46—47
sticking, 317
wear testing and, 276—277
Frictional heating, 64
Frictional oscillation, see Stick-slip
Frictional vibration, see Stick-slip
Friction polymer, 56
Fuel economy, 302
Full crown, 506
Full-film EHL, 139—151
Full film lubrication, 327, 333, 524
Full journal bearings, 441—443
stability of, 443
Full-scale tests, 657
Function, 651—654
Furnace roll bearings, 528

G

Gages
pressure, 658
strain, 38
Gamma distribution, 653—654
Gap
interfacial, 3
map of, 10, 11
Gas, see also specific types, 361
absolute viscosity of, 292
absorbed, 17
density of, 292
equation of state of, 292
kinematic viscosity of, 292
mean free path of, 292
nature of, 291
"perfect", 291
properties of, 291—300
solubility of, 241—243
sonic velocity of, 292
specific heat of, 292

specific volume of, 298
temperature of, 292
viscosity of, 292—298
Gas chromatography, 238
Gaseous lubricants, 52—53
Gas turbine engines, 505
Gaussian height distribution, 11—12
GDMS, see Glow discharge mass spectroscopy
Gear couplings, 565—573
oils for, 579
Gear lubrication, 543—544
hypoid, 517
Gear pump, 399—400
Gears, see also specific types, 279
bevel, 541, 543, 546, 549
breakage of, 559
burning of, 563
center distance of, 541
corrosion of, 563
double helical, 541
elastohydrodynamic lubrication and, 157—158
enclosed, 549—551
face width of, 541
helical, see Helical gears
helix angle of, 542
herringbone, 549
hypoid, 541, 546
loads on, 544
material for, 542
open, 390—391, 545, 552, 555
open installation of, 391
pitting of, 559
planet, 521, 522
plastic flow of, 561
scoring of, 561—563
scratching of, 561
size of, 544
spalling of, 561
speed of, 544
spur, 539, 543, 546, 549
temperature of, 544
tooth size of, 542
types of, 541, 544
wear of, 558—563
worm, 541, 546, 550—551
Gel permeation chromatography (GPC), 248
Geometry
contact, 517—519
errors in, 51
machine, 65
Glasses, 329
"Glass transition", 59
Glow discharge mass spectroscopy (GDMS), 23
Gold, 17, 28
Gouging wear, 176
Gow-Joule effect, 612
GPC, see Gel permeation chromatography
Grain boundaries, 18
Grain distortion, 487, 488
Graphite, 269—272, 282—284, 286, 327, 329,
505, 542

carbon, 474
Graphite composite brushes, 194
Graphite fluoride, 273
Grease, see also specific types, 56, 255, 525—528, 546, 574
    aluminum, 259, 306
    as coupling lubricant, 574—575
    barium, 259
    bulk, 392—393
    calcium (lime), 257
    calcium 12-hydroxystearate, 257—258
    centralized system of, 380—390, 530
    consistency of, 525
    defined, 255
    dropping point of, 264
    failure and, 535
    handling of, 392—393
    lime (calcium), 257
    lithium, 255—257
    lithium 12-hydroxystearate, 255
    nonsoap, 255
    penetration of, 264
    selection of, 264—267
    sodium, 258, 262
    sodium-aluminum, 258
    sodium-calcium, 258
    stiffness of, 525
    viscosity of base oil of, 574
Grease cups, 379
Grease gun, 380
Grid couplings, 569, 573
Grinding, 13, 7, 352—355
    debris of, 465
    mechanics of, 353—355
Guidance of motion, 646
Gyroscopic forces, 513

H

Halide crystals, 29
Halogenated hydrocarbons, 528
Handling ease, 260
Hardening
    case, 171
    electron beam, 637
    flame, 637
    laser, 637
    spark, 632, 638
    work, 173
Hard facings, 635
Hardness, 56, 57
    abrasive, 175
    hot, 517, 518
    indentation, 202
    material, 517
    short-term hot, 517
    standards for, 524
    surface, 51, 628
Hazardous fluids, 589
HDPE, see High density polyethylene

Header tank, 405
Heads of seals, 581
Health factors, 368
Heat, see also entries beginning with Thermal, 357
    of adsorption, 210
    conduction of, 78
    specific, 244, 292
    vaporization, 244—245
Heat capacity of lubricants, 414
Heaters, 404
Heating
    frictional, 64
    induction, 637
    oil reservoirs, 398
Heavy gear wear, 559
Height, 7
Height distributin, 14
    Gaussian, 11—12
    non-Gaussian, 12—13
Helical gears, 541, 543, 546, 549
    double, 541
Helix angle of gears, 542
Herringbone gears, 549
Hertzian conditions, 181
Hertzian contact, 164
Hertzian inlet film, 141
Hertzian stresses, 167, 175
Hexagonal crystal structure, 628
High density polyethylene (HDPE), 190
High-misalignment applications, 579
High-performanc couplings, 571—573
High pressure application, 367—368
High-pressure, high-shear capillary viscosities, 235
High-slip friction and EHL, 150—151
High-speed applications, 578
High speed and seals, 588
High speed steels (HSS), 346, 352
High-temperature coatings, 279
High-torque, high-misalignment applications, 579
Hindered phenols, 304
Horseshoe-shaped constriction, 147
Hot hardness, 518
    short-term, 517
Hot-working lubricants, 329—333
HSS, see High speed steels
Hybrid seals, 614
Hydrocarbons, 17, 52, 66, 310
    aromatic, see Aromatic hydrocarbons
    halogenated, 528
    naphthenic, 53
    nonmethane, 53
    olefinic, 54
    paraffinic, 53
    synthetic, 528
Hydrodynamic bearings, 155—156
Hydrodynamic designs, 298—299
Hydrodynamic films, 497
Hydrodynamic seals, 612—614
Hydrogen embrittlement, 633
Hydrostatic bearings, 113—116
    defined, 105

power consumption of, 107
Hydrostatic designs, 298—299
Hydrostatic extrusion, 324
Hydrostatic seals, 614—615
Hypoid gears, 541, 546
    lubricants for, 517

I

Ideal elastic, plastic solid, 59
Imbalance of rotor, 420
Impact, 175, 648
Impedance, 127, 129
Implantation of ions, 274, 635
Impression die forging, 319—320
Inactive oils, 357
Inclined plane, 37
Indentation, 320, 324
    hardness of, 202
Induction heating, 637
Inertia
    effects of, 418—420
    pad, 445—447
"Infant death" of system, 654
Infinite slider bearings, 69—71
Influence coefficients, 97—98
Information transmission, 646
Infrared (IR) instruments, 658
Inhomogeneous deformation, 320, 322, 324
Inlet film, 141, 142, 146
Inlet shear, 141
Inner ring of tapered bore, 522
Inorganic coatings, 636
Input-output relations, 646
    mathematical, 648
Instability of flow, 417
Installation
    centralized lubrication systems, 389—390
    open gear, 391
    seals, 604, 607, 610
Instationary loading, 121
Insulation, 64
Interacting surfaces, 164
Interaction factor, 65
Interaction problems, 306—307
Interface pressures, 317, 320, 327
Interface shear, 317
    strength of, 329
Interfacial gap, 3
Interfacial wear on smooth surfaces, 189—190
Interference fit, 522
Interlocking of asperities, 35
Internal clearance, 499, 519—521
    radial, 500, 501—503
Internal combustion engines, 578
Internal friction, 165
Internal seals, 581
International Organization for Standardization (ISO),
        6—7, 497
    ten-point height of, 7

Inversion, 96—97
Inverted emulsion, 365
Ion bombardment, 169
Ion implantation, 274, 635
Ion microscope, 18
Ion plating 178
Ion-scattering spectroscopy (ISS), 23
IR, see Infrared
Iron, 17
Iron retainers, 505
Iron-silicon bronze, 505
ISO, see International Organization for
        Standardization
ISS, see Ion-scattering spectroscopy

J

Jamming of debris, 481
Jiggling, 43
Joint ring, 594
Journal bearings
    compliant hydrodynamic, 155—156
    cylindrical, 126—136
    fixed-type, 435—443
    full, 441—443
    hydrodynamic, 155—156
    hydrostatic, 113—116
    pivoted-pad, 443—453
    short, 72—73
Journal scoring, 479

K

Ketones, 306
Kiln car wheels, 528
Kinematic friction, 42—43
Kinematics, 132, 341—342
    rolling bearings, 512—513
    surface reactions, 215—216
Kinematic viscosity, 413
    gases, 292

L

Labyrinth seals, 617
Lamellar solid-metal composites, 280—282
Lamellar solids, 269—273
Laminar concentric annular flow, 616
Laminar flow, 414, 616
    equation for, 106
Land tool wear, 347
Lantern ring, 608
Laser alloying, 638
Laser hardening, 637
Lay, 9
    to surface, 8
Layer-lattice compounds, 327
Layer of oxide, 9, 17

LDPE, see Low density polyethylene
Lead-base babbitts, 490
    bearings of, 469
Lead-copper alloys, 471
    bearings of, 490, 491
Lead corrosion, 490
Leaded bronzes, 469, 505
    bearings of, 490
Leakage, 597
    seals, 587, 598—604
LEED, see Low energy electron diffraction
Length in roughness sampling, 6
Life
    cutting tool, 345—349
    seal, 587—588, 598—604
    tool, 346—347
    wear, 66
Lift pockets, 117
Lifts, 105, 116—119
Light gear wear, 558—559
Lime (calcium) grease, 257
Limiting reduction, 323
Linear expansion coefficient, 175
Line contacts, 139—144, 163
    theoretical, 506
Line strainer, 401
Linings, 477
Lip contact load, 599
Lip formation, 175
Lip packing, 605
Lip seals
    dual, 598
    radial, 597—605
Liquid lubricants, 53—56, 209, 229—254
Lithium 12-hydroxystearate grease, 255
Lithium grease, 255—257
Load, 51, 216
    applied, 40—41
    asperity, 153
    defined, 105—106
    film thickness vs., 140
    gear, 544
    lip contact, 599
    transition, 50
Load-bearing characteristics, 56—59
Load-carrying capacity, 50, 55, 57, 66, 72
''Load coefficient'', 105
Load-deflection exponent, 512
Load distribution, 510—512
Loading
    instationary, 121
    non-stationary, 121
    seal, 584
Load orientation of pad, 444
Load ratings of bearings, 515—516
Load ratio, 153
Load-supporting area, 49, 59
Longitudinal asperities, 155
Long-term temperature resistance, 517
Loose wear debris, 628
Low density polyethylene (LDPE), 186, 188, 190

Low energy electron diffraction (LEED), 20, 27
Low friction components, 32
Low-slip friction and EHL, 147—150
Low-speed applications, 578
Low-temperature pumpability, 260
Low-temperature vicosities, 232
Lubricity, 54

M

Machinability, 362
Machine components and EHL, 156—160
Machine geometry, 65
Machine tool paint, 365
Machining
    economics of, 351—352
    orthogonal, 335—336
Machining chips, 465
Magnesium alloys, 640
Magnesium oxides, 29, 263
Magnetic chip detectors, 660
Magnetic seals, 584
Main bearings, 136
Manganese steels, 173, 625
Maps, 129
    gap, 10, 11
Mass spectroscopy, 23
Material losses
    tribo-induced, 651
    wear-induced, 651, 660
Materials, see also specific materials
    forming of, 646
    hardness of, 517
Mathematical input-output relations, 648
Mathematical models, 209
    contact fatigue, 514
Mathematical representation of friction, 317
Matrix inversion, 96—97
Maximum film pressure, 134
Maximum shear stress, 196
Maximum temperature, 66
Mean free path, 299
    gases, 292
Mean free time, 165
Mechanical face seals, 581—591, 595
Mechanical force-feed lubricators, 379
Mechanically cleaning filter, 402
Mechanical shaft couplings, 565—579
Mechanical stability, 266
Mechanical systems
    lubricated, 645
    reliability of, 651—654
    wear-induced failure of, 654
Mechanical wear, 477—481
Mechanical work transmission, 646
Mechanics
    cutting, 335—344
    grinding, 353—355
Melting, 13
    surface, 39

Metal alkylphenate sulfides, 313
Metal alkylsalicylates, 313
Metal bellows seals, 588
Metal dialkyldithiocarbamates, 310
Metal dialkyldithiophosphates, 309
Metal films, 273—274
Metal-lamellar solid composites, 280—282
Metallic bearing materials, 467—472
Metallurgical compatibility, 203, 204
Metallurgical solubility, 204
Metallurgy of surfaces, 25—26
Metals, see also specific metals
  babbitt, see Babbitts
  bearing, 64
  forming process for, 50
  noble, 28
  porous, 108
  removal of, 335—356
Metal salts, 273
Metal sulfides, 273
Metal sulfonates, 312—313
Metal transfer, 320
Metalworking, 324—325
Mica-filled epoxy, 194
Micro-arc, 180
Microbial action, 371—378
Microbial control of working coolants, 374—375
micro-EHL, 154—155
Micronized crystals, 170
Microprobe, 23
"Micro-rheodynamic" lubrication, 60
Microscopy, 17—18
  electron, see Electron microscopy
  field ion, 18
  scanning electron, 3, 23
  transmission electron, 170
Micro-slip, 43
Mild extreme pressure oil, 50
Mile High Stadium, Denver, Colorado, 105
Milling, 13, 255
Mineral fatty oil blend, 358
Mineral oil, 53, 525—528
  straight, 50, 51, 357
  sulfo-chlorinated, 359
  sulfurized, 358—359
Minimum film thickness, 134, 146
Misalignment, 43, 51, 488, 497, 506, 517—520,
    535, 559, 579
  components, 659
Mist application, 366
Mist systems
  centralized oil, 530
  oil, 532
  orifice oil, 387—389
Mixed film, 333
Mixed lubrication, 327
Mixed metal sulfides, 273
Mixed surface textures, 13—15
Mixing valve, 407
Mobility, 127, 129, 133—134
Moderate gear wear, 559

Moes diagram, 144
Molybdenum, 632
Molybdenum disulfide, 269—272, 279, 327, 330
Monitoring
  component, 659
  condition, 658—660
  friction-induced energy loss, 660
  lubriction, 659
  operating variable, 658—659
  wear-induced material loss, 660
Motion
  guidance of, 646
  one-dimensional, 126—128
  radial, 126
  sliding, 565
  stick-slip, 43—47
  two-dimensional, 129—132
Mounted-bearing units, 497
Mountings
  bearing, 521—523
  outer ring, 523
Multidisciplinary approach, 645
Mutual solubility, 203

N

Naphthalene, 305
Naphthenic hydrocarbons, 53
Naphthenic oils, 260
Narrow bearing, 72
"Narrow-section" formulas, 125
National Lubricating Grease Institute (NLGI), 546
Natural frequency of bearings, 81—83
NBS, see Nuclear back scattering spectroscopy
Needle roller bearings, 524
Neutral metal sulfonates, 312
Newtonian model, 147
Nickel plate, 633
Nitride coatings, 638
Nitriding, 637
Nitrogen, 52
  dissolved, 53
Nitrogen dioxide, 53
NLGI, see National Lubricating Grease Institute
Noble metals, 28
Nodular-graphite iron retainers, 505
Noise fields, 660
Noncohesive wear, 165
Noncontinuous contact, 49
Non-Gaussian height distribution, 12—13
Nonlubricated (dry) couplings, 565
Nonmethane hydrocarbons, 53
Non-Newtonian models, 150
Nonplanar surfaces, 126
Nonreversing lubrication system, 381—383
Nonrotating bearings, 126
Nonsoap grease, 255
Nonsoap thickeners, 259
Nonsoluble combinations, 628
Non-stationary loading, 121

Normal distribution, 653
Normal running, 654
No-slip condition, 69
Nuclear back scattering spectroscopy (NBS), 23
Nylon, 53, 190, 195

O

Oil, see also specific types, 528
    active, 357
    base, see Base oil
    blends of, 358
    centralized lubrication systems and, 390
    centrifuging of, 403
    compounded, 545
    coupling lubricant, 575—577
    cutting, 357—359
    emulsified, 359
    engine, 307, 308
    extreme-pressure, 64, 545
    filtration of, 401
    flow rate of, 399
    gear coupling, 579
    inactive, 357
    mild extreme pressure, 50
    mineral, see Mineral oil
    mineral fatty blends of, 358
    naphthenic, 260
    paraffinic, 260
    petroleum, 260
    pressure control for, 404—405
    settling of, 400—401
    soluble, 359
    spectrographic analysis of, 660
    straight fatty, 357
    straight mineral, 50, 51, 357
    strong extreme-pressure, 50
    suction conditions of, 399
    sulfo-chlorinated mineral, 359
    sulfurized fatty mineral blends of, 358
    sulfurized mineral, 358—359
    supply pressure of, 399
    synthetic base, 308
    temperature in, 51, 404
    universal engine, 307, 308
    viscosity of, 399
Oil additives, 301—315
Oil coolers, 403—404
    plate, 403
    radiator, 404
    shell and tube, 403
Oil flow lubrication, 579
Oil fog systems, 530
Oiliness, 54, 55, 57
Oil lifts, 105, 116—119
Oil mist systems, 532
    centralized, 530
    orifice, 387—389
Oil phase, 260
Oil pumps, 399—400

    protection of, 399
Oil reservoirs, 395—398
    capacity of, 395
    heaters for, 404
    heating of, 398
Oil systems
    centralized, 380—389, 530
    circulating, 530, 532
    cleanliness of, 409
    force-fed, 554
    orifice, 387
    piping for, 408—409
    splash, 547
Olefinic hydrocarbons, 54
Olefin oligomers, 528
Olefins, 310
Oligomers, 528
One-dimensional motion, 126—128
Open gear, 552, 555
Open gear compounds, 545
Open gear installations, 391
Open gear spray systems, 390—391
Operating conditions, 65
Operating principles of oil and grease centralized
        systems, 380—389
Operating variables, 645, 648
    monitoring of, 658—659
Operational performance, 277—288
Operational severity, 50—51
Optical viscometers, 235
Orbit computation, 133—134
Organic coatings, 636
Organic copolymers, 305
Organic esters, 528
Organic thickeners, 259
Orientation, 18
Orifice oil lubrication system, 387—389
Orifices, 108—109
O-ring seals, 605, 611
    dynamic, 612
Orthogonal machining, 335—336
Oscillation, frictional, see Stick-slip
Osculation, 507
Osculation clearance, 506
Outer ring mountings, 523
Outflow rate, 122
Outgassing of elastomers, 588
Output-input relations, 646
    mathematical, 648
Overbased metal sulfonates, 312—313
Overheating, 50
Oxidation, 259
    inhibitors of, 258, 303—304
    stability of, 247—249, 265, 286
Oxide films, 64, 169, 637
    formation of, 628
Oxide layer, 9, 17
Oxides, 32, 35, 325, 329
Oxygen, 52
    dissolved, 53

P

Packing, see also specific types
  automatic, 605, 608—610
  compression, 605—608
  cup, 608
  flange, 608
  lip, 605
  squeeze, 605, 610—612
  U-ring, 608
  V-ring, 609—610
Packing seals, 605—612
Pads
  centrally pivoted, 431—434
  contour of (preloading), 444—445, 521, 584
  inertia of, 445—447
  load orientation of, 444
  pivoted, 431—434, 444
Paint on machine tools, 365
Pallets, 105
PAN, see Phenyl alpha naphthylamine
Paraffinic hydrocarbons, 53
Paraffinic oils, 260
Partial bearings, 435—441
Partial crown, 506
Partial-film EHL, 152—155
Partial-film lubrication, 524
Partial pressure, 291
Particles
  formation of, 181—182
  spherical, 171
PCTFE, see Polychlorotrifluoroethylene
Peaks, 21
Peeling, 34
Penetration of greases, 264
Percussive mode, 175
''Perfect gas'', 291
Performance
  dynamic, 648, 652
  thrust bearing, 427—431
Permeable bearings, 69
Permeation chromatography, 248
Petroff equation, 73
Petroleum oils, 260
Phases, 18
Phenols, 305
  hindered, 304
Phenothiazine, 311
Phenyl alpha naphthylamine (PAN), 311
Phosphate esters, 528
Phosphating, 178, 631, 636
  steel, 333
Phosphosulfurized pinene, 312
Photoemission spectroscopy, 22—23
Physical properties of lubricants, 229
Physical sizes, 62
Physical vapor deposition (PVD), 634
Pickup of die, 324, 328
Pinene, 312
Pipes, 408—409
Piston rings, 281, 583, 591, 594

floating, 611
Pitting, 171, 218, 222, 481, 482, 560
  gear, 559
Pivoted-pad journal bearings, 443—453
Pivoted pads, 431—434, 444
Pivoted-pad (fixed-land) thrust bearing, 425—427
Planar bearings, 121—126
Plane, 37
Planet gear bearings, 521, 522
Planning of centralized lubrication systems, 389—390
Plasma-sprayed coatings, 281
Plastic bearings, 473—474
Plastic coatings, 636
Plastic flow, 169, 562
  gear, 561
  shear stress of, 59
  strength of, 56
Plasticity
  criterion for, 58
  index of, 8, 58
Plastics, 325
  deformation of, 25, 171
Plastic solids, 60
  ideal elastic, 59
Plate oil coolers, 403
Plates, 633
Platings
  composite, 633
  ion, 178
  porous chrome, 632
Plowing, 59
Pneumatically controlled diaphragm valve, 405
Pocket pressure, 106
Pockets of lift, 117
Point contact film shape, 147
Point contacts, 144—147, 163
  spin roll, 149
Poiseuille's law, 230
Poisson's ratios, 8
Polyacetal, 195
Polyacrylates, 305
Polyamine amide imidazoline, 314
Polychlorotrifluoroethylene (PCTFE), 186
Polyesters, 528
Polyethylene, 195
  high density, 190
  low density, 186, 188, 189
  ultrahigh molecular weight, 186
Polyglycol ethers, 528
Polyimides, 473
Polyisobutylene, 314—315
Polymeric films, 333
Polymers, see also specific types, 275—276, 325
  chlorinated, 305
  composites of, 280
  degradation of, 236—237
  friction, 56
  selection of, 195—198

transfer of, 186
unfilled, 185—190
wax alkylated naphthalene, 305
wax alkylated phenol, 305
Polymethacrylates, 194, 305
Polymethacrylic acid esters, 314
Polyol esters, 528
Polyphenyl ethers, 528
Polypropylene, 190
Polysiloxanes, 305
Polytetrafluoroethylene (PTFE), 39, 186, 189, 190,
    275, 279, 280, 473
Porous bushings, 472
Porous chrome plating, 632
Porous metals, 108
Porous surface coatings, 638
Positive rubbing contact seals, 581—612
Pour point depressants, 305
Power consumption of hydrostatic bearing, 107
Power dissipation, 134
Power loss calculation, 98—99
Precoat, 363
Predictions
    film thickness, 135
    wear, 201
Preloading, 444—445, 521, 584
Preservatives, 372—373
Pressure, 599
    ambient, 415
    average, 417
    breakaway, 117
    cavitation, 415
    contact, 40—41, 506, 648
    extreme, see Extreme pressure
    high, 367—368
    interface, 317, 320, 327
    limits of, 586—587
    maximum film, 134
    partial, 291
    pocket, 106
    supply, 399
    thermal, 175
    vapor, 237—239
    viscosity vs., 543
Pressure control for oil, 404—405
Pressure distribution, 140, 147
    film shape vs., 142—144
Pressure feeding grease cups, 379
Pressure gages, 658
Pressure gradient, 73
Pressure profile factor, 585
Pressure ratio, 127, 129
Pressure spike, 143
Pressure tank, 407
Pressure-velocity (PV) limit, 195, 196
Pressure-viscosity coefficient, 60
Pressure-viscosity relationships, 140, 234—235
Pressurized seals, 614
Probabilistic function, 652
Profile analyzers, 3, 9
Progressive lubrication system, 381—386

Protection ranges, 66
PTFE, see Polytetrafluoroethylene
Pumpability at low-temperature, 260
Pumping rings, 588
Pumps, 399—400
Pure surface textures, 13—15
PV, see Pressure-velocity
PVD, see Physical vapor deposition

Q

Quality control, 32
    of concentrates, 372
Quenching connection, 588

R

Radial ball bearings, 508
Radial internal clearance of bearings, 500—503
Radial lip seals, 597—605
Radial motion, 126
Radiation, 648
Radiator oil coolers, 404
Radioactive tracer methods, 660
Rail lubrication, 168
Rayleigh distribution, 653
Reaction films, 273
Reaction products, 56
Reactivity, 212
Real contact area, 56
Real load-supporting area, 59
Reciprocating compressors, 578
Recirculating lubricant-coolant systems, 328
Reconstruction, 28
Recovery, 34
Rectangular section, 123—124
Reduced Reynolds number, 417
Reduction, 323
Reinforcing fibers, 275
Relative wear resistance, 174
    steels, 167
Relaxation, 95—96
Reliability, 515
    mechanical systems, 651—654
    system, 654
Relief valves, 404
Relubrication, 579
Removal of metal, 335—356
Replica of surface, 659
Reservoir heater, 404
Reservoirs of oil, 395—398
Residual stress, 180—181, 504
Resistance
    abrasion, 171—175
    fatigue, 274, 504
    long-term temperature, 517
    score, 463, 464
    temper, 518
    wear, 167, 174

Resonance, 420
Restorative materials, 194
Restrictors of flow, 107—109
Retained austenite, 505
Retainers
    bearing, 279
    materials for, 505
    nodular-graphite iron, 505
Reversing lubrication system, 383—386
Reynolds equation, 69, 70, 72, 121, 414
Reynolds number, 417
Rheodynamic boundary film formation, 60—61
Rheological model, 147
Rigid couplings, 565
Rigid rollers, 139
Ring compression test, 329
Ring extrusion, 612
Ring mountings, 523
Rings
    annular, 125
    contracting split, 594
    expanding split, 594
    floating piston, 611
    inner, 522
    joint, 594
    lantern, 608
    piston, see Piston rings
    pumping, 588
    second step joint, 594
    segmented, 597
    split, 594
    upsetting of, 319
Ring seals, 591—597
Ring support, 519—521
RMS standard deviation of roughness, 152
R&O, see Rust and oxidation
Rod seals, 594
Roller bearing lubrication, 524—532
Roller bearings
    backup, 521
    clearance of, 501—503
    concave, 496
    cylindrical, 495—497, 509, 524
    EHL and, 156
    endurance of, 514—515
    friction in, 513
    furnace, 528
    kinematics of, 512—513
    needle, 524
    radial internal clearance of, 501—503
    rigid, 139
    spherical, 496, 497, 510
    standards for, 497—499
    symmetrical, 496
    tapered, 495—497, 523
    thrust, 511
Roller crowning, 506—508
Roll flattening, 322
Roll formation wear, 190
Rolling, 320—322, 648
Rolling friction, 43

Roll point contact, 149
Roll stability test, 266
Roll steel wear, 179
Rotary seals, 611
    heads of, 581
Rotational viscometers, 230
    concentric cylinder, 231
Rotor imbalance, 420
Roughness
    cartography of, 9—11
    coefficient of friction and, 41
    defined, 6
    grades of, 6, 7
    measurement of, 3—9
    parameter of, 11
    rms standard deviation of, 152
    sampling length of, 6
    surface, see Surface roughness
Roughness-width cutoff, 6, 9
Rough surfaces, 333
    deformation wear on, 186—189
Rubbed films, 277
Rubber, 190
Rubber bearings, 624
Rubbing contact seals, 581—612
Run-in, 14, 51—52
Running, 654
Rust, 302
Rust and oxidation (R&O) inhibitors, 544—545
Rust inhibitors, 262, 307

S

Safety, 368, 371
Salts, 273
Sampling length, 6
Scanning electron microscope (SEM), 3, 23
Score resistance, 463, 464
Scoring, 563
    gear, 561—563
    journal, 479
    limit of, 50
Scratching, 562
    gear, 561
Screw dislocations, 25, 26
Screw pump, 400
Scuffing limit, 50
Seal balance, 584—586, 588
Seal faces
    energy dissipation between, 584
    temperature at, 587
Seal heads, 581
Sealing dam dimensions, 594
Seals, see also specific types, 279, 574
    automatic secondary, 583
    buffered bushing, 617
    circumferential, 594—597
    clearance, 581, 612—617
    compression, 583
    contact, 581

controlled clearance, 612—616
double, 589
dual lip, 598
dynamic, 581
externally pressurized, 614
external operation of, 582
ferromagnetic, 618
fixed-geometry clearance, 617
floating bushing, 615—616
flushable, 497
freeze, 617—618
high speed and, 588
hybrid, 614
hydrodynamic, 612—614
hydrostatic, 614—615
installation of, 604, 607, 610
internal, 581
labyrinth, 617
leakage of, 587, 598—604
life of, 587—588, 598—604
loading of, 584
magnetic, 584
mechanical face, 581—591, 595
metal bellows, 588
O-ring, see O-ring seals
packing, 605—612
positive rubbing contact, 581—612
radial lip, 597—605
ring, 591—597
rod, 594
rotary, 611
secondary, 583—584
self-activated, 614
semistatic, 583
slinger, 618
specialized control of fluid in, 617—619
split ring, 594
static, 581
swell additives for, 306
tandem, 589
temperature extremes for, 588
Secondary seals, 583—584
Secondary tensile stresses, 319, 322, 324
Second step joint ring, 594
Section
broad, 125
circular, 121—123
conic, 126
elliptical, 123
equilateral triangular, 125
narrow, 125
rectangular, 123—124
spherical, 126
square, 125
Segmented ring, 597
Segregation, 28
Seizure, 36
Selection
cutting fluid, 361—365
dry bearings materials, 285—286
lubricant, 328, 529—530, 577—579, 656—658

polymer, 195—198
Selective etching, 632
Self-activated seals, 614
Self-aligning ball bearings, 495
Self-lubricating composites, 275, 279—285
SEM, see Scanning electron microscope
Semistatic seals, 583
Semisynthetics, 360
Separations at surface, 59
Servo valves, 108
Settling of oil, 400—401
Severe wear, 165—169
Severity, 50—51
"Severity parameter", 51
Shaft couplings, 565—579
Shaft speed, 604
Shape, 123
convergent film, 414
film, 142—144, 147, 414
point contact film, 147
Shear, 59
inlet, 141
interface, 317
Shear angle, 357
Shear modulus, 149
elastic, 151
Shear rates, 230, 231, 236, 291
Shear strain in chip formation, 340—341
Shear strength, 196
interface, 329
Shear stress, 230
maximum, 196
plastic flow, 59
viscosity of, 151
yield, 151
Shear-viscosity relationships, 235—237
Sheet metalworking, 324—325
Shell and tube oil coolers, 403
Short-bearing relations, 127—128
Short journal bearing, 72—73
Short-term hot hardness, 517
Shrink fit, 522
Side effects, 365
Silicate esters, 528
Silicones, 305, 528
Silicon-iron bronze, 505
Siliconizing, 638
Silver, 17, 28
Silver bearings, 472
Single acting thrust bearings, 109—110
Single line spring return lubrication system, 381
Single-traversal wear, 197
Sintered porous surface coatings, 638
Size
effects of, 63, 64
gear, 544
gear tooth, 542
physical, 62
Skewness, 8
Slab upsetting, 319
Sleeve, 522

Slider bearing, 69—71
Slide-roll ratio, 51
Sliding, 648
  smooth, 165
  well-lubricated, 61
Sliding compatibility, 204
Sliding distance, 52
Sliding electrical contacts, 279
Sliding friction, 35—36
  coefficient of, 46—47
Sliding motion, 565
Sliding speed and coefficient of friction, 39
Sliding velocity, 216
Sliding wear, 170
  severe, 166—169
Slinger seals, 618
Slip, 43, 512, 513
  forward, 329
Slotted washers, 584
Sludge, 575
Slurries, 589
Slush pan method, 555
Smearing wear, 168
Smooth surface interfacial wear, 189—190
SOAP, see Spectrographic oil analysis procedure
Soap, soap grease, see Grease
Sodium-aluminum grease, 258
Sodium-calcium grease, 258
Sodium grease, 258, 262
Soft metal films, 273—274
Soft nitride coatings, 638
Solid contaminants, 465
Solid-like boundary films, 63
Solid lubricants, 53, 547
Solids, see also specific solids, 361
  ideal elastic, plastic, 59
  lamellar, 269—273, 280—282
  plastic, 60
Solid-state bonding, 26
Solid wall bearings, 474—475
Solubility
  gas, 241—243
  metallurgical, 204
  mutual, 203
Soluble oils, 359
Sonic velocity of gases, 292
Sound, 299—300
Spalling, 218
  gear, 561
Spark hardening, 632, 638
Special deoxidized steels, 348—349
Specialized control of fluid in seals, 617—619
Specifications of lubrication, 249—252
Specific cutting energy, 342—343
Specific film thickness, 218
Specific gravity, 239
Specific heat, 244
  of gases, 292
Specific volume of gases, 298
Spectographic oil analysis procedure (SOAP), 660
Spectroscopy

atomic absorption, 248
electron, see Electron spectroscopy
glow discharge mass, 23
ion-scattering, 23
nuclear back scattering, 23
X-ray photoemission, 22—23
Speed, 51
  bearing, 516—517
  gear, 544
  seals and, 588
  shaft, 604
  sliding, 39
  sound, 299—300
  vibration, 420
Speed limits, 586—587
Spherical contact, 144, 163—164
Spherical particles, 171
Spherical roller bearings, 496, 497, 510
Spherical section, 126
Spikes, 143
Spin, 648
Spindle couplings, 571
Spin roll point contact, 149
Spiral bevel gears, 549
Spiral failure, 611
Splash system, 547
Split rings, 594
Split sleeve adapters, 522
Spragging, 445
Sprayed coatings, 634—635
Spray systems, 390—391
Spring-loaded relief valve, 404
Spring return lubrication system, 381
Springs, see also specific types, 582, 584, 609
Spur gears, 539, 543, 546, 549
Sputtering, 275, 635
Square section, 125
Squeal, 43
Squeeze films, 69, 121—138
Squeeze packing, 605, 610—612
Squeeze rate, 121
Stability
  bearings, 85—87, 443
  dimensional, 505
  full journal bearings, 443
  mechanical, 266
  oxidation, 247—249, 265, 286
  roll, 266
  thermal, 246—247, 269
Stable life of lubricants and temperature, 249
Staining testing, 329
Stainless steel, 640
Standard couplings, 570—571
Standard deviation, 152
Standards
  cleanliness, 524
  hardness, 524
  roller bearing, 497—499
Starting coefficient of friction, 42
Starting rate and coefficient of friction, 40
Starvation, 142

Starved contacts, 146
Static capacity, 514
Static friction, 42—43
    coefficient of, 46—47, 72
Static seals, 581
Stationary seal heads, 581
Steady-state wear, 197
Stearic acid, 9
Steel grid couplings, 569, 573
Steel-on-steel lubricant system, 66
Steels, 17
    austenitic manganese, 173, 625
    high speed, 344, 346, 352
    phosphating of, 333
    relative wear resistance of, 167
    roll, 179
    special deoxidized, 348—349
    stainless, 640
Sticking friction, 317
Stick-slip, 43—47, 598
Stiffness
    film, 110
    grease, 525
Stochastic theory, 153
Straight bevel gear drives, 549
Straight fatty oils, 357
Straight mineral oils, 50, 51, 357
Strain, 340—341
Strainers, 401
Strain gages, 38
Stratified surface textures, 13—15
Streamers, 72
Strength, 463, 503—504
    fatigue, 463—465, 477
    interface shear, 329
    plastic flow, 56
    shear, 196, 329
    yield, 36
Stress
    concentrations of, 506, 520
    contact, 503, 508, 513
    Hertzian, 167, 175
    maximum shear, 196
    plastic flow shear, 59
    residual, 180—181, 504
    secondary tensile, 319, 322, 324
    shear, 151, 196, 230
    yield shear, 151
Stretching, 324
Stribeck curve, 655
Strip chart, 39, 45—47
Strong EP oil, 50
Structure, 651—654
    body-centered cubic, 18, 19
    crystalline, 25—26
    face-centered cubic, 18, 19
    hexagonal crystal, 628
    surface, 32
    system, 645, 651
    tribological system, 648—651
Stuffing box, 581

temperature of, 587
    throat of, 588
Subboundaries, 25
Subharmonic resonance, 420
Substrate, 633—639
Suction conditions of oil, 399
Suction pipes, 408
Sulfide coatings, 637
Sulfides, 273
Sulfo-chlorinated mineral oil, 359
Sulfur, 21
Sulfur dioxide, 52, 53
Sulfurized fats, 310
Sulfurized mineral oil, 358—359
Superalloys, 640
Supply pipes, 408
Supply pressure of oil, 399
Support area, 49
Surface
    adsorption on, 32—33
    apparent wavelength of, 8
    atomically clean, 61
    bead-blasted, 9
    chemistry of, 24, 27—29
    clamped, 31
    clean, 27, 40
    compound formation of, 28—29
    conformal, 163
    contaminants on, 21
    counterformal, 163
    crystalline structure of, 25—26
    damage to, 50, 56, 65
    drive, 31
    energy of, 27
    events of, 28
    fatigue of, 201, 207, 651
    finish of, 32
    formation of, 28—29
    fresh, 61
    hardness of, 51, 628
    integrity of, 355
    interacting, 164
    kinetics of reactions of, 215—216
    lay to, 8
    melting of, 39
    metallurgy of, 25—26
    nonplanar, 126
    properties of, 649
    replica of, 659
    rough, 186—189
    separations of, 59
    smooth, 189—190
    structure of, 25—26, 32
    technological, 32
    temperatures of, 209, 215
    topography of, 186, 187
    traction, 31
    virgin, 328
    wavelength of, 8
Surface asperities, 189
Surface coatings, 328

sintered porous, 638
Surface films, 51, 56
Surface finish, 44
Surface roughness, 56, 57, 152, 327, 333
   cartography of, 9—11
   coefficient of friction and, 41
   measurement of, 3—9
Surface tension, 245—246, 598
Surface textures, 8, 50
   mixed, 13—15
   pure, 13—15
   stratified, 13—15
Surface tools, 18—25
Surface treatments, 623—644
Swell additives, 306
Symmetrical rollers, 496
Synthetic base oils, 308
Synthetic fluids, 260, 360—361
Synthetic hydrocarbons, 528
Synthetic lubricants, 528, 546
System accessories for centralized lubrication systems, 389
Systematic lubricant selection, 656—658
System concept, 645—654
System ''infant death'', 654
System reliability, 654
Systems analysis, 645
System structure tribo-induced changes, 651

T

Tackiness additives, 263, 306
Tandem seal, 589
Tanks
   header, 405
   pressure, 407
Tapered bore inner ring, 522
Tapered roller bearings, 495—497, 511, 523
Tapping, 43
Taylor number, 417
Taylor tool life equation, 346—347
Taylor vortex flow, 74
Technical function of tribological systems, 646—648
Technological surfaces, 32
Temperature, 215—216, 599
   bearing, 517
   coefficient of friction and, 39—40
   corrosive wear and, 632
   density and, 240
   distribution of, 77
   EHL and, 147
   extremes of for seals, 588
   of gases, 292
   gear, 544
   interacting surfaces, 164
   limits on, 527, 586—587
   long-term resistance to, 517
   maximum calculated, 66
   maximum conjunction, 66

oil, 51
   resistance to, 517
   seal, 588
   seal faces, 587
   stable life of lubricants and, 249
   stuffing box, 587
   surface, 164, 209, 215
   tool face, 343
   transition, 50, 205
Temperature control, 231, 588
   for oil, 404
Temperature gradients, 417, 648
Temperature-viscosity relationships, 231—234, 260
Temper resistance, 518
Ten-point height, 7
Tensile stresses, 319, 322, 324
Tension on surface, 245—246, 598
Testing, 371
   abrasion resistance, 171—175
   bench, 657
   Falex, 267
   fatigue, 17
   full-scale, 657
   lubricant, 328, 657
   ring compression, 329
   roll stability, 266
   staining, 329
   Timken, 266
   tribo-technical components and, 657
   twist compression, 329
   wear, 276—277
   wheel bearing, 266
Textures of surface, 8, 13—15, 50
THD, see Thermohydrodynamic
Theoretical line contact, 506
Thermal conductivity, 64, 244
Thermal damage, 487—488
Thermal effects, see also Heat, 417—418
Thermal fatigue, 488
Thermal insulation, 64
Thermal parameters, 65
Thermal pressure, 175
Thermal properties, 243—246, 466—467
Thermal stability, 246—247, 269
Thermal treatments, 637
Thermocouples, 658
Thermohydrodynamic (THD) theory, 414
Thermostatic valve, 406
Thickeners, 255—259
Thickness of film, see Film thickness
Thin film lubricants, 277—279
Three-body abrasion, 625, 627
Three-dimensional cutting, 344
Throat of stuffing box, 588
Thrust bearings, 109—113, 424—435, 495—497
   compliant hydrodynamic, 155—156
   double-acting, 110—113
   fixed-land (pivoted-pad), 425—427
   fixed-type, 424—425
   flexible-type, 434—435
   performance charts for, 427—431

single acting, 109—110
tapered roller, 511
Timken test, 266
Tin babbitts, 482, 487
bearings of, 467—469
Tin bronzes, 469
Tip curvature, 51
Titanium, 17
alloys of, 350, 640
Tools, see also specific types
analytical surface, 18—25
coated, 349
cutting, 345—349, 357
face temperature of, 343
life of, 345—349
machine, 365
materials for, 345—346
Taylor equation for life of, 346—347
wear of, 345—349
Tooth size on gears, 542
Topography of surface, 186, 187
Torque, 579
fluctuating, 578
Torsional failure, 611
Tracer methods, 660
Traction surfaces, 31
Transducers, 658
Transfer
metal, 320
polymer, 186
Transfer element, 169
Transient behavior of films, 121
Transition, 50—51
flow, 417
glass, 59
load of, 50
temperature of, 50, 205
Transmission
of information, 646
of mechanical work, 646
Transmission electron microscopy, 170
Transport of debris, 188
Transverse asperity collision, 155
Triangular section, 125
Tricresyl phosphate, 309
Tri-metal bearings, 475
Tube drawing, 322—323
Turbine engines, 505
Turbulence, 73—77
Turbulent concentric flow, 616
Turbulent dissipation, 418
Turbulent flow, 417
Turbulent lubricating film, 75
Turning, 13
Twist compression test, 329
Two-body abrasion, 625, 627
Two-dimensional motion, 129—132

U

UHMWPE, see Ultrahigh molecular weight
polyethylene
Ultrahigh molecular weight polyethylene
(UHMWPE), 186
Unbalance excitation frequency, 447—453
Undercoatings, 635
Unfilled polymers, 185—190
Universal engine oils, 307, 308
Unsaturated fatty acids, 306
Upsetting
of cylinder, 317—319
of ring, 319
of slab, 319
U-ring packing, 608

V

Vacuum equipment, 36
Valves
check, 117
diaphragm, 405
direct-operated diaphragm, 405
mixing, 407
pneumatically controlled diaphragm, 405
servo, 108
spring-loaded relief, 404
thermostatic, 406
Van der Waals forces, 26, 169, 170
Vapor, 52
coatings deposited by, 634
deposition of, 634
Vaporization heat, 244—245
Vapor pressure, 237—239
Velocity, 51
sliding, 216
sonic, 292
Velocity transducers, 658
VI, see Viscosity index
Vibrating crystal viscometers, 235
Vibration, 648
of components, 659
critical speed of, 420
frequency of, 168
frictional, see Stick-slip
Virgin surfaces, 328
Visco-elastic response, 39
Viscometers, 234
capillary, 230, 234
concentric cylinder rotational, 231
cone and plate, 231
failure of, 234
falling weight, 234
optical, 235
rotational, 230, 231
vibrating crystal, 235
Viscoseal, 618—619
Viscosity, 229—231, 260, 291, 579
absolute, 292

apparent, 265—266
average, 413—414
base oil of grease, 574
boundary film, 63
effective, 413
gases, 292—298
high-pressure, high-shear capillary, 235
kinematic, 292, 413
low-temperature, 232
modifiers of, 304—305, 306
oil, 399
pressure vs., 543
ranges of, 548
shear stress, 151
Viscosity grade of lubricants, 51
Viscosity index (VI), 232, 260, 304
Viscosity-pressure relationships, 140, 234—235
Viscosity-shear relationships, 235—237
Viscosity-temperature relationships, 231—234, 260
Volume
specific, 298
wear, 49
Volumetric outflow rate, 122
Vortex flow, 74
V-ring packing, 609—610

W

Walther equation, 231, 232
Washers, 584
Water
absorbed, 17
dissolved, 53
effect of, 218—219
quality of, 363
Water vapor, see Vapor
Wavelength, 8
Wave springs, 584
Waviness, 9
defined, 6
Wax, 305
Wax alkylated naphthalene polymers, 305
Wax alkylated phenol polymers, 305
Wear, see also Erosion
abrasive, see Abrasion
adhesive, see Adhesion
anticipated, 204
balance between adhesive and chemical, 212—216
bearing, 467
boundary lubrication and, 49—50
catastrophic, 56
chemical, 632
coefficients of, 49, 50, 65, 202—205, 209, 216
cohesive, 165
corrosive, see Corrosion
crater, 347—348
cutting, 181
cutting tool, 345—349
deformation, see Deformation
delamination theory of, 170—171
depth of, 49
die, 320
dusting, 282
electrical, 180
electrochemical, 222—223
erosive, see Erosion
fatigue, 61, 192, 218
fretting, 176—178, 523, 536, 631—632
gear, 558—563
gouging, 176
heavy gear, 559
impact, 175, 648
inhibitors of, 30., 306, 309
interfacial, 189—190
light gear, 558—559
lubricated, 209—225
materials losses induced by, 651, 660
mechanical, 477—481
mechanical failure induced by, 654
mechanisms of, 61—62
moderate gear, 559
modes of, 624—633
noncohesive, 165
predictions of, 201
relative resistance to, 167, 174
resistance to, 167, 174
roll formations and, 190
roll steels and, 179
severe, 165
severe sliding, 166—169
single-traversal, 197
sliding, 166—170
smearing, 168
steady-state, 197
surface fatigue, 201, 207
testing of and friction, 276—277
wear land tool, 347
Wear debris, 182, 660
loose, 628
Wear land tool wear, 347
Wear life, 66
Wear protection ranges, 66
Wear rates, 187, 287
coefficient of friction and, 41—42
Wear-resistant coatings, 623—644
Wear volume, 49
Wedge action, 69
Weibull distribution, 653
Weld, 64
Wetting additives, 360
Wheel bearing test, 266
Wheels, 528
Width of gear face, 541
Wiping, 488—489
damage from, 483
Wire drawing, 322—323
Work hardening, 173
Working coolant microbial control, 374—375
Worm gears, 541, 546, 550—551

X

XPS, see X-ray photoemission spectroscopy
X-ray photoemission spectroscopy (XPS or ESCA),
    22—23

Y

Yawing, 44
Yield shear stress, 151

Yield strength, 36
Young's moduli, 8

Z

ZDTP, see Zinc dithiophosphate
Zinc dialkyldithiophosphates, 303
Zinc diisopropyldithiophosphate, 56
Zinc dithiophosphate (ZDTP), 55, 303, 306, 309
Zinc oxides, 258, 263